Edited by
Takayuki Shioiri, Kunisuke Izawa,
and Toshiro Konoike

Pharmaceutical Process Chemistry

Related Titles

Yasuda, N. (ed.)

The Art of Process Chemistry

2010
ISBN: 978-3-527-32470-5

Fischer, Janos and Ganellin, C. Robin (eds.)

Analogue-based Drug Discovery II

2010
ISBN: 978-3-527-32549-8

Chorghade, M. S. (ed.)

Drug Discovery and Development

2 Volume Set

2007
ISBN: 978-0-471-39846-2

Wesselingh, J. A., Kiil, S., Vigild, M. E.

Design and Development of Biological, Chemical, Food and Pharmaceutical Products

2007
ISBN: 978-0-470-06154-1

Pollak, P.

Fine Chemicals

The Industry and the Business

2007
ISBN: 978-0-470-05075-0

Edited by Takayuki Shioiri, Kunisuke Izawa,
and Toshiro Konoike

Pharmaceutical Process Chemistry

WILEY-VCH Verlag GmbH & Co. KGaA

The Editors

Prof. Dr. Takayuki Shioiri
1-18-12 Minamigaoka, Nisshin
Aichi 470-0114
Japan

Dr. Kunisuke Izawa
Intermediate Chemicals Dept.
Ajinomoto Co., Inc.
1-15-1 Kyobashi, Chuo-ku
Tokyo 104-8315
Japan

Dr. Toshiro Konoike
CMC Development Laboratories
Shionogi & Co., Ltd.
1-3 Kuise Terajima 2-chome
Amagasaki, Hyogo 660-0813
Japan

All books published by Wiley-VCH are carefully produced. Nevertheless, authors, editors, and publisher do not warrant the information contained in these books, including this book, to be free of errors. Readers are advised to keep in mind that statements, data, illustrations, procedural details or other items may inadvertently be inaccurate.

Library of Congress Card No.: applied for

British Library Cataloguing-in-Publication Data
A catalogue record for this book is available from the British Library.

Bibliographic information published by the Deutsche Nationalbibliothek
The Deutsche Nationalbibliothek lists this publication in the Deutsche Nationalbibliografie; detailed bibliographic data are available on the Internet at <http://dnb.d-nb.de>.

© 2011 WILEY-VCH Verlag GmbH & Co. KGaA, Boschstr. 12, 69469 Weinheim, Germany

All rights reserved (including those of translation into other languages). No part of this book may be reproduced in any form – by photoprinting, microfilm, or any other means – nor transmitted or translated into a machine language without written permission from the publishers. Registered names, trademarks, etc. used in this book, even when not specifically marked as such, are not to be considered unprotected by law.

Composition   Laserwords Private Ltd., Chennai, India
Printing and Binding   Fabulous Printers Pte. Ltd., Singapore
Cover Design   Adam Design, Weinheim

Printed in Singapore
Printed on acid-free paper

ISBN: 978-3-527-32650-1

Contents

Preface *XVII*
List of Contributors *XXI*

1	**From Milligrams to Tons: The Importance of Synthesis and Process Research in the Development of New Drugs** *1*	
	Martin Karpf	
1.1	Introduction *1*	
1.2	The Synthetic Development of the Monoamine Oxidase-B Inhibitor Lazabemide™ *6*	
1.3	The Synthetic Development of the Lipase Inhibitor Tetrahydrolipstatin (Xenical™) *6*	
1.4	The Synthetic Development of the HIV Protease Inhibitor Saquinavir (Invirase™) *13*	
1.5	The Synthetic Development of the Influenza Neuraminidase Inhibitor Oseltamivir Phosphate (Tamiflu™) *16*	
1.5.1	Introduction *16*	
1.5.2	The Development of the Current Technical Synthesis of Oseltamivir Phosphate *18*	
1.5.3	The Search for Alternative Routes to Oseltamivir Phosphate *23*	
1.5.3.1	The Development of Azide-Free Transformations of the Key Epoxide Intermediate to Oseltamivir Phosphate *23*	
1.5.3.2	The Development of Alternative Syntheses for Oseltamivir Phosphate *27*	
	References *36*	
2	**Design of Dynamic Salt Catalysts Based on Acid–Base Combination Chemistry** *39*	
	Kazuaki Ishihara	
2.1	Introduction *39*	
2.2	Dehydrative Condensation Catalysts *41*	

2.2.1	Esterification Catalysts 41
2.2.2	Dehydrative Cyclocondensation Catalysts 43
2.3	Asymmetric Mannich-Type Catalysts 50
	References 56

3	**Asymmetric Oxidation with Hydrogen Peroxide, an Effective and Versatile Oxidant** 59
	Tsutomu Katsuki
3.1	Introduction 59
3.2	Asymmetric Epoxidation 60
3.2.1	Asymmetric Epoxidation with Synthetic Enzymes or Organocatalysts 60
3.2.2	Metal-Catalyzed Asymmetric Epoxidation of Unfunctionalized Olefins 62
3.2.3	Metal-Catalyzed Asymmetric Epoxidation of Allylic Alcohols 67
3.3	Asymmetric Oxidation of Sulfides 67
3.3.1	Metal–Salen-Catalyzed Oxidation 68
3.3.2	Metal–Schiff Base-Catalyzed Oxidation 68
3.3.3	Metal–ONNO–Tetradentate Ligand-Catalyzed Oxidation (Including *cis*-β Metal–Salen-Catalyzed Oxidation) 69
3.3.4	Miscellaneous 72
3.4	Conclusion 73
	References 74

4	**Development of Palladium Catalysts for Chemoselective Hydrogenation** 77
	Hironao Sajiki and Yasunari Monguchi
4.1	Catalyst Poisons and Chemoselective Heterogeneous Catalysts 77
4.1.1	Background 77
4.1.2	Chemoselective Inhibition of the Hydrogenolysis for *O*-Benzyl Protective Groups by the Addition of a Nitrogen-Containing Base 77
4.1.3	Pd/C(en) Complex as a Heterogeneous Chemoselective Hydrogenation Catalyst 81
4.1.4	Pd/C (Ph$_2$S) Complex as a Heterogeneous Chemoselective Hydrogenation Catalyst 85
4.2	Catalyst Supports and Chemoselective Heterogeneous Catalysts 90
4.2.1	Pd/Fib as a Silk-Fibroin-Supported Chemoselective Hydrogenation Catalyst 90
4.2.2	Pd-PEI as a Partial Hydrogenation Catalyst of Alkynes to Alkenes 93
4.3	Summary 96
	Acknowledgment 97
	References 97

5	**Silicon-Based Carbon–Carbon Bond Formation by Transition Metal Catalysis** *101*
	Yoshiaki Nakao and Tamejiro Hiyama
5.1	Introduction *101*
5.2	Cross-Coupling Reactions *102*
5.2.1	Brief Assessment of Early Stage Protocols *102*
5.2.2	Cross-Coupling Reactions Using Tetraorganosilanes through Intramolecular Activation *103*
5.2.3	Cross-Coupling Reactions Using Organosilanolates *106*
5.2.4	Other Tetraorganosilicon Compounds for Cross-Coupling Chemistry *108*
5.2.5	New Types of Electrophiles for Silicon-Based Cross-Coupling *111*
5.3	Carbonyl Addition Reaction *114*
5.3.1	Rhodium-Catalyzed Reactions *114*
5.3.2	Nickel-Catalyzed Reactions *115*
5.3.3	Palladium-Catalyzed Reactions *117*
5.3.4	Copper-Catalyzed Reactions *120*
5.3.5	Silver-Catalyzed Reactions *121*
5.4	Recent Developments in Catalytic Preparation of Organosilanes *121*
	References *123*
6	**Direct Reductive Amination with Amine Boranes** *127*
	Karl Matos and Elizabeth R. Burkhardt
6.1	Introduction *127*
6.2	Types of Amine Boranes *128*
6.2.1	Alkylamine Boranes *128*
6.2.2	Aromatic Amine Boranes *129*
6.2.2.1	Pyridine borane *130*
6.2.2.2	2-Picoline borane *131*
6.2.2.3	5-Ethyl-2-methylpyridine borane *131*
6.3	Comparison to Sodium Triacetoxyborohydride (STAB) *134*
6.4	Primary Amine Synthesis *135*
6.5	Stereoselective Reductive Amination *137*
6.6	Reaction Solvents *138*
6.7	Reaction Workup *138*
6.8	Conclusion *141*
	References *141*
7	**Industrial Synthesis of Perfluorinated Building Blocks by Liquid-Phase Direct Fluorination** *145*
	Takashi Okazoe
7.1	Introduction *145*
7.2	History of Direct Fluorination *146*
7.3	Synthetic Methods Using Perfluorinated Acyl Fluorides for Industrially Important Perfluorinated Monomers *149*

7.3.1	Direct Application of Liquid-Phase Fluorination	149
7.3.2	The PERFECT Method	150
7.4	Synthesis of Perfluorinated Building Blocks by the PERFECT Method	152
7.4.1	Perfluorinated Acyl Fluorides	152
7.4.2	Synthesis of Perfluorinated Ketones by the PERFECT Method	154
7.5	Conclusion	156
	References	157

8	**Cross-Linked Enzyme Aggregates as Industrial Biocatalysts**	**159**
	Roger A. Sheldon	
8.1	Introduction	159
8.2	Cross-Linked Enzyme Aggregates	160
8.2.1	Cross-Linking Agents	160
8.2.2	Protocols for CLEA Preparation	161
8.2.3	Advantages of CLEAs	163
8.2.4	Multi-CLEAs and Combi-CLEAs	164
8.3	CLEAs from Hydrolases	164
8.3.1	Lipase and Esterase CLEAs	165
8.3.2	Protease CLEAs	168
8.3.3	Amidase CLEAs	170
8.3.4	Nitrilases	171
8.3.5	Glycosidases	172
8.4	Oxidoreductases	172
8.4.1	Oxidases	172
8.4.2	Peroxidases	173
8.5	Lyases	174
8.5.1	Nitrile Hydratases	174
8.5.2	C–C Bond Forming Lyases	174
8.6	Combi-CLEAs and Cascade Processes	175
8.7	Reactor Design	176
8.7.1	Membrane Slurry Reactor	177
8.7.2	CLEAs in Microchannel Reactors	177
8.8	Conclusions and Prospects	178
	References	178

9	**Application of Whole-Cell Biocatalysts in the Manufacture of Fine Chemicals**	**183**
	Michael Schwarm	
9.1	Introduction: Early Applications of Biocatalysis for Amino Acid Manufacture at Evonik Degussa	183
9.2	Hydantoinase Biocatalysts	187
9.3	Amino Acid Dehydrogenase Biocatalysts	191
9.4	Alcohol Dehydrogenase Biocatalysts	195

9.5	Summary 203
	Acknowledgments 204
	References 204

10	**Process Development of Amrubicin Hydrochloride, an Anthracycline Anticancer Drug** 207
	Kazuhiko Takahashi and Mitsuharu Hanada
10.1	Introduction 207
10.2	Original Synthetic Route for Amrubicin 208
10.3	Amrubicin Bulk Production Synthetic Method 210
10.3.1	Safe Synthetic Method of 9-Aminoketone 211
10.3.2	Stereoselective Introduction of 7-Hydroxy Group 213
10.3.3	Polymorphism Study of Amrubicin Hydrochloride 215
10.3.4	Stability of Amrubicin Hydrochloride with Reference to Moisture 216
10.3.4.1	Amrubicin Hydrochloride Moisture Adsorption 217
10.3.4.2	Stability in Various Water Contents 217
10.3.4.3	Establishment of Drying Method 217
10.4	Conclusion 219
	References 219

11	**Process Development of HIV Integrase Inhibitor S-1360** 221
	Toshiro Konoike and Sumio Shimizu
11.1	Introduction 221
11.2	Discovery of Integrase Inhibitor S-1360 221
11.2.1	Discovery Route of S-1360 222
11.3	Synthesis of Two Starting Materials for S-1360 225
11.3.1	Two One-Step Syntheses of Benzylfuryl Methyl Ketone **2** 225
11.3.1.1	Friedel–Crafts Alkylation by Anhydrous $ZnCl_2$ in Dichloromethane 226
11.3.1.2	Friedel–Crafts Alkylation Using Aqueous $ZnCl_2$ 226
11.3.2	Two Synthetic Methods to Triazole Ester **3** 228
11.3.2.1	Ring Construction Method 228
11.3.2.2	Ring Modification Method 228
11.4	Process Chemistry of S-1360 and Scale-Up of THP Route 229
11.4.1	Protection of Triazole **3** by the Tetrahydropyranyl (THP) Group and Claisen Condensation 229
11.4.2	Deprotection of the THP Group and Purification of API Deprotection of the THP Group 230
11.4.2.1	Purification of API 231
11.4.2.2	Quality Assurance and Productivity 232
11.5	Process Development of S-1360 and Commercial Route by Methoxyisopropyl (MIP) Protection 233
11.5.1	MIP Route 234
11.5.2	Further Improvement of Productivity 235

11.6	Summary and Outlook	235
	Acknowledgments	237
	References	237

12 An Efficient Synthesis of the Protein Kinase Cβ Inhibitor JTT-010 239
Takashi Inaba

12.1	Introduction	239
12.2	Synthetic Strategies	240
12.3	Key Intermediate Synthesis	240
12.3.1	Optical Resolution	240
12.3.2	Enzymatic Chiral Induction	242
12.3.3	C–H Bond Activation by a Chiral Catalyst	243
12.3.4	Formal [3 + 2] Cycloaddition Using Chiral Cyclopropane	244
12.4	Replacement of the Hydroxyl Group of **1** with an Amino Group	250
12.5	Construction of JTT-010	251
12.5.1	Stepwise Maleimide Construction	251
12.5.2	Convergent Coupling Reaction to JTT-010	251
12.6	Conclusion	253
	References	254

13 Process Development of Oral Carbapenem Tebipenem Pivoxil, TBPM-PI 257
Takao Abe and Masataka Kitamura

13.1	Introduction	257
13.2	Discovery of TBPM-PI	257
13.3	Synthetic Process of Side Chain on the C2-Position of TBPM, TAT	260
13.3.1	Original Synthetic Process of TAT Starting from Benzhydrylamine	260
13.3.2	Practical Synthetic Process of TAT from Benzylamine	261
13.3.3	Industrial Synthetic Process of TAT: Back to Classic Bunte's Salt	263
13.4	Synthetic Process of TBPM-PI from 4-Nitrobenzyl $(1R,5R,6S)$-2-diphenylphosphoryloxy-6-[(R)-1-hydroxyethyl]-1-methyl-1-carbapen-2-em-3-carboxylate, MAP	265
13.4.1	Synthesis of PNB Ester of TBPM, L-188	265
13.4.2	Synthesis of TBPM-4H$_2$O	266
13.4.3	Prodrug Esterification: Synthesis of TBPM Hexetil, LJC11,143	267
13.4.4	Synthesis of TBPM-PI	269
13.5	Summary and Outlook	270
	Acknowledgments	271
	References	271

14	**Some Progress in Organic Synthesis of Pharmaceuticals in China** *273*	
	Delong Liu and Wanbin Zhang	
14.1	Introduction *273*	
14.2	Industrial Synthesis of Chinese Herbal Medicines *274*	
14.2.1	Industrial Synthesis of Berberine *274*	
14.2.2	Industrial Synthesis of D,L-Tetrahydropalmatine (THP) *277*	
14.3	New Agents Derived from Chinese Herbal Medicines *281*	
14.3.1	Bifendate and Bicyclol *281*	
14.3.1.1	Bifendate *281*	
14.3.1.2	Bicyclol *284*	
14.3.2	Qinghaosu *285*	
14.3.2.1	Synthesis *285*	
14.3.2.2	Fixed-Dose Riamet/Coartem *287*	
14.4	Process Chemistry for L-Ascorbic Acid and Biotin *291*	
14.4.1	Two Steps Fermentation Method for the Preparation of L-Ascorbic Acid (Vitamin C) *291*	
14.4.2	Total Synthesis of Biotin (Vitamin B7) *294*	
14.5	Conclusion and Perspectives *298*	
	Abbreviations *299*	
	References *299*	
15	**The Use of Continuous Processing to Make AZD 4407 Intermediates** *303*	
	Andrew S. Wells	
15.1	Green Chemistry and the Drive for Sustainability *303*	
15.2	Advantages of Chemistry in Continuous-Flow Reactors *304*	
15.3	Introduction to AZD 4407 *305*	
15.4	Comparison of the Synthetic Routes Used to Prepare AZD 4407 *305*	
15.5	Conversion of Batch to a Flow Process *308*	
15.5.1	Preparation of Synthons Used in the Microreactor Study *308*	
15.5.1.1	Disulfide Synthesis *308*	
15.5.1.2	(S)-2-Methyl Tetrahydropyran-4-one *309*	
15.5.1.3	Thiophene Synthon *310*	
15.5.2	*n*-Hexyl Lithium Versus *n*-Butyl Lithium *310*	
15.5.3	Batch Reactions: Lithiation Reaction and Disulfide Linking *311*	
15.5.4	Microreactor Description *311*	
15.5.5	Lithiation Reaction in Flow Mode *312*	
15.5.6	Coupling Reaction with Disulfide in Flow Mode *313*	
15.5.7	Linking the Lithiation and Reaction with Disulfide in Flow Mode *313*	
15.6	Conclusions *316*	
	Acknowledgments *317*	
	References *317*	

16	**Sustainable Processes Based on Enzymes Enabling 100% Yield and 100% ee Concepts** *321*	
	Oliver May	
16.1	Introduction *321*	
16.2	Asymmetric Synthesis *322*	
16.2.1	C–C Bond Formations *322*	
16.2.2	Production of Chiral Alcohols by Enzymatic Reduction of Ketones *325*	
16.2.3	Reduction of Activated C=C Bonds *326*	
16.2.4	Reductive Amination of α-Keto-Acids for the Production of α-Amino Acids *328*	
16.2.5	Transamination Reactions *330*	
16.3	Enzymatic Desymmetrization *332*	
16.4	Enzymatic Deracemization *335*	
16.5	Dynamic Kinetic Resolution *337*	
16.6	Summary and Outlook *338*	
	References *339*	
17	**Development of a Novel Synthetic Method for RNA Oligomers** *345*	
	Tadaaki Ohgi and Junichi Yano	
17.1	Introduction *345*	
17.2	Synthesis of CEM Amidites *349*	
17.3	Synthesis of RNA Oligomers from CEM Amidites *350*	
	Acknowledgments *360*	
	References *360*	
18	**Process Research with Explosive Reactions** *363*	
	Hiromu Kawakubo	
18.1	Introduction *363*	
18.2	Safety Evaluation of an Explosive Chemical Process *363*	
18.3	Standard Procedures for Risk Assessment *365*	
18.3.1	CHETAH (Chemical Thermodynamic and Energy Release Evaluation Program) Calculation *365*	
18.3.2	DSC (Differential Scanning Calorimetry) *365*	
18.3.3	DTA (Differential Thermal Analysis) and TG (Thermogravimetry) *367*	
18.3.4	Impact Sensitivity Test *368*	
18.3.5	Friction Sensitivity Test *368*	
18.3.6	Pressure Vessel Test *369*	
18.3.7	Steel Pipe Test *371*	
18.4	Safety Evaluation of Nitroacetic Acid Ethyl Ester *371*	
18.4.1	Various Safety Evaluations of Nitroacetic Acid Ethyl Ester *373*	
18.4.2	Synthesis of Nitroacetic Acid Ethyl Ester and Its Risk Assessment *374*	

18.5	Development of an Efficient Method for the Synthesis of Nitrobenzene Derivatives *376*	
	References *380*	

19	**Scientific Strategy for Optical Resolution by Salt Crystallization: New Methodologies for Controlling Crystal Shape, Crystallization, and Chirality of Diastereomeric Salt** *381*	
	Rumiko Sakurai and Kenichi Sakai	
19.1	Introduction *381*	
19.2	Control of Crystal Shape: Crystal Habit Modification *382*	
19.2.1	Significance of Crystal Shape in Industrial-Scale Production *382*	
19.2.2	Effective Additive for Controlling Crystal Shape *384*	
19.2.3	Mechanism of Crystal Habit Modification *385*	
19.3	Control of Crystallization: Concept of Space Filler *387*	
19.3.1	Evaluation of Molecular Size *387*	
19.3.2	Concept of Space Filler *388*	
19.3.3	Resolution of MMT *388*	
19.3.4	Crystal Structures of the Salts *390*	
19.4	Control of Chirality: Dielectrically Controlled Optical Resolution (DCR) *391*	
19.4.1	DCR *391*	
19.4.1.1	DCR in Resolution of (*RS*)-ACL with (*S*)-TPA *391*	
19.4.1.2	DCR in Resolution of (*RS*)-PTE with (*S*)-MA *394*	
19.5	Conclusion and Prospect *397*	
	References *397*	

20	**Development of New Drug and Crystal Polymorphs** *401*	
	Mitsuhisa Yamano	
20.1	Introduction *401*	
20.2	Scope of Crystal Polymorphs *402*	
20.3	Late-Appearing Polymorphs *402*	
20.4	Late-Appearing Polymorphs as a Process Research Issue *403*	
20.5	Drug Substance Form Selection *404*	
20.6	Polymorph Screening *405*	
20.7	Thermodynamically Stable Polymorphs *406*	
20.7.1	One-Component System *406*	
20.7.2	Multicomponent System *406*	
20.8	Polymorph Control *409*	
20.8.1	Nucleation and Seeding *411*	
20.8.2	Ostwald's Stage Rule *411*	
20.8.3	Unintentional Seeding *416*	
20.9	Primary Nucleation *416*	
20.10	Summary *417*	
	Acknowledgments *418*	
	References *418*	

21		**Development of LIPOzymes Based on Biomembrane Process Chemistry** *421*
		Hiroshi Umakoshi, Toshinori Shimanouchi, and Ryoichi Kuboi
21.1		Introduction *421*
21.2		From "Process Chemistry" to "Biomembrane Process Chemistry" *422*
21.3		Recognition (Separation) Function of Liposomes *425*
21.4		LIPOzyme: Liposome with Enzyme-Like Activity? *428*
21.4.1		Break-Down Type LIPOzyme *428*
21.4.2		Build-Up Type LIPOzyme *431*
21.5		Biomembrane Interference *433*
21.6		Summary *438*
		Acknowledgments *439*
		References *439*
22		**Matching Chemistry with Chemical Engineering for Optimum Design and Performance of Pharmaceutical Processing** *443*
		Amit V. Mahulkar, Parag R. Gogate, and Aniruddha B. Pandit
22.1		Concept of Molecule to Money *443*
22.2		Steps Involved in Bringing Molecule to Market *444*
22.2.1		Synthesis in Lab *444*
22.2.2		Kilo Lab *445*
22.2.3		Pilot Plant *446*
22.2.4		Full-Scale Plant *447*
22.3		Interrelation in Each Step and Concept of Unit Operations *448*
22.4		Unit Operations *449*
22.4.1		Mixing *449*
22.4.1.1		Utility of Mixing *449*
22.4.1.2		Types of Equipments *449*
22.4.1.3		Selection Criteria *451*
22.4.2		Crystallization *452*
22.4.2.1		Utility of Crystallization *452*
22.4.2.2		Types of Equipments *452*
22.4.2.3		Selection Criteria *453*
22.4.3		Filtration and Centrifugation *454*
22.4.3.1		Utility of Filtration *454*
22.4.3.2		Types of Equipments *455*
22.4.3.3		Selection Criteria for Filtration *457*
22.4.4		Centrifugation *458*
22.4.4.1		Utility of Centrifugation *458*
22.4.4.2		Types of Equipments *458*
22.4.4.3		Selection Criteria *460*
22.4.5		Drying *460*
22.4.5.1		Utility of Drying *460*
22.4.5.2		Types of Equipments *461*

22.4.5.3	Selection Criteria 463
22.5	Scale-up Problems 464
22.6	Optimization and Intensification of Unit Operations 464
22.6.1	Step I: Establishing Material and Energy Balance across All the Equipments 465
22.6.2	Step II: Preliminary Evaluation of All the Major Equipments 465
22.6.3	Step III: Analysis of All the Operations and Equipments at Relatively Intricate Levels 465
22.6.4	Step IV: Mathematical Modeling of Individual Equipment and Subsequently the Entire Plant 465
22.7	Summary 466
	References 466
23	**The Integration of Safety, Health, and Environmental Considerations into Process Development** 469
	Wesley White, Vyv Coombe, and Jonathan Moseley
23.1	Introduction 469
23.1.1	The SHE Triggers Model 469
23.1.2	Introduction to AZD4619 473
23.2	Process Safety 474
23.2.1	Assessment of Chemical Reaction Hazards 475
23.2.2	Assessment of Operational Hazards 475
23.2.3	Basis of Safety 475
23.2.4	Process Safety Triggers 476
23.2.4.1	Early Delivery 476
23.2.4.2	Later Process Safety Triggers 476
23.2.4.3	Technology Transfer 477
23.2.5	Application of Process Safety to AZD4619 477
23.2.5.1	General Hazards of Aqueous Diazotization and the Basis of Safety 477
23.2.5.2	Additions of Acetone, Water, Hydrochloric Acid, and Acrylic Acid 478
23.2.5.3	Addition of Sodium Nitrite 478
23.3	Health 479
23.3.1	Introduction 479
23.3.2	Health Triggers 479
23.3.2.1	Early Delivery 479
23.3.2.2	Synthetic Route Evaluation 480
23.3.2.3	Process Design 480
23.3.2.4	Process Optimization and Understanding 481
23.3.2.5	Technology Transfer 481
23.3.3	Application of Health Triggers to AZD4619 482
23.4	Environment 482
23.4.1	Introduction 482
23.4.2	Environment Triggers 482

23.4.2.1	Early Delivery and Synthetic Route Evaluation	*482*
23.4.2.2	Process Design	*484*
23.4.2.3	Process Optimization and Understanding	*484*
23.4.2.4	Technology Transfer	*485*
23.4.3	Application of Environmental Triggers to AZD4619	*485*
23.5	The Use of Risk Assessment	*486*
23.6	Conclusion	*487*
	Acknowledgments	*488*
	References	*488*

Index *489*

Preface

Pharmaceutical Process Chemistry is a very important field bridging medicinal chemistry and the industrial and commercial production of medicines. Although medicinal chemistry research is essential for finding novel innovative medicines and active pharmaceutical ingredients (APIs), neither medicines nor API can be made available to the world without active progress in pharmaceutical process chemistry. Further growth in this field should promote rapid progress in the development of new medicines.

The main purpose of medicinal chemistry is the discovery of pharmacologically active molecules at even a milligram scale while pharmaceutical process chemistry is concerned with the scale-up process for the production of useful molecules from milligrams to kilograms or even tons. Thus, pharmaceutical process chemistry has the following notable features:

1) selection of inexpensive and easily available starting materials in large quantities;
2) utilization of inexpensive catalysts and/or reagents and solvents;
3) establishment of robust and speedy procedures for producing drug candidates and API with high quality;
4) development of methods to produce drug candidates and API in an economical, convenient, and efficient manner;
5) avoidance of dangerous procedures and hazardous reagents;
6) selection of safer and environmentally friendly processes;
7) reduction of wastes.

Although synthetic & bioorganic chemistry currently plays a central role in pharmaceutical process chemistry, pharmaceutical process chemistry holds the hub position in relation to various other sciences including separation techniques, analytical science, chemical engineering, environmental science, pharmaceutical engineering, regulatory science, intellectual property, and so on, as shown in the figure.

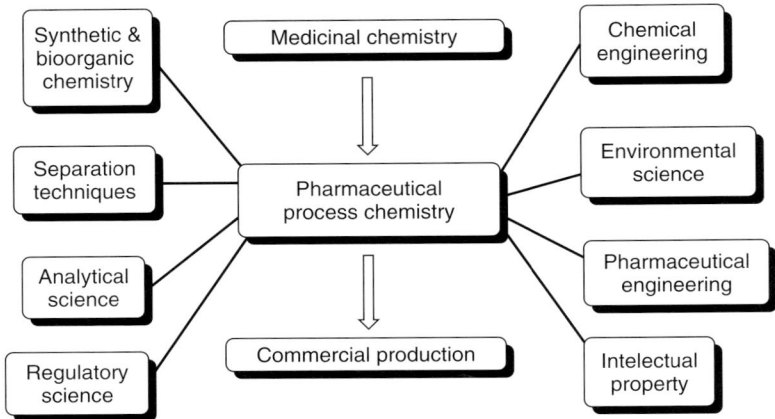

This book intends to cover the broad area of pharmaceutical process chemistry, and should be useful not only for process chemists but also for academics and students. Although there has been a recent tendency in medicines to proceed "From Small Molecules to Macromolecules," small synthetic molecules still have an important role in therapeutics. Thus, the introductory chapter addresses the importance of pharmaceutical process chemistry based on synthetic organic chemistry in the development of new small-molecule API. Various forefront synthetic methodologies that will be useful in the future growth of pharmaceutical process chemistry are then discussed. These methods may be applicable to process chemistry and may be useful for the production of new drug candidates and API, either as they stand or after the development of a suitable device for large-scale synthesis. The third part of this book addresses useful industrial synthetic and biochemical technologies and the fourth part discusses practical processes for the production of new medicines and API. The methods discussed should provide useful information for both process and academic chemists who are working on their own problems. The final part addresses aspects of this field, other than synthesis, that are useful for the production of new medicines: treatment of explosive reactions, crystal engineering, use of biomembranes, safety and environmental considerations, and so on.

All of the authors were selected from among the leaders in each area of specialization, and many of them have been invited to speak at symposia organized by the Japanese Society for Process Chemistry (JSPC) (*http://www.jspc-home.com/process/index.html*).

JSPC was founded in 2001 after several symposia and discussions between process chemistry directors in industry and university professors. JSPC aims to promote and encourage the development of process chemistry in various directions among members. JSPC has regular symposia twice a year in the summer and winter. Members of JSPC are affiliated with industry, especially the pharmaceutical industry, manufacturers of intermediates, universities, and so on. The number of participants in symposia continues to increase yearly. JSPC believes that it can play an important role in advancing and improving process chemistry.

Finally, we would like to express our gratitude to all of the contributors to this volume for their hard and ingenious work, and the reviewers of each chapter for their valuable comments and critical feedback. We want to thank the members of JSPC for their support and encouragement. We also want to thank the staff at Wiley for their outstanding editorial management.

Takayuki Shioiri
Kunisuke Izawa
Toshiro Konoike
(On behalf of JSPC)

List of Contributors

Takao Abe
Medicinal Chemistry
Research Laboratory
CMC Research Laboratories
Pharmaceutical Research Center
Meiji Seika Kaisha, Ltd.
760 Morooka-cho
Yokohama 222-8567
Japan

Elizabeth R. Burkhardt
BASF Corporation
1424 Mars-Evans City Road
Evans City, PA 16033
USA

Vyv Coombe
AstraZeneca
Brixham Environmental
Laboratory
Freshwater Quarry
Brixham
Devon TQ5 8BA
United Kingdom

Parag R. Gogate
Institute of Chemical Technology
Chemical Engineering
Department
N. P. Marg
Mumbai 400 019
India

Mitsuharu Hanada
Pharmacology Research Labs
Dainippon Sumitomo
Pharma Co., Ltd.
3-1-98 Kasugade-naka
Konohana-Ku
Osaka 554-0022
Japan

Tamejiro Hiyama
Kyoto University
Department of Material
Chemistry
Graduate School of Engineering
Nishikyo-Ku
Katsura
Kyoto 615-8510
Japan

Takashi Inaba
Central Pharmaceutical
Research Institute
Japan Tobacco Inc.
1-1 Murasaki-cho
Takatsuki
Osaka 569-1125
Japan

Kazuaki Ishihara
Nagoya University
Graduate School of Engineering
Nagoya 464-8603
Japan

Pharmaceutical Process Chemistry. Edited by Takayuki Shioiri, Kunisuke Izawa, and Toshiro Konoike
Copyright © 2011 WILEY-VCH Verlag GmbH & Co. KGaA, Weinheim
ISBN: 978-3-527-32650-1

List of Contributors

Martin Karpf
F. Hoffmann-La Roche Ltd
Synthesis and Process Research
pRED, Therapeutic Modalities
Grenzacherstrasse 124
4070 Basel
Switzerland

Tsutomu Katsuki
Kyushu University
Department of Chemistry
Faculty of Science
Graduate School
6-10-1 Hakozaki
Higashi-Ku
Fukuoka 812-8581
Japan

Hiromu Kawakubo
Asahi Kasei Chemicals
Corporation
New Business Group
Explosive Division
1-105 Kanda
Jinbo-Cho
Chiyoda-Ku
Tokyo 101-8101
Japan

Masataka Kitamura
Medicinal Chemistry
Research Laboratory
CMC Research Laboratories
Pharmaceutical Research Center
Meiji Seika Kaisha, Ltd.
760 Morooka-cho
Yokohama 222-8567
Japan

Toshiro Konoike
CMC Development Laboratories
Shionogi & Co., Ltd.
1-3 Kuise-terajima 2-chome
Amagasaki 660-0813
Japan

Ryoichi Kuboi
Osaka University
Division of Chemical Engineering
Graduate School of Engineering
Science
1-3 Machikaneyama-cho
Toyonaka, Osaka 560-8531
Japan

Delong Liu
Shanghai Jiao Tong University
School of Pharmacy
800 Dongchuan Road
Shanghai 200240
China

Amit V. Mahulkar
Institute of Chemical Technology
Chemical Engineering
Department
N. P. Marg
Mumbai 400 019
India

Karl Matos
BASF Corporation
1424 Mars-Evans City Road
Evans City, PA 16033
USA

Oliver May
DSM Innovative Synthesis B.V.
PO Box 18
6160 MD Geleen
The Netherlands

Yasunari Monguchi
Gifu Pharmaceutical University
Laboratory of Organic Chemistry
1-25-4 Daigaku-nishi
Gifu 501-1196
Japan

Jonathan Moseley
AstraZeneca
Avlon Works
Bristol BS10 7ZE
United Kingdom

Yoshiaki Nakao
Kyoto University
Department of Material
Chemistry
Graduate School of Engineering
Nishikyo-Ku
Katsura
Kyoto 615-8510
Japan

Tadaaki Ohgi
Nippon Shinyaku Co., Ltd
Discovery Research Laboratories
3-14-1 Sakura
Tsukuba City
Ibaraki 305-0003
Japan

Takashi Okazoe
AGC Chemicals
Asahi Glass Co., Ltd.
Business Management
General Division
1-12-1 Yuraku-cho
Chiyoda-ku
Tokyo 100-8405
Japan

Aniruddha B. Pandit
Institute of Chemical Technology
Chemical Engineering
Department
N. P. Marg
Mumbai 400 019
India

Hironao Sajiki
Gifu Pharmaceutical University
Laboratory of Organic Chemistry
1-25-4 Daigaku-nishi
Gifu 501-1196
Japan

Kenichi Sakai
Toray Fine Chemicals Co. Ltd.
Technology Development
Division
Nagoya
Aichi 455-8502
Japan

Rumiko Sakurai
Iwaki Meisei University
Faculty of Pharmacy
Iwaki
Fukushima 970-8044
Japan

Michael Schwarm
Evonik Degussa GmbH
Business Line Exclusive
Synthesis & Amino Acids
Rodenbacher Chaussee 4
63457 Hanau-Wolfgang
Germany

Roger A. Sheldon
Delft University of Technology
Biocatalysis and Organic
Chemistry
Julianalaan 136
2628 BL Delft
The Netherlands

Toshinori Shimanouchi
Osaka University
Division of Chemical Engineering
Graduate School of Engineering Science
1-3 Machikaneyama-cho
Toyonaka
Osaka 560-8531
Japan

Sumio Shimizu
CMC Development Laboratories
Shionogi & Co., Ltd.
1-3 Kuise-terajima 2-chome
Amagasaki 660-0813
Japan

Kazuhiko Takahashi
Process Chemistry R&D Labs
Dainippon Sumitomo
Pharma Co., Ltd.
1-5-51 Ebie
Fukushima-ku
Osaka 553-0001
Japan

Hiroshi Umakoshi
Osaka University
Division of Chemical Engineering
Graduate School of Engineering Science
1-3 Machikaneyama-cho
Toyonaka
Osaka 560-8531
Japan

Andrew S. Wells
AstraZeneca
Global Process R&D
42/2/2.0 Bakewell Road
Loughborough
Leicestershire LE11 5RH
United Kingdom

Wesley White
AstraZeneca
Milford 2-06
Macclesfield Works
Silk Road Business Park
Charter Way
Macclesfield
Cheshire SK10 2NA
United Kingdom

Mitsuhisa Yamano
Takeda Pharmaceutical
Company Limited
Chemical Development
Laboratories
CMC Center
17-85 Jusohonmachi 2-Chome
Osaka 532-8686
Japan

Junichi Yano
Nippon Shinyaku Co, Ltd.
Discovery Research Laboratories
3-14-1 Sakura
Tsukuba City
Ibaraki 305-0003
Japan

Wanbin Zhang
Shanghai Jiao Tong University
School of Chemistry and
Chemical Technology
800 Dongchuan Road
Shanghai 200240
China

1
From Milligrams to Tons: The Importance of Synthesis and Process Research in the Development of New Drugs
Martin Karpf

1.1
Introduction

Synthetic chemistry plays a key role in the multidisciplinary development process of new small molecule pharmaceuticals. In this context, organic synthesis is not only the essential tool to find potential drug candidate molecules but is also in charge of the subsequent creation, exploration, and evaluation of short, efficient, safe, reproducible, scalable, ecological but still economical syntheses for the selected clinical candidates. This second activity generally named *synthesis and process research* or just *process research* is the indispensable link between discovery chemistry and technical development heading toward future large-scale industrial production. In addition to solving the gradually rising synthetic problems associated with the ever increasing structural complexity of new potential drug molecules, the resulting synthesis has to show technical potential and has, particularly, to take into account the basic requirements and limitations of a prospective technical process.

In this chapter, the role and importance of synthesis and process research in the development process of new drugs from discovery chemistry (medicinal chemistry) synthesis up to the technical route will be outlined and exemplified with specific examples also taking into account large-scale production requirements. The chapter concentrates on the synthetic strategies and tactics applied to drug candidates in order to create efficient chemical syntheses with technical potential suitable for further technical optimization aiming at the large-scale industrial production of new pharmaceuticals.

Owing to the permanently changing environment of the pharmaceutical industry and the tremendous advancements of science, neither general rules nor final or permanent principles and recipes for the successful transformation of a synthetic process from milligrams to tons or from discovery chemistry to production can be provided. Synthetic organic chemists know too well that molecules frequently behave incalculably, and that they are usually hard to control and therefore deserve individual treatment.

Pharmaceutical Process Chemistry. Edited by Takayuki Shioiri, Kunisuke Izawa, and Toshiro Konoike
Copyright © 2011 WILEY-VCH Verlag GmbH & Co. KGaA, Weinheim
ISBN: 978-3-527-32650-1

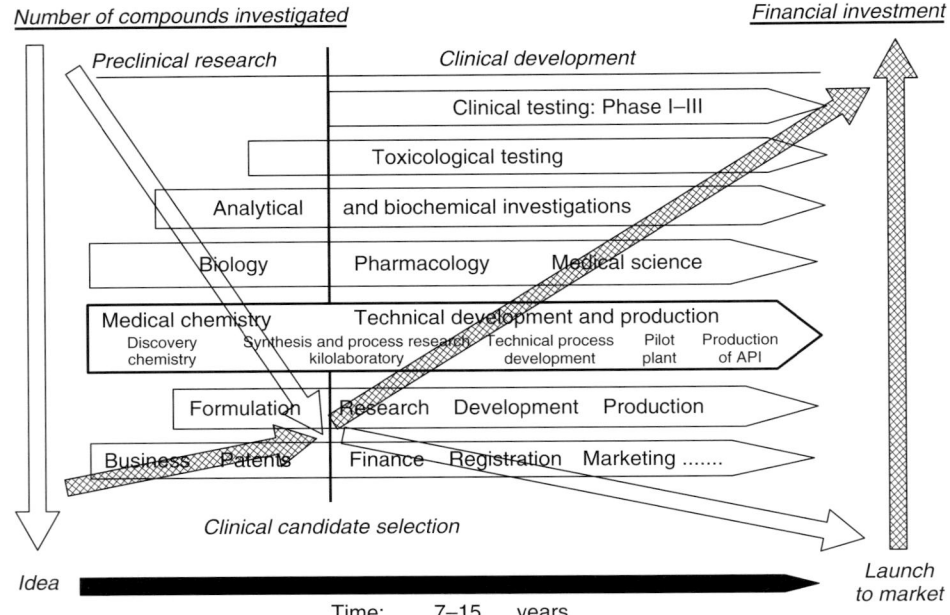

Figure 1.1 The drug development process (overview).

After a short introduction to the environment of chemists working in a synthesis and process research environment, the topic will be illustrated by four specific examples of innovative pharmaceuticals on their way to production for the market.

The classical development process of a new drug is sketched as a very coarse overview in Figure 1.1 and starts with the idea of which disease to treat and – in the best case – ends up with the introduction of a new pharmaceutical to the market. This multifaceted process easily takes 7–15 years, requires financial investments of up to $1–$2 billion and starts with a high number of up to many thousand compounds to be tested parallely by sophisticated methods such as high-throughput screening to finally come up with one or two clinical candidates to be further evaluated. In this overall research and development process, synthesis and process research – together with a scale-up or a kilolaboratory – represents a central activity at the important borderline linking preclinical research with clinical development at the stage of the clinical candidate selection. At this point in time, the so-called "clinical candidate" molecule emerging from discovery chemistry is selected, and first larger, sometimes up to kilogram amounts are immediately required to start clinical development and all related activities mentioned in Figure 1.1 including extended toxicological programs, analytical and biochemical investigations, as well as formulation research and development to find the appropriate pharmaceutical dosage form for the potential drug.

The role of synthesis and process research at this stage is twofold; namely, first to support the scale-up or kilolaboratory in troubleshooting and scaling up of the discovery chemistry route to allow for an initial small-scale production of the new drug candidate as quickly as possible, material urgently required to start the clinical development activities mentioned in Figure 1.1.

If the project continues – but the attrition rate at this point is still high – this activity is followed by a partial or full synthetic redesign to finally identify a synthetic route with technical potential to be handed over to the technical development department, which then has the task to transform the new synthesis into an efficient production process that will later be used for the manufacture of the active pharmaceutical ingredient (API) in commercial amounts.

An important question that regularly comes up is: why is synthesis and process research needed since there already exists a synthesis established by discovery chemistry about the way to find the clinical candidate? The answer to this question is emblematized in Figure 1.2 and relates to the different synthetic strategies applied by discovery chemistry on the one hand against synthesis and process research on the other hand.

The goal of discovery chemistry – the task of medicinal chemists – is to synthesize as many new compounds as quickly as possible, which will then be tested by biologists against the chosen biological target. Therefore, the synthetic strategy of discovery chemistry is an overall diversity-oriented process, allowing for finding

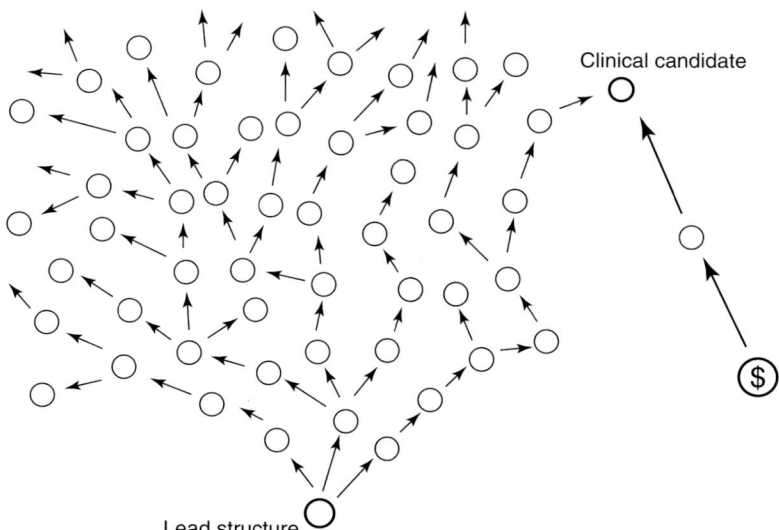

Figure 1.2 Synthetic strategies used by drug discovery versus synthesis and process research.

Prof. Ryoji Noyori
Nobel Laureate 2001

"Chemical synthesis with practical elegance"

Key requirements:

Absolute efficiency using perfect chemical reactions
▶ 100% selectivity & 100% yield
Economical processes ▶ No unwanted wastes
Environmentally friendly ▶ Resource and energy-saving

"The need for efficient and practical synthesis remains one of the greatest intellectual challenges with which chemists are faced in the 21st Century"

Figure 1.3 Chemical synthesis with practical elegance [1].

access to a large number of new compounds as quickly as possible. Starting from a hit compound obtained by high-throughput screening, automated parallel chemistry, or related techniques, lead structures are selected, which will further be optimized regarding the key parameters using multidimensional optimization to finally reach viable clinical candidates, which meet the criteria set regarding activity, selectivity, toxicology, safety, and so on. Despite all the modern and rational methods of contemporary drug discovery, a large number of compounds still have to be synthesized and tested.

After the identification of a new clinical candidate, the situation regarding the synthetic strategy changes entirely from diversity to target orientation. The task now is to create for the selected clinical candidate molecule, a specific synthesis with the potential to be later technically developed by the technical development department into a large-scale production process.

In this context, it is motivating for synthetic organic chemists working in the field of synthesis and process research and related departments to recall the definition of a "synthesis with practical elegance" introduced by Prof. Noyori, Nobel laureate, 2001 (Figure 1.3). Although the key requirements for a "synthesis with practical elegance" are highly challenging regarding efficiency, environmental impact, and economy, it is the obligation of responsible chemists to intensively and persistently strive for these goals.

To illustrate the synthetic development process for new pharmaceuticals at F. Hoffmann-La Roche Ltd, Basel, Switzerland, the four examples shown in Figure 1.4 will be discussed and commented, examples already patented and/or published for which syntheses with technical potential were created. Three of them (Xenical™, Saquinavir™, Tamiflu™) finally reached the market place. Tempium, a monoamine oxidase type B (MAO-B) inhibitor, which finally was dropped, but is included in the discussion since substantial amounts have already been produced according to the final one-step process starting off from a nine-step discovery synthesis. Taking into account the fundamental differences of the synthetic strategies, it is

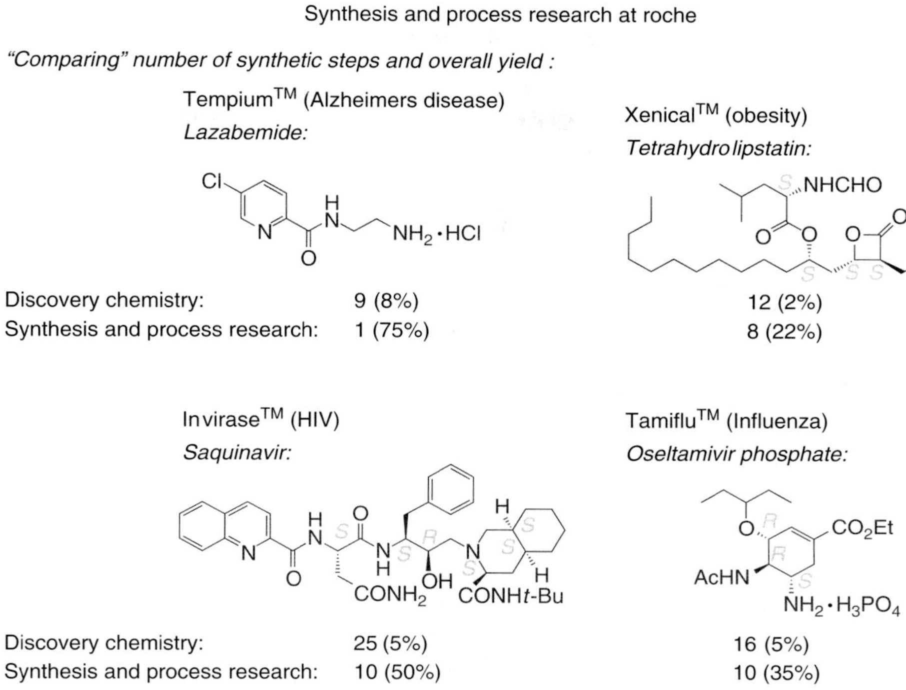

Synthesis and process research at roche

"Comparing" number of synthetic steps and overall yield :

Tempium™ (Alzheimers disease)
Lazabemide:

Discovery chemistry:	9 (8%)
Synthesis and process research:	1 (75%)

Xenical™ (obesity)
Tetrahydrolipstatin:

Discovery chemistry:	12 (2%)
Synthesis and process research:	8 (22%)

Invirase™ (HIV)
Saquinavir:

Discovery chemistry:	25 (5%)
Synthesis and process research:	10 (50%)

Tamiflu™ (Influenza)
Oseltamivir phosphate:

Discovery chemistry:	16 (5%)
Synthesis and process research:	10 (35%)

Figure 1.4 Selected examples of synthesis and process research at F. Hoffmann-La Roche Ltd, Basel, Switzerland.

very important to stress that a "comparison" of syntheses between the discovery chemistry route and the resulting synthesis created by synthesis and process research only with regard to the number of steps and overall yields is not admissible, since the key tasks, goals, and strategies of both areas are fundamentally different. As discussed above, the key task of discovery chemistry is to synthesize in a diversity-oriented manner – and as fast as possible – small amounts of new, biologically active molecules to be tested as potential clinical candidates. To this end, all available and sophisticated synthetic methods and separation techniques of modern organic chemistry should be applied. In contrast and after clinical candidate selection, it is the distinguished task of synthesis and process research chemists to create and evaluate in a target-oriented manner, a synthesis with technical potential for the selected clinical candidate to subsequently be technically developed for large-scale production.

Since a detailed discussion on the chemistry of all these four projects is out of scope of this review, the focus has been on the strategic principals and solutions, which has finally led in all four cases to practical technical solutions. Additional details are easily accessible through the chemical literature and through patents referred to in the corresponding schemes.

1.2
The Synthetic Development of the Monoamine Oxidase-B Inhibitor Lazabemide™

Lazabemide, an MAO-B inhibitor developed for the treatment of Alzheimer's disease, was first synthesized following a classical pathway starting from cheap "aldehyde collidine" (5-ethyl-2-methylpyridine). The key intermediate was 5-chloro-2-picolinic acid that was further converted to a wide variety of the corresponding amides to finally select the 2-aminoethyl-amide as the clinical candidate. The synthesis completed in a nine linear sequence and about 2–8% overall yield in the discovery chemistry stage, as shown in Scheme 1.1.

This classical approach required about 26 kg of the starting material to produce 1 kg of the active substance. An additional immense challenge for the kilolaboratory was the barely selective permanganate oxidation of the starting material at the first step, which was confronted with the cumbersome filtration of a large amount of manganese dioxide.

Although a troubleshooting of the discovery chemistry route enhanced the overall yield to about 10%, it was not essential for further scaling up. Efforts on the search for an alternative synthesis allowed the catalysis group of synthesis and process research to identify commercially available 2,5-dichloropyridine as an ideal starting material, which underwent Pd-catalyzed Sonogashira reaction with various acetylenes to afford the corresponding acetylenic pyridines in a highly efficient and selective manner. Although a permanganate oxidation was still required to gain access to the 5-chloro-picolinic acid intermediate, the overall yield of this scalable four-step process was already improved to 58% and only 1.1 kg of the starting material was required.

Finally, a direct, one-step amido carbonylation process was introduced using the same starting material, which provided the API in one step and 75% yield. The reaction was later developed to the 100 kg scale, requiring only 0.8 kg of the starting material to obtain 1 kg of the API.

As shown in Scheme 1.1, the reaction with an excess of ethylenediamine and carbon monoxide at 10 bar using only 0.1 mol% of the Pd(0) precursor and a phosphine ligand in toluene at reflux led after appropriate workup directly to the pure API in 75% yield. Interestingly, ethylenediamine used in excess acted as a reagent and as the base without deactivating the catalytically active Pd(0) species.

Although a one-step solution looks very favorable from a synthesis and process research chemist's point of view, the potential issues regarding the registration of such a short approach with the health authorities should not be neglected.

1.3
The Synthetic Development of the Lipase Inhibitor Tetrahydrolipstatin (Xenical™)

The second case to be discussed concerns tetrahydrolipstatin, a very potent and irreversible inhibitor of pancreatic lipase found and developed at Roche for the

1.3 The Synthetic Development of the Lipase Inhibitor Tetrahydrolipstatin (Xenical™)

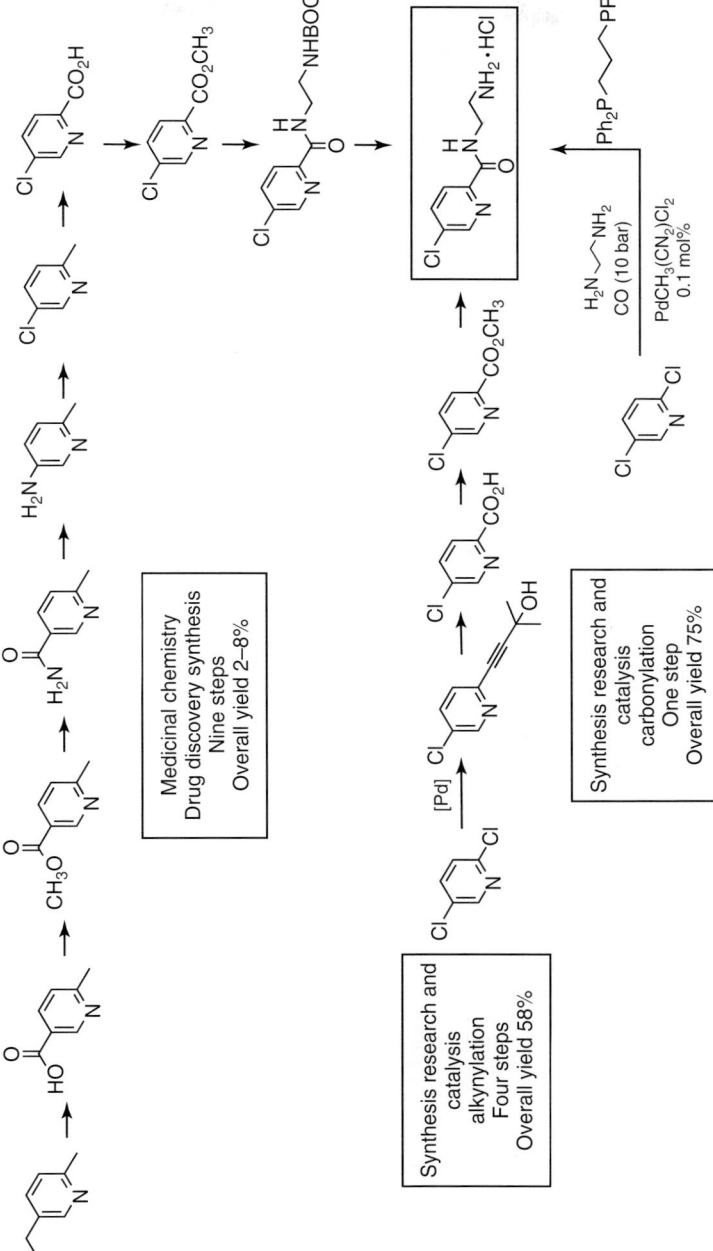

Scheme 1.1 The synthetic development of the monoamine oxidase-B inhibitor, lazabemide [3].

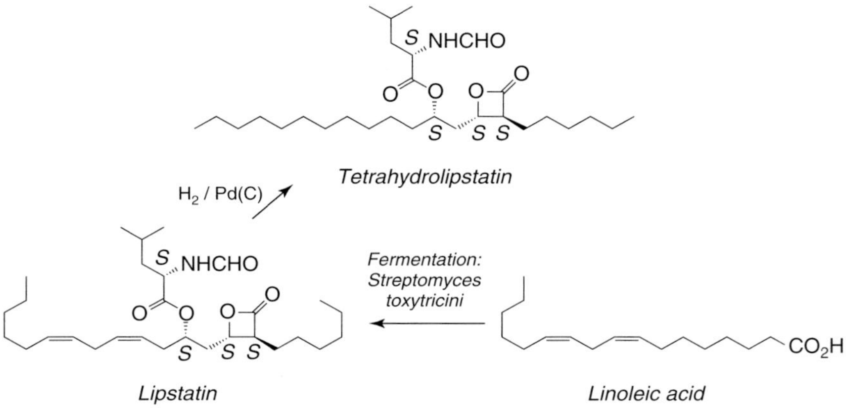

Scheme 1.2 Tetrahydrolipstatin [4].

treatment of overweight is shown in Scheme 1.2. This molecule represents the tetrahydro derivative of the natural product lipstatin, a secondary metabolite produced by *Streptomyces toxytricini*. Tetrahydrolipstatin as well as its natural counterpart lipstatin effectively inhibit the hydrolysis of triglycerides in food, thereby reducing digestion and uptake of dietary fats.

Although it finally became possible to produce lipstatin by fermentation using a mutant strain and employing linoleic acid as an auxiliary material, the search for an organic chemical synthesis with technical potential was first allowing for the production of the API at many a hundred ton scale.

From a synthetic standpoint, two intriguing structural features have to be synthetically controlled, also at large scale, namely a rather labile and quite reactive trans-substituted β-lactone moiety embedded in a C21 aliphatic chain and four stereocenters. Only one of them is commercially available in high optical purity in the form of the amino acid, leucine. Therefore the three stereocenters embedded in the aliphatic C21 chain had to be created by stereoselective synthesis.

From a stability standpoint it was essential to introduce the rather labile β-lactone moiety toward the end of the synthesis. Therefore, all practical syntheses of tetrahydrolipstatin proceeded through a common intermediate, the 2S,3S,5R-configured α-hexyl-β-hydroxy-δ-benzyloxy acid, shown in Scheme 1.3.

This key intermediate was then converted to the final product tetrahydrolipstatin by β-lactone formation, followed by debenzylation and introduction of the *N*-formyl leucine side chain under Mitsunobu conditions with complete inversion at the reacting stereo center.

The control of configuration of the three stereocenters in the chain was first achieved using two independent stereoselective transformations; namely, first an enantioselective hydrogenation of the starting β-keto-ester followed by O-benzylation and reduction to the aldehyde. Second, the aldehyde was condensed at low temperature in a Mukayama-type antialdol manner according to

1.3 The Synthetic Development of the Lipase Inhibitor Tetrahydrolipstatin (Xenical™)

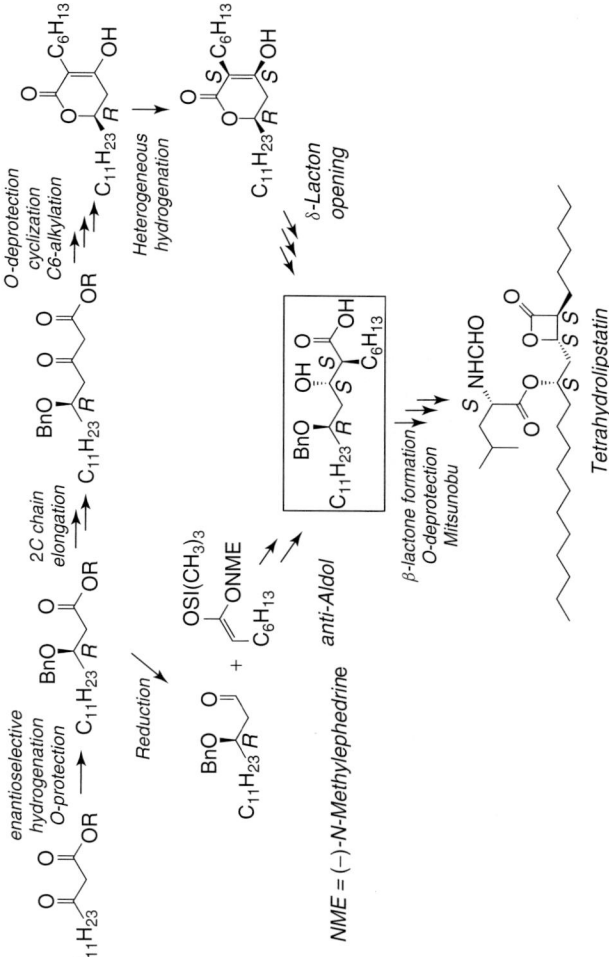

Scheme 1.3 Early access to tetrahydrolipstatin [6, 7].

Gennari and Scolastico with the (trimethylsilyl) TMS-ketene acetal bearing the N-methylephedrine auxiliary group to induce the two additional stereo centers in the required absolute configuration.

From a technical standpoint, this 12-step discovery route, however, is clearly not deemed for future technical purposes, not only due to the large number of transformations involved including low temperature steps, but also due to the large number of chromatographic purification steps required and leading to a modest overall yield. Nevertheless, in a heroic effort, the kilolaboratory succeeded in producing kilogram amounts of tetrahydrolipstatin according to gradually troubleshooted versions of this initial approach, since the material was urgently needed for starting clinical development.

In order to take advantage of a preexisting stereo center in 3-position of the optically pure β-hydroxy ester to induce additional stereo centers in a diastereoselective manner, synthesis and process research first investigated the path shown on the right section of Scheme 1.3. Starting from the optically pure O-benzylated β-hydroxy ester a two-carbon chain elongation using Masamune's protocol followed by debenzylation, cyclization, and alkylation directly led to the dihydropyrone which now, by heterogeneous hydrogenation over Ra-Ni was transformed to the β-hydroxy-δ-lactone with high diastereoselective induction of the two additional stereo centers. Subsequent opening of the resulting β-hydroxy-δ-lactone unit provided access to the key intermediate, the optically pure α-hexyl-β-hydroxy-δ-benzyloxy acid.

From a technical perspective, this procedure not only had the advantage to be devoid of low temperature reactions but also had the additional advantage that the cyclic intermediates are crystalline compounds and therefore easy to purify by recrystallization. However, this approach – although without chromatographic purifications involved – is even more demanding than the troubleshot discovery route regarding the number of steps involved, already requiring five steps for the access to the dihydropyrone key intermediate.

The problem was finally solved in a very pragmatic way when it was found that access to the racemic dihydropyrone requires only one step that is achievable by addition of the double anion of 2-hexyl-methyl-acetoacetate to laurinic aldehyde followed by spontaneous cyclization at workup as summarized in Scheme 1.4.

Since the diastereoselectivity argument also holds true for the racemic dihydropyrone, the current technical synthesis of tetrahydrolipstatin proceeds through racemates followed by classical optical resolution. Diastereoselective hydrogenation provided the still racemic, but all-cis β-hydroxy-δ-lactone in 90% yield.

Tetrahydropyranyl (THP) protection of the β-hydroxy group followed by lactone ring opening through basic hydrolysis provided the sodium salt of the corresponding acid, which allowed for the selective O-benzylation in δ-position providing the key intermediate as a pure racemate. Optical resolution using economical (−)-phenyl-ethylamine was performed with high yield furnishing the desired α-hexyl-β-hydroxy-δ-benzyloxy acid in optically pure form ready for the final sequence including β-lactone formation, hydrogenolysis, and introduction of

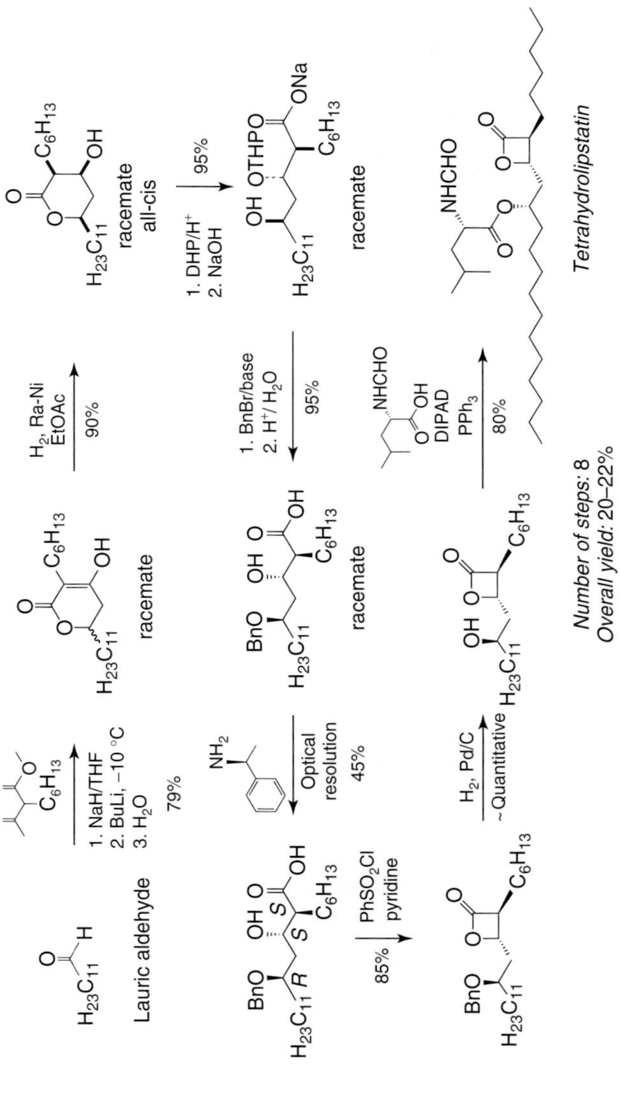

1.3 The Synthetic Development of the Lipase Inhibitor Tetrahydrolipstatin (Xenical™)

Scheme 1.4 The technical synthesis of tetrahydrolipstatin [7].

Number of steps: 8
Overall yield: 20–22%

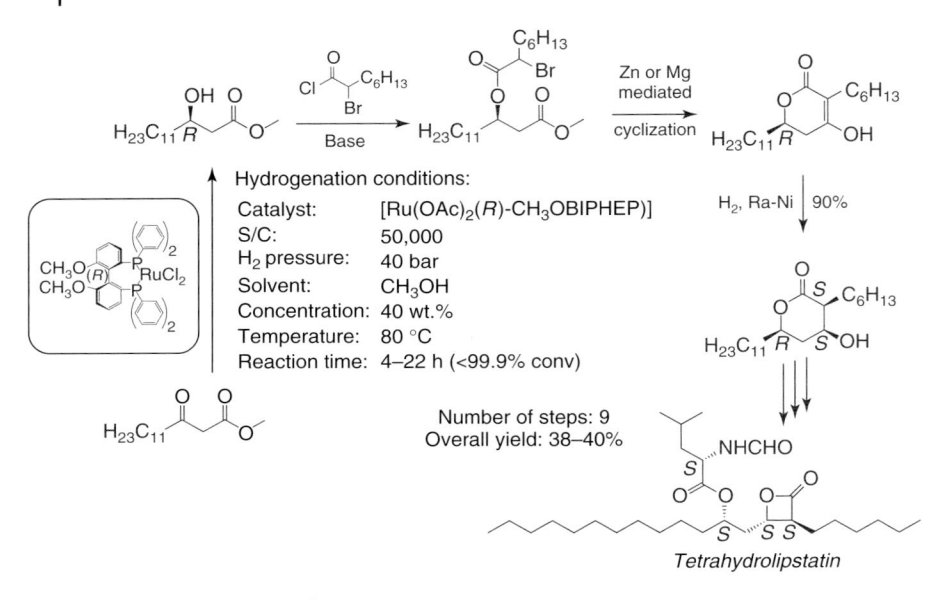

Scheme 1.5 Synthesis of tetrahydrolipstatin via enantioselective access to the optically pure α-hexyl-β-hydroxy-δ-lactone [8].

the N-formyl leucine side chain under Mitsunobu-type conditions with complete inversion of configuration.

This eight-step synthesis was developed by technical development to a large-scale process by which many hundred tons of the active principle have been synthesized so far.

Evidently, optical resolution at an intermediate or even at the late stage of a technical synthesis is clearly not the preferred option regarding efficiency, in the case that the undesired enantiomer cannot be recycled and has to be discarded. Therefore, various routes toward an efficient enantioselective access to the optically active α-hexyl-β-hydroxy-δ-lactone were investigated. A successful approach as shown in Scheme 1.5 is based on earlier findings and started with the O-acylation of the previously mentioned optically pure β-hydroxy ester with α-bromo caprylic acid chloride followed by Zn- or Mg-mediated Reformatzky-type ring closure. Raney nickel hydrogenation then provided the optically active β-hydroxy-δ-lactone moiety ready for the transformation to tetrahydrolipstatin in analogy to the transformation shown above, but without requiring optical resolution.

To this end, a high-yielding and efficient enantioselective hydrogenation procedure for the β-keto ester starting material in order to obtain the optically pure β-hydroxy ester was required. This method was developed by the catalysis group of synthesis and process research and uses a ruthenium catalyst with the Roche owned (R)-CH$_3$OBIPHEP-ligand and applies the conditions summarized in Scheme 1.5 providing the optically pure β-hydroxy ester in almost quantitative yield and a high optical purity of more than 99% ee.

In summary, this enantioselective protocol provided access to tetrahydrolipstatin with a nearly doubled overall yield compared to the racemic approach shown in Scheme 1.4.

1.4
The Synthetic Development of the HIV Protease Inhibitor Saquinavir (Invirase™)

Saquinavir (Figure 1.5) represents a very potent HIV protease inhibitor found and developed at Roche for the treatment of AIDS. The speedy development of this peptidomimetic molecule allowed entering the market first in 1995, followed by Merck's Indinavir™.

A coarse retrosynthetic view presented in Scheme 1.6 depicts the essential starting materials required to be finally combined to the API. Quinaldic acid as well as L-asparagine representing the left-hand part are commercially available in large amounts. Both the isoster subunit, assumed to be mainly responsible for the inhibitory activity by mimicking a peptide bond, as well as the decahydroamide part are derived from another abundant amino acid, namely L-phenylalanine by diastereoselective reactions.

Despite enormous efforts put into the evaluation of alternative enantioselective approaches to the decahydroamide building block, the diastereoselectivity based route evolved as the most efficient strategic one. The overview on the discovery chemistry synthesis presented in Scheme 1.7 summarizes the enormous effort of discovery chemistry in the initial access to the key intermediates, the so-called phthaloyl epoxide and the decahydroamide both derived from L-phenylalanine. Combination of these key fragments by epoxide ring opening was then followed by the stepwise introduction of pentafluorophenol activated L-asparagine and N-hydroxy-succinimidyl activated quinaldic acid to obtain – after 25 steps and about the same number of chromatographic purifications – the active compound in an overall yield of about 5%. It is important to mention at this point that discovery chemistry synthesized several hundred compounds of related structure and a kilolaboratory was able to produce several 10 kg of the API applying a gradually developing route, material urgently required to start clinical development.

After stepwise improvement, a synthesis with technical potential was handed over to the chemists of the technical development department in order to develop the large-scale technical process shown in Scheme 1.8.

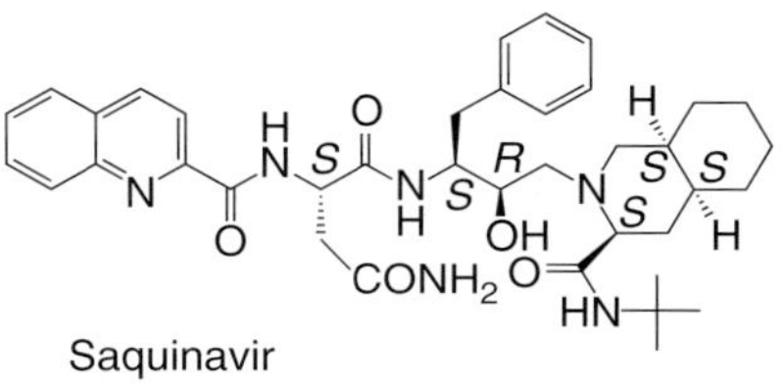

Saquinavir
Invirase™

Figure 1.5 The structure of Saquinavir (Invirase™).

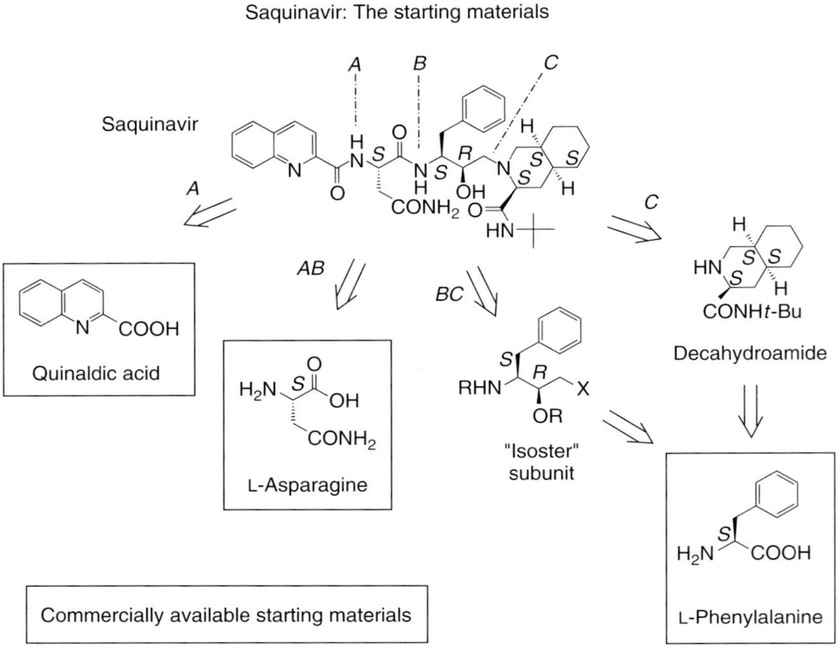

Scheme 1.6 Starting materials for Saquinavir.

The same primary starting material, L-phenylalanine, was used to propagate configuration and induce additional stereo centers. The chlorohydrine, which was employed as the phthaloyl epoxide equivalent as well as the decahydroamide were both accessible in only three steps. Coupling of these building blocks followed by N-deprotection gave access to the so-called amino alcohol building block, which was joined in the last step with quinargine, the coupling product of quinaldic acid and asparagine efficiently obtained via the mixed anhydride of pivalic acid.

The contribution of synthesis and process research to the process regarding ecological factors such as materials and solvents required is summarized in Table 1.1. The amount of materials and solvents hypothetically required to produce 1 metric ton of Saquinavir according to the discovery chemistry synthesis would be huge (calculation based on the experimental description of the discovery chemistry publication presented in Scheme 1.7), whereas these amounts were already considerably reduced by troubleshooting and optimization work toward a first scalable 16-step route providing additional active material required for the further clinical development program. The technically developed process based on the 10-step synthesis uses considerably less materials and solvents. However, it is important to stress that the main contribution of synthesis and process research concerns the evaluation of a short, efficient 10-step synthesis with technical potential. The key contribution to the reduction of materials involved is mainly

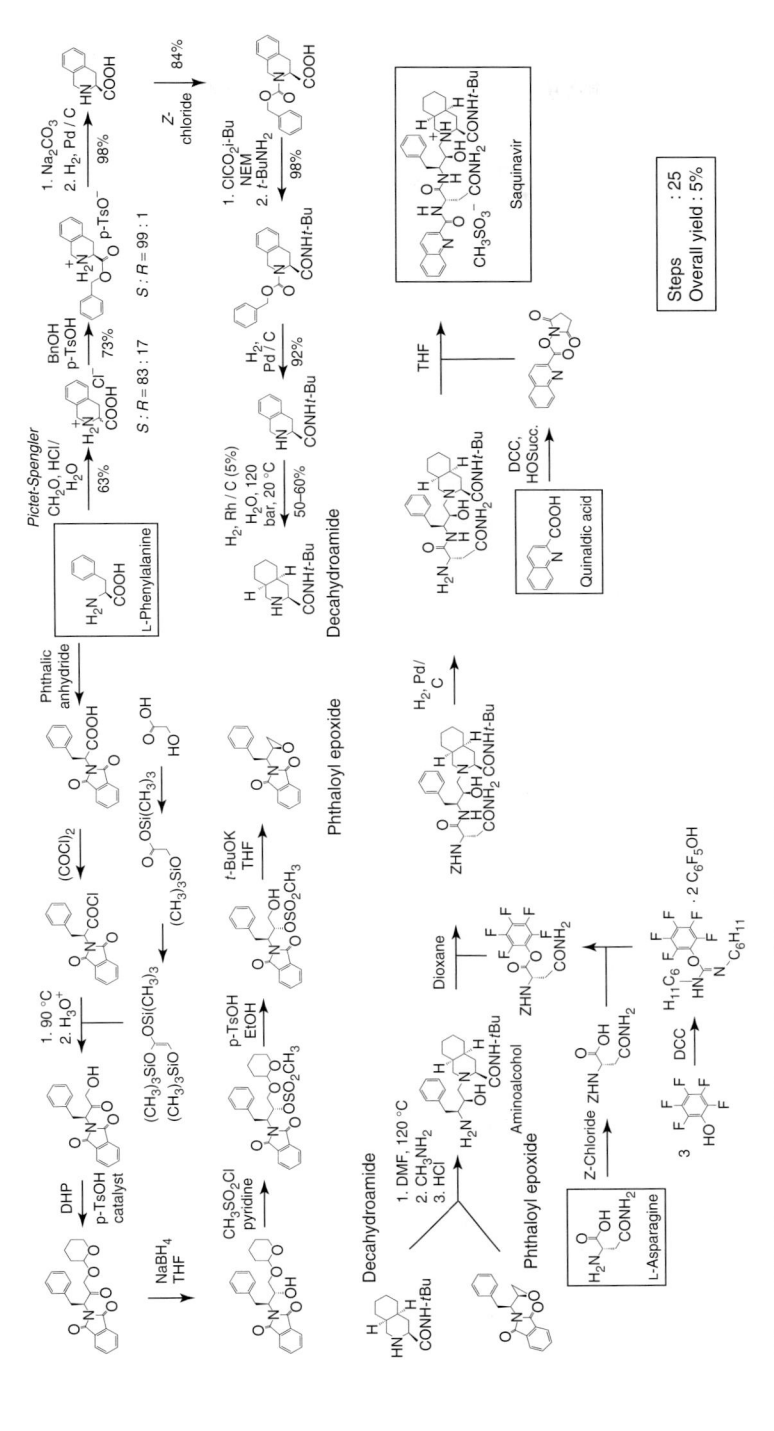

1.4 The Synthetic Development of the HIV Protease Inhibitor Saquinavir (Invirase ™)

Scheme 1.7 Discovery chemistry synthesis of Saquinavir [9, 10].

Table 1.1 Contribution of synthesis and process to the production route of Saquinavir.

Synthesis	Steps	Overall yield (%)	Required for 1 ton active drug	
			Reagents (tons)	Solvents
Discovery route	25	5	700	176
Troubleshooting	25	20	88	23
Scalable synthesis	16	26	80	17
Commercial synthesis	10	50	13	3

based upon the work of technical development chemists by streamlining and optimizing all process steps to the optimum.

1.5
The Synthetic Development of the Influenza Neuraminidase Inhibitor Oseltamivir Phosphate (Tamiflu™)

1.5.1
Introduction

The Roche/Gilead influenza neuraminidase inhibitor oseltamivir phosphate (Tamiflu™) (Figure 1.6), a trisubstituted cyclohexene ethyl carboxylate, is the orally available prodrug of the corresponding acid, which in turn is a very selective and potent inhibitor of influenza neuraminidase at nanomolar concentrations with an ideal half-life of about 3 h. The highly water-soluble phosphate salt is now used for the oral treatment and prevention of influenza virus infections, a disease that affects several million people each winter and providentially, the compound is also active against the H5N1 bird flu as well as the H1N1 swine flu virus that spreadduring 2009.

The inhibitor was found at Gilead Sciences, California, and a codevelopment contract was signed with Roche in 1996 followed by one of the fastest development programs culminating, after only three years of chemical and clinical development, in the launch of Tamiflu in as early as November 1999.

The ambitious program was triggered by the competitive situation since GlaxoSmithKline was concurrently developing their neuraminidase inhibitor zanamivir (Relenza™) (Figure 1.6). Even though their heterocyclic guanidino substituted dihydropyrane carboxylic acid derivative is also a very potent inhibitor of influenza neuraminidase, the compound shows low oral bioavailability and a short half-life

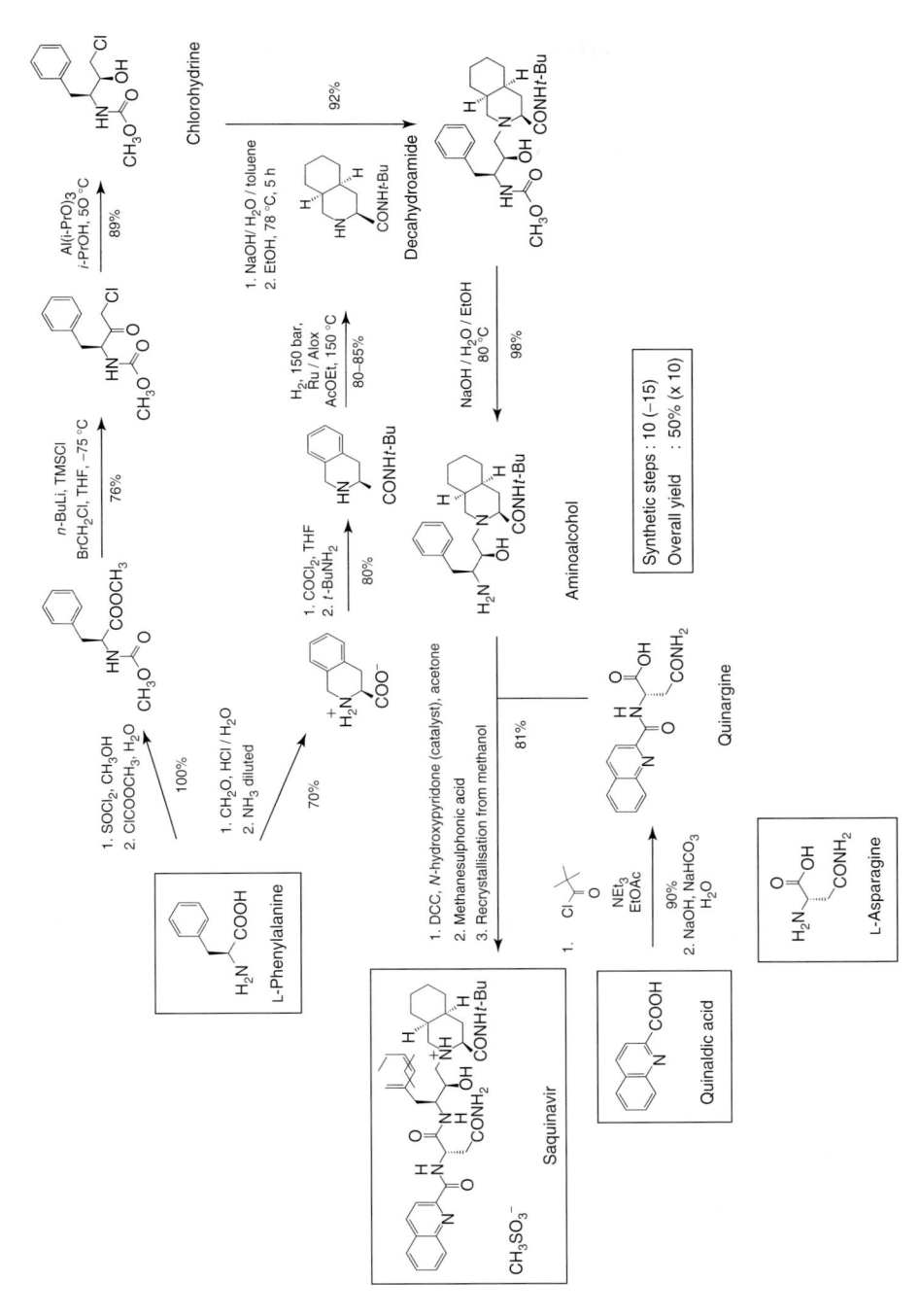

Scheme 1.8 Commercial synthesis of Saquinavir [1].

Oseltamivir phosphate Tamiflu™		Zanamivir Relenza™		Peramivir Rapiacta™ (Japan)	

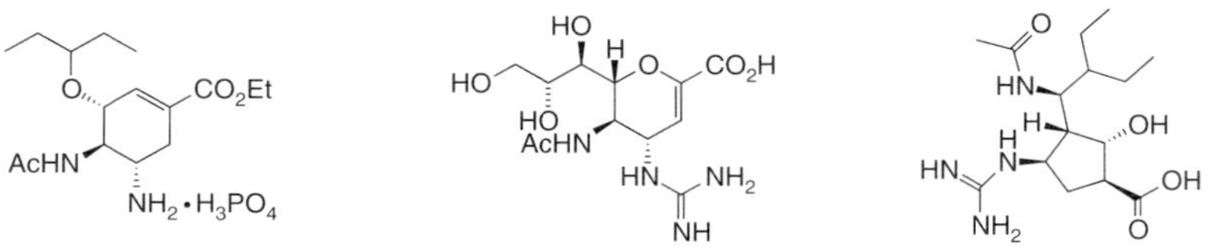

Use	Oral treatment and prevention of influenza virus infections	Use	Topical treatment of influenza virus infections by application via disk inhaler	Use	Treatment of influenza virus infections by intravenous application
Originator:	Gilead Sciences, California	Originator:	Biota holdings, Australia	Originator:	BioCryst
Licensee:	F. Hoffmann-La Roche Ltd.	Licensee:	Glaxo SmithKline	Licensee:	Shionogi and others
Launch:	November 1999	Launch:	July 1999	Launch:	October 2009 (USA, for emergency use) January 2010 (Japan)

Figure 1.6 Marketed anti-influenza neuraminidase inhibitors.

allowing only for its topical application via disk inhaler technology compared to an easy to administer capsule for oseltamivir phosphate. Zanamivir originates from the laboratories of Monash University in Australia and was licensed to GlaxoSmithKline via Biota Holding.

A third compound peramivir (Figure 1.6), found at BioCryst Pharmaceuticals, Inc. entered the US market in October 2009, when the FDA authorized its emergency use as an intravenous antiviral for certain patients and was introduced in January 2010 in Japan by the Shionogi & Co., Ltd under the trade name Rapiacta™.

The influenza neuraminidase represents a viral surface protein with the important role of cleaving the sialic acid end groups of the glycoproteins present on the surface of the infected cell. According to current knowledge this cleavage process allows the newly formed viral particles to escape from the "sialic acid glue" of the infected cells' surface and to infect new host cells. Inhibition of this cleavage process schematically depicted in Scheme 1.9 leads to the aggregation of the emerging viral particles on the surface of the destroyed cell, thereby efficiently stopping the infective cycle. Oseltamivir-free acid as well as zanamivir are thought to be effective mimics of the postulated oxonium-type transition state of this cleavage process.

1.5.2
The Development of the Current Technical Synthesis of Oseltamivir Phosphate

To illustrate the general remarks provided in the introduction regarding the task and role of synthesis and process research, the discussion starts with an overview of the Gilead discovery chemistry route as a typical example demonstrating the

1.5 Synthetic Development of Oseltamivir Phosphate (Tamiflu™)

Scheme 1.9 The role of the neuraminidase in the life cycle of the influenza virus.

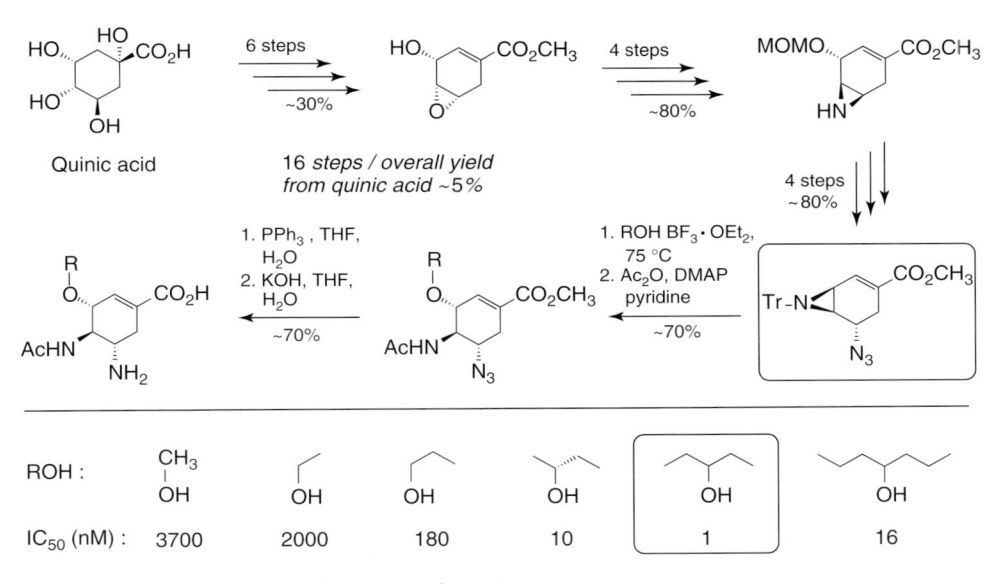

Scheme 1.10 The Gilead drug discovery synthesis [12, 13].

different synthetic strategies of drug discovery on the one hand compared to synthesis and process research on the other.

For discovery chemistry, the trityl aziridine azide was synthesized as the branching intermediate allowing for a fast transformation to a variety of potential drug candidates by regio- and stereoselective opening of the aziridine ring at the allylic position using various hydroxy components under Lewis acid catalysis followed by N-acetylation, azide reduction, and saponification. As indicated with a small selection of derivatives in Scheme 1.10 variation of the ether side chain led to a tremendous effect on activity, starting with the methoxy derivative still in the micromolar range passing through the ethyl and propyl derivative and finally arriving at the nanomolar activity range with the 3-pentyloxy derivative as the most active derivative.

This discovery chemistry route obviously was hardly amenable to scale-up due to a number of issues starting with (−)-quinic acid, a compound that is scarcely available in larger amounts. The access to the hydroxy-epoxide took six known steps, another four steps to the aziridine, and four additional steps to reach the branching trityl aziridine azide. All together about 16 steps required with an overall yield of roughly 10% including numerous chromatographic purifications.

After choosing the 3-pentyl-ether derivative as the most active inhibitor and the ethyl ester as the ideal prodrug, synthesis and process research activities were already initiated at Gilead Sciences for making oseltamivir phosphate available at least in kilogram amounts.

The first scalable synthesis shown in Scheme 1.11 was based on the elegant and early introduction of the 3-pentyl-ether side chain achieved by the regioselective reductive opening of the 3-pentanone ketal intermediate directly followed by the base-induced epoxide ring closure leading to the key precursor epoxide. This approach still required (−)-quinic acid as the starting material, which was easily converted to the acetonide mesylate. However, the dehydration step turned out to be particularly problematic both regarding yield and regioselectivity. Purification and isolation of the required cyclohexene intermediate became possible by the selective transformation of the accompanying but undesired '1,6'-double bond isomer by Pd-catalyzed allylic substitution of the mesylate group and subsequent extraction of the resulting pyrrolidino derivative into the acidic aqueous phase upon workup. Trans-ketalization of the resulting crystalline acetonide purified by crystallization then led to the oily key ketal ready to be used for the reductive ketal opening step followed by epoxide formation.

The subsequent transformation of the key precursor epoxide to the drug substance essentially represents the transformation of an epoxide to a 1,2-diamine derivative, a transformation involving azide reagents and intermediates. The sequence started with the epoxide ring opening using sodium azide at 65 °C followed by the direct transformation to the aziridine via a Staudinger phosphine imine using the extremely irritating and barely available reagent trimethylphosphine but with the advantage of allowing the removal of the formed trimethylphosphine oxide by extraction into the aqueous phase.

Opening of the aziridine ring with sodium azide under slightly acidic conditions at 85 °C led to the amino-azide intermediate, but with the hazard of forming hydrazoic acid, a low boiling liquid with a known tendency to detonate bond and therefore requiring stringent safety measures even for small-scale production.

Acetylation and reduction of the azido group with Lindlar's catalyst led after phosphate salt formation to the drug substance in an overall yield lower than the discovery synthesis but already with the advantage to proceed without chromatographic purifications and therefore allowing for the production of the first kilogram quantities directly required to continue clinical and biological investigations.

With the prospect of an upcoming multihundred-ton production for oseltamivir phosphate, a far reaching development of the process by Gilead, Roche, and also by third parties was critical to ensure the future market supply of the API. Owing to the ambitious timelines set by management, intensive troubleshooting of this

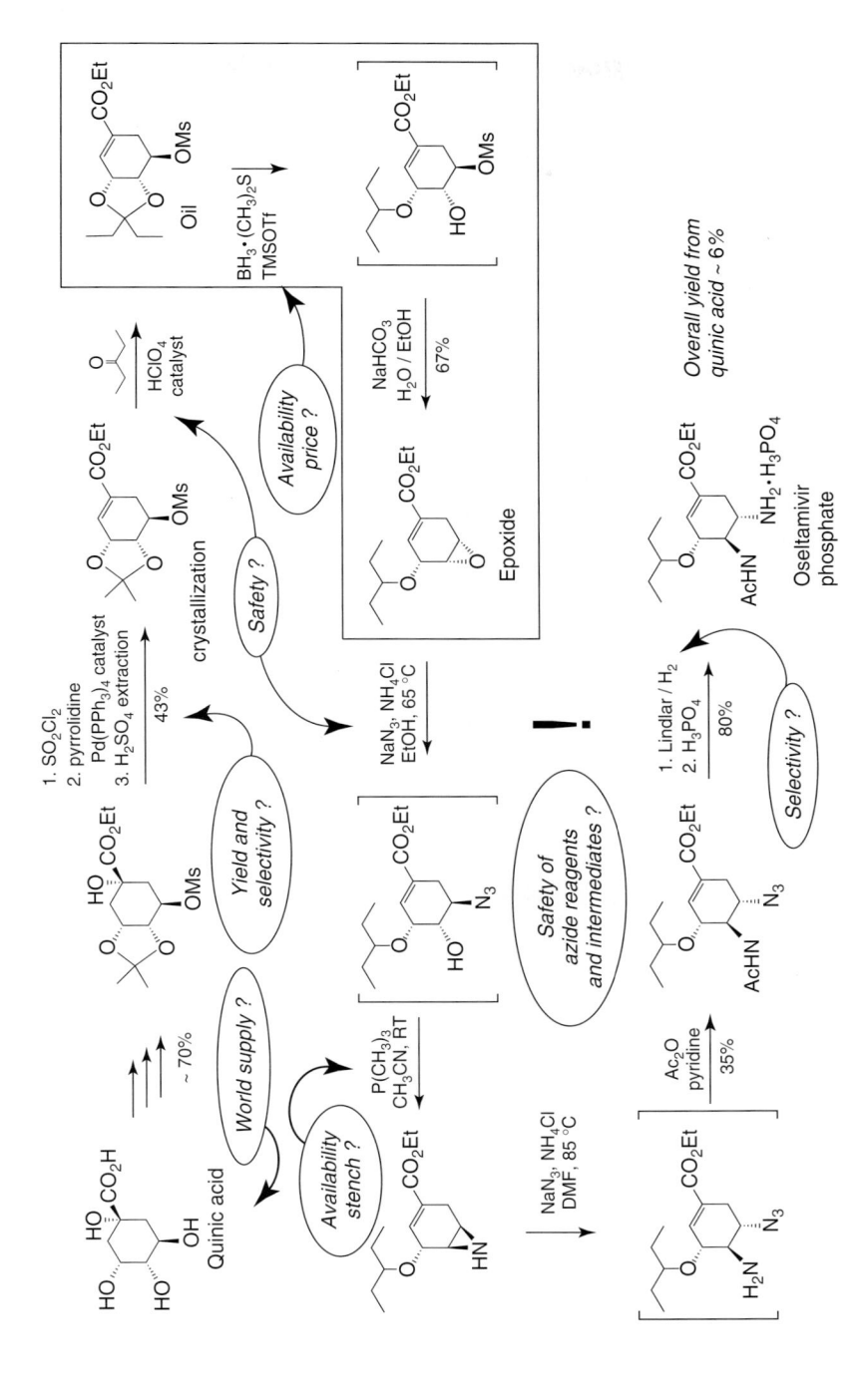

Scheme 1.11 Gilead scalable synthesis of the prodrug [14].

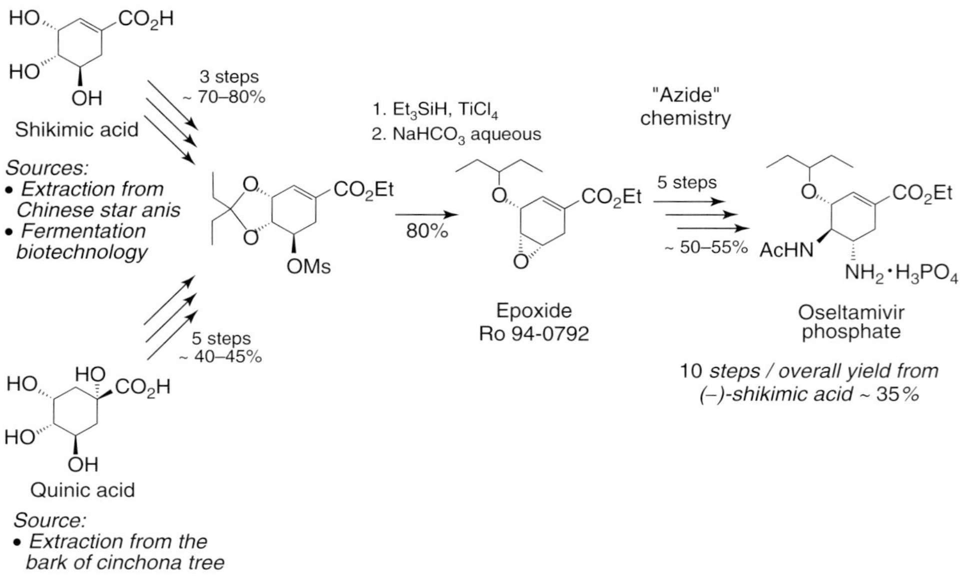

Scheme 1.12 Current commercial synthesis of oseltamivir phosphate [14–16].

synthesis finally led to the current industrial synthesis shown in Scheme 1.12, which preferentially starts from (−)-shikimic acid, since this raw material already contains the cyclohexene 1,2-double bond and therefore allows a much more efficient access to the key ketal intermediate.

For the reductive ketal opening a new cheap and very selective reagent combination triethylsilane/$TiCl_4$ was introduced. The "azide chemistry" was sourced out and developed by specialists in the field together with companies dealing with this type of chemistry in a safe way for many decades. This process now allows for the safe transformation of the key epoxide intermediate to the drug substance in an overall yield of 50–55% resulting in an overall yield of about 35% from (−)-shikimic acid, a route that now secures – together with the support of worldwide partners – a yearly production of several hundred metric tons for pandemic stockpiling.

Although (−)-shikimic acid was only available in research quantities at a very high price at the start of this project, it is now obtainable in ton quantities either by extraction from Chinese star anise or by fermentation using a genetically engineered *Escherichia coli* strain. Thus, the production of oseltamivir phosphate represents a noteworthy example for the proficient combination of biotechnology and synthetic organic chemistry.

In contrast, the overall yield to the epoxide intermediate starting from (−)-quinic acid never exceeded 50% due to the selectivity problem encountered in the dehydration step. Furthermore, the world supply of (−)-quinic acid is very limited since it is just a by-product of the extraction of quinine for tonic waters from the bark of the African cinchona tree.

1.5.3
The Search for Alternative Routes to Oseltamivir Phosphate

Although the main problems of the synthesis were solved and the supply of the API for clinical studies and its launch were secured, the price and availability of the starting material and the required outsourcing of the "azide chemistry" prompted synthesis and process research on the one hand to establish "azide-free" transformations of the key precursor epoxide in order to establish an independent, safe, and efficient alternative route amenable to a risk-free large-scale production. On the other hand, a search for shikimic acid independent new syntheses departing from cheap and abundant starting materials was initiated.

1.5.3.1 The Development of Azide-Free Transformations of the Key Epoxide Intermediate to Oseltamivir Phosphate

The first step was to identify an appropriate nitrogen nucleophile for replacing azide and the appropriate conditions for the epoxide ring opening. After an extensive search summarized in Scheme 1.13, allylamine was identified as the best substitute and magnesium bromide diethyletherate as a cheap catalyst which was new and not yet described in the chemical literature for this purpose.

After an intensive search for the most effective way to transform the key epoxide building block into oseltamivir phosphate replacing azide by allylamine, the conditions described in Scheme 1.14 were elaborated. The sequence starts with the regio- and stereoselective opening of the epoxide with allylamine in a nearly quantitative yield followed by Pd-catalyzed deallylation leading to the amino alcohol by acidic workup. Ethanolamine was shown to speed up the deallylation reaction considerably although the role of this promoter is not yet fully understood.

The transformation of the amino alcohol to the aminoallylamine was accomplished by a "near to one-pot" protocol discussed below including a reaction cascade without the need to isolate the intermediates encountered.

Selectivity in the N-acetylation of the aminoallylamine intermediate was achieved under acidic conditions, namely through transient protonation of the more basic secondary amino function followed by deallylation and phosphate salt formation leading to the drug substance in an overall yield from the epoxide of up to 40%.

This azide-free reaction sequence now compares well with the azide route concerning the number of steps and the number of intermediates isolated.

The reaction cascade or "domino" sequence mentioned above and depicted in Scheme 1.15 includes six consecutive reaction steps. The sequence started with the benzaldehyde imine formation allowing for the subsequent O-mesylation. After filtering off the triethylamine hydrochloride formed, the resulting iminomesylate was heated with 3 equiv. of allylamine in a Büchi autoclave for 16 h. Tracking the sequence by analytics revealed a first trans-imination to form the amino mesylate, which underwent fast ring closure to the aziridine releasing methanesulfonic acid triggering the aziridine ring opening. Interestingly, after completion of the reaction, the product found in the autoclave was not yet the desired aminoallylamine

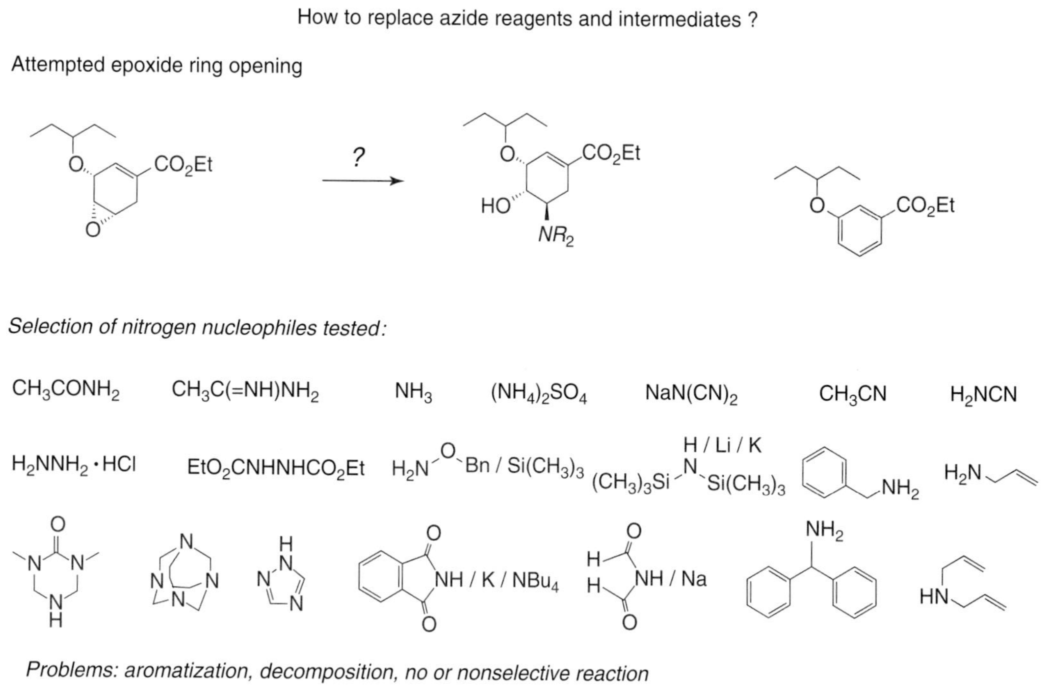

How to replace azide reagents and intermediates?

Attempted epoxide ring opening

Selection of nitrogen nucleophiles tested:

CH₃CONH₂   CH₃C(=NH)NH₂   NH₃   (NH₄)₂SO₄   NaN(CN)₂   CH₃CN   H₂NCN

H₂NNH₂·HCl   EtO₂CNHNHCO₂Et   H₂N–O–Bn / Si(CH₃)₃   (CH₃)₃Si–N(H/Li/K)–Si(CH₃)₃

Problems: aromatization, decomposition, no or nonselective reaction

Scheme 1.13 Replacing azide reagents and intermediates.

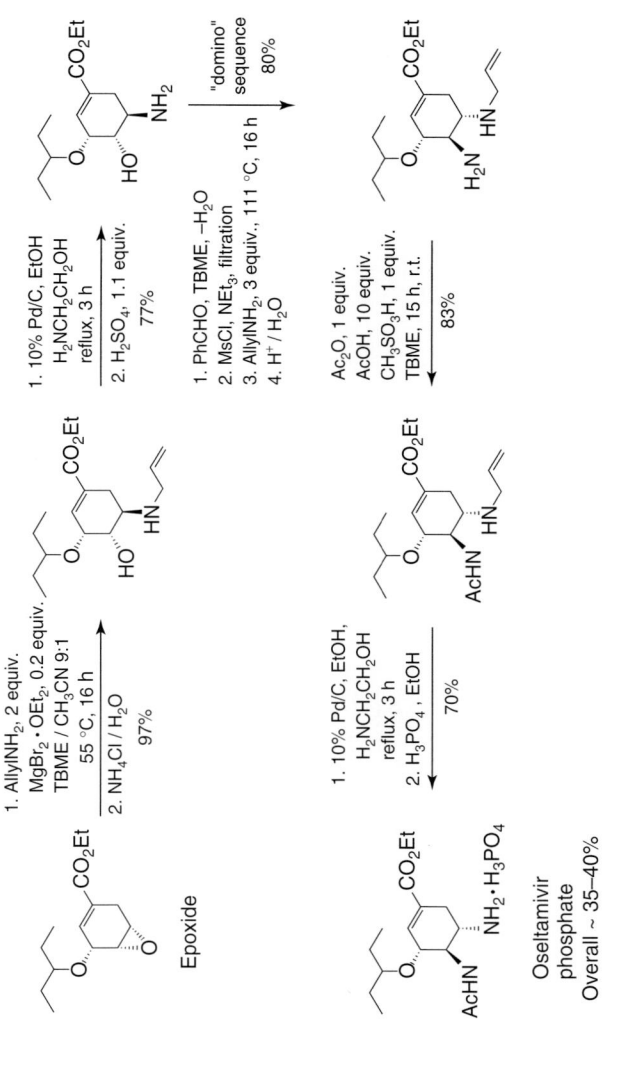

Scheme 1.14 The allylamine promoted azide-free synthesis of oseltamivir phosphate [17].

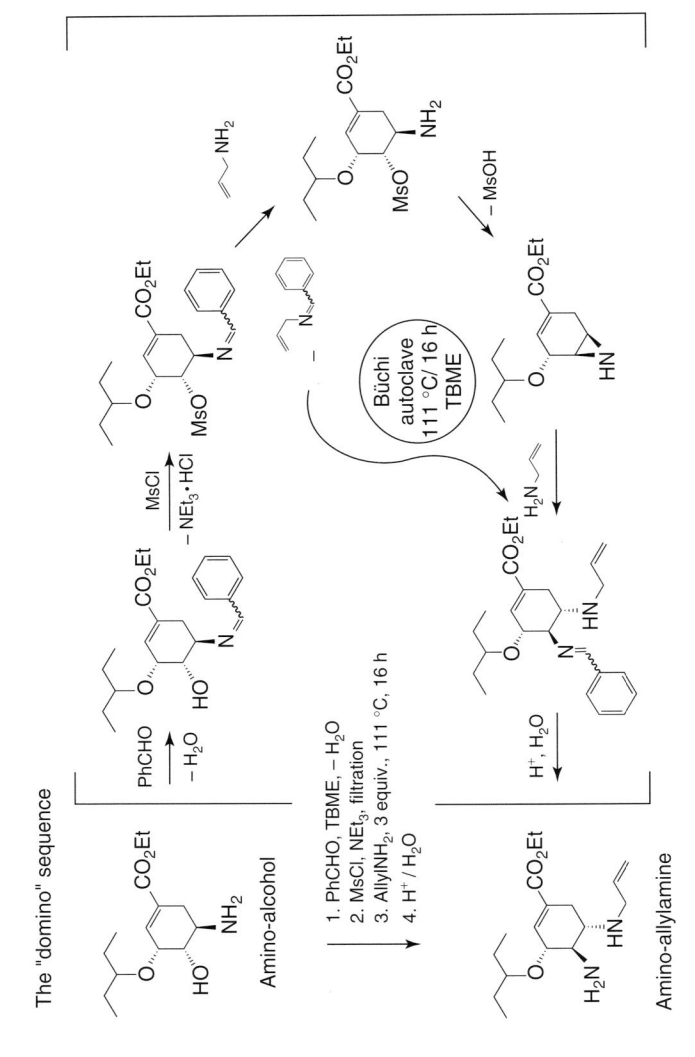

Scheme 1.15 The reaction cascade introducing the 5-allylamino functionality [17].

but the corresponding benzaldehyde imine interpreted as the product of a second trans-imination process. Acidic hydrolysis then led to the required aminoallylamine.

This reaction sequence demonstrates effectively the value of reaction cascades as an ideal tool for process improvement, which in this case allowed for transforming the amino alcohol to the aminoallylamine including five chemical transformations but requiring only one isolation and purification step.

This azide-free process compares well with the azide protocol also concerning the overall yield, which was improved up to 50% by technical development.

A related approach for the azide-free transformation of the key epoxide intermediate to oseltamivir phosphate employing *t*-butylamine to promote the epoxide ring opening and the less volatile diallylamine for the succeeding introduction of the 5-amino functionality is shown in Scheme 1.16. This route provides the API in an overall yield of about 60% starting from the epoxide.

1.5.3.2 The Development of Alternative Syntheses for Oseltamivir Phosphate

Since at an early stage of the synthetic development of oseltamivir phosphate, the commercial availability of (−)-shikimic acid in multihundred tons to secure the large-scale production of the API was still under exploration, the evaluation of new and different approaches to the API and potentially independent on (−)-shikimic acid as the raw material was deemed very important. Therefore, synthesis and process research was given the task to also evaluate alternative routes.

Figure 1.7 summarizes the major synthetic challenges of new syntheses of oseltamivir phosphate independent on (−)-shikimic acid. These include, primarily, the installation of the cyclohexene ring with the 1,2-double bond and three stereogenic centers of the required absolute configuration. Concurrently, the formation of the 4,5-amino substituents as well as the formation of the 3-pentylether side chain has to be efficiently controlled. Diels–Alder approaches as well as routes based upon suitably substituted aromatic rings and their transformations are as conceivable as novel ring constructions or even starting from suitable abundant chiral pool materials.

The key problems:

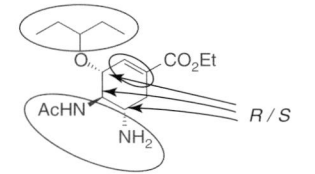

- Efficient induction of three stereogenic centers
- Regioselective introduction of the 1,2-double bond
- Introduction of the 4,5-amino substituents
- Formation of the 3-pentylether

Conceivable solutions and starting materials:

- Diels–Alder approaches
- Starting from aromatic rings and transformation
- Ring construction and transformation
- Starting from chiral pool

Figure 1.7 The evaluation of shikimic acid independent syntheses.

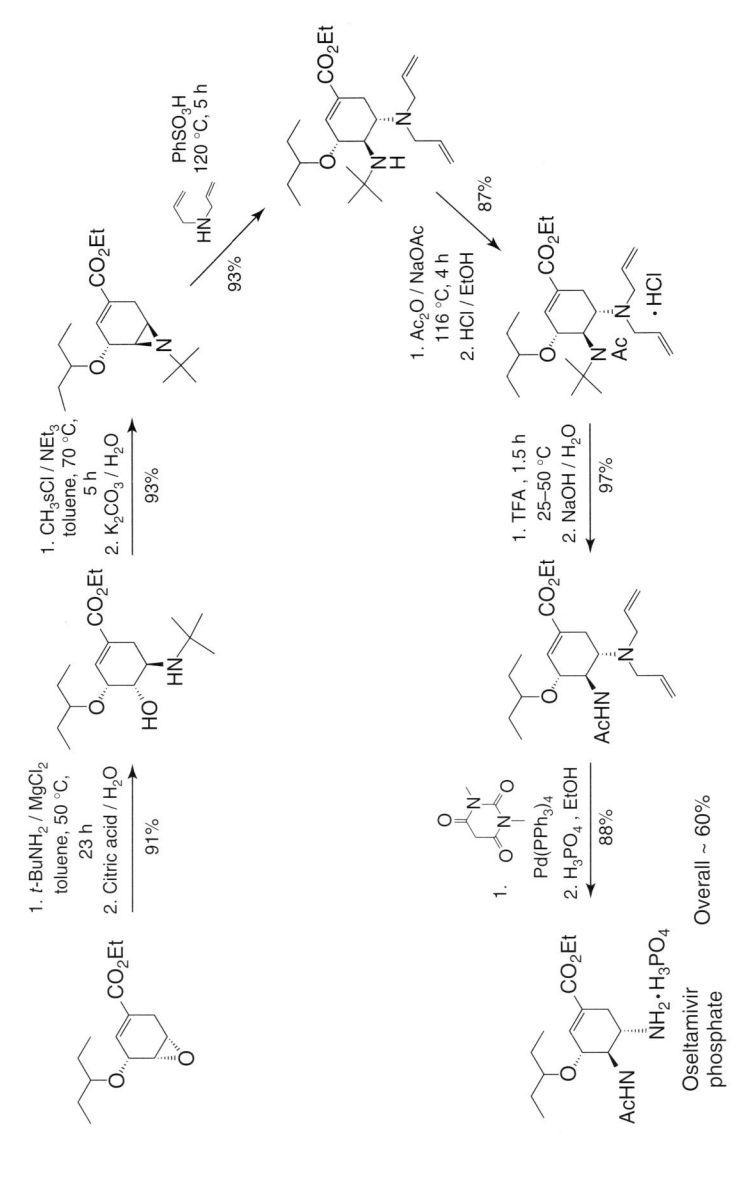

Scheme 1.16 The azide-free t-butylamine-diallylamine transformation [18].

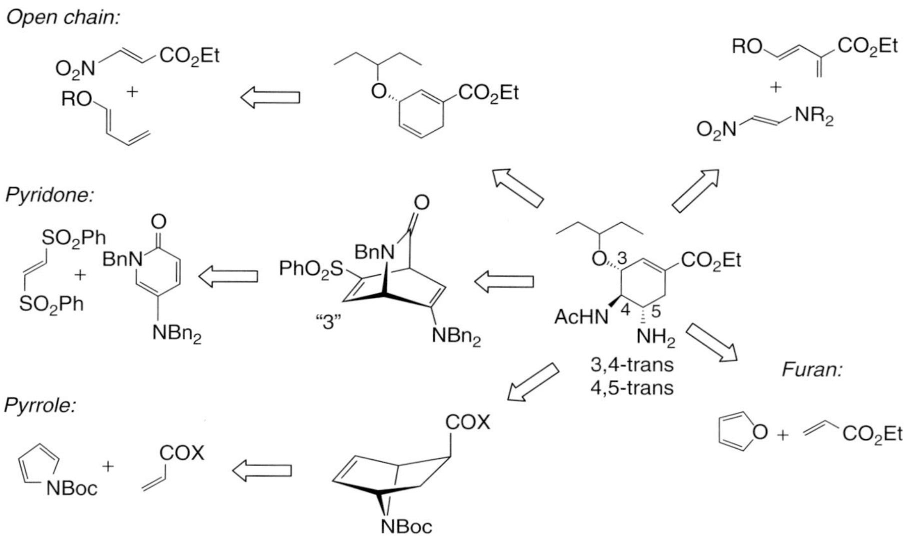

1.5 Synthetic Development of Oseltamivir Phosphate (Tamiflu™)

Scheme 1.17 The evaluation of Diels–Alder concepts.

Dealing with a cyclohexene derivative, obviously much effort was devoted to Diels–Alder approaches as shown in Scheme 1.17, summarizing the Diels–Alder concepts evaluated. Several "open-chain" concepts tested with the goal to introduce the two amino functions directly with the Diels–Alder reaction at a very early stage of the synthesis had to be abandoned mainly due to the instability of the corresponding dienophiles or the dienes, some of them representing quite unstable and hardly accessible compounds.

Attempts toward a 1,4-cyclohexadiene with the option to attack the more electron-rich nonconjugated 4,5-double bond with a nitrene or its equivalent were stopped due to the same reasons.

The "pyridone" Diels–Alder concept was based on the [4 + 2] cycloaddition of the perbenzylated 5-amino-2-pyridone to ethylene diphenyl disulfone. This reaction not only proceeded with high yield but also was also followed by the selective elimination of one of the sulfonyl groups. After exo selective sodium cyanoborohydride reduction of the enamine double bond, access to the bicyclic vinyl sulfone intermediate depicted in Scheme 1.17 was opened. This compound was expected to be ideally suited for conjugate nucleophilic addition at the prospective position "3." Although addition at this position was achievable, the concept had to be abandoned due to the stereoselectivity problem since it was not possible to reach the required 3,4-trans 4,5-trans configuration.

Since acrylic systems are not known or only sluggishly to react with Boc pyrrole, the "pyrrole" approach was also abandoned in favor of a classical furan Diels–Alder chemistry, starting with the very cheap and abundant starting materials furan and ethyl acrylate.

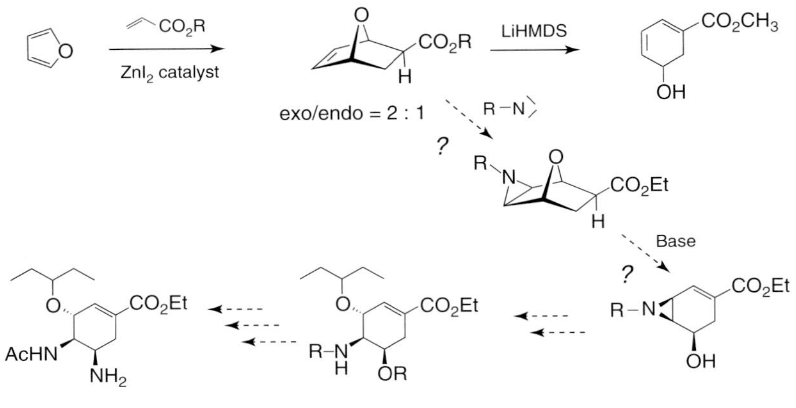

Scheme 1.18 The furan Diels–Alder/nitrene addition concept [19].

The "furan" concept is based on known investigations describing the zinc iodide catalyzed Diels–Alder reaction of furan and acrylates leading preferentially to the exo bicyclic isomer as shown in Scheme 1.18. Base-induced eliminative opening of the oxabicyclic system led to rather reactive 2,4-cyclohexadienols. As a variation of this protocol, we envisaged to first form an aziridine ring by nitrene addition or an equivalent protocol prior to the eliminative opening of the bicyclic system. This approach would lead to a cyclohexene-aziridine intermediate that should facilitate the regio- and stereoselective introduction of the 3-pentylether side chain as already known from the discovery chemistry approach (cf. Scheme 1.10). Further manipulations including the introduction of the amino function in 5-position were planned in order to reach the desired target.

As shown in Scheme 1.19, it was first possible to improve the exo selectivity of the Lewis acid catalyzed Diels–Alder reaction by up to 9 : 1, by replacing zinc iodide by the cheap zinc chloride and driving the reaction to the thermodynamic equilibrium. Second, an enzymatic resolution step allowed obtaining the pure R-isomer of the exo oxabicyclic intermediate after removal of the remaining parts of the endo isomer by distillation.

As a nitrene equivalent, one of the most stable, safe, and commercially available azides, namely *Shioiri's reagent* diphenylphosphoryl azide (DPPA) was applied, which at somewhat above room temperature added in an exo manner to the oxabicyclic system leading to a mixture of regioisomeric triazoles. The exo configuration of both isomers was clearly indicated by ^{1}H NMR coupling constants, and for the major isomer it was confirmed by X-ray analysis. After the thermal extrusion of nitrogen from the triazole mixture occurring at about 70 °C followed by trans-esterification at phosphorus (advantageous for a later step), the endo isomer, surprisingly, was isolated as determined by X-ray analysis of the corresponding acid.

This formal but still unexplained exo to endo "inversion," however, paved the way toward a very short and effective synthetic completion since the endo aziridine smoothly underwent eliminative ring opening followed by direct

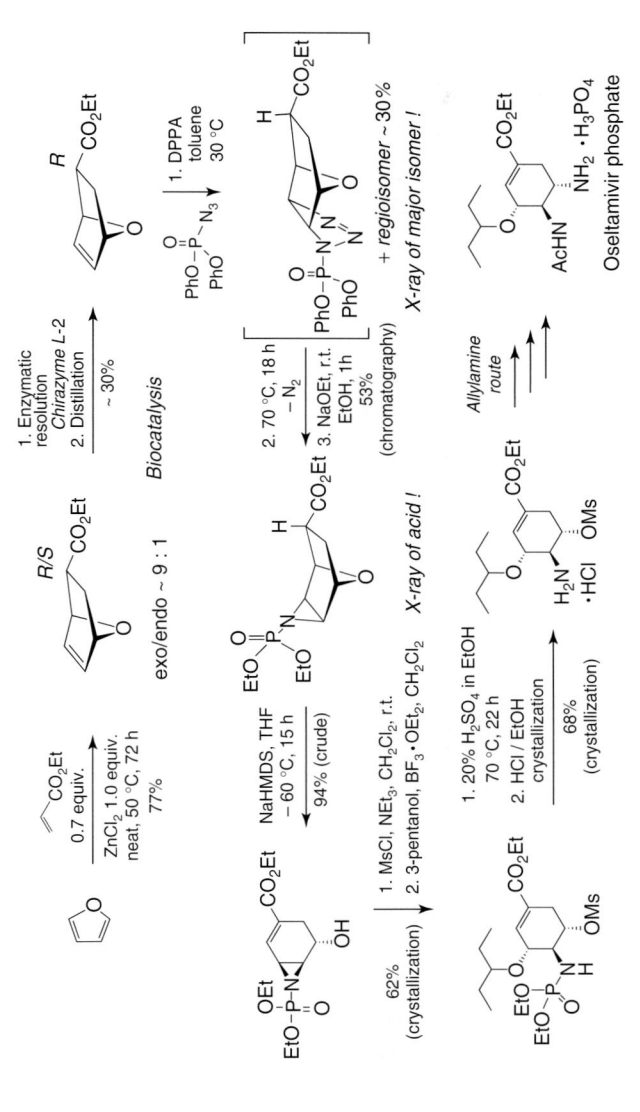

Scheme 1.19 The furan Diels–Alder/diphenylphosphoryl azide approach [15, 20].

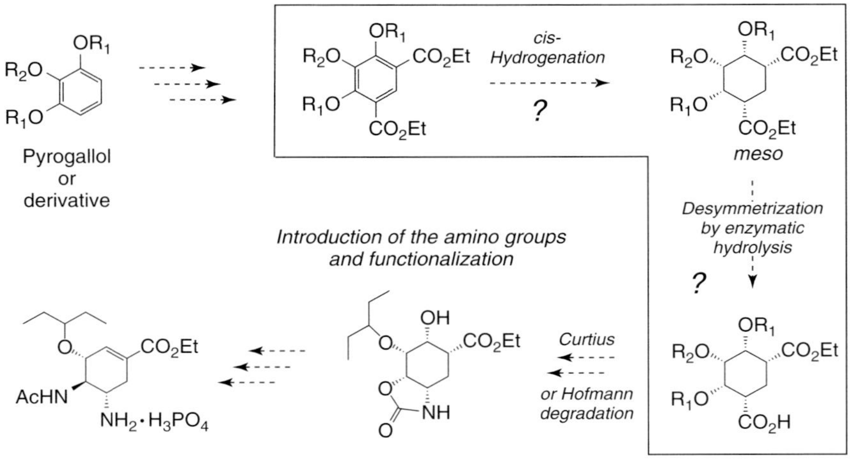

Scheme 1.20 The desymmetrization concept.

O-mesylation and regio- and stereoselective introduction of the 3-pentyl-ether side chain. Trans-esterification to the diethoxy phosphoryl compound – as mentioned above – was essential for the acidic cleavage of the N,P-bond leading to the amino mesylate isolated as the crystalline hydrochloride. With the specific configuration of this last intermediate, the transformation to the drug substance applying an analogous azide-free "allylamine" protocol as described in Schemes 1.14 and 1.15 became feasible, leading to the optically pure drug substance.

Taking advantage of a desymmetrization protocol over racemate cleavage regarding effectiveness, the synthetic desymmetrization concept sketched in Scheme 1.20 is based on two key steps. First, the possible all-cis hydrogenation of the iso-phthalic diester derivative will expectantly proceed by the use of a pyrogallol-type starting material, leading to an all-cis *meso*-diester. The meso ester will have the potential to undergo an enantioselective enzymatic desymmetrization, giving optimistically an optically pure acid–ester intermediate ready for the introduction of the required nitrogen functionalities in 4- and 5-positions via Hoffmanns- or Curtius-type degradation.

The result of this investigation is presented in Schemes 1.21 and 1.22. The synthesis started from cheap dimethoxyphenol with a sequence of high-yielding reactions. After effective 3-pentylether formation followed by dibromination and Pd-catalyzed double ethoxy carbonylation, the hydrogenation of the appropriately substituted iso-phthalic diester – although at somewhat elevated pressure and temperature – indeed led to the desired *meso*-diester.

Nearly quantitative and highly selective cleavage of the methyl ether groups using *in situ* generated trimethylsilyl iodide then led to the meso dihydroxy intermediate ready for smooth desymmetrization investigated by the biocatalysis group. Using cheap pig liver enzyme, it was possible to obtain the desired (+)-monoacid in quantitative yield.

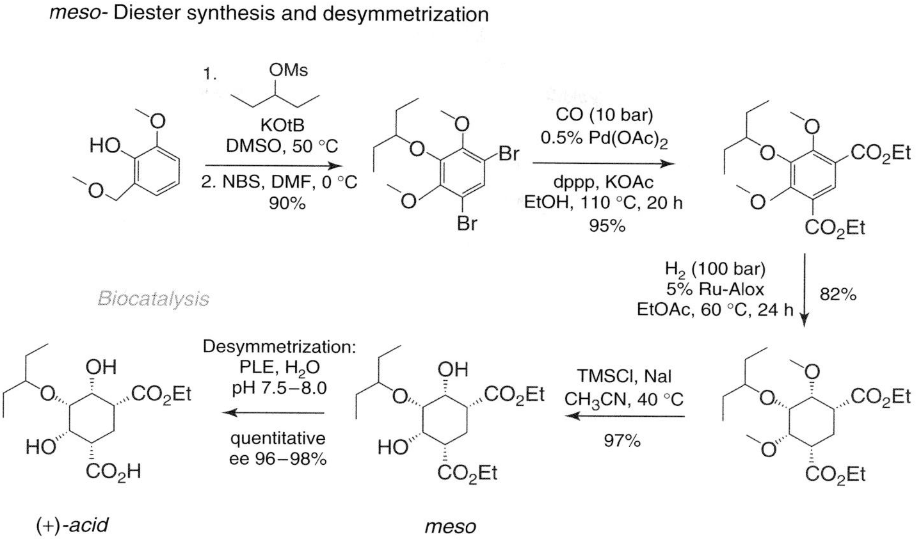

1.5 Synthetic Development of Oseltamivir Phosphate (Tamiflu™)

meso- Diester synthesis and desymmetrization

Scheme 1.21 *meso*-Diester synthesis and desymmetrization.

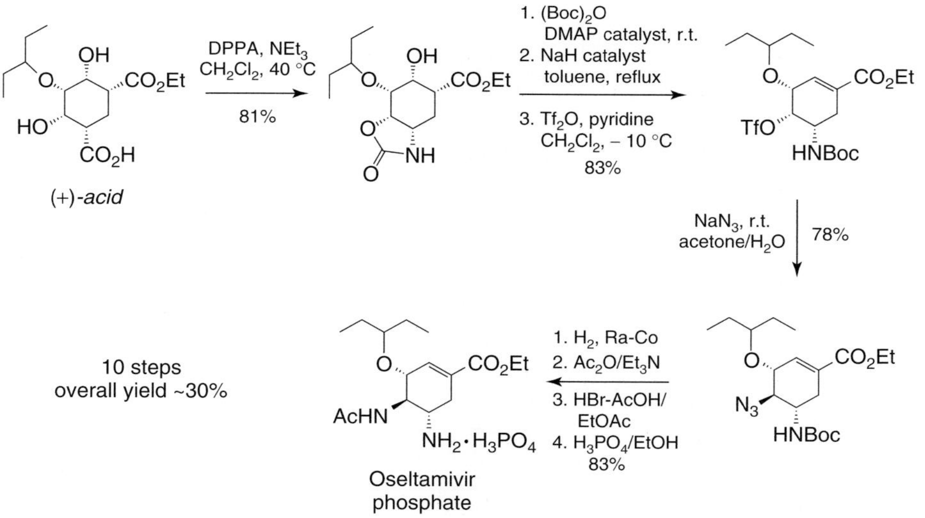

Scheme 1.22 Introducing the amino functions [21].

The conversion of this optically pure key intermediate to the drug substance was straightforward as shown in Scheme 1.22 starting with a Curtius-type degradation of the β-hydroxy-acid allowing introducing the 5-amino group with direct formation of the oxazolidinone. The subsequent reaction cascade takes advantage of the special configuration of this intermediate. The *N*-Boc protected intermediate was

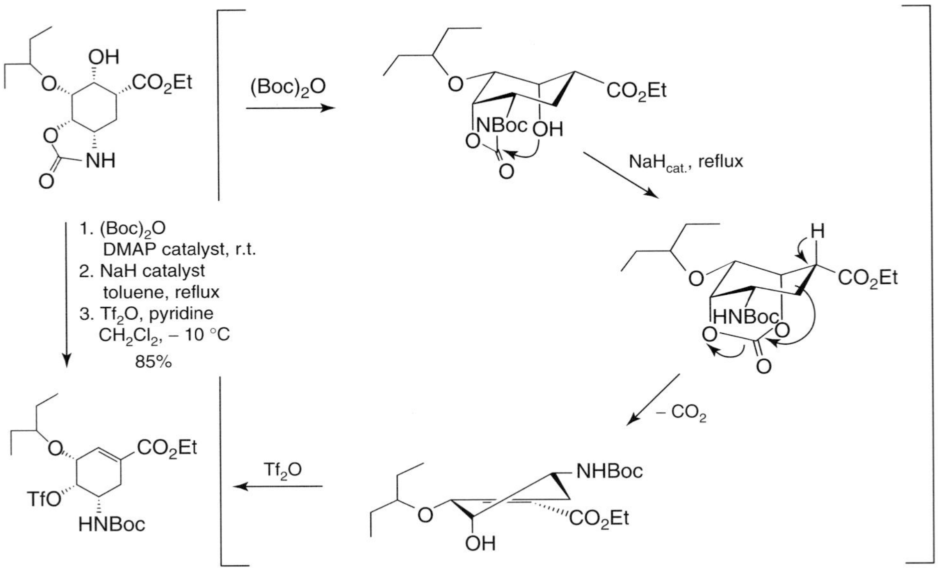

Scheme 1.23 Introducing the amino functions: the oxazolidinone shortcut [21].

treated with catalytic amounts of strong base triggering the effective and selective formation of the 1,2-double bond and – at the same time – the cleavage of the oxazolidinone system, followed by the formation of a triflate leaving group in an overall yield of 83%.

Inversion at position 4 so far still uses sodium azide but at room temperature and under neutral conditions, avoiding the formation of hydrazoic acid and sets the stage for the final sequence, namely, azide reduction, N-acetylation, Boc deprotection, and phosphate salt formation. The overall yield starting from a cheap aromatic starting material accounts for 30%, a result that compares favorably with the – already well developed – shikimic acid route.

To achieve this overall result, substantial fine tuning of all the reaction sequences is a clear requirement. A typical example for this process – the oxazolidinone transformation with concomitant introduction of the 1,2-double bond – is illustrated in Scheme 1.23. The transformation consists of a multistep cascade starting with N-Boc protection of the oxazolidinone intermediate followed by sodium hydride promoted deprotonation. An intramolecular attack on the oxazolidinone formed a strained cyclic carbonate stable enough to be isolated, but substantially activated for the subsequent carbon dioxide fragmentation process, directly providing the hydroxy compound ready for activation.

Several groups have published new approaches to oseltamivir phosphate also from alternative sources than (−)-shikimic acid, claiming a high need for routes independent of (−)-shikimic acid often on the basis of unfounded arguments regarding the availability of this acid as a technical starting compound as well as the potential risks involved in handling azide chemistry on an industrial

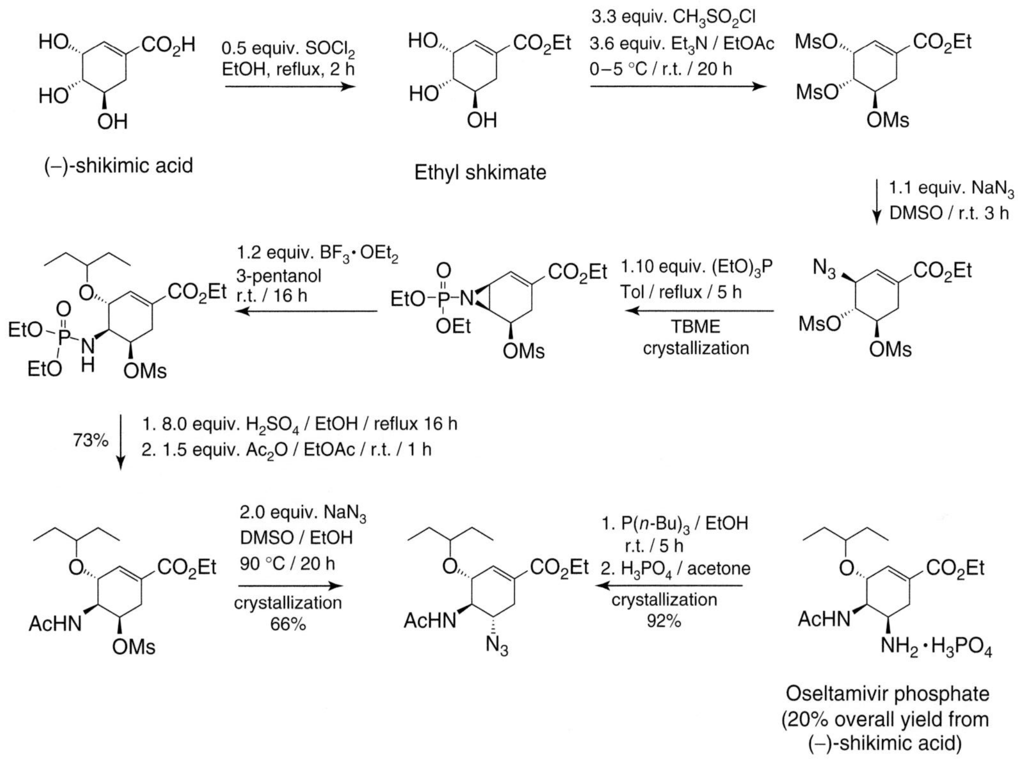

Scheme 1.24 Synthesis of oseltamivir phosphate via the O-trimesylate of ethyl shikimate starting from (−)-shikimic acid [22].

scale. Although commercial sources of large quantities of (−)-shikimic acid were unobtainable at the outset of this project, also prompting us to create and evaluate shikimic acid independent routes as described above, it became widely available in multihundred-ton amounts through this endeavor.

With large amounts of (−)-shikimic acid on hand and the constructive experience with partners performing azide chemistry on a bulk scale, the search for even shorter routes than the current commercial synthesis starting from this now abundant material finally led us to the protocol shown in Scheme 1.24 proceeding via the O-trimesylate of ethyl shikimate, obtained in high yield from (−)-shikimic acid by way of ethyl shikimate. Subsequent mesylation and regio- and stereoselective substitution of the allylic O-mesylate group with sodium azide at room temperature under nonacidic conditions led to the azide intermediate. Subsequent treatment thereof with triethyl phosphite in toluene at reflux produced the aziridine intermediate, which underwent regio- and stereoselective ring opening at the allylic position. The N,P-bond cleavage afforded the last mesylate, which by azide substitution under neutral conditions furnished the penultimate intermediate of the current commercial route of oseltamivir phosphate. This eight-step route provides the API in 20% overall yield from (−)-shikimic acid, already at a technically undeveloped stage.

The successful synthetic development of new small molecule drugs depends primarily and most vitally on dedicated people willing to cooperate and form teams starting with discovery chemistry and proceeding through synthesis and process research, kilolaboratory, and technical development departments. Only through thorough discussions and with high mutual respect for and acceptance of the individual competencies among the responsible chemists involved, the speedy evaluation and development of routes from milligrams to tons is attainable.

References

1. Noyori, R. (2001) *Adv. Synth. Catal.*, **343**, 1.
2. Imhof, R. and Kybuz, E. (1986) *J. Fluorine CH*, Patent DE 3530046.
3. Salone, M. and Vogt, P. (1990) EP 385210.
4. Weibel, E.K., Hadvary, P., Hochuli, E., Kupfer, E. and Lengsfeld, H. (1987) *J. Antibiotics*, **40**, 1081.
5. Weber, W. *et al.* (1997) EP 803576.
6. Barbier, P., Schneider, F. and Widmer, U. (1987) *Helv. Chim. Acta*, **70**, 1412.
7. Zutter, U. and Karpf, M. (1991) EP 443449.
8. Birk, R., Karpf, M., Püntener, K., Scalone, M., Schwindt, M., and Zutter, U. (2006) *Chimia*, **60**, 561; (2007) *Org. Proc. Res. & Dev.*, **11**, 524.
9. Roberts, N.A., Martin, J.A., Kinchington, D., Broadhurst, A.V., Craig, J.C., Duncan, I.B., Galpin, S.A., Handa, B.K., Kay, J., Krohn, A., Lambert, R.W., Merrett, J.H., Mills, J.S., Parkes, K.E.B. Redshaw, S., Ritchie, A.J., Taylor, D.L., Thomas, G.J. and Machin, P.J. (1990) *Science*, **248**, 358.
10. Parkes, K.E.B., Bushnell, D.J., Crackett, P.H., Dunsdon, S.J., Freeman, A.C., Gunn, M.P., Hopkins, R.A., Lambert, R.W. and Martin, J.A. (1994) *J. Org. Chem.*, **59**, 3656.
11. Hilpert, H., Goehring, W., Gokhale, S., Hilpert, H., Roessler, F., Schlageter, M. and Vogt, P. (1996) *Chimia*, **50**, 532.
12. Kim, C.U., Lew, W., Williams, M.A., Liu, H., Zhang, L., Swaminathan, S.,

13. Bischofberger, N., Chen, M.S., Mendel, D.B., Tai, C.Y., Laver, W.G. and Stevens, R.C. (1997) *J. Am. Chem. Soc.*, **119**, 681.
13. Kim, C.U., Lew, W., Williams, M.A., Wu, H., Zhang, L., Chen, X., Escarpe, P.A., Mendel, D.B, Laver, W.G. and Stevens, R.C. (1998) *J. Med. Chem.*, **41**, 2451.
14. Rohloff, J.C., Kent, K.M., Postich, M.J., Becker, M.W., Chapman, H.H., Kelly, D.E., Lew, W., Louie, M.S., McGee, L.R., Prisbe, E.J., Schultze, L.M., Richard, H., Yu, R.H. and Zhang, L. (1998) *J. Org. Chem.*, **63**, 4545.
15. Karpf, M., Abrecht, S., Cordon Federspiel, M., Estermann, H., Fischer, R., Karpf, M., Mair, H.-J., Oberhauser, T., Rimmler, G., Trussardi, R. and Zutter, U. (2007) *Chimia*, **61**, 93.
16. Chandran, S.S., Yi, J., Draths, K.M., Daeniken, R.V., Weber, W. and Frost, J.W. (2003) *Biotechnol. Prog.*, **19**, 808.
17. Karpf, M. and Trussardi, R. (2001) *J. Org. Chem.*, **66**, 2044.
18. Harrington, P., Brown, J.D., Foderaro, T. and Hughes, R.C. (2004) *Org. Proc. Res. & Dev.*, **8**, 86.
19. Brion, F. (1982) *Tetrahedron Lett.*, **23**, 5299.
20. Abrecht, S. *et al.* (2000) EP 20557.
21. Zutter, U., Iding, H., Spurr, P. and Wirz, B. (2008) *J. Org. Chem.*, **73**, 4895.
22. Karpf, M. and Trussardi, R. (2009) *Angew. Chem. Int. Ed.*, **121**, 5760.

2
Design of Dynamic Salt Catalysts Based on Acid–Base Combination Chemistry

Kazuaki Ishihara

2.1
Introduction

The acidity of catalysts can be controlled based on acid–base combination chemistry [1, 2]. For example, if *p*-toluenesulfonic acid is too strong as an acid catalyst, pyridinium *p*-toluenesulfonate may be chosen. *p*-Toluenesulfonic acid is a classical single-molecule catalyst, but, because of the equilibration between pyridinium *p*-toluenesulfonate and free pyridine, and *p*-toluenesulfonic acid, pyridinium *p*-toluenesulfonate behaves as dynamic complexes in solution. Enzymes also function through dynamic complexation between acidic molecules and basic molecules *in vivo*. It is possible to introduce enzymatic functions such as induced-fit, molecular recognition, cooperative effect, and allosteric effect to artificial salt catalysts due to their conformational and dynamic flexibility. The key to designing dynamic complexes is nonbonding interactions such as hydrogen-bonding-, electrostatic-, ionic-, $\pi-\pi$-, cation-π cation-*n*-, hydrophobic-, and hydrophilic interactions. Recently, a new type of catalysts that function through dynamic aggregation between acids and bases has been developed by the author and his coworkers. In this chapter, the potential of dynamic salt complexes as highly functional catalysts is demonstrated through two examples.

Scheme 2.1 shows the equilibrium in a solution of an equimolar mixture of a sulfonic acid and a tertiary amine. In this case, a 1 : 1 complex is expected to be the major species.

Scheme 2.2 shows the equilibrium in a solution of an equimolar mixture of a sulfonic acid and a secondary amine. In this case, linear or cyclic $n:n$ salt complexes may be formed because secondary ammonium cations have two acidic protons. The selectivity of complexation can be controlled by steric repulsions or other nonbonding interactions.

Scheme 2.3 shows the equilibrium in a solution of an equimolar mixture of a sulfonic acid and a primary amine. In this case, more complicated $n:n$ salt complexes may be formed because primary ammonium cations have three acidic protons. Therefore, it is not easy to control the selectivity of complexation.

Pharmaceutical Process Chemistry. Edited by Takayuki Shioiri, Kunisuke Izawa, and Toshiro Konoike
Copyright © 2011 WILEY-VCH Verlag GmbH & Co. KGaA, Weinheim
ISBN: 978-3-527-32650-1

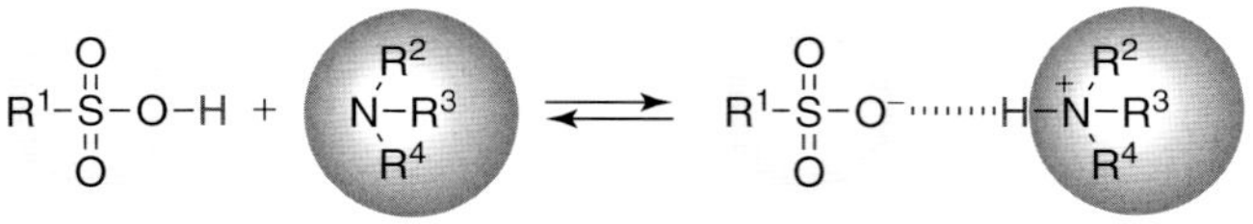

Scheme 2.1 Dynamic ammonium salts of a sulfonic acid with a tertiary amine.

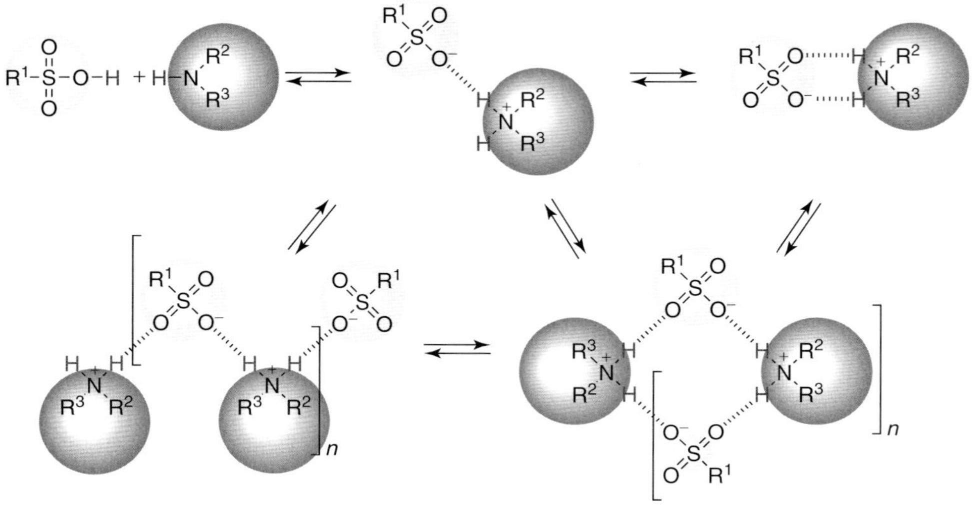

Scheme 2.2 Dynamic ammonium salts of a sulfonic acid with a secondary amine.

Nevertheless, a primary ammonium cation may be useful as a building block to construct a three-dimensional structure.

Thus, dynamic salt complex catalysts can be prepared *in situ* from acidic and basic small molecules, which are elaborately designed based on acid–base combination chemistry, as shown in Schemes 2.1–2.3.

In Section 2.2, dynamic salt complexes of sulfonic acids with secondary amines that function as dehydrative condensation catalysts are described. These salt catalysts are aggregated like a reverse micelle in nonpolar hydrocarbons, and their catalytic activities are much higher than those of the corresponding sulfonic acids themselves due to local hydrophobic function around the ammonium protons of their salts.

In Section 2.3, dynamic salt complexes of chiral disulfonic acids with tertiary amines that function as asymmetric Mannich-type reaction catalysts are described. Interestingly, although ammonium cations are achiral, high enantioselectivity is induced through the chiral counteranion. Chiral dynamic salt complexes can be rapidly optimized to induce high enantioselectivity by a combinatorial approach.

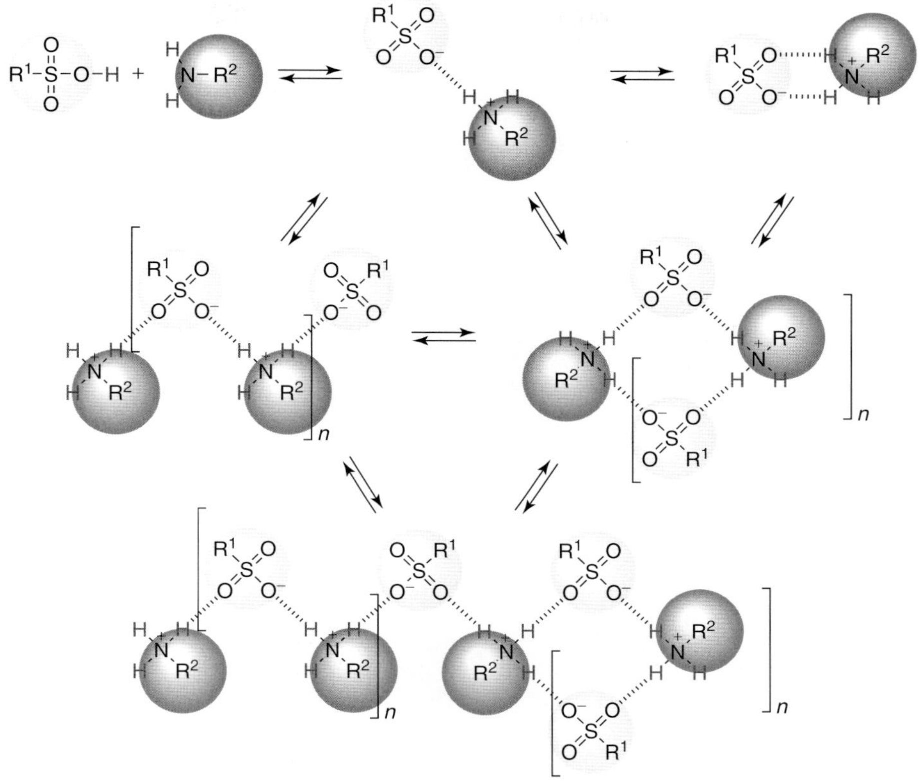

Scheme 2.3 Dynamic ammonium salts of a sulfonic acid with a primary amine.

2.2
Dehydrative Condensation Catalysts

2.2.1
Esterification Catalysts

Acid-catalyzed dehydration reactions such as the ester condensation reaction are some of the most fundamental reactions, and more environmentally benign alternatives are desired by the chemical industry [3]. Recently, some ammonium salts were developed as mild Brønsted acid dehydration catalysts (Scheme 2.4). Tanabe and coworkers reported N,N-diphenylammonium triflate (**1**) and pentafluoroanilinium triflate (**2**), which efficiently catalyzed the ester condensation reaction of carboxylic acids with equimolar amounts of alcohols [4]. Years later, it was reported that bulky diarylammonium pentafluorobenzenesulfonates **3a** and **4a**, which are much milder Brønsted acids than the corresponding ammonium triflates, were extremely active ester condensation catalysts [5].

$$R^1CO_2H + R^2OH \underset{}{\overset{\text{Catalyst HX}}{\rightleftharpoons}} R^1CO_2R^2 + H_2O$$

Scheme 2.4 N,N-Diarylammonium arenesulfonates.

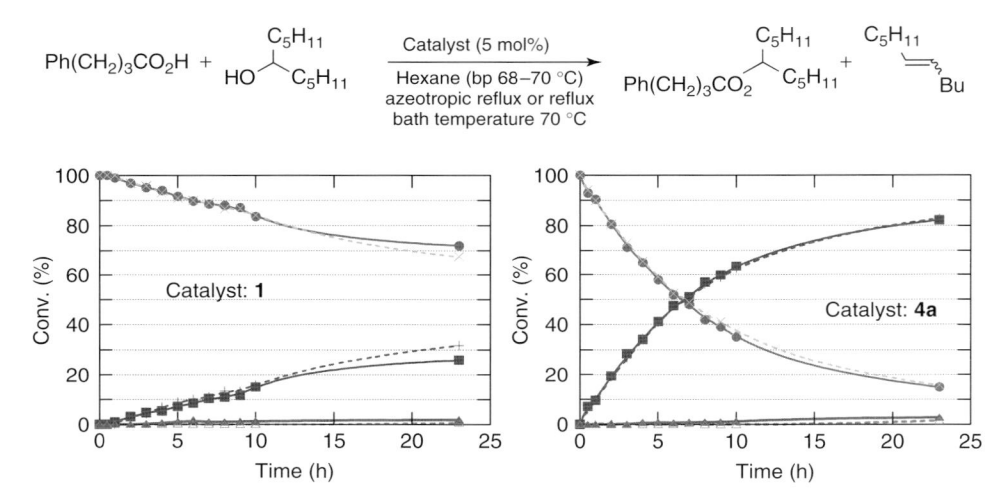

1: Ph$_2$N$^+$H$_2$ $^-$OTf

2: (C$_6$F$_5$)NH$_3^+$ $^-$OTf (pentafluorophenyl ammonium triflate)

3a: R = C$_6$F$_5$
3b: R = Tol

4a: R = C$_6$F$_5$
4b: R = Tol

$$Ph(CH_2)_3CO_2H + HO\text{-}CH(C_5H_{11})_2 \xrightarrow[\text{azeotropic reflux or reflux bath temperature 70 °C}]{\text{Catalyst (5 mol\%)} \atop \text{Hexane (bp 68--70 °C)}} Ph(CH_2)_3CO_2\text{-}CH(C_5H_{11})_2 + \text{5-undecene}$$

Scheme 2.5 Ester condensation of 4-phenylbutyric acid with 6-undecanol. The catalytic activities of **1** and **4a** under reflux conditions without the removal of water (solid lines) and under azeotropic reflux conditions (dotted lines) were compared. The proportions of 6-undecanol (circles and ×), 6-undecyl 4-phenylbutyrate (squares and +), and 5-undecene (filled triangles and open triangles) in the reaction mixture over time were evaluated by ^{1}H NMR analysis.

Comparative experiments using catalysts **1** and **4a** were performed using the ester condensation reaction of 4-phenylbutyric acid with 6-undecanol in hexane under reflux conditions without the removal of water and under azeotropic reflux conditions with the removal of water (Scheme 2.5). While the reaction catalyzed by **1** was slightly decelerated under reflux conditions without the removal of water, the reaction catalyzed by **4a** proceeded efficiently without the influence of water [5].

Interestingly, ammonium salt **3a** (lozenges) exhibits higher catalytic activity than C$_6$F$_5$SO$_3$H (circles) for the dehydrative ester condensation reactions of carboxylic acids with an equimolar amount of alcohols under the reaction conditions without the removal of water, despite the weaker acidity of **3a** (Scheme 2.6) [5b]. In contrast,

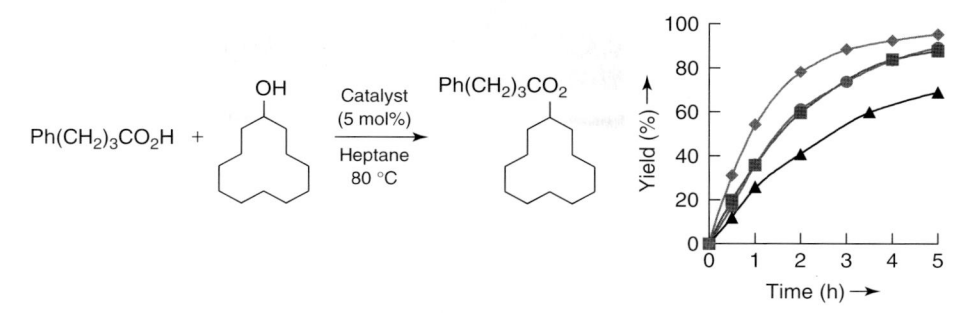

Scheme 2.6 Ester condensation reaction of 4-phenylbutyric acid with an equimolar amount of cyclododecanol catalyzed by **3a** (lozenges), **4a** (squares), $C_6F_5SO_3H$ (circles), and **1** (triangles) [5b].

ammonium salt **4a** (squares) shows catalytic activity similar to that of $C_6F_5SO_3H$ (circles), and ammonium salt **1** (triangles) shows catalytic activity lower than that of $C_6F_5SO_3H$ (circles). These phenomena can be interpreted by assuming that the hydrophobic environment created around the ammonium protons in anilinium arenesulfonates effectively promotes the dehydrative condensation reaction and the steric bulkiness of arenesulfonates suppresses the dehydrative elimination of secondary alcohols to produce alkenes.

2.2.2 Dehydrative Cyclocondensation Catalysts

Bulky N,N-diarylammonium pentafluorobenzenesulfonates promoted the dehydrative cyclization of 1,3,5-triketones to γ-pyrones much more effectively than the ester condensation reaction, since 1,3,5-triketones are generally less polar than carboxylic acids and alcohols [5d, 6]. The local hydrophobic environment created around the ammonium protons in ammonium sulfonates is the key to the unusual acceleration of dehydration reactions. The author and coworkers investigated the relationship between the catalytic activity and the steric and/or stereoelectronic factors of N,N-diarylammonium arenesulfonate catalysts for the dehydrative cyclization of 1,3,5-triketones, and discussed the microscopic hydrophobic environment created in aggregated ammonium sulfonates, taking their X-ray single-crystal structures into account.

The catalytic activities of **4a** and $C_6F_5SO_3H$ in the dehydrative cyclization of 4,6-dimethylnonan-3,5,7-trione (**5a**) were compared under heating without the removal of water and under azeotropic reflux conditions with the removal of water (Scheme 2.7). While the reaction catalyzed by $C_6F_5SO_3H$ at 80 °C without the removal of water gave γ-pyrone **6a** in 74% yield after 8 h (circles, graph a), the reaction under azeotropic reflux conditions in cyclohexane (bp 80.7 °C) with the removal of water gave **6a** in 96% yield after the same reaction time (squares, graph a). The reaction catalyzed by $C_6F_5SO_3H$ was considerably decelerated without

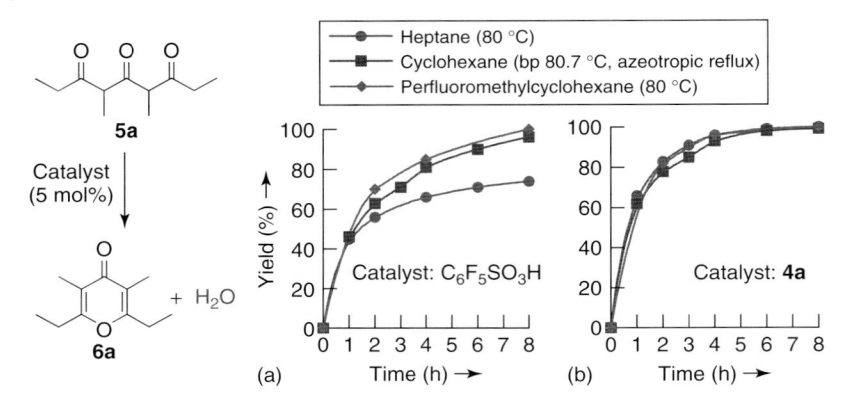

Scheme 2.7 Dehydrative cyclization of **5a** in solvent. The yield of γ-pyrone **6a** was evaluated by HPLC analysis.

the removal of water. However, the reaction proceeded more rapidly in perfluoromethylcyclohexane, which is a water-repellent solvent, without the influence of water, and gave **6a** quantitatively after 8 h (lozenges, graph a) [7]. Fluorous media appear to release the water produced from the active site of the catalyst. In contrast, the reaction catalyzed by **4a** at 80 °C gave **6a** quantitatively regardless of the above three conditions (graph b). More importantly, **4a** exhibited much higher catalytic activity than $C_6F_5SO_3H$ under heating conditions without the removal of water despite the weaker acidity of **4a**. These experimental results show that the use of **4a** (as the use of a Dean–Stark apparatus or hydrophobic perfluoromethylcyclohexane) gives the same rate-accelerating effect.

The solvent effect for the dehydrative cyclization of **5a** catalyzed by **4a** or $C_6F_5SO_3H$ without the removal of water was also investigated (Table 2.1). When the reaction catalyzed by **4a** was conducted in heptane at 80 °C for 8 h, **6a** was obtained in quantitative yield (entry 1). On the other hand, this reaction gave **6a** in 56–64% yield in more polar solvents such as toluene, 1,4-dioxane, and propionitrile (EtCN) (entries 2–4). Similarly, the reaction catalyzed by $C_6F_5SO_3H$ proceeded better in heptane than in toluene, 1,4-dioxane, or EtCN (entry 1 versus entries 2–4). Interestingly, **4a** and $C_6F_5SO_3H$ had similar catalytic activities in toluene, 1,4-dioxane, and EtCN (entries 2–4), while **4a** showed much higher catalytic activity than $C_6F_5SO_3H$ in heptane (entry 1). In less polar solvents such as heptane, **4a** should form a stable ion pair, in which the ammonium cation was tightly surrounded by two bulky *N*-aryl groups and an *S*-pentafluorophenyl group. It is conceivable that due to the hydrophobic environment created by these bulky aryl groups, the generated water was rapidly released from the active site of **4a** and the hydrophobic wall prevented polar water from gaining access to the active site of **4a**. On the other hand, it was suggested that **4a** could not form a stable ion pair in polar solvents such as toluene, 1,4-dioxane, and EtCN, and no rate-accelerating effect was exhibited.

The catalytic activities of $C_6F_5SO_3H$, TsOH and their *N,N*-diarylammonium salts for the dehydrative cyclization of 4,6,9-trimethyldecan-3,5,7-trione (**5b**) to

2.2 Dehydrative Condensation Catalysts

Table 2.1 Dehydrative cyclization of **5a** to **6a** catalyzed by **4a** or $C_6F_5SO_3H$.[a]

$$5a \xrightarrow[\text{Heptane, 80 °C, 8 h}]{\text{Catalyst (5 mol\%)}} 6a + H_2O$$

Entry	Solvent	Yield (%) of 6a[b]	
		4a	$C_6F_5SO_3H$
1	Heptane	100	74
2	Toluene	56	55
3	1,4-Dioxane	64	58
4	EtCN	64	61

[a] Reactions were carried out with 0.2 mmol of 5a and 5 mol% of catalyst in 4 ml of solvent at 80 °C for 8 h without the removal of generated water.
[b] Determined by HPLC analysis.

γ-pyrone **6b** were examined in heptane at 80 °C without the removal of water (Scheme 2.8). Compound **5b** has a methyl group at its 9-position, and is less reactive than **5a** because of the steric hindrance of the methyl group. As in the ester condensation reactions, N,N-diarylammonium tosylates showed slightly lower catalytic activities than TsOH in the dehydrative cyclization of **5b** (graph c, specific acid catalysis) [8]. In contrast, the catalytic activities of N,N-diarylammonium pentafluorobenzenesulfonates strongly depended on the structures of the N,N-diarylamines (graph d, general acid catalysis) [8]. TsOH ($pK_a(CD_3CO_2D) = 8.5$) is a stronger acid than $C_6F_5SO_3H$ ($pK_a(CD_3CO_2D) = 11.1$) [5, 9, 10]. Compared to $C_6F_5SO_3H$ (circles), N,N-diphenylammonium pentafluorobenzenesulfonate ($[Ph_2N^+H_2][^-O_3SC_6F_5]$) showed lower catalytic activity due to its weaker acidity and slight hydrophobicity (triangles), while **3a** (lozenges) and N,N-dimesitylammonium pentafluorobenzenesulfonate ($[Mes_2N^+H_2][^-O_3SC_6F_5]$) (squares) exhibited significantly higher catalytic activities than $C_6F_5SO_3H$. In particular, the most bulky catalyst **3a** had the highest catalytic activity due to its efficient creation of local hydrophobic environment. Very interestingly, **3a** showed higher catalytic activity than TsOH despite the much weaker acidity of **3a** (circles, graph c versus lozenges, graph d). These experimental results suggested that two bulky N-aryl groups and an S-pentafluorophenyl group, which surrounded the active site ($^+NH_2$) of **3a**, synergistically accelerated the dehydration reactions. On the basis of the results in Schemes 2.7 and 2.8 and Table 2.1, the rate-accelerating effect on the **3a**-catalyzed dehydration reaction could be attributed to the local hydrophobic environment in **3a**. A similar tendency was observed in the ester condensation reaction of carboxylic acids with alcohols as well as the dehydrative cyclization of 1,3,5-triketones [5b]. However, the latter reaction was promoted much more effectively than the former.

X-ray single-crystal structures of the N,N-diarylammonium sulfonates suggest that a hydrophobic environment in their aggregates may play a role. Crystal

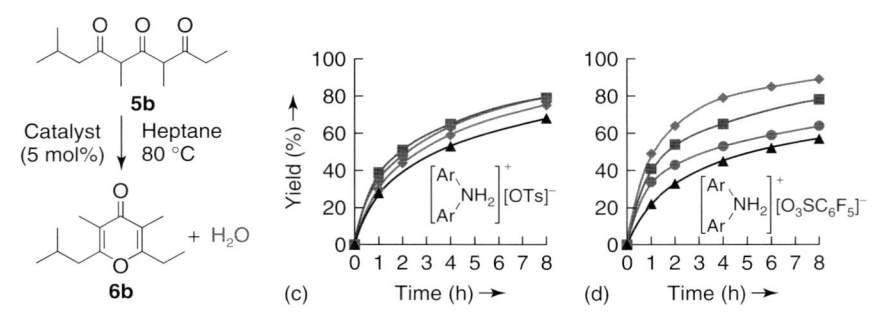

Scheme 2.8 Dehydrative cyclization of **5b** catalyzed by N,N-diarylammonium tosylates (graph c) and N,N-diarylammonium pentafluorobenzenesulfonates (graph d). The yield of γ-pyrone **6b** was evaluated by HPLC analysis. Lozenges, [(2,6-Ph$_2$C$_6$H$_3$)MesN$^+$H$_2$]; squares, [Mes$_2$N$^+$H$_2$]; circles, H$^+$; triangles, [Ph$_2$N$^+$H$_2$].

Scheme 2.9 Formation of **7** by crystallization of **3a**.

7 was obtained by the recrystallization of **3a**, which was a 1:1 molar mixture of N-(2,6-diphenylphenyl)-N-mesitylamine and C$_6$F$_5$SO$_3$H in CHCl$_3$–hexane (Scheme 2.9). Surprisingly, X-ray crystallographic analysis revealed that **7** is a supramolecular complex composed of two diarylammonium cations, four pentafluorobenzenesulfonate anions, and two oxonium cations (Figure 2.1).

Two ammonium cations and two oxonium cations in **7** are surrounded by 12 hydrophobic aryl groups, like reverse micelles. Furthermore, the cyclic ion pair is thermodynamically and conformationally stabilized by not only four "HN$^+$–H•••O=SO$_2^-$" and six "H$_2$O$^+$–H•••O=SO$_2^-$" intermolecular hydrogen bonds but also two intermolecular π–π interactions between mesityl groups and pentafluorophenyl groups, two intermolecular π–π interactions between phenyl groups and pentafluorophenyl groups and two intramolecular π–π interactions between mesityl groups and phenyl groups: the distance between the mesityl and pentafluorophenyl groups is 3.6–3.8 Å and that between the mesityl and phenyl groups is 3.0–3.6 Å. The extremely high catalytic activity of **3a** in the ester condensation and dehydrative cyclization of 1,3,5-triketones may be ascribed to the hydrophobic environment around ammonium protons in **7**, which includes carboxylic acids or 1,3,5-triketones in place of water.

2.2 Dehydrative Condensation Catalysts | 47

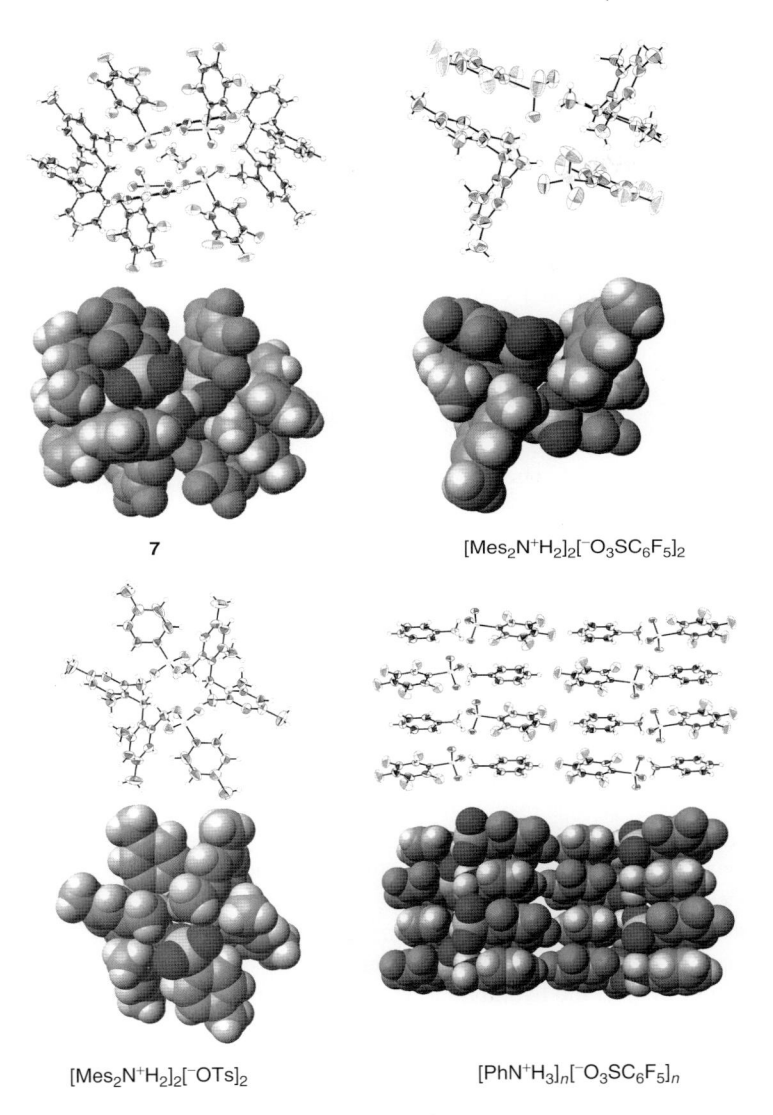

Figure 2.1 X-ray single-crystal structures of **7**, [Mes$_2$N$^+$H$_2$]$_2$[$^-$O$_3$SC$_6$F$_5$]$_2$, [Mes$_2$N$^+$H$_2$]$_2$[$^-$OTs]$_2$, and [PhN$^+$H$_3$]$_n$[$^-$O$_3$SC$_6$F$_5$]$_n$. Upper, ORTEP drawing; lower, space-filling drawing.

When crystal **7** was used as a catalyst instead of a 1 : 1 molar mixture of C$_6$F$_5$SO$_3$H and *N*-(2,6-diphenylphenyl)-*N*-mesitylamine, the ester condensation reaction of 4-phenylbutyric acid with cyclododecanol proceeded more slowly. An equilibrium mixture of **7** and *N*-(2,6-diphenylphenyl)-*N*-mesitylamine probably exists in a 1 : 1 molar solution of C$_6$F$_5$SO$_3$H and *N*-(2,6-diphenylphenyl)-*N*-mesitylamine in heptane. The above experimental results suggest that the ratio of **7** in a 1 : 1 molar solution of $C_6F_5SO_3H$ and N-(2,6-diphenylphenyl)-N-mesitylamine is much higher than that in a 2 : 1 molar solution.

The X-ray single-crystal structure of $[Mes_2N^+H_2][^-O_3SC_6F_5]$ is a dimeric complex composed of two N,N-dimesitylammonium cations and two pentafluorobenzenesulfonate anions (Figure 2.1). The ammonium cation moiety is surrounded by six aryl groups, and the cyclic ion pair is also stabilized by two intermolecular $\pi-\pi$ interactions. The distance between the N-mesityl and pentafluorophenyl groups is 3.5–3.6 Å. N,N-Dimesitylammonium tosylate ($[Mes_2N^+H_2][^-OTs]$) also forms a complex composed of two N,N-dimesitylammonium cations and two p-toluenesulfonate anions. In contrast to 7 and $[Mes_2N^+H_2]_2[^-O_3SC_6F_5]_2$, $[Mes_2N^+H_2]_2[^-OTs]_2$ does not exhibit intermolecular $\pi-\pi$ interactions between the N-mesityl and tolyl groups. In contrast to the N,N-diarylammonium sulfonates, anilinium pentafluorobenzenesulfonate did not form a cyclic ion pair structure. Therefore, N,N-diarylamine structures are important for the formation of cyclic ion pairs in which the ammonium cation moieties are surrounded by aryl groups.

Aggregated cyclic ion pairs 7 and $[Mes_2N^+H_2]_2[^-O_3SC_6F_5]_2$ may be similar to real active species in the dehydration reactions such as the ester condensation and dehydrative cyclization of 1,3,5-triketones. They are stabilized by not only intermolecular hydrogen bondings but also intermolecular $\pi-\pi$ interactions between the mesityl and pentafluorophenyl groups. Moreover, the use of a less polar solvent such as heptane also promotes tight aggregation between diarylamines and $C_6F_5SO_3H$, which is less acidic and less polar than TsOH. In contrast, $[Mes_2N^+H_2][^-OTs]^-$ is less active as a dehydration catalyst because of the instability of the cyclic ion pair, which is due to the absence of intermolecular $\pi-\pi$ interaction and the more acidic and more polar nature of TsOH. Therefore, it seems that the catalytic activity of $[Mes_2N^+H_2][^-OTs]$ is mainly due to its strong acidity. The formation of a stable cyclic ion pair, in which the ammonium protons are located in the local hydrophobic environment, is crucial for the excellent catalytic activity.

Water molecules produced at the active site of bulky N,N-diarylammonium pentafluorobenzenesulfonates are easily exchanged for less polar substrates such as carboxylic acids and 1,3,5-triketones. Once water molecules are released from the ammonium cation moiety, the hydrophobic wall prevents polar water molecules from gaining access to the active site of the catalysts, leading to the inhibition of inactivation of the catalysts by water. In contrast, less polar substrates can easily approach the active site through the hydrophobic wall and be efficiently activated. Thus, bulky N,N-diarylammonium pentafluorobenzenesulfonates exhibit remarkable catalytic activities for the dehydration reactions without any loss of catalytic activities even under conditions without the removal of water. In contrast, sulfonic acids interact with water more strongly than with less polar substrates. Therefore, sulfonic acids are inactivated under conditions without the removal of water.

A proposed mechanism for the dehydrative cyclization of 1,3,5-triketones is shown in Scheme 2.10. Bulky ammonium catalyst 3a should coordinate more preferentially with the carbonyl oxygen at the 5-position of 1,3,5-triketones 5 than with that at the 1-position, to avoid the steric hindrance of the R group (step 1). However, enol intermediate 9 is less stable than 8 because of steric

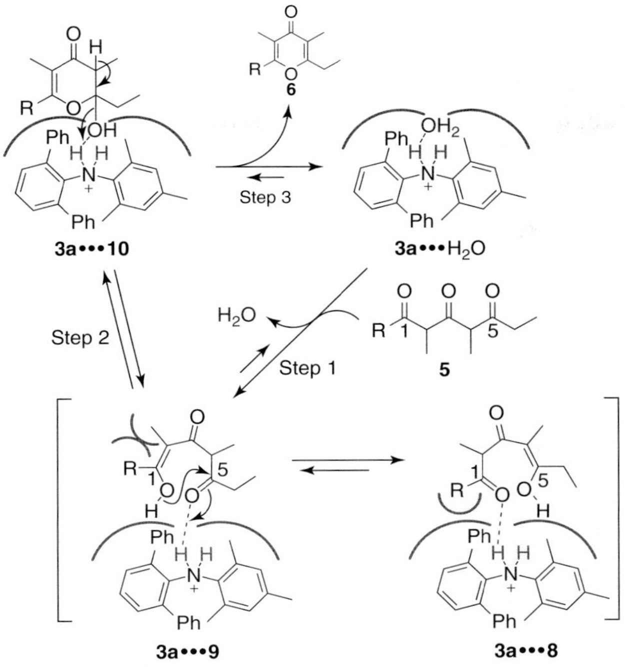

Scheme 2.10 Proposed mechanism of the dehydrative cyclization of 1,3,5-triketones **5**. $C_6F_5SO_3^-$ is omitted for clarity.

hindrance between the R group and methyl group. Therefore, **5b**, which has an isobutyl group, is less reactive than **5a**. Compound **9** is reversibly converted to cyclic hemiacetal **10** (step 2). Coordination of **3a** with the hydroxy group of **10** should make the hydrophilic hemiacetal moiety of **10** labile in the hydrophobic environment, and the dehydration of **10** (step 3) is promoted. Dehydration is also promoted by the steric hindrance of **3a**, and thus **10** is easily converted to the corresponding γ-pyrones **6**. The generated water is rapidly released from the active site of **3a** and easily exchanged for less polar **5**, due to the hydrophobic environment (step 1). Therefore, compound **3a** exhibited remarkable catalytic activities without being affected by the generated water.

In conclusion, bulky *N,N*-diarylammonium pentafluorobenzenesulfonates show an unusual rate-accelerating effect for dehydration reactions such as the ester condensation and the dehydrative cyclization of 1,3,5-triketones. In particular, the most bulky and hydrophobic catalyst **3a** shows much higher catalytic activity than $C_6F_5SO_3H$ even though **3a** has weaker acidity. It is conceivable that the local hydrophobic environment created by the tight aggregation of **3a** in less polar solvent efficiently promotes dehydrative reactions. In addition, **3a** remains as a catalyst in less polar solvents even in the presence of water generated by the condensation reaction. On the other hand, $C_6F_5SO_3H$ can easily be transferred to an aqueous layer generated by the dehydrative condensation. The X-ray crystallographic analysis of

7, which may be the real active species, suggests that stabilization of the cyclic ion pair by intermolecular $\pi-\pi$ interactions between hydrophobic bulky aryl groups is crucial for the creation of the hydrophobic environment.

2.3
Asymmetric Mannich-Type Catalysts

A chiral organic salt which consists of a Brønsted acid and a Brønsted base is one of the most promising catalysts in modern asymmetric syntheses [1, 2, 11, 12]. In general, acid–base combined salts have several advantages over single-molecule catalysts, with regard to the flexibility in the design of their dynamic complexes. Chiral ammonium salts of chiral amines with achiral Brønsted acids are typical examples of these organocatalysts with enantioselective function [13]. 2,2′-Disubstituted 1,1′-binaphthyl is one of the most popular chiral auxiliaries of asymmetric catalysts [14, 15]. However, bulky substituents at the 3,3′-positions of 1,1′-binaphthyl are often required to achieve high enantioselectivity in asymmetric catalyses. In sharp contrast, chiral 1,1′-binaphthyl-2,2′-disulfonic acid (BINSA, **11**) [16] is a promising chiral Brønsted acid, since both the Brønsted acidity and bulkiness can easily be controlled by complexation with achiral amines without substitutions at the 3,3′-positions in a binaphthyl skeleton (Scheme 2.11). However, despite this potential, there had been no reports on the application of chiral **11** to asymmetric catalyses since the first synthesis of *rac*-**11** in 1928 by Barber and Smiles [16]. The author and coworkers developed a practical synthesis of chiral **11** from inexpensive 1,1′-bi(2-naphthol) (BINOL) and efficient enantioselective catalysis in direct Mannich-type reactions using **11**–2,6-diarylpyridine (**12**) combined salts as tailor-made chiral Brønsted acid–base organocatalysts *in situ* [17].

The method used to prepare (*R*)-**11** from (*R*)-BINOL via the oxidation of dithiol (*R*)-**15** is shown in Scheme 2.12 [18].[1)] Thermolysis in the Newman–Kwart rearrangement of (*R*)-**13** to (*R*)-**14** was dramatically improved by using a microwave technique at a milder temperature (200 °C) [18, 19].[2)] The oxidation of thiols (RSH) to sulfonic acids (RSO$_3$H) is usually accompanied by the generation of disulfides (RS–SR) via intermolecular reactions [20]. In particular, it seems that dithiol (*R*)-**15** bearing two SH groups at the 2,2′-positions in a binaphthyl skeleton may be suitable for the formation of an oxidative S–S bond, intramolecularly [18]. However, the oxidation of (*R*)-**13** proceeded smoothly in 82% yield without epimerization under 7 atm of O$_2$/KOH in HMPA (hexamethylphosphoric triamide). The chemical structure of the potassium salt of (*R*)-**11** was ascertained by X-ray diffraction analysis

---

1) Compound **15** is commercially available from International Laboratory USA, Interchim, and WaterstoneTech.
2) In the original literature in Refs. [18], the Newman–Kwart rearrangement of **13** to **14** was examined at 285–400 °C. Moreover, during the preparation of this manuscript, a 1–400 kg-scale synthesis of aryl *S*-thiocarbamates from aryl *O*-thiocarbamates in microwave reactors was reported by Moseley and coworkers. The industrial synthesis of **11** should offer a great advantage.

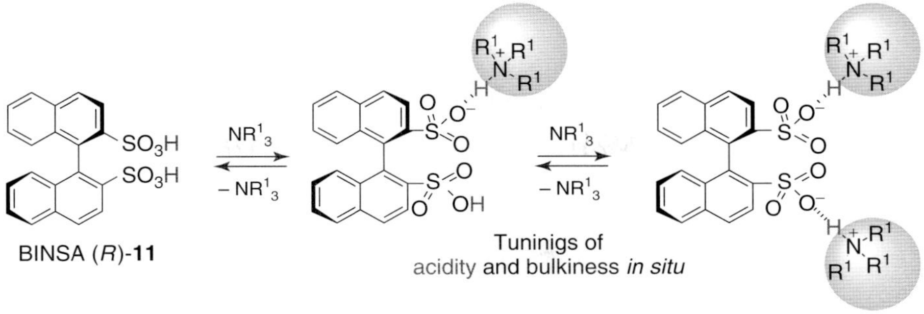

Scheme 2.11 Dynamic complexation of BINSA (R)-**11** with a tertiary amine.

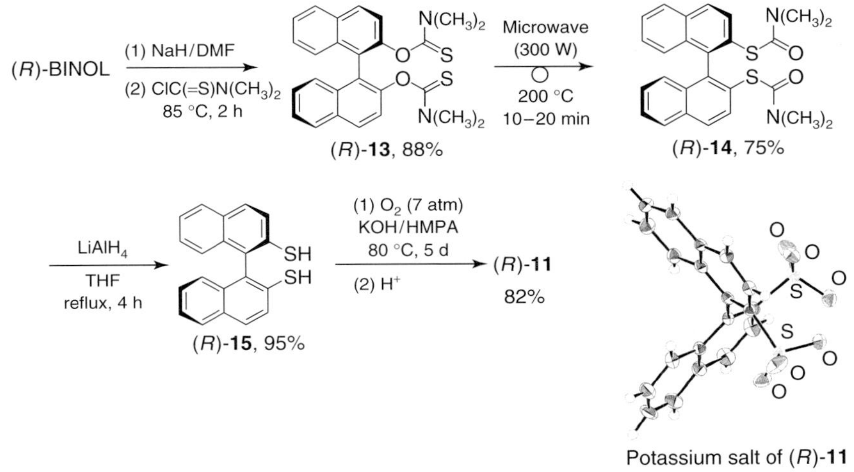

Scheme 2.12 Asymmetric synthesis of (R)-**11** and the X-ray structure of potassium salt of (R)-**11** (K$^+$ is omitted).

of its single crystal. Compound (R)-**11** was isolated by ion-exchange. Thus, (R)-**11** could be prepared in 51% yield over five steps from (R)-BINOL, or in 82% yield in one step from commercially available (R)-**15** [21].[3)]

The enantioselective direct Mannich-type reaction [15, 22] was examined using (R)-**11** as a chiral Brønsted acid catalyst (Table 2.2). Since the reaction between N-Cbz-phenylaldimine (**16a**) and acetylacetone (**17a**) proceeded without catalysts in dichloromethane at 0 °C, the slow addition of **17a** was the key to preventing the achiral pathway. However, despite such care, the enantioselectivity of **18a** was low (17% ee) when 5 mol% of (R)-**11** was used (entry 1). Next, chiral (R)-**11**–achiral amine combined salts prepared *in situ* were examined as chiral Brønsted acid–base

3) 3,3′-Bis[4-(β-naphthyl)phenyl]-substituted chiral binaphthol phosphoric acid catalyst, which was proved to be the most effective in the direct Mannich-type reaction by Terada and coworkers, was prepared from (R)-BINOL in 31% yield over seven steps. See [15].

Table 2.2 Ammonium salts of (R)-11 as tailor-made catalysts.[a]

$$\underset{\substack{\text{16a (R = Cbz)}\\\text{16b (R = Boc)}}}{\text{Ph}\overset{N^{\nearrow R}}{\underset{H}{\diagup\!\!\!\diagdown}}} + \underset{\substack{\text{17a}\\\text{(1.1 equiv.)}}}{\text{Ac-Ac}} \xrightarrow[\text{CH}_2\text{Cl}_2,\ 0\ ^\circ\text{C, 30 min}]{\substack{(R)\text{-11 (5 mol\%)}\\\text{amine (10 mol\%)}}} \underset{\substack{\text{18a (R = Cbz)}\\\text{18b (R = Boc)}}}{\text{Ph}\overset{\text{NHR}}{\underset{(R)}{\diagup\!\!\!\diagdown}}\text{Ac}\!\!-\!\!\text{Ac}}$$

Entry	16	Amine	Yield (%)	ee (%)
1	16a	–	81	17
2	16a	C_5H_5N	8	5
3	16a	$2\text{-Ph-}C_5H_4N$	11	10
4	16a	$2,6\text{-}(CH_3)_2\text{-}C_5H_3N$	19	0
5	16a	$2,6\text{-}t\text{-Bu}_2\text{-}C_5H_3N$	32	76
6	16a	$2,6\text{-Ph}_2\text{-}C_5H_3N$ (12a)	74	92
7	16b	12a	83	85
8[b]	16a	12a	85	8 (S)

[a] Acetylacetone **17a** was added at 0 °C over 1 h, and the resultant mixture was stirred for 30 min.
[b] (S)-1,1′-Binaphthyl-2,2′-dicarboxylic acid was used instead of (R)-**11**. After being stirred at 0 °C for 30 min, the reaction mixture was further stirred at room temperature for 4 h.

catalysts (Scheme 2.11). Some preliminary results using (R)-**11** (5 mol%)–amines (10 mol%) suggested that pyridines with weak Brønsted basicity would be better Brønsted bases, while trialkylamines with strong Brønsted basicity were much less active, and anilines caused side reactions such as the Friedel–Crafts reaction. Even then, pyridine, 2-phenylpyridine, and 2,6-lutidine also gave (R)-**18a** in low yield due to the insolubility of the corresponding salts (entries 2–4). In sharp contrast, 2,6-di-*tert*-butylpyridine improved the enantioselectivity up to 76% ee (entry 5). Moreover, (R)-**11** with 2,6-diphenylpyridine (**12a**), which led to a homogeneous catalyst *in situ*, was found to be highly effective, and (R)-**18a** was obtained in 74% yield with 92% ee (entry 6). N-Boc-phenylaldimine (**16b**), which had been reported by Terada and coworkers using pioneering chiral phosphoric acids (Scheme 2.13) [15, 21], was compatible with the present reaction conditions using (R)-**11** (5 mol%) and **12a** (10 mol%), and the corresponding adduct (R)-**18b** was obtained in 83% yield with 85% ee (entry 7). In their catalytic enantioselective reactions, acetylacetone [15a] was the sole nucleophile. Moreover, N-Boc protection in aldimines is essential for achieving high enantioselectivities. (S)-BINSA (5 mol%) and **12a** (10 mol%) showed low catalytic activity and low enantioselectivity (8% ee) under the same conditions as in entry 6 (entry 8) [14j].

The molar ratio of **12a** (0–15 mol%) to (R)-**11** (5 mol%) was optimized for the above direct Mannich-type reaction of **17a** with **16a** (Table 2.3). Interestingly, the enantioselectivities of **18a** were dramatically improved when a more than 1 : 0.75

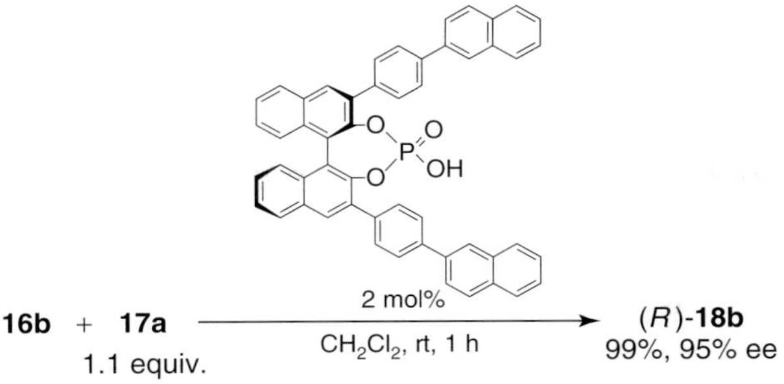

Scheme 2.13 Example of the enantioselective direct Mannich reaction using a single-molecule catalyst.

16b + 17a (1.1 equiv.) → (R)-18b, 99%, 95% ee
2 mol% catalyst, CH$_2$Cl$_2$, rt, 1 h

Table 2.3 Effect of the ratio of (R)-**11** : **12a**.

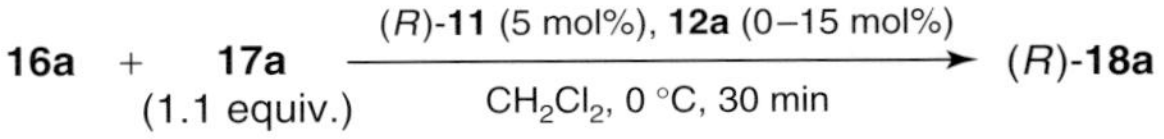

16a + 17a (1.1 equiv.) → (R)-**18a**
(R)-**11** (5 mol%), **12a** (0–15 mol%), CH$_2$Cl$_2$, 0 °C, 30 min

Entry	11 : 12a	Yield (%)	ee (%)	Entry	11 : 12a	Yield (%)	ee (%)
1	1 : 0	81	17	6	1 : 1.5	84	90
2	1 : 0.25	82	17	7	1 : 2	74	92
3	1 : 0.5	83	34	8	1 : 2.5	76	95
4	1 : 0.75	81	79	9	1 : 3	68	86
5	1 : 1	82	84	–	–	–	–

ratio of (R)-**11** : **12a** was examined (entries 4–9 vs entries 1–3).[4] As a result, a 1 : 1.5−1 : 2.5 ratio of (R)-**11** : **12a** was effective for achieving both a high yield and a high enantioselectivity (entries 6–8). Probably, the wide range of suitable ratios for (R)-**11** : **12a** is due to the dynamic structure of the catalysts (Scheme 2.11).

Fortunately, (R)-**18a** was obtained in 91% yield with 90% ee with the use of 1 mol% of (R)-**11** and 2 mol% of **12a** in the presence of 1.7 equiv. of MgSO$_4$, which would prevent the decomposition of **16a** (1.5 equiv.) due to adventitious moisture (Table 2.4, entry 1). Under these optimized conditions, N-Boc-Mannich product (R)-**18b** was obtained in 99% yield with 84% ee (entry 2). From **17a** and a variety of N-Cbz-arylaldimines bearing electron-donating or electron-withdrawing groups in the aryl or heteroaryl moiety, the corresponding adducts (**18c–j**) were obtained in excellent yields (92–99%) and with high enantioselectivities (89–98% ee) (entries 3–8). When other diketones such as 3,5-heptanedione (**17b**) and 1,3-diphenylpropane-1,3-dione (**17c**) were made to react with **16a**, **18i** and **18j** were obtained with 95 and 84% ee, respectively (entries 9 and 10).

4) Probably, (R)-**11**·**12** rather than (R)-**11**·(**12**)$_2$ may be the active species *in situ*, since there should be a dynamic equilibrium among (R)-**11**, (R)-**11**·**12**, and (R)-**11**·(**12**)$_2$ even if there is a large amount of **12** relative to (R)-**11** (Scheme 2.11 and Table 2.3).

Table 2.4 Catalytic enantioselective direct Mannich-type reaction.

Entry	16 (R, Ar)	17 (R')	18	Yield (%)	ee (%)
1	16a (Cbz, Ph)	17a (Me)	18a	91	90 (R-(−))
2	16b (Boc, Ph)	17a (Me)	18b	99	84 (R-(+))
3	16c (Cbz, o-CH$_3$C$_6$H$_4$)	17a (Me)	18c	99	96 ((+))
4	16d (Cbz, m-CH$_3$C$_6$H$_4$)	17a (Me)	18d	99	89 ((+))
5	16e (Cbz, p-CH$_3$OC$_6$H$_4$)	17a (Me)	18e	95	96 ((+))
6	16f (Cbz, p-BrC$_6$H$_4$)	17a (Me)	18f	92	98 (R-(−))
7	16g (Cbz, 1-Naph)	17a (Me)	18g	99	96 ((−))
8	16h (Cbz, 3-Thionyl)	17a (Me)	18h	98	98 ((+))
9	16a (Cbz, Ph)	17b (Et)	18i	95	95 ((−))
10	16a (Cbz, Ph)	17c (Ph)	18j	>99	84 ((−))

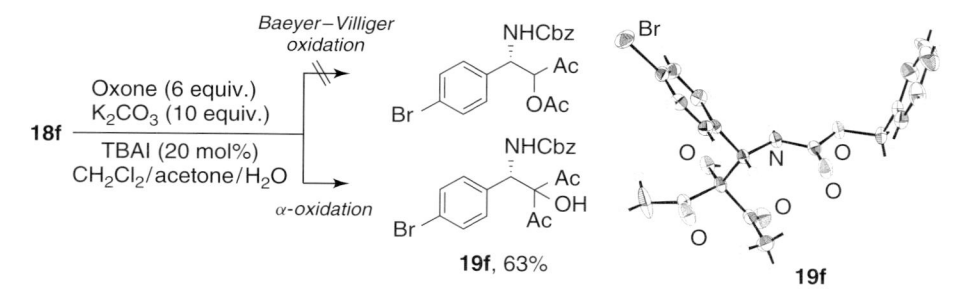

Scheme 2.14 Unexpected oxidation of **18f** and X-ray analysis of **19f**.

The absolute stereochemistry of the products **18a** and **18b** was determined by following Terada's procedure, which includes Baeyer–Villiger oxidation [15a]. However, unexpected tertiary alcohols (**19**) [23] were obtained exclusively instead of the Baeyer–Villiger products when Mannich adducts (**18**) were oxidized under the same reaction conditions as reported by Terada. The chemical structure and absolute conformation of **19f** were determined by X-ray analysis (Scheme 2.14).

Moreover, cyclic 1,3-diketone **17d** could also be used, and the corresponding adduct **18k** with a quaternary carbon center was obtained in 98% yield with a *syn/anti* diastereomer ratio of 83/17 and high enantioselectivity (91 and 96% ee, respectively) (Scheme 2.15).

2.3 Asymmetric Mannich-Type Catalysts

Scheme 2.15 Enantio- and diastereoselective direct Mannich-type reaction.

Scheme 2.16 Enantio- and diastereoselective direct Mannich-type reaction between **16a** and 1,3-ketoamide (**20**).

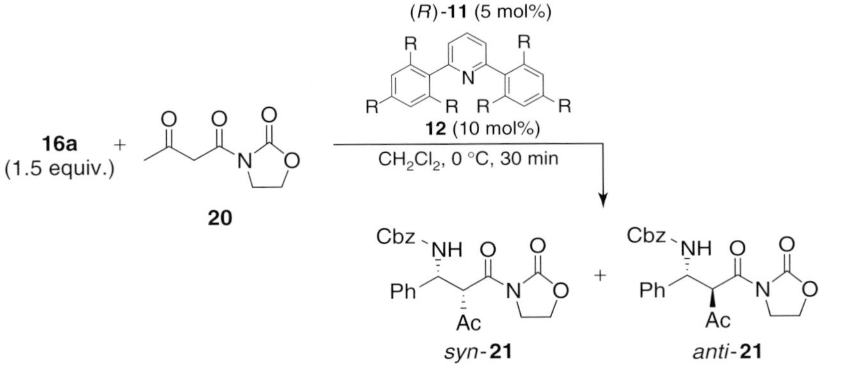

12a (R = H): 86% (dr = 53:47), 72% ee (*syn*) and 20% ee (*anti*)
12b (R = CH$_3$): 81% (dr = 60:40), 93% ee (*syn*) and 90% ee (*anti*)

A suitable chiral ammonium salt was easily tailor-made for a ketoester equivalent such as 3-acetoacetyl-2-oxazolidinone (**20**) (Scheme 2.16). Chiral ammonium salt (*R*)-**11**·**12a**$_2$, which was optimized for the reaction of diketones **17** with **16**, was not effective, and the desired product **21** was obtained in 86% yield with low diastereo- and enantioselectivities. In contrast, the enantioselectivity of **21** increased to 93% ee when 2,6-dimesitylpyridine (**12b**) was used in place of **12a**. In this way, tailor-made salts of (*R*)-**11** and **12** made it possible to avoid preparing single-molecule catalysts in advance and offered a quick solution to this type of optimization problem.

Compound **21** could easily be transformed to β-amino carbonyl compound **22** via deprotection of the oxazolidinone moiety without the loss of enantioselectivity (Scheme 2.17).

In summary, BINSA (*R*)-**11** is a highly effective chiral Brønsted acid that can be combined with an achiral Brønsted base. The combination of the achiral bulky 2,6-diarylpyridine **12** with the simple disulfonic acid (*R*)-**11** circumvents the trouble of having to build bulky substituents at the 3,3′-positions, as is normally required in analogous binaphthyl phosphoric acid catalysts. In the presence of 1 mol% of

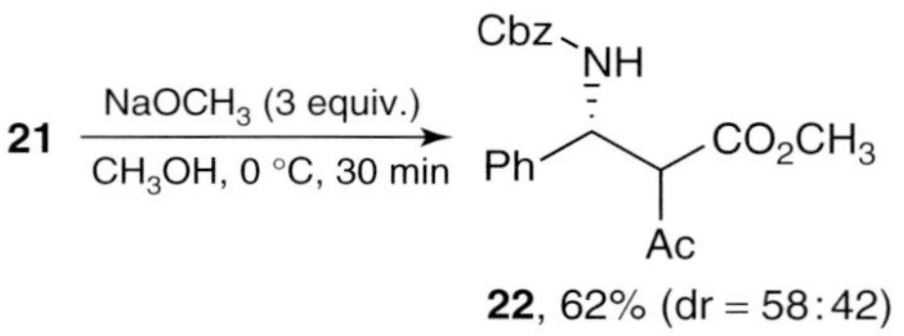

Scheme 2.17 Deprotection of the oxazolidinone moiety.

(R)-**11** and 2 mol% of **12**, highly enantioselective direct Mannich-type reactions of a variety of 1,3-diketones and a 1,3-ketoester equivalent with arylaldimines proceed smoothly with high enantioselectivities. BINSA is a powerful chiral auxiliary like BINOL, BINAP (2,2′-bis(diphenylphosphino)-1,1′-binaphthalene), and BINAM (2,2′-diamino-1,1′-binaphthalene), and is expected to trigger a new frontier in acid–base chemistry in asymmetric catalyses [24].

References

1. For excellent reviews in acid– base chemistry: (a) Shibasaki, M. and Yoshikawa, N. (2002) *Chem. Rev.*, **102**, 2187; (b) List, B. (2002) *Tetrahedron*, **58**, 5573; (c) Maruoka, K. and Ooi, T. (2003) *Chem. Rev.*, **103**, 3013; (d) Notz, W., Tanaka, F., and Barbas, C.F. III (2004) *Acc. Chem. Res.*, **37**, 580; (e) Tian, S.-K., Chen, Y., Hang, J., Tang, L., Mcdaid, P., and Deng, L. (2004) *Acc. Chem. Res.*, **37**, 621; (f) Lelais, G. and MacMillan, D.W.C. (2006) *Aldrichim. Acta*, **39**, 79.
2. Ishihara, K., Sakakura, A., and Hatano, M. (2007) *Synlett*, 686.
3. Ishihara, K. (2009) *Tetrahedron*, **65**, 1085–1109.
4. (a) Wakasugi, K., Misaki, T., Yamada, K., and Tanabe, Y. (2000) *Tetrahedron Lett.*, **41**, 5249–5252; (b) Funatomi, T., Wakasugi, K., Misaki, T., and Tanabe, Y. (2006) *Green Chem.*, **8**, 1022–1027.
5. (a) Ishihara, K., Nakagawa, S., and Sakakura, A. (2005) *J. Am. Chem. Soc.*, **127**, 4168–4169; (b) Sakakura, A., Nakagawa, S., and Ishihara, K. (2006) *Tetrahedron*, **62**, 422–433; (c) Sakakura, A., Nakagawa, S., and Ishihara, K. (2007) *Nat. Protoc.*, **2**, 1746–1751. (d) Sakakura, A., Watanabe, H., Nakagawa, S., and Ishihara, K. (2007) *Chem. Asian J.*, **2**, 477–483.
6. For examples, of other catalytic methods for the dehydrative cyclization of 1,3,5-triketones, see: (a) TsOH (ca. 60% yield): Asami, T., Yoshida, S., and Takahashi, N. (1986) *Agric. Biol. Chem.*, **50**, 469–474; (b) 5% methanolic H_2SO_4 (50% yield): Harris, T.M., Murphy, G.P., and Poje, A.J. (1976) *J. Am. Chem. Soc.*, **98**, 7733–7741; (c) 0.5 M HCl (50% yield): Dorman, L.C. (1967) *J. Org. Chem.*, **32**, 4105–4107; d) conc. H_2SO_4 (59–91% yield): Light, R.J. and Hauser, C.R. (1960) *J. Org. Chem.*, **25**, 538–546; (e) 45% HBr or polyphosphoric acid (no data on yields): O'Sullivan, W.I. and Hauser, C.R. (1960) *J. Org. Chem.*, **25**, 1110–1114.
7. For esterification in fluorous media, see: Gacem, B. and Jenner, G. (2003) *Tetrahedron Lett.*, **44**, 1391–1393.
8. Clayden, J., Greeves, N., Warren, S., and Wothers, P. (2001) *Organic Chemistry*, Oxford University Press, Oxford.
9. Rode, B.M., Engelbrechit, A., and Schantl, J. (1973) *Z. Physik. Chem. (Leipzig)*, **253** (1–2), 17–24.
10. Habel, W. and Sartori, P. (1982) *J. Fluorine Chem.*, **20**, 559–572.
11. For an excellent textbook: Berkessel, H. and Gröger, H. (eds) (2005) *Asymmetric Organocatalysis*, WILEY-VCH Verlag GmbH, Weinheim.

12. Reviews for chiral Brønsted acid catalysts: (a) Pihko, P.M. (2004) *Angew. Chem., Int. Ed.*, **43**, 2062; (b) Pihko, P.M. (2005) *Lett. Org. Chem.*, **2**, 398; (c) Akiyama, T., Itoh, J., and Fuchibe, K. (2006) *Adv. Synth. Catal.*, **348**, 999; (d) Taylor, M.S. and Jacobsen, E.N. (2006) *Angew. Chem., Int. Ed.*, **45**, 1520; (e) Connon, S.J. (2006) *Angew. Chem., Int. Ed.*, **45**, 3909; (f) Akiyama, T. (2007) *Chem. Rev.*, **107**, 5744.
13. (a) Corey, E.J. and Grogan, M.J. (1999) *Org. Lett.*, **1**, 157–160; (b) List, B., Lerner, R.A., and Barbas, C.F. III (2000) *J. Am. Chem. Soc.*, **122**, 2395–2396; (c) Hodous, B.L. and Fu, G.C. (2002) *J. Am. Chem. Soc.*, **124**, 10006–10007; (d) Huang, J. and Corey, E.J. (2004) *Org. Lett.*, **6**, 5027–5029; (e) Ouellet, S.G., Tuttle, J.B., and MacMillan, D.W.C. (2005) *J. Am. Chem. Soc.*, **127**, 32–33; (f) Wiskur, S.L. and Fu, G.C. (2005) *J. Am. Chem. Soc.*, **127**, 6176–6177; (g) Nugent, B.M., Yoder, R.A., and Johnston, J.N. (2004) *J. Am. Chem. Soc.*, **126**, 3418; (h) Ishihara, K. and Nakano, K. (2005) *J. Am. Chem. Soc.*, **127**, 10504; (i) Hoffmann, S., Seayad, A.M., and List, B. (2005) *Angew. Chem., Int. Ed.*, **44**, 7424; (j) Sakakura, A., Suzuki, K., Nakano, K., and Ishihara, K. (2006) *Org. Lett.*, **8**, 2229; (k) Sakakura, A., Suzuki, K., and Ishihara, K. (2006) *Adv. Synth. Catal.*, **348**, 2457–2465; (l) Kano, T., Tanaka, Y., and Maruoka, K. (2006) *Org. Lett.*, **8**, 2687; (m) Itoh, J., Fuchibe, K., and Akiyama, T. (2006) *Angew. Chem., Int. Ed.*, **45**, 4796; (n) Rueping, M. and Azap, C. (2006) *Angew. Chem., Int. Ed.*, **45**, 7832; (o) Mayer, S. and List, B. (2006) *Angew. Chem., Int. Ed.*, **45**, 4193; (p) Martin, N.J.A. and List, B. (2006) *J. Am. Chem. Soc.*, **128**, 13368; (q) Ishihara, K. and Nakano, K. (2007) *J. Am. Chem. Soc.*, **129**, 8930; (r) Ishihara, K., Nakano, K., and Akakura, M. (2008) *Org. Lett.*, **10**, 2893–2896.
14. Organocatalyses with chiral 3,3'-disubstituted binaphthyl compounds: (a) McDougal, N.T. and Schaus, S.E. (2003) *J. Am. Chem. Soc.*, **125**, 12094; (b) Akiyama, T., Itoh, J., Yokota, K., and Fuchibe, K. (2004) *Angew. Chem., Int. Ed.*, **43**, 1566; (c) Matsui, K., Takizawa, S., and Sasai, H. (2005) *J. Am. Chem. Soc.*, **127**, 3680; (d) Rueping, M., Sugiono, E., Azap, C., Theissmann, T., and Bolte, M. (2005) *Org. Lett.*, **7**, 3781; (e) Rowland, G.B., Zhang, H., Rowland, E.B., Chennamadhavuni, S., Wang, Y., and Antilla, J.C. (2005) *J. Am. Chem. Soc.*, **127**, 15696; (f) Liu, H., Cun, L.-F., Mi, A.-Q., Jiang, Y.-Z., and Gong, L.-Z. (2006) *Org. Lett.*, **8**, 6023; (g) Storer, R.I., Carrera, D.E., Ni, Y., and MacMillan, D.W.C. (2006) *J. Am. Chem. Soc.*, **128**, 84; (h) Nakashima, D. and Yamamoto, H. (2006) *J. Am. Chem. Soc.*, **128**, 9626; (i) Kang, Q., Zhao, Z.-A., and You, S.-L. (2007) *J. Am. Chem. Soc.*, **129**, 1484; (j) Hashimoto, T. and Maruoka, K. (2007) *J. Am. Chem. Soc.*, **129**, 10054; (k) Guo, Q.-S., Du, D.-M., and Xu, J. (2008) *Angew. Chem., Int. Ed.*, **47**, 759.
15. (a) Uraguchi, D. and Terada, M. (2004) *J. Am. Chem. Soc.*, **126**, 5356; (b) Terada, M., Sorimachi, K., and Uraguchi, D. (2006) *Synlett*, 133; (c) Gridnev, I.D., Kouchi, M., Sorimachi, K., and Terada, M. (2007) *Tetrahedron Lett.*, **48**, 497.
16. (a) Barber, H.J. and Smiles, S. (1928) *J. Chem. Soc.*, 1141; (b) Armarego, W.L.F. and Turner, E.E. (1957) *J. Chem. Soc.*, 13; (c) Takahashi, K. and Fukishi, K. (2005) Japan Patent 2005132815.
17. Hatano, M., Maki, T., Moriyama, K., Arinobe, M., and Ishihara, K. (2008) *J. Am. Chem. Soc.*, **130**, 16858–16860.
18. (a) Fabbri, D., Delogu, G., and De Lucchi, O. (1993) *J. Org. Chem.*, **58**, 1748; (b) Bandarage, U.K., Simpson, J., Smith, R.A.J., and Weavers, R.T. (1994) *Tetrahedron*, **50**, 3463.
19. (a) Moseley, J.D., Lenden, P., Lockwood, M., Ruda, K., Sherlock, J.-P., Thomson, A.D., and Gilday, J.P. (2008) *Org. Process Res. Dev.*, **12**, 30; (b) Gilday, J.P., Lenden, P., Moseley, J.D., and Cox, B.G. (2008) *J. Org. Chem.*, **73**, 3130.
20. (a) Wallace, T.J. and Schriesheim, A. (1965) *Tetrahedron*, **21**, 2271; (b) Agami, C., Prince, B., and Puchot, C. (1990) *Synth. Commun.*, **20**, 3289.
21. Also see: Wipf, P. and Jung, J.-K. (2000) *J. Org. Chem.*, **65**, 6319.

22. (a) Poulsen, T.B., Alemparte, C., Saaby, S., Bella, M., and Jørgensen, K.A. (2005) *Angew. Chem., Int. Ed.*, **44**, 2896; (b) Lou, S., Taoka, B., Ting, A., and Schaus, S.E. (2005) *J. Am. Chem. Soc.*, **127**, 11256; (c) Tillman, A.L., Ye, J., and Dixon, D.J. (2006) *Chem. Commun.*, 1191; (d) Rueping, M., Sugiono, F., and Schoepke, F.R. (2007) *Synlett*, 1441; (e) Ting, A. and Schaus, S.E. (2007) *Eur. J. Org. Chem.*, 5797.
23. House, H.O. and Gannon, W.F. (1958) *J. Org. Chem.*, **23**, 879.
24. García-García, P., Lay, F., García-García, P., Rabalakos, C., and List, B. (2009) *Angew. Chem. Int. Ed.*, **48**, 4363–4366.

3
Asymmetric Oxidation with Hydrogen Peroxide, an Effective and Versatile Oxidant

Tsutomu Katsuki

3.1
Introduction

Oxygen functionalities such as hydroxy and epoxy groups can be found in many biologically active compounds, and it is well known that their biological activities largely depend on the location and stereochemistry of the functional groups. Thus, development of an efficient method for introducing oxygen functionalities with strict stereo- and regiocontrol has long been a challenging research target in organic synthesis and many stereoselective functionalization methods have been developed to date. Among the methods of preparation of enantioenriched oxygen functionalities, asymmetric oxidation reactions are the most potent, straightforward, and widely used, as judged by the enormous number of references to hydroxylation, epoxidation, dihydroxylation, aminohydroxylation, and so on, in the literature [1]. Today, high enantioselectivity has been achieved in many oxidation reactions, but the atom efficiencies of most oxidations are still below satisfactory levels. On the other hand, resource depletion and environment deterioration are rapidly increasing global problems. Considering the importance of oxidation reaction as a tool for chemical transformation, enhancement of its atom efficiency and ecological sustainability is an urgent issue for realizing green sustainable chemistry. The atom efficiency of an oxidation reaction primarily depends on the active oxygen content of the oxidant used, and the ecological sustainability is significantly related to the chemical and physiological properties of the coproduct derived from the oxidant [2]. Although a variety of oxidants are available for catalytic oxidation today, the oxidants that meet the above criteria are very limited. The active oxygen content of most oxidants is lower than 30% and the coproducts derived from those oxidants are generally not compatible with the environment. Only two oxidants, molecular oxygen and hydrogen peroxide, satisfy the criteria. The active oxygen content of molecular oxygen is 50% when one of the two oxygen atoms is used for oxidation, and 100% when the two oxygen atoms are used, whereas that of hydrogen peroxide is 47% (Scheme 3.1). In addition to the high active oxygen content of these oxidants, the coproduct derived from the oxidants is only water, if the reaction selectivity is complete.

Pharmaceutical Process Chemistry. Edited by Takayuki Shioiri, Kunisuke Izawa, and Toshiro Konoike
Copyright © 2011 WILEY-VCH Verlag GmbH & Co. KGaA, Weinheim
ISBN: 978-3-527-32650-1

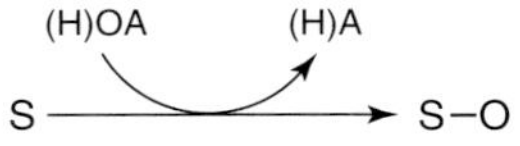

(H)OA   (H)A   A = active group, (H)A = coproduct,
S ⟶ S–O   active oxygen content (AOC) (%) = [O]/[(H)OA] × 100,
                     H_2O_2 : AOC = 47%, coproduct = H_2O

Scheme 3.1

Thus, enormous effort has been directed to the development of asymmetric oxidation catalysts that induce high asymmetry with the use of an ecofriendly oxidant, molecular oxygen, or hydrogen peroxide. Efficient catalysts for asymmetric aerobic oxidation have been developed in this decade and they have been successfully used for dehydrogenation-type asymmetric oxidation such as oxidative kinetic resolution of racemic alcohols, desymmetrization of *meso*-diols, and oxidative coupling of 2-naphthols [2b, 3], while oxygen atom transfer–type asymmetric oxidation using molecular oxygen continues to be rare in the literature [4, 5]. On the other hand, development of catalysts for asymmetric oxidation using aqueous hydrogen peroxide which is commercially available, inexpensive, and easy-to-handle has been rapidly growing and various excellent oxygen atom–transfer oxidation reactions have been reported. In this chapter, the development of asymmetric oxygen atom transfer reactions, in particular epoxidation and oxidation of sulfides, using hydrogen peroxide as the terminal oxidant is described.

3.2
Asymmetric Epoxidation

3.2.1
Asymmetric Epoxidation with Synthetic Enzymes or Organocatalysts

The pK_a of hydrogen peroxide ($pK_a = 11.6$) is less than that of water ($pK_a = 15.7$). In 1976, considering this nature of hydrogen peroxide, Wynberg *et al.* reported the first asymmetric epoxidation using aqueous hydrogen peroxide under basic conditions in the presence of a cinchona alkaloid–derived phase transfer catalyst, albeit with modest enantioselectivity [6]. Subsequently, Juliá *et al.* reported highly enantioselective epoxidation of chalcones using aqueous hydrogen peroxide in a triphasic toluene–water–poly amino acid system [7]. Moreover, Shioiri *et al.* also reported that *N*-(*p*-iodobenzyl)cinchona alkaloid salt serves as a phase transfer catalyst for highly enantioselective epoxidation of α,β-unsaturated ketones [8]. Although high enantioselectivity has been achieved with these methods, good substrates were largely limited to chalcone derivatives.

Recently, some chiral organic compounds including phase transfer reagents have been found to show unique and diverse asymmetric catalysis. Among them, asymmetric organocatalysis via iminium salt and enamine intermediates has recently received great attention due to its wide applicability [9]. As one of such applications, Jørgensen *et al.* reported enantioselective epoxidation of α,β-unsaturated aldehydes using aqueous hydrogen peroxide as the oxidant and a chiral pyrrolidine derivative **1** as the catalyst. This reaction can be successfully applied to both

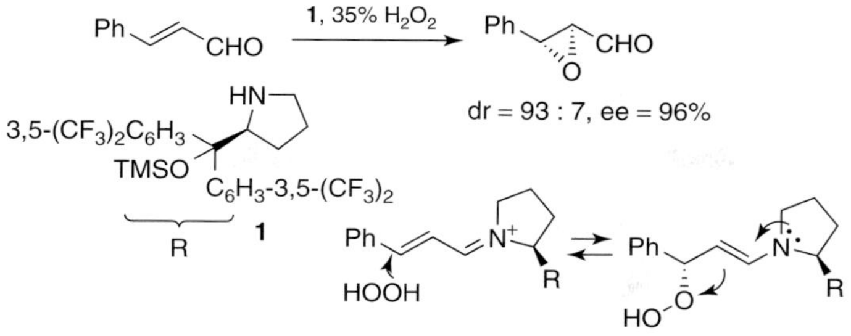

Scheme 3.2

β-aryl- and β-alkyl-substituted α,β-unsaturated aldehydes with high enantioselectivity (Scheme 3.2) [10a]. Since the peroxide addition produces an enamine intermediate and the rotation around the $C_{\alpha-\beta}$ bond is easy, geometrical isomerization occurs to some extent (<10%), even when the substrate is an E-enal. Córdova et al. have, independently, reported that a slightly different chiral pyrrolidine derivative catalyzes the epoxidation of α,β-unsaturated aldehydes using aqueous hydrogen peroxide with a similar level of enantioselectivity [10b].

A unique new concept, asymmetric counteranion-directed catalysis, proposed by Wang and List enabled asymmetric epoxidation of α,β-unsaturated aldehydes using t-butyl hydroperoxide. The epoxidation proceeds in the presence of an achiral ammonium chiral phosphate salt with high enantioselectivity (up to 96% ee) [11]. They further reported a method for epoxidizing cyclic enones using aqueous hydrogen peroxide with excellent enantioselectivity in the presence of a chiral diamine/chiral phosphate salt **2** as the catalyst (Scheme 3.3) [12].

The substrates of epoxidation using phase transfer catalysts or organocatalysts are mainly limited to α,β-unsaturated carbonyl compounds and their derivatives. On the other hand, in 1984, Curci et al. reported a seminal study on asymmetric epoxidation of simple olefins using a chiral ketone in the presence of potassium peroxomonosulfate, though enantioselectivity was only modest [13]. Later, enantioselectivity was remarkably improved by two groups. Yang et al. reported that a binaphthyl-1,2-dicarboxylic acid–derived ketone showed good enantioselectivity up to 87% ee [14], and Shi et al. reported that a fructose-derived ketone **3** showed high to excellent enantioselectivity (80–95% ee) in the epoxidation of various olefins [15]. The active species of this epoxidation is a dioxirane, and Shu and Shi

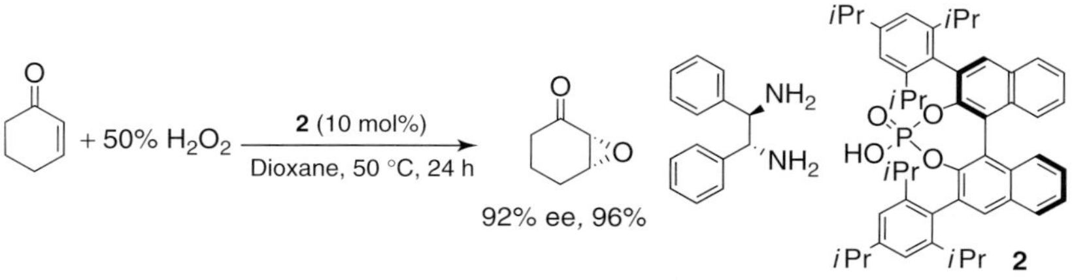

Scheme 3.3

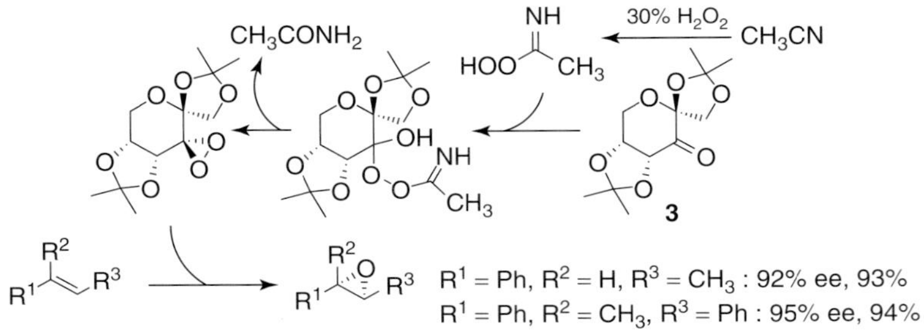

$R^1 = Ph$, $R^2 = H$, $R^3 = CH_3$: 92% ee, 93%
$R^1 = Ph$, $R^2 = CH_3$, $R^3 = Ph$: 95% ee, 94%

Scheme 3.4

reported that the dioxirane could be prepared *in situ* from aqueous hydrogen peroxide and acetonitrile and undergo epoxidation without eroding enantioselectivity (Scheme 3.4) [16]. In this reaction, hydrogen peroxide first reacts with acetonitrile to give peroxyimidic acid that converts the ketone to the dioxirane.

Oxaziridinium salt is a synthetic equivalent of dioxirane, and chiral iminium salts have also been used as a catalyst for asymmetric epoxidation. Oxone and peracid are usually used as the terminal oxidant. Recently, Page *et al.* reported epoxidation using aqueous hydrogen peroxide and obtained moderate to good enantioselectivity [17].

3.2.2
Metal-Catalyzed Asymmetric Epoxidation of Unfunctionalized Olefins

Since the late 1970s, the mechanism of the oxygen atom–transfer reactions including epoxidation catalyzed by cytochrome P-450, a typical oxidizing enzyme, has been extensively studied. The consensus is that a Fe(III)OOH species derived from molecular oxygen or hydrogen peroxide is converted via a push–pull mechanism to an active Fe(IV)=O species that undergoes oxygen atom transfer reactions [18, 19]. On the basis of this mechanism, Groves *et al.* reported Fe(porphyrin)-catalyzed epoxidation using iodosylmesitylene, which directly converts a Fe(III)porphyrin complex to an active Fe=O species [20]. In 1985, Kochi *et al.* reported metallosalen-catalyzed epoxidation using iodosylbenzene (salen = ethylene-1,2- bis(salicylideneiminato)) as the oxidant [21]. Subsequent to this, Jacobsen *et al.* and Katsuki *et al.* reported asymmetric epoxidation using Mn(III)–salen complex as the catalyst in the presence of a terminal oxidant such as iodosylbenzene or sodium hypochlorite [22]. Metalloporphyrin and metallosalen complexes are isoelectronic, and Berkessel *et al.* synthesized a bio-inspired Mn–salalen complex (salalen = salen/salan hybrid = half-reduced salen, salan = fully reduced salen) that has a (4-imidazoyl)methyl substituent at C7 and achieved asymmetric epoxidation using aqueous 1% hydrogen peroxide, though the enantioselectivity was moderate (up to 64% ee) [23]. Subsequent to these reports, Pietikäinen and Katsuki *et al.* independently reported asymmetric epoxidation using aqueous hydrogen peroxide in the presence of *N*-methylimidazole [24]. Although 30% hydrogen peroxide was available for this epoxidation, the enantioselectivity was only slightly improved.

On the other hand, the collaborative research of the Moro-Oka and Katsuki groups on the X-ray structures of diastereomeric Mn–salen complexes disclosed that the conformation of Mn–salen complexes, which has been presumed to play an important role in asymmetric induction, is controlled mainly by two factors: the orientation of the substituents at the ethylene part and the OH–π interaction between the salen and the apical ligands [25]. On the basis of structural analysis of the Mn–salen complexes, Shitama and Katsuki designed a new type of Mn–salen complex **4** bearing a (4-imidazoyl)methyl substituent at the ethylene part and achieved highly enantioselective epoxidation (up to 98% ee) using 30% hydrogen peroxide (Scheme 3.5) [26]. However, the epoxidation of acyclic *cis*-olefins is nonstereospecific.

Bleomycin, that is, a glycopeptide antibiotic, forms a Fe(II) complex, activates molecular oxygen, and oxidatively cleaves DNA. Ohno *et al.* reported that the Fe(III) complex of a pyridine model ligand-6 (PYML-6) of bleomycin catalyzes epoxidation of some olefins using aqueous hydrogen peroxide. Although enantioselectivity and yield are modest, the report is interesting in terms of development of asymmetric oxidation catalysis of abundant iron [27].

Nishiyama *et al.* have reported that the Ru(pybox)(pyridinedicarboxylate) complex serves as a catalyst for asymmetric epoxidation using iodosylbenzene as oxidant [28]. On the basis of the modification of the complex by ligand tuning, Beller *et al.* revealed that Ru(Ph_2pyboxazine)(pyridinedicarboxylate) catalyzes epoxidation using hydrogen peroxide as the oxidant in an enantioselective manner (up to 84% ee), though the substrate scope is narrow [29]. Beller *et al.* also reported that the $FeCl_3$-(*N*-tosyl-1,2-diphenylthylenediamine) **5** complex is a good catalyst for the asymmetric epoxidation of bulky *trans*-stilbene derivatives (Scheme 3.6) [30].

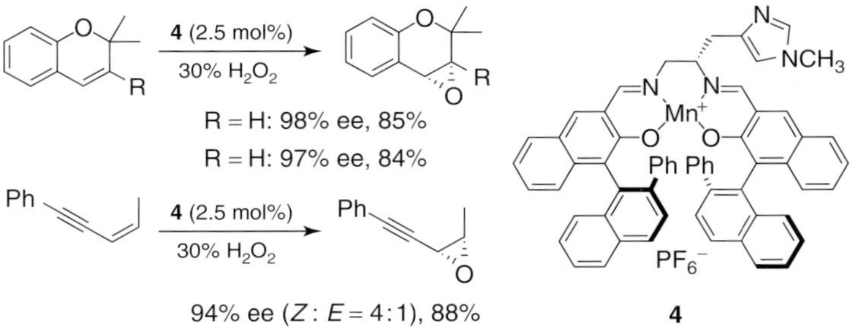
4 (2.5 mol%)
30% H2O2
R
R = H: 98% ee, 85%
R = H: 97% ee, 84%
Ph
4 (2.5 mol%)
30% H2O2
94% ee (Z : E = 4 : 1), 88%
NCH3
Mn+
Ph Ph
PF6−
4

Scheme 3.5

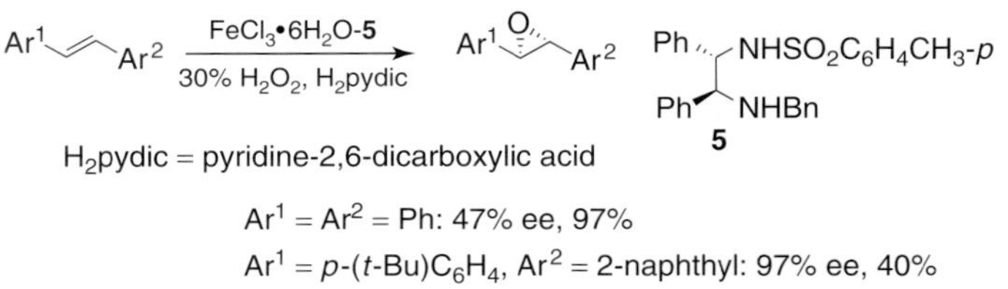
FeCl3•6H2O-5
30% H2O2, H2pydic
Ar1
Ar2
Ph
NHSO2C6H4CH3-p
NHBn
5
H2pydic = pyridine-2,6-dicarboxylic acid
Ar1 = Ar2 = Ph: 47% ee, 97%
Ar1 = p-(t-Bu)C6H4, Ar2 = 2-naphthyl: 97% ee, 40%

Scheme 3.6

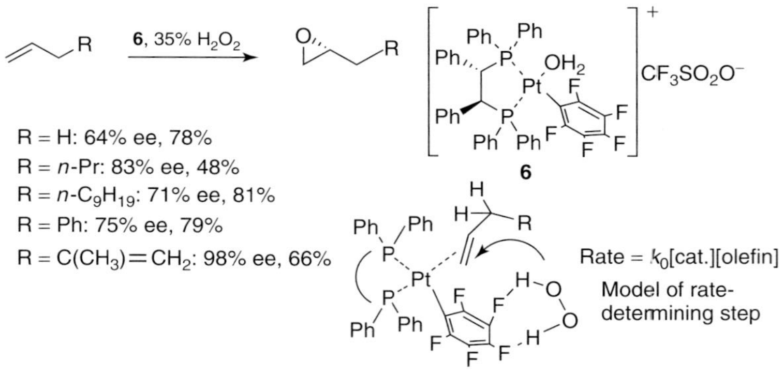

R = H: 64% ee, 78%
R = n-Pr: 83% ee, 48%
R = n-C$_9$H$_{19}$: 71% ee, 81%
R = Ph: 75% ee, 79%
R = C(CH$_3$)=CH$_2$: 98% ee, 66%

Rate = k_0[cat.][olefin]
Model of rate-determining step

Scheme 3.7

In most metal-catalyzed epoxidation using hydrogen peroxide, it is activated on a metal ion and an oxidation event occurs on the metal center. Recently, Strukul et al. reported an unprecedented epoxidation catalysis of cationic pentafluorophenylplatinum(II)(diphosphine) complex **6**, for which hydrogen peroxide is activated by hydrogen bond formation between hydrogen peroxide and the m- and p-fluoro substituents (Scheme 3.7) [31]. Although substrates are limited to terminal olefins without any allylic substituent, the reaction of simple terminal alkenes showed good to high enantioselectivity (63–79% ee) and that of 1,4-diene or 3-aryl-1-propenes showed good to excellent enantioselectivity (63–98% ee). Epoxidation of styrene did not proceed. This substrate scope has been explained by the aforementioned unique mechanism of hydrogen peroxide activation: a platinum-bound olefin directing two allylic hydrogen atoms toward the bulky diphosphine ligand attacks the activated hydrogen peroxide in an intramolecular fashion [32]. Thus, the presence of an allylic substituent retards the desired epoxidation.

The diperoxo(oxo)molybdenum(IV) complex bearing (S)-lactic acid piperidineamide as a chiral ligand has been used for the epoxidation of alkenes and moderate enantioselectivity (49% ee) is achieved, though the reaction is stoichiometric [33]. Later, Saito and Katsuki have revealed that a (aR,R)-di-µ-oxo titanium–salen complex serves as the efficient catalyst for oxidation of sulfides [34]; this has been inferred to proceed through a peroxo-titanium–salen species from the NMR study (see below) [35]. However, the titanium–salen complex did not catalyze epoxidation of olefins under the same conditions.

In the course of this study, another new (aR,R)-di-µ-oxo titanium complex was isolated and found to show unique epoxidation catalysis using a urea–hydrogen peroxide adduct [36]. X-ray analysis of the complex disclosed that it carries a salalen ligand that has one imino nitrogen atom and one amino nitrogen atom, and coordinates to a titanium ion in a cis-β configuration. The metal-coordinated amine is a kind of quaternary ammonium ion and a good hydrogen bond donor, and it has been inferred that the amine proton forms a hydrogen bond with the peroxo species to activate it toward epoxidation. Moreover, (aR,S)-(di-µ-oxo)titanium–salalen complex **7**, a diastereomer of (aR,R)-salalen complex, has been found to show much

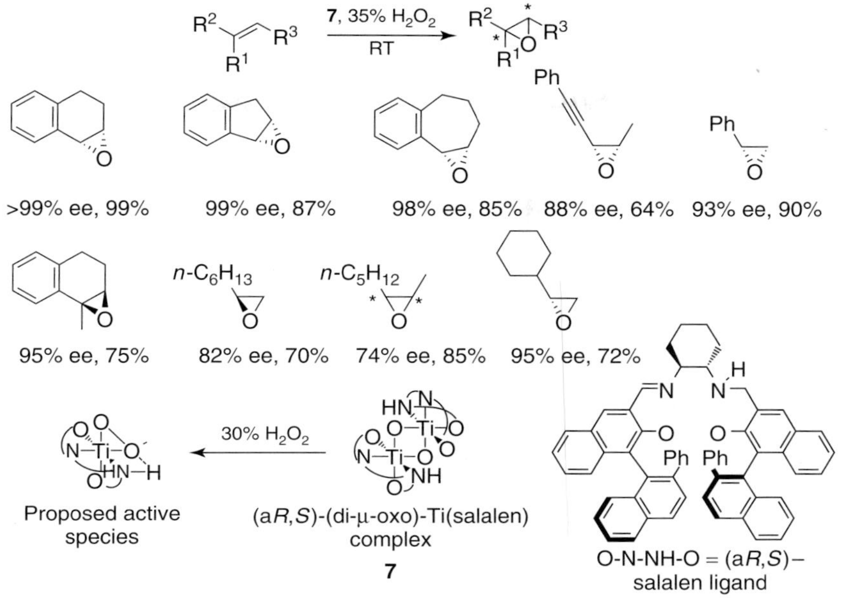

Scheme 3.8

better epoxidation catalysis. The epoxidation of a wide range of olefins, tri-, cis-disubstituted, and terminal olefins, using aqueous hydrogen peroxide proceeds with high enantioselectivity in good yields (Scheme 3.8). Even less reactive non-conjugated terminal and cis-disubstituted olefins can be epoxidized with good to excellent enantioselectivity in satisfactory yields [37]. It is noteworthy that this epoxidation (Matsumoto–Sawada–Katsuki method) is stereospecific and can be carried out with 1 equiv. of hydrogen peroxide.

Although the (di-µ-oxo)titanium complex **7** shows excellent epoxidation catalysis, the salalen ligand includes two chiral binaphthyl units and its synthesis needs a rather long step. This makes the epoxidation with **7** less practical. On the basis of the X-ray structure of **7**, the two pairs of interligand nonbonded interactions – the hydrogen bonds between the amino proton and the phenoxo oxygen atom and the CH–π interaction between the α-methylene proton of the amino group and the naphthyl substituent at C3 (or C3′) – are very likely to contribute to the stabilization of **7**. Moreover, the proposed mechanism of activation of the putative peroxo-titanium species suggests that the presence of a hydrogen bond donor close to the peroxo species is necessary for the epoxidation catalysis. A metal–salan complex bearing an aryl group at C3 and C3′ satisfies these two criteria. In addition, metal–salan complexes can be prepared by reduction of the corresponding metal–salen complexes that can be synthesized in a few steps from salicyl aldehydes, diamines, and metal ion sources [38]. As expected, (di-µ-oxo)titanium(salan) complex **8a** catalyzed the epoxidation using aqueous hydrogen peroxide, although with slightly inferior enantioselectivity and yield to those obtained with **7**.

After the tuning of the 3 and 3′-substituents, (di-μ-oxo)titanium(salan) complex **8b** that has an *ortho*-methoxyphenyl group at C3 and C3′ was found to show significantly improved enantioselectivity and yield [39] that are almost identical with those obtained with **7** (Scheme 3.9). However, titanium–salan complexes are less stable to the epoxidation conditions than titanium–salalen complex and it partly decomposes during the epoxidation. Although the decomposition product of complex **8b** does not affect enantioselectivity, it makes the epoxidation medium acidic and some acid-sensitive epoxides such as indene oxide decompose under the epoxidation conditions. However, the addition of a small amount of phosphate buffer (pH = 7.4–8.0) prevents the decomposition of the epoxides [40]. It is noteworthy that this reaction can be carried out on a mole scale. (Caution! The reaction is exothermic and care must be taken to strictly keep the temperature at about room temperature by adding hydrogen peroxide portionwise.) Although high enantioselectivity (about 80–90% ee) has been obtained for epoxidation of styrenes, the titanium(salan) complex derived from proline was found to show significantly improved enantioselectivity (96–98% ee) (Scheme 3.10) [41].

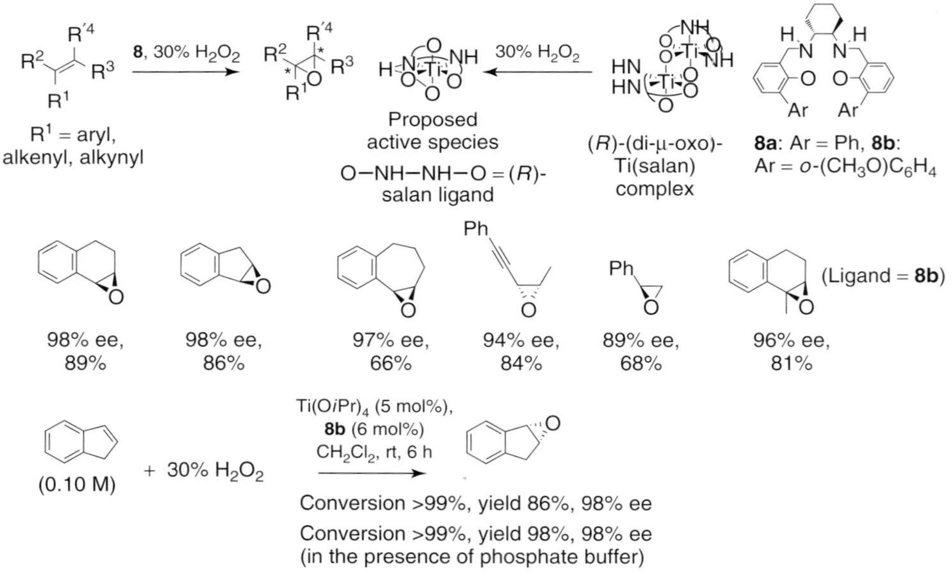

Scheme 3.9

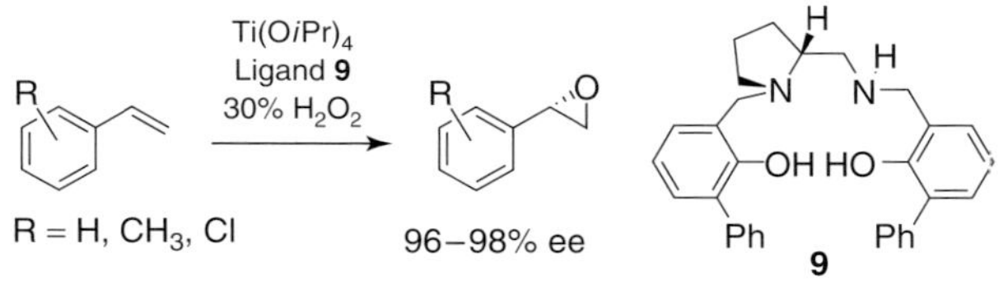

Scheme 3.10

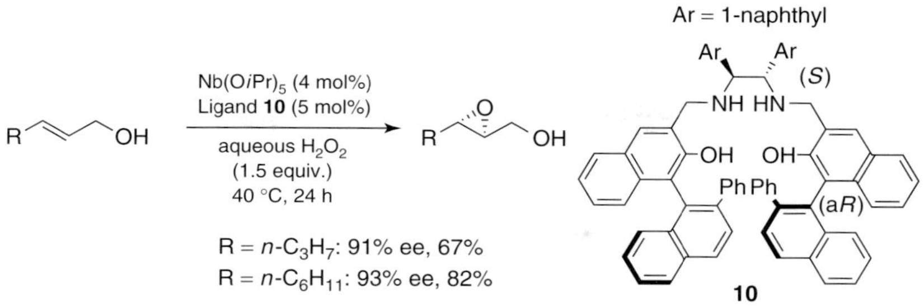

Scheme 3.11

$R\diagdown\!\!\diagdown\!\!\diagup OH$
Nb(O*i*Pr)$_5$ (4 mol%)
Ligand **10** (5 mol%)
aqueous H$_2$O$_2$ (1.5 equiv.)
40 °C, 24 h
→ R—(epoxide)—OH

R = *n*-C$_3$H$_7$: 91% ee, 67%
R = *n*-C$_6$H$_{11}$: 93% ee, 82%

Ar = 1-naphthyl
(Ligand **10**, (a*R*), (*S*))

3.2.3
Metal-Catalyzed Asymmetric Epoxidation of Allylic Alcohols

Optically active 2,3-epoxy alcohols react stereo- and regioselectively with a variety of nucleophiles to give various useful compounds. Therefore, much effort has been directed toward the development of metal-catalyzed asymmetric epoxidation of allylic alcohols [42]. In 1980, Katsuki and Sharpless reported a highly enantioselective and practical epoxidation using titanium/tartrate/*t*-butyl hydroperoxide (TBHP) system, which has been widely used in organic synthesis for construction of various optically active compounds and has made a large contribution to the development of asymmetric synthesis [43]. Recently, Yamamoto *et al.* reported excellent epoxidation catalysis of a vanadium-bis(hydroxamic acid) complex: high enantioselectivity and low catalyst loading [44]. Despite these advancements, there is no report on the method for epoxidation of allylic alcohols using hydrogen peroxide. In 2008, Egami and Katsuki revealed that (μ-oxo)Nb–salan complex catalyzes epoxidation of allylic alcohols using urea–hydrogen peroxide adduct with good to high enantioselectivity [45]. Use of aqueous hydrogen peroxide deteriorates enantioselectivity and yield. Although the complex is a μ-oxo dimer, a genuine active species has been suggested to be a monomeric Nb–salan **10** complex based on the study of a linear relationship between the enantiomeric excesses of product and catalyst. In agreement with the suggestion, the complex prepared *in situ* from Nb(O*i*Pr)$_5$ and salan ligand catalyzes highly enantioselective epoxidation using aqueous hydrogen peroxide in a mixed solvent of dichloromethane and brine (Scheme 3.11) [45b].

3.3
Asymmetric Oxidation of Sulfides

Most sulfoxides are biologically active and they have broad pharmaceutical application. Moreover, they have been widely used as chiral auxiliaries [46]. Thus, development of enantioenriched synthesis of sulfoxides has been a growing research interest. Among the many methods for synthesizing optically active sulfoxides reported to date, enantioselective oxidation of sulfides is the most straightforward and practical. This section deals mainly with enantioselective oxidation of sulfides using hydrogen peroxide as the terminal oxidant.

3.3.1
Metal–Salen-Catalyzed Oxidation

In 1986, Fujita *et al.* reported that the chiral oxo vanadium(IV)–salen complex catalyzes oxidation of sulfides using *t*-butyl hydrogen peroxide as an oxidant, albeit with modest enantioselectivity [47]. To the best of our knowledge, this is the first report on asymmetric reactions using an optically active metallosalen complex as a catalyst. They also reported that the titanium–salen complex catalyzes oxidation of sulfides with a slightly better enantioselectivity (up to 63% ee) [48]. Subsequent to these reports, Jacobsen *et al.* reported that manganese–salen complexes catalyze oxidation of sulfides using aqueous hydrogen peroxide in acetonitrile with moderate enantioselectivity [49]. It is noteworthy that enantioselectivity is significantly affected by the electronic nature of the 5, 5'-substituent of the complex: the presence of an electron-donating group at C5 and 5' of the complex increases enantioselectivity (up to 68% ee).

Miyazaki and Katsuki reported that the niobium–salen complex catalyzes oxidation of various sulfides using urea–hydrogen peroxide adduct with good enantioselectivity (77–86% ee) [50]. A positive nonlinear relationship between the enantiomeric excesses of the catalyst and the product has been observed in this reaction.

3.3.2
Metal–Schiff Base-Catalyzed Oxidation

In 1995, Bolm and Bienewald reported the vanadium-catalyzed asymmetric oxidation of sulfides, which employs aqueous hydrogen peroxide as the oxidant. Although enantioselectivity was moderate to high (53–85%), this method has several advantages: high catalytic activity, mild reaction conditions, ease of preparation of the ligand **11a** and the catalyst, and use of aqueous hydrogen peroxide [51]. Owing to these advantages, many improvements have been made with regard to ligands and protocols [52]. Anson *et al.* reported that Schiff base ligand **11b** derived from 3,5-iodosalicyl aldehyde and (S)-*t*-leucinol induces high asymmetry in the oxidation of alkyl aryl sulfides (up to 97% ee) (Scheme 3.12) [53]. For example, oxidation of methyl phenyl sulfide proceeds with 90% ee.

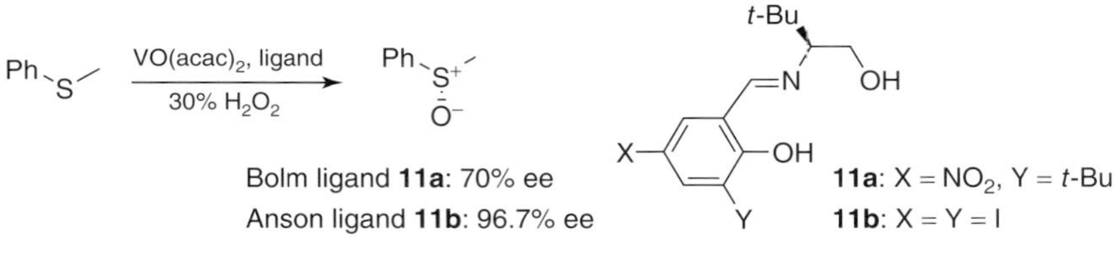

Ph–S–Ph $\xrightarrow{\text{VO(acac)}_2,\text{ ligand}}_{30\%\ H_2O_2}$ Ph–S$^+$(O$^-$)–Ph

Bolm ligand **11a**: 70% ee
Anson ligand **11b**: 96.7% ee

11a: X = NO$_2$, Y = *t*-Bu
11b: X = Y = I

Scheme 3.12

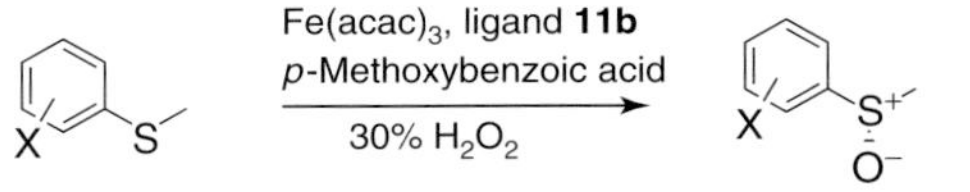

Fe(acac)₃, ligand **11b**
p-Methoxybenzoic acid
30% H₂O₂

X = H: 90% ee, 63%; X = 4-CH₃: 92% ee, 63%;
X = 4-Br: 94% ee, 59%

Scheme 3.13

It is known that sulfoxides are often further oxidized to sulfones under the conditions used for the oxidation of sulfides, and the oxidation occurs in an enantiomer-differentiating manner [54]. The enantiomeric excesses of sulfoxides usually increase as the oxidation of the sulfoxides proceeds. After screening reaction temperature and solvent, Jackson et al. revealed that tandem oxidation of aryl alkyl sulfides provides highly enantioenriched sulfoxides in good yields [55].

Legros and Bolm recently reported that iron complexes bearing Schiff base ligands also serve as a catalyst for oxidation of alkyl aryl sulfides. Moderate to good enantioselectivity (27–90%) has been obtained with the complex bearing the ligand **11b** modified by Anson [56]. Subsequent to this report, however, they discovered that the addition of a half equivalent of carboxylic acid relative to iron ion remarkably improves the enantioselectivity of the oxidation of alkyl aryl sulfides (66–94%), while the enantioselectivity of benzyl methyl sulfide is modest (Scheme 3.13) [57].

3.3.3
Metal−ONNO−Tetradentate Ligand-Catalyzed Oxidation (Including cis-β Metal−Salen-Catalyzed Oxidation)

As mentioned before, Jacobsen et al. have reported asymmetric sulfoxidation using a Mn−salen complex/H_2O_2 system [49] but it was moderately enantioselective. Although an oxo manganese(VII) species has been postulated as an active species for asymmetric epoxidation using a Mn−salen complex/H_2O_2 system [58], the mechanism of the sulfoxidation is unclear. In 2001, Saito and Katsuki reported asymmetric sulfoxidation catalysis of di-μ-oxo-titanium−salen complex **12** using the urea−hydrogen peroxide adduct [34a]. The oxidation is highly enantioselective and it can be applied to the oxidation of a variety of sulfides: high enantioselectivity has been obtained in oxidation of dialkyl sulfides (>91%), 1,3-dithiolanes (79–95%), 1,3-dithianes (>95%) [34b], and racemic oxathiane ($k_{S/R}$ = >30) [59], as well as alkyl aryl sulfides (>91%) (Scheme 3.14) [34a]. Although most methods for oxidation of sulfides are more or less accompanied by overoxidation, it is noteworthy that no overoxidation to sulfone or disulfoxide has been observed in this oxidation [34a]. On the basis of the NMR analysis, it has been inferred that a peroxo-titanium-(cis-β-salen) intermediate participates in this oxidation [35]. The use of aqueous hydrogen peroxide instead of urea−hydrogen peroxide adduct diminished enantioselectivity [34a]. For example, the oxidation of methyl phenyl sulfide with aqueous hydrogen peroxide was 76% ee at 25 °C. The decrease in enantioselectivity has been considered to be due to the ring opening of the

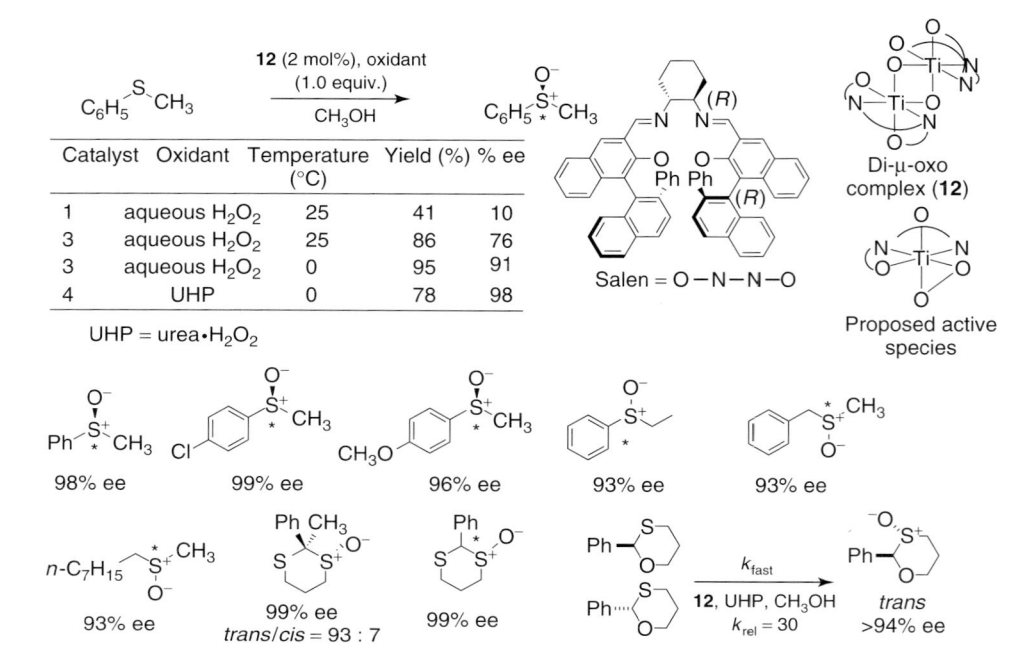

Scheme 3.14

Catalyst	Oxidant	Temperature (°C)	Yield (%)	% ee
1	aqueous H_2O_2	25	41	10
3	aqueous H_2O_2	25	86	76
3	aqueous H_2O_2	0	95	91
4	UHP	0	78	98

UHP = urea·H_2O_2

peroxo-titanium species by water to give the hydroxo-hydroperoxo-titanium species that undergoes lower enantioselective oxidation. However, it has been found that the enantioselectivity of the oxidation using aqueous hydrogen peroxide is significantly improved by lowering the reaction temperature, though it is still inferior to that obtained with urea–hydrogen peroxide adduct. For example, the oxidation of methyl phenyl sulfide with aqueous hydrogen peroxide at 0 °C showed 91% ee (B. Saito and T. Katsuki, unpublished results). However, different from the oxidation using the urea–hydrogen peroxide adduct, the formation of a trace amount (<5%) of the sulfone has been observed in this oxidation.

If the aforementioned ring-opening step is reversible and the reversed (ring-closing) reaction can be accelerated somehow, the lifetime of the hydroxo-hydroperoxo-titanium species becomes short and aqueous hydrogen peroxide is expected to be usable as the terminal oxidant for the oxidation. In the opening of the peroxo-titanium species, an anionic hydroxo ligand is generated. If the quadrivalent titanium is replaced with a tervalent metal, the ring-opening reaction should lead to a hydroperoxo species that carries a neutral dissociable ligand such as an aqua ligand and its reversed reaction should be facilitated [60]. Indeed, chiral aluminum–salalen complex **13** is a good catalyst for oxidation of sulfides using aqueous hydrogen peroxides in the presence of a small amount of a phosphate buffer (Scheme 3.15). Under this condition, the resulting sulfoxides are slowly oxidized with moderate enantiomer differentiation, and the enantiomeric excesses of the sulfoxides increase, as the reaction proceeds. The active species of this

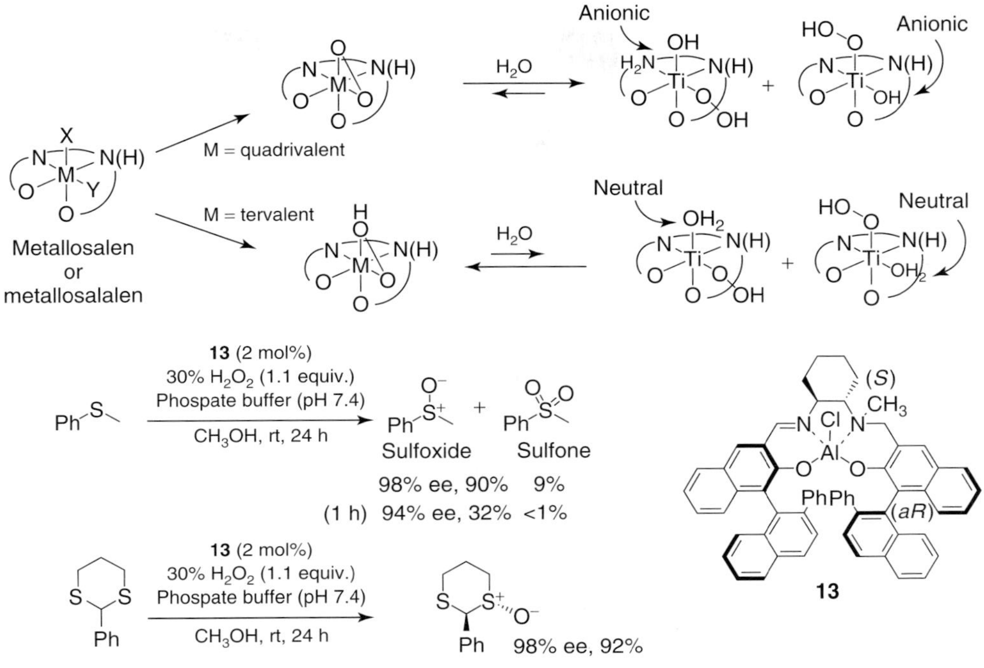

3.3 Asymmetric Oxidation of Sulfides | 71

Scheme 3.15

reaction has been considered to be a cis-$\beta\eta_2$-hydroperoxoaluminum species. This oxidation can be successfully applied to the transformation of alkyl aryl sulfides and 1,3-dithianes with enantioselectivity of greater than 90%. However, the oxidation of benzyl methyl sulfide is 80% ee. In general, the stereochemistry of oxidation reactions is significantly influenced by the properties of the solvent used, but this oxidation is not greatly affected by solvent and substrate concentration. This fact indicates the possibility of performing the oxidation under solvent-free conditions, if catalyst loading can be reduced to the utmost extent and substrate and product serve as the solvent. Indeed, the oxidation proceeds under solvent-free conditions with high enantioselectivity and the turnover number of the catalyst amounts to 46 000 at most [61]. (Caution! The oxidation is an exothermic reaction, and the reaction vessel must be cooled at 0 or −10 °C and hydrogen peroxide must be added portionwise, when the oxidation is carried out under solvent-free conditions.)

Zhu *et al.* reported that the oxovanadium–salan complex catalyzes oxidation of sulfides using aqueous hydrogen peroxide. The reaction is more than 80% enantioselective toward several alkyl aryl sulfides at 0 °C, and no oxidation of the sulfoxides proceeds under this condition [62]. However, oxidation of sulfoxides occurs at 25 °C with enantiomer differentiation, and the oxidation of racemic sulfoxide provides enantioenriched sulfoxides at the expense (of about 70%) of starting sulfoxides. It is noteworthy that N,N′-dimethylated salan ligands show only modest enantioselectivity.

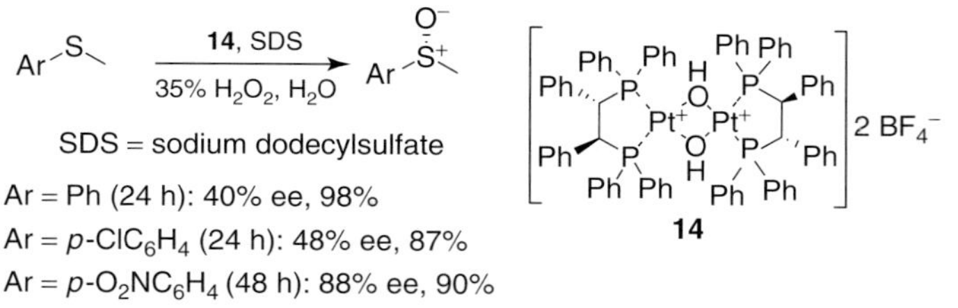

$Ar-S\diagdown$ $\xrightarrow[35\%\ H_2O_2,\ H_2O]{14,\ SDS}$ $Ar-\overset{O^-}{\underset{}{S^+}}\diagdown$

SDS = sodium dodecylsulfate

Ar = Ph (24 h): 40% ee, 98%
Ar = p-ClC$_6$H$_4$ (24 h): 48% ee, 87%
Ar = p-O$_2$NC$_6$H$_4$ (48 h): 88% ee, 90%

14 2 BF$_4^-$

Scheme 3.16

Bryliakov and Talsi recently reported oxidation of sulfides using the titanium–salan complex as a catalyst. The asymmetric catalysis of the titanium complex is similar to that of the above oxovanadium–salan complex [63].

Although many methods for asymmetric oxidation of sulfides using hydrogen peroxide have been developed till date, most of them use a biphasic solvent (organic solvent/water) system. Scarso and Strukul revealed that platinum-mediated asymmetric oxidation of alkyl aryl sulfides can be carried out in water in the presence of sodium dodecyl sulfate, a surfactant [64]. Enantioselectivity of the reactions is moderate, except that oxidation of some substrates, such as methyl p-nitrophenyl sulfide, bearing an electron-withdrawing group is highly enantioselective (88% ee) (Scheme 3.16).

Egami and Katsuki disclosed that iron–salan complex **15** catalyzes oxidation of sulfides using aqueous hydrogen peroxide at 20 °C in water even in the absence of surfactant with high enantioselectivity [65]. This oxidation can be successfully applied to not only alkyl aryl sulfides but also dialkyl sulfides with high enantioselectivity. The iron–salan complex is insoluble to water, but mixing it with sulfide in water gives a biphasic liquid system. It has been considered that the formation of the biphasic liquid system by coordination of sulfide to **15** promotes the oxidation of sulfides in water (Scheme 3.17). As with many sulfoxidation reactions, formation of sulfones has also been observed in this oxidation. Although the degree of the enantiomer differentiation in the oxidation is small (2.5–4.5), the enantiomeric excesses of the sulfoxides slightly increase as the oxidation proceeds. Lowering the reaction temperature to 0 °C enhances enantioselectivity and significantly suppresses oxidation of sulfoxides [65b]. Moreover, the oxidation at 0 °C is tolerant of an acid-sensitive functional group such as a benzylic acetal. Further lowering the temperature to −10 °C deteriorates enantioselectivity.

3.3.4
Miscellaneous

Page et al. reported that camphorsulfonylimines, in particular, dimethoxycamphorsulfonylimines, serve as catalysts for oxidation of sulfides using aqueous hydrogen peroxide, and high enantioselectivity greater than 80% has been obtained in the oxidation of tert-butyl methyl sulfide and 2-substituted 1,3-dithianes [66]. Although there are two possible catalytic cycles for this oxidation, via a

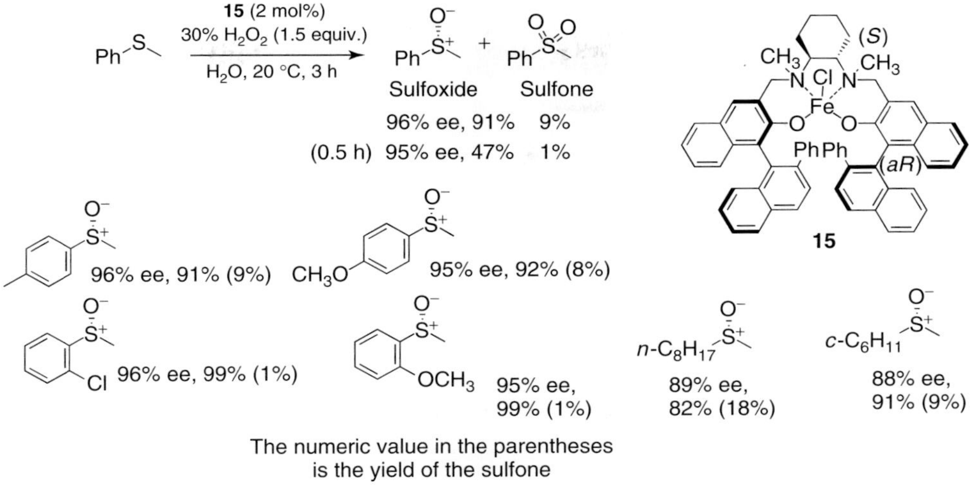

Scheme 3.17

hydroperoxyamine or an oxaziridine intermediate, the former catalytic cycle is more likely, because the oxidation using the oxaziridine intermediate separately prepared shows the enantioselectivity slightly different from that obtained in this oxidation.

Microsomal flavin adenine dinucleotide–containing monooxygenase metabolizes various xenobiotic substrates. As a biomimetic approach [67] toward flavin-based oxidation catalysis, Shinkai and Toda have reported asymmetric oxidation using facially chiral flavinophan as catalyst in the presence of aqueous hydrogen peroxide [68]. The reaction of methyl phenyl sulfide gives the sulfoxide of 65% ee.

3.4
Conclusion

The past few decades have witnessed remarkable progress in asymmetric oxidation using hydrogen peroxide. Highly enantioselective and high-yielding epoxidation and oxidation of sulfides under mild reaction conditions have been achieved to date by introduction of various chiral catalysts, which can generate an active hydrogen peroxide–derived species in a chiral steric environment. Moreover, understanding of the reaction mechanisms and of the mechanisms of asymmetric induction has also been deepened through these studies. Although there are still many issues to be solved before developing practical oxidation methods for industrial use, tremendous knowledge of the oxidation reactions, accumulated through decades of oxidation research, should ensure satisfactory development of oxidation reactions using hydrogen peroxide and we will certainly witness the development in the coming decades.

References

1. (a) Trost, B.M. and Fleming, I. (eds) (1991) *Comprehensive Organic Synthesis*, vol. 7, Pergamon Press, Oxford; (b) Bäckval, J.-E. (ed.) (2004) *Modern Oxidation Methods*, Wiley-VCH Verlag GmbH, Weinheim; (c) Beller, M. and Bolm, C. (eds) (2004) *Transition Metals for Organic Synthesis*, 2nd edn, vol. 2, Wiley-VCH Verlag GmbH, Weinheim.
2. (a) ten Brink,G.-J., Arends, I.W.C.E., and Sheldon, R.A. (2004) *Chem. Rev.*, **104**, 4105; (b) Irie, R. and Katsuki, T. (2004) *Chem. Rec.*, **4**, 96; (c) Kaczorowaka, K., Kolarska, Z., Mitka, K., and Kowalski, P. (2005) *Tetrahedron*, **61**, 8315.
3. (a) Wills, M. (2008) *Angew. Chem. Int. Ed.*, **47**, 4264; (b) Imada, Y. and Naota, T. (2007) *Chem. Rec.*, **7**, 354; (c) Schultz, M.J. and Sigman, M.S. (2006) *Tetrahedron*, **62**, 8227.
4. (a) Mukaiyama, T., Yamada, T., Nagata, T., and Imagawa, K. (1993) *Chem. Lett.*, 327; (b) Mukaiyama, T. and Yamada, T. (1995) *Bull. Chem. Soc. Jpn.*, **68**, 17; (c) Nagata, T., Imagawa, K., Yamada, T., and Mukaiyama, T. (1995) *Bull. Chem. Soc. Jpn.*, **68**, 3241.
5. (a) Groves, J.T. and Quinn, R. (1985) *J. Am. Chem. Soc.*, **107**, 5790; (b) Lai, T.-S., Zhang, R., Cheung, K.-K., Kwong, H.-L., and Che, C.-M. (1998) *Chem. Commun.*, 1583; (c) Döbler, C., Mehltretter, G., and Beller, M. (1999) *Angew. Chem. Int. Ed. Engl.*, **38**, 3026; (d) Einhorn, C., Einhorn, J., Marcadal-Abbadi, C., and Pierre, J.-L. (1999) *J. Org. Chem.*, **64**, 4542; (e) Shen, J. and Tan, C.-H. (2008) *Org. Biomol. Chem.*, **6**, 4096.
6. (a) Helder, R., Hummelen, J.C., Laane, R.W.P.M., Wiering, J.S., and Wynberg, H. (1976) *Tetrahedron Lett.*, **17**, 1831; (b) Wynberg, H. and Greijdanus, B. (1978) *Chem. Commun.*, 427.
7. (a) Juliá, S., Masana, J., and Vega, J.C. (1980) *Angew. Chem. Int. Ed.*, **19**, 829; (b) Colonna, S., Morinari, H., Banfi, S., Juliá, S., Masana, J., and Alvarez, A. (1983) *Tetrahedron Lett.*, **39**, 1635.
8. Arai, S., Tsuge, H., and Shioiri, T. (1998) *Tetrahedron Lett.*, **39**, 7563.
9. (a) MacMillan, D.W.C. (2008) *Nature*, **455**, 304; (b) Mielgo, A. and Palomo, C. (2008) *Chem. Asian J.*, **3**, 922.
10. (a) Marigo, M., Franzén, J., Poulsen, T.B., Zhuang, W., and Jørgensen, K.A. (2005) *J. Am. Chem. Soc.*, **127**, 6964; (b) Sundén, H., Ibrahem, I., and Córdova, A. (2006) *Tetrahedron Lett.*, **47**, 99.
11. Wang, X. and List, B. (2008) *Angew. Chem. Int. Ed.*, **47**, 1119.
12. Wang, X., Reisinger, C.M., and List, B. (2008) *J. Am. Chem. Soc.*, **130**, 6070.
13. Curci, R., Fiorentino, M., and Serio, M.R. (1984) *Chem. Commun.*, 155.
14. Yang, D., Yip, Y.-C., Tang, M.-W., Wong, M.-K., Zheng, J.-H., and Cheung, K.-K. (1996) *J. Am. Chem. Soc.*, **118**, 491.
15. Zu, Y., Wang, Z.-X., and Shi, Y. (1996) *J. Am. Chem. Soc.*, **118**, 9806.
16. Shu, L. and Shi, Y. (2001) *Tetrahedron*, **57**, 5213.
17. (a) Page, P.C.B., Parker, P., Rassias, G.A., Buckley, B.R., and Bethell, D. (2008) *Adv. Synth. Catal.*, **350**, 1867; (b) Page, P.C.B., Buckley, B.R., and Blacker, A.J. (2004) *Org. Lett.*, **6**, 1543.
18. (a) Groves, J.T., Nemo, T.E., and Myers, R.S. (1979) *J. Am. Chem. Soc.*, **101**, 1032; (b) Groves, J.T. and Myers, R.S. (1983) *J. Am. Chem. Soc.*, **105**, 5791.
19. For other studies on the oxidation catalysis of cytochrome p-450, see: (a) Yuan, L.C. and Bruice, T.C. (1986) *J. Am. Chem. Soc.*, **108**, 1643; (b) Battioni, P., Renaud, J.P., Bartoli, J.F., Artiles, M.R., Fort, M., and Mansuy, D. (1988) *J. Am. Chem. Soc.*, **110**, 8462; (c) Yamaguchi, K., Watanabe, Y., and Morishima, I. (1993) *J. Am. Chem. Soc.*, **115**, 4058; (d) Machii, K., Watanabe, Y., and Morishima, I. (1995) *J. Am. Chem. Soc.*, **117**, 6691; (e) Ozaki, S., Inaba, Y., and Watanabe, Y. (1998) *J. Am. Chem. Soc.*, **120**, 8020; (f) Nam, W., Lee, H.J., Oh, S.-Y., Kim, C., and Jang, H.G. (2000) *J. Inorg. Biochem.*, **80**, 219; (g) Watanabe, Y. and Ueno, T. (2003) *Bull. Chem. Soc. Jpn.*, **76**, 1309, references cited therein.

20. Groves, J.T. and Myers, R.S. (1983) *J. Am. Chem. Soc.*, **105**, 5791.
21. (a) Samsel, E.G., Srinivasan, K., and Kochi, J.K. (1985) *J. Am. Chem. Soc.*, **107**, 7606; (b) Srinivasan, K., Michaud, P., and Kochi, J.K. (1986) *J. Am. Chem. Soc.*, **108**, 2309.
22. (a) Zhang, W., Loebach, J.L., Wilson, S.R., and Jacobsen, E.N. (1990) *J. Am. Chem. Soc.*, **112**, 2801; (b) Irie, R., Noda, K., Ito, Y., Matsumoto, N., and Katsuki, T. (1990) *Tetrahedron Lett.*, **31**, 7345.
23. (a) Schwenkreis, T. and Berkessel, A. (1993) *Tetrahedron Lett.*, **34**, 4785; (b) Berkessel, A., Frauenkron, M., Schwenkreis, T., Steinmetz, A., Baum, G., and Fenske, D. (1996) *J. Mol. Cat. A*, **113**, 321.
24. (a) Pietikäinen, P. (1994) *Tetrahedron Lett.*, **35**, 941; (b) Pietikäinen, P. (1998) *Tetrahedron*, **54**, 4319; (c) Irie, R., Hosoya, N., and Katsuki, T. (1994) *Synlett*, 255.
25. Hashihayata, T., Punniyamurthy, T., Irie, R., Katsuki, T., Akita, M., and Moro-oka, Y. (1999) *Tetrahedron*, **55**, 14599.
26. Shitama, H. and Katsuki, T. (2006) *Tetrahedron Lett.*, **47**, 3203.
27. Kaku, Y., Otsuka, M., and Ohno, M. (1989) *Chem. Lett.*, 611.
28. Nishiyama, H., Shimada, T., Itoh, H., Sugiyama, H., and Motoyama, Y. (1997) *Chem. Commun.*, 1863.
29. Tse, M.K., Döbler, C., Bhor, S., Klawonn, M.K., Mägerlein, W., Hugl, H., and Beller, M. (2004) *Angew. Chem. Int. Ed.*, **43**, 5255.
30. Gelalcha, F.G., Bitterlich, B., Anilkumar, G., Tse, M.K., and Beller, M. (2007) *Angew. Chem. Int. Ed.*, **46**, 7293.
31. Colladon, M., Scarso, A., Sgarbossa, P., Michellin, R.A., and Strukul, G. (2006) *J. Am. Chem. Soc.*, **128**, 14006.
32. Colladon, M., Scarso, A., Sgarbossa, P., Michellin, R.A., and Strukul, G. (2007) *J. Am. Chem. Soc.*, **129**, 7680.
33. Schurig, V., Hintzer, K., Leyrer, U., Mark, C., Pitchen, C., and Kagan, H.B. (1989) *J. Organometal. Chem.*, **370**, 81.
34. (a) Saito, B. and Katsuki, T. (2001) *Tetrahedron Lett.*, **42**, 3873; (b) Tanaka, T., Saito, B., and Katsuki, T. (2002) *Tetrahedron Lett.*, **43**, 3259.
35. Saito, B. and Katsuki, T. (2001) *Tetrahedron Lett.*, **42**, 8333.
36. Matsumoto, K., Sawada, Y., Saito, B., Sakai, K., and Katsuki, T. (2005) *Angew. Chem. Int. Ed.*, **44**, 4935.
37. Sawada, Y., Matsumoto, K., and Katsuki, T. (2007) *Angew. Chem. Int. Ed.*, **46**, 4559.
38. Sawada, Y., Matsumoto, K., Kondo, S., Watanabe, H., Ozawa, T., Suzuki, K., Saito, B., and Katsuki, T. (2006) *Angew. Chem. Int. Ed.*, **45**, 3478.
39. Matsumoto, K., Sawada, Y., and Katsuki, T. (2006) *Synlett*, 3545.
40. Shimada, Y., Kondo, S., Ohara, Y., Matsumoto, K., and Katsuki, T. (2007) *Synlett*, 2445.
41. Matsumoto, K., Oguma, T., and Katsuki, T. (2009) *Angew. Chem. Int. Ed.*, **48**, 7432.
42. For seminal studies on metal-catalyzed epoxidation of allylic alcohols, see: (a) Sheng, M.N. and Zajacek, J.G. (1970) *J. Org. Chem.*, **35**, 1839; (b) Sharpless, K.B. and Michaelson, R.C. (1973) *J. Am. Chem. Soc.*, **95**, 6136; For seminal studies on the asymmetric version of the epoxidation, see: (c) Yamada, S.-I., Mashiko, T., and Terashima, S. (1977) *J. Am. Chem. Soc.*, **99**, 1988; (d) Michaelson, R.C., Palermo, R.E., and Sharpless, K.B. (1977) *J. Am. Chem. Soc.*, **99**, 1990.
43. Katsuki, T. and Sharpless, K.B. (1980) *J. Am. Chem. Soc.*, **102**, 5974.
44. (a) Zhang, W., Basak, A., Kosugi, Y., Hoshino, Y., and Yamamoto, H. (2005) *Angew. Chem. Int. Ed.*, **44**, 4389; (b) Li, Z., Zhang, W., and Yamamoto, H. (2008) *Angew. Chem. Int. Ed.*, **47**, 7520.
45. (a) Egami, H. and Katsuki, T. (2008) *Angew. Chem. Int. Ed.*, **47**, 5171; (b) Egami, H., Oguma, T., and Katsuki, T. (2010) *J. Am. Chem. Soc.*, **132**, 5886.
46. (a) Bolm, C., Muñiz, K., and Hildebrand, J.P. (1999) in *Comprehensive Asymmetric Catalysis*, vol. 2 (eds E.N., Jacobsen, A., Pfaltz, and H., Yamamoto), Springer-Verlag, Berlin, p. 697; (b) Kagan, H. (2000) in *Catalytic Asymmetric Synthesis*, 2nd edn (ed.

47. Nakajima, K., Kojima, M., and Fujita, J. (1986) *Chem. Lett.*, 1483.
48. Sasaki, C., Nakajima, K., Kojima, M., and Fujita, J. (1991) *Bull. Chem. Soc. Jpn.*, **64**, 1318.
49. Palucki, M., Hanson, P., and Jacobsen, E.N. (1992) *Tetrahedron Lett.*, **33**, 7111.
50. Miyazaki, T. and Katsuki, T. (2003) *Synlett*, 1046.
51. Bolm, C. and Bienewald, F. (1995) *Angew. Chem. Int. Ed.*, **34**, 2640.
52. (a) Vetter, A.H. and Berkessel, A. (1998) *Tetrahedrn Lett.*, **39**, 1741; (b) Skarzewski, J., Ostrycharz, E., and Siedlecka, R. (1999) *Tetrahedron: Asymmetry*, **10**, 3457; (c) Karpyshev, N.N., Yakovleva, O.D., Talsi, E.P., Bryliakov, K.P., Tolstikova, O.V., and Tolstikov, A.G. (2000) *J. Mol. Catal. A.*, **157**, 91; (d) Ohta, C., Shimizu, H., Kondo, A., and Katsuki, T. (2002) *Synlett*, 161; (e) Jeong, Y.C., Choi, S., Hwang, Y.D., and Ahn, K.H. (2004) *Tetrahedron Lett.*, **45**, 9249.
53. Pelotier, B., Anson, M.S., Campbell, I.B., Macdonald, S.J.F., Priem, G., and Jackson, R.F.W. (2002) *Synlett*, 1055.
54. Komatsu, N., Hashizume, M., Sugita, T., and Uemura, S. (1993) *J. Org. Chem.*, **58**, 7624.
55. Drago, C., Caggiano, L., and Jackson, R.F.W. (2005) *Angew. Chem. Int. Ed.*, **44**, 7221.
56. Bolm, C. and Legros, J. (2003) *Angew. Chem. Int. Ed.*, **42**, 5487.
57. (a) Legros, J. and Bolm, C. (2004) *Angew. Chem. Int. Ed.*, **43**, 4225; (b) Legros, J. and Bolm, C. (2005) *Chem. Eur. J.*, **11**, 1086.
58. Katsuki, T. (1995) *Coord. Chem. Rev.*, **140**, 189.
59. Saito, B. and Katsuki, T. (2003) *Chirality*, **15**, 24.
60. (a) Yamaguchi, T., Matsumoto, K., Saito, B., and Katsuki, T. (2007) *Angew. Chem. Int. Ed.*, **46**, 4729; (b) Matsumoto, K., Yamaguchi, T., Fujisaki, J., Saito, B., and Katsuki, T. (2008) *Chem. Asian J.*, **3**, 351.
61. Matsumoto, K., Yamaguchi, T., and Katsuki, T. (2008) *Chem. Commun.*, 1704.
62. Sun, J., Zhu, C., Dai, Z., Yang, M., Pan, Y., and Hu, H. (2004) *J. Org. Chem.*, **69**, 8500.
63. Bryliakov, K.P. and Talsi, E.P. (2008) *Eur. J. Org. Chem.*, 3369.
64. Scarso, A. and Strukul, G. (2005) *Adv. Synth. Catal.*, **347**, 1227.
65. (a) Egami, H. and Katsuki, T. (2007) *J. Am. Chem. Soc.*, **129**, 8940–8941; (b) Egami, H. and Katsuki, T. (2008) *Synlett*, 1453.
66. (a) Page, P.C.B., Heer, J.P., Bethell, D., and Lund, B.L. (1999) *Phosphorus, Sulfur Silicon*, **153–154**, 247; (b) Page, P.C.B., Heer, J.P., Bethell, D., Collington, E.W., and Andrews, D.M. (1994) *Tetrahedron Lett.*, **35**, 9629.
67. For the review of organocatalysis of flavins, see: Imada, Y. and Naota, T. (2007) *Chem. Rec.*, **7**, 354.
68. Shinkai, S., Yamaguchi, T., Manabe, O., and Toda, F. (1988) *J. Chem. Soc., Chem. Commun.*, 1399.

4
Development of Palladium Catalysts for Chemoselective Hydrogenation
Hironao Sajiki and Yasunari Monguchi

4.1
Catalyst Poisons and Chemoselective Heterogeneous Catalysts

4.1.1
Background

Catalytic hydrogenation using a heterogeneous catalyst has been a powerful tool for functional group transformation in both laboratory and industrial scale reaction processes. Especially, Pd on charcoal (Pd/C), the most frequently used heterogeneous hydrogenation catalyst [1], has many advantages over homogeneous catalysts, such as stability, ease of separation, recyclability, and cost performance [2]. However, the high catalytic activity of Pd/C makes it difficult to achieve chemoselective hydrogenation among some reducible functionalities.

4.1.2
Chemoselective Inhibition of the Hydrogenolysis for *O*-Benzyl Protective Groups by the Addition of a Nitrogen-Containing Base

Functional group manipulation is fundamental in synthetic organic chemistry. Therefore, the development of new chemoselective transformations continues to be of great importance in synthetic and process chemistries [2, 3]. A number of chemoselective hydrogenation methods that minimize undesired reductions have been reported; for example, platinum metal sulfides [4] or ZnX_2–Pd/C (or Pt/C) systems [5] have been used for the selective hydrogenation of nitro groups in the presence of aromatic halides to obtain haloaniline derivatives. A recent study reported that a polymer-incarcerated platinum species worked for the selective reduction of alkenes, alkynes, and aromatic nitro groups while leaving benzyl ethers and aromatic halides untouched [6]. Other methods for the chemoselective hydrogenation of aromatic nitro groups using vanadium-promoted Raney nickel [7], Ir/C–MgO [8], and Pt/Al_2O_3–$P(OPh)_3$ [8] were also reported to give aniline derivatives that bear aromatic chlorides and aromatic carbonyl groups. Semihydrogenations of alkynes to alkenes were also achieved using nickel [9] or

---

Pharmaceutical Process Chemistry. Edited by Takayuki Shioiri, Kunisuke Izawa, and Toshiro Konoike
Copyright © 2011 WILEY-VCH Verlag GmbH & Co. KGaA, Weinheim
ISBN: 978-3-527-32650-1

gold [10] nanoparticles. However, only a few catalyst systems, such as the Lindlar and Rosenmund methods, are practically useful [11].

There has been extensive interest in controlling the benzyl ether hydrogenolysis during the transformation of an organic compound containing multiple reducible functional groups [12]. Although benzyl ethers are widely used as protective groups of alcohols chiefly because of their stability under a variety of reaction conditions, low cost, ease of formation, and removal by mild catalytic hydrogenolysis, the lack of chemoselectivity between the benzyl groups and other reducible functional groups toward Pd-catalyzed hydrogenolysis has been a serious problem [13]. During our effort to overcome these problems, we found that the addition of a nitrogen-containing base, such as ammonia, pyridine, or ammonium acetate, to the Pd/C-catalyzed hydrogenation system as a mild catalyst poison selectively suppressed the hydrogenolysis of the aliphatic benzyl ether with smooth hydrogenation of other reducible functionalities, such as an alkyne, alkene, azide, nitro, benzyl ester, and N-Cbz [14].

Our first attempts to achieve the hydrogenolysis of the O-benzyl protective group were carried out by making 3-benzyloxy-1-phenyl-1-propene (1) react in the presence of an additive (0.5 equiv.) and a Pd catalyst (10% of the weight of the substrate) in CH_3OH (Table 4.1). The 5% Pd/C or Pd-black-catalyzed hydrogenolysis of the O-benzyl protective group was entirely blocked by the addition of ammonia, triethylamine, pyridine, and ammonium acetate, and the corresponding 1-benzyloxy-3-phenylpropane (2a) was quantitatively obtained (entries 3–7).

Table 4.1 Effect of the nitrogen-containing base on Pd-catalyzed hydrogenolysis of the aliphatic O-benzyl ether 1.

$$PhCH=CHCH_2OBn \xrightarrow[\text{Additive (0.5 equiv.)}]{\text{Catalyst, } H_2, CH_3OH} Ph(CH_2)_3OBn + Ph(CH_2)_3OH$$
$$\mathbf{1} \qquad\qquad\qquad\qquad\qquad \mathbf{2a} \qquad\quad \mathbf{2b}$$

Entry	Catalyst	Additive	Yield (%)[a]	
			2a[b]	2b
1	5% Pd/C	None	0	100[c]
2	Pd black	None	0	98
3	5% Pd/C	NH_3	98	0
4	Pd black	NH_3	99	0
5	5% Pd/C	Et_3N	97	0
6	5% Pd/C	Pyridine	98	0
7	5% Pd/C	NH_4OAc	97	0
8	5% Pd/C	NH_4Cl	0	98

[a] Isolated yield unless otherwise noted.
[b] Though reactions were completed within 15 min, these were allowed to stand for 15 h to prove the selectivity of the inhibition.
[c] By ^1H NMR.

Although these hydrogenations of the alkene moiety of **1** were completed within 15 min in each case, the benzyl group was not removed even after 15 h. On the other hand, when ammonium chloride was used as an additive, no inhibition was observed, and debenzylated 3-phenylpropanol (**2b**) was formed (entry 8) analogous to the case of no additive (entries 1 and 2).

The reaction profile on the hydrogenation of **1** using 5% Pd/C (a) or 5% Pd/C plus ammonia (b) is shown in Figure 4.1. Under 5% Pd/C-catalyzed hydrogenation conditions, the alkene moiety of **1** could be rapidly reduced prior to the hydrogenolysis of the O-benzyl group (Figure 4.1a). Cinnamyl alcohol was never detected in the reaction mixture indicating that the hydrogenation of the alkene moiety of **1** occurred prior to the debenzylation. The ability of the catalyst for the hydrogenation of the alkene moiety seems to be completely retained in spite of the addition of ammonia, which is seen in the fact that the rates for the hydrogenation of **1** into **2a** are not affected by the presence of ammonia. The addition of ammonia, however, entirely blocked the hydrogenolysis of the benzyl group.

To explore the generality of the chemoselective hydrogenation of benzyl ether derivatives, a number of substrates were investigated using 5% Pd/C as a catalyst in the presence of 0.5 equiv. of ammonia or ammonium acetate. As shown in Table 4.2, the addition of a nitrogen-containing base selectively and thoroughly suppressed the hydrogenolysis of the aliphatic O-benzyl group with smooth hydrogenation of the alkene (entry 2), azido (entry 3), N-Cbz (entries 4–6), nitro (entry 7), and benzyl ester (entry 8) functionalities. Control experiments indicated that under 5% Pd/C hydrogenolysis conditions without the additive, the benzyl group was smoothly deprotected to the corresponding alcohol in excellent yields (entry 1).

Although the benzyl group of a phenolic ether was unfortunately and easily deprotected in the presence of ammonia, methylamine, or pyridine (Table 4.3, entries 2–4) [15, 16], we found a significant difference in the inhibitory effect

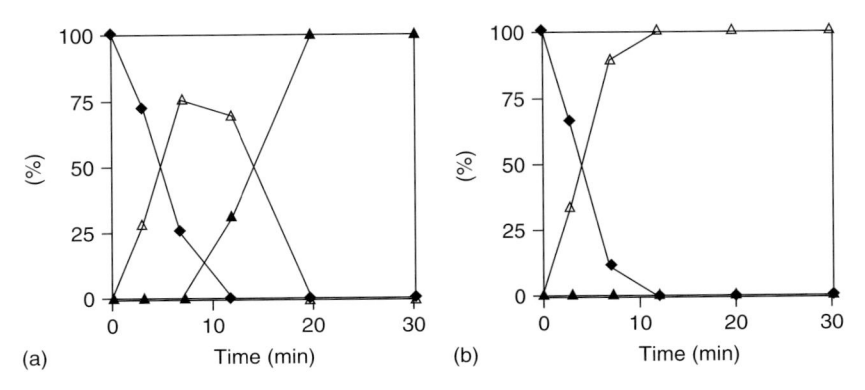

Figure 4.1 Reaction profile of the hydrogenation of 3-benzyloxy-1-phenyl-1-propene (**1**) using 5% Pd/C (a) or 5% Pd/C plus ammonia (b). ♦, 3-benzyloxy-1-phenyl-1-propene (**1**); △, 1-benzyloxy-3-phenylpropane (**2a**); ▲, 3-phenyl-1-propanol (**2b**).

Table 4.2 Chemoselective hydrogenation of aliphatic benzyl ether derivatives.[a]

Entry	Substrate	Additive	Product	Yield (%)[b]
1	(dienyl)-OBn	None	(alkyl)-OH	93
2	(dienyl)-OBn	NH_3	(alkyl)-OBn	93
3	$BnO(CH_2)_2N_3$	NH_3	$BnO(CH_2)_2NH_2$	95
4	Cbz-Ser(Bn)-OH	NH_3	H-Ser(Bn)-OH	97
5	BnO-(cyclohexyl)-NCbz	NH_3	BnO-(cyclohexyl)-NH	97
6	BnO-(cyclohexyl)-NCbz	NH_4OAc	BnO-(cyclohexyl)-NH	95
7	O_2N-(C$_6H_4$)-$CO_2(CH_2)_2OBn$	NH_4OAc	H_2N-(C$_6H_4$)-$CO_2(CH_2)_2OBn$	90
8	Boc-Ser(Bn)-OBn	NH_4OAc	Boc-Ser(Bn)-OH	96

[a]The reaction was carried out using 1.0 mmol of the substrate with 5% Pd/C (10% of the weight of the substrate) and an additive (0.5 equiv.) in CH_3OH (2 ml) under hydrogen atmosphere (balloon). Although most of the chemoselective hydrogenations were completed within 3 h, the benzyl group remained intact even after 15–24 h.
[b]Isolated yield.

on the hydrogenolysis of the aromatic O-benzyl ether of N-Boc-O-benzyltyrosine methyl ester (3) depending upon the nitrogen-containing bases employed as an additive. Upon employment of ethylenediamine, a 1,2-ambident amine, the hydrogenolysis of the O-benzyl protective group of 3 was dramatically suppressed (entry 5), although 1,4-butanediamine did not indicate any significant suppressive effect under identical reaction conditions (entry 7). This fact suggests that the length of the C–C chain between the two nitrogen atoms plays a crucial role in the inhibitory effect. Consequently, the structural requisite for the nitrogen-containing base would be the 1,2-ambident base moiety which possesses the ability to form a five-membered cyclic complex chelated with Pd. Much better results were obtained using an aromatic 1,2-ambident nitrogen-containing base (entries 8–10), and the hydrogenolysis of the O-benzyl protective group of 3 was completely blocked by the addition of 2,2′-dipyridyl or 1,10-phenanthroline. Another critical property of the additives for the successful inhibition of the hydrogenolysis may be their affinity for the Pd metal based on the π-electrons of the aromatic ring [17].

4.1 Catalyst Poisons and Chemoselective Heterogeneous Catalysts

Table 4.3 Conversion (%) of **3** to **4** after a 2 h hydrogenolysis with 5% Pd/C in the presence of a nitrogen-containing base.[a]

3 (BocHN-CH(CO₂CH₃)-CH₂-C₆H₄-OBn) → 4 (BocHN-CH(CO₂CH₃)-CH₂-C₆H₄-OH)

Conditions: 5% Pd/C, H_2, CH_3OH, Nitrogen-containing base (0.5 equiv.)

Entry	Amine or base	Conversion (%)
1	None	100
2	NH_3	100
3	CH_3NH_2	100
4	Pyridine	100
5	Ethylenediamine	28
6	1,3-Propanediamine	41
7	1,4-Butanediamine	95
8	o-Phenylenediamine	19
9	2,2′-Dipyridyl	0
10	1,10-Phenanthroline	0

[a] The reaction was carried out using 0.10 mmol of **3** and a nitrogen-containing base or amine (0.05 mmol) with 5% Pd/C (3.9 mg) in CH_3OH (1 ml) under a hydrogen atmosphere (balloon) for 2 h.

4.1.3
Pd/C(en) Complex as a Heterogeneous Chemoselective Hydrogenation Catalyst

As noted in the preceding section, we documented that the addition of a nitrogen-containing base to a Pd/C-catalyzed reduction system selectively suppressed the hydrogenolysis of an aliphatic benzyl ether with smooth hydrogenation of the other reducible functionalities [14]. We also found that the catalyst activity of Pd/C toward the phenolic benzyl ether of **3** was time-dependently suppressed by the addition of ethylenediamine, 1,3-propanediamine, or 1,4-butanediamine (Figure 4.2). The slope of the graphs time-dependently decreased and finally reached a plateau, which represent the tangible catalyst inactivation in the ethylenediamine and 1,3-propanediamine cases.

Furthermore, the catalyst activity was also time-dependently suppressed by the preliminary processing with ethylenediamine without the substrate. The hydrogenolysis activity of 5% Pd/C toward the benzyl protective group of **5** was analyzed in relation to the pretreatment time with ethylenediamine (Figure 4.3). After processing with ethylenediamine for 30 h, the 5% Pd/C was no longer

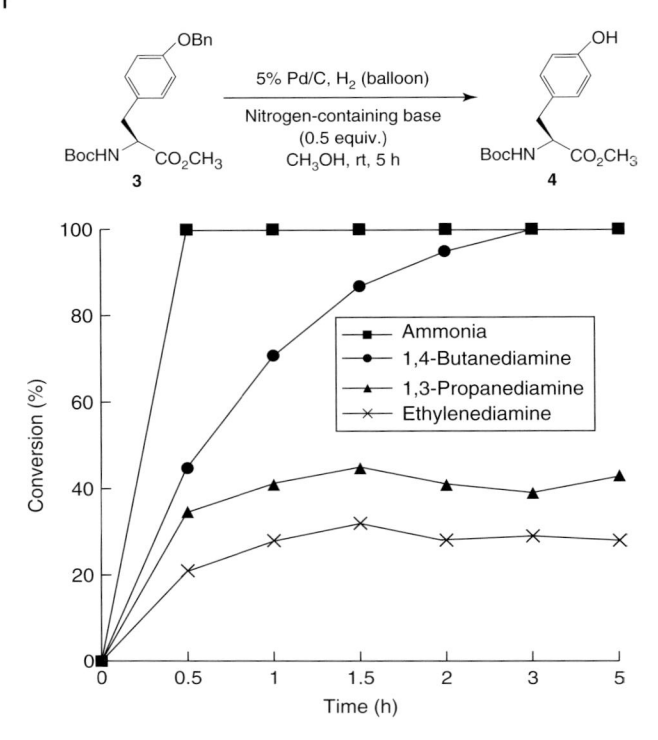

Figure 4.2 Time-course study on the Pd/C-catalyzed hydrogenolysis of phenolic benzyl ether (**3**) in the presence of a nitrogen-containing base.

catalytically active for the hydrogenolysis of even the phenolic benzyl ether. On the basis of these results, isolation of the pretreated 5% Pd/C catalyst with ethylenediamine was attempted.

Thus, a suspension of commercial Pd/C and large excess (about 70 equiv. vs Pd metal of Pd/C) of ethylenediamine in CH_3OH was stirred for 48 h at room temperature, and then filtered, vigorously washed with CH_3OH and ether, and finally dried under a vacuum for 48 h at room temperature. The isolated catalyst consisted of approximately an equal mole ratio of ethylenediamine and Pd metal.[1] The isolated Pd on the activated carbon–ethylenediamine complex [Pd/C(en)] is nonpyrophoric and can be stored in a general reagent bottle for a long period at ambient temperature, and also completely lost its catalytic activity for the hydrogenolysis of **1**, **3**, and **5**. Since the treatment of ethylenediamine with Pd black or 5% Pd on alumina powder instead of Pd/C did not form such a selective catalyst, the carbon support of Pd/C(en) seems to play a crucial role as an electron-accepting species [18a–d] or as a polymer support [18e–h] for the zero-valent Pd complex. The x-ray photoelectron

---

1) The 5% Pd/C(en) catalyst contains 1.35–1.75% of nitrogen.

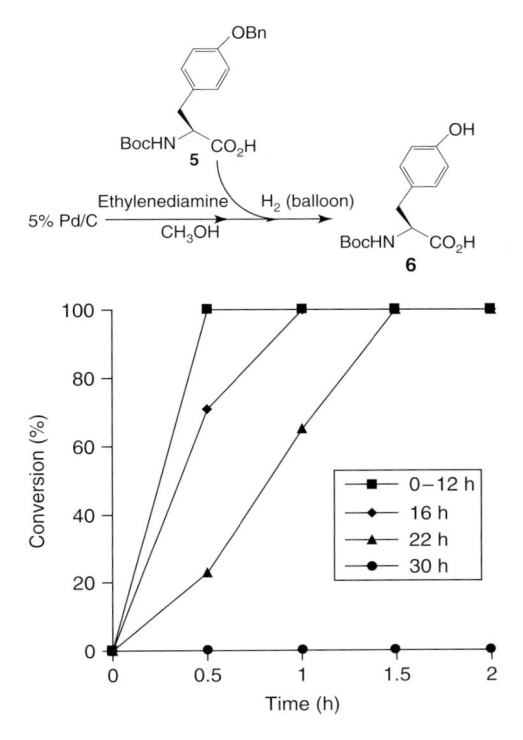

Figure 4.3 Effect of the pretreatment time of Pd/C with ethylenediamine on the hydrogenolysis of **5**.

spectroscopy (XPS) of Pd/C(en) was measured in order to gain insight into the surface interaction step of Pd metal. Although the core binding energy of a commercial Pd/C catalyst (Aldrich) is 335.4 and 337.1 eV corresponds to that of Pd(0) on a carbon support [16b], the observed Pd $3d_5$ peaks in the Pd/C(en) catalyst shifted to significantly higher energies of 335.8 and 338.4 eV. Furthermore, N 1s (399.80 eV) was observed in the Pd/C(en), while no significant N 1s peak was found in the Pd/C [19].

To explore the scope of the Pd/C(en) catalyst, the hydrogenation of a number of substrates was investigated (Table 4.4). The catalyst activity toward a wide variety of reducible functionalities, such as an alkene, alkyne, benzyl ester, nitro, azide, aromatic ketone, and aldehyde, was retained. On the other hand, the hydrogenation of benzyl ether (entries 1–7) [19], Cbz (entries 8–11) [19, 20], aromatic nitrile (entries 12 and 13) [21], benzyl alcohol (entries 13–15) [21, 22], and epoxide (entries 16–18) [23] was suppressed, with appropriate optimization of the reaction conditions (e.g., solvent exchange and addition of a base) in some cases. Although it has been established that O-TBDMS (t-butyldimethylsilyl) protective groups are believed to be stable toward Pd/C-catalyzed hydrogenation conditions [12, 24], the frequent and unexpected loss of the TBDMS protective group from a variety of hydroxyl functions occurred under neutral and mild Pd/C-catalyzed hydrogenation conditions in CH_3OH [25, 26]. On the other hand, the undesirable

Table 4.4 Chemoselective hydrogenation in the presence of O-benzyl or N-Cbz protective group using Pd/C(en) catalyst.[a]

Entry	Substrate	Time (h)	Product	Yield (%)[b]
1	PhCH=CHCH$_2$OBn	0.25	Ph(CH$_2$)$_3$OBn	95
2	(prenyl-type structure)–OBn	24	(saturated)–OBn	91
3	N$_3$(CH$_2$)$_2$OBn	6	NH$_2$(CH$_2$)$_2$OBn	93
4	Boc–Ser(Bn)–OBn	0.33	Boc–Ser(Bn)–OH	91
5	O$_2$N–C$_6$H$_4$–CO$_2$(CH$_2$)$_2$OBn	2	H$_2$N–C$_6$H$_4$–CO$_2$(CH$_2$)$_2$OBn	98
6	BnO,CH$_3$O-C$_6$H$_3$–CH=CH$_2$	2	BnO,CH$_3$O-C$_6$H$_3$–Et	96
7[c]	BnO–C$_6$H$_4$–CH$_2$CO$_2$Bn	2	BnO–C$_6$H$_4$–CH$_2$CO$_2$H	94
8[d]	diallyl-N-Cbz	17	dipropyl-N-Cbz	92
9[d]	HC≡C–CH$_2$NHCbz	4	CH$_3$CH$_2$CH$_2$NHCbz	83
10[c]	Cbz–Pro–OBn	6	Cbz–Pro–OH	76
11[d]	N$_3$(CH$_2$)$_4$NHCbz	2	H$_2$N(CH$_2$)$_4$NHCbz	99
12[d]	NC–C$_6$H$_4$–O–CH$_2$CH=CH$_2$	24	NC–C$_6$H$_4$–O–CH$_2$CH$_2$CH$_3$	96
13[d]	NC–C$_6$H$_4$–CH(OH)–C≡C–Ph	24	NC–C$_6$H$_4$–CH(OH)–CH$_2$CH$_2$–Ph	83
14	Ph–C(O)–CH$_2$CH$_2$–C(O)–CH$_3$	24	Ph–CH(OH)–CH$_2$CH$_2$–C(O)–CH$_3$	91
15	2-(allyloxy)benzaldehyde	24	2-(propyloxy)benzyl alcohol	94
16[d]	allyl–O–CH$_2$–epoxide	3	propyl–O–CH$_2$–epoxide	92
17[d]	prenyl–O–C(O)–epoxide–OH	12	saturated–O–C(O)–epoxide–OH	86

Table 4.4 (continued)

Entry	Substrate	Time (h)	Substrate	Yield (%)
18[d,e]	4-O$_2$N-C$_6$H$_4$-CO-O-(CH$_2$)$_3$-epoxide	18	4-H$_2$N-C$_6$H$_4$-CO-O-(CH$_2$)$_3$-epoxide	100
19	PhCH=CHCH$_2$OTBDMS	24	Ph(CH$_2$)$_3$OTBDMS	93
20	cyclohexenyl-CH$_2$OTBDMS	24	cyclohexyl-CH$_2$OTBDMS	92

[a] Unless otherwise specified, the reaction was carried out using 0.2 mmol of the substrate in CH$_3$OH (1 ml) with Pd/C(en) (10% of the weight of the substrate) under a hydrogen atmosphere (balloon) for the given reaction time.
[b] Isolated yield.
[c] Reaction was performed in the presence of DMAP (4-dimethylaminopyridine) or Dabco (1.4 equiv. vs substrate).
[d] Reaction was performed with THF (tetrahydrofuran) or dioxane as the solvent.
[e] Reaction was performed under 5 atm of hydrogen.

problem was perfectly overcome and the chemoselective hydrogenation of reducible functionalities leaving the TBDMS protective group intact was achieved using Pd/C(en) instead of Pd/C (entries 19 and 20) [25, 27].[2)]

4.1.4
Pd/C (Ph$_2$S) Complex as a Heterogeneous Chemoselective Hydrogenation Catalyst

As mentioned in Section 4.1.2, we reported that the addition of a nitrogen-containing base to a Pd/C-catalyzed hydrogenation system selectively suppressed the hydrogenolysis of benzyl ether in the presence of other reducible functionalities [14, 15]. While the use of Pd/C(en) further achieved an efficient chemoselectivity [19–23], [25], development of a novel catalyst possessing a different catalyst activity will reinforce the versatility of synthetic processes. In order to establish a catalytic hydrogenation system possessing a distinct chemoselectivity, we focused on sulfur-containing compounds with the expectation that they should be quite strong catalyst poisons compared to nitrogenous compounds. Several sulfur-containing catalyst poisons have been reported; for

2) While TIPS (triisopropylsilyl) and TBDPS (t-butyldiphenylsilyl) ethers were quite stable under the hydrogenation conditions in CH$_3$OH, THF, AcOEt, and CH$_3$CN, the TBDMS and TES protective groups were readily cleaved in CH$_3$OH [25, 26]. Consequently, the Pd/C-catalyzed hydrogenation in CH$_3$OH can be applied as a convenient and neutral deprotection method of TBDMS and TES (triethylsilyl) protective groups in the presence of other protective groups. In contrast, the TBDMS ether was not deprotected under the hydrogenation conditions in AcOEt and CH$_3$CN at all, and the TES ether was stable in CH$_3$CN. Thus, the solvent-mediated chemoselective hydrogenation of reducible functionalities, such as alkene, alkyne, and benzyl ethers, as distinguished from the TBDMS and TES ethers, can also be achieved under Pd/C-catalyzed hydrogenation conditions in AcOEt or CH$_3$CN as the solvent.

example, quinoline-S is used for the Rosenmund reduction in order to avoid the overreduction of aldehydes to the corresponding alcohols [28], thiophene, or butylmercaptan is used for the selective reduction of an alkene-tolerating O-benzyl protective groups [29], and platinum sulfide (PtS) catalyzes the conversion from halonitrobenzenes to haloanilines without dehalogenation [30]. During our effort to establish a chemoselective hydrogenation method using Pd/C by the addition of sulfur-containing compounds, we found that the addition of diphenylsulfide (Ph_2S) moderately depressed the catalyst activity of Pd/C to furnish a distinguishing chemoselectivity upon hydrogenation [31]. With the use of commercial 10% Pd/C in the absence of an additive, the hydrogenation of the alkene and aromatic ketone readily proceeded to quantitatively give the fully hydrogenated 1,3-diphenylpropane (**10**) (Table 4.5, entry 1), and a selective partial reduction was achieved using Pd/C(en) to afford the corresponding benzyl alcohol (**9**) without hydrogenolysis (entry 2) [22]. The addition of 0.01 equiv. of diphenylsulfone and diphenylsulfoxide could not depress the hydrogenation of the aromatic ketone (entries 3 and 4). However, diphenyldisulfide and thiophenol completely deactivated the 10% Pd/C, and only the starting material (**7**) was afforded (entries 8 and 9). On the other hand, we found that the addition of diphenylsulfide (0.01–0.1 equiv.) completely blocked the hydrogenation of only the aromatic ketone while hydrogenation of the alkene smoothly and chemoselectively proceeded (entries 6 and 7) [3]. Reduction of the diphenylsulfide use to 0.001 equiv. did not exert a sufficient suppressing effect on the Pd/C-catalyzed hydrogenation of the aromatic ketone (entry 5) [31]. This method

Table 4.5 Assessment of additives for chemoselective hydrogenation between alkene and aromatic ketone using chalcone (7) as the substrate.

Entry	Additive	7	:	8	:	9	:	10[a]
1	None	0	:	0	:	0	:	100
2	None[b]	0	:	0	:	100	:	0
3	Ph_2SO_2	0	:	0	:	100	:	0
4	Ph_2SO	0	:	93	:	7	:	0
5	Ph_2S (0.001 equiv.)	0	:	94	:	6	:	0
6	Ph_2S	0	:	100	:	0	:	0
7	Ph_2S (0.1 equiv.)	0	:	100	:	0	:	0
8	Ph_2S_2	100	:	0	:	0	:	0
9	PhSH	100	:	0	:	0	:	0

[a] By ^{1}H NMR.
[b] 10% Pd/C(en) was used as the catalyst instead of 10% Pd/C.

is totally applicable to the complete chemoselective hydrogenation of alkene, alkyne, and azide functionalities in the presence of other reducible functional groups.

In spite of such unique chemoselectivities, this method requires the removal of Ph_2S from the reaction mixture by silica gel column chromatography. Therefore, the development of a Pd/C–Ph_2S complex, $Pd/C(Ph_2S)$, by immobilization of Ph_2S to Pd/C as a heterogeneous and chemoselective hydrogenation catalyst possessing a chemoselectivity similar to the addition of Ph_2S to the Pd/C-catalyzed hydrogenation system is strongly desired. Detailed investigations led us to establish the preparation method of $Pd/C(Ph_2S)$. The suspension of 10% Pd/C and Ph_2S (2.0 equiv. vs Pd metal of Pd/C) in CH_3OH was stirred for five days at room temperature under an argon atmosphere, and then the resulting black powder was collected by filtration, successively washed with CH_3OH and ether, and finally dried under reduced pressure to give a nonpyrophoric and stable complex [32]. The elemental analysis of the complex indicated that the ratio of Pd metal and Ph_2S was 1 : 1.27. From the XRD (X-ray diffraction) analysis [33], it was observed that the (0, 0, 2) face of the activated carbon of this complex was shifted to a lower angle compared to that of a commercial Pd/C (Figure 4.4). This means that a certain amount of Ph_2S penetrates into the graphite layers of the activated carbon [34]. Furthermore, EPMA (electron probe microanalyzer) [35] data (Figure 4.5) showed that Ph_2S and the Pd metal are not evenly distributed on the carbon support. The XPS spectrum of Pd 3d of $Pd/C(Ph_2S)$ indicates a certain and partial interaction of Ph_2S with the Pd metal. One of the binding energy peaks of the Pd metal in $Pd/C(Ph_2S)$ was observed at 337.7 eV, which is obviously higher than that of the Pd(0) standard (335.1–335.5 eV) (Figure 4.6) [36], suggesting that the peak should be derived from the Pd–S complex, and a portion of the Pd metal is strongly bound to Ph_2S. Furthermore, a comparison of the effective Pd metal surface area, determined by measurement of the CO absorbability, between Pd/C and $Pd/C(Ph_2S)$ indicated that 47% of the active site on the Pd surface of $Pd/C(Ph_2S)$

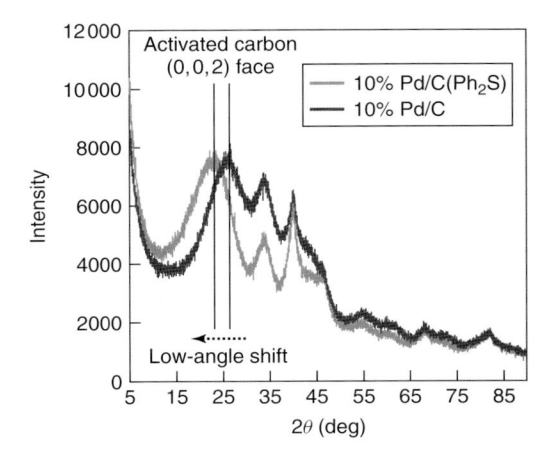

Figure 4.4 XRD analysis of $Pd/C(Ph_2S)$.

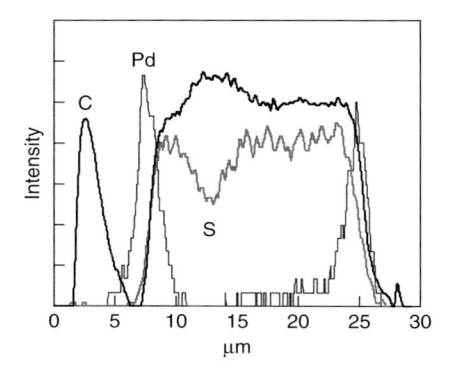

Figure 4.5 EPMA analysis of Pd/C(Ph$_2$S).

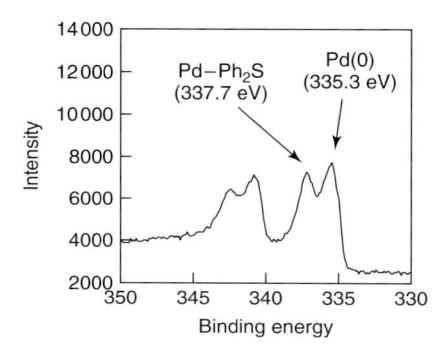

Figure 4.6 XPS analysis of Pd/C(Ph$_2$S).

was covered by Ph$_2$S. These results indicated that about one half of the fresh Pd(0) surface (active sites) of the original Pd/C still remained intact and the other half had interacted with Ph$_2$S by the Pd–S bond based upon the moderately occluded sulfur lone pair. Therefore, Pd/C(Ph$_2$S) should possess a very weak catalyst activity and was expected to produce chemoselective hydrogenation among some reducible functionalities [32].

Pd/C(Ph$_2$S) possesses an excellent chemoselectivity for hydrogenation as follows: alkene (Table 4.6, entries 2–6 and 11), alkyne (entries 1 and 7), nitro (entries 8 and 9), and azide (entries 10 and 11) functionalities were smoothly hydrogenated, while aromatic carbonyl (entries 1 and 10) and chloride (entries 2, 3, and 8), benzyl ester (entries 4, 5, and 9), and N-Cbz (entries 6 and 7) functionalities remained intact. It is noteworthy that the aromatic carbonyls could remain intact under Pd/C(Ph$_2$S)-catalyzed hydrogenation conditions (entries 1 and 10), while the aromatic carbonyl groups could be very smoothly hydrogenolyzed to the methylene compounds under the commercial Pd/C-catalyzed conditions (e.g., Table 4.5, entry 1) and hydrogenated to the benzyl alcohols using Pd/C(en) (e.g., Table 4.5, entry 2).

Table 4.6 Chemoselective hydrogenation of alkene or alkyne in the presence of aromatic ketone using Pd/C(Ph$_2$S).

Substrate $\xrightarrow[\text{CH}_3\text{OH, 24 h}]{\text{10\% Pd/C(Ph}_2\text{S)} \atop \text{H}_2 \text{ (balloon)}}$ Product

Entry	Substrate	Product	Yield (%)
1	Ph-C≡C-C(O)-Ph (alkyne ketone)	Ph-CH$_2$-CH$_2$-C(O)-Ph	94
2	4-Cl-C$_6$H$_4$-CH=CH$_2$	4-Cl-C$_6$H$_4$-CH$_2$CH$_3$	100
3	4-Cl-C$_6$H$_4$-CH=CH-C(O)NH$_2$	4-Cl-C$_6$H$_4$-CH$_2$CH$_2$-C(O)NH$_2$	100
4	CH$_2$=C(CH$_3$)-CO$_2$Bn	(CH$_3$)$_2$CH-CO$_2$Bn	100
5	4-(BnO$_2$C)-C$_6$H$_4$-CH=CH$_2$	4-(BnO$_2$C)-C$_6$H$_4$-CH$_2$CH$_3$	99
6	(CH$_2$=CHCH$_2$)$_2$N-Cbz	(CH$_3$CH$_2$CH$_2$)$_2$N-Cbz	98
7	4-(HC≡C)-C$_6$H$_4$-NHCbz	4-(CH$_3$CH$_2$)-C$_6$H$_4$-NHCbz	94
8	4-Cl-C$_6$H$_4$-NO$_2$	4-Cl-C$_6$H$_4$-NH$_2$	91
9	4-O$_2$N-C$_6$H$_4$-CO$_2$Bn	4-H$_2$N-C$_6$H$_4$-CO$_2$Bn	97
10	4-(PhC(O))-C$_6$H$_4$-N$_3$	4-(PhC(O))-C$_6$H$_4$-NH$_2$	99
11	4-N$_3$-C$_6$H$_4$-CH=CH-CO$_2$H	4-H$_2$N-C$_6$H$_4$-CH$_2$CH$_2$-CO$_2$H	95

Table 4.7 Reuse test of Pd/C(Ph$_2$S).

Ph−CH=CH−CO$_2$Bn (**11**) $\xrightarrow[\text{CH}_3\text{OH, 24 h}]{\text{10\% Pd/C(Ph}_2\text{S)} \text{ H}_2 \text{ (balloon)}}$ Ph−CH$_2$−CH$_2$−CO$_2$Bn (**12**) + Ph−CH$_2$−CH$_2$−CO$_2$H (**13**)

Run	Ratio (11:12:13)	Yield (%) of 12[a]
1	0:100:0	100
2	0:100:0	92
3	0:100:0	97
4	0:100:0	99
5	0:100:0	99

[a] Isolated yield.

Pd/C(Ph$_2$S) is entirely reusable for the chemoselective hydrogenation of benzyl cinnamate at least until the fifth run without any significant loss of the selectivity and yield (Table 4.7) [32].[3)]

4.2
Catalyst Supports and Chemoselective Heterogeneous Catalysts

4.2.1
Pd/Fib as a Silk-Fibroin-Supported Chemoselective Hydrogenation Catalyst

The silk secreted from the silk gland of the silkworm *bombix mori* is composed of two principal protein components, fibroin and sericin, characterized by their solubility and stability in hot water, which enable the industrial degumming of the silk threads. Recently, fibroin and sericin are considered as useful biomaterials accounting for a wide variety of interesting properties, such as in food, cosmetics, medical, and biological materials, not to mention fibers. In particular, fibroin has an approximate 370 kDa molecular mass and mainly consists of glycine, alanine, serine, tyrosine, and very few residues of sulfur amino acids [37], which can be a strong catalyst poison of transition metal catalysts. Many of the electron-rich functional groups in the protein easily bind with the metal surface, leading to the bioconjugate. During 1956–1962 [38], Akabori and coworkers reported a silk-supported zero-valent Pd catalyst for asymmetric hydrogenation. According to their reported procedure [38b], silk-fibroin was boiled for 8 min in 0.1 N AcOH containing PdCl$_2$ and the resulting chelate was reduced under hydrogen (80 kg cm^{-2}) in an autoclave.

3) Leached Pd and sulfur were not detected (below measurable limits: <1 ppm for Pd and <20 ppm for sulfur) after the second run of the reuse test.

Based upon this preparation method [38], proteins of the silk-supported Pd catalyst were most likely denatured under the drastic reaction conditions for the reason that the proteins were exposed to the strongly acidic conditions as a result of liberation of HCl from $PdCl_2$ in boiling 0.1 N AcOH. This could be due to the fact that the silk Pd-catalyzed asymmetric hydrogenations are inefficient in as much as the reproducibility of the method was invariably poor [38]. We developed the preparation method of a highly dispersed Pd catalyst supported on silk-fibroin by an HCl and hydrogen-free mild method at room temperature that is active and chemoselective in the heterogeneous phase hydrogenation [39]. The Pd deposition onto the fibroin was initiated by incipient wetness impregnation with a rust-colored CH_3OH solution of $Pd/C(OAc)_2$ (Figure 4.7a,b). The mixture was sonicated [39d] for 12 h (30 ± 3 °C) at room temperature under atmospheric conditions and the colorless fibroin gradually changed to black, while the solution completely changed to colorless-clear (Figure 4.7c–e). After simple filtration, the obtained Pd/Fib was stable at room temperature under atmospheric conditions and nonpyrophoric (Figure 4.7f). The quantitative analysis of the formation of acetic acid and formaldehyde was carried out to gain insight into the reduction of $Pd(OAc)_2$. For the reason that 90% of acetic acid and 70% of formaldehyde were determined from the filtered clear solution of Figure 4.7e, CH_3OH contributed to the reduction of $Pd(OAc)_2$ to Pd(0), which was strongly absorbed and coordinated by the silk-fibroin (Scheme 4.1). SEM images (Figure 4.8) showed a high dispersion of amorphous Pd clusters on the smooth fibroin surface. The variation in the Pd content made a difference only in the density of the Pd particles. This homogeneous

Figure 4.7 Preparation of Pd/Fib catalyst: (a) fibroin fiber; (b) $Pd(OAc)_2$ in CH_3OH; (c) fibroin in solution (0 h); (d) fibroin in solution (3 h); (e) Pd/Fib in colorless CH_3OH; (f) 2.5% Pd/Fib.

4 Development of Palladium Catalysts for Chemoselective Hydrogenation

$$\text{Pd(OAc)}_2 \xrightarrow[\text{CH}_3\text{OH, rt}]{\text{Fibroin}} \text{Pd(OAc)}_2/\text{Fib} \xrightarrow[\text{CH}_3\text{OH}]{\text{rt}} \text{Pd(0)/Fib} + \text{HCHO} + 2\text{AcOH}$$

Scheme 4.1 Mechanism of Pd/Fib generation.

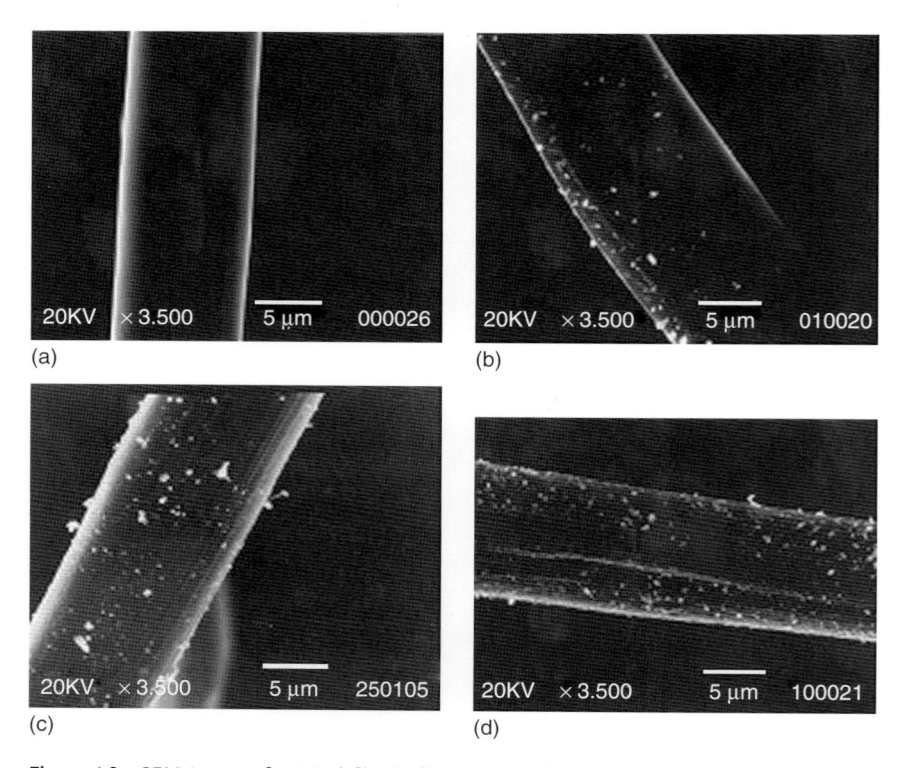

Figure 4.8 SEM image of original fibroin fiber (a), 1% (b), 2.5% (c), and 10% Pd/Fib (d).

dispersion was attributed to an interaction between Pd(OAc)$_2$ and the amino acids of the silk-fibroin, thus increasing the resistance to the growth of the Pd cluster.

As in the case of 10% Pd/C(Ph$_2$S) in Section 4.1.4, 2.5% Pd/Fib provides an excellent chemoselectivity for the hydrogenation of alkanes, alkynes, and azides in the presence of other reducible functionalities including aromatic carbonyl and halide, N-Cbz, and benzyl ester which are easily hydrogenated under even Pd/C(en)-catalyzed hydrogenation conditions [39]. While the 10% Pd/C(Ph$_2$S)-catalyzed selective reduction of the nitro group was successful, the 2.5% Pd/Fib-catalyzed hydrogenation of nitro compounds led to the formation of a complex mixture based on the nucleophilic coupling of the resulted amine and the intermediate nitroso and/or imine derivatives. Hence, it seems difficult to achieve a good selectivity in the Pd/Fib-catalyzed hydrogenation of nitro groups.

4.2.2
Pd-PEI as a Partial Hydrogenation Catalyst of Alkynes to Alkenes

The selective partial hydrogenation of alkynes to *cis*-alkenes represents an important class of chemical transformations that have found extensive use in the construction of enormous functional materials [40]. This important transformation has mainly been accomplished under ordinary hydrogen pressure at room temperature using heterogeneous Pd catalysts (Lindlar catalyst in conjunction with quinoline [41] and Pdc [42]), Ni catalysts (low-active Raney Ni [43], P-1 Ni [44], P-2 Ni [43m, 45], and Nic [46]), Au nanoparticles [47], or homogeneous Rh and Cr complexes [48]. In these cases, except for the Lindlar catalyst, the narrow substrate applicability, concomitant use with a basic additive, low *cis–trans* selectivity, pyrophoric property, and/or operational complexity of the catalysts are/is undesirable as a general synthetic methodology and these are the reasons why such catalysts have never been a general selective catalyst for the partial hydrogenation of an alkyne to a *cis*-alkene. On the other hand, the Lindlar catalyst is widely applicable and used as a selective catalyst, although the pretreatment using environmentally harmful $Pb(OAc)_2$ and the simultaneous use with quinoline are necessary to achieve the chemo- and geometrically selective partial hydrogenation. Furthermore, the smooth overreduction to the corresponding alkanes efficiently proceeded under Lindlar's hydrogenation conditions using monosubstituted (terminal) alkynes as substrates. Therefore, Lindlar's method is only applicable to the selective chemical transformation of disubstituted alkynes to *cis*-alkenes [49]. Furthermore, an efficient method for the partial hydrogenation of both internal and terminal alkynes using Ni nanoparticles was reported, although the use adjustment of the *in situ* generated molecular hydrogen source (Li powder and EtOH or *i*-PrOH) is essential [50]. Thus, it has been very important to develop a truly selective and environmentally benign catalyst.

We have developed a polyethyleneimine (PEI)-, branched polymer with a 25 000 average MW, supported catalyst (Pd/PEI) for the practical and partial (selective) hydrogenation of disubstituted alkynes to *cis*-alkenes and an efficient partial hydrogenation method (controlled by the catalyst activity) that overcomes these serious and long-standing problems. The 5% Pd/PEI complex catalyst was prepared by the direct introduction of $Pd(OAc)_2$ in the CH_3OH solution of deaerated PEI under an Ar atmosphere (Figure 4.9). The resulting rust-colored solution was stirred under an H_2 atmosphere at room temperature for 24 h. The solution gradually changed to black, indicating the formation of the zero-valent Pd species. After concentration *in vacuo*, the obtained black gummy 5% Pd/PEI is stable in a capped vial and is nonpyrophoric. The Pd deposition (5 wt %) onto the PEI was based on the incipient ratio of $Pd(OAc)_2$ and PEI [51].

The Pd/PEI catalyst is applicable to the selective transformation of various disubstituted alkynes to *cis*-alkanes (Table 4.8, entries 1–5). Even for the substrate bearing a conjugated ketone on the alkyne, no serious overreduction was observed, while a significant *cis–trans* isomerization of the methyl styryl ketone accompanied

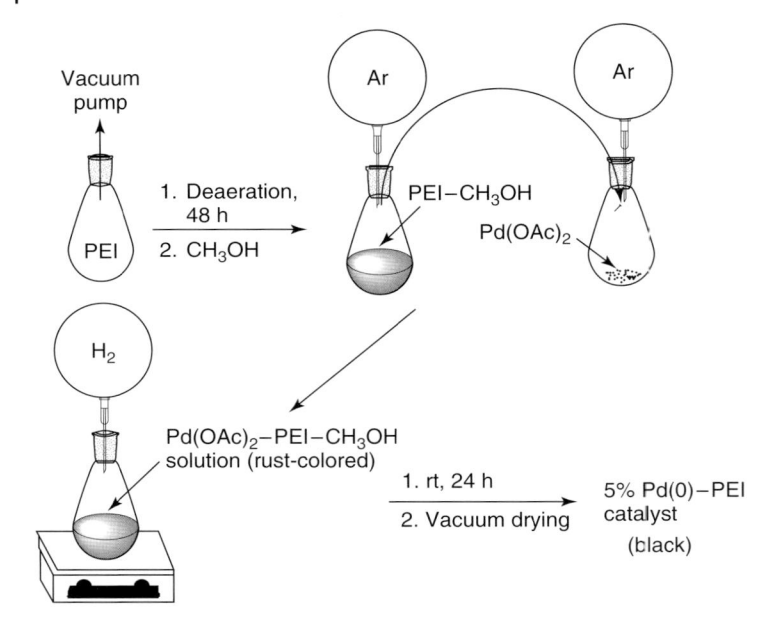

Figure 4.9 Preparation of Pd/PEI catalyst.

the partial hydrogenation on the basis of Pd-catalyzed keto–enol tautomerism (entry 3, *cis–trans* ratio = 36 : 58). It is noteworthy that monosubstituted (terminal) alkynes were also found to be viable substrates (entries 6–12), although solvent tuning (a mixture consisting of CH_3OH and more coordinating solvents such as dioxane) was necessary to obtain a better selectivity. Furthermore, other reducible functionalities, such as the *N*-Cbz protective group (entry 6), benzyl ester (entry 9), benzyl ether (entry 10), and tert-butyldimethylsilyl (TBS/TBDMS) ether (entry 12) could be universally tolerated under the hydrogenation conditions [52].

The partial hydrogenation was strongly influenced by the addition of acetic acid [19]. While *cis*-stilbene was obtained by the Pd/PEI-catalyzed partial hydrogenation of diphenylalkyne (entry 1), the drastic overreduction (about 80%) to 1,2-diphenylethane was observed by the addition of acetic acid (1 equiv. vs substrate) to the reaction mixture.

A postulated mechanistic image for the Pd/PEI-catalyzed partial hydrogenation is shown in Figure 4.10. The mild and multicoordinating nitrogen atoms of PEI with Pd(0) metal produce the Pd/PEI complex possessing a gentle wire-gauze-like equilibrium structure. Furthermore, enormous amounts of 1,4-dioxane as a part of the solvent could patch the open-seam of the Pd–PEI complex as a mild and bidentate oxygen ligand. On the other hand, the addition of acetic acid to the reaction mixture causes destruction of the mild coordinate bonds. It is speculated that the PEI gauze keeps alkenes possessing sp^2-carbons well away from strongly occluded Pd(0) metal, although the spatially compact alkynes can easily access the catalytically active sites.

Table 4.8 Pd/PEI-catalyzed partial hydrogenation of alkynes.

$$R\text{—}\!\!\!\equiv\!\!\!\text{—}R' \quad \xrightarrow[\substack{CH_3OH + \text{Dioxane,} \\ \text{rt, 24 h}}]{\substack{5\% \text{ Pd(0)–PEI} \\ (10 \text{ wt \%}), H_2}} \quad \begin{array}{c} R \quad R' \\ \diagup\!=\!\diagdown \\ H \quad H \end{array}$$

$$R\text{—}\!\!\!\equiv\!\!\!\text{—}H \quad \longrightarrow \quad \begin{array}{c} R \quad H \\ \diagup\!=\!\diagdown \\ H \quad H \end{array}$$

Entry	Substrate	CH_3OH : dioxane	Selectivity (%)[a]
1	Ph—≡—Ph	1:1	98[b]
2	Ph—≡—CO$_2$H	1:1	100[c,d]
3	Ph—≡—COCH$_3$	1:1	94[e]
4	(structure)	1:1	100
5	(structure with OH)	1:1	100
6	≡—C$_6$H$_4$—NHCbz	1:4	93
7	(structure)	0:1	83
8	≡—SPh	1:1	98[f]
9	≡—(CH$_2$)$_3$—CO$_2$Bn	1:4	100
10	≡—OBn	1:4	96
11	(estradiol-ethynyl structure)	4:1	100[c]
12	(Ph, OTBS alkyne structure)	1:0	100

[a] The alkane selectivity (%) versus the generation of the fully hydrogenated alkane was determined by ^{1}H NMR.
[b] One percent of the trans isomer was generated.
[c] K_2CO_3 (1 equiv.) was used as an additive.
[d] Four percent of the trans isomer was generated.
[e] The product was obtained as a cis–trans (36 : 58) mixture due to the significant isomerization.
[f] AcOEt was used instead of dioxane.

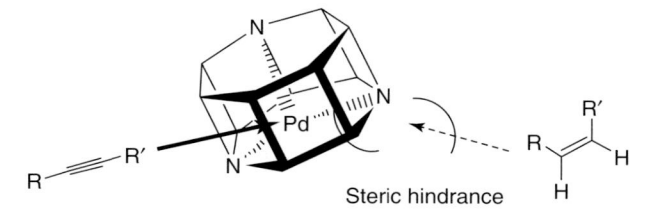

Figure 4.10 Postulated mechanistic image for the chemoselectivity.

4.3 Summary

In this chapter, we introduced the development of a way to suppress the Pd/C-catalyzed hydrogenation of benzyl ethers by the addition of nitrogen-containing bases and four kinds of practical and novel heterogeneous catalysts, as well as the catalyst properties and significance of the chemoselective hydrogenation in synthetic and process chemistries. A summarized overall comparison of the catalyst activities toward hydrogenation among Pd/C, Pd/C plus nitrogen-containing base, Pd/C(en), Pd/C(Ph$_2$S), Pd/Fib, and Pd/PEI is shown in Figure 4.11. By using these catalyst systems as the synthetic situation demands, a wide variety of chemoselective hydrogenation methods and the reduction of synthetic processes, which meet the criteria of green chemistry, become available. Pd/C(en), Pd/C(Ph$_2$S), Pd/Fib, and Pd/PEI are commercially available from Wako Pure Chemical Industries and the N.E. Chemcat Corporation.

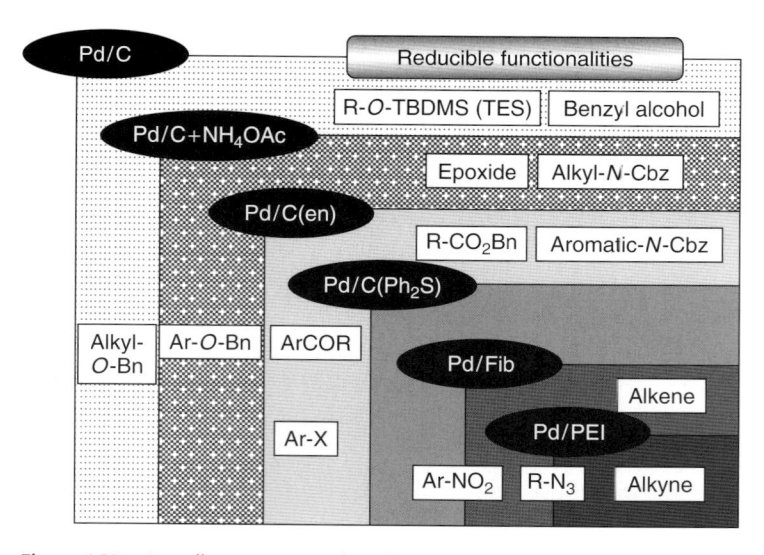

Figure 4.11 Overall comparison of catalyst activities.

Acknowledgment

We are grateful to Wako Pure Chemical Industries and the N.E. Chemcat Corporation for support. We also thank Drs Kazuyuki Hattori, Takashi Ikawa, Akinori Mori, Shigeki Mori, and many graduates of our laboratory in Gifu Pharmaceutical University for substantial contributions to the development of practical catalysts, and Ms Chie Morishige for preparing this manuscript for publication.

References

1. Burke, S. D. and Danheiser, R. L. (eds) (1999) *Handbook of Reagents for Organic Synthesis*, Oxidising and Reducing Agents, Vol. 2, John Wiley & Sons, Inc., New York, p. 280.
2. For reviews, see (a) Rylander, P.N. (1985) *Hydrogenation Methods*, Academic Press, New York;   (b) Rylander, P.N. (1979) *Catalytic Hydrogenation in Organic Synthesis*, Academic Press, New York;   (c) Freifelder, M. (1971) *Practical Catalytic Hydrogenation Techniques and Applications*, Wiley-Interscience, New York;   (d) Siegel, S. (1991) in *Comprehensive Organic Synthesis*, vol. 8 (eds B. M. Trost and I. Fleming), Pergamon Press, Oxford, p. 417;   (e) Hudlicky, M. (1996) *Reductions in Organic Chemistry*, 2nd edn, American Chemical Society, Washington, DC;   (f) Nishimura, S. (2001) *Handbook of Heterogeneous Catalytic Hydrogenation for Organic Synthesis*, Wiley-Interscience, New York.
3. Larock, R.C. (1999) *Comprehensive Organic Transformations*, 2nd edn, Wiley VCH Verlag GmbH, New York.
4. Dovell, S.F. and Greenfield, H. (1965) *J. Am. Chem. Soc.*, **87**, 2767.
5. Wu, G., Huang, M., Richards, M., Poirier, M., Wen, X., and Draper, R.W. (2003) *Synthesis*, 1657.
6. Miyazaki, Y., Hagio, H., and Kobayashi, S. (2006) *Org. Biomol. Chem.*, **4**, 2529.
7. Studer, M., Neto, S., and Blaser, H.-U. (2000) *Top. Catal.*, **13**, 205.
8. Hoogenraad, M., van der Linden, J.B., and Smith, A.A. (2004) *Org. Process. Res. Dev.*, **8**, 469.
9. Benzyl ether could survive in the semihydrogenation conditions using nickel nanoparticles, see Alonso, F., Osante, I., and Yus, M. (2007) *Tetrahedron*, **63**, 93.
10. Segura, Y., López, N., and Pérez-Ramírez, J. (2007) *J. Catal.*, **247**, 383.
11. (a) Lindlar, H. (1952) *Helv. Chim. Acta*, **35**, 446;   (b) Lindlar, H. and Dubuis, R. (1973) *Org. Synth. Coll.*, **5**, 880;   (c) Rosenmund, K.W. (1918) *Ber. Dtsch. Chem. Ges.*, **51**, 585;   (d) Rosenmund, K. and Zetzsche, F. (1921) *Ber. Dtsch. Chim. Ges.*, **54B**, 425.
12. Wuts, P.G.M. and Greene, T.W. (2006) *Greene's Protective Groups in Organic Synthesis*, 4th edn, John Wiley-Intersciences, New York.
13. Several and pioneering chemoselective hydrogenation methods of alkenes without hydrogenolysis of the aliphatic O-benzyl groups were previously reported, however, such methods are unsatisfactory for general applicability, see: (a) Bindra, J.S. and Grodski, A. (1978) *J. Org. Chem.*, **16**, 3240;   (b) Czech, B.P. and Bartsh, R.A. (1984) *J. Org. Chem.*, **49**, 4076;   (c) Ghosh, A.K. and Krishnan, K. (1998) *Tetrahedron Lett.*, **39**, 947;   (d) Misiti, D., Zappia, G., and Monache, G.D. (1999) *Synthesis*, 873.
14. (a) Sajiki, H. (1995) *Tetrahedron Lett.*, **36**, 3465;   (b) Sajiki, H., Kuno, H., and Hirota, K. (1998) *Tetrahedron Lett.*, **39**, 7127;   (c) Sajiki, H. and Hirota, K. (1998) *Tetrahedron*, **54**, 13981.
15. The problem could also be solved by employing an electron-rich 4-methoxybenzyl (MPM) group instead of the benzyl group as a protective group and using a Pd/C and pyridine combination as a catalyst, see: Sajiki, H.,

Kuno, H., and Hirota, K. (1997) *Tetrahedron Lett.*, **38**, 399.
16. (a) McPhee, W.D. and Erickson, E.S. (1946) *J. Am. Chem. Soc.*, **68**, 624; (b) A similar over-reduction of aromatic substrates has been documented for the Rosenmund's reduction using a Pd catalyst with suitable catalyst poisons (modifiers), see: Maier, W.F., Chettle, S.J., Rai, R.S., and Thomas, G. (1986) *J. Am. Chem. Soc.*, **108**, 2608.
17. Gaund, M.J., Yu, J., and Spencer, J.B. (1998) *J. Org. Chem.*, **63**, 4172.
18. The zerovalent Pd complexes that consist of an electron-rich d^{10} metal center with an electron-donating nitrogen ligand should (A) be stabilized by an electron-accepting species or (B) be supported by a polymer-backbone. (A) For examples: (a) Cavell, K.J., Stufkens, D.J., and Vrieze, K. (1980) *Inorg. Chim. Acta*, **47**, 67; (b) Sustmann, R., Lau, J., and Zipp, M. (1986) *Tetrahedron Lett.*, **105**, 5207; (c) Sustmann, R., Lau, J., and Zipp, M. (1986) *Recl. Trav. Chim. Pays– Bas*, **105**, 356; (d) van Asselt, R., Elsevier, C.J., Smeets, W.J.J., and Spek, A.L. (1994) *Inorg. Chem.*, **33**, 1521; (e) Klein, R.A., Witte, P., van Belzen, R., Fraanje, J., Goubitz, K., Numan, M., Schenk, H., Ernsting, J.M., and Elsevier, C.J. (1998) *Eur. J. Inorg. Chem.*, **1998**, 319; and references therein; (B) For examples: (f) Card, R.J. and Neckers, D.C. (1977) *J. Am. Chem. Soc.*, **99**, 7733 (g) Card, R.J. and Neckers, D.C. (1978) *Inorg. Chem.*, **17**, 2345; (h) Card, R.J., Liesner, C.E., and Neckers, D.C. (1979) *J. Org. Chem.*, **44**, 1095.
19. Sajiki, H., Hattori, K., and Hirota, K. (1998) *J. Org. Chem.*, **63**, 7990.
20. Hattori, K., Sajiki, H., and Hirota, K. (2000) *Tetrahedron*, **56**, 8433.
21. Maegawa, T., Fujita, Y., Sakurai, A., Sato, M., Oono, K., and Sajiki, H. (2007) *Chem. Pharm. Bull.*, **55**, 837.
22. (a) Sajiki, H., Hattori, K., and Hirota, K. (1998) *J. Chem. Soc., Perkin Trans. 1*, 4043; (b) Hattori, K., Sajiki, H., and Hirota, K. (2001) *Tetrahedron*, **57**, 4817.
23. Sajiki, H., Hattori, K., and Hirota, K. (2000) *Chem. Eur. J.*, **6**, 2200.
24. Corey, E.J. and Venkateswarlu, A. (1972) *J. Am. Chem. Soc.*, **94**, 6190.
25. (a) Hattori, K., Sajiki, H., and Hirota, K. (2000) *Tetrahedron Lett.*, **41**, 5711; (b) Hattori, K., Sajiki, H., and Hirota, K. (2001) *Tetrahedron*, **57**, 2109.
26. Ikawa, T., Sajiki, H., and Hirota, K. (2004) *Tetrahedron*, **60**, 6189.
27. See: Ikawa, T., Hattori, K., Sajiki, H., and Hirota, K. (2004) *Tetrahedron*, **60**, 6901.
28. (a) Rosenmund, K.W. (1918) *Ber. Dtsch. Chem. Ges.*, **51**, 585; (b) Rosenmund, K.W. and Zetzsche, F. (1921) *Ber. Dtsch. Chem. Ges.*, **54**, 425; (c) Mosettig, E. and Mozingo, R. (1948) *Org. React.*, **4**, 362.
29. Oelschläger, H. (1960) *Arch. Pharm.*, **293**, 442.
30. Dovell, F.S. and Greenfield, H. (1965) *J. Am. Chem. Soc.*, **87**, 2767.
31. (a) Mori, A., Miyakawa, Y., Ohashi, E., Haga, T., Maegawa, T., and Sajiki, H. (2006) *Org. Lett.*, **8**, 3279; (b) Mori, A., Mizusaki, T., Miyakawa, Y., Ohashi, E., Haga, T., Maegawa, T., Monguchi, Y., and Sajiki, H. (2006) *Tetrahedron*, **62**, 11925.
32. Mori, A., Mizusaki, T., Kawase, M., Haga, T., Maegawa, T., Monguchi, Y., Takao, S., Takagi, Y., and Sajiki, H. (2008) *Adv. Synth. Catal.*, **350**, 406.
33. Fagherazzi, G., Canton, P., Riello, P., Pinna, F., and Pernicone, N. (2000) *Catal. Lett.*, **64**, 119.
34. (a) Warren, B.E. (1941) *Phys. Rev.*, **59**, 693; (b) Suzuki, T. and Kaneko, K. (1988) *Carbon*, **26**, 743; (c) Rodriguez-Reinoso, F., Martin-Martinez, J.M., Prado-Burguete, C., and McEnaney, B. (1987) *J. Phys. Chem.*, **91**, 515.
35. Li, X.Z., Deng, Y., Yan, Y.J., Huang, H.F., Yu, W.Z., and Li, C.X. (2001) *J. New Energy*, **6**, 80.
36. Maier, W.F., Chettle, S.J., Rai, R.S. and Thomas, G. (1986) *J. Am. Chem. Soc.*, **108**, 2608.
37. (a) Lucas, F. (1966) *Nature*, **210**, 952; (b) Mita, K., Ichimura, S., and James, T.C. (1994) *J. Mol. Evol.*, **38**, 583.
38. (a) Akabori, S., Sakurai, S., Izumi, Y., and Fujii, Y. (1956) *Nature*, **178**, 323; (b) Izumi, Y. (1959) *Bull. Chem. Soc., Jpn.*, **32**, 932; (c) Izumi, Y. (1959) *Bull. Chem.*

Soc., Jpn., **32**, 936; (d) Izumi, Y. (1959) *Bull. Chem. Soc., Jpn.*, **32**, 942; (e) Akamatsu, A., Izumi, Y., and Akabori, S. (1961) *Bull. Chem. Soc., Jpn.*, **34**, 1067; (f) Akamatsu, A., Izumi, Y., and Akabori, S. (1961) *Bull. Chem. Soc. Jpn.*, **35**, 1706.

39. (a) Sajiki, H., Ikawa, T., and Hirota, K. (2003) *Tetrahedron Lett.*, **44**, 171; (b) Sajiki, H., Ikawa, T., and Hirota, K. (2003) *Tetrahedron Lett.*, **44**, 4837; (c) Ikawa, T., Sajiki, H., and Hirota, K. (2005) *Tetrahedron*, **61**, 2217; (d) Kitamura, Y., Tanaka, A., Sato, M., Oono, K., Ikawa, T., Maegawa, T., Monguchi, Y., and Sajiki, H. (2007) *Synth. Commun.*, **37**, 4381.

40. (a) Campbell, K.N. and Campbell, B.K. (1942) *Chem. Rev.*, **30**, 145; (b) Crombie, L. (1952) *Quart. Rev. (Lond.)*, **6**, 101; (c) Burwell, R.L. Jr. (1957) *Chem. Rev.*, **57**, 895; (d) Wells, P.B. (1964) *Chem. Ind. (Lond.)*, 1742; (e) Bond, G.C. and Wells, P.B. (1964) *Adv. Catal.*, **15**, 205; (f) Marvell, E.N. and Li, T. (1973) *Synthesis*, 457.

41. Lindlar, H. (1952) *Helv. Chim. Acta*, **35**, 446.

42. Brunet, J.-J. and Caubere, P. (1984) *J. Org. Chem.*, **49**, 4058.

43. (a) Campbell, K.N. and Eby, L.T. (1941) *J. Am. Chem. Soc.*, **63**, 216; (b) Ahmad, K. and Strong, F.M. (1948) *J. Am. Chem. Soc.*, **70**, 1699; (c) Max, R.A. and Deatherage, F.E. (1951) *J. Am. Oil Chem. Soc.*, **28**, 110; (d) Fusari, S.A., Greenlee, K.W., and Brown, J.B. (1951) *J. Am. Oil Chem. Soc.*, **28**, 416; (e) Howton, D.R. and Davis, R.H. (1951) *J. Org. Chem.*, **16**, 1405; (f) Huber, W.F. (1951) *J. Am. Chem. Soc.*, **73**, 2730; (g) Oroshnik, W., Karmas, G., and Mebane, A.D. (1952) *J. Am. Chem. Soc.*, **74**, 295; (h) Khan, N.A. (1952) *J. Am. Chem. Soc.*, **74**, 3018; (i) Oroshnik, W., Karmas, G., and Mebane, A.D. (1952) *J. Am. Chem. Soc.*, **74**, 3807; (j) Elsner, B.B. and Paul, P.F.M. (1953) *J. Chem. Soc.*, 3156; (k) Oroshnik, W. and Mebane, A.D. (1954) *J. Am. Chem. Soc.*, **76**, 5719; (l) Knight, J.A. and Diamond, J.H. (1959) *J. Org. Chem.*, **24**, 400; (m) Brown, C.A. and Ahuja, V.K. (1973) *J. Chem. Soc., Chem. Commun.*, 553.

44. Nitta, Y., Imanaka, T., and Teranishi, S. (1981) *Bull. Chem. Soc. Jpn.*, **54**, 3579.

45. (a) Oroshnik, W., Karmas, G., and Mebane, A.D. (1952) *J. Am. Chem. Soc.*, **74**, 3807; (b) Oroshnik, W. and Mebane, A.D. (1954) *J. Am. Chem. Soc.*, **76**, 5719.

46. (a) Brunet, J.-J., Gallois, P., and Caubere, P. (1980) *J. Org. Chem.*, **45**, 1937; (b) Gallois, P., Brunet, J.-J., and Caubere, P. (1980) *J. Org. Chem.*, **45**, 1946.

47. Segura, Y., López, N., and Pérez-Ramíez, J. (2007) *J. Catal.*, **247**, 383.

48. (a) Schrock, R.R. and Osborn, J.A. (1985) *J. Am. Chem. Soc.*, **98**, 2143; (b) Sodeoka, M. and Shibasaki, M. (1985) *J. Org. Chem.*, **50**, 1147.

49. Lindlar, H. and Dubuis, R. (1973) *Org. Synth., Coll.*, **5**, 880.

50. Alons, F., Osante, I., and Yus, M. (2006) *Adv. Synth. Catal.*, **348**, 305.

51. (a) Sajiki, H., Mori, S., Ohkubo, T., Ikawa, T., Kume, A., Maegawa, T., and Monguchi, Y. (2008) *Chem. Eur. J.*, **14**, 5109; (b) Mori, S., Ohkubo, T., Ikawa, T., Kume, A., Maegawa, T., Monguchi, Y., and Sajiki, H. (2009) *J. Mol. Catal. A: Chem.*, **307**, 77.

52. The smooth cleavage of an *N*-Cbz protective group during the hydrogenation of an alkene function using the Lindlar catalyst without quinoline conditions was reported, see: Ghosh, A.K. and Krishnan, K. (1998) *Tetrahedron Lett.*, **39**, 947.

5
Silicon-Based Carbon–Carbon Bond Formation by Transition Metal Catalysis
Yoshiaki Nakao and Tamejiro Hiyama

5.1
Introduction

Carbon–carbon bond forming reactions constitute the most important class of transformations in organic synthesis for constructing the frameworks of organic molecules. Organometallic reagents play a key role in this process, because carbon–metal bonds are generally highly polarized and react with carbon electrophiles to make new C–C bonds. In modern organic synthesis, target molecules have become very diverse and complex, with a variety of functional groups and stereocenters. Accordingly, there is a growing demand for organometallic reagents that exhibit favorable chemoselectivity and stereoselectivity. The stability of the organometallic reagents is also an important issue for maintaining reagent purity. Further, ready availability of reagents with the correct stoichiometry is essential for reproducibility and reliability. The use of organosilicon compounds has received much attention in organic synthesis because with respect to these criteria, they offer many advantages over other groups of compounds such as organometallics and organometalloid reagents.

Many types of organosilicon reagents are readily available. A wide range of organosilicon compounds are commercially available, and these compounds are generally inexpensive because of the natural abundance of silicon. Organosilicon reagents are, in general, highly stable owing to the presence of less polarized C–Si bonds. On the other hand, this makes these reagents less reactive toward electrophiles than the other main group, metallic nucleophiles. One can expect high chemoselectivity associated with C–C bond forming reactions using silane-based reagents because of the inherent nature of the organosilicon compounds. However, the trade-off has made these reagents less familiar to the synthetic community. The exceptions are allylsilanes and silyl enol ethers, both of which are sufficiently reactive because of their highly electron-rich double bond.

This chapter focuses on ways to use transition metal catalysis to make organosilicon reagents highly useful for C–C bond forming reactions. Transmetalation is an elemental step that delivers a reacting organic group from silicon to a transition metal center. It allows organosilicon reagents to participate in a variety of C–C

Pharmaceutical Process Chemistry. Edited by Takayuki Shioiri, Kunisuke Izawa, and Toshiro Konoike
Copyright © 2011 WILEY-VCH Verlag GmbH & Co. KGaA, Weinheim
ISBN: 978-3-527-32650-1

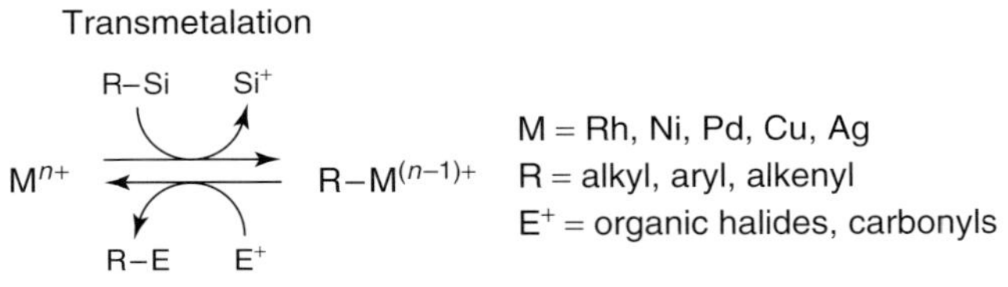

Transmetalation

$$M^{n+} + R\text{-Si} \rightleftarrows R\text{-M}^{(n-1)+} + \text{Si}^+$$
$$R\text{-M}^{(n-1)+} + E^+ \rightarrow R\text{-E}$$

M = Rh, Ni, Pd, Cu, Ag
R = alkyl, aryl, alkenyl
E^+ = organic halides, carbonyls

Scheme 5.1 General scheme for transmetalation from silicon to transition metals.

bond forming reactions in the presence of a catalytic amount of transition metals. The resulting organometallic compounds perform various C–C bond forming events in a highly chemo- and stereoselective manner. Thus, a highly selective overall transformation can be envisioned if an efficient protocol is developed for the transmetalation step (Scheme 5.1). We review successful examples reported mainly in the last five years, particularly focusing on cross-coupling [1] and carbonyl addition reactions.

5.2
Cross-Coupling Reactions

5.2.1
Brief Assessment of Early Stage Protocols

In 1988, Hiyama and Hatanaka reported for the first time that organotrimethylsilanes undergo cross-coupling reactions with aryl and alkenyl halides in the presence of a palladium catalyst and tris(dimethylamino)sulfonium difluorotrimethylsilane (TASF) (Scheme 5.2) [2]. Later, they demonstrated that fluorine-substituted tetracoordinate alkenyl [3] and arylsilanes [4] showed a reactivity much higher than that of the trimethylsilyl reagents and could significantly expand the scope of organic groups transferable by the silicon-based cross-coupling reaction (Scheme 5.3). Pentacoordinate silicates generated *in situ* by the nucleophilic attack of a fluoride ion on silicon are supposed to be sufficiently reactive to participate in transmetalation with palladium(II) species. The fluorine substituent apparently enhances the Lewis acidity of the silicon center and so promotes the formation of pentacoordinate silicates. A similar effect of electron-withdrawing heteroatom substituents

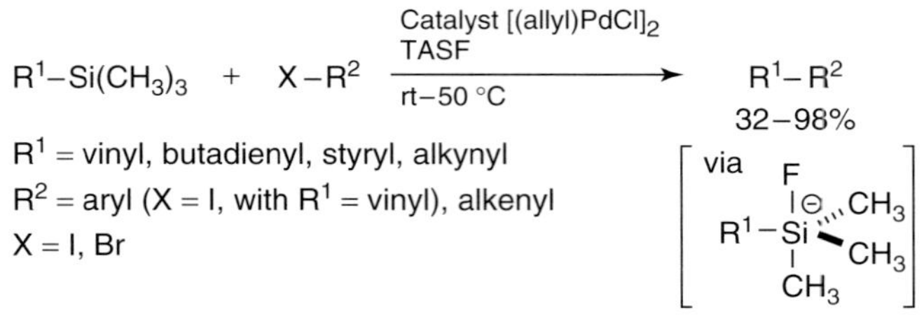

$$R^1\text{-Si(CH}_3)_3 + X\text{-}R^2 \xrightarrow[\text{rt–50 °C}]{\text{Catalyst [(allyl)PdCl]}_2, \text{TASF}} R^1\text{-}R^2 \quad 32\text{–}98\%$$

R^1 = vinyl, butadienyl, styryl, alkynyl
R^2 = aryl (X = I, with R^1 = vinyl), alkenyl
X = I, Br

via pentacoordinate fluorosilicate $[R^1\text{-Si}(CH_3)_2(CH_3)F]^-$

Scheme 5.2 Cross-coupling reactions using organotrimethylsilanes.

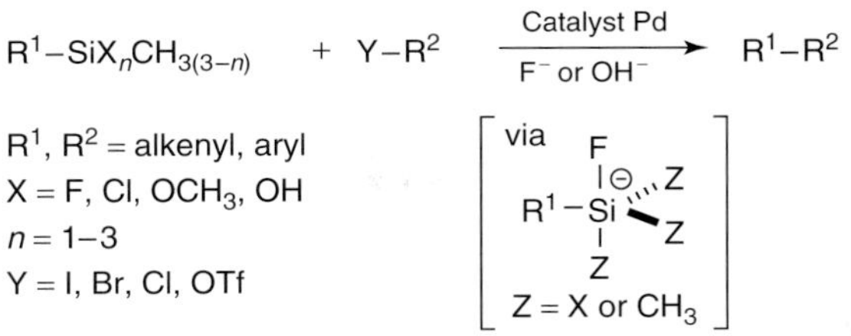

$R^1-SiX_nCH_{3(3-n)}$ + Y–R² $\xrightarrow[\text{F}^-\text{ or OH}^-]{\text{Catalyst Pd}}$ R¹–R²

R¹, R² = alkenyl, aryl
X = F, Cl, OCH₃, OH
n = 1–3
Y = I, Br, Cl, OTf

via

$$R^1-\underset{\underset{Z}{|}}{\overset{\overset{F}{|\ominus}}{Si}}\overset{Z}{\underset{Z}{\diagdown}}$$

Z = X or CH₃

Scheme 5.3 Cross-coupling reactions using halogenated organosilanes.

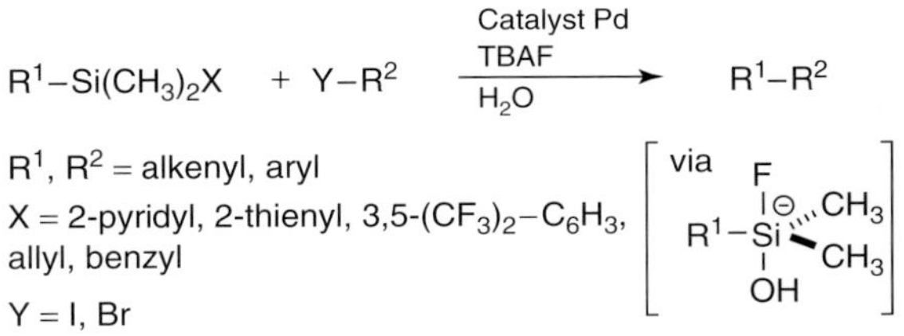

$R^1-Si(CH_3)_2X$ + Y–R² $\xrightarrow[\text{H}_2\text{O}]{\substack{\text{Catalyst Pd}\\ \text{TBAF}}}$ R¹–R²

R¹, R² = alkenyl, aryl
X = 2-pyridyl, 2-thienyl, 3,5-(CF₃)₂–C₆H₃, allyl, benzyl
Y = I, Br

via

$$R^1-\underset{\underset{OH}{|}}{\overset{\overset{F}{|\ominus}}{Si}}\overset{CH_3}{\underset{CH_3}{\diagdown}}$$

Scheme 5.4 Cross-coupling reactions using tetraorganosilanes as masked silanols.

has subsequently been revealed for the chloro [5], alkoxy [6], and hydroxy [7] groups. Fluoride-free conditions using NaOH [8] or KOSi(CH₃)₃ [9] have also been found effective in forming pentacoordinate silicates with heteroatom-substituted organosilicon reagents. These original protocols, however, suffer from the sensitivity of the silicon reagents toward acids, bases, and even moisture in some cases. The use of highly nucleophilic fluoride or hydroxide as an activator makes the transformation less chemoselective, especially for substrates with a base-sensitive functionality and for common silyl ethers that are invariably used in the synthesis of complex natural products. Whereas the use of tetraorganosilicon reagents as "masked silanols" [10] has solved the stability issue, the use of fluoride activators is still essential for the *in situ* generation of the silanols that are considered to be the reactive species (Scheme 5.4).

5.2.2
Cross-Coupling Reactions Using Tetraorganosilanes through Intramolecular Activation

The use of stable organosilicon compounds, especially tetraorganosilanes under mild activation, is highly desirable in cross-coupling chemistry. However, tetraorganosilanes are generally reluctant to undergo transmetalation, because the less electrophilic silicon center is not well motivated to promote the formation of reactive pentacoordinate silicates even with fluoride activation. However, this is not the case when a metal alkoxide is located at an appropriate intramolecular position, as is often seen in the Brook rearrangement [11]. Indeed, efficient transmetalation of alkenyl(trimethyl)silanes with copper(I) has been demonstrated: a palladium-catalyzed cross-coupling reaction occurs with the aid of a proximal

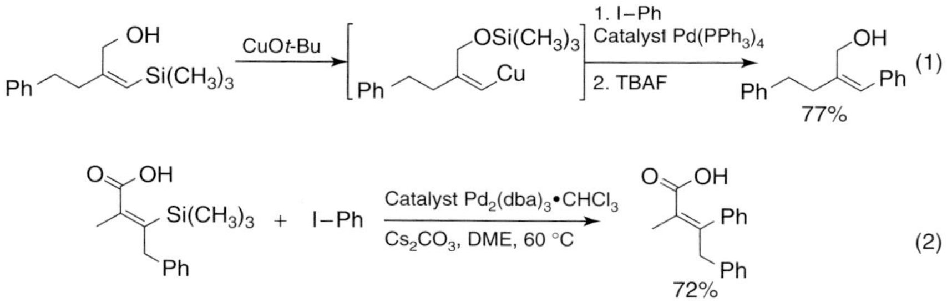

Scheme 5.5 Seminal discoveries of transmetalation assisted by intramolecular activation.

hydroxy group (Scheme 5.5, Eq. (1)) [12]. A similar effect using a carboxylate group has also been reported (Scheme 5.5, Eq. (2)) [13]. These two examples clearly demonstrate that intramolecular attack of a nucleophilic oxygen is highly effective for transmetalation from silicon to a transition metal. This effect is presumably due to the high oxophilicity of silicon and it eliminates the need for strong nucleophiles such as fluoride for silicate formation.

These observations as well as the related reagent design for carbonyl allylation and benzylation [14] have led to the use of alkenyl- and aryl[2-(hydroxymethyl)phenyl]dimethylsilanes as novel silicon-based cross-coupling reagents [15]. These silicon reagents have many advantages over conventional ones: excellent chemostability based on the tetraorganosilicon structure, the hydroxy-containing activating group being free from a transferring organic group, and recoverable and reusable cyclic silyl ether as a silyl residue. The alkenylsilanes undergo palladium-catalyzed cross-coupling in the presence of K_2CO_3, a mild and inexpensive base. An excellent chemoselectivity that is tolerant toward a range of functional groups including common silyl protection is observed (Scheme 5.6) [16].

The coupling reactions of the arylsilanes also proceed smoothly with the addition of a copper(I) cocatalyst, suggesting that sequential transmetalation from silicon to copper and then to palladium may be possible (Scheme 5.7) [17].

The high reactivity of the intramolecularly activated silicon reagents is particularly exemplified by reactions with stannyl- or pinacolatoboryl-substituted

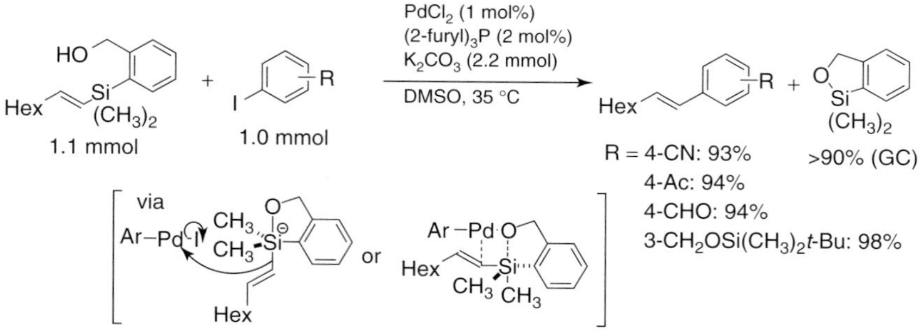

Scheme 5.6 Cross-coupling reactions using alkenyl[2-(hydroxymethyl)phenyl]dimethylsilanes.

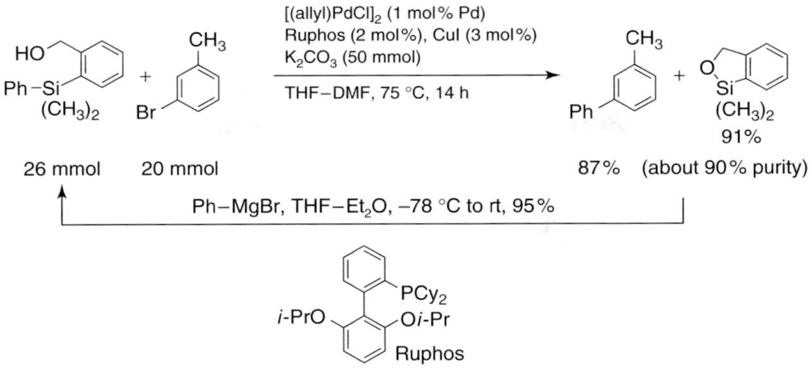

Scheme 5.7 Cross-coupling reactions using aryl[2-(hydroxymethyl)phenyl]dimethylsilanes.

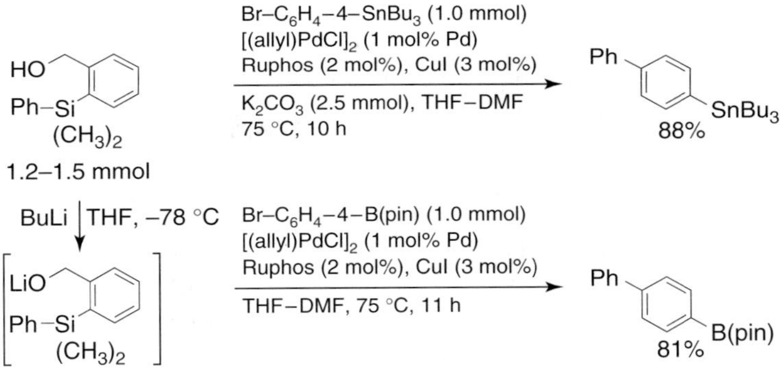

Scheme 5.8 Cross-coupling reactions of phenyl[2-(hydroxymethyl)phenyl]dimethylsilanes with metalated aryl bromides.

bromobenzenes. Here, under appropriate reaction conditions, the C–Si bond is activated preferentially over other C–metal bonds believed to be much more reactive to yield metalated biaryls (Scheme 5.8) [18].

The cross-coupling reactions with allylic and benzylic carbonates proceed in the absence of an external base, because the oxidative addition of the electrophiles gives palladium alkoxides [19] that probably act as bases to promote transmetalation. The protocol allows a silicon-based approach to diarylmethanes which are found in many biologically active substances (Scheme 5.9).

Even the highly strong C(sp^3)–Si bonds of tetraorganosilicon compounds can be activated by intramolecular activation to undergo fluoride-free alkyl cross-coupling reactions [20] with alkyltriorganosilicon reagents (Scheme 5.10) [21]. In this case, *gem*-dimethyls at the benzylic position are essential for a smooth reaction. Otherwise, oxidation occurs at the benzylic carbon, yielding a benzaldehyde derivative. Isopropyl groups on silicon allow the discrimination of primary and secondary alkyls.

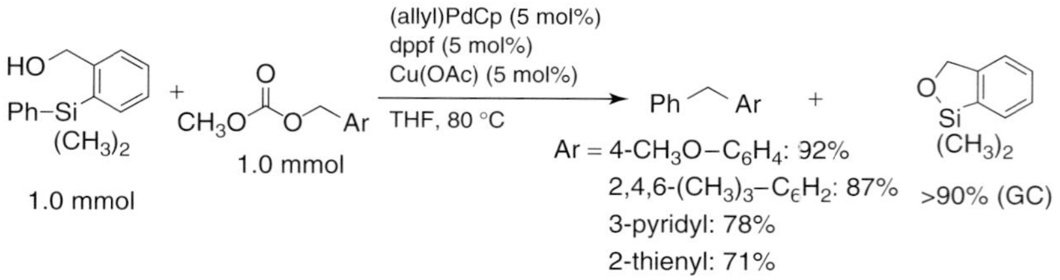

Scheme 5.9 Cross-coupling reactions of aryl[2-(hydroxymethyl)phenyl]dimethylsilanes with arylmethyl carbonates.

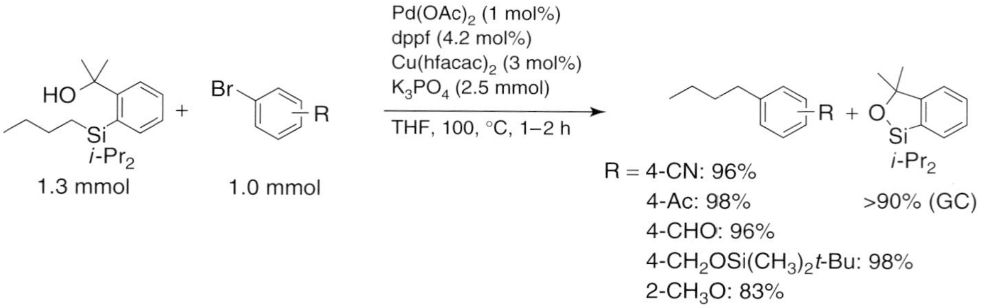

Ar = 4-CH$_3$O-C$_6$H$_4$: 92%
2,4,6-(CH$_3$)$_3$-C$_6$H$_2$: 87%
3-pyridyl: 78%
2-thienyl: 71%

>90% (GC)

R = 4-CN: 96%
4-Ac: 98%
4-CHO: 96%
4-CH$_2$OSi(CH$_3$)$_2$t-Bu: 98%
2-CH$_3$O: 83%

>90% (GC)

Scheme 5.10 Cross-coupling reactions using alkyl[2-(hydroxyprop-2-yl)phenyl]diisopropylsilanes.

The free benzylic hydroxy group is essential for activating the C–Si bonds of the silicon reagents; with O-protected reagents, no cross-coupling reaction takes place. Taking advantage of this simple switch in reagent reactivity, electrophiles having the silyl group with O-protection undergo a cross-coupling reaction, with the protected silyl group remaining totally intact. The resulting products can participate in the subsequent cross-coupling event upon O-deprotection. By simply repeating these operations using a variety of halogenated arylsilanes, oligoarenes such as oligothiophenes can be assembled in a highly efficient manner (Scheme 5.11) [22].

A similar strategy employing o-hydroxybenzyl-substituted silicon reagents generated *in situ* has also been demonstrated (Scheme 5.12) [23]. An amide carbonyl can also serve as an intramolecular activator to facilitate the activation of CH$_3$–Si bonds, allowing palladium-catalyzed oxidative methylation of electron-deficient olefins (Scheme 5.13) [24].

5.2.3
Cross-Coupling Reactions Using Organosilanolates

Organosilanols have received increasing attention in the last decade as alternatives to the highly reactive and relatively stable organosilicon reagents for cross-coupling chemistry [25]. The silicon reagents participate in the cross-coupling reaction using KOSi(CH$_3$)$_3$ as a base in the absence of a fluoride activator [9].

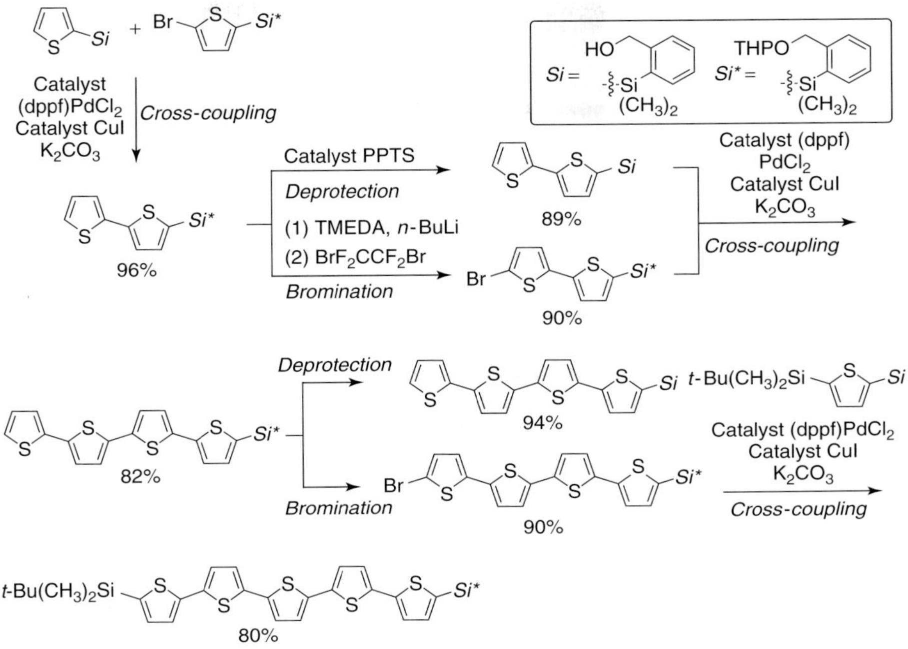

Scheme 5.11 Oligothiophene synthesis through iterative cross-coupling using protected and unprotected aryl[2-(hydroxymethyl)phenyl]dimethylsilanes.

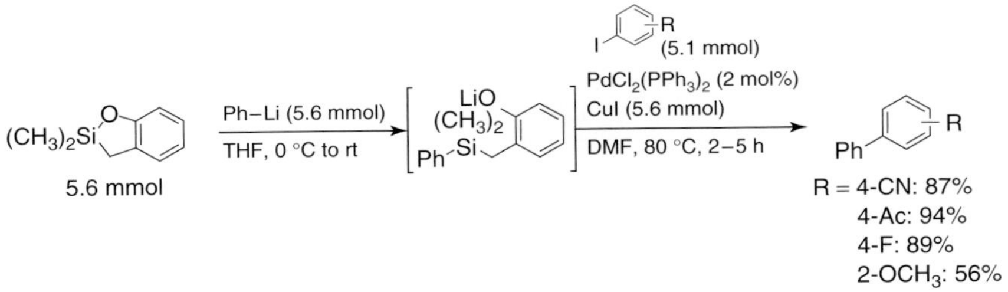

Scheme 5.12 Cross-coupling reactions using aryl(2-hydroxybenzyl)dimethylsilanes.

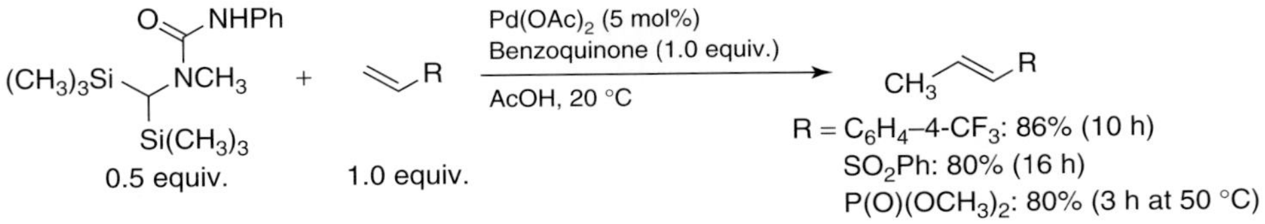

Scheme 5.13 Oxidative methylation of olefins using N-bis(trimethylsilyl)methyl-substituted ureas.

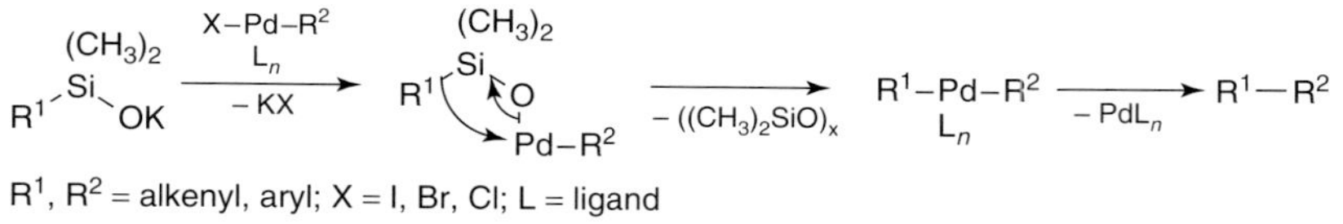

R^1, R^2 = alkenyl, aryl; X = I, Br, Cl; L = ligand

Scheme 5.14 Suggested Mechanism for transmetalation of organosilanolates through palladium silanolates.

Denmark and Sweis extensively studied the mechanism of transmetalation in fluoride-free conditions and proposed palladium silanolate as a key intermediate, which may undergo intramolecular transmetalation in a manner different from the pentacoordinate silicates (Scheme 5.14) [26].

This novel paradigm for transmetalation has led the research group to develop a protocol using preformed organosilanolate as a coupling partner by irreversible deprotonation of the corresponding organosilanol. Most silanolate salts are in a bench-stable powdery form that can be isolated, stored, and handled easily. The use of silanolate salts eliminates the need for an external base, thus suppressing both the undesirable dehydration of silanols to form siloxanes and the protodesilylation possibly caused by conjugate acid derived from the base. Various alkenyl- [27] and arylsilanolates [28] including heteroarylsilanolates [29] undergo palladium-catalyzed cross-coupling reactions under mild conditions (Scheme 5.15, Eqs. (1–5)). Simple vinyldimethylsilanolate can be generated *in situ* from commercially available tetramethyldivinyldisiloxane by a reaction with $KOSi(CH_3)_3$. This process is reversible; the equilibrium in a polar DMF solvent favors vinyldimethylsilanolate, which participates in the vinylation of aryl halides in the presence of a palladium catalyst (Scheme 5.15, Eq. (3)) [30]. Crotylsilanolates also react with aryl halides with γ-selectivity [31] to provide a wide variety of 3-arylbut-1-enes (Scheme 5.15, Eq. (6)) [32].

The protocol using alkenylsilanolates generated *in situ* can be applied to the total synthesis of the polyene macrolide RK-397 (Scheme 5.16) [33]. A benzyldimethylsilyl group remains intact under the reaction conditions for the silanolate coupling. It subsequently undergoes a second coupling reaction with another alkenyl iodide by conventional fluoride activation to assemble the stereochemically well-defined tetraene structure required to access the target molecule. This example provides a good demonstration of the power of silicon-based cross-coupling technology.

5.2.4
Other Tetraorganosilicon Compounds for Cross-Coupling Chemistry

Because of the importance and ubiquitous nature of substituted 2-pyridyl groups in pharmaceuticals and materials, the introduction of cross-coupling reactions has gained much attention. Whereas 2-pyridylboronic acids are too unstable to be readily hydrolyzed [34], stable 2-triorgano(2-pyridyl)silanes have been shown to undergo a palladium-catalyzed cross-coupling reaction with aryl iodides in the presence of a stoichiometric amount of copper [35] or silver salts [36] (Scheme 5.17).

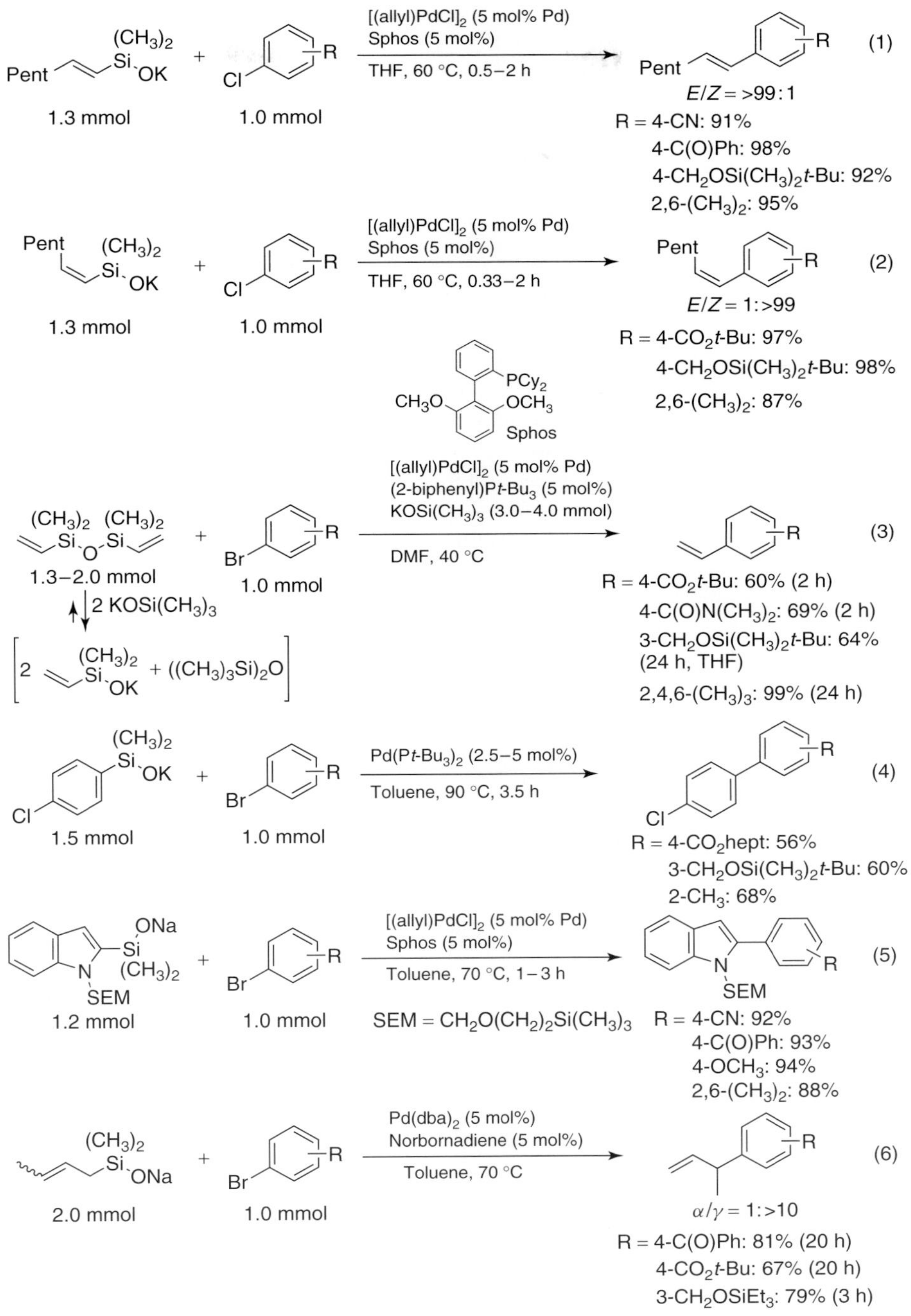

Scheme 5.15 Cross-coupling reactions using organosilanolates.

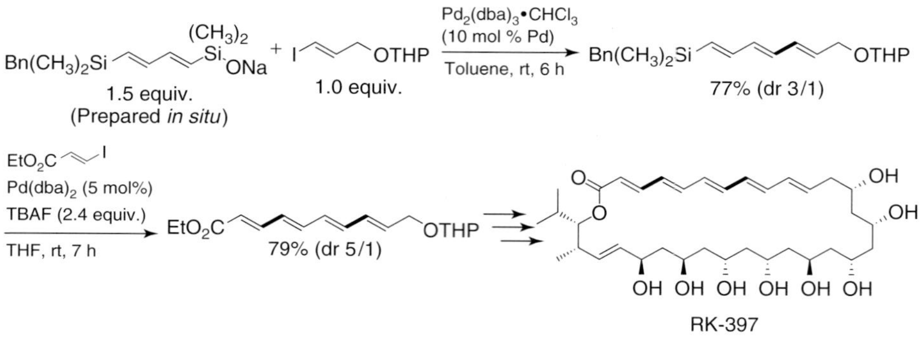

Scheme 5.16 Sequential silicon-based cross-coupling reactions in total synthesis of RK-397.

A palladium-catalyzed cross-coupling reaction of silyl enol ethers (Scheme 5.18) [37] and α-silylnitriles (Scheme 5.19) [38] with aryl halides is a useful protocol for accessing α-arylalkyl ketones and carboxylic acid derivatives.

Carbamoylsilanes undergo cross-coupling with a range of electrophiles in the presence or absence of palladium catalysts to yield the respective amides (Scheme 5.20) [39]. Simple reaction conditions without base activators may be useful for introducing a carbamoyl group.

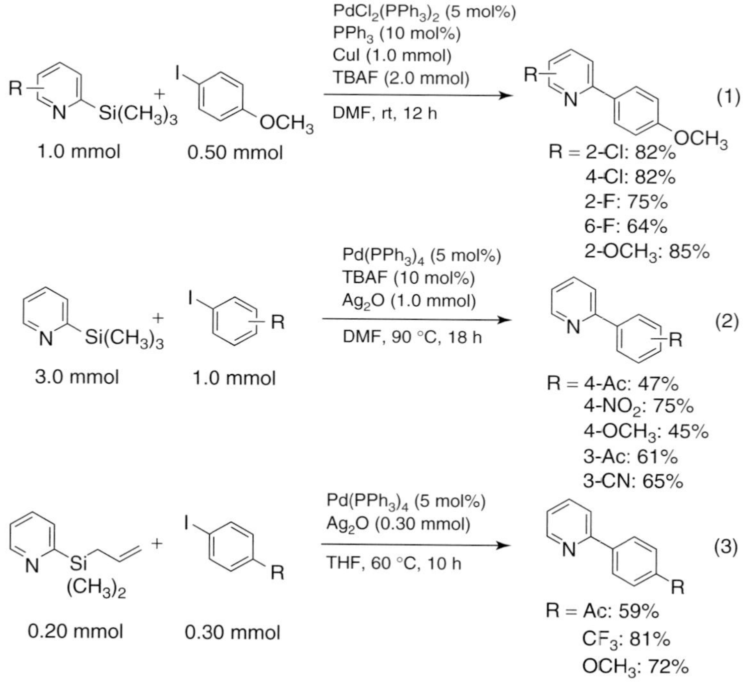

Scheme 5.17 Cross-coupling reactions using triorgano(2-pyridyl)silanes.

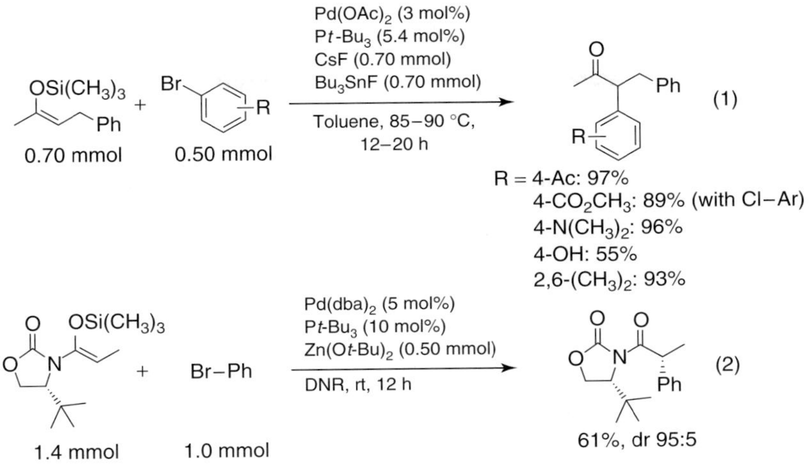

Scheme 5.18 Cross-coupling reactions using silyl enolates.

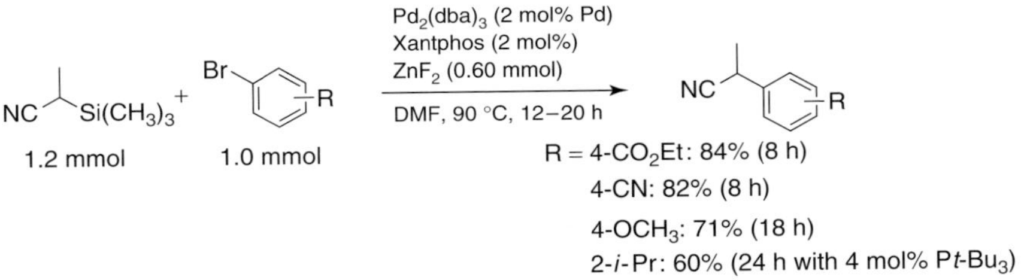

Scheme 5.19 Cross-coupling reactions using α-silylnitriles.

5.2.5
New Types of Electrophiles for Silicon-Based Cross-Coupling

Recently performed extensive studies of metal-catalyzed cross-coupling reactions have expanded the scope of the substrates significantly by a sophisticated design of the transition metal complexes with diverse ligands and detailed mechanistic studies. Some of the newly developed systems have also been applied to the silicon-based protocol. For example, aryl sulfonates that are inexpensive and readily available from the corresponding phenol derivatives serve as aryl electrophiles for a coupling reaction with aryl(trimethoxy)silanes (Scheme 5.21) [40]. The use of bulky and highly electron-donating phosphorus ligands further promotes the oxidative addition of the less-reactive C–O bond.

Successful cross-coupling reactions of aryl(trimethoxy)silanes [41] and aryl(trifluoro)silanes [42] with alkyl halides have also been achieved (Scheme 5.22). Nickel catalysts are particularly effective for secondary alkyl halides. It has been suggested that the mechanism governing the coupling reaction involves radical intermediates. The use of an optically active diamine ligand allows dynamic

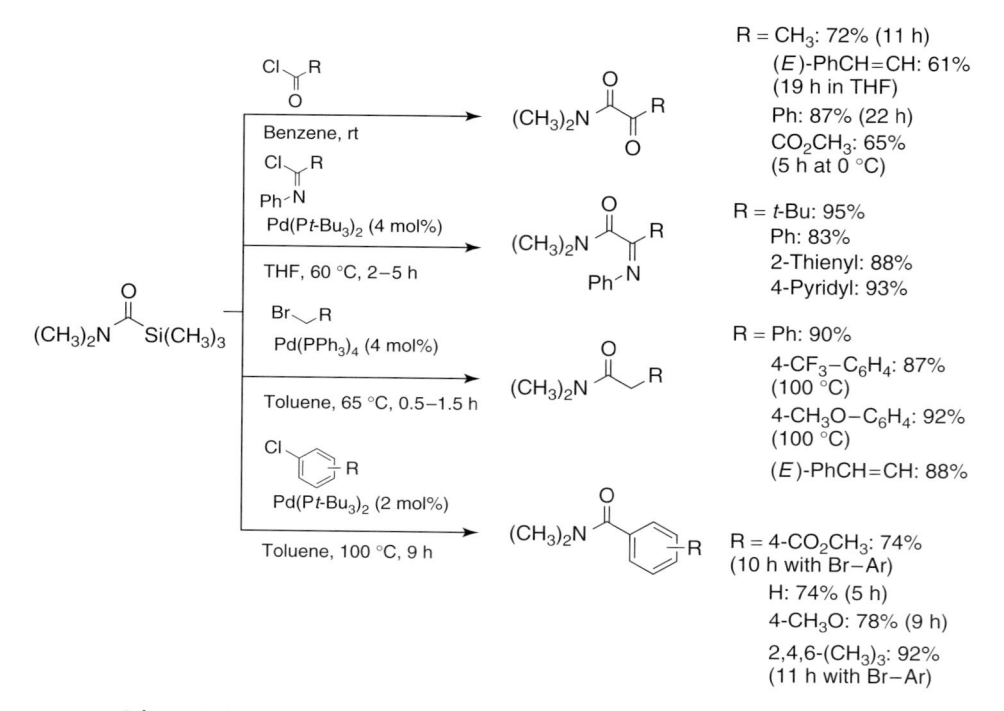

Scheme 5.20 Cross-coupling reactions using carbamoylsilanes.

Scheme 5.21 Silicon-based cross-coupling reactions with aryl sulfonates.

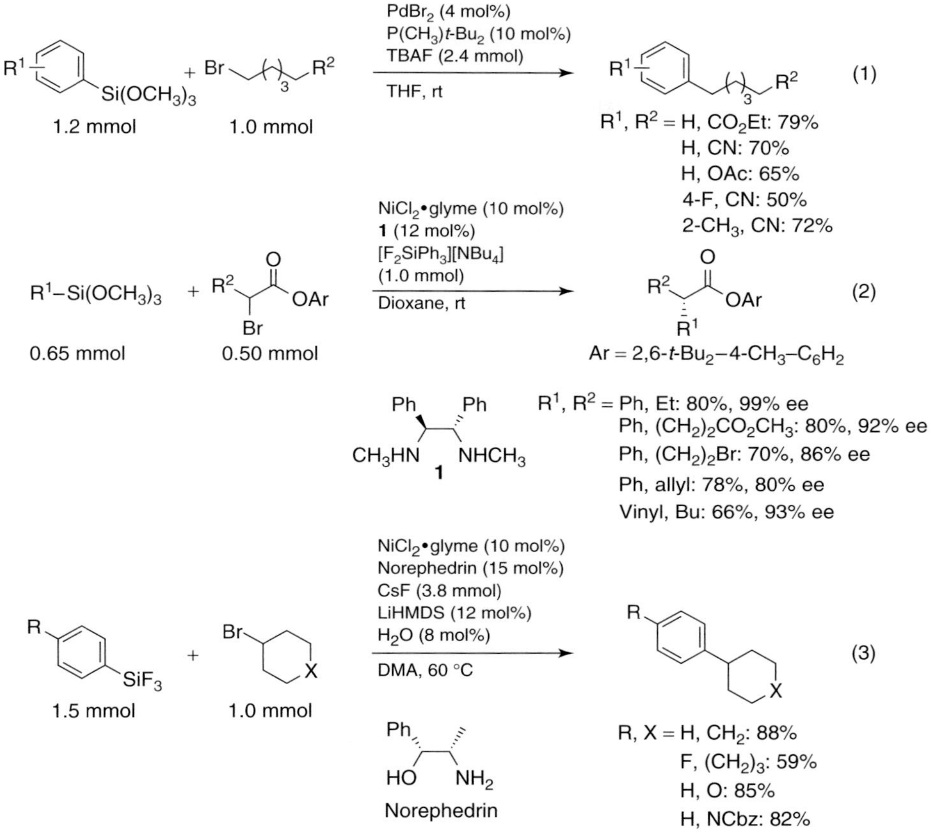

Scheme 5.22 Silicon-based cross-coupling reactions with alkyl halides.

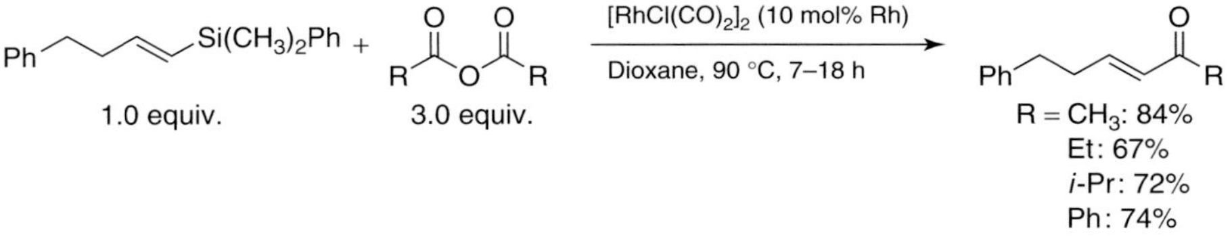

Scheme 5.23 Silicon-based cross-coupling reactions with acid anhydrides.

kinetic resolution of α-bromocarboxylates to yield α-arylcarboxylates with high enantioselectivity (Scheme 5.22, Eq. (2)) [41b].

Acid anhydrides serve as coupling partners for alkenyldimethylphenylsilanes in the presence of a rhodium catalyst to provide α,β-unsaturated ketones (Scheme 5.23) [43]. It is worth noting that the highly stable tetraorganosilicon reagents, which are readily available through alkyne hydrosilylation, undergo transmetalation without any activators.

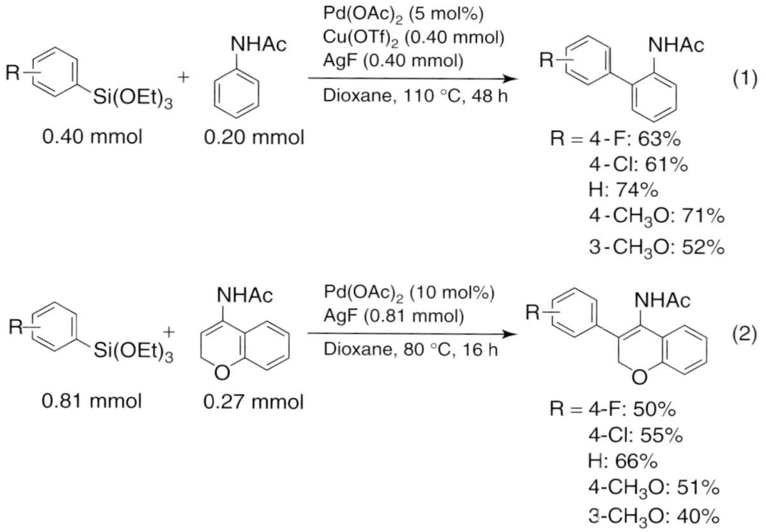

Scheme 5.24 Silicon-based C–H arylation reactions.

Transition metal–catalyzed activations of unreactive C–H bonds and their functionalizations have received increasing attention in the last decade [44]. Related transformations using organosilicon nucleophiles have appeared as a promising alternative to conventional cross-coupling reactions requiring prefunctionalized electrophilic coupling partners. For example, palladium-catalyzed C–H arylation of acetanilides [45] and enamides [46] proceeds with aryltrimethoxysilanes (Scheme 5.24). A plausible catalytic cycle comprises the following steps in sequence: electrophilic palladation at the ortho position of the electron-rich aromatic ring and double bond, transmetalation with the arylsilane reagents, reductive elimination, and finally oxidation of the palladium catalyst.

5.3
Carbonyl Addition Reaction

5.3.1
Rhodium-Catalyzed Reactions

In 1997, Miyaura reported that organoboronic acids undergo 1,4-addition reactions with the aid of a rhodium(I) catalyst [47]. This novel reaction found many applications in organic synthesis soon after enantioselective carbonyl addition reactions (including 1,2-addition to aldehydes and imines) were reported. This was because the corresponding enantio- and chemoselectivities were now sufficiently high for these reagents to be useful in organic synthesis [48]. The use of organosilicon reagents for rhodium-catalyzed carbonyl addition reactions has subsequently

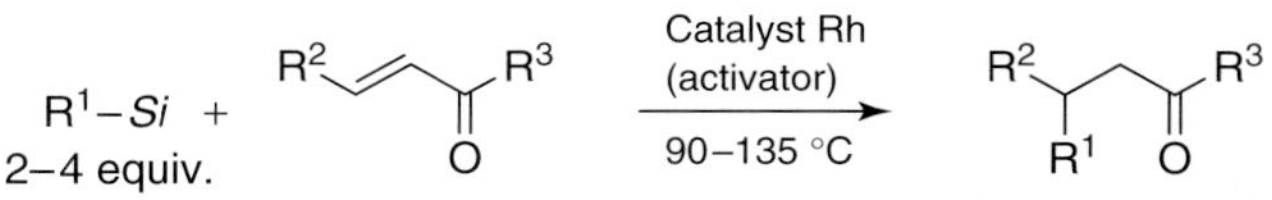

Scheme 5.25 Rhodium-catalyzed 1,4-addition reactions using organosilicon reagents.

R^1–Si + 2–4 equiv.

R^1–Si (activator):
Ar–Si(OEt)$_3$, alkenyl–Si(OEt)$_3$ (TBAF at 70 °C), Ar–SiEt(OH)$_2$ (1.0 equiv.),
(Ar–SiCH$_3$O)$_n$, Ar$_2$SiCl$_2$ (NaF)

been investigated. Initial studies focused on alkoxysilanes [49], halosilanes [50], silanols [51], and silicones [52]. Compared with the boron reagents, the silicon reagents required relatively harsh reaction conditions and nucleophilic activators. In some cases, this might have been true because the C–Si bonds of the silicon reagents were less reactive than those of the boron-based reagents (Scheme 5.25).

Alkenyl- and aryl[2-(hydroxymethyl)phenyl]dimethylsilanes have recently been found to undergo rhodium-catalyzed 1,4-addition reactions under mild conditions without any activators (Scheme 5.26, Eq. (1)) [53]. High enantioselectivities are achieved with Rh/chiral diene catalysts, enabling silicon-based access to the nitrogen-containing heterocycles (Scheme 5.26, Eqs. (2) and 3)). The silicon-based protocol is compatible with α-substituted vinyl nucleophiles such as propen-2-ylsilane, thereby facilitating the enantioselective synthesis of highly substituted allylsilanes through 1,4-addition reactions with β-silyl-substituted enones (Scheme 5.26, Eq. (4)) [54]. The corresponding transformations with α-substituted vinylboronic acids suffer from the inherent instability of the borane reagents [55].

The propen-2-ylsilane reagent also undergoes enantioselective 1,2-addition to imines in the presence of a similar Rh/chiral diene catalyst to give chiral allylic amines with a high percentage enantiomeric excess (Scheme 5.27) [56].

On the basis of the kinetic resolution observed with the addition of phenyl[2-(1-hydroxyethyl)phenyl]dimethylsilane to cyclohexen-2-one in the presence of the Rh/chiral diene catalyst, a mechanism involving transmetalation through equatorial R rather than axial R from a pentacoordinate silicate is suggested (Scheme 5.28).

5.3.2
Nickel-Catalyzed Reactions

Three-component coupling of aryl[2-(hydroxymethyl)phenyl]dimethylsilanes, 1,3-dienes, and carbonyl electrophiles takes place with a high diastereoselectivity in the presence of a nickel/carbene catalyst (Scheme 5.29) [57]. Aryl transfer from pentacoordinate silicates to a π-allylnickel species generated through the reaction of 1,3-dienes, carbonyl, and nickel(0) [58] is a plausible reaction mechanism.

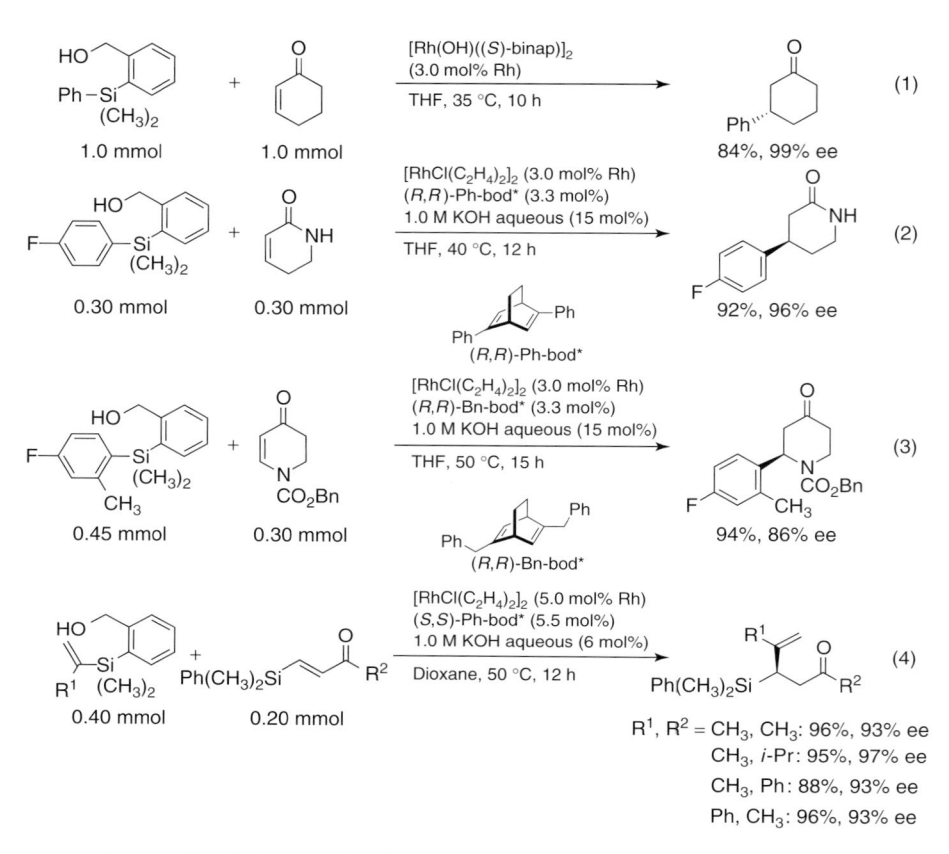

Scheme 5.26 Rhodium-catalyzed enantioselective 1,4-addition reactions using organo-[2-(hydroxymethyl)phenyl]dimethylsilanes.

Scheme 5.27 Rhodium-catalyzed enantioselective 1,2-addition reactions of alkenyl[2-(hydroxymethyl)phenyl]dimethylsilanes to imines.

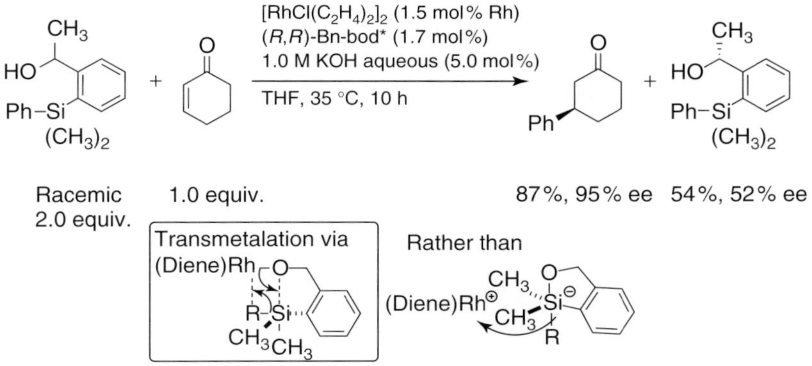

Scheme 5.28 Plausible mechanism for transmetalation of organo[2-(hydroxy-methyl)phenyl-]dimethylsilanes to rhodium(I) based on kinetic resolution of phenyl[2-(hydroxyethyl)phenyl-]dimethylsilane.

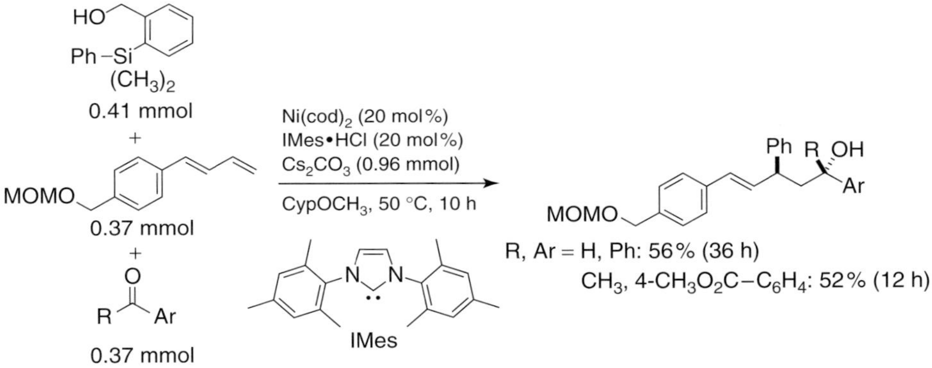

Scheme 5.29 Nickel-catalyzed three-component coupling of organo[2-(hydroxymethyl)phenyl-]dimethylsilanes, 1,3-dienes, and carbonyls.

5.3.3
Palladium-Catalyzed Reactions

Transmetalation from the silicon of aryl(trialkoxy)silanes [59] and aryl(trifluoro)silanes [60] to cationic palladium(II) species proceeds in the presence of ZnF_2 as an activator, and the resulting aryl palladium species undergoes an enantioselective 1,4-addition reaction across the Michael acceptors (Scheme 5.30).

Palladium(II) catalysts are also effective for enantioselective Mannich-type reactions with silyl enol ethers [61] and tetraallylsilane [62]; in both cases, the reaction involves the respective organopalladium species as an intermediate that reacts with imines (Scheme 5.31) [63].

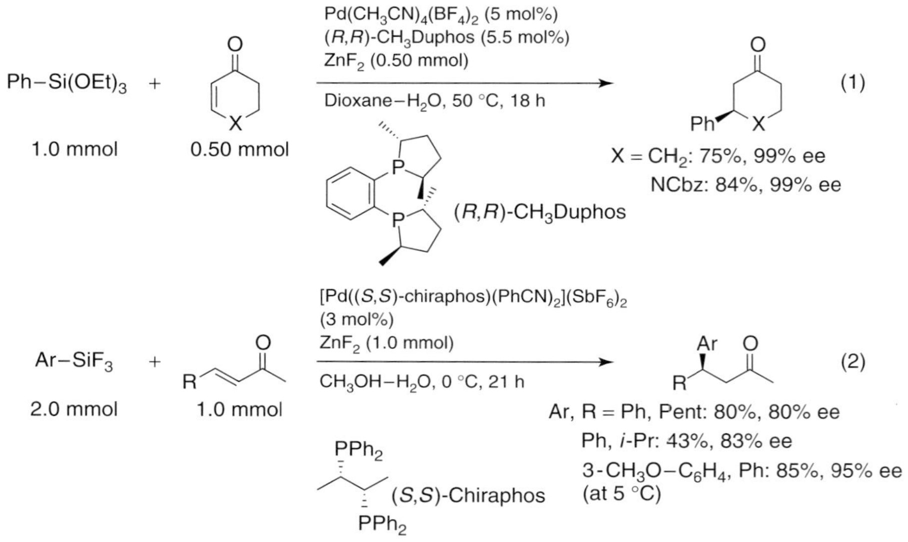

Scheme 5.30

Ph–Si(OEt)$_3$ (1.0 mmol) + cyclohexenone derivative (X) (0.50 mmol)

Conditions: Pd(CH$_3$CN)$_4$(BF$_4$)$_2$ (5 mol%), (R,R)-CH$_3$Duphos (5.5 mol%), ZnF$_2$ (0.50 mmol), Dioxane–H$_2$O, 50 °C, 18 h

(1)

X = CH$_2$: 75%, 99% ee
NCbz: 84%, 99% ee

(R,R)-CH$_3$Duphos

Ar–SiF$_3$ (2.0 mmol) + R-enone (1.0 mmol)

Conditions: [Pd((S,S)-chiraphos)(PhCN)$_2$](SbF$_6$)$_2$ (3 mol%), ZnF$_2$ (1.0 mmol), CH$_3$OH–H$_2$O, 0 °C, 21 h

(2)

Ar, R = Ph, Pent: 80%, 80% ee
Ph, i-Pr: 43%, 83% ee
3-CH$_3$O–C$_6$H$_4$, Ph: 85%, 95% ee
(at 5 °C)

(S,S)-Chiraphos

Scheme 5.30 Palladium-catalyzed enantioselective 1,4-addition reactions using aryl(triethoxy)silanes and aryl(trifluoro)silanes.

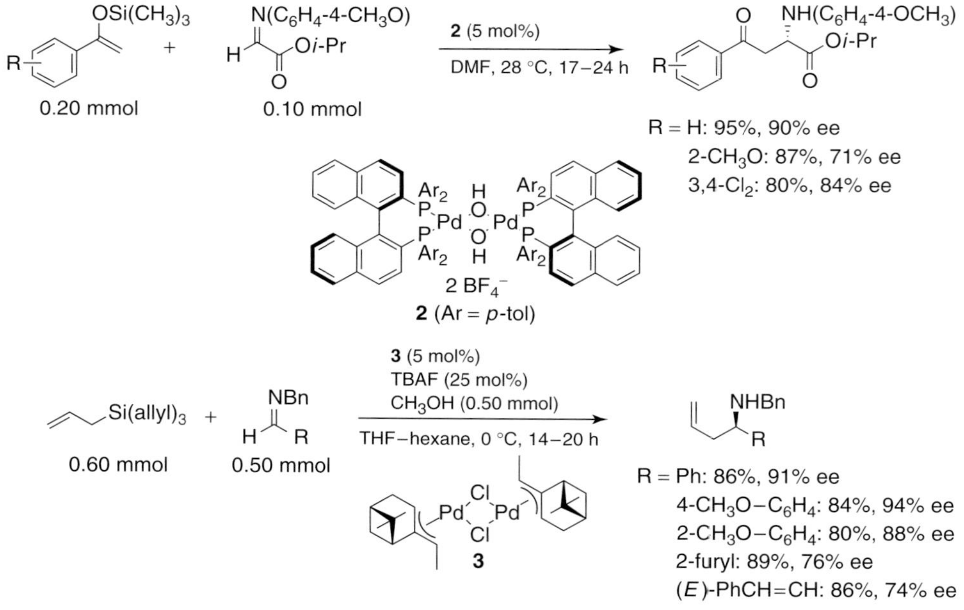

Scheme 5.31

R–C$_6$H$_4$–C(OSi(CH$_3$)$_3$)=CH$_2$ (0.20 mmol) + HN=CH–C(=O)Oi-Pr with N(C$_6$H$_4$-4-CH$_3$O) (0.10 mmol)

Conditions: 2 (5 mol%), DMF, 28 °C, 17–24 h

Product: R–C$_6$H$_4$–CO–CH$_2$–CH(NH(C$_6$H$_4$-4-OCH$_3$))–C(=O)Oi-Pr

R = H: 95%, 90% ee
2-CH$_3$O: 87%, 71% ee
3,4-Cl$_2$: 80%, 84% ee

Catalyst 2 (Ar = p-tol), 2 BF$_4^-$

Allyl–Si(allyl)$_3$ (0.60 mmol) + HN(NBn)=CH–R (0.50 mmol)

Conditions: 3 (5 mol%), TBAF (25 mol%), CH$_3$OH (0.50 mmol), THF–hexane, 0 °C, 14–20 h

Product: CH$_2$=CH–CH$_2$–CH(NHBn)–R

R = Ph: 86%, 91% ee
4-CH$_3$O–C$_6$H$_4$: 84%, 94% ee
2-CH$_3$O–C$_6$H$_4$: 80%, 88% ee
2-furyl: 89%, 76% ee
(E)-PhCH=CH: 86%, 74% ee

Catalyst 3

Scheme 5.31 Palladium-catalyzed enantioselective Mannich-type reactions.

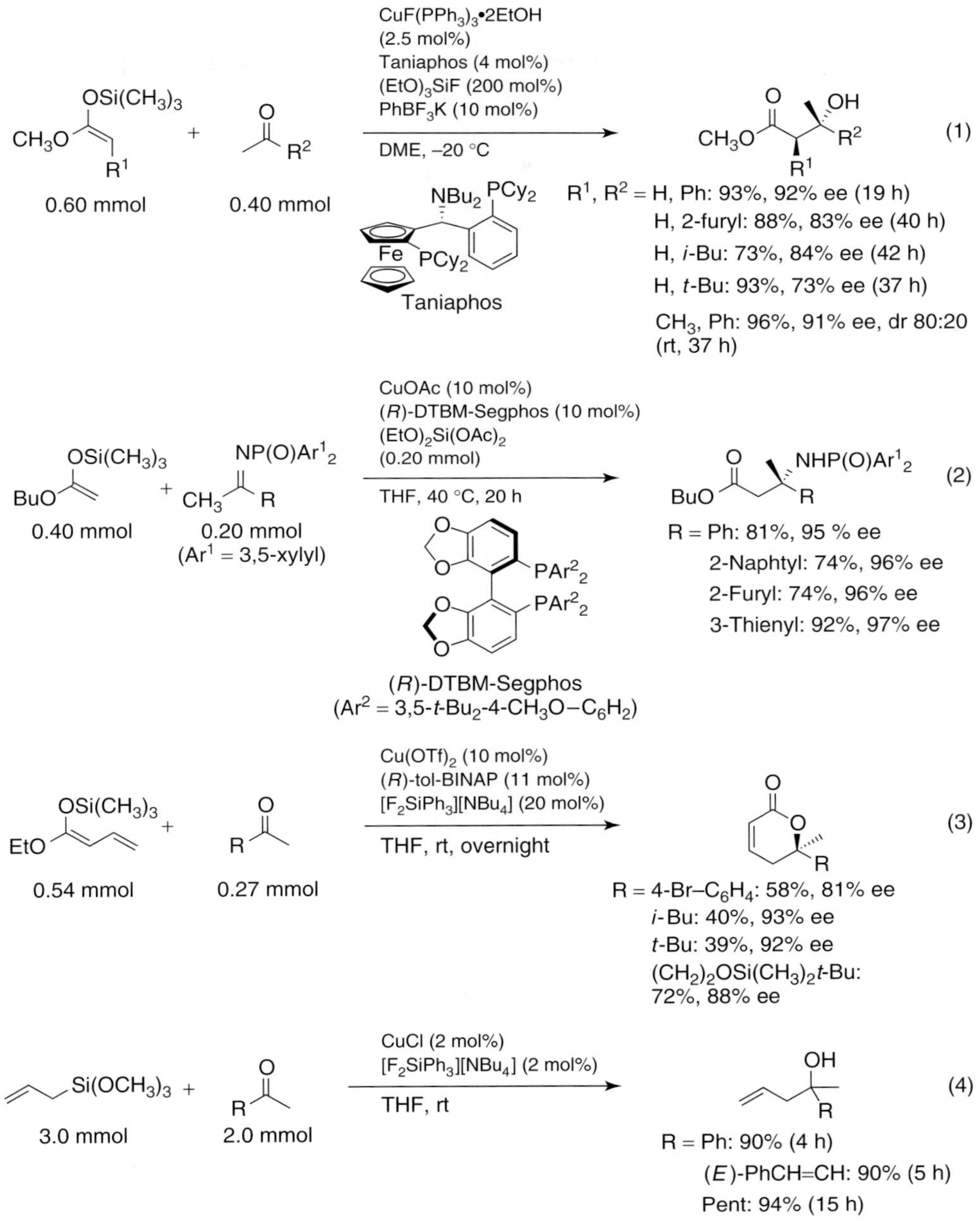

5.3 Carbonyl Addition Reaction

Scheme 5.32 Copper-catalyzed carbonyl addition reactions using organosilicon reagents.

Reaction (1):
- CH$_3$O–C(OSi(CH$_3$)$_3$)=CHR^1 (0.60 mmol) + R^{2}C(O)H (0.40 mmol)
- CuF(PPh$_3$)$_3$·2EtOH (2.5 mol%), Taniaphos (4 mol%), (EtO)$_3$SiF (200 mol%), PhBF$_3$K (10 mol%), DME, –20 °C
- Product: CH$_3$O–C(O)–CR^1–CH(OH)R^2

Taniaphos: NBu$_2$, PCy$_2$, Fe, PCy$_2$

R^1, R^2 = H, Ph: 93%, 92% ee (19 h)
H, 2-furyl: 88%, 83% ee (40 h)
H, i-Bu: 73%, 84% ee (42 h)
H, t-Bu: 93%, 73% ee (37 h)
CH$_3$, Ph: 96%, 91% ee, dr 80:20 (rt, 37 h)

Reaction (2):
- BuO–C(OSi(CH$_3$)$_3$)=CH$_2$ (0.40 mmol) + CH$_3$–CR=NP(O)Ar$^1{}_2$ (0.20 mmol) (Ar^1 = 3,5-xylyl)
- CuOAc (10 mol%), (R)-DTBM-Segphos (10 mol%), (EtO)$_2$Si(OAc)$_2$ (0.20 mmol), THF, 40 °C, 20 h
- Product: BuO–C(O)–CH$_2$–CR(NHP(O)Ar$^1{}_2$)

R = Ph: 81%, 95% ee
2-Naphtyl: 74%, 96% ee
2-Furyl: 74%, 96% ee
3-Thienyl: 92%, 97% ee

(R)-DTBM-Segphos (Ar^2 = 3,5-t-Bu$_2$-4-CH$_3$O–C$_6$H$_2$)

Reaction (3):
- EtO–C(OSi(CH$_3$)$_3$)=CH–CH=CH$_2$ (0.54 mmol) + RC(O)H (0.27 mmol)
- Cu(OTf)$_2$ (10 mol%), (R)-tol-BINAP (11 mol%), [F$_2$SiPh$_3$][NBu$_4$] (20 mol%), THF, rt, overnight
- Product: dihydropyranone with R substituent

R = 4-Br–C$_6$H$_4$: 58%, 81% ee
i-Bu: 40%, 93% ee
t-Bu: 39%, 92% ee
(CH$_2$)$_2$OSi(CH$_3$)$_2$t-Bu: 72%, 88% ee

Reaction (4):
- CH$_2$=CH–CH$_2$–Si(OCH$_3$)$_3$ (3.0 mmol) + RC(O)H (2.0 mmol)
- CuCl (2 mol%), [F$_2$SiPh$_3$][NBu$_4$] (2 mol%), THF, rt
- Product: CH$_2$=CH–CH$_2$–C(OH)(H)R

R = Ph: 90% (4 h)
(E)-PhCH=CH: 90% (5 h)
Pent: 94% (15 h)

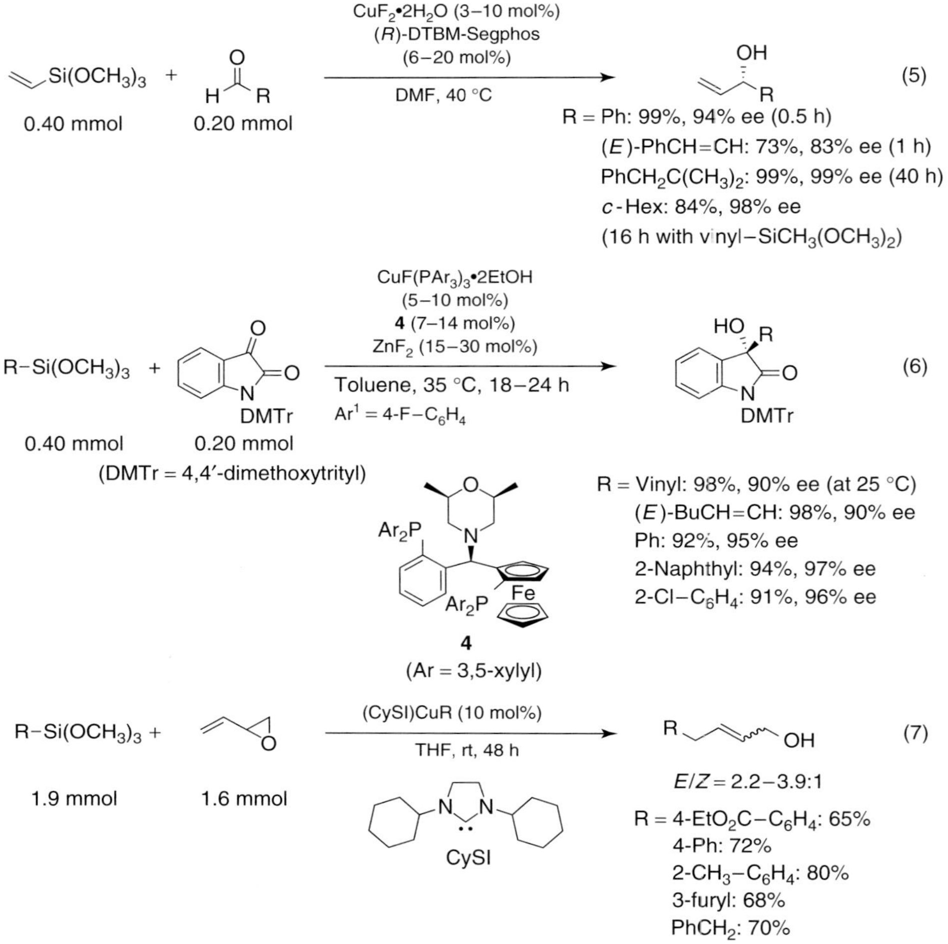

Scheme 5.32 (Continued)

Reaction (5):

≈Si(OCH₃)₃ + H-C(=O)-R → [CuF₂·2H₂O (3–10 mol%), (R)-DTBM-Segphos (6–20 mol%), DMF, 40 °C] → CH₂=CH-CH(OH)-R

0.40 mmol, 0.20 mmol

R = Ph: 99%, 94% ee (0.5 h)
(E)-PhCH=CH: 73%, 83% ee (1 h)
PhCH₂C(CH₃)₂: 99%, 99% ee (40 h)
c-Hex: 84%, 98% ee
(16 h with vinyl–SiCH₃(OCH₃)₂)

Reaction (6):

R–Si(OCH₃)₃ + isatin-DMTr → [CuF(PAr₃)₃·2EtOH (5–10 mol%), 4 (7–14 mol%), ZnF₂ (15–30 mol%), Toluene, 35 °C, 18–24 h, Ar¹ = 4-F–C₆H₄] → 3-hydroxy-3-R-oxindole-DMTr

0.40 mmol, 0.20 mmol
(DMTr = 4,4′-dimethoxytrityl)

Ligand 4: Ar₂P-substituted ferrocene with morpholine
(Ar = 3,5-xylyl)

R = Vinyl: 98%, 90% ee (at 25 °C)
(E)-BuCH=CH: 98%, 90% ee
Ph: 92%, 95% ee
2-Naphthyl: 94%, 97% ee
2-Cl–C₆H₄: 91%, 96% ee

Reaction (7):

R–Si(OCH₃)₃ + glycidyl (epoxide) → [(CySI)CuR (10 mol%), THF, rt, 48 h] → R-CH=CH-CH₂-OH

1.9 mmol, 1.6 mmol

CySI = N,N'-dicyclohexyl imidazol-2-ylidene

E/Z = 2.2–3.9:1
R = 4-EtO₂C–C₆H₄: 65%
4-Ph: 72%
2-CH₃–C₆H₄: 80%
3-furyl: 68%
PhCH₂: 70%

5.3.4
Copper-Catalyzed Reactions

The use of copper catalysts for silicon-based carbonyl addition reactions has generated significant interest because copper catalysts are generally inexpensive. Moreover, the organocopper intermediates generated by transmetalation with enol- [64], dienol- [65], allyl- [66], alkenyl-, and arylsilanes [67] are highly reactive and participate in 1,2-addition reactions of various electrophiles including ketones and ketimines as well as aldehydes and epoxides. Some of these transformations are achieved with a high degree of enantioselection by using a range of chiral bisphosphine ligands (Scheme 5.32).

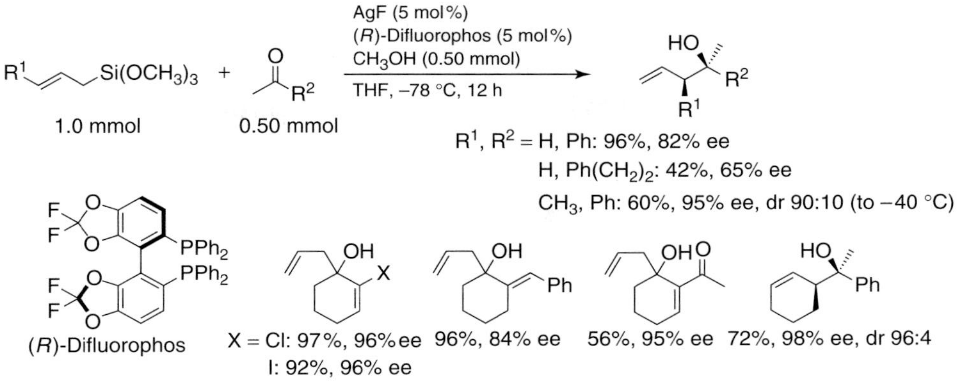

Scheme 5.33 Silver-catalyzed enantioselective allylation of ketones using allyl(trimethoxy)silanes.

5.3.5
Silver-Catalyzed Reactions

Chiral silver complexes have been found to be effective catalysts for enantioselective aldol reactions and the allylation of aldehydes [68] and ketones [69] (Scheme 5.33). An allyl silver species that possibly reacts with the carbonyl electrophiles through a six-membered transition state is proposed.

5.4
Recent Developments in Catalytic Preparation of Organosilanes

Organosilicon reagents are conventionally synthesized by the reaction of highly nucleophilic organolithium or organomagnesium reagents with silicon electrophiles, typically chlorosilanes. Extensive developments in chemo- and

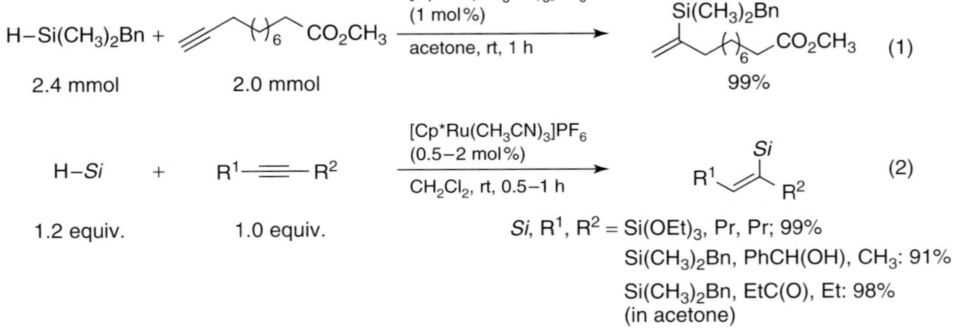

Scheme 5.34 Ruthenium-catalyzed hydrosilylation of alkynes.

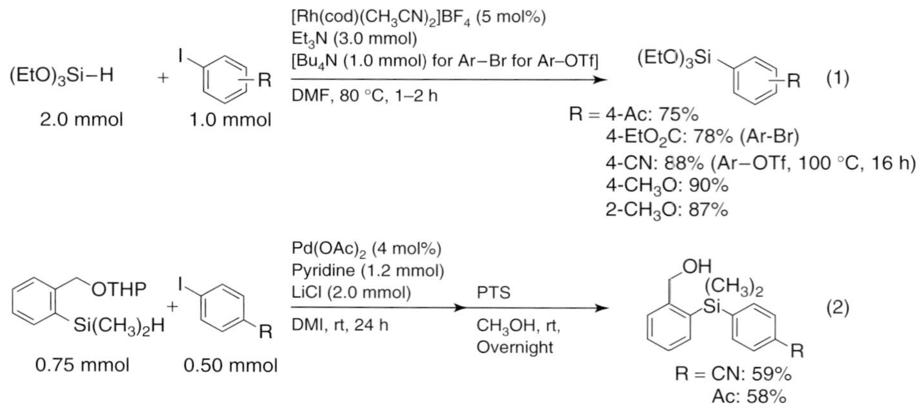

Scheme 5.35 Transition metal–catalyzed silylation of aryl halides with hydrosilanes.

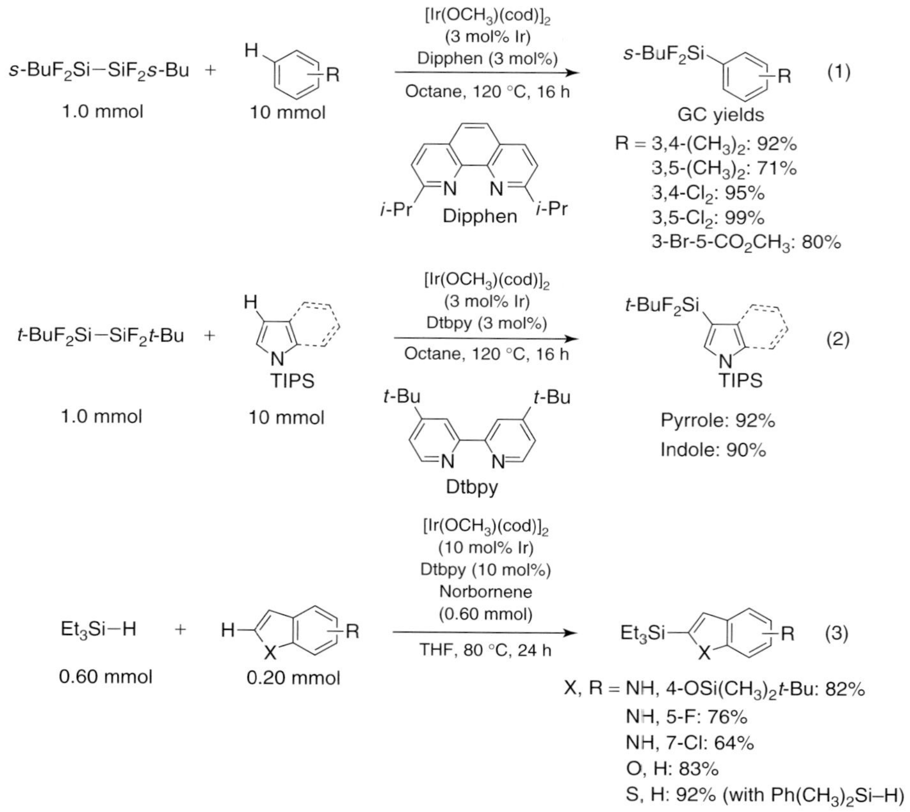

Scheme 5.36 Iridium-catalyzed silylation of C–H bonds.

stereoselective C–C bond forming reactions based on organosilicon reagents have led to increasing demands to prepare such organosilanes with functional groups that do not tolerate conventional synthetic conditions. Here again, transition metal catalysis plays a key role by providing chemoselective protocols for installing a silyl group. Whereas platinum- or rhodium-catalyzed alkyne hydrosilylation chemistry has long been the standard for accessing functionalized alkenylsilanes [70], ruthenium catalysis [71] has recently been reported to allow unique stereo- and regioselection for this transformation, which is complementary to reactions of the conventional catalysis scheme (Scheme 5.34).

Metal-catalyzed silylation of aryl halides with hydrosilanes has been developed as a chemoselective and atom economical way to access arylsilane reagents having a range of functional groups (Scheme 5.35). Rhodium- [72] and palladium-catalyzed [73] reactions are particularly useful for directly preparing arylsilane reagents having reactive silyl groups.

An ideal route to organosilicon compounds is direct C–H silylation by transition metal catalysis. Such transformations have been developed with ruthenium [74], rhodium [75], iridium [76], nickel [77], and platinum [78] catalysts. Particularly useful are the iridium-catalyzed protocols that allow regio- and chemoselective installation of the alkyldifluorosilyl groups that participate in the cross-coupling chemistry (Scheme 5.36).

References

1. Reviews up to 2004, see: (a) Hiyama, T. (1998) in *Metal-Catalyzed Cross-Coupling Reactions* (eds F. Diederich and P.J. Stang), Wiley-VCH Verlag GmbH, Weinheim, p. 421; (b) Hiyama, T. and Shirakawa, E. (2002) *Top. Curr. Chem.*, **219**, 61; (c) Denmark, S.E. and Sweis, R.F. (2004) in *Metal-Catalyzed Cross-Coupling Reactions*, 2nd edn (eds F. Diederich and P.J. Stang), Wiley-VCH Verlag GmbH, Weinheim, p. 163; (d) Tsuji, J. (2004) *Palladium Reagents and Catalysts*, John Wiley & Sons, Ltd, Chichester, p. 338.
2. Hatanaka, Y. and Hiyama, T. (1988) *J. Org. Chem.*, **53**, 918.
3. Hatanaka, Y. and Hiyama, T. (1989) *J. Org. Chem.*, **54**, 265.
4. Hatanaka, Y., Fukushima, S., and Hiyama, T. (1989) *Chem. Lett.*, 1711.
5. Hatanaka, Y., Goda, K., Okahara, Y., and Hiyama, T. (1994) *Tetrahedron*, **50**, 8301.
6. Tamao, K., Kobayashi, K., and Ito, Y. (1989) *Tetrahedron Lett.*, **30**, 6051.
7. (a) Hirabayashi, K., Kawashima, J., Nishihara, Y., Mori, A., and Hiyama, T. (1999) *Org. Lett.*, **1**, 299; (b) Denmark, S.E. and Wehrli, D. (2000) *Org. Lett.*, **2**, 565.
8. Hagiwara, E., Gouda, K., Hatanaka, Y., and Hiyama, T. (1997) *Tetrahedron Lett.*, **38**, 439.
9. Denmark, S.E. and Sweis, R.F. (2001) *J. Am. Chem. Soc.*, **123**, 6439.
10. (a) Denmark, S.E. and Choi, J.Y. (1999) *J. Am. Chem. Soc.*, **121**, 5821; (b) Denmark, S.E., Wehrli, D., and Choi, J.Y. (2000) *Org. Lett.*, **2**, 2491; (c) Itami, K., Nokami, T., and Yoshida, J. (2001) *J. Am. Chem. Soc.*, **123**, 5600; (d) Hosoi, K., Nozaki, K., and Hiyama, T. (2002) *Chem. Lett.*, 138; (e) Katayama, H., Nagao, M., Moriguchi, R., and Ozawa, F. (2003) *J. Organomet. Chem.*, **676**, 49; (f) Trost, B.M., Machacek, M.R., and Ball, Z.T. (2003) *Org. Lett.*, **5**, 1895; (g) Anderson, J.C. and Munday, R.H. (2004) *J. Org. Chem.*, **69**, 8971.

11. For reviews, see: (a) Brook, A.G. (1974) *Acc. Chem. Res.*, **7**, 77; (b) Jankowski, P., Raubo, P., and Wicha, J. (1994) *Synlett*, 985; (c) Moser, W.H. (2001) *Tetrahedron*, **57**, 2065.
12. Taguchi, H., Ghoroku, K., Tadaki, M., Tsubouchi, A., and Takeda, T. (2002) *J. Org. Chem.*, **67**, 8450.
13. Shindo, M., Matsumoto, K., and Shishido, K. (2005) *Synlett*, 176.
14. Hudrlik, P.F., Arango, J.O., Hijji, Y.M., Okoro, C.O., and Hudrlik, A.M. (2000) *Can. J. Chem.*, **78**, 1421.
15. Nakao, Y., Imanaka, H., Sahoo, A.K., Yada, A., and Hiyama, T. (2005) *J. Am. Chem. Soc.*, **127**, 6952.
16. Nakao, Y., Imanaka, H., Chen, J., Yada, A., and Hiyama, T. (2007) *J. Organomet. Chem.*, **692**, 585.
17. Nakao, Y., Sahoo, A.K., Yada, A., Chen, J., and Hiyama, T. (2006) *Sci. Technol. Adv. Mater.*, **7**, 536.
18. Chen, J., Tanaka, M., Sahoo, A.K., Takeda, M., Yada, A., Nakao, Y., and Hiyama, T. (2010) *Bull. Chem. Soc. Jpn.*, **83**, 554.
19. (a) Guibe, F. and Saintmleux, Y. (1981) *Tetrahedron Lett.*, **22**, 3591; (b) Tsuji, J., Shimizu, I., Minami, I., Ohashi, Y., Sugiura, T., and Takahashi, K. (1985) *J. Org. Chem.*, **50**, 1523.
20. For alkyl cross-coupling with alkyl-trifluorosilanes, see: (a) Matsuhashi, H., Kuroboshi, M., Hatanaka, Y., and Hiyama, T. (1994) *Tetrahedron Lett.*, **35**, 6507; (b) Matsuhashi, H., Asai, S., Hirabayashi, K., Mori, A., and Hiyama, T. (1997) *Bull. Chem. Soc. Jpn.*, **70**, 437.
21. Nakao, Y., Takeda, M., Matsumoto, T., and Hiyama, T. (2010) *Angew. Chem. Int. Ed.*, **49**, 4447.
22. Nakao, Y., Chen, J., Tanaka, M., and Hiyama, T. (2007) *J. Am. Chem. Soc.*, **129**, 11694.
23. Son, E.C., Tsuji, H., Saeki, T., and Tamao, K. (2006) *Bull. Chem. Soc. Jpn.*, **79**, 492.
24. Rauf, W. and Brown, J.M. (2008) *Angew. Chem. Int. Ed.*, **47**, 4228.
25. Denmark, S.E. and Sweis, R.F. (2002) *Acc. Chem. Res.*, **35**, 835.
26. Denmark, S.E. and Sweis, R.F. (2004) *J. Am. Chem. Soc.*, **126**, 4876.
27. Denmark, S.E. and Kallemeyn, J.M. (2006) *J. Am. Chem. Soc.*, **128**, 15958.
28. (a) Denmark, S.E., Smith, R.C., and Tymonko, S.A. (2007) *Tetrahedron*, **63**, 5730; (b) Denmark, S.E., Smith, R.C., Chang, W.T.T., and Muhuhi, J.M. (2009) *J. Am. Chem. Soc.*, **131**, 3104.
29. Denmark, S.E., Baird, J.D., and Regens, C.S. (2008) *J. Org. Chem.*, **73**, 1440.
30. Denmark, S.E. and Butler, C.R. (2008) *J. Am. Chem. Soc.*, **130**, 3690.
31. For similar γ-selective cross-coupling using crotyltrifluorosilane, see: Hatanaka, Y., Ebina, Y., and Hiyama, T. (1991) *J. Am. Chem. Soc.*, **113**, 7075.
32. Denmark, S.E. and Werner, N.S. (2008) *J. Am. Chem. Soc.*, **130**, 16382.
33. Denmark, S.E. and Fujimori, S. (2005) *J. Am. Chem. Soc.*, **127**, 8971.
34. Tyrrell, E. and Brookes, P. (2003) *Synthesis*, 469.
35. Pierrat, P., Gros, P., and Fort, Y. (2005) *Org. Lett.*, **7**, 697.
36. (a) Nokami, T., Tomida, Y., Kamei, T., Itami, K., and Yoshida, J. (2006) *Org. Lett.*, **8**, 729; (b) Napier, S., Marcuccio, S.M., Tye, H., and Whittaker, M. (2008) *Tetrahedron Lett.*, **49**, 6314.
37. (a) Kuwajima, I. and Urabe, H. (1982) *J. Am. Chem. Soc.*, **104**, 6831; (b) Liu, X.X. and Hartwig, J.F. (2004) *J. Am. Chem. Soc.*, **126**, 5182; (c) Su, W.P., Raders, S., Verkade, J.G., Liao, X.B., and Hartwig, J.F. (2006) *Angew. Chem. Int. Ed.*, **45**, 5852.
38. Wu, L.Y. and Hartwig, J.F. (2005) *J. Am. Chem. Soc.*, **127**, 15824.
39. (a) Cunico, R.F. and Maity, M.C. (2002) *Org. Lett.*, **4**, 4357; (b) Cunico, R.F. and Maity, B.C. (2003) *Org. Lett.*, **5**, 4947; (c) Chen, J.X. and Cunico, R.F. (2004) *J. Org. Chem.*, **69**, 5509; (d) Cunico, R.F. and Pandey, R.K. (2005) *J. Org. Chem.*, **70**, 5344; (e) Cunico, R.F. and Pandey, R.K. (2005) *J. Org. Chem.*, **70**, 9048.
40. (a) Zhang, L. and Wu, J. (2008) *J. Am. Chem. Soc.*, **130**, 12250; (b) Zhang, L., Qing, J., Yang, P.Y., and Wu, J. (2008) *Org. Lett.*, **10**, 4971; (c) So, C.M., Lee, H.W., Lau, C.P., and Kwong, F.Y. (2009) *Org. Lett.*, **11**, 317.
41. (a) Lee, J.Y. and Fu, G.C. (2003) *J. Am. Chem. Soc.*, **125**, 5616; (b) Dai, X.,

Strotman, N.A., and Fu, G.C. (2008) *J. Am. Chem. Soc.*, **130**, 3302.
42. Strotman, N.A., Sommer, S., and Fu, G.C. (2007) *Angew. Chem. Int. Ed.*, **46**, 3556.
43. Yamane, M., Uera, K., and Narasaka, K. (2004) *Chem. Lett.*, **33**, 424.
44. For reviews, see: (a) Kakiuchi, F. and Chatani, N. (2003) *Adv. Synth. Catal.*, **345**, 1077; (b) Kakiuchi, F. and Kochi, T. (2008) *Synthesis*, 3013.
45. Yang, S.D., Li, B.J., Wan, X.B., and Shi, Z.J. (2007) *J. Am. Chem. Soc.*, **129**, 6066.
46. Zhou, H., Xu, Y.H., Chang, W.J., and Loh, T.P. (2009) *Angew. Chem. Int. Ed.*, **48**, 5355.
47. (a) Sakai, M., Hayashi, H., and Miyaura, N. (1997) *Organometallics*, **16**, 4229; For palladium-catalyzed reactions, see: (b) Cho, C.S., Motofusa, S.-I., Ohe, K., Uemura, S., and Shim, S.C. (1995) *J. Org. Chem.*, **60**, 883.
48. For a review, see: Hayashi, T. and Yamasaki, K. (2003) *Chem. Rev.*, **103**, 2829.
49. (a) Oi, S., Honma, Y., and Inoue, Y. (2002) *Org. Lett.*, **4**, 667; (b) Murata, M., Shimazaki, R., Ishikura, M., Watanabe, S., and Masuda, Y. (2002) *Synthesis*, 717; (c) Oi, S., Taira, A., Honma, Y., and Inoue, Y. (2003) *Org. Lett.*, **5**, 97; (d) Otomaru, Y. and Hayashi, T. (2004) *Tetrahedron-Asymmetry*, **15**, 2647; (e) Sanada, T., Kato, T., Mitani, M., and Mori, A. (2006) *Adv. Synth. Catal.*, **348**, 51; (f) Oi, S., Taira, A., Honma, Y., Sato, T., and Inoue, Y. (2006) *Tetrahedron-Asymmetry*, **17**, 598; (g) Hargrave, J.D., Herbert, J., Bish, G., and Frost, C.G. (2006) *Org. Biomol. Chem.*, **4**, 3235.
50. Huang, T.S. and Li, C.J. (2001) *Chem. Commun.*, 2348.
51. (a) Mori, A., Danda, Y., Fujii, T., Hirabayashi, K., and Osakada, K. (2001) *J. Am. Chem. Soc.*, **123**, 10774; (b) Fujii, T., Koike, T., Mori, A., and Osakada, K. (2002) *Synlett*, 298.
52. Koike, T., Du, X.L., Mori, A., and Osakada, K. (2002) *Synlett*, 301.
53. Nakao, Y., Chen, J., Imanaka, H., Hiyama, T., Ichikawa, Y., Duan, W.L., Shintani, R., and Hayashi, T. (2007) *J. Am. Chem. Soc.*, **129**, 9137.
54. Shintani, R., Ichikawa, Y., Hayashi, T., Chen, J., Nakao, Y., and Hiyama, T. (2007) *Org. Lett.*, **9**, 4643.
55. Peyroux, E., Berthiol, F., Doucet, H., and Santelli, M. (2004) *Eur. J. Org. Chem.*, 1075.
56. Nakao, Y., Takeda, M., Chen, J., Hiyama, T., Ichikawa, Y., Shintani, R., and Hayashi, T. (2008) *Chem. Lett.*, **37**, 290.
57. (a) Saito, N., Yamazaki, T., and Sato, Y. (2008) *Tetrahedron Lett.*, **49**, 5073; (b) Saito, N., Yamazaki, T., and Sato, Y. (2009) *Chem. Lett.*, **38**, 594.
58. Ogoshi, S., Tonomori, K., Oka, M., and Kurosawa, H. (2006) *J. Am. Chem. Soc.*, **128**, 7077.
59. Nishikata, T., Yamamoto, Y., and Miyaura, N. (2004) *Organometallics*, **23**, 4317.
60. Nishikata, T., Yamamoto, Y., Gridnev, I.D., and Miyaura, N. (2005) *Organometallics*, **24**, 5025.
61. Hagiwara, E., Fujii, A., and Sodeoka, M. (1998) *J. Am. Chem. Soc.*, **120**, 2474.
62. (a) Nakamura, K., Nakamura, H., and Yamamoto, Y. (1999) *J. Org. Chem.*, **64**, 2614; (b) Fernandes, R.A. and Yamamoto, Y. (2004) *J. Org. Chem.*, **69**, 735.
63. For a palladium enolate intermediate, see: Fujii, A., Hagiwara, E., and Sodeoka, M. (1999) *J. Am. Chem. Soc.*, **121**, 5450.
64. (a) Oisaki, K., Suto, Y., Kanai, M., and Shibasaki, M. (2003) *J. Am. Chem. Soc.*, **125**, 5644; (b) Oisaki, K., Zhao, D., Kanai, M., and Shibasaki, M. (2006) *J. Am. Chem. Soc.*, **128**, 7164; (c) Suto, Y., Kanai, M., and Shibasaki, M. (2007) *J. Am. Chem. Soc.*, **129**, 500.
65. (a) Kruger, J. and Carreira, E.M. (1998) *J. Am. Chem. Soc.*, **120**, 837; (b) Pagenkopf, B.L., Kruger, J., Stojanovic, A., and Carreira, E.M. (1998) *Angew. Chem. Int. Ed.*, **37**, 3124; (c) Moreau, X., Bazan-Tejeda, B., and Campagne, J.M. (2005) *J. Am. Chem. Soc.*, **127**, 7288.
66. Yamasaki, S., Fujii, K., Wada, R., Kanai, M., and Shibasaki, M. (2002) *J. Am. Chem. Soc.*, **124**, 6536.

67. (a) Tomita, D., Wada, R., Kanai, M., and Shibasaki, M. (2005) *J. Am. Chem. Soc.*, **127**, 4138; (b) Motoki, R., Tomita, D., Kanai, M., and Shibasaki, M. (2006) *Tetrahedron Lett.*, **47**, 8083; (c) Tomita, D., Yamatsugu, K., Kanai, M., and Shibasaki, M. (2009) *J. Am. Chem. Soc.*, **131**, 6946; (d) Herron, J.R., Russo, V., Valente, E.J., and Ball, Z.T. (2009) *Chem. Eur. J.*, **15**, 8713.
68. Yanagisawa, A., Kageyama, H., Nakatsuka, Y., Asakawa, K., Matsumoto, Y., and Yamamoto, H. (1999) *Angew. Chem. Int. Ed.*, **38**, 3701.
69. Wadamoto, M. and Yamamoto, H. (2005) *J. Am. Chem. Soc.*, **127**, 14556.
70. For a review, see: Trost, B.M. and Ball, Z.T. (2005) *Synthesis*, 853.
71. (a) Kawanami, Y., Sonoda, Y., Mori, T., and Yamamoto, K. (2002) *Org. Lett.*, **4**, 2825; (b) Trost, B.M. and Ball, Z.T. (2005) *J. Am. Chem. Soc.*, **127**, 17644.
72. Murata, M., Yamasaki, H., Ueta, T., Nagata, M., Ishikura, M., Watanabe, S., and Masuda, Y. (2007) *Tetrahedron*, **63**, 4087.
73. Iizuka, M. and Kondo, Y. (2008) *Eur. J. Org. Chem.*, 1161.
74. Kakiuchi, F., Igi, K., Matsumoto, M., Hayamizu, T., Chatani, N., and Murai, S. (2002) *Chem. Lett.*, 396; (b) Kakiuchi, F., Matsumoto, M., Tsuchiya, K., Igi, K., Hayamizu, T., Chatani, N., and Murai, S. (2003) *J. Organomet. Chem.*, **686**, 134; (c) Ihara, H. and Suginome, M. (2009) *J. Am. Chem. Soc.*, **131**, 7502.
75. (a) Sakakura, T., Sodeyama, T., Tokunaga, Y., and Tanaka, M. (1987) *Chem. Lett.*, 2211; (b) Ezbiansky, M.K., Djurovich, P.I., LaForest, M., Sinning, D.J., Zayes, R., and Berry, D.H. (1998) *Organometallics*, **17**, 1455; (c) Tobisu, M., Ano, Y., and Chatani, N., (2008) *Chem. Asia. J.*, **3**, 1585.
76. (a) Ishiyama, T., Sato, K., Nishio, Y., and Miyaura, N. (2003) *Angew. Chem. Int. Ed.*, **42**, 5346; (b) Ishiyama, T., Sato, K., Nishio, Y., Saiki, T., and Miyaura, N. (2005) *Chem. Commun.*, 5065; (c) Saiki, T., Nishio, Y., Ishiyama, T., and Miyaura, N. (2006) *Organometallics*, **25**, 6068; (d) Lu, B. and Falck, J.R. (2008) *Angew. Chem. Int. Ed.*, **47**, 7508.
77. Ishikawa, M., Okazaki, S., Naka, A., and Sakamoto, H. (1992) *Organometallics*, **11**, 4135.
78. (a) Uchimaru, Y., Brandl, P., Tanaka, M., and Goto, M. (1993) *J. Chem. Soc., Chem. Commun.*, 744; (b) Williams, N.A., Uchimaru, Y., and Tanaka, M. (1995) *J. Chem. Soc., Chem. Commun.*, 2045; (c) Tsukada, N. and Hartwig, J.F. (2005) *J. Am. Chem. Soc.*, **127**, 5022; (d) Murata, M., Fukuyama, N., Wada, J., Watanabe, S., and Masuda, Y. (2007) *Chem. Lett.*, **36**, 910.

6
Direct Reductive Amination with Amine Boranes
Karl Matos and Elizabeth R. Burkhardt

6.1
Introduction

Many current drugs and active pharmaceutical intermediates contain amine functionalities. Direct reductive amination of aldehydes and ketones is an elegant and powerful tool for the transformation of carbonyl compounds into structurally diverse amines [1]. In this important carbon–nitrogen bond forming process, aldehydes or ketones react with ammonia, primary or secondary amines in the presence of a reducing agent to produce primary, secondary, and tertiary amines, respectively [2]. An advantage of this efficient process is that the intermediate imine or iminium ion is not isolated prior to reduction. Since water is a coproduct during the *in situ* formation of the imine or iminium ion, the reducing agent must therefore be stable in the presence of water. Some preparation methods utilize a drying agent, such as molecular sieves [3] or $TiCl_4$ [4] to bind water. In addition, the reduction of the ketone or aldehyde must be slower than imine formation and reduction to avoid the creation of alcohol by-products.

Besides the formation of water in the condensation process, imines are generally formed under acidic catalysis, further limiting the types of reducing agents capable of withstanding aqueous acidic conditions. Acetic acid is commonly used for ketone amination due to favorable protonation of the hydroxyamine and loss of water. However, for aliphatic aldehyde reductions, use of acetic acid may increase the undesirable aldehyde reduction.

In order to successfully carry out a direct reductive amination, the choice of the reducing agent is critical. Three acid-stable reducing agents are sodium cyanoborohydride [5], sodium triacyloxyborohydride [6], and amine borane [7]. Industrial use of sodium cyanoborohydride is unattractive due to the toxicity, sluggish rate of reaction, and the large volumes needed. Sodium triacetoxyborohydride (STAB), although quite selective for reductive amination, has only one hydride and is insoluble in most common organic solvents. In addition, STAB is often used in excess, is sluggish with less reactive aromatic and substituted ketones, and cannot be utilized in protic solvents such as methanol, which favors imine formation. Catalytic hydrogenation has limited use with compounds containing double bonds, triple

Pharmaceutical Process Chemistry. Edited by Takayuki Shioiri, Kunisuke Izawa, and Toshiro Konoike
Copyright © 2011 WILEY-VCH Verlag GmbH & Co. KGaA, Weinheim
ISBN: 978-3-527-32650-1

bonds, or hydrogen-sensitive functionalities in the substrate [8]. Other reagents are nonselective in their reducing properties or are decomposed in acidic media precluding an *in situ* direct reductive amination.

6.2
Types of Amine Boranes

A number of amine boranes are effective reagents to carry out direct reductive aminations due to their acid stability [9] and functional group compatibility [10]. However, not all amine boranes are suitable, since sterically hindered amine boranes and aniline-derived amine boranes are reactive toward acid and protic media [11]. Advantages of the amine boranes described here are (i) the effective amination of aldehydes and ketones with weakly basic aniline substrates; (ii) reactions that can be conducted in polar protic solvents, such as alcohol which favors imine formation; (iii) that water can be used as a solvent or cosolvent; and (iv) reactions may even be conducted without a solvent.

The purpose of this review is to highlight the uses of amine boranes for direct reductive amination. Other selective reducing agents for reductive amination, such as STAB, have been recently reviewed [6] and are not discussed in this article.

6.2.1
Alkylamine Boranes

Amine boranes, derived from primary or secondary amines, are more appropriate for reduction of preformed imines because incorporation of the amine from the borane complex to reduction products often occurs.

tert-Butylamine borane [12] (TBAB) and dimethylamine borane [13] (DMAB) are commercially available reagents used primarily in reduction of preformed imines, enamines, or Schiff bases. TBAB was the reagent of choice to reduce multikilogram quantities of the preformed imine **2** shown using methanesulfonic acid as an activator (Scheme 6.1) [14]. The nitro group did not suffer from reduction with amine boranes, whereas when using sodium borohydride, **2** gave some by-products having azo- and azoxy-functional groups, along with the desired amine **3**. The methylbenzylamine product **3** was isolated as the HCl salt in 87% yield. In this procedure, the reduction product was obtained consistently with high yield and purity.

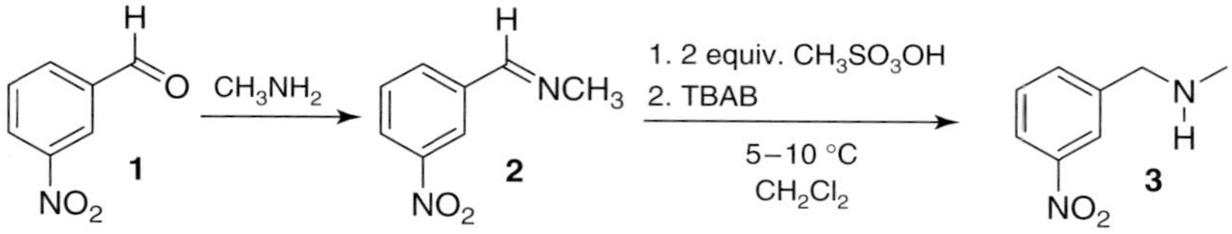

Scheme 6.1

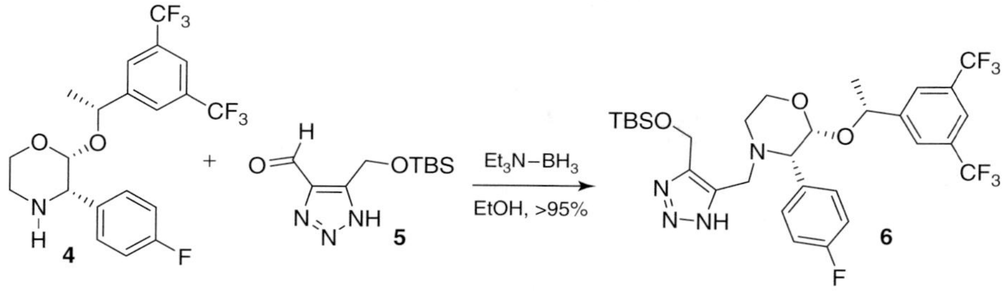

Scheme 6.2

TBAB has also been used in the reductive alkylation of amine-containing glycopeptide antibiotics with aldehydes or ketones [15]. The syntheses of glycopeptide antibiotics had previously used the more toxic sodium cyanoborohydride as the reducing agent, but TBAB was a preferable reagent for larger scale applications.

Tertiary amines or aromatic amines do not get incorporated into the imine intermediate, therefore amine boranes prepared from these types of amines are advantageous in direct reductive amination. Although, triethylamine borane [16] (TEAB) has been used infrequently in reductive aminations, Merck's exploratory route toward the synthesis of Aprepitant [17], applied TEAB in ethanol for the reductive amination which gave a 95% yield as shown in Scheme 6.2. However, in the final route, STAB was used because the product **6** formed amine borane complexes as impurities.

Recently, Scott and coworkers reported the synthesis of a hepatitis C polymerase inhibitor [18]. The key step involved a convergent coupling via reductive alkylation using trimethylamine borane (TMAB). They screened several amine boranes (BH_3 complexed with dimethylamine or *t*-butylamine) but TMAB gave the best results. The optimal conditions on a large scale involved the addition of dihydropyrone and TMAB to solid aldehyde. They obtained 7.3 kg of desired material, which on recrystallization from $CH_3OH/EtOAc$ resulted in 5.4 kg of product in >98% purity.

In summary, alkylamine boranes are suitable reducing reagents for reductive amination because of their stability under the reaction conditions, high yields, high product purity, and easy distillative separation of low-boiling amine by-product during workup.

6.2.2
Aromatic Amine Boranes

Aromatic amine borane complexes, derived from pyridine, 2-picoline, or 5-ethyl-2-methylpyridine will be the focus of the remainder of this review on direct reductive amination. Boranes derived from pyridine compounds are stable in the presence of carboxylic acid and even hydrochloric acid for the duration of the reductive amination.

6.2.2.1 Pyridine borane

Pyridine borane (PYB) has been commercially available for more than 20 years in ton quantities. PYB is soluble in most common solvents, even slightly soluble in water, but is insoluble in hexane. Storage of PYB between 15 and 35 °C is necessary to prevent freezing (mp 10–11 °C) or undesirable decomposition at elevated temperature [19]. Although freezing of PYB is not hazardous, thawing by means of external heating of containers could lead to serious unintended decomposition and fire.

Reductive aminations using PYB [3, 20] have been intensively studied. PYB has been demonstrated in successful reductive aminations to prepare several drug intermediates at a large scale [21]. Researchers at AstraZeneca synthesized AZD3409, a novel antiproliferative agent with potential application in the treatment of breast cancer and other tumors [22]. The synthesis of key aniline intermediate **9** required a reductive amination between thiolactol **7** and aniline derivative **8** as shown in Scheme 6.3. They used PYB in a 100-kg scale reductive amination in isopropanol with HCl as a protic accelerator. The substituted aniline was taken a step farther before isolation of desired product in 76% yield.

In the reaction of Scheme 6.4, PYB was used instead of the discovery method utilizing sodium cyanoborohydride [23]. Parker used factorial experimental design to optimize the reductive amination and obtained the desired amine product **12** in 85% yield. The process was scaled to 1–2 kg in dichloromethane instead of methanol to avoid formation of a dimethyl acetal of **12** as a by-product.

In another example, proteins were methylated successfully in the presence of formaldehyde with amine boranes [24]. PYB gave higher methylation than other amine boranes and the methylation was most effective in the pH range of 7–9. Lower methyl incorporation was observed at pH 5.

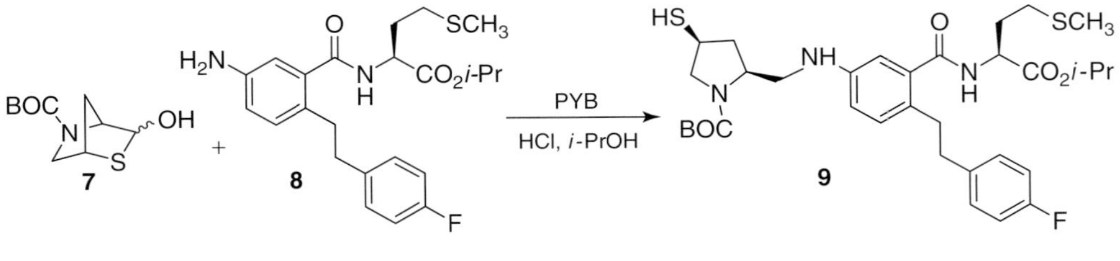

Scheme 6.3

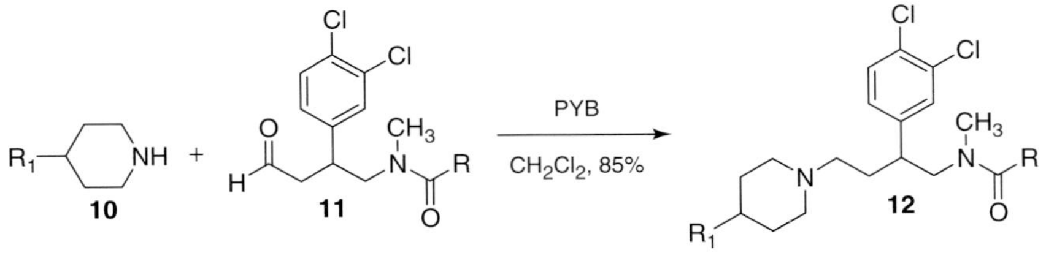

Scheme 6.4

Another promising use of PYB is in reductive amination of aldehyde-containing organotrifluoroborates to provide valuable amine-containing organoboron reagents for use in Suzuki–Miyaura C–C bond coupling reactions [25].

Examples of reductive amination in polymers using PYB extend this method to bind oligosaccharides to the surface of a β-alanine-poly(lysine)dendrimer [26] and to immobilize proteins on agarose beads [27].

6.2.2.2 2-Picoline borane

2-Picoline borane [28] (PICB) was recently introduced for reductive aminations of aldehydes and ketones in either methanol, water, or without solvent. PICB, a commercially available stable solid (mp 44–45 °C), is soluble in typical organic solvents. To support the "green chemistry mindset," [29] reductions were successfully demonstrated without solvent as a means to increase volume productivity and decrease solvent use. Previously, reductive aminations with PYB or STAB were conducted at 0.25–0.5 M concentration, so a substantial increase in volume productivity can be achieved by conducting reactions without solvent or in high concentration. For most reductions in methanol, water, or neat, the yields of aminated products were high. However, since the imine or iminium formed during the reaction is a solid in some cases, the reaction mixture must be either warmed to melt the intermediate or solvent added to achieve stirring and efficient heat transfer.

Another application of PICB to the synthesis of alkoxyamine derivatives was reported by Kawase and coworkers [30]. They found the reduction of oxime ethers proceeds rapidly in the presence of aqueous HCl in CH_3OH–AcOH to give alkoxyamines. The method was extended to one-pot synthesis of alkoxyamine derivatives from aldehydes and ketones.

During analysis of air samples containing carbonyl compounds, a method was developed based on PICB-mediated reduction of the dinitrophenylhydrazones formed from aldehydes trapped during air sampling [31]. This method is superior to former methods because only the C=N bond is reduced.

In summary, PICB was shown to be extremely useful in direct reductive amination.

6.2.2.3 5-Ethyl-2-methylpyridine borane

5-Ethyl-2-methylpyridine borane (PEMB), a stable liquid, was introduced in 2008 for reductive amination at a commercial scale. BASF recently reported the synthesis and reductive amination studies with PEMB, Scheme 6.5 [32]. Further work is elaborated here on the scope of the reductive amination and reaction workup. The results of aldehyde and ketone amination are listed in Table 6.1.

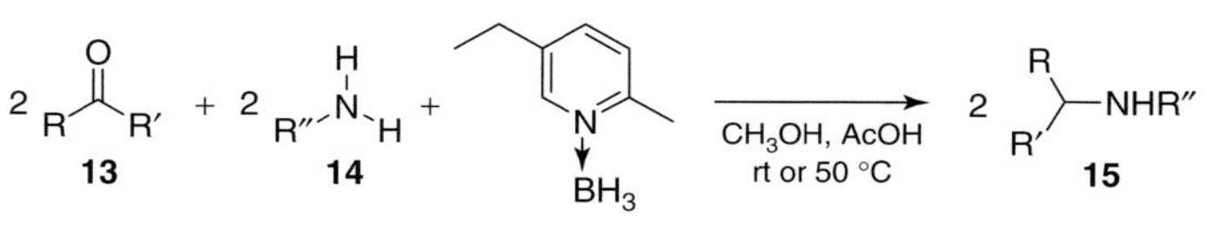

Scheme 6.5

Table 6.1 Reductive amination of aldehydes and ketones; comparison of PEMB to PICB and PYB.

Entry	Aldehyde or ketone/amine pairs	% yield[a] with PEMB in CH$_3$OH	% yield[a] with PEMB neat	% yield with PICB in CH$_3$OH	% yield with PICB neat	% yield with PYB in CH$_3$OH [17][b]
1	Benzaldehyde/aniline	99	99, 81[b]	95	99	93[c]
2	Benzaldehyde/benzylamine	87[d]	92[d]	76	NA	96
3	Benzaldehyde/di-n-propylamine	0[e]	96	NA	48[f]	50[g]
4	Hexanal/aniline w/o AcOH	94	92	82	77	91
5	Pivaldehyde/n-propylamine	99	100	NA	NA	NA
6	Cyclohexanone/aniline	92, 85[b]	93	95	94	71, 93[c]
7	Cyclohexanone/benzylamine	98[h], 83[d]	70[b]	73	78	96
8	3-Methyl-2-butanone/aniline	81	97	NA	NA	NA
9	3-Methyl-2-butanone/n-propylamine	100	100	NA	NA	NA
10	Pinacolone/aniline	63	96	NA	NA	NA
11	Pinacolone/n-propylamine	94	92	NA	NA	NA
12	2-Pentanone/aniline	74	94	89	NA	27
13	2-Pentanone/benzylamine	84	83	79[i]	63[i]	80
14	Acetophenone/aniline	70	98	NA	NA	NA
15	Acetophenone/benzylamine	62[j], 76[k]	70	72	87	10, 69[c]
16	Acetophenone/n-propylamine	69	97	NA	NA	NA
17	2-Acetylnaphthone/benzylamine	66[j], 70[k]	NA	NA	NA	NA

[a] GC yield.
[b] Amine isolated as HCl salt.
[c] Reaction of petroleum ether with acetic acid [17a].
[d] Amine isolated as dialkylammonium acetate.
[e] No reductive amination product, benzylalcohol generated.
[f] 24 h reaction time with 3 equiv. amine.
[g] Di-n-butylamine.
[h] Ratio of ketone, amine, PEMB was 1 : 1 : 0.39.
[i] Similar ketone.
[j] Percent conversion in 1 h 50 °C.
[k] Percent conversion in 6 h 50 °C with 1 : 1 : 0.5 ratio, PEMB consumed.

To increase the economic advantage of this process, the reagent ratio was aligned to use two of the three hydrides available in PEMB, thus a 2:2:1 ratio for the carbonyl, amine, and PEMB was successfully used in all examples. In fact, in many aldehyde aminations, 20–30% of the PEMB remained after consumption of substrates. Comparative literature results for reductive amination with PICB [8] and PYB [33] are also listed where a reagent ratio of 1:1:1 had been used for the carbonyl, amine, and amine borane reagents. Since the presence of carboxylic acid is beneficial for reductive amination, 2 equiv. of acetic acid relative to the carbonyl compound was used in all reductions, except entry 4 with hexanal which easily forms an imine. Most reactions were conducted at 20–25 °C by a slow addition of PEMB to the stirred mixture of amine and carbonyl substrate under nitrogen. For the less reactive ketones (entries 8–16), reactions were conducted at 50 °C to increase the reductive amination rate.

The reductive aminations with PEMB compare quite well with the results of other aromatic amine borane complexes considering the lower amount of PEMB used, as compared to PYB and PICB examples. The unoptimized yields were higher using PEMB than PYB in some examples, notably, the neat reductive amination of benzaldehyde with di-n-butylamine (entry 3), amination of hexanal without acetic acid present (entry 4), and the reductive amination of 2-pentanone with weakly basic aniline (entry 12).

PEMB was very selective toward reductive amination over carbonyl reduction. Competitive reduction of the ketone or aldehyde was not observed in any of the reactions except for entry 3 where the reduction of benzaldehyde preferably occurs in methanol and benzyl alcohol was obtained as the sole product. However, for this substrate combination, solvent-less reduction with PEMB gave a high yield (96%) of desired tertiary amine without competitive benzaldehyde reduction. Using PICB, Kikugawa *et al.* observed only a 48% yield of trialkylamine in 24 h while using an excess of di-n-propylamine without solvent and a 3% yield of trialkylamine and 77% benzyl alcohol with water as the solvent.

The reductive amination of aldehydes using 0.6 equiv. (relative to aldehyde) PEMB occurred quite rapidly. For example, amination occurred in 2 h or less for benzaldehyde. For hexanal, the reductive amination without acetic acid was complete within 5 min after the PEMB addition. Similar fast reactions times have been observed with PICB and PYB in solvent. The cyclohexanone aminations were typically complete at ambient temperature within 1 h after the PEMB addition. For the reductive aminations without solvent, PEMB performed considerably faster than PICB. Kikugawa *et al.* [28] reported reaction times of 3.5 and 15.5 h for entries 5 and 6 respectively, with PICB. Presumably, the liquid state of PEMB gives it a strong kinetic advantage over solid PICB in solvent-less reductive aminations.

Because acyclic and aromatic ketones, especially acetophenone, only slowly undergo reductive amination, the reactions in Entries 8–16 were conducted at 50 °C to increase reaction rate. In contrast to PICB amination of acetophenone with benzyl amine, which took 72 h at ambient temperature (both neat or in methanol), PEMB amination of acetophenone by benzylamine at 50 °C was complete in 4 h (neat), and 62% complete in 1 h in methanol. Advantages of reductive amination with neat PEMB are rapid reactions and high reactor utilization as well as decreased amount of waste (solvent).

Maximizing the utilization of complexed borane in PEMB was studied in the reductive amination because ^{11}B NMR spectroscopy indicated only borate (18 ppm) and unreacted PEMB (−12 ppm) after consumption of aldehyde or cyclohexanone. A reduction using 16% over stoichiometric (1.16 hydride relative to cyclohexanone) proceeded to 98% completion by GC analysis (Table 6.1, entry 7). Owing to slow methanolysis during the course of the reaction, a slight excess of borane reagent is needed (6% PEMB methanolysis in 4 h at 30 °C). Methanolysis occurs more rapidly at 50 °C, therefore 0.6 equiv. of PEMB was insufficient to complete some ketone aminations.

6.3
Comparison to Sodium Triacetoxyborohydride (STAB)

STAB has been demonstrated by Abdel-Magid et al. and others for direct reductive amination most often in 1,2-dichloromethane or tetrahydrofuran (THF). Typically, STAB is used in excess (from 1.4 to 4 equiv.) and the carbonyl compound is the limiting reagent with a slight excess of amine (5–10%). STAB is a hygroscopic solid with low solubility in most solvents and cannot be used in alcohol or aqueous solvent. An advantage of amine boranes is the ability to use protic solvents for reductive aminations. Table 6.2 lists the literature yields for reductive amination with STAB for comparison to our results using PEMB. Aldehyde aminations

Table 6.2 Reductive amination of aldehydes and ketones; comparison of PEMB to sodium triacetoxyborohydride.

Entry	Aldehyde or ketone/amine pairs	Yield[a] with PEMB in CH$_3$OH	Yield[a] with PEMB neat	Yield with STAB [31][a]
1	Benzaldehyde/aniline	99	99, 81[b]	88–95
2	Benzaldehyde/di-n-propylamine	0[c]	96	64[d]
3	Cyclohexanone/aniline	92	93, 85[b]	71
4	Cyclohexanone/benzylamine	83[e]	70[e]	88
5	2-Pentanone/aniline	74	94	90
6	2-Pentanone/benzylamine	84	83	81[d]
7	Acetophenone/benzylamine	62[f]	70	55[g]

[a] GC yield.
[b] Amine isolated as HCl salt.
[c] No reductive amination product, benzylalcohol generated.
[d] BASF result.
[e] Amine isolated as ammonium acetate salt.
[f] Conversion in 1 h at 50 °C.
[g] 240 h reaction time.

with STAB typically take 1–24 h at ambient temperature. However, ketone amination with STAB typically requires at least 24 h at ambient temperatures and for acetophenone, the reaction required 10 days at ambient temperature. One limitation of STAB is that reaction mixtures should not be heated above 50 °C due to STAB self-reduction and decomposition.

The results using PEMB are comparable to those reported for STAB and superior for several entries. Reductive amination of ketones with PEMB at 50 °C is considerably faster than that with STAB. For example, reductive amination of 2-pentanone with benzyl amine using STAB took 24 h at ambient temperature to reach 81% completion versus 84% in 2 h at 50 °C for the same reduction with PEMB (Table 6.2, entry 6). Reductive amination of acetophenone with benzyl amine using PEMB was also quicker, 62% conversion in 1 h at 50 °C in methanol, and 70% conversion in 4 h at 50 °C for the reaction without solvent as compared to 55% conversion with STAB after 240 h at ambient temperature.

6.4
Primary Amine Synthesis

The synthesis of primary amines from ammonia or ammonium salts is fraught with difficulties from poor functional group tolerance using transition metal catalysts to over-alkylation of the amine giving secondary-amine products along with the primary amine. For example, attempted primary amination of a glucose-derived aldehyde with sodium cyanoborohydride resulted in dialkylamine formation [34]. Reductive amination with ammonium trifluoroacetate or ammonium acetate with STAB as the reducing agent in THF produced 80% dialkylated amine in 48 h [35]. Another method for primary amination used hydroxylamine as an amine source and LAH as a reductant [36]. Scheme 6.6 shows primary amination with ammonium hydroxide using hydrogenation over Pd/C. This method was successful in the synthesis of intermediate **17** toward preparation of MN-447, thus decreasing the route by three steps [37]. The reaction conditions required 10% Pd/C, 8 equiv. of ammonium hydroxide and atmospheric hydrogen pressure in dioxane over 22 h. While this palladium-catalyzed reductive amination was successful, the long reaction time and low concentration (0.11 M) detract from the advantage of a shorter route. STAB, NaCNBH$_3$, and NaBH$_4$ were not successful in this amination, but amine boranes were not tested.

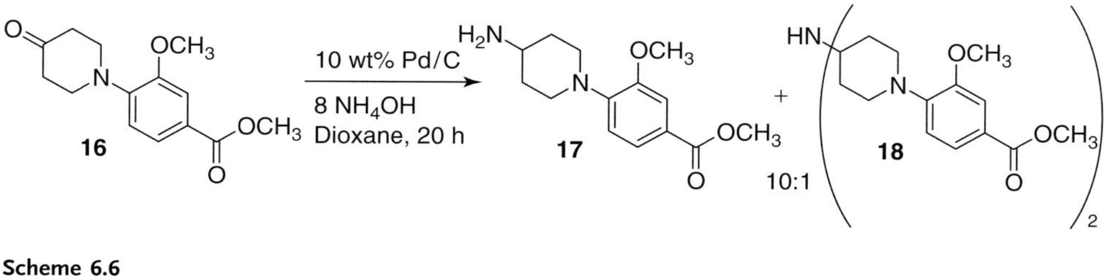

Scheme 6.6

Table 6.3 Reductive amination using ammonium acetate.

Entry	Ketone[a]	20 (%)	21 (%)	% Conversion at 2 h rt
1	α-Tetralone	94	6	12
2	Acetophenone	93	7	30
3	2-Methylcyclohexanone[b]	94	6	90[b]
4	3-Methyl-2-butanone	93	7	100
5	Cyclohexanone	85	14	100
6	2-Butanone	52	48	100
7	Cyclohexanone[c]	49	50	100

[a] Reagent ratio of 2:10:1 for ketone, NH$_4$OAc, PEMB.
[b] Reduction was conducted at 50 °C.
[c] Reagent ratio of 2:2:1 for ketone, NH$_4$OAc, PEMB.

Ammonium acetate is insoluble in THF, 1,2-dichloroethane or acetonitrile, but soluble in methanol, so direct reductive amination using amine boranes may provide the desired primary amine, prompting our investigation with PEMB. Using a 2:2:1 ratio of cyclohexanone, ammonium acetate, and PEMB in methanol with acetic acid, a rapid reduction was observed, but unfortunately the product mixture was 49% cyclohexylamine and 50% dicyclohexylamine (relative product ratio); see Table 6.3, entry 7. Better results were obtained using a 2:10:1 ratio of ketone, ammonium acetate, and PEMB in methanol with acetic acid at room temperature (Table 6.3). Two of the three hydrides on PEMB are utilized in this excellent synthesis of moderate to high yields of primary amine. Dialkylamine products were observed, although the amount of dialkylamine was lower for sterically hindered ketones. The rate of reduction was slower with aromatic ketones (Table 6.3), showing low conversion at 2-h post-PEMB addition. 3-Methyl-2-butanone, cyclohexanone, and 2-butanone were fully aminated in about 2 h at room temperature. In comparison, 99% conversion of acetophenone was observed after 24 h (Scheme 6.8).

The amination of 2-methylcyclohexanone was conducted at 50 °C. The diastereomer ratio for 2-methylcyclohexylbenzylamine was 1.3:1.0 favoring the *cis*-diastereomer. Reductive amination of 2-methylcyclohexanone by Ganem with 9-BBN-NaCN delivered a ratio of 2.3:1.0 [38]. The diastereomer ratio for the dialkylamine derived from acetophenone was 1.4:1.0 favoring the ((R),(R);(S),(S)) enantiomer pair. A reductive amination of acetophenone using ammonium acetate and stoichiometric amount (relative to acetophenone) of PEMB in methanol/acetic

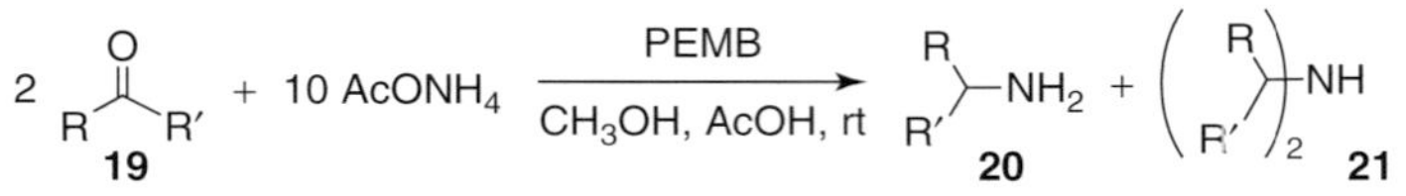

Scheme 6.7

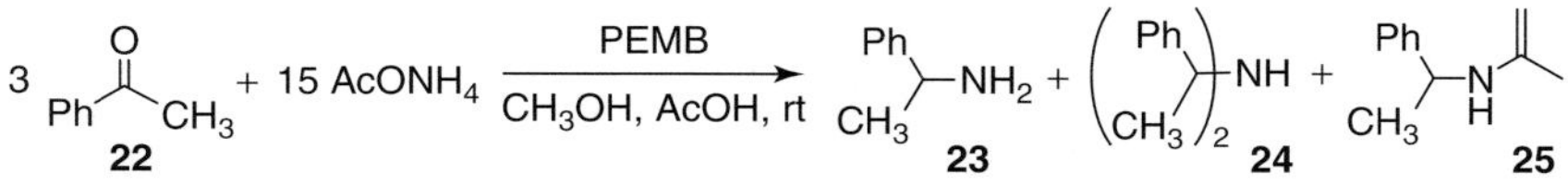

[Scheme 6.8 structures: 3 Ph-C(O)-CH₃ (22) + 15 AcONH₄ → (PEMB, CH₃OH, AcOH, rt) → Ph-CH(CH₃)-NH₂ (23) + (Ph-CH(CH₃))₂NH (24) + Ph-C(CH₃)=N-H acetyl (25)]

Scheme 6.8 Amination with stoichiometric PEMB.

acid for 4 h resulted in 47% conversion to monoalkylamine **23** and dialkylamine **24** in a ratio of 86 : 14. The dialkylamine was a mixture of ((R),(R);(S),(S)) and ((R),(S);(S),(R)) in a diastereomeric ratio of 2.2 : 1.0. The presence of enamine intermediate **25** enroute to dialkylamine **24** was observed at 7% (Scheme 6.8).

The rate for reductive amination of tetralone was low at ambient temperature, to 18% conversion in 5 h, and continuing to 66% conversion in 66 h (left stirring over weekend). In addition to the primary and secondary amines, the product mixture also contained a small amount of N-tetrahydronaphthyl-N-dihydronaphthylamine (6% enamine product).

Further optimization using PEMB shows promise for direct conversion of a ketone to primary amine synthesis by using easy-to-handle ammonium acetate.

6.5
Stereoselective Reductive Amination

Hutchins studied stereoselective reduction of substituted cyclohexyl imine systems with borohydride reducing agents and TBAB [39]. However, examples of amine boranes in stereoselective reductive aminations are scarce. Recently, Cai and coworkers used TBAB in a stereoselective reduction of oxime **26** to synthesize allergy and asthmatic drug candidate, S-5751 (Scheme 6.9) [38]. Oximes are usually formed in aqueous media; therefore, amine borane complexes were ideal for this reduction because of their stability in aqueous acidic solution up to pH 5. Aqueous $TiCl_3$ was buffered with sodium acetate and oxime **26** was added to this solution at 0 °C followed by reduction with TBAB (Scheme 6.9). The desired amino ester product **27** was isolated by crystallization in 85% yield with selectivities higher than >98% de.

We briefly explored the diastereoselectivity for the amination of 2-methylcyclohexanone with benzyl amine with various amine boranes. Using PEMB, the diastereomer ratio for the product, 2-methylcyclohexylbenzylamine, was 3.2 : 1 favoring the *cis*-diastereomer regardless of reaction temperature (comparison of 35 °C versus 50 °C). This reductive amination with STAB delivered a ratio of 6 : 1 in our lab as compared to 8 : 1 found by McGill [40]. Using a preformed imine, reduction by TBAB gave a 9 : 1 ratio of *cis*- to *trans*-diastereomers [41].

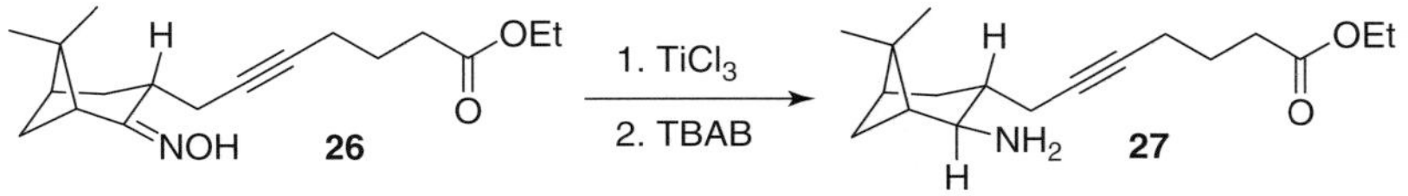

[Scheme 6.9: oxime **26** with NOH group → (1. TiCl₃, 2. TBAB) → amine **27** with NH₂]

Scheme 6.9

6.6
Reaction Solvents

Amine boranes are miscible in most organic solvents, thus, reductive amination can be conducted in any suitable solvent for the substrates and product, including polar protic solvents. Because of the strong interaction of the nitrogen in pyridine compounds with Lewis acidic borane, amine boranes such as PEMB are relatively stable toward hydrolysis (5% per day in water/THF at rt) and methanolysis (6% per day in methanol at 22 °C, 27% in 24 h at 30 °C in methanol/acetic acid, 10:1). DiMare et al. [31] observed that reductive aminations with PYB were significantly faster in methanol than ethers or chlorinated hydrocarbon solvents mainly due to facile imine and imminium ion formation in methanol. Another advantage of methanol as the solvent is that methyl borate can be distilled from the reaction mixture as the methyl borate/methanol azeotrope (bp 53 °C).

Reductive amination in water can be challenging because of the reversible nature of imine formation, especially with ketones and amines possessing high water solubility [42]. In any event, reductive amination in water was successfully demonstrated by Kikugawa et al. using PICB [28], obtaining slightly lower yields of amines from the aqueous system compared to reactions in methanol or neat. We conducted one reductive amination of benzaldehyde and benzylamine with PEMB in water with acetic acid. The isolated yield of the dibenzylammonium acetate salt was 69%. The lower yield compared to the amination without solvent (Table 6.1, entry 2) can be attributed to loss of product to the aqueous solvent. PEMB is immiscible in water causing the reaction mixture to consist of the substrates and amine borane in a concentrated hydrophobic organic layer above the aqueous layer. The aqueous layer may facilitate heat transfer during exothermic reactions of the neat organic components and be beneficial in that respect.

6.7
Reaction Workup

Residual or excess amine borane in the reaction mixture should be quenched before isolation of the desired amine product. Several methods are available for this transformation (e.g., Pd-catalyzed methanolysis [43] or refluxing with acid). Refluxing the reaction mixture with excess acetic acid or HCl effectively deactivates residual amine borane. For example, PEMB is completely quenched in methanol (0.23 M solution) with 10 v/v% acetic acid is completely quenched after 7 h at reflux (Figure 6.1).

In reductive amination via borane amines, isolation of the product amine can be complicated by the presence of the original borane carrier amine. If the carrier amine has a significantly different boiling point from the product amine, the reduction product can be isolated by distillation. For example, *tert*-butylamine from TBAB and triethylamine from TEAB have relatively low boiling points.

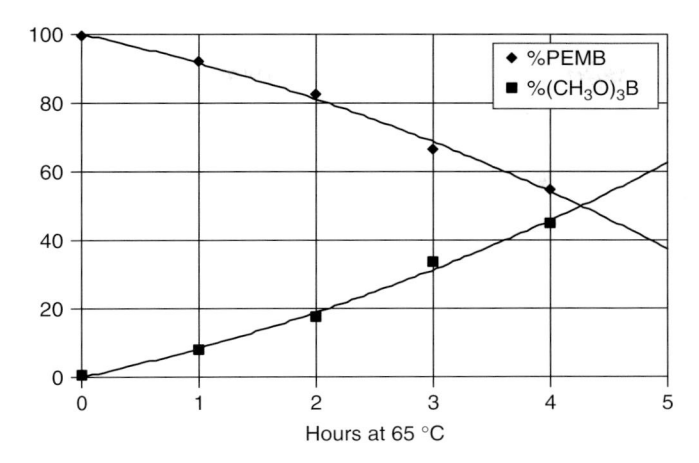

Figure 6.1 Methanolysis of 5-ethyl-2-methylpyridine in acetic acid/methanol measured by ^{11}B NMR spectroscopy.

Separation via acidification is also very attractive when using aromatic amine boranes because protonated pyridine bases have a pK_a between 5 and 7 while dialkylamines and alkylamines have a basicity similar to trialkylamines with a pK_a for the protonated form around 10. For a simulated separations example, a mixture of dicyclohexylamine and 5-ethyl-2-methylpyridine (8 and 4 mmol respectively) was acidified with HCl in water (enough HCl to protonate only the dicyclohexylamine). The solid dicyclohexylammonium hydrochloride was not water soluble, so it was filtered and rinsed with diethyl ether (86% yield of recovered dicyclohexylammonium hydrochloride, 100% pure by ^{1}H NMR spectroscopy). In this unoptimized example, dicyclohexylammonium salt from a PEMB-mediated reductive amination was isolated in 62% yield. This demonstrates the feasibility of amine separation via pK_a difference.

The carboxylic acid used to assist imine formation is also valuable during amine separation from the reaction mixture. While some dialkylamine carboxylates are solids, a number of PEMH$^+$ carboxylate$^-$ pairs are ionic liquids with a viscosity lower than PEMB (3.2 cst at 20 °C for PEMBH$^+$acetate$^-$ versus 7.0 cst for PEMB). In addition, 5-ethyl-2-methylpyridinium benzoate, acetate, propionate, 2-methylbutyrate, and pivalate are ionic liquids, completely miscible in hydrocarbons such as hexanes, ethers, and methanol. To test amine separation using this concept, a mixture similar to a reductive amination reaction solution was prepared with 5-ethyl-2-methylpyridine (0.3 g, 2.5 mmol), dicyclohexylamine (0.9 g, 5 mmol), and benzoic acid (0.9 g, 7.5 mmol) in 15 ml of methanol (Scheme 6.9). The slurry was filtered to get 1.01 g of Cy$_2$NH$_2$-benzoate salt (67% isolated yield of first crop, >99% pure by ^{1}H NMR). The methanol was stripped from the filtrate to give a gummy solid of the mixed ammonium salts. Rinsing the gummy salts with (2 × 5 ml) hexanes removed the PEMH-benzoate from the Cy$_2$NH$_2$-benzoate

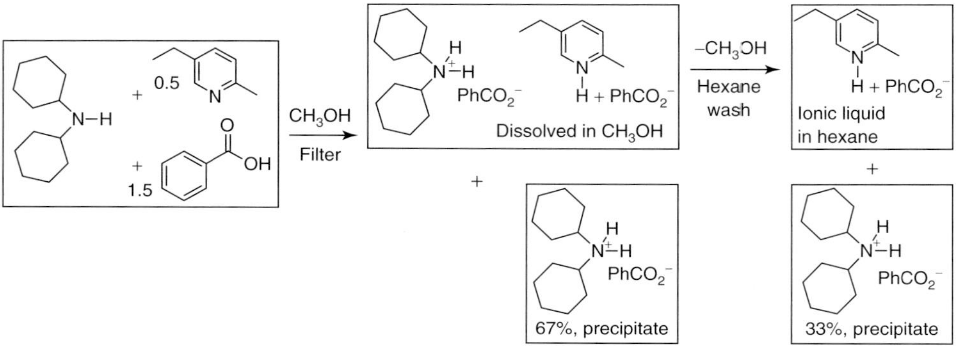

Scheme 6.10 Separation of dicyclohexylamine.

(0.54 g, 98% pure, 100% recovery crops 1 and 2). PEMH-benzoate (0.5 g, 83%) was recovered by stripping the hexanes (Scheme 6.11).

Furthermore, isolation of desired amine product from reductive amination mixtures was successful as demonstrated by the following example. Benzylamine, benzaldehyde, glacial acetic acid were combined in a flask under N_2 and stirred at room temperature followed by PEMB addition. After completion of the reaction, it was refluxed with methanol and concentrated HCl to quench the excess borane. The mixture was concentrated under vacuum. The resulting white needles were filtered, washed with chilled heptanes, and dried to obtain pure dibenzylammonium acetate in 87% yield, demonstrating the efficiency of this protocol in isolating the desired amine product.

A more difficult separation of 5-ethyl-2-methyl-pyridine from substituted aniline reductive amination products was expected due to the similar basicity of the two amines.

For demonstration purposes, a reductive amination was conducted with cyclohexanone and aniline in methanol using 6 M HCl instead of acetic acid (Scheme 6.11). The reaction was more exothermic (than when using acetic acid) and reached 100% conversion in 1 h. Stripping the methanol gave isolated amine salts as a mixture of cyclohexylaniline–HCl and PEMHCl (10 g, 98% yield corrected for PEMHCl by-product). Owing to the high solubility of PEMHCl in alcohol solvent, washing of the cyclohexylaniline–HCl/PEMHCl mixture with isopropanol successfully accomplished a separation. Half of the isolated salts (5 g) were slurried in 10 ml of isopropanol to dissolve PEMHCl, and filtered to leave white cyclohexylaniline–HCl salt (3.41 g, 85% yield, 99% pure by ^{1}H NMR).

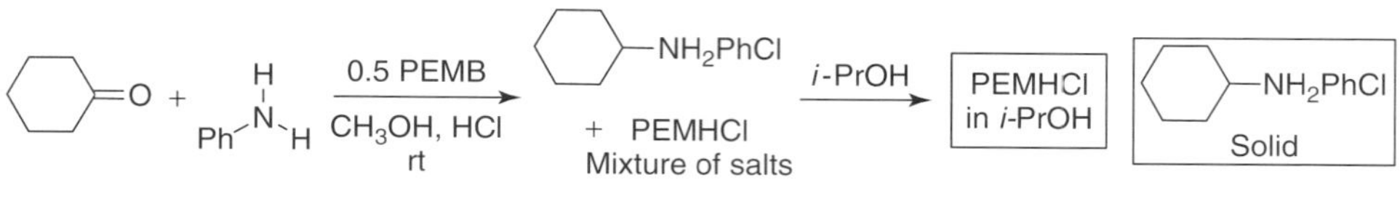

Scheme 6.11 Separation of cyclohexylaniline.

In summary, the desired amine can be separated from 5-ethyl-2-methylpyridine by selective choice of amine salts. The miscibility of several PEMHX ionic liquids in organic solvents including hexanes or heptanes was observed. This property contrasts with the low solubility of dialkylammonium salts. Therefore, the desired amine salt can be washed with hydrocarbon solvent to remove PEMH-carboxylate. In several model examples and reductive amination mixtures, solubility differences allowed washing of the desired amine salt with solvent to successfully separate the desired product from 5-ethyl-2-methylpyridine.

6.8
Conclusion

Amine borane reducing agents are very effective for reductive aminations of ketones and aldehydes. Especially with aromatic amine boranes, two or even three of the hydrides are effectively utilized maximizing the atom economy of the reagent. Both PICB and PEMB are viable replacements for PYB in reductive aminations.

Reaction workup for the isolation of aliphatic secondary or tertiary amines can be accomplished via distillation, acid/base workup, or crystallization of the ammonium carboxylate salt and rinsing of the solids with hexanes to remove by-products, especially when using PEMB for the reduction. The high solubility of 5-ethyl-2-methylpyridinium hydrochloride in alcohols is advantageous for purification of ammonium hydrochloride salts. Furthermore, amine boranes enable a new method of monoamination reaction of ketones with ammonium acetate as an inexpensive, easy-to-handle nitrogen source.

References

1. (a) Rappoport, Z. (ed.) (1994) *The chemistry of enamines*, Patai Series: The Chemistry of Functional Groups (ed. Z. Rappoport), John Wiley and Sons, Inc., New York; (b) Hutchins, R.O., Hutchins, M.K. (1991) in *Comprehensive Organic Synthesis*, vol. 8 (ed. B.N. Trost), Pergamon, Oxford, p. 25; (c) Rubio-Pérez, L., Pérez-Flores, F.J., Sharma, P., Velasco, L., and Cabrera, A. (2009) *Org. Lett.*, **11**, 265.
2. Baxter, E.W. and Reitz, A.B. (2002) *Organic Reactions*, vol. 59, John Wiley & Sons, Inc., New York, p. 1.
3. (a) Bomann, M.D., Guch, I.C., and DiMare, M. (1995) *J. Org. Chem.*, **60**, 5995; (b) Peterson, M.A. (2002) *Synthetic Commun.*, **32**, 443.
4. Whitesell, J.K. (1991) in *Comprehensive Organic Synthesis*, vol. 6 (ed. E. Winterfeldt), Pergamon Press, Oxford, p. 705.
5. (a) Borch, R.F. and Hassid, A.I. (1972) *J. Org. Chem.*, **37**, 1673; (b) Borch, R.F., Bernstein, M.D., and Durst, H.D. (1971) *J. Am. Chem. Soc.*, **93**, 2897.
6. (a) Abdel-Magid, A.F., Carson, K.G., Harris, B.D., Maryanoff, C.A., and Shah, R.D. (2006) *Org. Proc. Res. Dev.*, **10**, 971; (b) Gribble, G.W. (2006) *Org. Proc. Res. Dev.*, **10**, 1062; (c) Abdel-Magid, A.F. and Maryanoff, C.A. (1996) in *Reductions in Organic Synthesis: Recent Advances and Practical Applications* (ed. A.F. Abdel-Magid), ACS Simposium Series, 641, American Chemical Society, Washington, DC, p. 201.
7. (a) Lane, C.F. (1973) *Aldrichim. Acta*, **6**, 51; (b) Hutchins, R.O., Learn, K.,

Nazer, B., Pytlewski, D., and Pelter, A. (1984) *Org. Prep. Proc. Int.*, **16**, 335.

8. (a) de Vries, J.G. and Elsevier, D.J. (2007) *The Handbook of Homogeneous Hydrogenation*, Wiley-VCH Verlag GmbH, Weinheim; (b) Nishimura, S. (2001) *Handbook of Heterogeneous Catalytic Hydrogenation for Organic Synthesis*, John Wiley & Sons, Inc., New York; (c) Byun, E., Hong, B., De Castro, K.A., Lim, M., and Rhee, H. (2007) *J. Org. Chem.*, **72**, 9815; (d) Tararov, V.I. and Boerner, A. (2005) *Synlett*, **2**, 203; (d) Robichaud, A. and Ajjou, A.N. (2006) *Tetrahedron Lett.*, **47**, 3633.
9. Rablen, P.R. (1997) *J. Am. Chem. Soc.*, **119**, 8350.
10. Matos, K., Pichlmair, S., and Burkhardt, E.R. (2007) *Chim. Oggi/Chem. Today*, **25**, 17.
11. Brown, H.C., Kanth, J.V.B., Dalvi, P.V., and Zaidlewicz, M. (1999) *J. Org. Chem.*, **64**, 6263.
12. (a) Hutchins, R.O., Su, W.Y., Sivakumar, R., Cistone, F., and Stercho, Y.P. (1983) *J. Org. Chem.*, **48**, 3412; (b) Singh, S.K., Dev, I.K., Duch, D.S., Ferone, R., Smith, G.K., Freisheim, J.H., and Hynes, J.B. (1991) *J. Med. Chem.*, **34**, 606; (c) Singh, S.K., Singer, S.C., Ferone, R., Waters, K.A., Mullin, R.J., and Hynes, J.B. (1992) *J. Med. Chem.*, **35**, 2002; (d) Gala, D., Dahanukar, V.H., Eckert, J.M., Lucas, B.S., Schumacher, D.P., and Zavialov, I.A. (2004) *Org. Proc. Res. Dev.*, **8**, 754.
13. Billman, J.H. and McDowell, J.W. (1961) *J. Org. Chem.*, **26**, 1437.
14. Connolly, T.J., Constantinescu, A., Lane, T.S., Matchett, M., McGarry, P., and Paperna, M. (2005) *Org. Process Res. Dev.*, **9**, 837.
15. Berglund, R.A., Lockwood, N.A., Magadanz, H.E., and Zheng, H. (1999) US Patent 5,952,466.
16. (a) Taylor, E.C., Palmer, D.C., George, T.J., Fletcher, S.R., Tseng, C.P., Harrington, P.J., Beardsley, G.P., Dumas, D.J., Rosowsky, A., and Wick, M. (1983) *J. Org. Chem.*, **48**, 4852; (b) Taylor, E.C., Yoon, C., and Hamby, J.M. (1994) *J. Org. Chem.*, **59**, 7092; (c) Gangjee, A., Devraj, R., and Queener, S.F. (1997) *J. Med. Chem.*, **40**, 470; (d) Hirokawa, Y., Horikawa, T., Noguchi, H., Yamamoto, K., and Kato, S. (2002) *Org. Process Res. Dev.*, **6**, 28; (e) Gangjee, A., Wang, Y., Queener, S.F., and Kisliuk, R.L. (2006) *J. Heterocycl. Chem.*, **43**, 1523.
17. Brands, K.M.J., Krska, S.W., Rosner, T., Conrad, K.M., Corley, E.G., Kaba, M., Larsen, R.D., Reamer, R.A., Sun, Y., and Tsay, F.-R. (2006) *Org. Process Res. Dev.*, **10**, 109.
18. Camp, D., Matthews, C.F., Neville, S.T., Rouns, M., Scott, R.W., and Truong, Y. (2006) *Org. Process Res. Dev.*, **10**, 814.
19. (a) Ryschkewitsch, G.E. and Birnbaum, E.R. (1959) *Inorg. Chem.*, **4**, 575; (b) Baldwin, R. and Washburn, R.M. (1961) *J. Org. Chem.*, **26**, 3549; (c) Brown, H.C. and Domash, L. (1956) *J. Am. Chem. Soc.*, **78**, 5384.
20. (a) Pelter, A. and Rosser, R.M. (1984) *J. Chem. Soc., Perkin Trans. I*, 717; (b) Moormann, A.E. (1993) *Synth. Commun.*, **23**, 789.
21. (a) Parker, J.S., Bowden, S.A., Firkin, C.R., Moseley, J.D., Murray, P.M., Welham, M.J., Wisedale, R., Young, M.J., and Moss, W.O. (2003) *Org. Proc. Res. Dev.*, **7**, 67; (b) Moseley, J.D., Moss, W.O., and Welham, M.J. (2001) *Org. Proc. Res. Dev.*, **5**, 491; (c) Bernstein, P.R., Dedinas, R.F., Russell, K., and Shenvi, A.B. (1999) WO Patent 0002859; (d) Hishitanki, Yasuhiro-Shionogi & Co., Ltd.; Itani, Hikaru-Shionogi & Co., Ltd.; Irie, Tadashi-Shionogi & Co., Ltd (1999) WO Patent 0032606, EP 1134222; (e) Smyser, T.E. and Confalone, P.N. (1996) DuPont Merck Pharmaceutical Co. US Patent 5,532,356. July 2; (f) Bergland, R.A., Lockwood, N.A., Magadanz, H.E., and Zheng, H. (1999) Hua, Eli Lilly and Co. US Patent 5,998,581, December 7.
22. Abbas, S., Ferris, L., Norton, A.K., Powell, L., Robinson, G.E., Siedlecki, P., Southworth, R.J., Stark, A., and Williams, E.G. (2008) *Org. Proc. Res. Dev.*, **12**, 202.
23. Parker, J.S., Bowden, S.A., Firkin, C.R., Moseley, J.D., Murray, P.M., Welham, M.J., Wisedale, R., Young, M.J., and Moss, W.O. (2003) *Org. Process Res. Dev.*, **7**, 67–67.

24. Cabacungan, J.C., Ahmed, A.I., and Feeney, R.E. (1982) *Analytical Biochemistry*, **124**, 272–272.
25. Molander, G.A. and Cooper, D.J. (2008) *J. Org. Chem.*, **73**, 3885.
26. Baigude, H., Katsuraya, K., Okuyama, K., Tokunaga, S., and Uryu, T. (2003) *Macromolecules*, **36**, 7100.
27. Stults, N.L., Asta, L.M., and Lee, Y.C. (1989) *Anal. Biochem.*, **180**, 114.
28. (a) Sato, S., Sakamoto, T., Miyazawa, E., and Kikugawa, Y. (2004) *Tetrahedron*, **60**, 7899–7906; (b) Kikugawa, Y. (2004) JP 2004256511, and (2005) JP 2005170917, June 30.
29. (a) Tanaka, K. (2003) *Solvent-free Organic Synthesis*, Wiley-VCH, Verlag GmbH, Weinheim, Germany; (b) Tanaka, K. and Toda, F. (2000) *Chem. Rev.*, **100**, 1025; (c) Metzger, J.O. (1998) *Angew. Chem., Int. Ed.*, **37**, 2975; (d) Walsh, P.J., Li, H., and de Parrodi, C.A. (2007) *Chem. Rev.*, **107**, 2503.
30. Kawase, Y., Yamagishi, T., Kutsuma, T., Ueda, K., Iwakuma, T., Nakata, T., and Yokomatsu, T. (2009) *Heterocycles*, **78**, 463.
31. Uchiyama, S., Inaba, Y., Matsumoto, M., and Suzuki, G. (2009) *Anal. Chem.*, **81**, 485.
32. Burkhardt, E.R. and Coleridge, B.M. (2008) *Tetrahedron Letters*, **49**, 5152.
33. Bomann, M.D., Guch, I.C., and DiMare, M. (1995) *J. Org. Chem.*, **60**, 5995.
34. Dunlap, N.K., Drake, J., Ward, A., Salyard, T.L.J., and Martin, L. (2008) *J. Org. Chem.*, **73**, 2928.
35. (a) Abdel-Magid, A.F., Carson, K.G., Harris, B.D., Maryanoff, C.A., and Shah, R.D. (1996) *J. Org. Chem.*, **61**, 3849; (b) Abdel-Magid, A.F. and Mehrman, S.J. (2006) *Org. Proc. Res. Dev.*, **10**, 971.
36. Yao, B., Ji, H., Cao, Y., Zhou, Y., Zhu, J., Lü, J., Li, Y., Chen, J., Zheng, C., Jiang, Y., Liang, R., and Tang, H. (2007) *J. Med. Chem.*, **50**, 5293.
37. Ishikawa, M., Tsushima, M., Kubota, D., Yanagisawa, Y., Hiraiwa, Y., Kojima, Y., Ajito, K., and Anzai, N. (2008) *Org. Proc. Res. Dev.*, **12**, 596.
38. Cai, D., Larsen, R., Journet, M., and Campos, K. (2002) Merck & Co., Inc., Assignee WO Patent 02 32892.
39. Hutchins, R.O., Su, W.Y., Sivakumar, R., Cistone, F., and Stercho, Y.P. (1983) *J. Org. Chem.*, **48**, 3412.
40. McGill, J.M., LaBell, E.S., and Williams, M. (1996) *Tetrahedron Lett.*, **37**, 3977.
41. Wrobel, J.E. and Ganem, B. (1981) *Tetrahedron Lett.*, **22**, 3447.
42. Hailes, H. (2007) *Org. Proc. Res. Dev.*, **11**, 114.
43. (a) Couturier, M., Tucker, J.L., Andresen, B.M., Dubé, P., and Negri, J.T. (2001) *Org. Lett.*, **3**, 465; (b) Couturier, M., Andresen, B.M., Jorgensen, J.B., Tucker, J.L., Busch, F.R., Brenek, S.J., Dubé, P., am Ende, D.J., and Negri, J.T. (2002) *Org. Proc. Res. Dev.*, **6**, 42.

7
Industrial Synthesis of Perfluorinated Building Blocks by Liquid-Phase Direct Fluorination

Takashi Okazoe

7.1
Introduction

Fluorine is sterically the smallest element next to hydrogen. Therefore, a living body cannot discriminate between fluorinated and nonfluorinated molecules. On the other hand, a carbon–fluorine bond is the strongest found in organic chemistry but with opposite polarization to C–H bonds because fluorine is the most electronegative element in the periodic table. Accordingly, the metabolism of the living body is disturbed at a certain stage of the metabolic sequences. This makes fluorine substitution attractive for the development of pharmaceuticals [1, 2]. However, there are not so many examples of drugs with a polyfluorinated structure except for those containing trifluoromethyl group, probably because the availability of polyfluorinated building blocks is quite restricted.

For the building block or synthon approach, the proper selection of polyfunctional fluorine-containing synthons is a key feature of synthetic schemes [3]. However, the reactivity of substrates and reagents drastically changes when fluorine atoms are introduced into the reacting sites [4]. This tendency becomes more obvious when the number of fluorine atoms increases.

A striking example is the reactivity of alkyl halides. Although S_N1 and S_N2 mechanisms operate when a few fluorine atoms are incorporated in relatively remote positions of aliphatic chains, perfluoroalkyl halides are usually resistant to these classical processes. This makes it difficult to connect perfluorinated building blocks through carbon–carbon bond formation with perfluorinated substrates.

There are only a few C–C bond forming reactions, which are applied in manufacturing perfluorinated compounds. One is telomerization of perfluorinated olefins [5, 6]. The resulting raw material is used for water and oil repellent agents. Hereby, only telomers with even carbon numbers are obtained; telomers with a specific value of *n* are rarely accessible in a selective manner (Scheme 7.1).

For the construction of compounds in a selective manner, carbon–oxygen bond formation is more important than carbon–carbon bond formation in industrial organofluorine chemistry. For example, perfluoro alkoxides can add to perfluoro

---

Pharmaceutical Process Chemistry. Edited by Takayuki Shioiri, Kunisuke Izawa, and Toshiro Konoike
Copyright © 2011 WILEY-VCH Verlag GmbH & Co. KGaA, Weinheim
ISBN: 978-3-527-32650-1

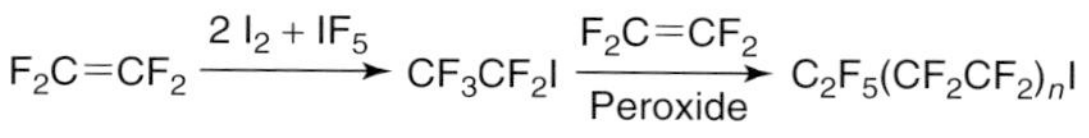

Scheme 7.1 Telomerization.

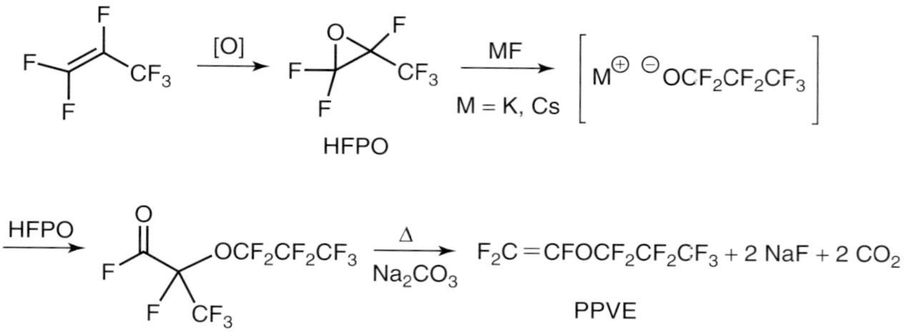

$F_2C=CF_2 \xrightarrow{2\,I_2 + IF_5} CF_3CF_2I \xrightarrow[\text{Peroxide}]{F_2C=CF_2} C_2F_5(CF_2CF_2)_nI$

Scheme 7.2 Synthesis of PPVE via HFPO.

epoxides. This reaction is applied to the synthesis of perfluorinated vinyl ethers as shown in Scheme 7.2. The alkoxides, usually prepared from perfluoroepoxides or perfluoroacyl fluorides, and a fluoride anion, react with hexafluoropropylene oxide (HFPO) to give C–O bond formation products [7]. The reaction may look curious to traditional organic chemists, because the perfluorinated alkoxides are not prepared from the corresponding alcohols and a base but from epoxides or acyl fluorides and a fluoride anion. Here, acyl fluorides are equivalent to the alkoxides. Moreover, the nucleophilic attack occurs at the sterically hindered side of the molecule due to the presence of a CF_3 substituent. In any event, this reaction is used on an industrial scale. However, this approach suffers from the problem that, at the moment, synthesis of starting acyl fluorides with complicated structures is difficult.

Therefore, an entirely new synthetic methodology for multifunctional fluorinated chemicals is required. Direct fluorination of substrates that are accessible by conventional organic synthesis has been considered a promising direction, because many C–C bond forming transformations are available for common substrates.

Reactions using CoF_3 [8, 9] or electrochemical fluorination (ECF) [10, 11] are direct fluorinations in a sense that they convert C–H bonds to C–F ones. However, they have serious drawbacks of low productivity and poor yields [12]. Thus, improvement of direct fluorination with elemental fluorine has been the target of one avenue of current synthetic research.

7.2
History of Direct Fluorination

Elemental fluorine was first prepared in small quantities by Henri Moissan in 1886 by electrolysis of anhydrous hydrogen fluoride [13]. This method is still used for the generation of fluorine by electrolysis of KF·2HF, not only in the laboratory but also on an industrial scale, particularly for the production of IF_5 for use as a raw material of telogen [5, 6], for water and oil repellent agents, and UF_6 from UF_4 in nuclear electricity generation [14].

Moissan himself was the first to carry out the reactions between neat fluorine and several organic compounds. However, only decomposed products resulted and, occasionally, explosions occurred [15]. During the early twentieth century, it was very difficult to control the reactions with fluorine because of their violent nature [16].

In order to modify the high reactivity of fluorine, the vapor-phase fluorination ("jet fluorination") apparatus for direct fluorination was developed by Bigelow in 1955 [17]. Jet fluorination seems to be suitable for the preparation of low–molecular weight fluorocarbons, and is used on a semitechnical scale for the production of perfluoropropane, which is used as a dry plasma etchant in the microelectronics industry [18].

The thermodynamics of fluorination reactions has been studied since the late 1950s [19, 20]. Reactions between hydrocarbons and fluorine are highly exothermic, because very strong C–F and H–F bonds are formed while the dissociation energy of fluorine is very low. Therefore, the propensity of fluorine to react by radical chain reactions is extremely high.

Consequently, exhaustive fluorination is generally regarded as a free radical process (Scheme 7.3) [15].

The overall reaction in which fluorine replaces hydrogen is highly exothermic (431 kJ mol^{-1}). In comparison, the central C–C bond strength in butane, for example, is 364 kJ mol^{-1}. Therefore, the overall reaction is exothermic enough to break carbon–carbon bonds. This may be the main reason for the common occurrence of explosions in direct fluorinations. However, the most important feature of this reaction sequence is that no individual step is sufficiently exothermic to break C–C bonds, with the exception of the termination step, which is exothermic

			ΔH_{25}(kJ mol^{-1})
Overall reaction	R–H + F$_2$	→ RF + HF	– 431
Initiation	F$_2$	→ 2 F$^\bullet$	+ 158
	R–H + F$_2$	→ R$^\bullet$ + HF + F$^\bullet$	+ 16
Propagation	R$^\bullet$ + F$_2$	→ RF + F$^\bullet$	– 289
	R–H + F$^\bullet$	→ R$^\bullet$ + HF	– 141
Termination	R$^\bullet$ + F$^\bullet$	→ RF	– 446
	R$^\bullet$ + R$^\bullet$	→ R–R	– 351

Scheme 7.3 Thermodynamic data for direct fluorination.

by 446 kJ mol^{-1}. Therefore, in the early stages of the fluorination, the reaction of F$_2$ rather than F radical is required. This means that a large excess of F$_2$ relative to the H to be replaced in a molecule should be maintained. Thus, the propagation step is reasonably mild if heat is removed rapidly. As the fluorination reaction proceeds, further fluorination of partly fluorinated substrates becomes increasingly difficult, because the carbon skeleton becomes increasingly protected by previously substituted fluorine atoms, whose nonbonding electron pairs inhibit further attack by incoming fluorine atoms. On the other hand, the introduction of F atoms makes C–C bonding stronger. For example, the C–C bond energy of perfluorobutane is as strong as 469 kJ mol^{-1}, higher than the 446 kJ mol^{-1} typically released in the termination step. Therefore, for complete fluorination, powerful fluorination with F radicals is required. This means that the presence of a radical generator is necessary throughout fluorination.

Another basic problem of direct fluorination involves kinetics. The rate of the reaction must be reduced so that the energy liberated from the reaction can be removed. This crucial aspect of control in direct fluorinations is usually accomplished by a dilution technique [21]. A new approach of this type to control fluorination reactions using gradient fluorination under low temperature conditions ("LaMar fluorination") was developed by Lagow and Margrave in 1970 [21, 22]. In this process, substrates are condensed at low temperature into a tube packed with copper turnings through which fluorine, initially highly diluted in either helium or nitrogen, is passed. The concentration of fluorine and the reaction temperature are slowly increased over a period of several days to permit perfluorination. This batch process requires relatively long reaction times to obtain perfluorinated material. Adcock adapted the concept to a flow version, inventing an aerosol fluorination process in the 1980s [23].

The use of a solvent inert to fluorine is preferable for reaction control. However, such an inert compound that could act as an appropriate solvent was unavailable for a long time. This problem was solved after other perfluorination methods such as ECF and indirect fluorination using CoF$_3$ were established.

In early 1990s, Scherer described a practical liquid-phase direct perfluorination method as part of a study on artificial blood substitutes [16, 24]. He disclosed that inverse addition of a substrate into an inert liquid saturated with undiluted 100% fluorine gas under UV irradiation should be employed. He pointed out that it is very important that the reactant is injected at a very slow constant rate into an inert fluorocarbon solvent saturated by fluorine, and that a large excess of F$_2$ relative to the hydrogen atoms to be replaced in a substrate should be maintained. This method is, however, only suitable for the perfluorination of partially fluorinated substrates, such as partially fluorinated ethers and amines, which are both soluble in a perfluorocarbon solvent and can withstand such vigorous reaction conditions.

At almost the same time, Exfluor Corporation claimed that various nonfluorinated ethers and esters are perfluorinated without the use of UV irradiation (Exfluor-Lagow method) [22, 25]. Their methods are based on essentially the same idea of inverse addition of substrates into an inert liquid that dissolves fluorine gas except that UV irradiation is not required. The Exfluor-Lagow method involves slow addition of both a nonfluorinated substrate and fluorine in excess into a vigorously stirred inert solvent such as a chlorofluorocarbon (CFC). Addition of a small quantity of a highly reactive hydrocarbon such as benzene, which reacts spontaneously with fluorine to produce a very high concentration of fluorine radicals that are much more reactive to the substrates, is effective for the completion of the reaction.

The Exfluor-Lagow liquid-phase elemental fluorine process gives products in high yields. However, reaction solvents are limited to now-regulated CFCs, particularly for direct fluorination of nonfluorinated substrates, which contain functional group(s) in the structure, because they have limited solubility to solvents other than regulated CFCs.

Partially fluorinated substrates are more stable toward the fluorination process than nonfluorinated substrates, since they are more soluble, and the presence of a polyfluorinated group significantly lowers the oxidation potential of the substrates. Therefore, it is easier to control the direct fluorination reaction. Consequently, perfluorinated compounds are generally produced in higher yields than the corresponding nonfluorinated compounds. This concept was applied in the preparation of perfluoroethers [26].

Considering these results, it seems that the direct fluorination method has reached an adequate level for monomer synthesis in industry.

7.3
Synthetic Methods Using Perfluorinated Acyl Fluorides for Industrially Important Perfluorinated Monomers

7.3.1
Direct Application of Liquid-Phase Fluorination

Perfluoroalkoxy copolymer (PFA: copolymer of tetrafluoroethylene (TFE) and perfluoro(propyl vinyl ether) (PPVE)) is one of the most important perfluorinated polymers. It is used both as a thermally and chemically resistant material in industrial use, and especially in recent medical, IT, and electronics applications [27, 28]. However, it has been prepared from very costly HFPO (Scheme 7.2) [7]. Thus, its resulting expense has been a disadvantage.

Although Lagow applied the Exfluor-Lagow elemental fluorine process only to simple molecules such as octyl octanoate, the method was considered to be applicable also to the synthesis of industrially important compounds such as perfluorinated alkyl vinyl ethers, providing that the corresponding hydrocarbon counterparts could be made. From such a viewpoint, Okazoe reexamined the synthesis of PPVE retrosynthetically (Scheme 7.4) [29].

PPVE would be made either by dechlorination of perfluorinated dichloroethyl ether **2** (Route I) [30] or by thermal elimination of perfluorocarboxylic acid derivatives **1a–c** (Route II) as in the industrial preparation. Compound **2** may be obtained

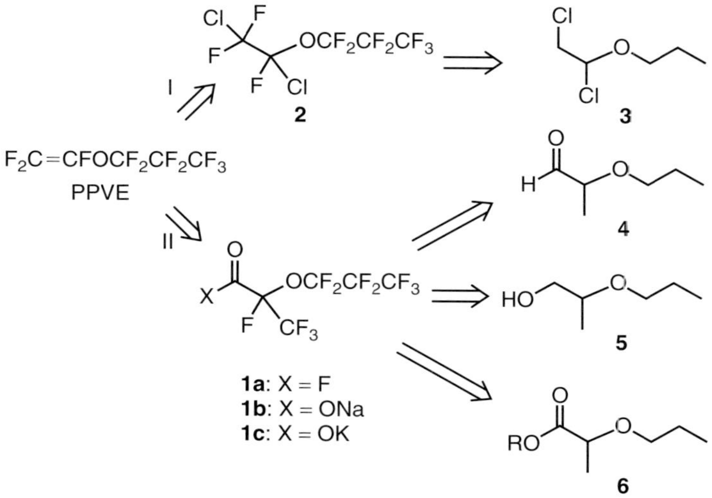

Scheme 7.4 Retrosynthetic analysis of PPVE.

by direct fluorination of the corresponding dichloroethyl ether **3**. Perfluoroacyl fluoride **1a** may be derived from aldehyde **4**, all of whose hydrogen atoms are to be substituted by fluorine atoms, or ester **6**, which would be converted to **1b** or **1c**. Alcohol **5** would also be a precursor of acid fluoride **1a** because its perfluorinated derivative would lead to the desired compound **1a**. Alkali salts of perfluorocarboxylic acid **1b** and **1c** would be derived from acid fluoride **1a**.

Among these candidates, compound **3** seemed the best at first sight, because it could be obtained from inexpensive propyl vinyl ether in only one step. Therefore, Route I was chosen for the first trial.

After chlorination of propyl vinyl ether [31], direct fluorination was carried out, basically applying Lagow's liquid-phase direct fluorination method. The fluorination was conducted with 20% F_2/N_2 and proceeded as expected, but the yield was only 40% at best: migration of chlorine atoms made the reaction complex. Furthermore, considerable amounts of products arising from C–C and C–O bond cleavage formed during the fluorination, probably because some reactions took place in the vapor phase. This suggested that substrate **3** was too volatile to be fluorinated in the liquid phase.

7.3.2
The PERFECT Method

Okazoe then returned to the retrosynthesis shown in Scheme 7.4 and decided to examine Route II. There were three possible precursors as discussed above. Among them, compound **5** was chosen because of easy preparation. However, direct fluorination of **5** seemed dangerous in view of the possible initial formation of an unstable hypofluorite. Protection of the OH group was essential in this route.

The protecting group in this case should be a perfluorinated group of high molecular weight in order to moderate the reactivity toward fluorination and to suppress volatility. Moreover, it should be removable after fluorination. Considering these requirements, the functional group $-COCF(CF_3)OCF_2CF_2CF_3$ seemed to be the most attractive as the protecting one, because it could be removed by thermolysis after perfluorination to give the same component **1a** (Scheme 7.5). In addition, this protecting agent has been available as the intermediate for PPVE in the conventional manufacturing process. Moreover, this seemed to suggest that the acyl fluoride **1a** could be recyclable.

Thus, before direct fluorination, the protection is carried out simply by mixing nonfluorinated alcohol **5** with perfluoroacyl fluoride **1a** and by removing the HF formed during the reaction, out of the reaction system, with a stream of nitrogen. Then, liquid-phase direct fluorination of the resulting partially fluorinated ester **7** should give perfluorinated ester **8**. Thermal fragmentation of **8** gives 2 molar amounts of acyl fluoride **1a**. The obtained acyl fluoride can be used either for further production or for further thermal fragmentation to give PPVE. Okazoe named this process "PERFECT," an abbreviation of PERFluorination of Esterified Compounds followed by Thermolysis [29].

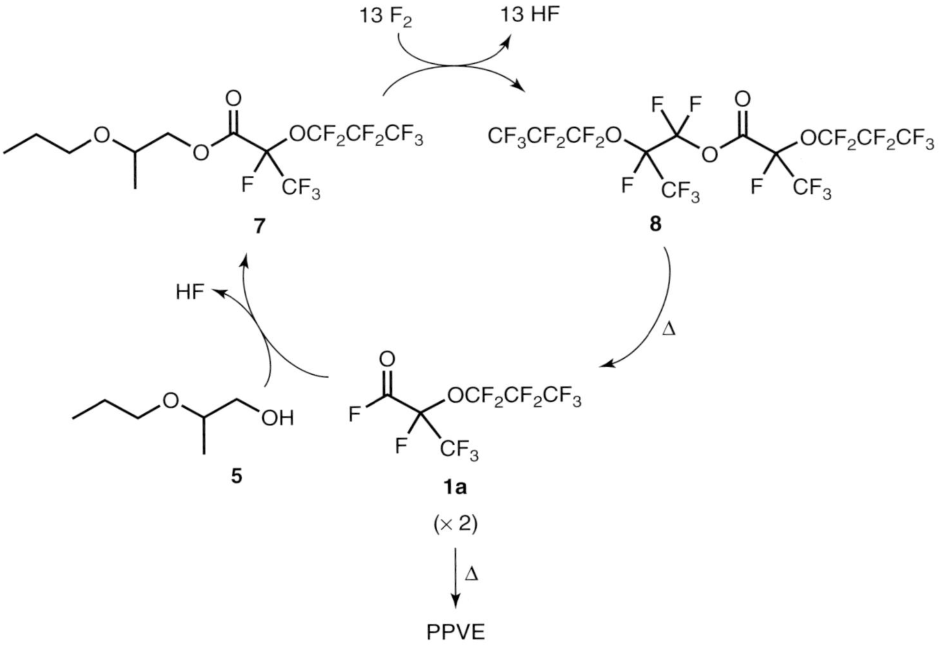

Scheme 7.5 The "PERFECT" process for PPVE.

7.4
Synthesis of Perfluorinated Building Blocks by the PERFECT Method

7.4.1
Perfluorinated Acyl Fluorides

Acyl fluorides are useful building blocks as they can be reduced to primary alcohols, or converted to other functional groups used as common building blocks in organic synthesis. Therefore, the synthesis of perfluorinated acyl fluorides at will allows the versatile preparation of desired perfluorinated building blocks.

In those cases where the desired perfluoroacyl fluoride is available, it can be multiplied by the method shown above. However, in many cases, the desired acyl fluorides are not always readily available. In such situations, a hydrocarbon counterpart alcohol **9** whose structure corresponds to desired acyl fluoride **13** is reacted with the available acyl fluoride **10** (Scheme 7.6). The resulting ester **11** is fluorinated to give the perfluoroester **12**, and the following thermal fragmentation gives the desired acyl fluoride **13** and recovered starting acyl fluoride **10**. Acyl fluorides **13** and **10** are separated by distillation. Once the desired acyl fluoride **13** is obtained, then it can be multiplied by synthesizing the homo ester from **9** and **13** applying the PERFECT process [32].

According to this methodology, various perfluoroacyl fluorides were synthesized from nonfluorinated primary alcohols, which, if required, can be prepared by organic synthesis. Examples of the results of direct fluorination using the PERFECT method are shown in Table 7.1 [32].

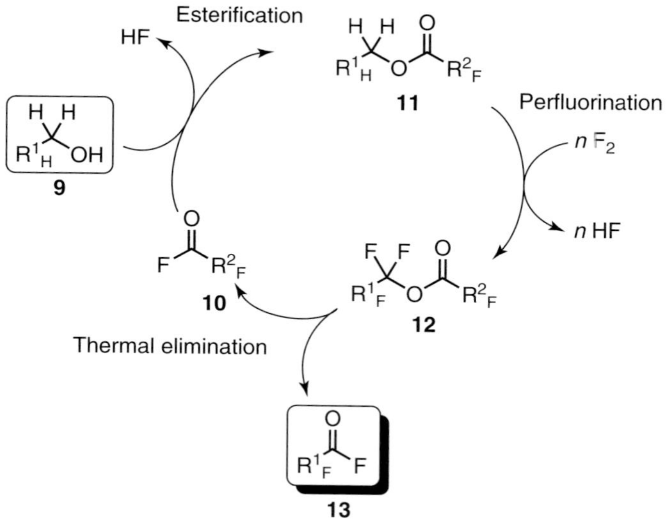

Scheme 7.6 The "PERFECT" process for a perfluorinated acyl fluoride.

7.4 Synthesis of Perfluorinated Building Blocks by the PERFECT Method

Table 7.1 Direct fluorination in the PERFECT process.

$$R_H\text{-O-CO-CF(CF}_3\text{)-OCF}_2\text{CF}_2\text{CF}_3 \xrightarrow[\text{R113}]{20\% \text{ F}_2/\text{N}_2} R_F\text{-O-CO-CF(CF}_3\text{)-OCF}_2\text{CF}_2\text{CF}_3$$

Run	Substrate	Product	Yield (%)
1	CH$_3$(CH$_2$)$_9$O-CH(CH$_3$)-CH$_2$-O-CO-Rf	CF$_3$(CF$_2$)$_9$O-CF(CF$_3$)-CF$_2$-O-CO-Rf	69
2	cyclohexyl-CH$_2$-O-CH(CH$_3$)-CH$_2$-O-CO-Rf	perfluorocyclohexyl-CF$_2$-O-CF(CF$_3$)-CF$_2$-O-CO-Rf	75
3	(methyl-dioxolane)-CH$_2$-O-CO-Rf	(perfluoro-methyl-dioxolane)-CF$_2$-O-CO-Rf	87
4	(dimethyl-dioxolane)-CH$_2$-O-CO-Rf	(perfluoro-dimethyl-dioxolane)-CF$_2$-O-CO-Rf	78
5	PhCH$_2$-O-CH(CH$_3$)-CH$_2$-O-CO-Rf	C$_6$F$_{11}$-CF$_2$-O-CF(CF$_3$)-CF$_2$-O-CO-Rf	21
6	PhCH$_2$-O-CH$_2$CH$_2$CH$_2$-O-CO-Rf	C$_6$F$_{11}$-CF$_2$-O-CF$_2$CF$_2$CF$_2$-O-CO-Rf	25
7	CH$_2$=CH-CH$_2$-O-CH$_2$CH$_2$CH$_2$-O-CO-Rf	CF$_3$-CF$_2$-CF$_2$-O-CF$_2$CF$_2$CF$_2$-O-CO-Rf	86

Rf: –CF(CF$_3$)OCF$_2$CF$_2$CF$_3$; R113: 1,1,2-trichlorotrifluoroethane.

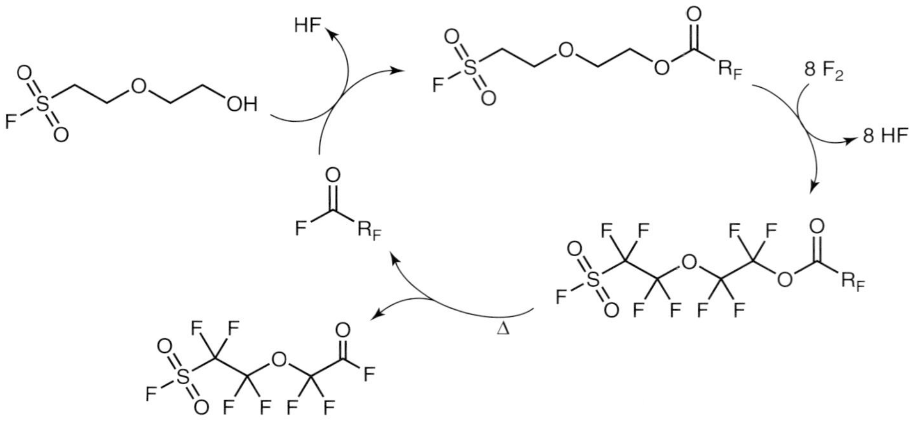

Scheme 7.7 The PERFECT process for synthesis of a precursor of monomers for perfluoroalkanesulfonic acid membrane.

Using the PERFECT process, direct fluorination of a partially fluorinated ester, which has an alkanesulfonyl fluoride functional group at one end, gave the desired precursor of a perfluorinated alkanesulfonic acid for use as an ion exchange membrane (Scheme 7.7) [33].

Direct fluorination of a partially fluorinated diester, which was prepared from a hydrocarbon diol and a perfluorinated acyl fluoride, followed by thermal fragmentation, gave a perfluorinated diacyl fluoride, which is a precursor of a perfluorinated carboxylic acid membrane monomer (Scheme 7.8) [34].

7.4.2
Synthesis of Perfluorinated Ketones by the PERFECT Method

Perfluorinated ketones are quite useful in order to introduce a perfluorinated building block by C–C bond formation. On the other hand, the synthesis of perfluoro ketones by direct fluorination [25, 35] is yet to be studied, because it has not been appropriate for large-scale synthesis, especially of small molecular weight ketones.

In the application of the PERFECT process for perfluoro ketone synthesis, a nonfluorinated secondary alcohol is employed as the starting material (Scheme 7.9) [36].

First, for a small hydrocarbon component with the backbone structure of the desired ketone, secondary alcohol **14** is selected and prepared by conventional organic synthesis when it is not commercially available. Then, it is coupled with a perfluorinated moiety, typically perfluoroacyl fluoride **10**, to make a high molecular weight substrate in the form of the partially fluorinated ester **15**. Perfluorination is achieved by liquid-phase direct fluorination with elemental fluorine to give the perfluorinated ester **16**. In the direct fluorination reaction of **15**, the vapor-phase

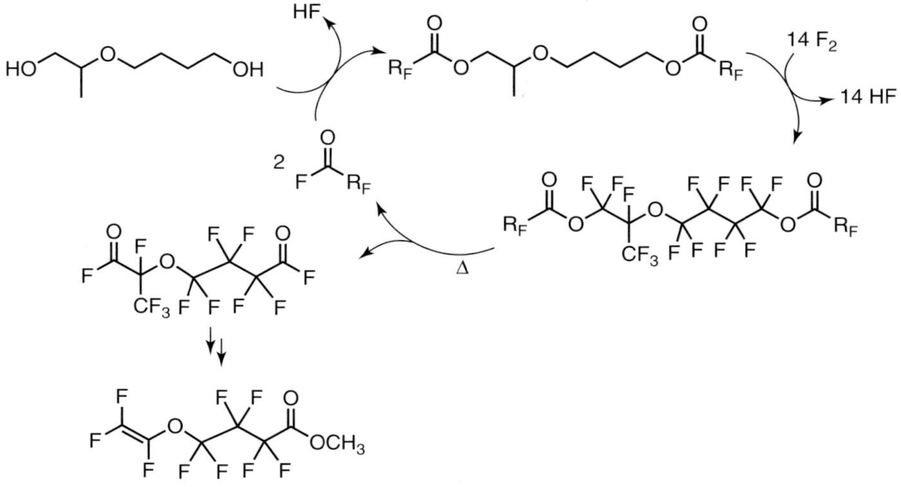

Scheme 7.8 The PERFECT process for the synthesis of a monomer for carboxylic acid membrane.

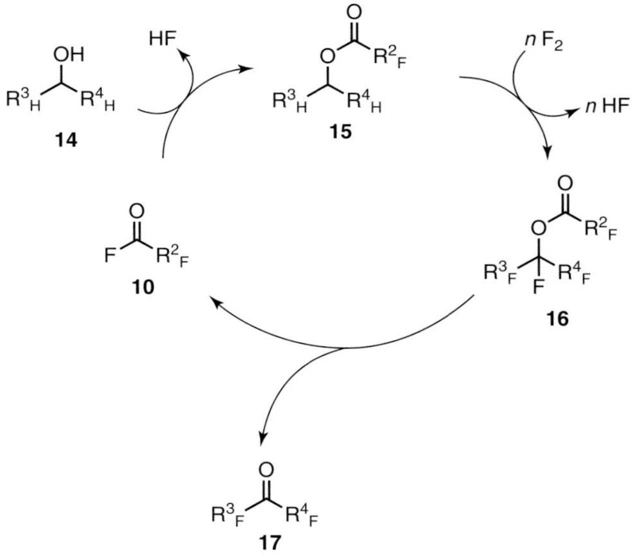

Scheme 7.9 The PERFECT process for a perfluorinated ketone.

reaction is suppressed since substrate **15** has a low vapor pressure and good solubility in a perfluorinated solvent. This approach has an advantage that various perfluorinated compounds other than CFCs can be used as the solvent. Typically, perfluoroacyl fluoride **10** itself is employed. This concept is also applied to the acyl fluoride synthesis (Scheme 7.6). Final thermal fragmentation regenerates the starting perfluoroacyl fluoride **10** and provides the desired perfluorinated ketone **17**. Separation of these products is readily achieved by distillation.

Application to the synthesis of a precursor for fluoropolymer resists for 157 nm microlithography was successfully performed by this process [36].

7.5
Conclusion

In addition to providing industrially important materials [37], the PERFECT perfluorination process can provide pharmaceutical building blocks with tailor-made functionalized materials. The raw materials are inexpensive hydrocarbons, and the synthesis from nonfluorinated components makes it possible to create entirely new fluorinated structure at will. Examples of the range of new fluorinated building blocks that can be synthesized from the PERFECT process are shown in Scheme 7.10 [38].

Theoretically, the only by-product of the PERFECT process is hydrogen, because HF, which is formed in the process, can be electrochemically converted to hydrogen and fluorine; the fluorine can be used again in the process. Moreover, the PERFECT process does not use any solvent other than the products or intermediates of the process itself. In that sense, it contributes to reduce the environmental burden associated with the production of industrial chemicals.

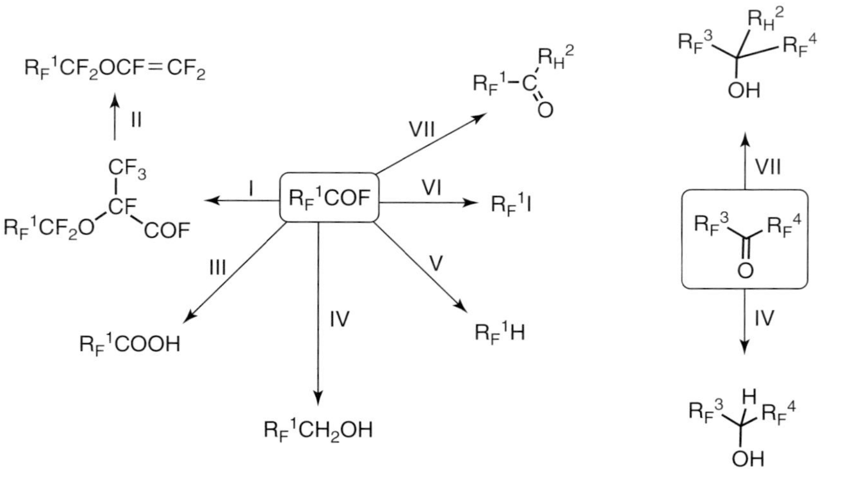

(I) $F_3C-\underset{O}{\overset{F}{C}}-CF_2$, MtIF; (II) Δ on Na_2CO_3 or glass beads; (III) H_2O; (IV) reduction;

(V) Δ in a solvent; (VI) KI; (VII) R_H^2MgX.

Scheme 7.10 Building blocks from the "PERFECT" products.

References

1. Filler, R., Kobayashi, Y., and Yagupolskii, L.M. (1993) *Organofluorine Compounds in Medicinal Chemistry and Biomedical Applications*, Elsevier, Amsterdam.
2. Hiyama, T. (2000) *Organofluorine Compounds, Chemistry and Applications*, Springer, Berlin, p. 137.
3. Soloshonok, V.A. (2005) *Fluorine-Containing Synthons*, American Chemical Society, Washington, DC.
4. Wakselman, C. (1995) in *Chemistry of Organic Fluorine Compounds II* (eds M. Hudlicky and A.E. Pavlath), American Chemical Society, Washington, DC, p. 446.
5. Haszeldine, R.N. (1949) *J. Chem. Soc.*, 2856.
6. Haszeldine, R.N. (1953) *J. Chem. Soc.*, 3761.
7. Millauer, H., Schwertfeger, W., and Siegemund, G. (1985) *Angew. Chem., Int. Ed. Engl.*, **24**, 161.
8. Fowler, R.D., Burford, W.B. III, Hamilton, J.M. Jr., Sweet, R.G., Weber, C.E., and Kasper, J.S. (1947) *Ind. Eng. Chem.*, **39**, 292.
9. Fowler, R.D., Burford, W.B. III, Hamilton, J.M. Jr., Sweet, R.G., Weber, C.E., and Kasper, J.S. (1951) in *Preparation, Properties and Technology of Fluorine and Organic Fluoro Compounds* (eds C. Slesser, and S.R. Schram), McGraw-Hill Book Co., Inc., New York, p. 349.
10. Simons, J.H. (1949) *J. Electrochem. Soc.*, **95**, 47.
11. Abe, T. and Nagase, S. (1982) in *Preparation, Properties and Industrial Applications of Organofluorine Compounds* (ed. R.E. Banks), Ellis Horwood, Chichester, p. 19.
12. Moldavskii, D.D, Bispen, T.A., Kaurova, G.I., and Furin, G.G. (1999) *J. Fluorine Chem.*, **94**, 157.
13. Flahaut, J. and Viel, C. (1986) *J. Fluorine Chem.*, **33**, 27.
14. Atherton, M.J. (2000) in *Fluorine Chemistry at the Millennium: Fascinated by Fluorine* (ed. R.E. Banks), Elsevier, Amsterdam, p. 1.
15. Hatchinson, J. and Sandford, G. (1997) *Top. Curr. Chem.*, **193**, 1.
16. Ono, T. (2003) *Chim. Oggi*, **21**, 39.
17. Tyczkowskyi, E.A. and Bigelow, L.A. (1955) *J. Am. Chem. Soc.*, **77**, 3007.
18. Banks, R.E. and Tatlow, J.C. (1986) *J. Fluorine Chem.*, **33**, 227.
19. Miller, W.T. and Dittman, A.L. (1956) *J. Am. Chem. Soc.*, **78**, 2793.
20. Sheppard, W.A and Sharts, C.M. (1969) *Organic Fluorine Chemistry*, W. A. Benjamin Inc., New York, p. 12.
21. Lagow, R. and Margrave, J. (1979) *Prog. Inorg. Chem.*, **26**, 161.
22. Lagow, R. (2000) in *Fluorine Chemistry at the Millennium: Fascinated by Fluorine* (ed. R.E. Banks), Elsevier, Amsterdam, p. 283.
23. Adcock, J.L., Horita, K., and Renk, E.B. (1981) *J. Am. Chem. Soc.*, **103**, 6937.
24. Scherer, K.V., Yamanouchi, K., and Ono, T. (1990) *J. Fluorine Chem.*, **50**, 47.
25. Bierschenk, T.R., Juhlke, T., Kawa, H., and Lagow, R.J. (1992) Exfluor research corporation. US Patent 5, 093, 432.
26. Badyal, J.P.S., Chambers, R.D., and Joel, A.K. (2000) *J. Fluorine Chem.*, **60**, 297.
27. Feiring, A.E. (1994) in *Organofluorine chemistry: Principles and Commercial Applications* (eds R.E. Banks, B.E. Smart, and J.C. Tatlow), Plenum Press, New York, p. 339.
28. Hintzer, K. and Löhr, G. (1997) in *Modern Fluoropolymers* (ed. J. Scheirs), John Wiley & Sons, Ltd, Chichester, p. 223.
29. Okazoe, T., Watanabe, K., Itoh, M., Shirakawa, D., Murofushi, H., Okamoto, H., and Tatematsu, S. (2001) *Adv. Synth. Catal.*, **343**, 215.
30. Navarrini, W., Tortelli, V., Russo, A., and Corti, S. (1999) *J. Fluorine Chem.*, **95**, 27.
31. Coker, J.N., Bjornson, A., Londergan, T.E., and Johnson, J.R. (1955) *J. Am. Chem. Soc.*, **77**, 5542.
32. Okazoe, T., Watanabe, K., Itoh, M., Shirakawa, D., and Tatematsu, S. (2001) *J. Fluorine Chem.*, **112**, 109.
33. Okazoe, T., Murotani, E., Watanabe, K., Itoh, M., Shirakawa, D., Kawahara, K.,

Kaneko, I., and Tatematsu, S. (2004) *J. Fluorine Chem.*, **125**, 1695.

34. Okazoe, T., Watanabe, K., Itoh, M., Shirakawa, D., Kawahara, K., and Tatematsu, S. (2005) *J. Fluorine Chem.*, **126**, 521.
35. Adcock, J.L. and Robin, M.L. (1983) *J. Org. Chem.*, **48**, 2437.
36. Okazoe, T., Watanabe, K., Itoh, M., Shirakawa, D., Takagi, H., Kawahara, K., and Tatematsu, S. (2007) *Bull. Chem. Soc. Jpn.*, **80**, 1611.
37. Okazoe, T. (2009) *Proc. Jpn. Acad., Ser. B*, **85**, 276.
38. Baasner, B., Hagemann, H., and Tatlow, J.C. (1999) *Methoden der Organischen Chemie (Houben-Weyl)*, 4th edn, vol. E10b, Georg Thieme Verlag, Stuttgart.

8
Cross-Linked Enzyme Aggregates as Industrial Biocatalysts
Roger A. Sheldon

8.1
Introduction

Biocatalysis has many benefits from the viewpoint of green and sustainable chemistry [1, 2]. Reactions are performed under mild conditions (ambient temperature and pressure at physiological pH) in water as solvent, in high chemo-, regio-, and enantioselectivities. Biocatalytic syntheses are generally more atom- and step economical, less energy intensive, and generate less waste than conventional organic syntheses. Furthermore, enzymes are derived from renewable resources and are "natural," biocompatible, and biodegradable. However, commercialization is often hampered by their lack of operational and storage stability, cumbersome recovery, and recycling and product contamination. These obstacles can generally be overcome by immobilization of the enzyme [3, 4], affording improved storage and operational stability and providing for its facile separation and reuse. Moreover, immobilized enzymes, in contrast to free enzymes that can penetrate the skin, are hypoallergenic.

Immobilization methods can be conveniently divided into three types [4]: binding to a carrier, encapsulation in an inorganic or organic polymeric matrix, or by cross-linking of the protein molecules. Binding to or encapsulation in a carrier inevitably leads to dilution of catalytic activity and, hence, lower productivities (kilogram products per kilogram enzyme) owing to the introduction of a large proportion (90–99% of the total) of noncatalytic mass. In contrast, immobilization via cross-linking of enzyme molecules with a bifunctional cross-linking agent is a carrier-free method and the resulting biocatalyst essentially comprises 100% active enzyme.

The technique of protein cross-linking, via reaction of glutaraldehyde with reactive NH_2 groups on the protein surface, was originally developed more than 40 years ago [5]. Such cross-linked enzymes (CLEs) are produced by mixing an aqueous solution of the enzyme with an aqueous solution of glutaraldehyde [6]. However, the CLEs exhibited low activity retention, poor reproducibility, low mechanical stability, and, owing to their gelatinous nature, were difficult to handle. Consequently, binding to a carrier became the most widely used methodology for enzyme immobilization.

Pharmaceutical Process Chemistry. Edited by Takayuki Shioiri, Kunisuke Izawa, and Toshiro Konoike
Copyright © 2011 WILEY-VCH Verlag GmbH & Co. KGaA, Weinheim
ISBN: 978-3-527-32650-1

The use of cross-linked enzyme crystals (CLECs) as industrial biocatalysts was introduced in the early 1990s and subsequently commercialized by Altus Biologics [7, 8]. The method was applicable to a broad range of enzymes and CLECs proved significantly more stable to denaturation by heat, organic solvents, and proteolysis than the corresponding soluble enzyme or lyophilized (freeze-dried) powder. Their operational stability, controllable particle size, and ease of recycling, coupled with their high catalyst and volumetric productivities, made them ideally suitable for industrial biocatalysis. However, CLECs have an inherent disadvantage: the need to crystallize the enzyme, a laborious procedure requiring enzyme of high purity, which translates to high costs.

Hence, several years ago we reasoned that perhaps crystallization could be replaced by precipitation of the enzyme from aqueous buffer, a simpler and less expensive method not requiring highly pure enzymes. This led us to develop a new class of immobilized enzymes which we called *cross-linked enzyme aggregates* (CLEA®s) [9, 10].

8.2
Cross-Linked Enzyme Aggregates

The addition of salts, or water miscible organic solvents, or nonionic polymers, to aqueous solutions of proteins leads to their precipitation as physical aggregates that are held together by noncovalent bonding without perturbation of their tertiary structure. Addition of water to this precipitate results in dissolution of the enzyme. In contrast, cross-linking of these physical aggregates by the reaction of reactive groups on the enzyme surface (e.g., free amino groups of lysine residues) with a bifunctional reagent such as glutaraldehyde renders the aggregates permanently insoluble while maintaining their preorganized superstructure, and hence their catalytic activity. Since precipitation is often used to purify enzymes, the CLEA methodology essentially combines purification and immobilization into a single unit operation.

8.2.1
Cross-Linking Agents

Glutaraldehyde is generally the cross-linking agent of choice as it is inexpensive and readily available in commercial quantities. It has been used for decades for cross-linking proteins [11, 12]. However, the chemistry is complex and not fully understood [11]. Cross-linking occurs via reaction of the free amino groups of lysine residues, on the surface of neighboring enzyme molecules, with oligomers or polymers resulting from aldol condensations of glutaraldehyde. This can involve both Schiff's base formation and Michael-type 1,4 addition to α, β-unsaturated aldehyde moieties [12] (Scheme 8.1). The exact mode of cross-linking is also dependent on the pH of the mixture [12].

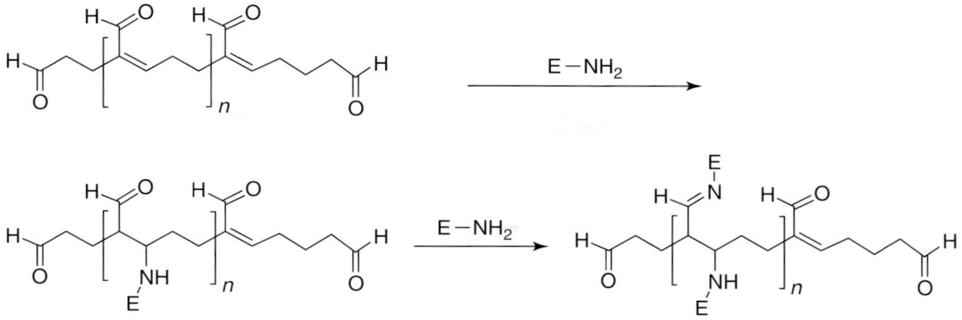

Scheme 8.1 Reaction of glutaraldehyde oligomers with enzyme amino groups.

Other dialdehydes, involving less complicated chemistry, can be used as cross-linkers. With some enzymes, for example, with nitrilases, we sometimes observed low or no retention of activity when glutaraldehyde was used as the cross-linker. In this case, we observed good activity retention using dextran polyaldehyde as a cross-linker [13], followed by reduction of the Schiff's base moieties with sodium borohydride to form irreversible amine linkages. Galactose dialdehyde, formed by galactose oxidase–catalyzed aerobic oxidation of galactose, has been proposed as a potential protein cross-linker [14].

Since every enzyme is a different molecule with a different number of accessible lysine residues, one would expect every enzyme to behave differently under cross-linking conditions. Indeed, with enzymes containing few or no accessible lysine residues, cross-linking may be ineffective and lead to CLEAs that are unstable toward leaching in aqueous media. One way to overcome this problem is to add polyamines, such as polyethyleneimine, which are then coimmobilized with the enzyme [15].

8.2.2
Protocols for CLEA Preparation

Optimization of the protocols for CLEA preparation, with regard to parameters such as temperature, pH, concentration, stirring rate, precipitant, additives, and cross-linking agent, is a relatively simple operation which lends itself to automation, for example, using 96-well plates [16]. The nature of the precipitant predictably has an important effect on the activity recovery and, hence, it is necessary to screen a number of water miscible organic solvents and polymers such as polyethylene glycols (PEGs). In the initial screening of precipitants, the amount of aggregates formed is determined and then they are redissolved in aqueous buffer and their activities are measured. However, it should be mentioned that the fact that the aggregates show a high activity on redissolution in buffer does not necessarily mean that they will retain this high activity after cross-linking. It could be that the aggregates contain the enzyme in an unfavorable conformation which will not be retained when they are redissolved in buffer but will be preserved when they are cross-linked. Hence, it is advisable to choose a few precipitants that give good yields of aggregates for further screening subsequent to cross-linking. The optimum precipitant may not be the one that ultimately gives the optimum CLEA.

The ratio of cross-linker to enzyme is obviously an important factor. If the ratio is too low, sufficient cross-linking does not occur and CLEAs are not formed, and if it is too high, too much cross-linking occurs resulting in a complete loss of the enzyme's flexibility which is necessary for its activity. Since every enzyme has a unique surface structure, containing varying numbers of lysine residues, for example, the optimum ratio has to be determined for each enzyme.

The ratio of cross-linker to enzyme is also important in determining the particle size of CLEAs [17]. Particle size is an important property from the viewpoint of large-scale applications since it directly affects mass transfer and filterability under operational conditions. A typical particle size of CLEAs is 10–50 μm and their filterability is generally sufficient for batch operation. However, for certain large-scale applications it may be necessary to perform the reaction in a continuous operation mode over a packed bed of biocatalyst. This will require relatively large particles in order to avoid a large pressure drop over the column. One approach to preparing CLEAs with increased particle size and mechanical stability is to encapsulate them in a polyvinyl alcohol matrix (the so-called Lentikats) [18]. The resulting dilution of activity observed with a penicillin amidase CLEA was an acceptable 40%.

CLEAs of α-chymotrypsin and *Mucor javanicus* lipase were immobilized in hierarchically ordered mesoporous silica, using a ship-in-the-bottle approach, to afford biocatalysts with good stability and activity [19]. Glucose oxidase CLEAs have been immobilized on the surface of carbon nanotubes for use in miniature biofuel cells [20]. The same group reported the immobilization of CLEAs in magnetic mesocellular carbon foam [21] and, together with magnetic nanoparticles, in mesocellular mesoporous silica [22]. Lipase CLEAs have also been embedded in the pores of microporous cellulose- or polytetrafluoroethylene (PTFE)-based membranes to afford biocatalytic membranes [23].

In another variation on the standard CLEA protocol, silica-CLEA nanocomposites were prepared by performing the cross-linking of the enzyme aggregates in the presence of alkoxysilanes as silica precursors (Scheme 8.2) [24]. The hydrophobicity of the resulting nanocomposites can be tuned by an appropriate choice of alkoxysilane and the minor dilution of catalytic activity is acceptable. This methodology

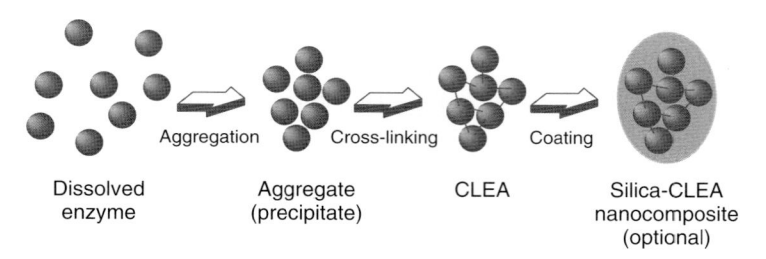

| Dissolved enzyme | Aggregate (precipitate) | CLEA | Silica-CLEA nanocomposite (optional) |

Scheme 8.2 The CLEA technology.

can also be applied to the synthesis of CLEA–polymer nanocomposites from other inorganic or organic polymer precursors. In addition to providing the possibility to prepare CLEAs with the desired hydrophobic or hydrophilic microenvironments, the methodology also allows for tuning of their particle size. We shall return to the question of robustness and particle size of CLEAs later when we discuss reactor design for implementation of CLEAs in production processes.

One of the important advantages of CLEAs is that they can be prepared from very crude enzyme abstracts, probably even extracts obtained directly from fermentation broth. However, sometimes it is difficult to achieve CLEA formation in cases where the protein concentration in the enzyme preparation is low. In such cases, CLEA formation can be promoted by the addition of a second protein, such as bovine serum albumin, as a so-called proteic feeder [25].

8.2.3
Advantages of CLEAs

CLEAs have many benefits in the context of industrial applications. There is no need for highly pure enzyme; they can easily be prepared from impure enzyme extracts. Since they are carrier-free they avoid the costs associated with the use of (often expensive) carriers. In industrial practice, these attributes translate to low costs and short time-to-market. Further cost benefits accrue from the fact that they exhibit high catalyst productivities (kilogram products per kilogram biocatalyst) and are easy to recover and recycle. Furthermore, they exhibit improved properties compared to the free enzymes that they are derived from. They generally exhibit much better storage and operational stability with regard to denaturation by heat, organic solvents, and autoproteolysis, and are stable toward leaching in aqueous media.

Since the costs of the precipitant and cross-linker are generally much lower than that of the free enzyme, the variable costs of CLEA production are largely determined by the price of the free enzyme, the labor costs, and the activity recovery. The price of the free enzyme differs for each enzyme, and labor costs are very much dependent on the scale of production. The key parameter is the activity recovery that should preferably be close to (or more than) 100%. We hasten to add, however, that the enzyme costs per kilogram of product are also determined by how many times the CLEA can be recycled.

With regard to activity recovery we emphasize that it is important, when reading the literature, to establish how a particular author is defining activity recovery. Activity recovery is simply activity recovered/activity charged expressed as a percentage and is independent of the weight of the CLEA. Thus, 100% activity recovery could be obtained with an increase or a decrease in weight of the CLEA compared to that of the free enzyme sample charged. *Specific activity*, on the other hand, is defined as an activity per unit weight of protein. The relevant comparison for calculating activity recovery is the activity of the free enzyme charged with that of the immobilized enzyme recovered, *under the same conditions*. This is relatively straightforward for a hydrolytic process in which the free enzyme is dissolved in water, for example, tributyrin hydrolysis for a lipase. However, comparison becomes more problematical for a synthesis reaction in nonaqueous media where the free enzyme is not soluble. In this case, many authors compare the specific activity of the immobilized lipase with that of the powder obtained by precipitation of the enzyme from aqueous solution with acetone. Another possibility is to compare it with freeze-dried enzyme.

The CLEA technology has broad scope and has been applied to an increasingly wide selection of enzymes (see below). Yet another advantage results from the possibility to coimmobilize two or more enzymes to provide CLEAs that are capable of catalyzing multiple biotransformations, independently or in sequence as catalytic cascade processes.

8.2.4
Multi-CLEAs and Combi-CLEAs

As noted above, additional proteins can be incorporated into CLEAs by coprecipitation and cross-linking. This notion also leads to the idea of preparing multipurpose CLEAs from crude enzyme extracts consisting of multiple enzymes. For example, Gupta and coworkers [26] prepared a CLEA from porcine pancreatic acetone powder extract which is known to be rich in several enzyme activities, particularly lipase, phospholipase A_2, and α-amylase. They further showed that all three enzyme activities were completely retained in the CLEA and the latter could be recycled three times without appreciable loss of activity. Similarly, the same group [27] prepared a multipurpose CLEA exhibiting pectinase, xylanase, and cellulose activities from the commercial preparation, Pectinex^{TM} Ultra SP-L, that is used in the food processing industry for hydrolyzing pectin. This multipurpose CLEA could be used for carrying out three independent reactions: the hydrolysis of polygalacturonic acid (pectinase activity), xylan, or carboxymethyl cellulose. With all three enzymes, the CLEA exhibited increased thermal stability compared to the free enzyme and it could be used three times without activity loss. We suggest that the name multi-CLEA is appropriate for such CLEAs that are made from heterogeneous populations of enzymes and can be used for performing different biotransformations independently.

On the other hand, we envisaged the deliberate coimmobilization of two or more enzymes in a single CLEA for the sole purpose of performing two or more biotransformations in sequence, that is, as multienzyme cascade processes. We have given the name combi-CLEAs to such catalysts and examples of cascade processes performed with such combi-CLEAs are discussed later.

8.3
CLEAs from Hydrolases

The majority of the CLEAs that have been reported to date involve hydrolases, mainly because they are the enzymes that have the most industrial applications but also because they are probably the simplest enzymes to work with.

8.3.1
Lipase and Esterase CLEAs

Lipases (E.C. 3.1.1.3) are widely used in industrial biotransformations and it is not surprising, therefore, that the preparation and application of lipase CLEAs have been widely studied [28–35]. We selected seven commercially available lipases for an investigation of the effect of various parameters, such as the precipitant and the addition of additives (e.g., surfactants and crown ethers), on the activities of the resulting CLEAs [28]. The activation of lipases by additives, such as surfactants and crown ethers, is well documented and is generally attributed to the lipase being induced to adopt a more active conformation [36]. We reasoned that cross-linking of enzyme aggregates, in the presence of such an additive, would "lock" the enzyme in this more favorable conformation. Since the additive is not covalently bonded to the enzyme, it can subsequently be washed from the CLEA. In this way, we prepared several hyperactive lipase CLEAs, in some cases exhibiting activities higher than the corresponding free enzyme [28]. The experimental procedure was further simplified by combining precipitation, in the presence or absence of additives, with cross-linking into a single operation [28].

Xu and coworkers [29] similarly prepared CLEAs from a range of commercially available lipases and investigated their performance in the esterification of lauric acid with n-propanol in a solvent-free system by comparing their activities with those of the corresponding acetone powders. The nature of the precipitant, as would be expected, had a profound effect on the activity recoveries and the specific activities of the CLEAs. For example, with *Candida antarctica* B lipase (CaLB) the CLEAs obtained using ammonium sulfate, acetone, and PEG 600 or PEG 200 as precipitants exhibited in excess of 200% activity recovery compared to the acetone powder directly prepared from the same solution of free enzyme. The PEG 200 precipitated CLEA showed the highest specific activity, 139% compared to acetone powder. Similarly, a *Candida rugosa* lipase CLEA, prepared by precipitation with PEG 200, exhibited 270% esterification activity relative to the corresponding acetone powder.

As noted above, particle size is an important property from the viewpoint of large-scale applications since it directly affects mass transfer and filterability under operational conditions. Yu and coworkers showed [30] that the nature and concentration of the precipitant, enzyme, and glutaraldehyde concentrations, and pH all play a role in determining the particle size of *Candida rugosa* lipase CLEAs. The most important of these parameters was the enzyme to glutaraldehyde ratio. Interestingly, the *C. rugosa* CLEAs, prepared using ammonium sulfate as precipitant, also showed about twofold increase in enantioselectivity in the kinetic resolution of racemic ibuprofen by esterification with 1-propanol [30].

Saxena and coworkers [31] prepared CLEAs from the alkaline and thermostable *Thermomyces lanuginosa* lipase. Efficient CLEA formation was observed using ammonium sulfate as precipitant along with a twofold increase in activity in the presence of the anionic surfactant, sodium dodecyl sulfate (SDS). The CLEA preparation was highly stable and could be used 10 times without appreciable loss of activity in the hydrolysis of olive oil in isopropanol.

Gupta and coworkers [32] prepared CLEAs from *Burkholderia cepacia* lipase (formerly known as *Pseudomonas cepacia lipase*) by precipitation with acetone, in the presence of bovine serum albumin as proteic feeder, and cross-linking with glutaraldehyde. They subsequently used the CLEAs as the catalyst in the transesterification of triglycerides from *Madhuca indica* oil, containing high free fatty acid content, with ethanol to afford biodiesel in the form of fatty acid ethyl esters. Triglycerides containing high amounts of free fatty acids are difficult to convert into biodiesel using the classical (alkali) chemical catalyst. The *B. cepacia* CLEAs afforded a 92% conversion in 2.5 h at 40 °C. The same group also reported [34] the successful use of *B. cepacia* lipase CLEAs in the resolution of racemic citronellol by enzymatic transesterification with vinyl acetate. They further showed, with the aid of scanning electron microscopy (SEM), that the morphology of the CLEA particles is dependent on the extent of cross-linking. Similarly, *B. cepacia* lipase CLEA was used by Kanerva *et al.* [35] for the highly enantioselective resolution of 1-phenylethanol (Scheme 8.3).

Candida antarctica lipase A (CaLA) is a rather exceptional lipase that is known to accept bulky tertiary alcohols as substrates [37] and, therefore, has considerable potential in organic synthesis. Özdemirhan and coworkers reported [33] that CaLA CLEA was an effective catalyst for the resolution of aromatic ring–fused cyclic tertiary alcohols by transesterification with vinyl acetate (Scheme 8.4).

CLEA preparations of the popular CaLB performed better than the industrial benchmark immobilized version of this enzyme: Novozyme 435 (CaLB immobilized on a macroporous acrylic resin) in water, but this superior activity could not

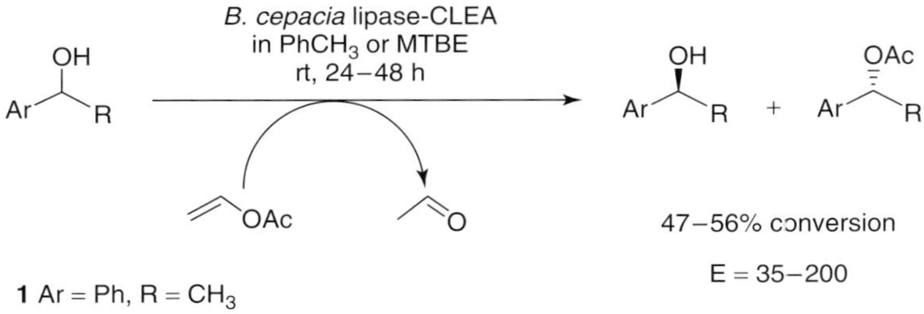

1 Ar = Ph, R = CH$_3$
2 Ar = 2-furyl, R = CH$_3$
3 Ar = Ph, R = CH$_2$NHCOPr

47–56% conversion
E = 35–200

Scheme 8.3 Resolution of 1-phenylethanol with *B. cepacia* lipase CLEA.

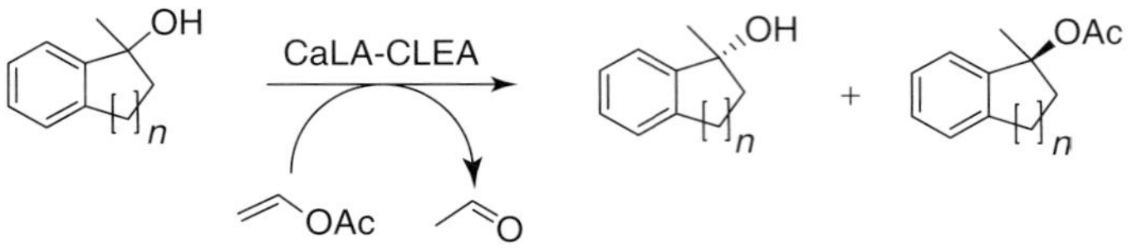

n = 1 or 2

Scheme 8.4 Resolution of tertiary benzylic alcohols with *C. antarctica* lipase A CLEA.

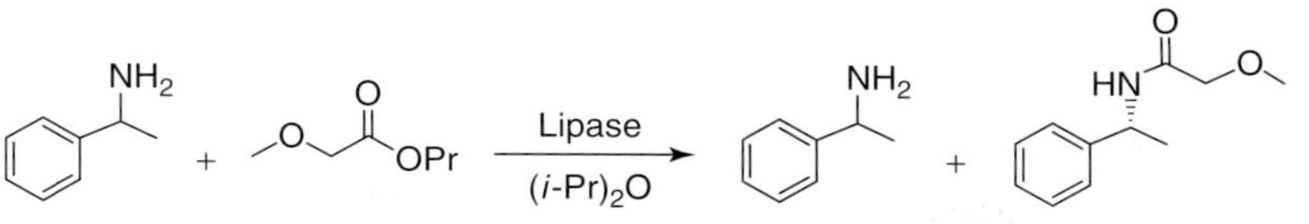

	Activity in H$_2$O (U g^{-1})	Activity in (i-Pr)$_2$O (U g^{-1})	Ratio
Novozyme 435	7300	250	29
CLEA-AM	38 000	50	760
CLEA-OM	31 000	1500	21

AM = aqueous media; OM = organic media

Scheme 8.5 Comparison of Novozyme 435 with CaLB CLEA in water and organic media.

be directly translated to organic media. However, modification of the protocol preparation, to produce a more lipophilic CLEA, afforded a dramatic improvement in the activity of CaLB CLEA in the enantioselective acylation of 1-phenethylamine in diisopropyl ether as solvent (Scheme 8.5) [24]. The optimized CaLB CLEAs displayed activities surpassing those of Novozyme 435 in both aqueous and organic media. We also note that enzyme can be leached from the surface of Novozyme 435 in aqueous media, whereas the CaLB CLEA, being covalently bound, is completely stable.

It was subsequently shown that CaLB CLEAs exhibit excellent activities in supercritical carbon dioxide and ionic liquids. Thus, CaLB CLEA displayed superior activity to that of Novozyme 435 in the kinetic resolution of 1-phenylethanol by acylation with vinyl acetate in scCO$_2$ [38–40]. Similarly, CaLB CLEA preparations were effective in the kinetic resolution of 1-phenylethanol and 1-phenylethylamine in the ionic liquids, [bmim] [NO$_3$] and [bmim] [N(CN)$_2$], whereas Novozyme 435 was completely inactive under these conditions [41].

In contrast to the extensive studies devoted to lipase CLEAs, there have been only sporadic reports of CLEAs from other esterases. Feruloyl esterases (E.C. 3.1.1.73) play a key role in the degradation of plant cell walls by catalyzing the hydrolysis of ferulate (4-hydroxy-3-methoxycinnamate) ester groups involved in the cross-linking of hemicellulose and lignin. Christakopoulos et al. [42] described the synthesis of a recombinant *Aspergillus niger* feruloyl esterase CLEA and its use in the enzymatic synthesis of various hydroxycinnamate esters [43] including the synthesis of esters of glycerol in ionic-liquid water mixtures [44]. Ju and coworkers [45] similarly reported the preparation of a CLEA from an *Aspergillus awamori* feruloyl esterase and studied its properties using differential scanning calorimetry and SEM.

Similarly, Sanchez-Ferreira and coworkers [46] prepared a CLEA from a recombinant acetyl xylan esterase from *Bacillus pumilus*, another enzyme which *in vivo* is involved in the degradation of lignocellulose. *In vitro* this enzyme catalyzes the hydrolysis of the acetate moiety in cephalosporin C and 7-aminocephalosporanic acid (7-ACA) to form advanced intermediates for the production of semisynthetic cephalosporins. They used a novel dispersing technology to obtain CLEAs with high activity and operational stability.

8.3.2
Protease CLEAs

We have prepared CLEAs from (chymo) trypsin (E.C. 3.4.21.4) [16] and from the *Bacillus licheniformis* alkaline protease (alcalase, E.C. 3.4.21.62, also known as *subtilisin Carlsberg*) an inexpensive enzyme used in laundry detergents. Alcalase has been widely used in organic synthesis, for example, in the resolution of (amino acid) esters [47], and amines [48] and peptide synthesis [49]. It is perhaps not surprising, therefore, that alcalase CLEA has been used in amino acid and peptide biotransformations [50–53]. For example, Quaedflieg et al. [50] reported a versatile and highly selective synthesis of α-carboxylic esters of N-protected amino acids and peptides (Scheme 8.6) using alcalase CLEA.

Alcalase CLEA was similarly used [51] in the synthesis of α-protected aspartic acid β-esters and glutamic acid γ-esters via selective α-hydrolysis of symmetrical aspartyl

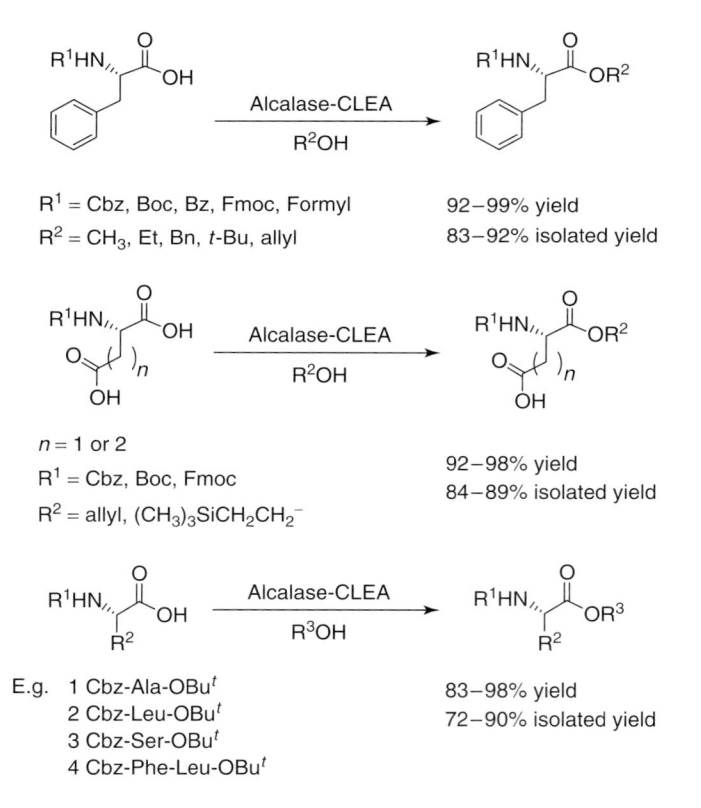

R^1 = Cbz, Boc, Bz, Fmoc, Formyl
R^2 = CH_3, Et, Bn, *t*-Bu, allyl

92–99% yield
83–92% isolated yield

n = 1 or 2
R^1 = Cbz, Boc, Fmoc
R^2 = allyl, $(CH_3)_3SiCH_2CH_2-$

92–98% yield
84–89% isolated yield

E.g. 1 Cbz-Ala-OBu^t
2 Cbz-Leu-OBu^t
3 Cbz-Ser-OBu^t
4 Cbz-Phe-Leu-OBu^t

83–98% yield
72–90% isolated yield

Scheme 8.6 Alcalase CLEA in esterification of amino acids.

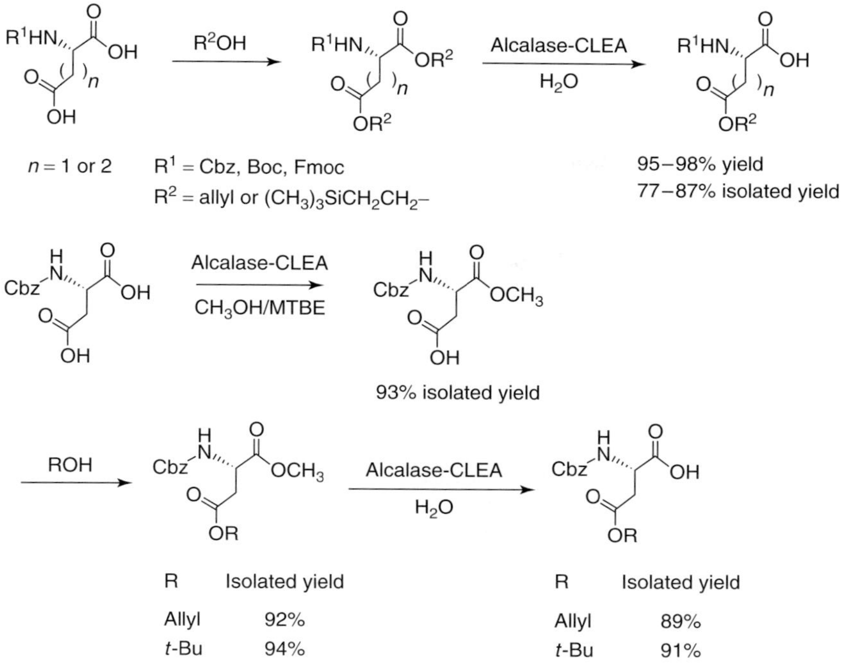

Scheme 8.7 Synthesis of monoesters of aspartic and glutamic acids using alcalase CLEA.

or glutamyl diesters or in a three-step protocol involving enzymatic formation of the α-methyl ester followed by chemical β-esterification and selective α-hydrolysis (Scheme 8.7).

The same group also described [52] the use of alcalase CLEA, under near anhydrous conditions, to catalyze the mild and cost-efficient synthesis of C-terminal aryl amides of amino acids and peptides by reaction of the corresponding free carboxylic acid, or the methyl or benzyl ester, with the aromatic amine (Scheme 8.8). The products were obtained in high chemical and enantio- and diastereomeric purities. In contrast to state-of-the-art chemical methods, no racemization was observed.

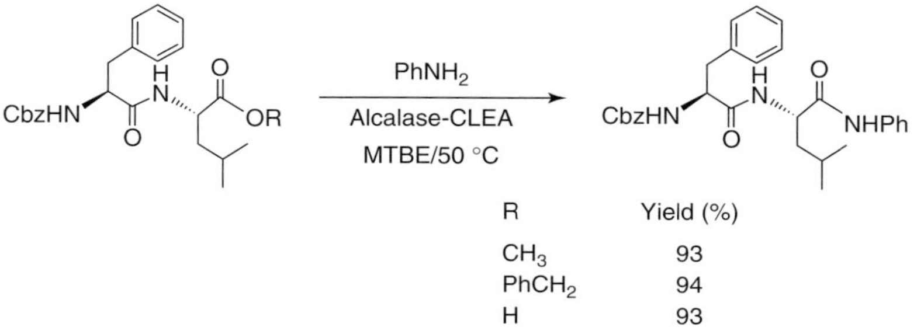

Scheme 8.8 Amidation of peptide C-terminal ester groups catalyzed by alcalase CLEA.

Eggen and Boeriu [53] similarly reported the use of Alcalase CLEA for the selective enzymatic hydrolysis and amidation of C-terminal esters or acids of peptides.

8.3.3
Amidase CLEAs

Our initial studies of CLEAs [9, 54] involved the preparation of CLEAs from penicillin G amidase (E.C. 3.5.1.11), an industrially important enzyme used in the synthesis of semisynthetic penicillin and cephalosporin antibiotics [55]. The free enzyme has limited thermal stability and a low tolerance to organic solvents which makes it an ideal candidate for stabilization as a CLEA. Indeed, a penicillin G amidase CLEA, prepared by precipitation with *tert*-butanol, and cross-linking with glutaraldehyde, proved to be an effective catalyst for the synthesis of ampicillin (Scheme 8.9) [9, 54]. Remarkably, the productivity of the CLEA was higher even than that of the free enzyme that it was made from and substantially higher than that of the CLEC. Not surprisingly, the productivity of the commercial catalyst was much lower, a reflection of the fact that it mainly consists of noncatalytic ballast in the form of the polyacrylamide carrier. Analogous to the corresponding CLECs, the penicillin G amidase CLEAs also maintained their high activity in organic solvents [56].

Similarly, a CLEA produced from a partially purified penicillin G amidase from a recombinant *Escherichia coli* strain was used by Illanes *et al.* [57] in the kinetically controlled synthesis of ampicillin in ethylene glycol–water (60 : 40 v/v). The same group [58] used a penicillin amidase CLEA for the synthesis of the semisynthetic cephalosporin, cephalexin in an aqueous medium (Scheme 8.10). They also recently

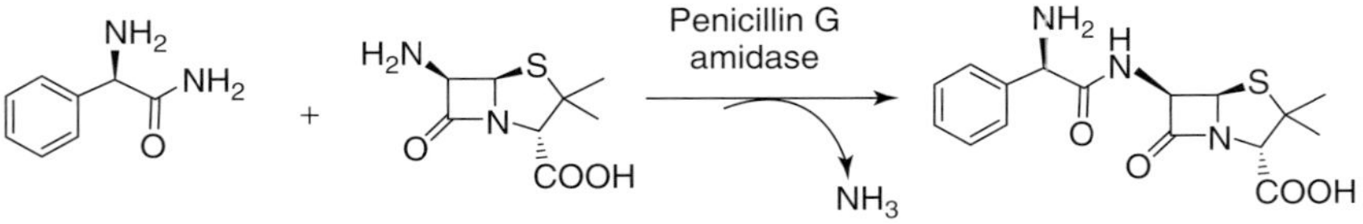

Ampicillin

Biocatalyst	Conversion (%)	S/H ratio	Relative productivity
Free enzyme	88	2.0	100
CLEC	72	0.71	39
T-CLEA	85	1.58	151
PGA-450	86	1.56	0.8

Scheme 8.9 Synthesis of ampicillin catalyzed by penicillin amidase CLEA.

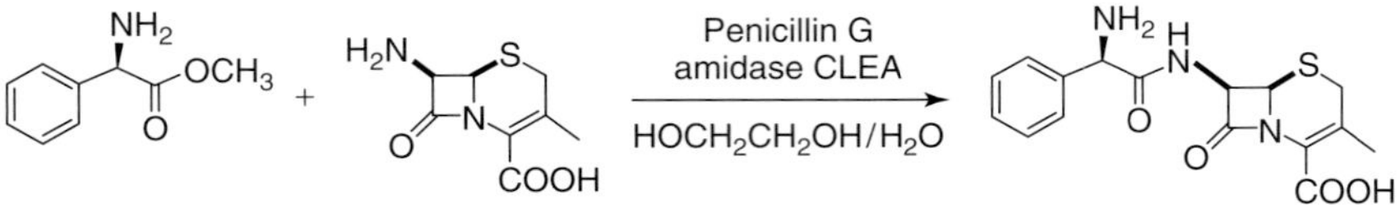

7-ADCA

Cephalexin

Scheme 8.10 Synthesis of cephalexin catalyzed by penicillin amidase CLEA.

reported a study of the influence of the degree of cross-linking (0.15 or 0.25) on the performance of penicillin amidase CLEAs in cephalexin synthesis in aqueous ethylene glycol [59]. The cross-linker to enzyme ratio had only a minor effect on the specific activity of the CLEA but significantly influenced the activity recovery, the thermal stability, and the productivity in cephalexin synthesis, all being higher at the lower glutaraldehyde to enzyme ratio.

Svedas and coworkers [60] recently reported a study of the effect of "aging" on the catalytic properties of penicillin amidase CLEAs in the hydrolysis of (R)-phenylglycine amide and the synthesis of ampicillin. The period of time between enzyme precipitation and cross-linking was found to influence the structural organization of the resulting CLEA, the "mature" CLEAs consisting of larger particles that were more effective in both the hydrolytic and synthetic process. They suggested that the aggregate size might regulate the extent of covalent modification and thereby the catalytic activity of the CLEAs.

We prepared a CLEA [61] from aminoacylase (EC 3.5.1.14) derived from an *Aspergillus* sp. which is the catalyst in the Evonik process for the manufacture of (S)-amino acids by highly enantioselective hydrolysis of N-acetyl amino acids (Scheme 8.11). The aminoacylase CLEA was an active and recyclable catalyst for this reaction. Interestingly, it was unable to catalyze the hydrolysis of simple esters, which is known to occur with the free enzyme preparation from which the CLEA was prepared. A plausible explanation for this observation is that the esterolytic activity is derived from a protein impurity in the free enzyme preparation. We conclude that it demonstrates the power of the CLEA methodology for combining enzyme purification and immobilization into a single operation.

8.3.4
Nitrilases

Nitrilases (EC 3.5.5.1) catalyze the hydrolysis of nitriles to the corresponding carboxylic acids and as such are potentially interesting catalysts for the enantioselective synthesis of carboxylic acids from readily available nitrile precursors. It

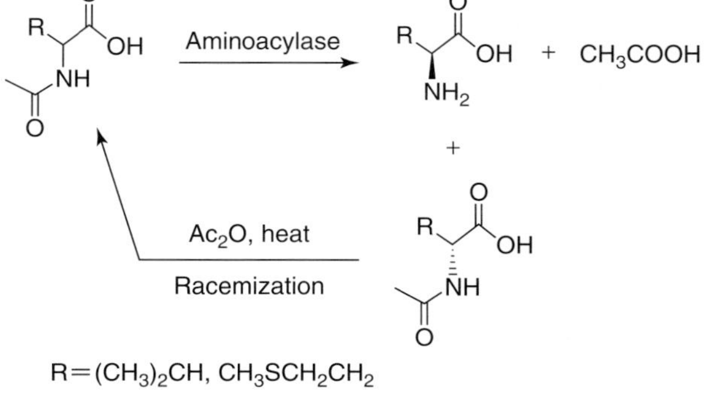

R=(CH$_3$)$_2$CH, CH$_3$SCH$_2$CH$_2$

Scheme 8.11 Resolution of N-acylamino acids catalyzed by aminoacylase.

is well known that nitrilases are rather sensitive enzymes owing to the presence of an oxygen-sensitive cysteine residue in the active site. As mentioned earlier, we experienced difficulties in obtaining CLEAs from nitrilases using the standard protocol of cross-linking with glutaraldehyde [13]. However, we were able to successfully prepare a nitrilase CLEA by using dextran polyaldehyde as the cross-linker [13]. On the other hand, Banerjee *et al.* [62] successfully prepared a CLEA from a recombinant nitrilase overexpressed in *E. coli* using the standard cross-linking with glutaraldehyde. We reasoned that the oxygen sensitivity of nitrilases could be suppressed by surrounding the enzyme molecules with a hydrophilic shell that would reduce the solubility of oxygen in the enzyme environment (NB: this could be the explanation for the positive effect of dextran polyaldehyde as a cross-linker). Indeed, we found that coaggregates of nitrilases with polyethyleneimine (PEI) are much more oxygen tolerant than the dissolved enzyme [63].

8.3.5
Glycosidases

Khare and coworkers [64] have described the preparation of a CLEA of the β-galactosidase from *Aspergillus oryzae* and applied them successfully in the synthesis of galacto-oligosaccharides. *In vivo* this enzyme catalyzes the hydrolysis of lactose and is used to alleviate the symptoms of lactose intolerance. It can also be used in the reverse direction to synthesize oligosaccharides by transgalactosylation. Similarly, Lopez-Munguia and coworkers [65] prepared a CLEA from *Bacillus subtilis* levansucrase and showed that it was an effective and robust catalyst for the synthesis of oligofructosides by transfructosylation. The latter are also of interest because of their nutraceutical properties.

8.4
Oxidoreductases

8.4.1
Oxidases

Recyclable CLEAs were prepared from a variety of oxidases: glucose oxidase (E.C.1.1.3.4), galactose oxidase (E.C. 1.1.3.9), and laccase (E.C. 1.10.3.2) [16, 24]. Laccase, in particular, has many potential applications; for example, in combination with the stable radical TEMPO (2,2,6,6-tetramethylpiperidine-1-oxyl) for the catalytic aerobic oxidation of starch to carboxy starch [24]. The latter is of interest as a biodegradable substitute for polyacrylates as a super water absorbent. However, the enzyme costs are too high, owing to the instability of the laccase under the reaction conditions, which is assumed to be a direct result of the oxidation of the surface of this heavily glycosylated enzyme. Cross-linking would be expected to increase the stability of the laccase by protecting reactive groups on the surface and a CLEA prepared from the laccase from *Coriolus versicolor* indeed exhibited improved stability and a better performance in starch oxidation [24]. Another potential application of laccases is in bioremediation of wastewater but, in order to be commercially viable, the enzyme costs must be very low and, hence, it is important to immobilize the laccase to improve stability and facilitate recycling. In this context, Agathos and coworkers [66] reported the use of laccase CLEAs in a perfusion basket reactor for the continuous removal of endocrine-disrupting chemicals such as bisphenol A, 4-nonyl phenol, and triclosan from urban wastewater. Tyrosinase (E.C. 1.14.18.1) is a copper-dependent oxidase with properties very similar to laccase. It catalyzes the ortho hydroxylation of phenols and has potential applications in the same areas as those for laccase. Aytar and Bakir [67] prepared a CLEA from mushroom tyrosinase with 100% activity recovery, via precipitation with ammonium sulfate and cross-linking with glutaraldehyde, and showed that it had enhanced thermal stability, both on storage and under operational conditions, compared to the free enzyme.

8.4.2
Peroxidases

An additional benefit of the CLEA technology is that it can stabilize the quaternary structures of multimeric enzymes, a structural feature often encountered with redox metalloenzymes. For example, the stability of CLEAs from two tetrameric catalases (E.C. 1.11.1.6) exhibited improved stability compared to the free enzyme [68]. Torres and coworkers [69] reported that they were able to increase the thermal stability of royal palm (*Roystonia regia*) peroxidase 5000-fold by preparing a CLEA from it. The peroxidase CLEA could be used several times in the decolorization of wastewater containing azo dyes.

We have recently reported [70, 71] the successful preparation of CLEAs from the heme-dependent chloroperoxidase (CPO; E.C.1.11.1.10) from *Caldariomyces fumago*. In vivo CPO catalyzes the oxidation of chloride ion, by hydrogen peroxide, to hypochlorite but in the absence of chloride ion it can catalyze various regio- and enantioselective oxygen transfer processes which endow it with enormous synthetic potential [72]. However, a major drawback of CPO is its low stability owing to the facile oxidative degradation of the porphyrin ring by hydrogen peroxide [73] even at relatively low peroxide concentrations. Coprecipitation with bovine serum albumin or pentaethylene hexamine was needed for optimum results, presumably because of the paucity of lysine residues on the surface and, hence, available for cross-linking. Under optimized conditions, an activity recovery of 68% was obtained and the CLEA exhibited enhanced thermal stability and tolerance toward hydrogen peroxide during sulfoxidation of anisole.

We previously reported [74] the design of a semisynthetic peroxidase by the addition of vanadate to the acid phosphatase, phytase. Recently, we described [75] the preparation, characterization, and performance of a CLEA of this semisynthetic peroxidase in sulfoxidations.

8.5
Lyases

8.5.1
Nitrile Hydratases

A class of lyases that has considerable industrial relevance comprises the nitrile hydratases (NHases; E.C. 4.2.1.84) [76, 77] that catalyze the addition of water to nitrile moieties. NHases are Fe- or Co-dependent metalloenzymes that usually consist of multimeric structures. They are generally used as whole-cell biocatalysts because the free enzymes have limited operational stability outside the cell, possibly owing to dissociation of tetramers resulting in deactivation. Hence, we reasoned that CLEA formation could have a beneficial effect by holding the catalytically active tetramer together, analogous to that observed with catalase (see above). This indeed proved to be the case: a CLEA prepared from a cell-free extract of an NHase isolated from an alkaliphilic bacterium showed excellent activity in the conversion of acrylonitrile to acrylamide and was active with a variety of aliphatic nitriles [78, 79]. Moreover, the NHase-CLEA could be recycled 36 times with little loss of activity.

8.5.2
C–C Bond Forming Lyases

The CLEA methodology has also been successfully applied to several C–C bond forming lyases, notably the *R*- and *S*-specific oxynitrilases (hydroxynitrile lyases; E.C. 4.1.2.10) which catalyze the enantioselective hydrocyanation of a wide range of aldehydes. For example, CLEAs prepared from the (*R*)-specific oxynitrilase from almonds, *Prunus amygdalis* (PaHNL) by cross-linking with glutaraldehyde [80] or dextran polyaldehyde [13] were highly effective catalysts for the hydrocyanation of aldehydes under microaqueous conditions and could be recycled several times without loss of activity. CLEAs were similarly prepared from the (*S*)-specific oxynitrilases from *Manihot esculenta* and *Hevea brasiliensis* [81, 82]. Because these oxynitrilase CLEAs perform exceptionally well in organic solvents, they can afford higher enantioselectivities than observed with the free enzymes owing to the essentially complete suppression of competing nonenzymatic hydrocyanation under these conditions [83]. For example, Roberge and coworkers [83] were able to obtain high enantioselectivities in the hydrocyanation of pyridine-3-aldehyde (Scheme 8.12). Such good results could not be obtained with the free enzyme or other immobilized forms. This is a difficult substrate for enantioselective hydrocyanation owing to the relatively facile nonenzymatic background reaction as a result of the electron attracting properties of the pyridine ring.

We recently reported [84] the preparation of a CLEA from the relatively unknown *R*-selective hydroxynitrile lyase from *Linus usitatissimum* (*Lu*HNL) and used it in the conversion of butanone to the *R*-cyanohydrin with 87% ee. Interestingly, addition of the 2-butanone substrate prior to precipitation of the enzyme improved the synthetic activity of the resulting CLEA.

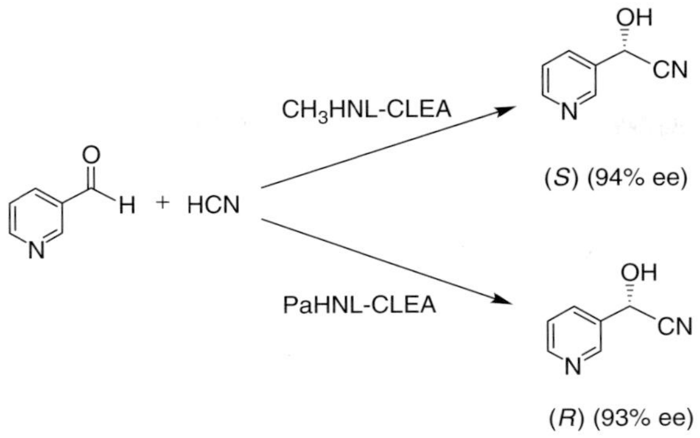

Scheme 8.12 Hydrocyanation of pyridine3-aldehyde catalyzed by hydroxynitrile lyase CLEAs.

CLEAs have also been successfully prepared from other C–C bond forming lyases, for example, pyruvate decarboxylase (E.C. 4.1.1.1) and deoxy-D-ribose phosphate aldolase (DERA, E.C. 4.1.2.4) [16].

8.6
Combi-CLEAs and Cascade Processes

Catalytic cascade processes [85] have numerous potential benefits: fewer unit operations, less reactor volume and higher volumetric and space–time yields, shorter cycle times, and less waste generation. Furthermore, by coupling steps together, unfavorable equilibria can be driven toward product. We have achieved this by immobilizing two or more enzymes in "combi-CLEAs," for example, containing catalase in combination with glucose oxidase or galactose oxidase, respectively [16].

We recently used a combi-CLEA containing the S-selective oxynitrilase from *M. esculenta* and a nonselective nitrilase from *Pseudomonas fluorescens* for the one-pot conversion of benzaldehyde to S-mandelic acid (Scheme 8.13) [86]. The enantioselectivity is provided by the oxynitrilase and *in situ* conversion by the nitrilase serves to drive the equilibrium of the first step toward product. This could, in principle, also be achieved by using an S-selective nitrilase in combination with nonenzymatic hydrocyanation (as we have previously shown with an

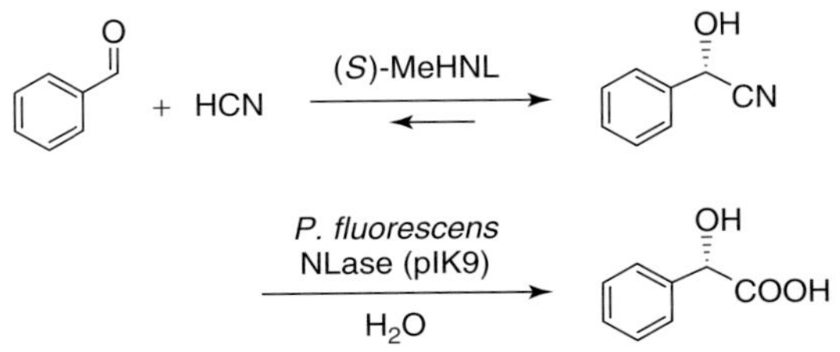

Scheme 8.13 Direct conversion of benzaldehyde to (S)-mandelic acid with a combi-CLEA.

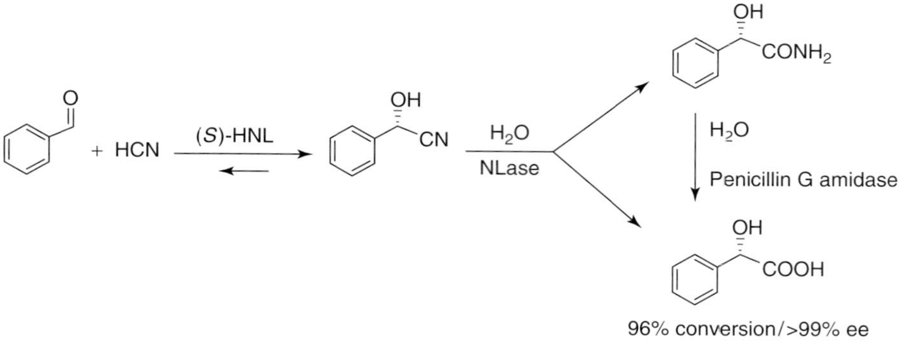

Scheme 8.14 A trienzymatic cascade process catalyzed by a triple-decker combi-CLEA.

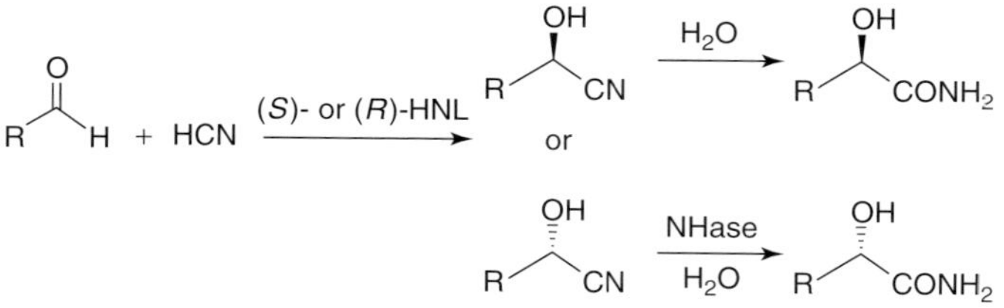

Scheme 8.15 Conversion of aldehydes to (S)-α-hydroxycarboxylic amides with a combi-CLEA.

R-nitrilase), but, unfortunately, there are no nitrilases that exhibit S-selectivity with mandelonitriles.

We later found that substantial amounts of the corresponding S-amide were also formed which led us to the idea of adding a third enzyme, penicillin G amidase, to catalyze the hydrolysis of the amide. In this way, using the same oxynitrilase/nitrilase combi-CLEA as shown in Scheme 8.13, in combination with an immobilized penicillin G amidase, S-mandelic acid was obtained as the sole product in high yield and enantiopurity (Chmura, A., manuscript in preparation). Subsequently, we showed that a combi-CLEA containing all three enzymes could be used to obtain S-mandelic acid (Scheme 8.14) in >99% ee at 96% conversion (Chmura, A., manuscript in preparation).

We also prepared a combi-CLEA from *M. esculenta* hydroxynitrile lyase and the alkaliphilic nitrile hydratase from *Nitriliruptor akaliphilus* (EC 4.2.1.84) and used it to catalyze the bienzymatic cascade process for the conversion of aldehydes to S-α-hydroxycarboxylic acid amides (Scheme 8.15) [87].

8.7
Reactor Design

The application of immobilized enzymes in batch processes generally involves recovery by filtration or centrifugation. Alternatively, the process can be adapted to continuous operation by employing the immobilized enzyme in a packed bed reactor. This generally requires fairly large particles in order to avoid a large pressure drop over the column. However, the use of large particles can lead to diffusion limitations, resulting in lower rates of reaction, and hence a compromise has to be found. One alternative is to use a fluidized bed which can contain very small particles which must, however, be relatively dense otherwise they will be blown out of the column.

8.7.1
Membrane Slurry Reactor

An interesting alternative is to use the CLEA in a membrane slurry reactor (MSR) [88]. The CLEA is retained in the reactor because it is too large to pass through the pores of the membrane. In contrast, the substrate and product can be pumped in and out of the reactor. CLEAs of small and/or broad particle size can be used, thus combining the advantages of high rates of reaction with ease of separation. This also enables better control of process conditions, eliminates downstream processing steps, and ensures a very efficient use of the biocatalyst. High catalyst loadings are possible, resulting in high space–time yields. In principle, any standard stirred tank reactor (continuous or batch) can be converted into a membrane slurry reactor by applying only minor changes, and many inexpensive, size-selective membranes are commercially available and can be used in the MSR.

Combination of the high volumetric and catalyst productivities and ease of preparation of CLEAs with the high catalyst loading capacity of an MSR and the retention of the catalyst within the reactor offers a very cost-effective system for performing continuous biotransformations on an industrial scale.

The practical utility of the MSR was demonstrated by performing the industrially important hydrolysis of penicillin G to 6-amino penicillanic acid (6-APA), the key intermediate in the synthesis of semisynthetic penicillins and cephalosporins, catalyzed by a penicillin amidase CLEA. The phenylacetic acid coproduct is an inhibitor and needs to be neutralized with a base. The conversion in time was monitored by measuring the amount of base consumed. The 6-APA product was isolated from the reactor effluent by crystallization at the isoelectric point (pH 4.3). Space–time yields up to 30.5 g l^{-1} h^{-1} were achieved using 10% catalyst loading at 20 °C compared to 18 g l^{-1} h^{-1} at 35 °C in current industrial processes based on batch operation. The MSR/CLEA system was allowed to operate over a period of two weeks without any loss in the catalyst efficiency or membrane fouling.

8.7.2
CLEAs in Microchannel Reactors

Microreactor technologies are rapidly gaining in popularity for the production of fine chemicals [89]. Key benefits are the extremely efficient heat exchange and mass transfer as a consequence of the much larger surface area to volume ratios compared to conventional reactors. Microchannel enzyme reactors can also enable the rapid screening and optimization of biotransformations. Consequently, there is a rapidly developing interest in the application of microchannel enzyme reactors, which generally require immobilization of the enzyme, either separately or attached to the microchannel surface [90, 91]. For example, Honda *et al.* [92] immobilized α-chymotrypsin CLEAs as an enzyme-polymer membrane on the inner surface of PTFE tubing of 500 μm diameter and 6 cm length by cross-linking with a mixture of formaldehyde and glutaraldehyde. The enzyme-immobilized microchannel reactor prepared in this way was tested in the hydrolysis of *N*-glutaryl-L-phenyl alanine and shown to be stable for 40 days of operation. The same groups [93] subsequently described the use of the same type of enzyme-immobilized microchannel reactor containing aminoacylase in an integrated microfluidic system for the continuous resolution of racemic *N*-acyl amino acids.

Littlechild and coworkers [94] used a different approach. They prepared CLEAs from a thermophilic L-aminoacylase from *Thermococcus litoralis*, which had been overexpressed in *E. coli*. The CLEAs were subsequently mixed with controlled pore glass and packed in capillary reactors fitted with a silica frit to contain them in the reactor. The CLEA microchannel reactor retained activity for at least two months during storage at 4 °C.

8.8
Conclusions and Prospects

The CLEA technology has many advantages in the context of industrial biotransformations. It is exquisitely simple and amenable to rapid optimization, which translates to low costs and short time-to-market. It is applicable to a wide variety of enzymes, including crude preparations and affording robust, recyclable catalysts that exhibit high retention of activity and improved properties, such as enhanced thermal stability, better tolerance to organic solvents, and are more resistant to proteolysis. The technique is applicable to the preparation of combi-CLEAs containing two or more enzymes, which can be advantageously used in catalytic cascade processes. The use of CLEAs immobilized in microchannel reactors has obvious potential in the rapid screening and optimization of biotransformations and, ultimately, in continuous production processes. In short, we believe that CLEAs will be widely applied in the future in industrial biotransformations and other areas requiring immobilized enzymes.

References

1. (a) Sheldon, R.A. and van Rantwijk, F. (2004) *Aust. J. Chem.*, **37**, 281; (b) Sheldon, R.A. (2000) *Pure Appl. Chem*, **72**, 1233.
2. Sheldon, R.A., Arends, I.W.C.E., and Hanefeld, U. (2007) *Green Chemistry and Catalysis*, Wiley-VCH Verlag GmbH, Weinheim.
3. (a) Hanefeld, U., Gardossi, L., and Magner, E. (2009) *Chem. Soc. Rev.*, **38**, 453; (b) Kallenberg, A.I., Van Rantwijk, F., and Sheldon, R.A. (2005) *Adv. Synth. Catal.*, **347**, 905; (c) Cao, L. (2005) *Curr. Opin. Chem. Biol.*, **9**, 217; (d) Bornsceuer, U.T. (2003) *Angew. Chem. Int. Ed.*, **42**, 3336; (e) Krajewska, B.

(2004) *Enz. Microb. Technol.*, **35**, 126.
4. Sheldon, R.A. (2007) *Adv. Synth. Catal.*, **349**, 1289.
5. Cao, L., van Langen, L.M., and Sheldon, R.A. (2003) *Curr. Opin. Biotechnol.*, **14**, 387, references cited therein.
6. (a) For more recent examples see: Khare, S.K. and Gupta, M.N. (1990) *Biotechnol. Bioeng.*, **35**, 94; (b) Tyagi, R., Batra, R., and Gupta, M.N. (1999) *Enz. Microb. Technol.*, **24**, 348.
7. St. Clair, N.L. and Navia, M.A. (1992) *J. Am. Chem. Soc.*, **114**, 7314.
8. (a) Margolin, A.L. (1996) *Tibtech*, **14**, 223; (b) Lalonde, J. (1997) *ChemTech*, **27** (2), 38; (c) Margolin, A.L. and Navia, M.A. (2001) *Angew. Chem. Int. Ed.*, **40**, 2204.
9. Cao, L., van Rantwijk, F., and Sheldon, R.A. (2000) *Org. Lett.*, **2**, 1361.
10. (a) For reviews see: Sheldon, R.A., Schoevaart, R., and van Langen, L.M. (2005) *Biocat. Biotrans.*, **23**, 141; (b) Sheldon, R.A., Sorgedrager, M., and Janssen, M.H.A. (2007) *Chim. Oggi, Chem. Today*, **25** (1), 48; (c) Sheldon, R.A. (2007) *Biochem. Soc. Trans.*, **35**, 1583; (d) Sheldon, R.A. and Patel, R.N. (eds) (2007) *Biocatalysis in the Pharmaceutical and Biotechnology Industries*, CRC Press, Boca Raton, pp. 351–362.
11. (a) Walt, D.R. and Agayn, V.L. (1994) *Trends Anal. Chem.*, **13**, 425; (b) Tashima, T., Imai, M., Kuroda, Y., Yagi, S., and Nakagawa, T. (1991) *J. Org. Chem.*, **56**, 694; (c) Migneault, I., Dartignuenave, C., Bertarand, M.J., and Waldron, K.C. (2004) *Biotechniques*, **37**, 790.
12. Wine, Y., Cohen-Hadar, N., Freeman, A., and Frolow, F. (2007) *Biotechnol. Bioeng*, **98**, 711.
13. Mateo, C., Palomo, J.M., van Langen, L.M., van Rantwijk, F., and Sheldon, R.A. (2004) *Biotechnol. Bioeng.*, **86**, 273.
14. Schoevaart, R. and Kieboom, T. (2002) *Carbohydr. Res.*, **337**, 899 and (2001), **334**, 1.
15. Lopez-Gallego, F., Betancor, L., Hidalgo, A., Alonso, N., Fernandez-Lafuenta, R., and Guisan, J.M. (2005) *Biomacromol*, **6**, 1639.
16. Schoevaart, R., Wolbers, M.W., Golubovic, M., Ottens, M., Kieboom, A.P.G., van Rantwijk, F., and van der Wielen, L.A.M. (2004) *Biotechnol. Bioeng.*, **87**, 754.
17. Yu, H.W., Chen, H., Wang, X., Yang, Y.Y., and Ching, C.B. (2006) *J. Mol. Catal. B: Enzym.*, **43**, 124.
18. Wilson, L., Illanes, A., Pessela, B.C.C., Abian, O., Fernandez_Lafuenta, R., and Guisan, J.M. (2004) *Biotechnol. Bioeng.*, **86**, 558.
19. Kim, M.I., Kim, J., Lee, J., Jia, H., Na, H.B., Youn, J.K., Kwak, J.H., Dohnalkova, A., Grate, J.W., Wang, P., Hyeon, T., Park, H.G., and Chang, H.N. (2007) *Biotechnol. Bioeng.*, **96**, 210.
20. Fischback, M.B., Youn, J.K., Zhao, X., Wang, P., Park, H.G., Chang, H.N., Kim, J., and Ha, S. (2006) *Electroanalysis*, **18**, 2016.
21. Lee, J., Lee, D., Oh, E., Kim, J., Kim, Y.-P., Jin, S., Kim, H.S., Hwang, Y., Kwak, J.H., Park, J.-G., Shin, C.-H., Kim, J., and Hyeon, T. (2005) *Angew. Chem. Int. Ed.*, **44**, 7427.
22. Kim, J., Lee, J., Na, H.B., Kim, B.C., Youn, J.K., Kwak, J.H., Moon, K., Lee, E., Kim, J., Park, J., Park, J., Dohnalkova, A., Park, H.G., Gu, M.B., Chang, H.N., Grate, J.W., and Hyeon, T. (2005) *Small*, **1**, 1203.
23. Hilal, N., Nigmatullin, R., and Alpatova, A. (2004) *J. Membr. Sci.*, **238**, 131.
24. Schoevaart, W.R.K., van Langen, L.M., van den Dool, R.T.M., and Boumans, J.W.L. (2006) WO Patent 2006/046865 A2, to CLEA Technologies.
25. Shah, S., Sharma, A., and Gupta, M.N. (2006) *Anal. Biochem.*, **351**, 207.
26. Dalal, S., Kapoor, M., and Gupta, M.N. (2007) *J. Mol. Datal. B: Enzym.*, **44**, 128.
27. Dala, S., Sharma, A., and Gupta, M.N. (2007) *Chem. Cent. J.*, 1.
28. (a) Lopez-Serrano, P., Cao, L., van Rantwijk, F., and Sheldon, R.A. (2002) *Biotechnol. Lett.*, **24**, 1379; (b) Lopez Serrano, P., Cao, L., van Rantwijk, F., and Sheldon, R.A. (2002) WO Patent 02/061067 A1, to Technische Universiteit Delft.
29. Prabhavathi, B.L.A., Guo, Z., and Xu, Z. (2009) *J. Am. Oil Chem. Soc.*, **86**, 637.
30. Yu, H.W., Chen, H., Wang, X., Yang, Y.Y., and Ching, C.B. (2006) *J. Mol. Catal. B: Enzym.*, **43**, 124–127.

31. Gupta, P., Dutt, K., Misra, S., Raghuwanshi, S., and Saxena, R.K. (2009) *Bioresou. Technol.*, **100**, 4074.
32. Kumari, V., Shah, S., and Gupta, M.N. (2007) *Energy Fuels*, **21**, 368.
33. Özdemirhan, D., Sezer, S., and Sönmez, Y. (2008) *Tetrahedron: Asymmetry*, **19**, 2717.
34. Majumder, A.B., Monsal, K., Singh, T.P., and Gupta, M.N. (2008) *Biocatal. Biotransform.*, **26**, 235.
35. Hara, P., Hanefeld, U., and Kanerva, L.T. (2008) *J. Mol. Catal. B: Enzym.*, **50**, 80.
36. Theil, F. (2000) *Tetrahedron*, **56**, 2905.
37. (a) Kourist, R., De Maria, P.D., and Bornscheuer, U.T. (2008) *Chem. Biol. Chem.*, **9**, 491; (b) Henke, E., Pleiss, J., and Bornscheuer, U.T. (2002) *Angew. Chem. Int. Ed.*, **41**, 3211.
38. Hobbs, H.R., Kondor, B., Stephenson, P., Sheldon, R.A., Thomas, N.R., and Poliakoff, M. (2006) *Green Chem.*, **8**, 816.
39. Matsuda, T., Tsuji, K., Kamitanaka, T., Harada, T., Nakamura, K., and Ikariya, T. (2005) *Chem. Lett.*, **34**, 1102.
40. Dijkstra, Z.J., Merchant, R., and Keurentjes, J.T.F. (2007) *J. Supercrit. Fluids*, **41**, 102.
41. Ruiz Toral, A., de los Rios, A.P., Hernandez, F.J., Janssen, M.H.A., Schoevaart, R., van Rantwijk, F., and Sheldon, R.A. (2007) *Enz. Microb. Technol.*, **40**, 1095.
42. Vafiadi, C., Topakas, E., and Christakopoulos, P. (2008) *J. Mol. Catal. B: Enzym.*, **54**, 35.
43. Vafiadi, C., Topakas, E., Alissandratos, A., Faulds, C.B., and Christakopoulos, P. (2008) *J. Biotechnol.*, **133**, 497.
44. Vafiadi, C., Topakas, E., Nahmias, V.R., Faulds, C.B., and Christakopoulos, P. (2009) *J. Biotechnol.*, **139**, 124.
45. Eid Fazary, A., Ismadji, S., and Ju, Y.-H. (2009) *Int. J. Biol. Macromol*, **44**, 240.
46. Montoro-Garcia, S., Gil-Ortiz, F., Navarro-Fernandez, J., Rubio, V., Garcia-Carmona, F., and Sanchez-Ferrer, A. (2009) *Bioresour. Technol.*, **101**, 331.
47. Miyazawa, T. (1999) *Amino Acids*, **16**, 191.
48. van Rantwijk, F. and Sheldon, R.A. (2004) *Tetrahedron*, **60**, 501.
49. Zhang, X.Z., Wang, X., Chen, S., Fu, X., Wu, X., and Li, C. (1996) *Enz. Microb. Technol.*, **19**, 538.
50. Nuijens, T., Cusan, C., Kruijtzer, J.A.W., Rijkers, D.T.S., Liskamp, R.M.J., and Quaedflieg, P.J.M. (2009) *Synthesis*, 809.
51. Nuijens, T., Kruijtzer, J.A.W., Cusan, C., Rijkers, D.T.S., Liskamp, R.M.J., and Quaedflieg, P.J.L.M. (2009) *Tetrahedron Lett.*, **50**, 2719.
52. Nuijens, T., Cusan, C., Kruijtzer, J.A.W., Rijkers, D.T.S., Liskamp, R.M.J., and Quaedflieg, P.J.L.M. (2009) *J. Org. Chem.*, **74**, 5145.
53. Eggen, I.F. and Boeriu, C.G. (2007) WO patent 2007/082890, and US Patent Appl. 2009053760 A1, to Organon.
54. Cao, L., van Langen, L.M., van Rantwijk, F., and Sheldon, R.A. (2001) *J. Mol. Catal. B: Enzym.*, **11**, 665.
55. Wegman, M.A., Janssen, M.H.A., van Rantwijk, F., and Sheldon, R.A. (2001) *Adv. Synth. Catal.*, **343**, 559.
56. van Langen, L.M., Oosthoek, N.H.P., van Rantwijk, F., and Sheldon, R.A. (2003) *Adv. Synth. Catal.*, **345**, 797.
57. Illanes, A., Wilson, L., Caballero, E., Fernandez-Lafuente, R., and Guisan, J.M. (2006) *Appl. Biochem. Biotechnol.*, **133** (3), 189.
58. (a) Illanes, A., Wilson, L., Altamirano, C., Cabrera, Z., Alvarez, L., and Aguirre, C. (2007) *Enz. Microb. Technol.*, **40**, 195; (b) Illanes, A., Wilson, L., and Aguirre, C. (2009) *Appl. Biochem. Biotechnol.*, **157**, 98.
59. Wilson, L., Illanes, A., Soler, L., and Henriquez, M.J. (2009) *Proc. Biochem.*, **44**, 322.
60. Pchelintsev, N.A., Youshko, M.I., and Svedas, V.K. (2009) *J. Mol. Catal. B: Enzym.*, **56**, 202.
61. Bode, M.L., van Rantwijk, F., and Sheldon, R.A. (2003) *Biotechnol. Bioeng.*, **84**, 710.
62. Kaul, P., Stolz, A., and Banerjee, U.C. (2007) *Adv. Synth. Catal.*, **349**, 2167.
63. Mateo, C., Fernandes, B., Van Rantwijk, F., Stolz, A., and Sheldon, R.A. (2006) *J. Mol. Catal. B: Enzym.*, **38**, 154.
64. Gauer, R., Pant, H., Jain, R., and Khare, S.K. (2006) *Food Chem.*, **97**, 426.
65. Ortiz-Soto, M.E., Rudino-Pinera, E., Rodriguez-Alegria, M.E., and

Lopez-Munguia, A. (2009) *BMC Biotechnol.*, **9**, 68, 2167.
66. (a) Cabana, H., Jone, J.P., and Agathos, S.N. (2007) *J. Biotechnol.*, **132**, 23; (b) Cabana, H., Jones, J.P., and Agathos, S.N. (2009) *Biotechnol. Bioeng.*, **102**, 1582.
67. Aytar, B.S. and Bakir, U. (2008) *Process Biochem.*, **43**, 125.
68. Wilson, L., Betancor, L., Fernandez-Lorente, G., Fuentes, M., Hidalgo, A., Guisan, J.M., Pessela, B.C.C., and Fernandez-Lafuente, R. (2004) *Biomacromolecules*, **5**, 814.
69. Grateron, C., Barosa, O., Rueda, N., Ortiz-lopez, C., and Torres, R. (2007) *J. Biotechnol.*, **131S**, S87.
70. Perez, D.I., Van Rantwijk, F., and Sheldon, R.A. (2009) *Adv. Synth. Catal.*, **351**, 2133.
71. See also Roberge, C., Amos, D., Pollard, D., and Devine, P. (2009) *J. Mol. Catal. B: Enzym.*, **56**, 41.
72. van Deurzen, M.P.J., van Rantwijk, F., and Sheldon, R.A. (1997) *Tetrahedron*, **53**, 13183.
73. Leak, D.J., Sheldon, R.A., Woodley, J.M., and Adlercreutz, P. (2009) *Biocat. Biotrans.*, **27**, 1.
74. van de Velde, F., Könemann, L., van Rantwijk, F., and Sheldon, R.A. (1998) *Chem. Commun.*, 1891.
75. Correia, I., Aksu, S., Adao, P., Pessoa, J.C., Sheldon, R.A., and Arends, I.W.C.E. (2008) *J. Inorg. Biochem.*, **102**, 318.
76. van Pelt, S., van Rantwijk, F., and Sheldon, R.A. (2008) *Chim. Oggi/Chem. Today*, **26** (3), 2.
77. DiCosimo, R. (2006) in *Biocatalysis in the Pharmaceutical and Biotechnology Industries* (ed. R. N. Patel), CRC Press, Boca Raton, pp. 1–26.
78. van Pelt, S., Quignard, S., Kubac, D., Sorokin, D.Y., van Rantwijk, F., and Sheldon, R.A. (2008) *Green Chem.*, **10**, 395.
79. See also: Kubac, D., Kaplan, O., Elisakova, V., Patek, M., Vejvoda, V., Slamova, K., Tothova, A., Lemaire, M., Gallienne, E., Lutz-Wahl, S., Fischer, L., Kuzma, M., Pelantova, H., van Pelt, Sander, S., Bolte, J., Kren, V., and Martinkova, L. (2008) *J. Mol. Catal. B: Enzym.*, **50**, 107.
80. van Langen, L.M., Selassa, R.P., van Rantwijk, F., and Sheldon, R.A. (2005) *Org. Lett.*, **7**, 327.
81. Cabirol, F.L., Hanefeld, U., and Sheldon, R.A. (2006) *Adv. Synth. Catal.*, **348**, 1645.
82. Chmura, A., van der Kraan, G.M., Kielar, F., van Langen, L.M., van Rantwijk, F., and Sheldon, R.A. (2006) *Adv. Synth. Catal.*, **348**, 1655.
83. Roberge, C., Fleitz, F., Pollard, D., and Devine, P. (2007) *Tetrahedron Lett.*, **48**, 1473.
84. Cabirol, F.L., Tan, P.L., Tay, B., Cheng, S., Hanefeld, U., and Sheldon, R.A. (2008) *Adv. Synth. Catal.*, **350**, 2329.
85. Bruggink, A., Schoevaart, R., and Kieboom, T. (2003) *Org. Proc. Res. Dev.*, **7**, 622.
86. Mateo, C., Chmura, A., Rustler, S., van Rantwijk, F., Stolz, A., and Sheldon, R.A. (2006) *Tetrahedron Asymmetry*, **17**, 320.
87. van Pelt, S., van Rantwijk, F., and Sheldon, R.A. (2009) *Adv. Synth. Catal.*, **351**, 397.
88. Sorgedrager, M.J., Verdoes, D., van der Meer, H., and Sheldon, R.A. (2008) *Chim. Oggi Chem. Today*, **26** (4), 24.
89. Roberge, D.M., Zimmermann, B., Rainone, F., Gottsponer, M., Eyholzer, M., and Kockmann, N. (2008) *Org. Proc. Res. Dev.*, **12**, 905.
90. Miyazaki, M. and Maeda, H. (2006) *Trends Biotechnol.*, **24**, 463, references cited therein.
91. Thomsen, M.S., Pölt, P., and Nidetzky, B. (2007) *Chem. Commun.*, 2527.
92. Honda, T., Miyazaki, M., Nakamura, H., and Maeda, H. (2005) *Chem. Commun.*, 5062.
93. Honda, T., Myazaki, M., Yamaguchi, Y., Nakamura, H., and Maeda, H. (2007) *Lab Chip*, **7**, 366.
94. Hickey, A.M., Marle, L., McCreedy, T., Watts, P., Greenway, G.M., and Littlechild, J.A. (2007) *Biochem. Soc. Trans.*, **35**, 1621.

9
Application of Whole-Cell Biocatalysts in the Manufacture of Fine Chemicals
*Michael Schwarm**

9.1
Introduction: Early Applications of Biocatalysis for Amino Acid Manufacture at Evonik Degussa

The development of industrially viable processes for the manufacture of enantiomerically pure compounds has received considerable interest in recent years, especially due to the growing importance of chiral building blocks in the synthesis of pharmaceutically active drugs. Many different technologies are well established for the preparation of such compounds; for example, chemical transformations of starting materials from the chiral pool, chemical resolution of racemic mixtures via diastereomeric salt pairs, asymmetric synthesis using chiral auxiliaries or catalysts, and modern biotechnology including fermentation and biocatalysis, which are becoming increasingly important.

At Evonik Degussa, enzymes have been successfully applied for the synthesis of amino acids for almost 30 years [1], starting with L-aminoacylase for the production of enantiomerically pure amino acids such as L-methionine. This process normally starts from racemic N-acetylamino acids which are readily available, for example, by Strecker synthesis and subsequent acetylation, by amidocarbonylation of the corresponding aldehydes, or by alkylation of acetamidomalonic ester and subsequent selective hydrolysis.

Treatment with commercially available L- or D-acylases in aqueous solution under neutral conditions then affords the desired enantiomerically pure L- or D-amino acids which can be isolated either by direct crystallization or by purification using ion exchange. The remaining N-acetyl-D- or N-acetyl-L-amino acids can be recycled easily either by chemical racemization using acetic anhydride or by treatment with an N-acetylamino acid racemase (Scheme 9.1). While most proteinogenic L-amino acids are nowadays available on a large scale through fermentation, L-methionine is still produced at industrial scale using the L-acylase process in connection with an enzyme membrane reactor [2]. However, due to the unusually broad substrate

*Dedicated to Prof. Karlheinz Drauz on the occasion of his 60th birthday in acknowledgment of his pioneering work on biocatalysis at Evonik Degussa.

Pharmaceutical Process Chemistry. Edited by Takayuki Shioiri, Kunisuke Izawa, and Toshiro Konoike
Copyright © 2011 WILEY-VCH Verlag GmbH & Co. KGaA, Weinheim
ISBN: 978-3-527-32650-1

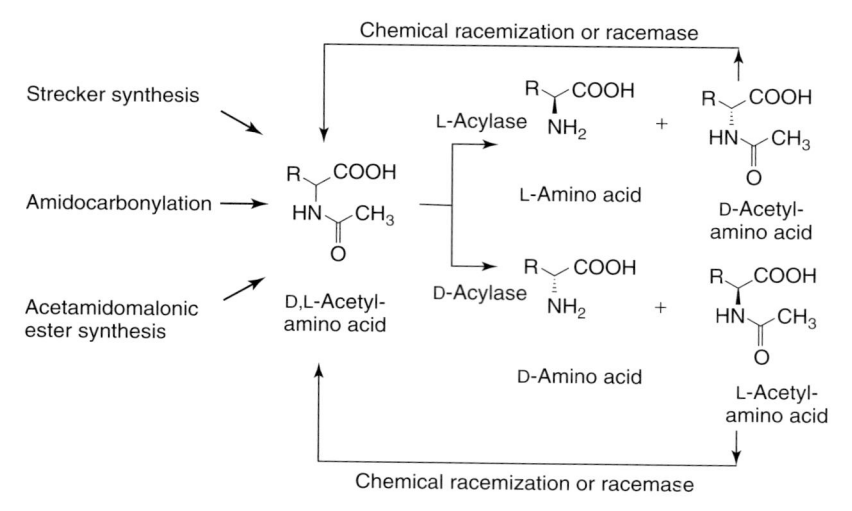

Scheme 9.1 Acylase process.

specificity especially of the L-acylase, these enzymes are particularly suitable for the manufacture of many other amino acids which are not available through fermentation. The robustness of both L- and D-acylases under process conditions allows for rapid scale-up in many cases, and therefore a great variety of different L- or D-amino acids have been produced at Evonik Degussa at a technical scale. Examples include D-serine, D-tryptophan, L- and D-3-(1′-naphthyl)alanine, and L- and D-3-(2′-naphthyl)alanine which have been manufactured using L-acylase, and again D-tryptophan and D-3-(4′-chlorophenyl)alanine, which have been produced using D-acylase (Scheme 9.2).

In the 1990s, the company also developed the application of resting whole cells of *Agrobacterium radiobacter* containing a D-hydantoinase and a D-carbamoylase for the

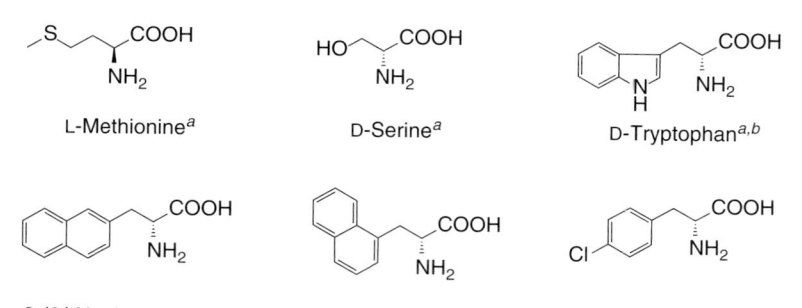

L-Methionine[a]   D-Serine[a]   D-Tryptophan[a,b]

D-3-(2′-Naphthyl)alanine[a]   D-3-(1′-Naphthyl)alanine[a]   D-3-(4′-Chlorophenyl)alanine[b]

[a] Prepared using L-acylase
[b] Prepared using D-acylase

Scheme 9.2 Enantiomerically pure amino acids manufactured using acylases.

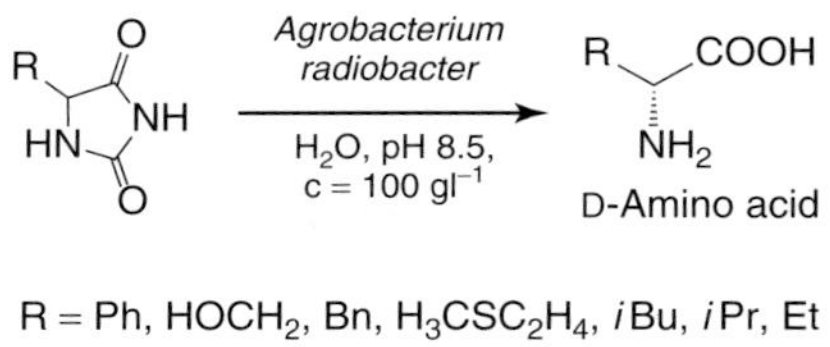

R = Ph, HOCH$_2$, Bn, H$_3$CSC$_2$H$_4$, iBu, iPr, Et

Scheme 9.3 Early D-hydantoinase whole-cell biocatalyst process.

preparation of a variety of D-amino acids (Scheme 9.3) [3]. The process starts from 5-substituted hydantoins which are readily available; for example by condensation of the respective aldehydes with hydantoin and subsequent catalytic hydrogenation or from the respective L-amino acids and racemization of the resulting hydantoins under alkaline conditions. The whole-cell biocatalyst was obtained from Debi Recordati and contained a D-hydantoinase, a D-carbamoylase, and most likely a hydantoin racemase. It was applied in aqueous solution at a slightly alkaline pH of about 8.5 to support the *in situ* racemization of the remaining L-hydantoin. Taking advantage of this dynamic resolution, this process allows for a high yield conversion of the hydantoin into the desired D-amino acid, without the need for a separate racemization of the undesired enantiomer as in the classical acylase process. The broad substrate specificity of the biocatalyst allows for the preparation of many different D-amino acids, such as D-phenylglycine, D-serine, D-phenylalanine, D-methionine, D-leucine, D-valine, and D-2-aminobutyric acid; D-phenylalanine has been produced at the ton scale with this process. However, the hydantoin of homophenylalanine was not accepted by this D-hydantoinase biocatalyst.

Alternatively, a commercial D-hydantoinase was used to convert certain racemic hydantoins into the corresponding *N*-carbamoyl-D-amino acids, followed by chemical deprotection using sodium nitrite under acidic conditions to obtain the free D-amino acids.

In parallel to these developments, production of L-*tert*-leucine (L-Tle) by enzymatic reductive amination was introduced at an industrial scale (Scheme 9.4). Owing to its bulky, hydrophobic side chain, this unusual amino acid has found widespread application both as chiral auxiliaries and catalysts in asymmetric synthesis and as a building block in new developmental drugs [4]. Introduction into new protease inhibitors for the treatment of viral diseases such as AIDS or hepatitis C has been particularly successful and resulted in a variety of drugs which are either already commercial or in advanced stages of clinical development [5]. The manufacturing process for L-Tle starts with trimethylpyruvate which is commercially available, but can also be obtained from pinacolin by treatment with chlorine, alkaline hydrolysis, and subsequent oxidation of the resulting 3,3-dimethyl-2-hydroxybutyric acid with either permanganate [6a] or aqueous hypochlorite solution in the presence of a ruthenium catalyst such as RuO$_2$·H$_2$O [6b]. Trimethylpyruvate is then converted into the desired chiral amino acid in very good yields and excellent optical purities by leucine dehydrogenase (LeuDH) from *Bacillus* sp. This transformation requires ammonia and, more importantly, NADH as a cofactor for the enzyme. As this cofactor is a rather expensive raw material, it needs to be regenerated and recycled

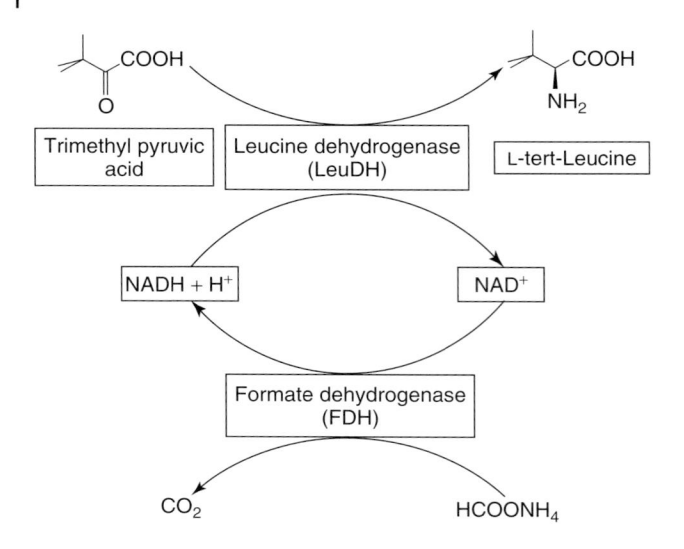

Scheme 9.4 Synthesis of L-*tert*-leucine.

in situ to make the process economically attractive. This regeneration is achieved by means of a second enzyme, formate dehydrogenase (FDH) from *Candida boidinii*, which irreversibly converts formate into carbon dioxide to regenerate NADH from NAD$^+$, thus closing the catalytic cycle. This process is now operated routinely at the ton scale at Evonik Rexim in Ham (France). It is also applicable for the production of a great variety of other bulky amino acids such as L-neopentylglycine [7].

Nowadays, the rapid progress of molecular biology has enabled researchers in both academia and industry to evolve and improve enzymes with respect to chemo- and enantioselectivity, temperature stability, pH sensitivity, and other properties. Optimized expression systems and high–cell density fermentation protocols have been developed to produce a desired enzyme efficiently and in large quantities. At the same time, high-throughput screening (HTS) technologies are available to identify the most suitable biocatalyst for a desired transformation as quickly as possible.

However, the use of isolated enzymes can be disadvantageous because of laborious isolation and purification of the enzymes and also often a reduced stability under process conditions. Furthermore, for biotransformations in the presence of more than one isolated enzyme (such as the dynamic resolution of hydantoins and redox processes), more than one fermentation run is required to produce the respective enzymes. Tailor-made whole-cell biocatalysts that contain all the desired enzymes in overexpressed form represent an economically attractive alternative, as they combine the production of all the desired enzymes for a specific biotransformation with only one fermentation run. Moreover, the application of whole cells also allows the use of cell-internal cofactor pools, greatly reducing or even eliminating the need for the addition of expensive external cofactors, for example, in keto acid-to-amino acid or ketone-to-alcohol transformations. Therefore, great efforts have been made at Evonik Degussa in recent years to develop such types of "designer bugs" which combine different tailor-made enzymes and, where necessary, also the required cofactor-regeneration system within the same cell for efficient and cost-attractive production of enantiomerically pure compounds, particularly amino acids and chiral alcohols.

9.2
Hydantoinase Biocatalysts

Recombinant hydantoinase biocatalysts have been the first examples of this successful approach. In the beginning, a D-hydantoinase, an L-carbamoylase, and a hydantoin racemase were combined in an *Escherichia coli* expression system. The enantioselectivity of the employed D-hydantoinase was only moderate, therefore the D-carbamoylamino acid was formed more quickly compared to the respective L-enantiomer. However, the employed L-carbamoylase was perfectly L-selective, converting only the L-carbamoylamino acid into the desired free L-amino acid. As this reaction is irreversible, the product was removed from the system of reaction equilibria. As the undesired D-carbamoylamino acid was reconverted to the respective D-hydantoin which racemizes under the reaction conditions, supported by the hydantoin racemase of the whole-cell biocatalyst, all the initial hydantoin was finally converted to the desired free L-amino acid (Scheme 9.5). However, because of the preferential D-selectivity of the employed hydantoinase and the necessary reconversions and racemizations resulting from that, this initial process suffered from low productivity.

The hydantoinase and the gene encoding it were therefore investigated in more detail. In the enzyme, four positions were identified at which certain amino acid exchanges result in an enhanced activity of the enzyme, while one other amino acid appeared to have a decisive influence on the enantioselectivity of the enzyme.

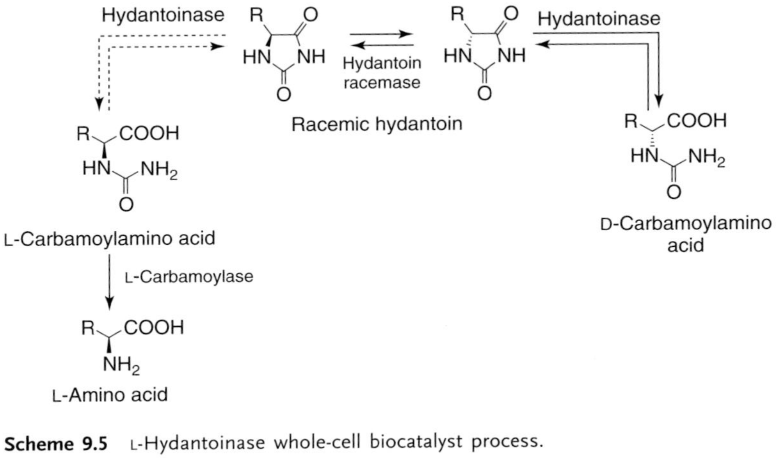

Scheme 9.5 L-Hydantoinase whole-cell biocatalyst process.

On the basis of this understanding, the hydantoinase gene was subjected to saturation mutagenesis (a special directed evolution method which modifies only a defined area of the protein-encoding gene in order to obtain protein variants with all the 20 possible amino acids at the selected position of the protein) at the position encoding the amino acid responsible for enantioselectivity. A great variety of different mutants was obtained from this, some of them with even increased D-selectivity, while others were obviously less D-selective. Some of these were selected and subjected to subsequent rounds of this directed evolution process (creation of gene variants by randomly mutating the DNA of the gene, then selection of the mutants possessing the desired properties by a suitable assay and finally DNA sequencing of the selected mutants to identify what has occurred to the DNA of the protein-encoding gene), and after the third cycle one mutant was identified which was more L- than D-selective. This understanding is based on HPLC analysis of crude reaction mixtures. The wild-type W produced more D- than L-carbamoylamino acid. Some mutants produced even higher quantities of the undesired D-enantiomer, while other mutants already showed an increase in the concentration of the desired L-carbamoylamino acid compared to the wild-type W. After the third cycle, the chiral HPLC analysis provided higher concentrations for the L-carbamoylamino acid than for the D-enantiomer for one mutant. This was apparently the first time that inversion of the enantioselectivity of an enzyme by directed evolution had been described in the literature [8].

This evolved gene was then transferred into the *E. coli* expression system which already contained the genes for the L-carbamoylase and the hydantoin racemase. The resulting whole-cell biocatalyst was then compared to the original one, taking production of L-methionine as an example. Concentration of the starting material D,L-methionine hydantoin dropped very rapidly from its initial value of 0.4 M. After 1 h, almost all the starting material had been consumed. At that time, the intermediate D(L)-carbamoylamino acid was still present in significant amounts, however, after 4–6 h more than 99% conversion could be achieved for the desired L-methionine, which is a remarkable improvement in comparison with the nonevolved original biocatalyst which only achieved 60–70% conversion in the same time. The optical purity of the formed L-methionine was excellent, with an enantiomeric excess of >99.5%.

This new biocatalyst has been applied successfully at an industrial scale for the production of L-norvaline (L-Nva) (Scheme 9.6) which is an important building

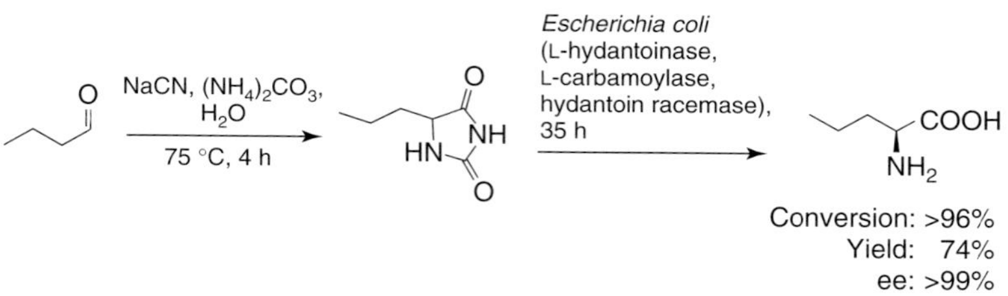

Scheme 9.6 Manufacture of L-norvaline.

block for commercial drugs as well as for drugs under clinical development [9]. L-Nva is in principle an excellent substrate for L-acylase, but preparation of the required starting material *N*-acetyl-D,L-norvaline is laborious and subsequent reracemization of the undesired D-enantiomer is required for an economic process. Compared to that, the new hydantoinase process offered significant advantages. Racemic norvaline hydantoin, the required starting material, is readily available in a Bucherer–Bergs reaction from butyric aldehyde, NaCN, and $(NH_4)_2CO_3$ in water at 75 °C over 4 h. Subsequent treatment with an optimized *E. coli* whole-cell biocatalyst containing an L-hydantoinase, an L-carbamoylase, and a hydantoin racemase afforded the desired L-Nva in >96% conversion after 35 h, starting from an initial 0.5 M concentration of the Nva hydantoin. The resulting solution of crude L-Nva was then ultrafiltered to remove the biocatalyst and subjected to ion exchange to remove salts and impurities. Concentration and crystallization afforded pure L-Nva in 74% yield (unoptimized) and an excellent enantiomeric excess of >99%. The removed biocatalyst was recycled for use in subsequent batches.

Similarly, this L-hydantoinase biocatalyst has proven highly valuable for the manufacture of L-ε-oxonorleucine acetal (L-ONA) (Scheme 9.7) which is another building block for developmental drugs [10]. This amino acid is a particularly challenging one to work with as it is highly water soluble and, at the same time, highly sensitive to acidic conditions due to its acetal moiety. Therefore, the normal acylase process which is so valuable for the production of many other amino acids could not be applied in this case, as it requires acidic pH shifts for separation of the free L-amino acid from the unconverted *N*-acetylamino acid, either by extraction and crystallization or by ion exchange. These disadvantages could be overcome by the hydantoinase process. The synthesis started with glutardialdehyde monoethyleneglycol acetal which can be obtained from glutardialdehyde and ethylene glycol following known procedures [11]. This monoaldehyde was then converted to the required hydantoin using hydrocyanic acid and ammonium carbonate, again following a Bucherer–Bergs protocol in analogy to established procedures [12]. Starting from 0.5 M of the hydantoin, subsequent treatment with the L-hydantoinase biocatalyst over 8 h then afforded the desired L-ONA in >99% conversion. The biocatalyst was then removed by ultracentrifugation and ultrafiltration. Subsequent charcoal treatment was employed to remove impurities, and then L-ONA was precipitated by addition of acetone and isolated in 69% yield. The enantiomeric excess was >99.4%.

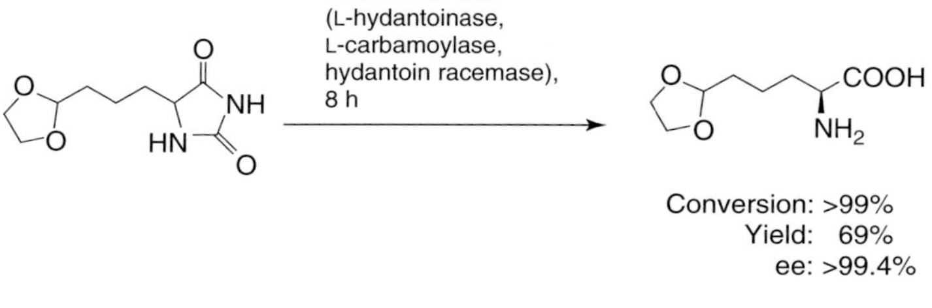

Scheme 9.7 Manufacture of L-ε-oxonorleucine acetal (L-ONA).

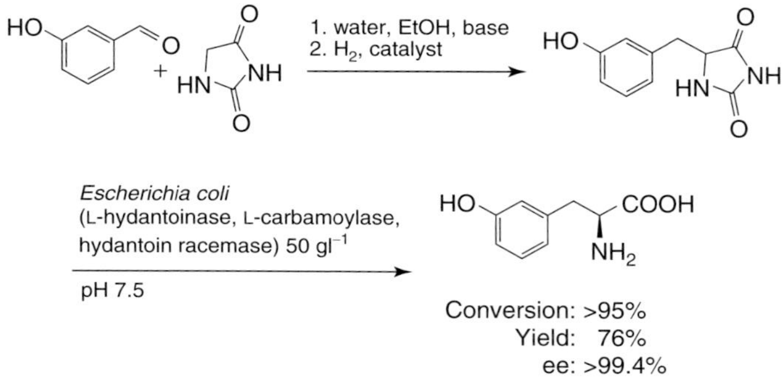

Scheme 9.8 Manufacture of L-*meta*-tyrosine.

In yet another example, the new hydantoinase biocatalyst has been employed for the preparation of production quantities of L-*m*-hydroxyphenylalanine (L-*meta*-tyrosine, L-*m*-Tyr) (Scheme 9.8) which has also been used as a building block for potential new drugs. Recently, preparation of L-*m*-Tyr using a classical kinetic resolution approach with alcalase as the resolving enzyme was described [13]; this provided the desired chiral amino acid in excellent optical purity (>99.5% ee). In contrast to that, the process using the hydantoinase/carbamoylase whole-cell biocatalyst is a dynamic resolution which generally offers the potential for significantly higher yields compared to kinetic resolutions. In this case, the starting hydantoin was prepared by condensation of commercially available *m*-hydroxybenzaldehyde and hydantoin in water/ethanol under base catalysis, followed by catalytic hydrogenation of the intermediate 3-hydroxybenzylidene hydantoin. The saturated 3-hydroxybenzylhydantoin was then suspended in water at a concentration of $62\,g\,l^{-1}$ and treated with the *E. coli* L-hydantoinase biocatalyst at pH 7.5 and a (wet) cell concentration of $50\,g\,l^{-1}$. After 24 h, the conversion was >95%, and after standard workup of heat treatment, acidification, addition of charcoal, filtration to remove the biocatalyst, neutralization of the filtrate, filtration of the crude amino acid, dissolution at acidic pH, charcoal treatment, and reprecipitation at neutral pH, L-*m*-Tyr was obtained in an unoptimized 76% yield and a very good optical purity of >99.4% ee.

Very similar biocatalysts were developed for the production of D-amino acids, using D-selective hydantoinases and D-carbamoylases. These whole-cell biocatalysts were shown to be very useful for the production of D-tryptophan and D-tyrosine, for example. The permanent *in situ* racemization, which is required for this dynamic resolution process, is ensured by means of a hydantoin racemase, which is also expressed in the whole-cell system. Thus, complete conversion of the racemic hydantoin to the desired D-amino acid is eventually achieved. This transformation has been run at a pilot scale for the manufacture of kilogram-scale quantities of D-tyrosine (Scheme 9.9). This process started from readily available natural L-tyrosine which reacts with sodium cyanate to give the *N*-carbamoylamino acid which was then cyclized in HCl, at elevated temperature, to the hydantoin. Treatment with aqueous ammonia for 6 h at pH 11 and 50 °C afforded the

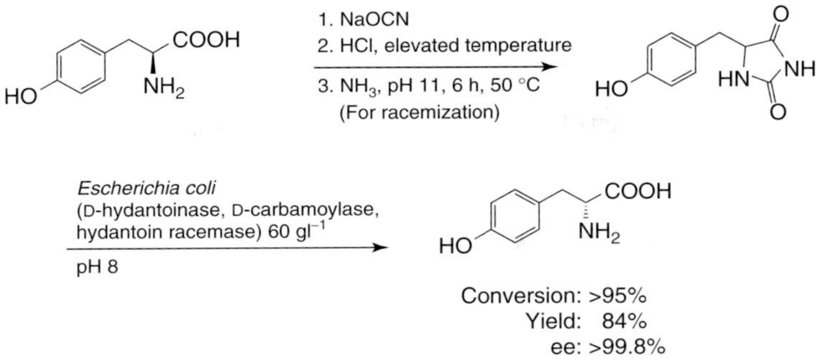

Scheme 9.9 Manufacture of D-tyrosine.

required racemic hydantoin. This was suspended in water at pH 8 at a concentration of 41 g l^{-1} and treated with the *E. coli* whole-cell biocatalyst containing a D-hydantoinase, a D-carbamoylase, and a hydantoin racemase at a biocatalyst concentration of 60 g l^{-1}. After 24 h, the conversion to the desired D-amino acid was >95%. After an easy and straightforward workup, D-Tyr was isolated in 84% yield and an excellent optical purity of >99.8% ee.

9.3
Amino Acid Dehydrogenase Biocatalysts

More recently, whole-cell biocatalysts were also developed for the reductive amination of α-keto acids to enantiomerically pure L-amino acids. As this transformation is not a resolution and the enzymes are perfectly L-specific, the enantiomeric purities of the L-amino acids produced have always been excellent, exceeding 99% ee. The first whole-cell catalysts for reductive amination [14] contained the gene encoding for LeuDH from *Bacillus cereus* and, in addition, a gene encoding for a mutant of the original FDH from *C. boidinii*. This mutant FDH had previously been shown to be more stable as compared to the wild-type enzyme [15]. However, LeuDH was known to be a highly active enzyme with a specific activity of 400 U mg^{-1}, while the FDH typically exhibited a specific activity of around 6 U mg^{-1} [14]. Therefore, it was decided to combine a medium–copy number plasmid containing the gene for the LeuDH with a high–copy number plasmid containing the FDH gene in the same host in order to end up with a biocatalyst with more balanced activities of the expressed enzymes.

This goal was achieved by Dr. Altenbuchner's research group at Stuttgart University with the construction of the new *E. coli* BW 3110 biocatalyst which exhibited activities of 2 U mg^{-1} for the LeuDH and up to 0.3 U mg^{-1} for FDH. Most likely, this can be explained by increased production of FDH due to the higher copy number vector for the gene encoding for FDH.

This LeuDH/FDH biocatalyst has the potential to reduce the costs of the reductive amination processes significantly due to the following reasons:

- Only one fermentation is required to produce the biocatalyst compared to two separate fermentations to produce the cells separately containing the LeuDH and the FDH, respectively.
- The biocatalyst is suitable for high–cell density fermentations.
- No laborious isolation and purification of the enzymes is required.
- No or only very little external cofactor is required, because this is already contained in the whole-cell biocatalyst.

The requirement for any amounts of external cofactor in addition to the intracellular quantities of NAD^+ was investigated in more detail for the example of conversion of trimethylpyruvic acid (TMP) to L-Tle [14]. Basically, the reaction proceeded without the need for external addition of cofactor as expected. However, at larger substrate concentrations above 500 mM, significant inhibition effects were observed, preventing the reaction from achieving full completion. This was in agreement with earlier reports for the L-Tle process using isolated enzymes [16]. For example, only 50% conversion of TMP was observed at 900 mM substrate concentration, 2.5 mM sodium formate, and without any addition of external NADH. In contrast, conversion was faster and went to almost 100% completion, when NADH was added to the reaction mixture at concentrations of 0.001 or 0.01 M, respectively.

In order to avoid this undesired addition of expensive external cofactor, a modified substrate feed concept was applied which had been developed earlier during the process development for L-Tle manufacture using the free enzymes [16]. In this process, the TMP concentration was always maintained below 500 mM by continuous addition following its conversion to L-Tle by the biocatalyst over a period of 12 h. The inhibition effects described previously could thus be avoided and after 24 h, >95% conversion of TMP to L-Tle was achieved with substrate inputs of up to 1 M and without any addition of external NADH. After separation of the biocatalyst cells by centrifugation and ultrafiltration, the crude reaction solution was subjected to ion exchange chromatography to afford the desired L-Tle in 84% isolated yield and, as expected, an excellent enantiomeric excess of >99% (Scheme 9.10).

This new whole-cell biocatalyst was also successfully applied to the synthesis of L-neopentylglycine [17]. Recently, this unusual bulky amino acid has also attracted increasing interest from research groups as a building block for new developmental drugs [18]. This process started from pivalaldehyde, which was condensed with hydantoin in water at elevated temperature and under ethanolamine catalysis to afford neopentylidene hydantoin. This was subsequently hydrolyzed with aqueous

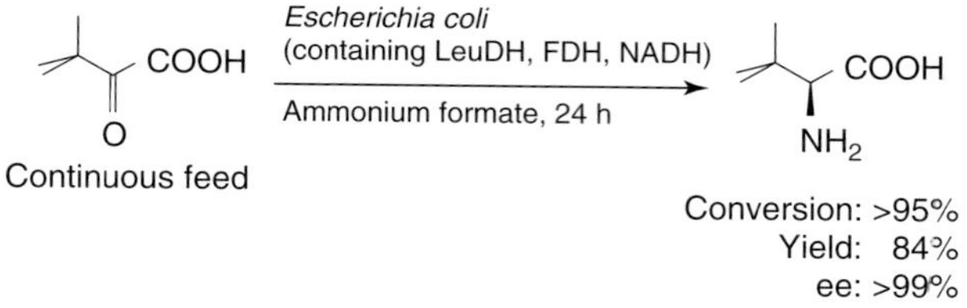

Scheme 9.10 Synthesis of L-*tert*-leucine using whole-cell biocatalysts.

sodium hydroxide to afford the required keto acid salt which was further purified by acidification and extraction of the free keto acid into methyl isobutyl ketone (MIBK). Finally, the required α-ketoneopentylglycine was isolated as its sodium salt after treatment with sodium methoxide. This starting material was then treated with the new *E. coli* whole-cell biocatalyst containing a LeuDH, an FDH, and NADH as the cofactor at a concentration of 66 g l^{-1} (wet cells), a temperature of 30 °C and pH 7. The aqueous reaction medium contained ammonium formate to regenerate the cofactor. The α-keto acid was added portionwise as described above for the synthesis of L-Tle, corresponding to a final substrate concentration of 88 g l^{-1}. After one day under these reaction conditions, >95% conversion and an enantioselectivity of >99% were achieved, without any addition of an external cofactor. After subsequent addition of 5% aqueous ammonia to pH 12 to dissolve the precipitated amino acid, removal of the biomass by centrifugation and ultrafiltration (molecular weight cutoff: 10 000 Da), neutralization, ion exchange using an acidic ion exchange resin (Amberlite 252C) and final elution, concentration, and crystallization, the desired L-neopentylglycine was obtained in 83% isolated yield [17].

However, the activity ratio between the LeuDH and the FDH was still somewhat unbalanced, and therefore a second generation of whole-cell biocatalysts was developed which now express a glucose dehydrogenase (GDH) from *Bacillus subtilis* in addition to the LeuDH. This GDH typically has a much higher specific activity of several 100 U mg^{-1} which is in favorable contrast to the relatively low specific activities reported for FDH [14, 19]. Thus, the amount of enzyme required for the production of a certain amount of amino acid is significantly reduced. This enzyme regenerates the cofactor (NADP$^+$ to NADPH + H$^+$ in this case) by converting glucose to gluconolactone which then irreversibly opens up to gluconic acid, thus shifting the reaction equilibria to the desired products of chiral L-amino acids (Scheme 9.11). As the cofactor was to be regenerated within the cells, ideally without any need for external addition of cofactor, the requirement of this GDH for the more expensive NADPH as compared to NADH was not relevant in this case.

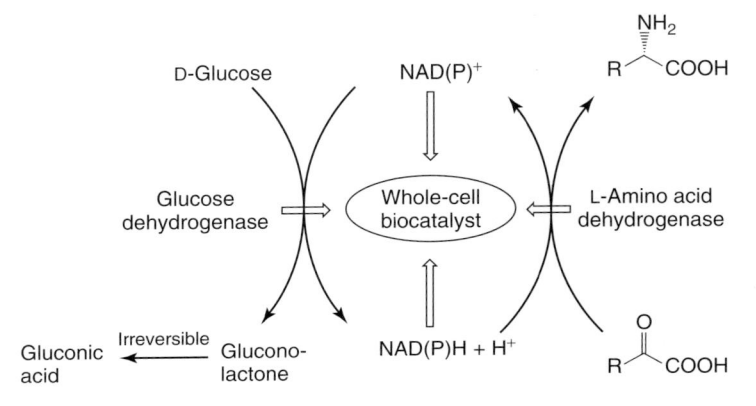

Scheme 9.11 Cofactor regeneration using glucose dehydrogenase.

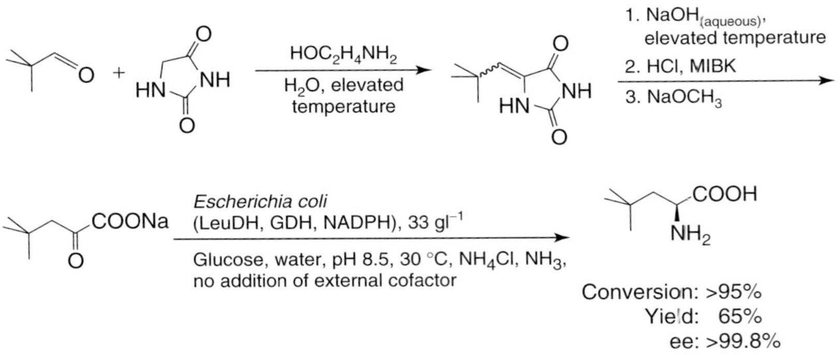

Scheme 9.12 Manufacture of L-neopentylglycine.

Subsequently, this new whole-cell biocatalyst was successfully applied at pilot scale for the production of multi-10 kg quantities of L-neopentylglycine (Scheme 9.12) under similar conditions as described above for the LeuDH whole-cell biocatalyst containing FDH. However, in this case the aqueous reaction medium contained glucose to regenerate the cofactor and was adjusted to pH 8.5 with ammonium chloride and aqueous ammonia. The concentration of the α-keto acid was adjusted to $42\,g\,l^{-1}$. After one day, >95% conversion was obtained, and after a similar workup as described above, L-neopentylglycine was obtained in an unoptimized 65% yield and an excellent optical purity of >99.8% ee. First pilot runs for the manufacture of L-Tle are in preparation.

Nevertheless, one downside of the current application of the whole-cell biocatalysts is the relatively inconvenient handling. The biocatalyst is isolated by ultracentrifugation and then added to the biotransformation reaction mixture as a wet cell mass. Moreover, the only possibility to store the biocatalyst for extended periods of time is by deep-freezing, which poses the risk of cell lysis by ice formation. Therefore, first experiments were made to obtain preparations of the biocatalyst which are more stable on storage and more convenient in application. One possibility could be spray drying which affords the option of using the biocatalyst as a dry powder that should be more stable and, in addition, can be added to the reaction mixture as easily as any other solid reagent. The LeuDH/GDH amino acid dehydrogenase biocatalyst was selected for the first experiments (Table 9.1). A spray dryer with disk atomizer was employed at a vacuum of 30 mbar, inlet temperatures in the range of 150–190 °C and outlet temperatures of 75–85 °C. The biomass concentration of the liquid feed was 18–30% based on dry matter,

Table 9.1 Activities of LeuDH/GDH biocatalyst preparations.

	Activity LeuDH	Activity GDH
Wet biomass (%)	100	100
Spray-dried biomass (%)	17	74

and no additives were used. Technically, the spray drying could be completed successfully resulting in the desired dry biocatalyst powder. However, the enzymes turned out to be insufficiently stable under the processing conditions. The activity of the LeuDH significantly dropped to 17% compared to the wet biomass starting material, while the GDH proved to be more resistant toward activity loss with a decrease to a favorable 74% with respect to the original value. These results are considered a promising start, but obviously require further optimization, before these spray-dried whole-cell biocatalysts can be applied economically at a commercial scale.

9.4
Alcohol Dehydrogenase Biocatalysts

As an extension of our portfolio of enantiomerically pure amino acids, we have now also entered the field of enantiomerically pure alcohols. Industrial production of these compounds has been well established for many years already, applying for example, wild-type microorganisms such as cheap baker's yeast (*Saccharomyces cerevisiae*), recombinant strains, or, alternatively, asymmetric catalytic hydrogenation with chiral metal catalysts. In order to be competitive or even superior to established technologies, a biocatalytic process has to be highly enantiospecific to deliver products of very good optical purity and, at the same time, be highly productive for good economics. This means that the process needs to run at high concentrations, preferably well above $100\,\mathrm{g\,l^{-1}}$, and with a processing time of only a few hours in order to keep the fixed costs as low as possible.

Baker's yeast is among the first whole-cell biocatalysts which have found technical application. It has already been known for a long time to contain many different alcohol dehydrogenases (ADHs), and it can be considered a "proof of principle" for nature's ability to catalyze asymmetric reactions such as conversion of ketones to chiral alcohols. However, quite a few disadvantages have prevented baker's yeast from being applied on a broader basis for the manufacture of enantiomerically pure alcohols. First, many competing (R)- and (S)-ADHs are available in the cells. In many cases, this leads to optical purities of <90%. Expression of the desired ADH is usually low, resulting in comparatively low productivity of the yeast cells. At the same time, only low substrate concentrations are tolerated in many cases, which means a relatively high catalyst loading and finally results in difficult catalyst removal and downstream processing. To overcome these downsides, more specific and more active biocatalysts have been highly desirable.

Numerous (R)- and (S)-specific ADHs from various microorganisms such as *Rhodococcus erythropolis* [20], *Thermoanaerobium brockii* [21], *Thermoanaerobium ethanolicus* [22], and *Lactobacillus brevis* [23] have already been identified and described in the literature. The enzymes show different, often complementary, activities and selectivities for substituted acetophenones, longer-chain arylalkyl ketones, and alkyl ketones. In our initial study [24], we explored the substrate specificity of (S)-ADH from *R. erythropolis*. This enzyme was selected because it consumes NADH as a cofactor, and this is considerably cheaper than the NADPH required by many other ADHs.

First, the enzyme was isolated from a crude extract of R. erythropolis, purified, and characterized. The enzyme protein contains 348 amino acids. The Zn-containing enzyme itself shows a tetrameric structure with a size of 36.2 kDa per subunit. After isolation of the corresponding gene, this was cloned into an E. coli K12 derivative which is a safety level 1 strain that can be easily handled in microbiological labs. This turned out to be an efficient expression system for the enzyme, producing more than 20% soluble protein with adjustable expression levels. After fermentation, the enzyme was isolated from the cells by ultrasonic disruption and centrifugation. The crude extract was subjected to a heat treatment, subsequent centrifugation, and further purification by ion exchange. The resulting partially purified (S)-ADH was then used for preparative studies, adding NADH as a cofactor, FDH from C. boidinii as the cofactor-regenerating enzyme, and sodium formate in a buffered solution.

(S)-ADH from R. erythropolis turned out to be a particularly useful enzyme having a broad substrate specificity (Scheme 9.13) and exhibiting high specific activities of >100 U mg^{-1} protein in most cases. Initial investigations for substrate specificities were carried out in dilute solutions at concentrations of about 1–10 mM, depending on the solubilities of the substrates in aqueous phosphate buffer solutions. Results are based on photometric assays [24]. When compared to acetophenone as the standard substrate with an enzyme activity set to 100%, para-substituted analogs were particularly well accepted, with relative enzyme activities ranging from 194 to 1333%. meta-Chloroacetophenone was also found to be an excellent substrate with relative activity of 2384%, whereas the corresponding ortho-derivative was somewhat less reactive with 85% activity relative to acetophenone. 2-Alkanones turned out to be very good substrates as well, with relative activities of 1096% for 2-heptanone, 2521% for 2-decanone, and 3328% for 2-octanone, respectively. Related β-keto esters are also readily converted with relative activities of 134% for methyl acetoacetate and 1020% for ethyl acetoacetate. Finally, acetone derivatives such as chloroacetone (119%) and phenoxyacetone (4180%) were analyzed to be good to excellent substrates as well.

Preparative experiments further demonstrated the usefulness of the new enzyme. When acetophenone, p-chloroacetophenone, ethyl acetoacetate, 2-heptanone, and 2-octanone reacted with the recombinant (S)-ADH, conversions of >95% and enantiomeric excesses of >99% were obtained in all cases (Scheme 9.14) [24]. This is a particular demonstration of the usefulness of the new recombinant (S)-ADH compared to the wild-type whole-cell biocatalyst, that is, R. erythropolis. In this case, the enantioselectivity was high only in the first few hours of the biotransformation, but dropped significantly with further progress of the reaction, although conversion rates remained as high as before. This was attributed to the probable presence of a second ADH in the wild-type biocatalyst. In contrast, the isolated, recombinant (S)-ADH afforded the desired products in high conversions and excellent enantioselectivities, independent of the reaction time.

Aromatic ketones:

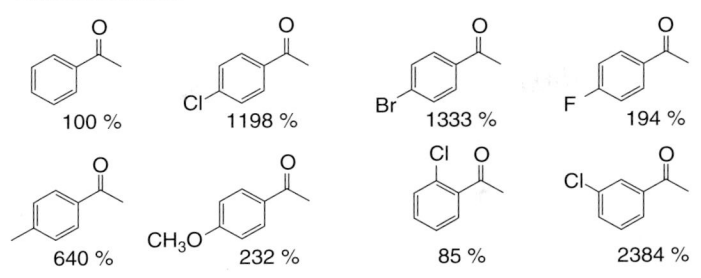

acetophenone 100 %	4'-Cl 1198 %	4'-Br 1333 %	4'-F 194 %
4'-Me 640 %	4'-CH₃O 232 %	2'-Cl 85 %	3'-Cl 2384 %

β-Keto esters:

methyl acetoacetate 134 %   ethyl acetoacetate 1020 %

2-Alkanones:

2-hexanone 1096 %   2-octanone 3328 %   2-decanone 2521 %

Acetone derivatives:

chloroacetone 119 %   phenoxyacetone 4180 %

Scheme 9.13 Substrate specificity of (S)-alcohol dehydrogenase from *Rhodococcus erythropolis* and relative activities.

These investigations convincingly demonstrated the usefulness of a pure, well-defined recombinant enzyme over crude enzyme preparations from wild-type microorganisms or an undeveloped wild-type whole-cell biocatalyst. In parallel, an (R)-alcohol dehydrogenase from *Lactobacillus kefir* had become available for the preparation of the corresponding (R)-alcohols [25].

In the next step, we wanted to overcome the disadvantage of employing two different enzymes plus an expensive external cofactor by combining these in one especially designed whole-cell biocatalyst, which would be able to regenerate the cofactor internally. In the beginning of this research and development program, the following goals were defined for the desired ADH whole-cell biocatalysts and the manufacturing processes for chiral alcohols:

- biocatalysts containing an (R)- or (S)-ADH, an NAD^+- or $NADP^+$-regenerating dehydrogenase and the required cofactor;
- high–cell density fermentations for the biocatalyst cells;
- strong enzyme expressions;

Product	Conversion (%)	ee (%)
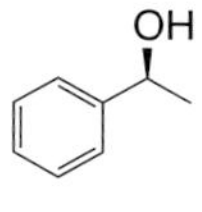 1-phenylethanol (OH on CH attached to phenyl)	>95	>99
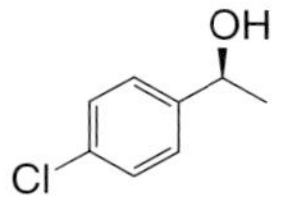 1-(4-chlorophenyl)ethanol	>95	>99
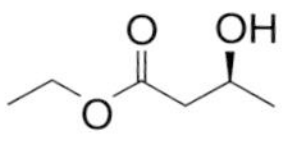 ethyl 3-hydroxybutanoate	>95	>99
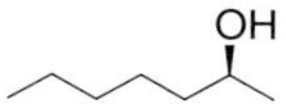 2-heptanol	>95	>99
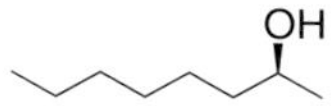 2-octanol	>95	>99

Scheme 9.14 Synthesis of chiral alcohols using (S)-alcohol dehydrogenase from R. erythropolis.

- high productivities (>100 g l^{-1} initial substrate concentrations);
- broad substrate specificity;
- no requirement for any external cofactor;
- high substrate conversion rates;
- high yields;
- excellent optical purities;
- easy workup;
- lowest possible costs.

As already described above, it was again decided to replace the FDH from *C. boidinii* by GDHs from *B. subtilis* or *Thermoplasma acidophilum*, due to much higher specific activities of these enzymes as compared to FDH.

These enzymes were then cloned into the established *E. coli* expression system. The specific activities of the ADHs could be improved to more than 100 U mg^{-1}, in certain cases more than 1000 U mg^{-1} protein. The strains are suitable for high–cell density fermentations, allowing the production of large biocatalyst quantities in relatively small fermentation vessels, and are stable without the addition of any antibiotics. Thus, a set of different tailor-made biocatalysts is available for the production of a broad variety of enantiomerically pure (R)- and (S)-alcohols. HTS techniques then allow the rapid identification of the most suitable biocatalyst for the manufacture of the desired chiral alcohol.

Subsequently, these new whole-cell biocatalysts were use-tested for the preparation of a great variety of enantiomerically pure (R)- and (S)-alcohols from the corresponding ketones (Scheme 9.15) [26–28].

9.4 Alcohol Dehydrogenase Biocatalysts

Product	Biocatalyst (g wet cells l^{-1})	Substrate (gl^{-1})	Time (h)	Conversion (%)	ee (%)	Reference
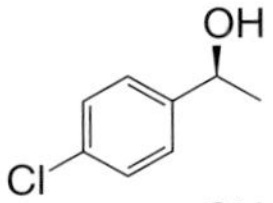 1-(4-chlorophenyl)ethanol	(S)-ADH/GDH 52 gl^{-1}	156	31	94	>99.8	27
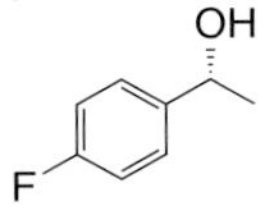 1-(4-fluorophenyl)ethanol	(R)-ADH/GDH 71 gl^{-1}	69	23	>95	>99	28
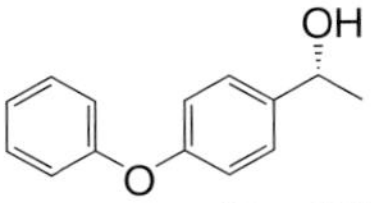 1-(4-phenoxyphenyl)ethanol	(R)-ADH/GDH 52 gl^{-1}	212	25	>95	>99.4	27
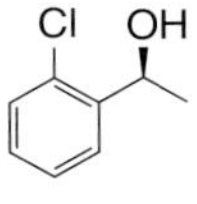 1-(2-chlorophenyl)ethanol	(S)-ADH/GDH 48 gl^{-1}	156	76	95	>99.4	27
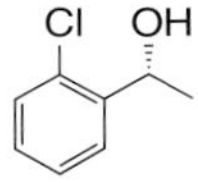 1-(2-chlorophenyl)ethanol	(R)-ADH/GDH 52 gl^{-1}	156	53	>95	90	26
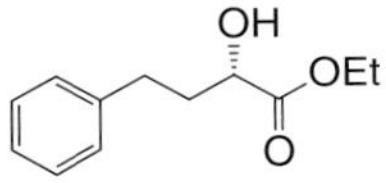 ethyl 3-hydroxy-5-phenylpentanoate	(S)-ADH/GDH 51 gl^{-1}	210	27	93	96	27
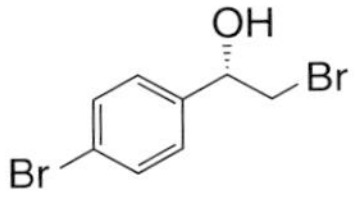 2-bromo-1-(4-bromophenyl)ethanol	(R)-ADH/GDH 52 gl^{-1}	140	26	94	97	27
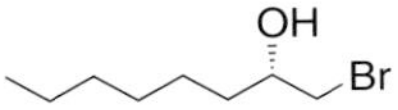 1-bromo-2-octanol	(R)-ADH/GDH 56 gl^{-1}	208	24	>95	>99	26

Scheme 9.15 Preparation of chiral alcohols using ADH whole-cell biocatalysts.

In all cases, the following advantages apply:
- no chemical side reactions are detected during the investigated biotransformations;
- biotransformation processes are robust and stable;
- external addition of cofactor is not needed;
- scale-up is easy;
- substrate concentrations are high;
- conversions are high;
- yields and product purities are high;
- enantioselectivities are high to excellent.

Downstream processing in most cases is straightforward following the standard procedure of acidification to inactivate the biocatalyst, addition of a filter aid material, filtration, extraction of the desired chiral alcohol with methyl *tert*-butyl ether (MTBE), evaporation of the solvent, and subsequent distillation of the alcohol [27].

(S)-1-(4′-Chlorophenyl)ethanol was prepared in this way from 4-chloroacetophenone using the new whole-cell biocatalyst containing (S)-ADH from R. erythropolis and a GDH from B. subtilis. No external cofactor had to be added during the reaction. The biotransformation was run in water containing glucose for the regeneration of the cofactor. Starting with a substrate concentration of $156\,g\,l^{-1}$, 94% conversion was achieved and the resulting (S)-alcohol was analyzed to have >99.8% ee [27].

Likewise, (R)-1-(4′-fluorophenyl)ethanol was obtained from 4-fluoroacetophenone using the (R)-selective whole-cell biocatalyst containing the (R)-ADH from L. kefir and the GDH from T. acidophilum (Scheme 9.16) [28]. Starting with a substrate concentration of $69\,g\,l^{-1}$, $71\,g\,l^{-1}$ biocatalyst (wet cells), and 1.18 equiv. of glucose to regenerate the cell-internal cofactor NADPH, >95% conversion was observed after 23 h. After the usual workup, the desired product was obtained in 87% yield and an excellent optical purity of >99% ee.

In preliminary lab trials, the biocatalyst was found to accept 3-fluoro- and 2-fluoroacetophenone as additional substrates, although the activities were somewhat reduced compared to acetophenone as a standard substrate, with 92% relative activity for the 3-F-isomer and 67% relative activity for the 2-F-isomer, respectively [28].

In a similar, but larger-scale (10 l) experiment, (R)-1-(4′-chlorophenyl)ethanol was prepared from 4-chloroacetophenone using the (R)-selective whole-cell biocatalyst. After 30 h at pH 6–7 and 25 °C and with 1.06 equiv. of D-glucose, conversion was 95%, starting from a substrate concentration of $156\,g\,l^{-1}$ (1.0 M) and a comparatively low biocatalyst loading of $25\,g\,l^{-1}$ (wet cells). Downstream processing as described above afforded the crude product in 91% yield, 95% purity, and an excellent enantiomeric excess of >99.8% [27].

Finally, technical applicability of the new (S)-ADH whole-cell biocatalyst was successfully demonstrated in the pilot production of a specifically substituted (S)-phenylethylalcohol which was required as a chiral building block for a new developmental drug. After receipt of the initial inquiry, possible options for the synthesis of the target molecule were evaluated, and biocatalytic reduction using whole cells quickly turned out to be the preferred option as compared to

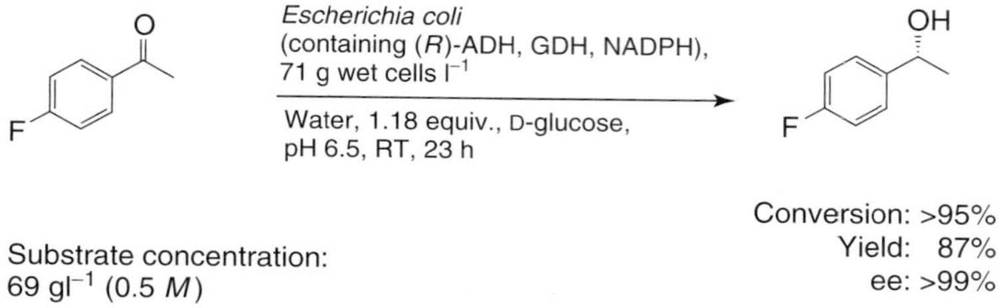

Scheme 9.16 Preparation of (R)-1-(4′-fluorophenyl)ethanol.

asymmetric catalytic hydrogenation in this specific case. In the following month, an HTS program was run to identify the most suitable (S)-ADH biocatalyst for the target molecule. Subsequently, chemical feasibility could be demonstrated at lab scale; based on that, a quotation was prepared and submitted to the customer. Within the next month, a sample was prepared that was representative of the future technical process and fully met all customer specifications. In parallel, the order for a pilot plant campaign was received from the customer. The next month was used for further process development and optimization, especially development of a suitable biotransformation, isolation, and purification process, and in parallel the required raw materials were procured. Once these had arrived on site, the pilot campaign was run. The biocatalyst was applied at a concentration of $50\,g\,l^{-1}$ (wet cells) in aqueous suspension at pH 6.5–7.0 and 22–24 °C in the presence of glucose to regenerate the cofactor. Initial substrate concentration was $>150\,g\,l^{-1}$. Final conversion was >98% and the isolated yield, although unoptimized, a favorable 80%. The product exhibited excellent optical purity with >99.4% ee.

In total, less than four months were required from receipt of the initial inquiry to successful completion of the first pilot campaign with an output of more than 100 kg of the required (S)-alcohol of excellent chemical quality and optical purity (Scheme 9.17).

In the meantime, this process has been run repeatedly at an industrial scale. This demonstrates that whole-cell biocatalysts can be very powerful tools in the rapid

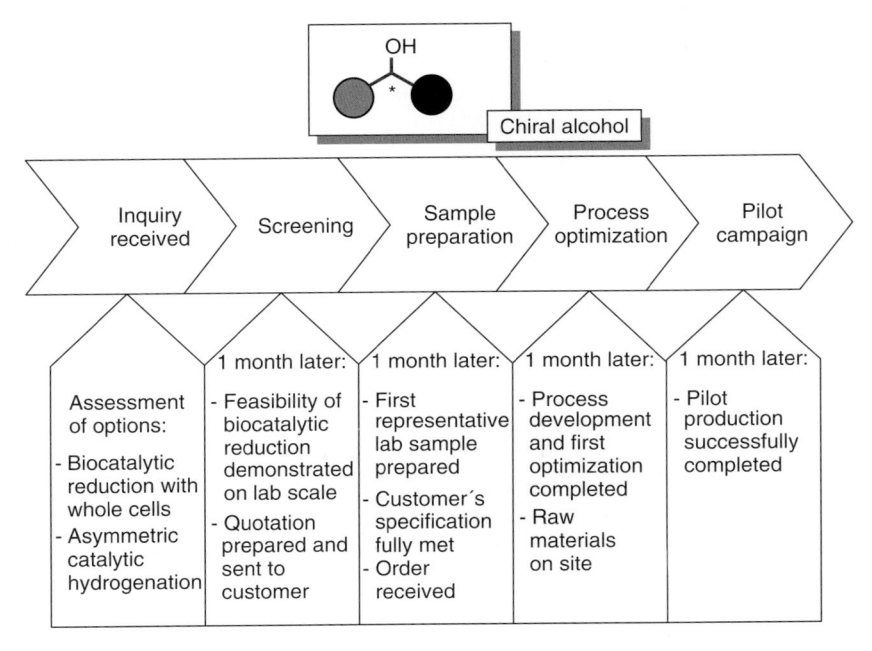

Scheme 9.17 From inquiry to delivery in less than four months.

large-scale preparation of chiral molecules, when all the required technologies are in place, such as

- well-characterized strain collection of complementary ready-to-use whole-cell biocatalysts with known substrate specificities and proven technical applicability;
- HTS for rapid identification of the most suitable biocatalyst;
- HPLC to determine optical purities of products;
- high–cell density fermentation for rapid and efficient production of large quantities of the required biocatalysts;
- ultracentrifugation and/or ultrafiltration equipment to remove the biocatalyst cells from the reaction mixture.

As for the amino acid dehydrogenase whole-cell biocatalysts (see above), first experiments were also made to obtain spray-dried preparations of the biocatalyst. Technically, spray drying could also be completed successfully using conditions as described above to result in the desired dry biocatalyst powder. In this case, the enzymes turned out to be more stable under the processing conditions compared with those enzymes contained in the LeuDH/GDH amino acid dehydrogenase biocatalyst. The activity of the (S)-ADH from R. erythropolis decreased to 68% compared to the wet biomass starting material, and the GDH from B. subtilis was again found to be more resistant toward activity loss with a very favorable residual activity of >80% with respect to the original value (Table 9.2). These results are considered suitable for potential technical application of these spray-dried whole-cell biocatalysts at commercial scale.

Recently, the ADH biocatalyst E. coli DSM 14459 containing the (R)-ADH from L. kefir and the GDH from T. acidophilum was also applied for the conversion of cinnamaldehyde to cinnamyl alcohol (Scheme 9.18) [19]. This compound was required as an aroma ingredient for the food industry. It is also available by chemical reduction of cinnamyl aldehyde; however, the "nature-identical" products obtained from these processes often suffer from the content of small amounts of by-products, for example, compounds with reduced C=C double bonds, which may lead to undesired odor properties. In addition, these chemical reduction technologies often require expensive reagents or catalysts and challenging reaction conditions such as high temperatures, partial exclusion of water or hazardous chemicals such as hydride reagents or gaseous hydrogen. These disadvantages can potentially be overcome by application of a biocatalytic reduction process and,

Table 9.2 Activities of (S)-ADH/GDH biocatalyst preparations.

	Activity (S)-ADH (from *Rhodococcus erythropolis*)	Activity GDH (from *Bacillus subtilis*)
Wet biomass (%)	100	100
Spray-dried biomass (%)	68	>80

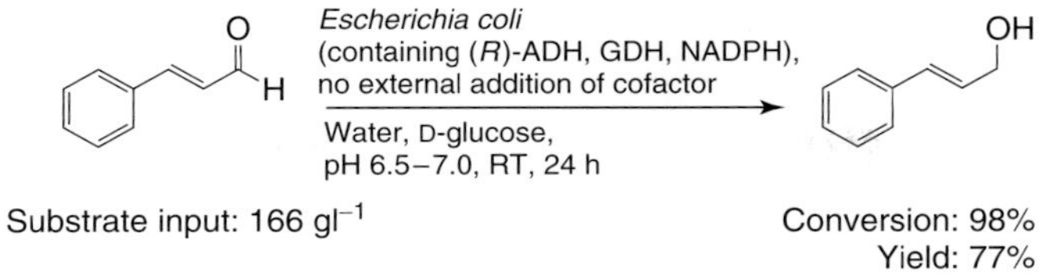

Scheme 9.18 Manufacture of cinnamyl alcohol using a whole-cell biocatalyst.

Substrate input: 166 g l^{-1}

Reaction conditions: *Escherichia coli* (containing (R)-ADH, GDH, NADPH), no external addition of cofactor; Water, D-glucose, pH 6.5–7.0, RT, 24 h.

Conversion: 98%
Yield: 77%

in addition, the resulting cinnamyl alcohol can be labeled "natural grade," if obtained from a biotechnological process, in contrast to the material manufactured by chemical synthesis.

Pleasingly, the (R)-ADH from *L. kefir* was found to be not only useful for the reduction of ketones, but also very suitable for the desired transformation of cinnamaldehyde to cinnamyl alcohol. In this case, the relative activity of the enzyme was 120% as compared to the standard reduction of acetophenone. The new production process was run at ambient temperature in an aqueous system at about neutral pH (6.5–7.0) without any cosolvent at a substrate input of up to 166 g l^{-1}. Natural glucose was added to regenerate the cofactor NADP$^+$. As expected, there was no need for any addition of external cofactor. Conversion was analyzed as 98% after 24 h. Isolation was performed by acidification to pH <3 with hydrochloric acid, addition of a filter aid, and subsequent removal of the biomass by filtration. The aqueous filtrate was extracted with ethyl acetate and the resulting organic solutions concentrated. Short-path distillation afforded the desired cinnamyl alcohol in 77% yield and ≥98% purity. Pilot campaigns were completed successfully, and technical quantities of the biocatalyst have already been supplied for future production.

9.5
Summary

As a result of a long-term strategic R&D program, whole-cell biocatalyst technology platforms have been developed for manufacture of the following substances:

- L- and D-amino acids using D- or L-hydantoinase biocatalysts;
- L-amino acids using LeuDH biocatalysts;
- (S)- and (R)-alcohols using (S)- and (R)-ADH biocatalysts.

More than 15 hydantoinase whole-cell biocatalysts, more than 15 redox whole-cell biocatalysts, and more than 5 whole-cell biocatalysts for other biotransformations, covering broad ranges of different substrates, are available for immediate application at a technical scale. These new whole-cell biocatalysts can be fine-tuned quickly for applications with new substrates. Therefore, biocatalyst and bioprocess development are no longer the rate-determining steps in the development of new manufacturing processes for valuable intermediates and building blocks for fine chemicals.

Acknowledgments

The author would like to thank Prof. Karlheinz Drauz, Prof. Andreas Bommarius, Dr Oliver May, Prof. Harald Groeger, Dr Stefan Verseck, Dr Francoise Chamouleau, Dr Chad Hagedorn, Dr Nicolas Orologas, Dr Stefan Buchholz, Dr Andreas Karau, Kai Boldt, Dr Juergen Roos, Dr Wolf-Ruediger Krahnert, Dr Jens Reinbold, Dr Matthijs ter Wiel, Dr Steffen Osswald, Dr Kai Doderer, and their respective lab teams for their enthusiasm and valuable contributions during the course of the respective research projects. Excellent cooperation from the scientific research groups of Prof. Werner Hummel and Prof. Maria-Regina Kula at the University of Düsseldorf (Germany), Prof. Christoph Syldatk and Dr Josef Altenbuchner at the University of Stuttgart (Germany), Prof. Klaus-Dieter Vorlop at the University of Braunschweig (Germany), Prof. Frances H. Arnold at the California Institute of Technology at Pasadena (USA), and the company BRAIN at Zwingenberg (Germany) is gratefully acknowledged. Dr Matthias Janik, Dr Kurt Guenther, Dr Franz-Rudolf Kunz, Stefan Merget, and their respective lab teams have contributed to the success of these projects through their outstanding analytical support. Kurt Klostermann, Jochen Lebert, Rolf Braun, Gerard Richet, and their teams in the respective production plants made these products possible using technical equipment. We also thank the German Ministry for Education and Technology for generous financial support during various R&D projects.

References

1. Drauz, K., Eils, S., and Schwarm, M. (2002) *Chim. Oggi/Chem. Today*, **20**, 15.
2. Bommarius, A.S., Schwarm, M., and Drauz, K. (2001) *Chimia*, **55**, 50.
3. Bommarius, A.S., Kottenhahn, M., Klenk, H., and Drauz, K. (1992) in *Microbial Reagents in Organic Synthesis* (ed. S. Servi), Kluwer, p. 161.
4. Bommarius, A.S., Schwarm, M., Stingl, K., Kottenhahn, M., Huthmacher, K., and Drauz, K. (1995) *Tetrahedron: Asymmetry*, **6**, 2851.
5. (a) Witherell, G. (2001) *Curr. Opin. Invest. Drugs*, **2**, 340; (b) Revill, P., Serradell, N., Bolós, J., and Rosa, E. (2007) *Drugs Fut.*, **32**, 788; (c) Prongay, A.J. et al. (2007) *J. Med. Chem.*, **50**, 2310; (d) Njoroge, F.G., Chen, K.X., Shih, N.-Y., and Piwinski, J.J. (2008) *Acc. Chem. Res.*, **41**, 50; (e) Arasappan, A. et al. (2009) *J. Med. Chem.*, **52**, 2806.
6. (a) Dickore, K., Engels, H.D., Kraetzer, H., and Merz, W. (1977) DE Patent 26 48 300; (b) Jackman, D.E. (1980) EP 0 011 207.
7. Krix, G., Bommarius, A.S., Drauz, K., Kottenhahn, M., Schwarm, M., and Kula, M.-R. (1997) *J. Biotechnol.*, **53**, 29.
8. May, O., Nguyen, P.T., and Arnold, F.H. (2000) *Nat. Biotechnol.*, **18**, 317.
9. (a) Mannhold, R. (1985) *Drugs Fut.*, **10**, 636; (b) Baltzer, S. and van Dorsselaer, V. (2005) WO Patent 2005073226; (c) Harada, N. and Hikota, M. (2008) JP 2008044933.
10. Patel, R. (2001) *Adv. Synth. Catal.*, **343**, 527.
11. Kikumoto, R. and Nakamura, A. (1973) JP 73 39,416.
12. (a) Taillades, J., Rousset, A., Lasperas, M., and Commeyras, A. (1986) *Bull. Soc. Chim. Fr.*, 650; (b) Rousset, A., Lasperas, M., Taillades, J., and Commeyras, A. (1980) *Tetrahedron*, **36**, 2649; (c) Greenstein, J.P. and Winitz, M. (1961) *Chemistry of the Amino Acids*,

vol. 1, John Wiley & Sons, Inc., New York, p. 698ff.
13. Humphrey, C.E., Furegati, M., Laumen, K., La Vecchia, L., Leutert, T., Müller-Hartwieg, J.C.D., and Vögtle, M. (2007) *Org. Proc. Res. Dev.*, **11**, 1069.
14. Menzel, A., Werner, H., Altenbuchner, J., and Groeger, H. (2004) *Eng. Life Sci.*, **4**, 573.
15. Slusarczyk, H., Felber, S., Kula, M.-R., and Pohl, M. (2000) *Eur. J. Biochem.*, **267**, 1280.
16. Kragl, U., Vasic-Racki, D., and Wandrey, C. (1996) *Bioprocess Eng.*, **14**, 291.
17. Groeger, H., May, O., Werner, H., Menzel, A., and Altenbuchner, J. (2006) *Org. Proc. Res. Dev.*, **10**, 666.
18. (a) Deziel, R. and Moss, N. (1994) WO Patent 9420528; (b) Wakimasu, M., Kikuchi, T., and Kawada, A. (1995) EP 655463; (c) Beaulieu, P.L., Deziel, R., Brunet, M.L., Moss, N., and Plante, R. (1998) US Patent 5,846,941; (d) Halbert, S.M., Michaud, E., Thompson, S.K., and Veber, D.F. (1999) WO Patent 9966925; (e) Emmanuel, M.J., Frye, L.L., Hickey, E.R., Liu, W., Morwick, T.M., Spero, D.M., Sun, S., Thomson, D.S., Ward, Y.D., and Young, E.R.R. (2001) WO Patent 2001019816.
19. Chamouleau, F., Hagedorn, C., May, O., and Groeger, H. (2007) *Flavour Fragr. J.*, **22**, 169.
20. (a) Hummel, W. and Gottwald, C. (1993) DOS Patent 4,209,022; (b) Zelinski, T., Peters, J., and Kula, M.-R. (1994) *J. Biotechnol.*, **33**, 283.
21. (a) Korkhin, Y., Kalb, A.J., Peretz, M., Bogin, O., Burstein, Y., and Frolow, F. (1998) *J. Mol. Biol*, **278**, 967; (b) Bogin, O., Peretz, M., and Burstein, Y. (1997) *Protein Sci.*, **6**, 450.
22. Holt, P.J., Williams, R.E., Jordan, K.N., Lowe, C.R., and Bruce, N.C. (2000) *FEMS Microbiol. Lett.*, **190**, 57.
23. Hummel, W. (1997) *Adv. Biochem. Eng. Biotechnol.*, **58**, 146.
24. Hummel, W., Abokitse, K., Drauz, K., Rollmann, C., and Groeger, H. (2003) *Adv. Synth. Catal.*, **345**, 153.
25. (a) Hummel W. and Kula M.-R. (1991) EP 456107; (b) Weckbecker, A. and Hummel, W. (2004) in *Microbial Enzymes and Biotransformations, Methods in Biotechnology*, vol. 17 (ed. J. L. Barredo), Humana Press, Totowa, p. 241.
26. Osswald, S., Doderer, K., Groeger, H., and Wienand, W. (2007) *Suppl. Focus Biocatal. Chim. Oggi/Chem. Today*, **25** (5), 16.
27. Groeger, H., Chamouleau, F., Orologas, N., Rollmann, C., Drauz, K., Hummel, W., Weckbecker, A., and May, O. (2006) *Angew. Chem. Int. Ed.*, **45**, 5677.
28. Groeger, H., Rollmann, C., Chamouleau, F., Sebastien, I., May, O., Wienand, W., and Drauz, K. (2007) *Adv. Synth. Catal.*, **349**, 709.

10
Process Development of Amrubicin Hydrochloride, an Anthracycline Anticancer Drug

Kazuhiko Takahashi and Mitsuharu Hanada

10.1
Introduction

Since the isolation of daunorubicin from *Streptomyces peucetius* in the 1950s, a number of anthracycline derivatives have been investigated for improved antitumor activity and reduced side effects. In terms of antitumor activity, only a few compounds have shown a clear advantage over doxorubicin, which is one of the most widely used antitumor agents in clinical oncology today. Most of these compounds, however, have been produced by fermentation or semisynthesis, and all of them have a hydroxyl group at the 9-position and an amino sugar moiety at the 7-position (Figure 10.1).

By adopting a different approach, we have been able to produce fully synthetic anthracycline derivatives that have antitumor activity superior to that of doxorubicin. Among these derivatives, amrubicin ((**1**), Calsed®) (formerly known as SM-5887, Figure 10.2) is the first completely synthetic anthracycline derivative in the world. Amrubicin is characterized by a 9-amino group and a simple sugar moiety at the 7-position and has exhibited in preclinical studies both greater efficacy in human tumor xenografts and less cardiotoxicity than doxorubicin [1–5].

A major pathway for anthracyclines' metabolism is the reduction of the C-13 carbonyl group to a hydroxyl group by carbonyl reductase, generally regarded as an inactivation. The 13-hydroxyl metabolites of daunorubicin, doxorubicin, and epirubicin are much less cytotoxic than the corresponding parent drugs. In contrast, amrubicinol, the 13-hydroxyl metabolite of amrubicin, is 5–100 times more active than amrubicin in inhibiting the growth of 17 human tumor cell lines [6]. In addition, amrubicinol at lower medium concentrations attains intracellular concentrations similar to those of amrubicin. Moreover, the level of amrubicinol after administration of amrubicin has been shown to be higher than that of doxorubicin in tumor tissues, but lower in normal tissues, with good correlation between the level of amrubicinol in tumors and the efficacy of amrubicin *in vivo* [7, 8]. These findings suggest that the antitumor activity of amrubicin is produced by selective disposition of amrubicinol in tumors.

Pharmaceutical Process Chemistry. Edited by Takayuki Shioiri, Kunisuke Izawa, and Toshiro Konoike
Copyright © 2011 WILEY-VCH Verlag GmbH & Co. KGaA, Weinheim
ISBN: 978-3-527-32650-1

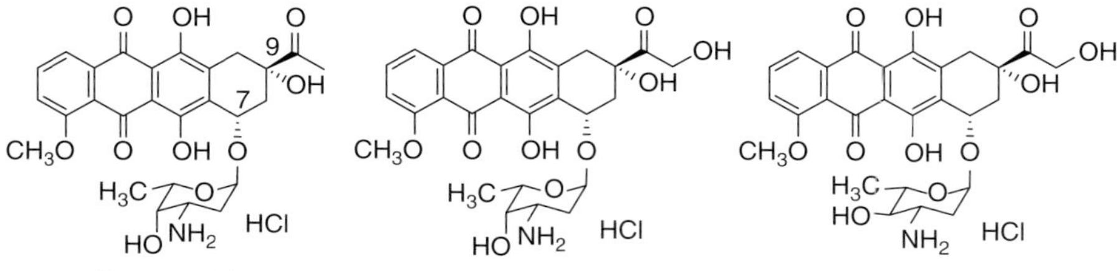

Figure 10.1 Anthracycline derivatives produced by fermentation or semisynthesis.

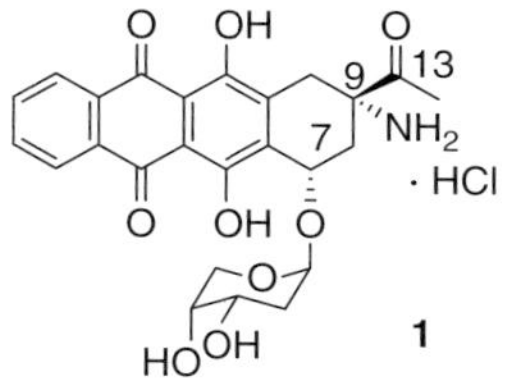

Figure 10.2 Amrubicin hydrochloride.

Both amrubicin and amrubicinol not only interact with deoxyribonucleic acid (DNA) by intercalation but also inhibit DNA topoisomerase II by stabilizing the cleavable complex, followed by double-stranded DNA breaks [9]. They induce apoptosis mediated by activation of caspase-3/7 preceding a loss of mitochondrial membrane potential [10].

Of 33 patients with extensive disease small-cell lung cancer (ED-SCLC) treated with amrubicin, 3 showed complete response and 22 had partial response, for an overall response rate of 75.8% [11]. In addition, amrubicin was active in the treatment of refractory or relapsed SCLC [12]. Consistent with the synergistic effects of amrubicin and cisplatin observed in preclinical studies [3, 13], the combination of amrubicin and cisplatin in previously untreated patients with ED-SCLC has demonstrated an overall response rate of 87.8% (36/41) [14]. In that clinical study, the median survival time was 13.6 months, and major toxicities were leucopenia and neutropenia, while non-hematological toxicities were relatively mild. Amrubicin is currently approved for the treatment of SCLC and non–small cell lung cancer (NSCLC) in Japan and clinical trials for its use in lung cancers are in progress in the United States.

This chapter outlines the process chemistry of amrubicin hydrochloride prepared by a fully synthetic method, distinct from the traditionally established fermentation or semisynthetic method used for preparation of anthracycline anticancer drugs.

10.2
Original Synthetic Route for Amrubicin

The original synthetic route for amrubicin hydrochloride is shown in Scheme 10.1 [1, 15]. Tetralone **2**, which is readily available from *p*-benzoquinone, was used

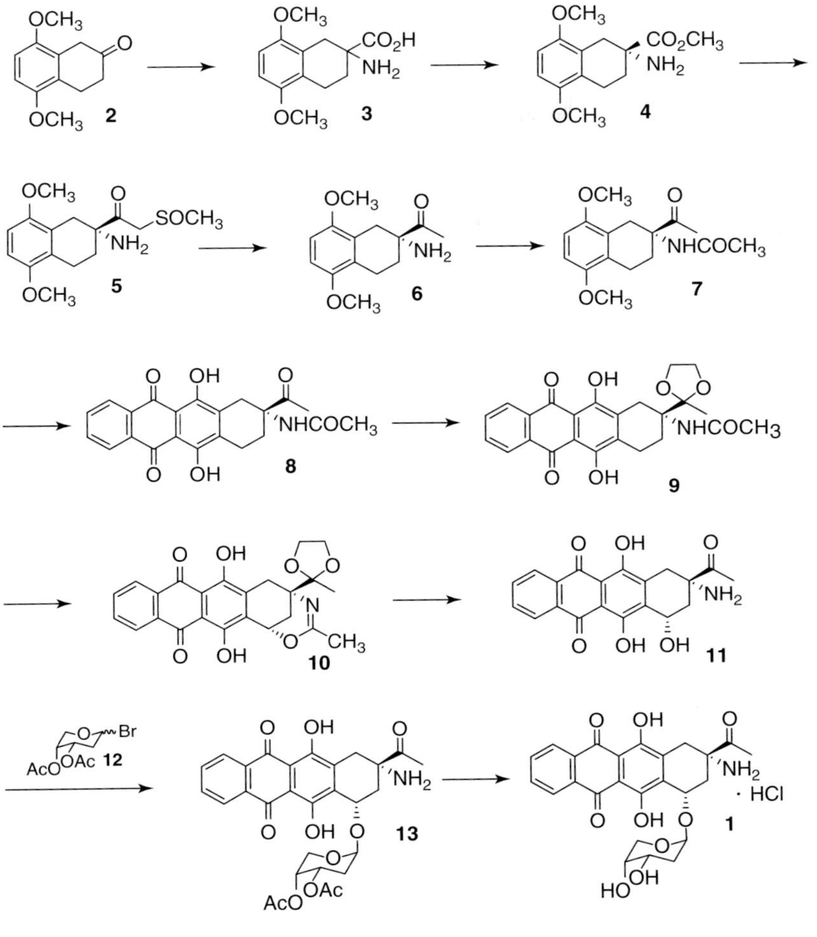

Scheme 10.1 Original synthetic route for amrubicin.

as a starting material for the synthesis of aglycon **11**. This tetralone **2** was converted to the amino acid **3** via hydantoin formation. Optical resolution of the methyl ester using D-(−)-mandelic acid gave the optically active amino ester **4**, and transformation of the methoxycarbonyl group to the acetyl group via methylsulfoxide gave the methylketone **6**. Friedel–Crafts condensation of the acetylated derivative **7** with phthalic anhydride gave an anthracycline structure; compound **8**. Protection of the ketone group by ketal formation followed by stereoselective oxidation at the 7-position using 1,3-dibromo-5,5-dimethylhydantoin (DDH) and 2,2′-azobisisobutyronitrile (AIBN) followed by continuously occurring intramolecular cyclization gave compound **10**. Deprotection and oxazine ring cleavage of compound **10** followed by crystallization gave the aglycon **11**. Glycosidation using the sugar moiety **12** followed by deprotection of the acetyl groups and salt formation gave amrubicin hydrochloride (**1**).

10.3
Amrubicin Bulk Production Synthetic Method

Amrubicin's bulk production route is described in Scheme 10.2. There have been many efforts to establish a synthetic method for the commercial production of amrubicin hydrochloride based on the above-mentioned original synthetic route. However, due to low solubility of the intermediates with anthracycline structure in organic solvents, usable solvents in the synthesis were very limited. In addition, most of the reaction intermediates are chemically unstable because of the unique structure of the labile amino group on the quaternary carbon center requiring a strict control of pH and temperature to handle these intermediates. Therefore, to establish a bulk production synthetic route for amrubicin, many modifications were necessary. The main changes in the synthesis are as follows:

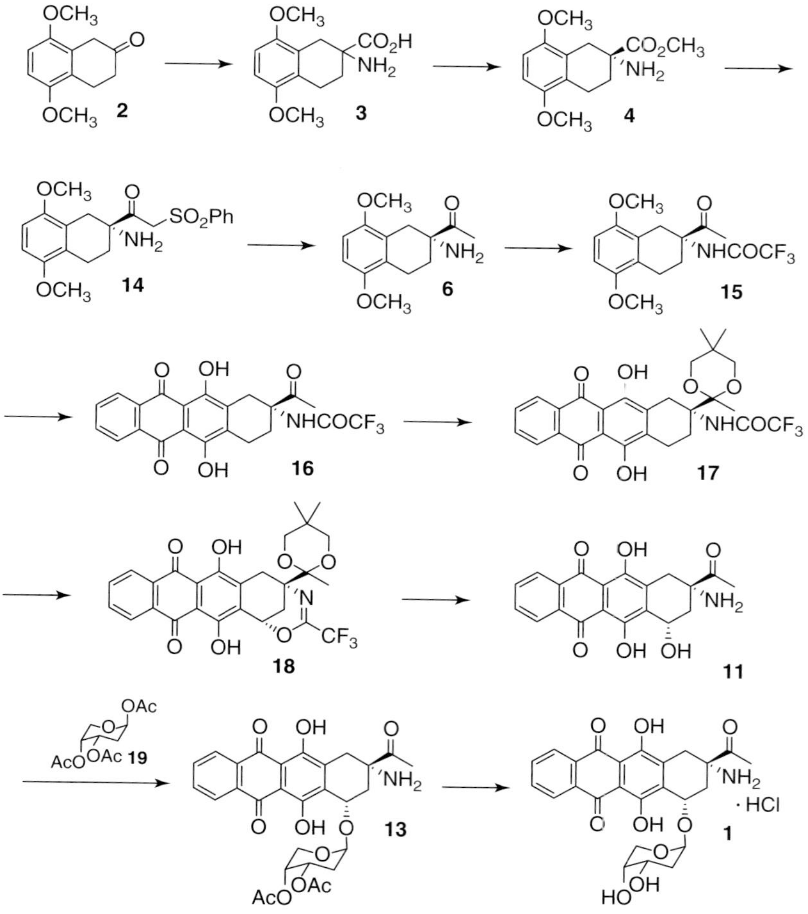

Scheme 10.2 Bulk production synthetic route for amrubicin.

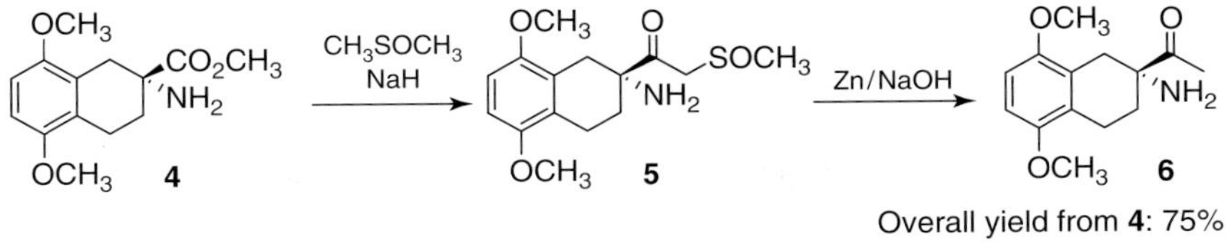

10.3 Amrubicin Bulk Production Synthetic Method

Scheme 10.3 Original method for preparation of ketone **6**.

Overall yield from **4**: 75%

- In the introduction of the methylketone, the use of dimethyl sulfoxide (DMSO) and sodium hydroxide was changed to the use of methyl phenyl sulfone.
- The protecting group for the 9-amino moiety was changed from an acetyl group to a trifluoroacetyl group.
- The protection reagents in the 13-keto group were changed from a combination of ethylene glycol/p-toluenesulfonic acid to a combination of neopentyl glycol/Amberlyst 15 resin.
- The sugar moiety precursor was changed from the bromoacetal **12** to the triacetoxy derivative **19**.

In this chapter, details of the following four synthetic improvements are described:

1) safe synthetic method for 9-aminoketone;
2) stereoselective introduction of 7-hydroxy group;
3) polymorphism study of amrubicin hydrochloride;
4) stability of amrubicin hydrochloride with reference to moisture.

10.3.1
Safe Synthetic Method of 9-Aminoketone

In the original amrubicin synthetic method, the DMSO–sodium hydride method was used to introduce a methyl group (Scheme 10.3). In this method, the methylsulfinylmethylide (dimsyl) anion, which was prepared from DMSO and sodium hydride, played the role of a nucleophile to the methoxycarbonyl group, although it is well known to be thermally unstable. This resulted in technical problems as reported in Chemical Engineering News [16, 17].

The sealed-cell differential scanning calorimetry (SC-DSC)[1] chart of the DMSO–sodium hydride mixture is shown in Figure 10.3. Exothermic decomposition started at 69 °C, and two sequential peaks were observed. The total heat of the two peaks was over 1000 J g^{-1}.

1) SC-DSC is a thermoanalytical technique in which the difference in the amount of heat required to increase the temperature of a sample and reference are measured as a function of temperature. The main application of DSC is in studying phase transitions, or exothermic decompositions.

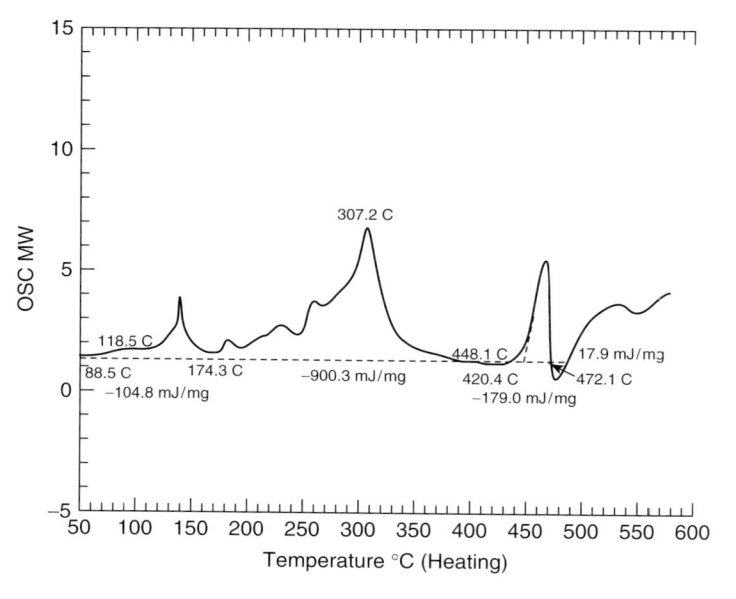

Figure 10.3 SC-DSC chart of DMSO–sodium hydride mixture.

Scheme 10.4 Transformation of ester **4** to phenylsulfone **14**.

In order to avoid using the DMSO–sodium hydride method, we chose the methyl phenyl sulfone method as shown in Scheme 10.4. According to the literature [18], n-butyl lithium was used as a base at −78 °C. However, combination of n-butyl lithium and a temperature as low as −78 °C requires special equipments, including a cryogenic system, which make this method costly. Alternative bases were then investigated, and lithium amide was found to be a suitable replacement for n-butyl lithium (Table 10.1).

To evaluate the safety of this new method, we used accelerating rate calorimetry (ARC).[2] The results of the ARC analysis were as follows: In the DMSO–sodium hydride method, five exothermic peaks were observed. At 57 °C, exothermic decomposition started, and it went up to 151.8 °C. The fifth peak went up to 376.5 °C adiabatically. On the other hand, the methyl phenyl sulfone method

---

2) ARC is well-known as one of the evaluation methods of the existence of explosive reactions. A sample is heated slowly, when a generation of heat exceeding 0.02 °C min−1 is detected, it automatically becomes an adiabatic system, and a self-exothermic state is recorded.

Table 10.1 Alternative base selection.

Base	n-BuLi	60% NaH	LiNH$_2$
Temperature (°C), time (h)	−78	60, 2	60, 2
Yield (%)	91	60	92

Table 10.2 Reagents molar ratio of reagents in lithium amide method.

Compound 4	CH$_3$SO$_2$Ph	LiNH$_2$	Yield (%)
1	2	4	97
1	1.5	3	70
1	2	2.5	46

showed three exothermic peaks, but no exothermic decomposition was observed below 140 °C. After optimization of the reagents' molecular ratio, the yield could be improved as shown in Table 10.2.

The obtained phenylsulfone could be easily reduced to the ketone **6** in zinc–acetic acid. The two-step overall yield was 83%, 8 points higher than that of the original DMSO method. As described above, we could establish an alternative synthetic process for introducing the aminoketone group via a phenylsulfone [19].

10.3.2
Stereoselective Introduction of 7-Hydroxy Group

In the second-generation manufacturing method, the introduction of a hydroxyl group in the 7-position was performed as described in Scheme 10.5. Protection of the ketone group in trifluoroacetamide derivative **16** using neopentyl glycol and p-toluenesulfonic acid, and oxidative bromination at the 7-position, followed by continuously occurring intramolecular cyclization using DDH and AIBN gave the bridging oxazine intermediate **18**. Acidic cleavage of both the ketal group and the oxazine ring of **18**, followed by neutralized crystallization in aqueous media gave the aglycon **11**. However, some issues had to be improved and are listed below:

1) Retardation of the reaction and de-trifluoroacetylation were observed in the step leading to the formation of the ketal **17**.
2) Epimerization of the 7-hydroxy group occurred in the hydrolysis step (about 2%).
3) Slow filtration of the aglycon **11** and the resulting highly wet filter cake caused low productivity due to a long process time for both, filtration and drying.
4) Overall yield from **16** to **11** was low at 37%.

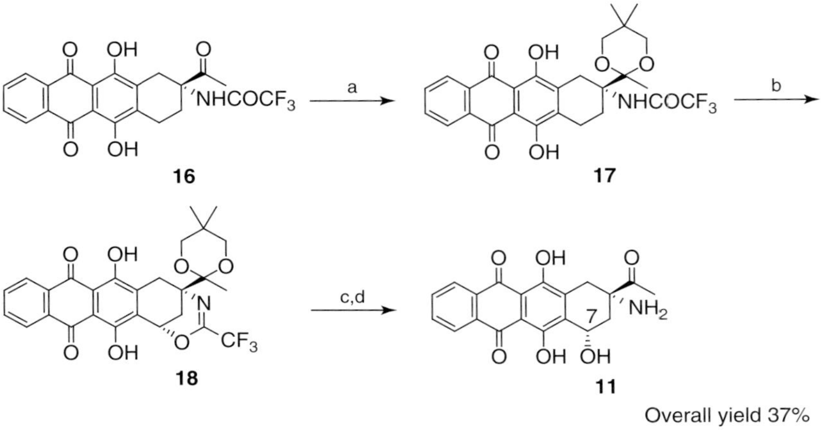

Scheme 10.5 Introduction of hydroxy group at 7-position.
(a) Neopentyl glycol/*p*-TsOH; (b) DDH/AIBN; (c) HCl;
(d) HCl/dioxane, then aqueous Na_2HPO_4.

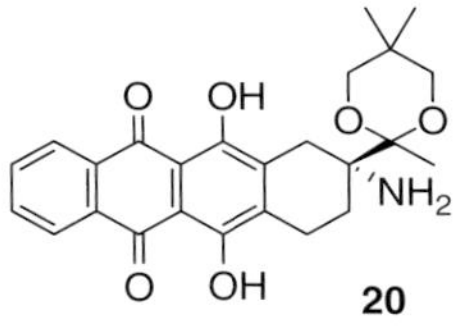

Figure 10.4 De-trifluoroacetylation by-product.

At the laboratory scale, ketal formation proceeded smoothly using *p*-toluenesulfonic acid as a catalyst but at the pilot scale, a retardation of the reaction was frequently observed. As a result, undesired de-trifluoroacetylation by-product **20** was formed (Figure 10.4). This amine by-product **20** interrupted the radical-type reaction such as bromination using DDH and AIBN, resulting in a need for silica gel column-chromatography purification prior to the next step. Many efforts were made to solve this problem, and Amberlyst resin–catalyzed ketal formation as reported in the literature [20] was viewed as suitable for this process. First, it was possible to select an acid with suitable strength for ketal formation, and second, the resin's capability to absorb moisture and amine could be beneficial for both accelerating the reaction rate and removing the amine by-product. Finally, the resin catalyst can be easily removed by simple filtration. As expected, the reaction proceeded smoothly, and the isolated ketal compound **17** did not contain de-trifluoroacetylation by-product **20**, either at the pilot scale or at the commercialization scale (Table 10.3) [21].

Epimerization of the 7-hydroxy group could be suppressed by changing the reaction solvent from 1,4-dioxane to *n*-butyl alcohol in the cleavage step. Slow filtration of the aglycon **11** resulted from the presence of very fine particles of hydrate crystal. This could be overcome by crystal form control to isolate **11** as the anhydrate crystal from the hydrophilic organic solvent. As a result, the overall yield from **16** to **11** could be improved from 37 to 53%.

Table 10.3 Comparison of ketal formation catalyst.

Catalyst	p-Toluenesulfonic acid	Amberlyst 15
Reaction time (h)	12	16
Remaining ketone **16** (%)	5–6	3.5–3.7
De-trifluoroacetylation **20** (%)	12–13	Not more than 0.1
Isolated yield (%)	75–80	90

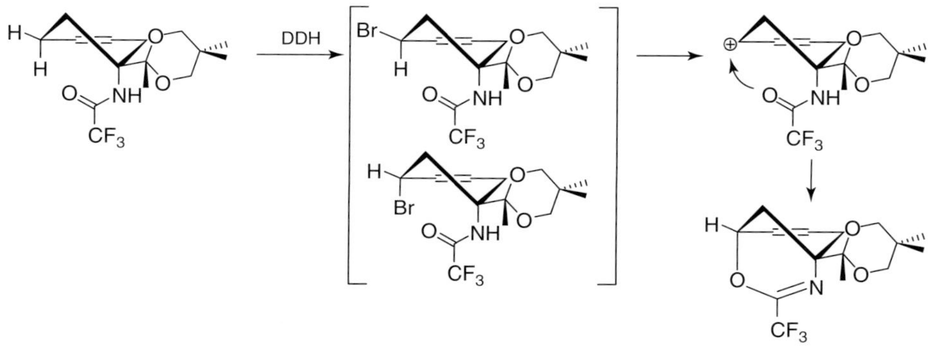

Scheme 10.6 Proposed mechanism for stereospecific cyclization.

In the oxidative cyclization step, the 7-hydroxy group was introduced stereospecifically on the α face. However, it is considered that the bromination of **17** proceeds on both faces, although continuous cyclization occurs stereospecifically via the benzylic cation intermediate as an S_N1 type reaction (Scheme 10.6) [22].

10.3.3
Polymorphism Study of Amrubicin Hydrochloride

As amrubicin hydrochloride is used for the injectable dosage form, its crystalline form does not affect bioavailability, but it affects drug substance stability. Specifically, this compound is thermally unstable due to the labile amino group at the 9-position affording deamination products. In the early development stage, amrubicin hydrochloride was isolated after crystallization at a temperature lower than 15 °C to afford α-form crystal. This α-form crystal, however, had low stability. Therefore, improvement of this instability was one of the most important issues at that time. After extensive study of the crystalline form of amrubicin hydrochloride, we found that a specific crystal form, the β-form, had excellent thermal stability. Powder X-ray diffraction (PXRD) of both crystal forms are shown in Figures 10.5 and 10.6.

The stable β-form of amrubicin hydrochloride can be produced by crystallization from a mixture of water and hydrophilic organic solvent at a temperature higher

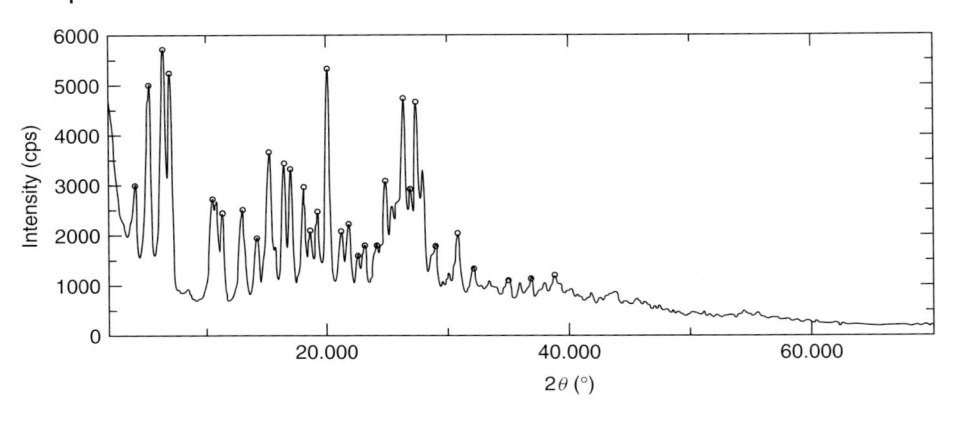

Figure 10.5 PXRD of α-form of amrubicin hydrochloride.

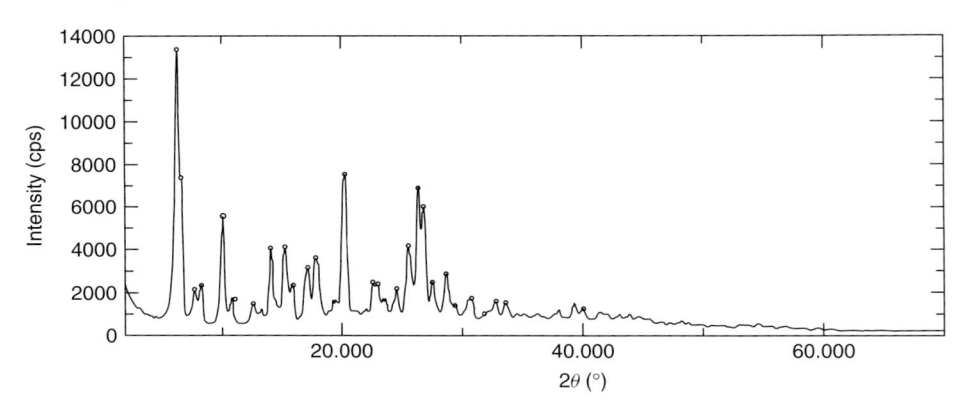

Figure 10.6 PXRD of β-form of amrubicin hydrochloride.

than 22 °C. A comparison of the stability of both crystal forms is shown in the table below (Table 10.4) [23].

10.3.4
Stability of Amrubicin Hydrochloride with Reference to Moisture

In the final step of amrubicin manufacturing, critical issues had to be considered. In order to remove the crystallization solvent from the wet filter cake, a drying method under reduced pressure with introduction of dry nitrogen gas was used. Although the drying time could subsequently be shortened, the content of a specific impurity, that is the desaccharification by-product (aglycon **11**), increased remarkably. We explored the reason for this decomposition, and examined all parameters related to the drying process. As a result, we found that the water content after drying was 2.5%, which was lower than the product preferred level (4–7%). These findings led us to investigate the stability of amrubicin hydrochloride with reference to moisture.

Table 10.4 Comparison of the stability of amrubicin hydrochloride crystal forms.

	α-form	β-form
Crystallization temperature	Lower than 15 °C	Higher than 22 °C
Stability after 24 h at 80 °C	7.8% degraded	No degradation products

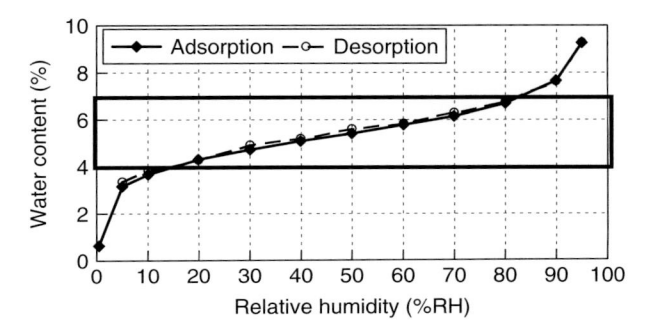

Figure 10.7 Moisture adsorption behavior.

10.3.4.1 Amrubicin Hydrochloride Moisture Adsorption

First, we measured the amrubicin hydrochloride moisture adsorption equilibrium. As shown in Figure 10.7, the moisture content in amrubicin hydrochloride changed depending on the relative humidity (RH):

- below 10% RH, the water content was less than 4%;
- above 80% RH, the water content was more than 7%.

10.3.4.2 Stability in Various Water Contents

In the next step, the relationship between stability and moisture content was investigated. The results of stability tests with varying water content at 30 °C are shown in Figure 10.8. At less than 3% water content, formation of the desaccharification by-product **11** was remarkable; on the other hand, at more than 8% water content, the level of deamination by-product **21** increased. It was therefore concluded that keeping the water content in the range of 3–8%, more preferably 4–7% is important not only for storage of amrubicin hydrochloride but also during its drying [24].

10.3.4.3 Establishment of Drying Method

To establish a drying method of amrubicin hydrochloride, the water content after various drying methods was tested (Table 10.5). A dry nitrogen gas through-flow drying method or a drying method under reduced pressure with introduction of dry nitrogen gas resulted in a moisture content of 1.3 or 2.5%, respectively. These drying conditions caused decomposition of amrubicin to desaccharification

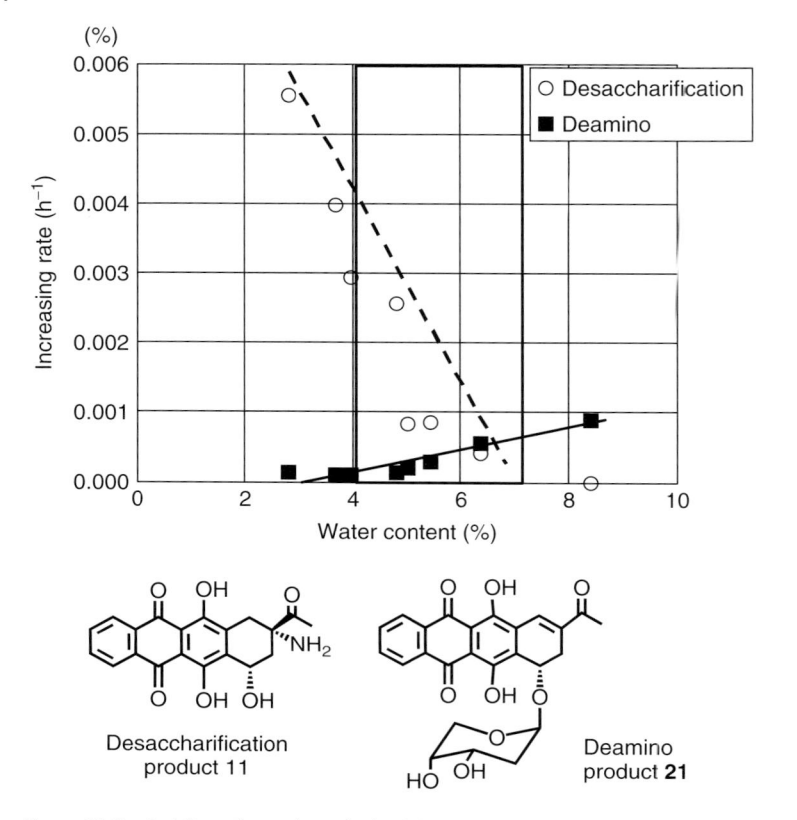

Figure 10.8 Stability of amrubicin hydrochloride with varying water content at 30 °C.

Table 10.5 Water contents after various drying conditions.

	Conditions	Water content (%)
Through-flow nitrogen	25 °C/14 h	1.3
Reduced pressure with introduction of dry nitrogen gas	2 kPa/17 h	2.5
Reduced pressure	2 kPa/28 h	3.4

by-product **11**. On the other hand, drying under reduced pressure at 2 kPa afforded a water content of 3.4% after 28 h. This drying method was selected for removing crystallization solvent from the wet filter cake.

The water content after primary drying was 3.4%, which is lower than the product preferred level (4–7%); therefore, the primary drug substance was exposed to a moistened nitrogen atmosphere in 40–60% RH to control water content at 4–7%. This could help control the final content of desaccharification and deamination by-products.

10.4
Conclusion

We have successfully established a bulk production method of amrubicin hydrochloride as a completely synthetic anthracycline derivative.

1) We established a safe synthetic method for production of aminoketone from carboxylic ester.
2) We established a stereoselective introduction method for the 7-hydroxy group.
3) We established a crystallization method to obtain the stable β form of amrubicin hydrochloride.
4) We clarified amrubicin hydrochloride water absorption properties and stability with regard to moisture and defined a method for handling this drug.

References

1. Ishizumi, K., Ohashi, N., and Tanno, N. (1987) *J. Org. Chem.*, **52**, 4477.
2. Morisada, S., Yanagi, Y., Noguchi, T., Kashiwazaki, Y., and Fukui, M. (1989) *Jpn. J. Cancer Res.*, **80**, 69.
3. Hanada, M., Noguchi, T., and Yamaoka, T. (2007) *Cancer Sci.*, **98**, 447.
4. Suzuki, T., Minamide, S., Iwasaki, T., Yamamoto, H., and Kanda, H. (1997) *Invest. New Drugs*, **15**, 219.
5. Noda, T., Watanabe, T., Kohda, A., Hosokawa, S., and Suzuki, T. (1998) *Invest. New Drugs*, **16**, 121.
6. Yamaoka, T., Hanada, M., Ichii, S., Morisada, S., Noguchi, T., and Yanagi, Y. (1998) *Jpn. J. Cancer Res.*, **89**, 1067.
7. Noguchi, T., Ichii, S., Morisada, S., Yamaoka, T., and Yanagi, Y. (1998) *Jpn. J. Cancer Res.*, **89**, 1061.
8. Noguchi, T., Ichii, S., Morisada, S., Yamaoka, T., and Yanagi, Y. (1998) *Jpn. J. Cancer Res.*, **89**, 1055.
9. Hanada, M., Mizuno, S., Fukushima, A., Saito, Y., Noguchi, T., and Yamaoka, T. (1998) *Jpn. J. Cancer Res.*, **89**, 1229.
10. Hanada, M., Noguchi, T., and Yamaoka, T. (2006) *Cancer Sci.*, **97**, 1396.
11. Yana, T., Negoro, S., Takada, M., Yokota, S., Takada, Y., Sugiura, T., Yamamoto, H., Sawa, T., Kawahara, M., Katakami, N., Ariyoshi, Y., and Fukuoka, M. (2007) *Invest. New Drugs*, **25**, 253.
12. Onoda, S., Masuda, N., Seto, T., Eguchi, K., Takiguchi, Y., Isobe, H., Okamoto, H., Ogura, T., Yokoyama, A., Seki, N., Asaka-Amano, Y., Harada, M., Tagawa, A., Kunikane, H., Yokoba, M., Uematsu, K., Kuriyama, T., Kuroiwa, Y., and Watanabe, K. (2006) *J. Clin. Oncol.*, **24**, 5448.
13. Yamauchi, S., Kudoh, S., Kimura, T., Hirata, K., and Yoshikawa, J. (2002) *Osaka City Med. J.*, **48**, 69.
14. Ohe, Y., Negoro, S., Matsui, K., Nakagawa, K., Sugiura, T., Takada, Y., Nishiwaki, Y., Yokota, S., Kawahara, M., Saijo, N., Fukuoka, M., and Ariyoshi, Y. (2005) *Annals. Oncol.*, **16**, 430.
15. Terashima, S., Kimura, Y., Ishizumi, K., and Tanno, N. (1985) Japanese Kokai Tokkyo Koho, 60188396.
16. French, F.A. (1966) *Chem. Eng. News*, **44** (15), 48.
17. Olson, G.L. (1966) *Chem. Eng. News*, **44** (24), 7.
18. Choudhry, S.C., Serico, L., and Cupano, J. (1989) *J. Org. Chem.*, **54**, 3755.
19. Kito, M., Dan, A., Shimago, K., and Sunagawa, M. (2000) Japanese Kokai Tokkyo Koho, 2000063336.
20. Constable, E.C. and Smith, D.R. (1997) *Tetrahedron*, **53**, 1715.
21. Kawano, H. and Shimago, K. (2002) Japanese Kokai Tokkyo Koho, 2002161078.

22. Yamamoto, K. and Shimago, K. (2003) Japanese Kokai Tokkyo Koho, 2003012620.
23. Shimago, K. and Uenishi, Y. (1999) PCT International Applications, WO Patent 9928331.
24. Takahashi, K., Fujimoto, K., Yamauchi, Y., and Okahashi, T. (2003) PCT International Applications, WO Patent 2003035660.

11
Process Development of HIV Integrase Inhibitor S-1360
Toshiro Konoike and Sumio Shimizu

11.1
Introduction

Acquired immunodeficiency syndrome (AIDS) was first reported in the United States in 1981, and two years later in 1983, the human immunodeficiency virus-1 (HIV-1) was identified as a pathogenic retrovirus causing AIDS. HIV infects human helper T cells, proliferates, and consequently destroys the immune functions, which causes AIDS after a certain incubation period. The replication cycle of HIV starts from the entry/fusion of HIV into a human T cell and ends with a mature HIV virion budding out of a T cell. Several enzymes and receptors are involved in the replication cycle, and they have become drug targets for treating HIV infection. Six classes of anti-HIV drugs are currently on the market [1]. The first antiretroviral drug is zidovudine (azidothymidine, AZT), which belongs to a class of nucleoside reverse transcriptase inhibitors. The other classes of anti-HIV drugs are nonnucleoside reverse transcriptase inhibitors, protease inhibitors, fusion inhibitors, entry coreceptor CCR5 (CC chemokine receptor 5) antagonists, and integrase inhibitors; nevirapine, saquinavir, enfuvirtide, maraviroc, and raltegravir are the first ones in each class, respectively.

Integration of viral DNA into the host DNA is a requisite process for HIV replication and the initial two steps of the integration reaction are carried out by the viral integrase encoded by HIV. Inhibition of integrase has become a novel drug target for treating HIV infection, and of late, integrase inhibitors have been extensively investigated. A prototype of HIV integrase inhibitors is a 1,3-diketo acid (DKA) that is known to inhibit the strand transfer reaction of HIV DNA into T-cell DNA by coordinating with two magnesium ions on the HIV integrase at the active site [2].

11.2
Discovery of Integrase Inhibitor S-1360

Shionogi and Merck have been pioneer companies in the R&D of integrase inhibitors (Figure 11.1). Screening of the Shionogi compound library was started

Pharmaceutical Process Chemistry. Edited by Takayuki Shioiri, Kunisuke Izawa, and Toshiro Konoike
Copyright © 2011 WILEY-VCH Verlag GmbH & Co. KGaA, Weinheim
ISBN: 978-3-527-32650-1

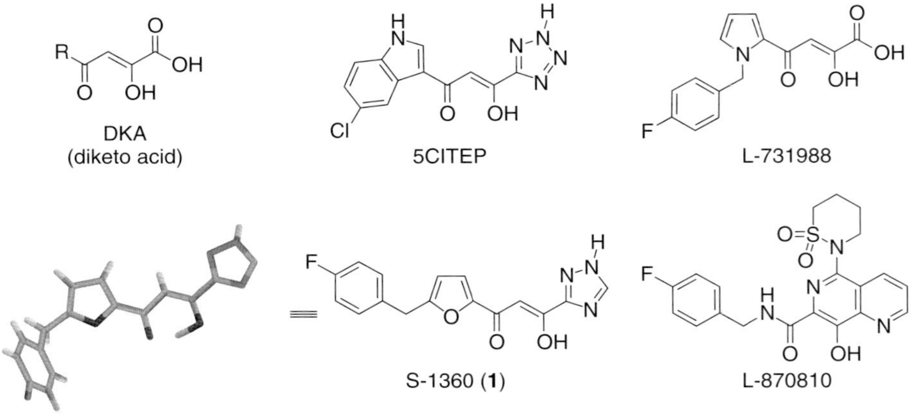

Figure 11.1 Leads of integrase inhibitor.

in 1995. From a seed compound, we synthesized a prototype integrase inhibitor and continued structure–activity relationship (SAR) studies. Among several compounds, 5CITEP [3] was found to give a cocrystal with an HIV integrase core domain, and a single crystalline X-ray diffraction study of the cocrystal showed the structure of the coordination site. Although 5CITEP was a weak inhibitor, it provided a platform for us to continue SAR studies. Eventually, we found S-1360 (**1**) to be a potent integrase inhibitor and selected it as a drug candidate for clinical trials. S-1360 is a 1,3-diketone bearing a benzylfuran ring at one end and a 1,2,4-triazole ring at the other end. The tetrazole of 5CITEP and the triazole of S-1360 are known as *bioisosteres of carboxylic acid*, and 5CITEP and S-1360 were recognized to be equivalent to a prototypical integrase inhibitor DKA. The NMR spectrum and the single crystal X-ray diffraction study of S-1360 showed that it exists in an enol form in solution as well as in a crystalline state. Its patent was filed in 1998 [4], and we began the GMP campaign 1 of the active pharmaceutical ingredients (APIs) of S-1360 in 1999. Phase 1 clinical trials started in 2000, making it the first inhibitor of HIV integrase to enter clinical testing. Clinical profiles of S-1360 were reported in 2002 [5].

Merck's prototypical pyrrole containing diketo acid inhibitor L-731988 was reported in 2000 [6]. Later, they discovered their clinical trial candidate L-870810 [7]. Thus, two candidates, S-1360 and L-870810, were being tested for efficacy and safety in clinical trials conducted concurrently around 2002.

11.2.1
Discovery Route of S-1360

The discovery route of S-1360 was linear and lengthy (Scheme 11.1). Two starting materials, 5-(4-fluorophenyl)-2-furyl methyl ketone **2** and ethyl 1,2,4-triazole-3-carboxylate **3**, were prepared by a long sequence of reactions. Protection of the triazole was done using the triphenymethyl (trityl, Tr-) group by

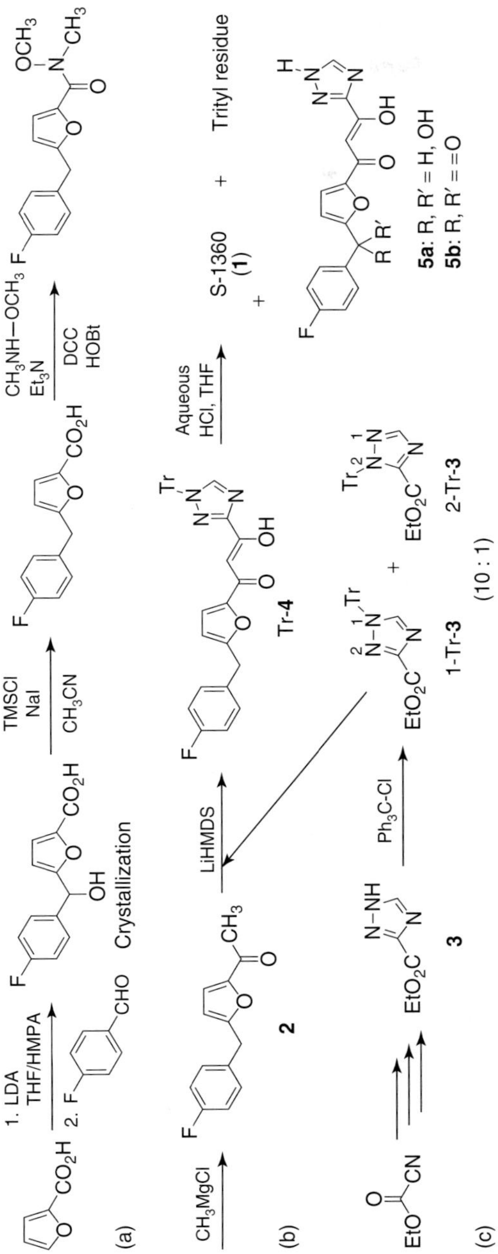

Scheme 11.1 Discovery route of S-1360 (1).

alkylation with trityl chloride (TrCl), which led to the production of a mixture of two regioisomers, 1-Tr-**3** and 2-Tr-**3**. The subsequent Claisen condensation [8] of **2** and 1-Tr-**3** was carried out with 2 equiv. of lithium hexamethyldisilazide (LiHMDS) under cryogenic conditions to give the final protected intermediate Tr-**4**.

In the early stage of the development, there were two requirements for the Claisen condensation to give **1**. One was that protection of **3** was essential for the Claisen condensation. Triazole **3** has a protic N–H bond in the triazole ring and was deprotonated under strong basic conditions for the Claisen condensation to give an anionic species of **3**, which reduced susceptibility of the ester group of **3** to nucleophilic attack of the lithium enolate of methyl ketone **2**; no condensation was observed without protection. Protection of the protic N–H proved to be crucial for the Claisen condensation (Scheme 11.2a).

The second requirement is related to the mode of the Claisen condensation. 1,3-Diketone analogous to S-1360 was conventionally prepared by Claisen

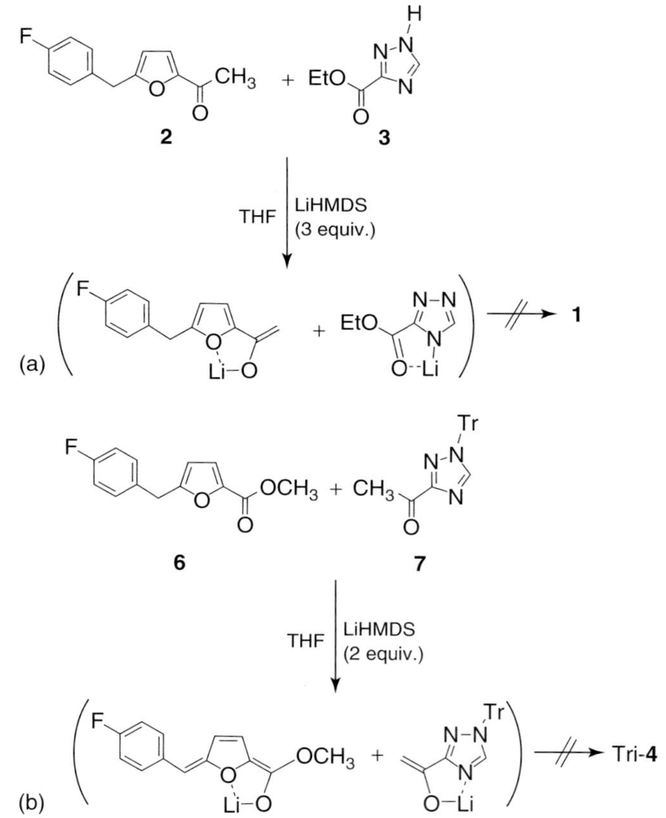

Scheme 11.2 Unsuccessful Claisen condensation (a) furyl methyl ketone and unprotected triazole ester. (b) furan ester and tritylated acetyltriazole.

condensation of a methyl ketone and an ester. Conceptually, two modes of C–C bond formation are feasible for S-1360. The results of the discovery route showed that the Claisen condensation of furyl methyl ketone **2** and the protected triazole ester 1-Tr-**3** proceeded uneventfully and gave the desired compound in a high yield (Scheme 11.1b). In contrast, the other mode of the Claisen condensation between furan ester **6** and acetyltriazole **7** led to recovery of the starting materials because of the formation of two lithium enolates which were inert to each other because of stabilization by conjugation and chelation (Scheme 11.2b).

Subsequent deprotection of the trityl group was carried out by treatment with aqueous HCl in tetrahydrofuran (THF). The isolation and purification of **1** were done by partitioning sodium enolate of **1** and trityl residue into two layers of aqueous sodium hydroxide and toluene, and subsequent extraction of **1** was achieved using ethyl acetate after acidifying the aqueous phase. Two by-products, benzylic alcohol **5a** and ketone **5b**, arose from air oxidation on the active methylene group of **1** in an aqueous alkaline solution during the partitioning, and consequently, repeated recrystallization of the extracts of **1** was required to remove the two by-products from **1**.

The entire synthetic reaction involved several drawbacks for large-scale preparation, such as expensive raw materials, toxic reagents, harsh or cryogenic reaction conditions, lack of selectivity, inefficient final deprotection, and tedious purification steps. Several kilograms of API **1** were prepared by this method for preclinical toxicity studies in spite of these technical shortcomings. Owing to a high medical need and the exciting prospect of a new drug for AIDS, research on S-1360 had been proceeding extremely rapidly. Thus, a practical synthesis was urgently required to support further toxicology and clinical trials. The details of the synthesis and manufacturing have been published elsewhere [9, 10] and the progress of the chemical development is reported in the following sections.

11.3
Synthesis of Two Starting Materials for S-1360

On the basis of the results from medicinal chemists, the remaining synthetic challenges were (i) concise syntheses of the two starting materials, (ii) selection of a suitable protective group for triazole, (iii) practical conditions for the Claisen condensation, and (iv) robust and effective final deprotection and purification. For efficient speed development and cost-effective manufacturing of **1**, we needed a low-cost and reliable synthetic method by which S-1360 could be prepared on a large scale. Described first are the syntheses of two starting materials **2** and **3**.

11.3.1
Two One-Step Syntheses of Benzylfuryl Methyl Ketone 2

Since the synthesis of **2** was quite long starting from expensive 2-furoic acid, we investigated an alternative synthesis of **2** from inexpensive raw materials and reagents. Furanes are known to undergo the Friedel–Crafts reactions [11] but there are few examples for furyl ketones [12]. The most straightforward Friedel–Crafts alkylation of 2-furyl methyl ketone by 4-fluorobenzyl chloride was attempted as both of them are commercially inexpensive raw materials for pharmaceuticals. Two one-pot processes for **2** were developed (Scheme 11.3) [10].

11.3.1.1 Friedel–Crafts Alkylation by Anhydrous ZnCl$_2$ in Dichloromethane

First, a variety of combinations of Lewis acids and solvents was surveyed for the Friedel–Crafts benzylation of 2-furyl methyl ketone. It was found that the desired ketone **2** could be obtained in a single step by treating a mixture of 2-furyl methyl ketone and 4-fluorobenzyl chloride with zinc chloride in several solvents. The reaction conditions were optimized; when 2-furyl methyl ketone and benzyl chloride (2 equiv.) were treated with anhydrous zinc chloride (1.5 equiv.) in refluxing dichloromethane, **2** (53%) was obtained along with a regioisomer **8** (1%) (Scheme 11.3a). A white, crystalline solid precipitated as the reaction proceeded, and it proved to be a zinc chloride complex of **2** without **8**. For synthetic purposes, this complex was collected by filtration and washed with dichloromethane. Subsequently, the resulting white solid was dissolved in water and extracted with toluene. The toluene extract contained zinc-free **2** as a sole product (42% yield), and this extract could be used for the reactions following concentration.

This method has a few merits, especially for expedited synthesis of **2** for speedy development of S-1360, and a few kilograms of **2** were synthesized by this procedure and supplied for the synthesis of **1** for nonclinical use. However, dichloromethane is an obvious drawback for industrial production due to environmental concerns and being a health hazard, we sought an alternative procedure for the Friedel–Crafts reaction.

11.3.1.2 Friedel–Crafts Alkylation Using Aqueous ZnCl$_2$

We surveyed a number of reaction conditions using anhydrous ZnCl$_2$ without a solvent and within a range of solvents including alcohols, ketones, carboxylic acids, nitromethane, and water. Water gave **2** in the highest yield and was an appropriate solvent to replace dichloromethane [13]. The Friedel–Crafts reaction in water is a two-layer system comprised of organic compounds and aqueous ZnCl$_2$ solution (Scheme 11.3). As water is the most favorable solvent for process development, the reaction conditions were optimized; the mixture of 2-furyl methyl ketone, 4-fluorobenzyl chloride (2 equiv.), and ZnCl$_2$ (1.05 equiv.) in two volumes of water was stirred at 85 °C for 6 h. The Friedel–Crafts reaction gave **2** (70%), accompanying ketone **8** (5%), and di-4-fluorobenzyl ether (5%). Isolation and purification of **2** required distillation under reduced pressure and subsequent crystallization from solvent mixture of 2-propanol and hexane at −10 °C. The desired ketone **2** was isolated in 43% yield.

Calorimetry of the Friedel–Crafts reaction revealed the total energy given off during the reaction in H$_2$O to be 23.7 kJ mol^{-1}, which meant an adiabatic temperature rise by 14 °C. Therefore, we concluded that the aqueous reaction was adequate for scale-up and optimized the conditions, which were adopted

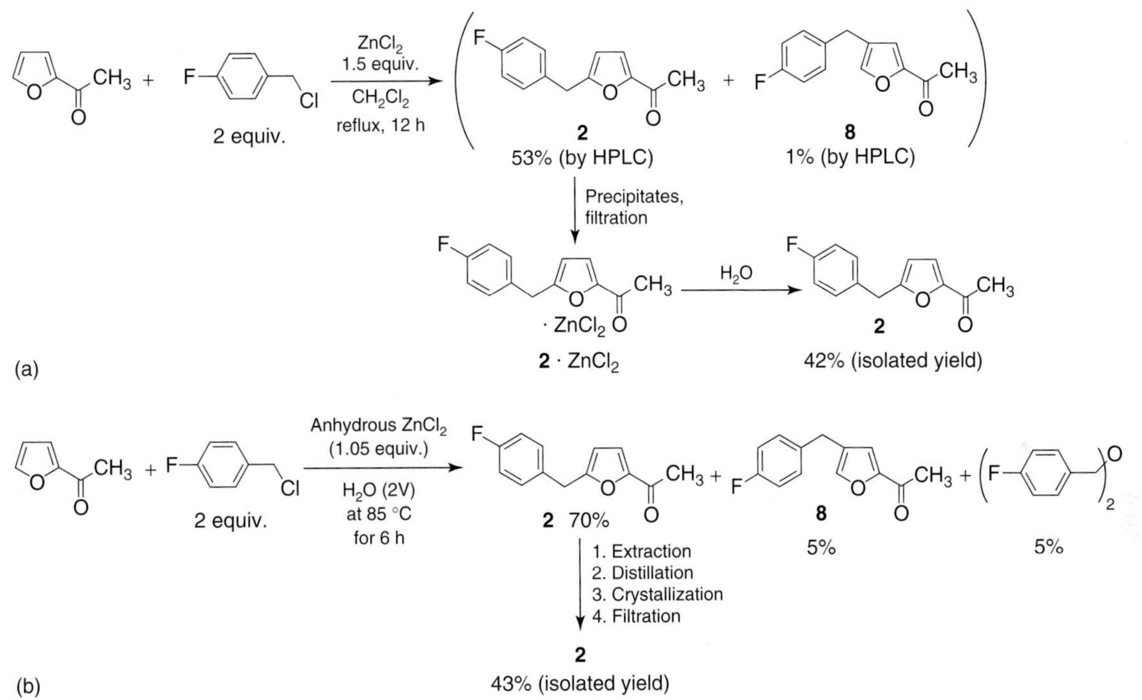

Scheme 11.3 Two synthetic methods of benzylfuryl methyl ketone. (a) Synthesis of **2** by anhydrous $ZnCl_2$ in dichloromethane. (b) Synthesis of **2** by aqueous $ZnCl_2$.

for large-scale synthesis. Hygroscopic anhydrous $ZnCl_2$ could be replaced by commercially available 50% aqueous $ZnCl_2$ which was inexpensive and safe to handle, and we obtained about 500 kg/lot of **2**.

11.3.2
Two Synthetic Methods to Triazole Ester 3

Triazole ester **3** was prepared in two ways, ring construction and ring modification. Both were satisfactory and we simply optimized several reactions for scale-up (Scheme 11.4).

11.3.2.1 Ring Construction Method
This method started from two raw materials, ethyl cyanoformate and hydrazine monohydrate (Scheme 11.4a). Ethyl cyanoformate was treated with hydrogen chloride in ethanol to give imidate, which was condensed with formic hydrazide to give **9**. Condensation product **9** was dehydrated by ethyl orthoformate under toluene refluxing conditions to give ethyl ester **3** [14]. This method was not applicable for synthesis of the corresponding methyl ester **10** because methyl cyanoformate was not available in large quantities. For this reason, ethyl ester **3** was chosen as the starting material for S-1360 and prepared in several hundred-kilogram quantities.

11.3.2.2 Ring Modification Method
3-Amino-1,2,4-triazole-5-carboxylic acid was chosen as a raw material for the ring modification method because it was a commercially available building block for agrochemicals. The 3-amino group was diazotized and subsequently reduced to give 1,2,4-triazole-3-carboxylic acid (Scheme 11.4b) [15]. Esterification by hydrogen chloride in ethanol or methanol gave the corresponding esters **3** and **10** in good yields, and this route enabled us to prepare methyl ester **10** as well as ethyl ester **3**.

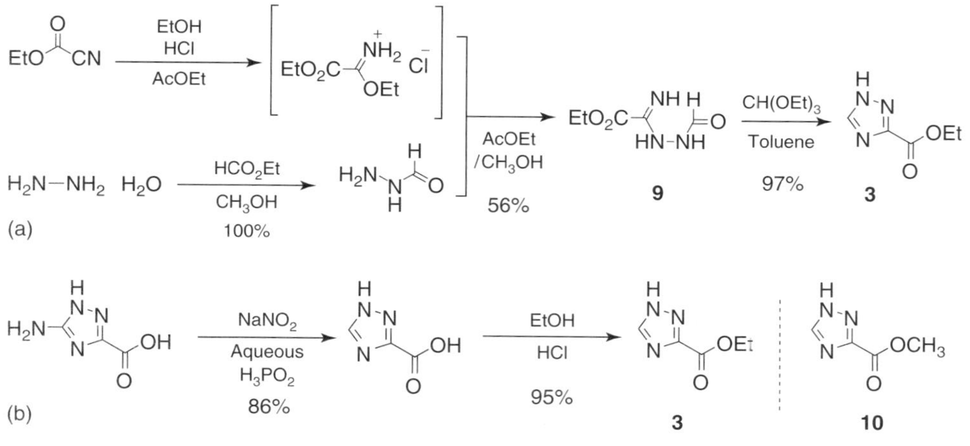

Scheme 11.4 Two synthetic methods of triazole esters. (a) Ring construction method. (b) Ring modification method.

In order to secure multisource vendors preparing **3** by two complimentary methods of ring construction and ring modification, several hundreds of kilograms of ethyl ester **3** was prepared by the ring modification method.

Several years later when we were preparing for campaign 3 of API of **1** on a ton scale for phase 2 clinical trials, inexpensive methyl ester **10** [15, 16] was found available for supply in metric tons because it had become a key building block for antiviral ribavirin, which is used for treatment of hepatitis C in combination therapy with pegylated interferon α [17]. However, we decided to proceed with the process development using the ethyl ester **3** in stock as long as the drug development could be continued until the proof of concept (POC) of S-1360 was established. Accordingly, we formulated a plan to substitute ethyl ester **3** by methyl ester **10** for the starting material in the case that further clinical development would continue.

11.4
Process Chemistry of S-1360 and Scale-Up of THP Route

11.4.1
Protection of Triazole 3 by the Tetrahydropyranyl (THP) Group and Claisen Condensation

Protective groups are often used in organic synthesis [18] although their usage is not always recommended because the introduction/removal of a protective group adds two more reactions to the total synthetic sequence. If protective groups have to be used, judicious choice is essential for efficient and robust processes to obtain APIs. Their usage often affects yields, qualities (contents, impurity profiles, residual solvents, etc.), productivities, and costs of APIs especially when the final reaction step is deprotection. Process chemists at three pharmaceutical companies in the United Kingdom reported a survey of reactions used for 128 drug candidates in the early development stages [19]. The study showed that the ratio of protection and deprotection occupied the largest portion (21%) of all reactions used; it is generally regarded as a necessary evil.

We first chose a tetrahydropyranyl (THP) protective group [20], which is commonly used for alcohol protection (Scheme 11.5). Preliminary experimental results showed that all reaction and purification steps proceeded uneventfully, and we further optimized all reaction conditions [9]. Protection of **3** by the THP group was done by treatment with 3,4-dihydropyran under acidic conditions, which initially gave a mixture of two regioisomers of THP-**3**. It was found that 2-THP-**3** isomerized to stable 1-THP-**3** to exclusively give 1-THP-**3** under protecting conditions upon prolonged reaction time. The next Claisen condensation of furyl methyl ketone **2** with 1-THP-**3** was carried out by lithium diisopropylamide (LDA) at a low temperature as in the discovery route. By screening various bases in place of expensive LDA, sodium methoxide in methanol was found to be most appropriate in terms of yield, cost and because it did not need cryogenic conditions. Further optimization of the conditions allowed us to run the protection efficiently and the

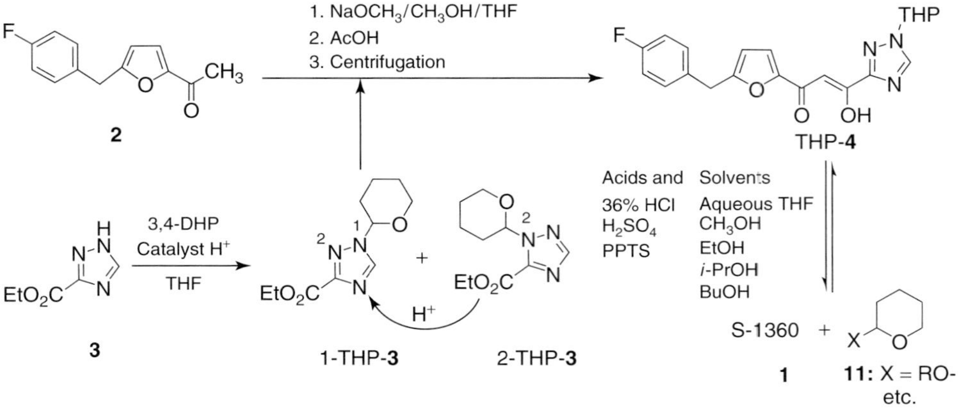

Scheme 11.5 THP protection, Claisen condensation, and deprotection.

Claisen condensation under mild conditions to give THP-4 in a high yield. THP-4 was crystallized from the solution by neutralizing with acetic acid (Scheme 11.5), and these three steps starting from protection to crystallization were carried out in a one-pot process. THP-4 was isolated by centrifugal filtration, washing, and drying (86% yield).

11.4.2
Deprotection of the THP Group and Purification of API Deprotection of the THP Group

The deprotection of THP-4 was carried out under a variety of acidic conditions commonly employed for THP-protected alcohols, generally giving **1** in a high yield (Scheme 11.5). However, as an appreciable amount of THP-4 remained in all of the reactions, we scrutinized conditions for deprotection using several acids, alcohols, and solvents under varying concentrations of THP-4 in order to find the conditions for achieving complete deprotection (Table 11.1). Methanol proved to be the most appropriate solvent to obtain **1** in a high yield, however, we observed an appreciable 1.6% of THP-4 (Table 11.1, entry 3), and could get sufficiently pure API with 0.1% of THP-4 only under impractical high dilution conditions (Table 11.1, entry 4).

As these results implied that an equilibrium occurs among **1**, THP-4, and THP-residue **11** (Scheme 11.5), a mixture resulting from the deprotection (Table 11.1, entry 3) was quenched and extracted with EtOAc, and the extracts were concentrated to give crude **1**. Crystalline **1**, obtained by adding CH_3OH to the concentrate, contained a significant amount of THP-4. Recrystallization of **1** from a number of solvents or solvent mixtures did not improve the ratio, and disappointingly, neither reprocessing nor reworking of **1** contaminated by THP-4 under deprotection conditions led to a satisfactory purity level of **1**. It was speculated from these results that the THP groups of THP-4 and **11** easily migrated to **1** and vice versa.

Table 11.1 Deprotection of THP-4 under acidic conditions.

Entry	Solvents				Acids		Conditions		Residual THP-4 (%)
		Volume		Volume		Equivalent	Temperature (°C)	h	
1	THF	20	H_2O	2	c-HCl	6	Reflux	1	14
2	THF	10	H_2O	1	c-HCl	3	Reflux	8	21
3	CH_3OH	20	–	–	H_2SO_4	Catalyst	Reflux	1.5	1.6
4	CH_3OH	100	–	–	H_2SO_4	Catalyst	Reflux	1.5	0.1
5	EtOH	20	–	–	H_2SO_4	Catalyst	Reflux	1.5	1.3
6	EtOH	20	–	–	PPTS	Catalyst	Reflux	1.5	2.4
7	EtOH	20	H_2O	5	c-HCl	6	Reflux	1.5	1.5
8	BuOH	20	–	–	H_2SO_4	Catalyst	Reflux	2	1.8
9	i-PrOH	20	–	–	H_2SO_4	Catalyst	100	1.5	4.1
10	i-PrOH	120	–	–	H_2SO_4	Catalyst	Reflux	1	0.7

11.4.2.1 Purification of API

In the course of investigation of the deprotection conditions, the hydrochloride salt of S-1360 (**1·HCl**) was found to crystallize out when THP-4 was treated with an excessive 5 equiv. of hydrochloric acid in methanol or 2-propanol although deprotection did not proceed to completion and THP-4 remained (0.9%). The resulting salt **1·HCl**, collected by filtration and washed with methanol or 2-propanol, proved to be free of contamination of THP-4 by HPLC, and the remaining THP-4 was removed into the mother liquor and washing methanol. When **1·HCl** was suspended in aqueous THF, it liberated HCl and **1** crystallized out to give free API **1** in a high yield (Scheme 11.6).

The THP route was scaled up and GMP campaign 1 was carried out. Every step up to the centrifugation of salt **1·HCl** apparently worked well. However, the obtained precipitate of **1·HCl** was an extremely fine crystalline powder, and its filtration was difficult and took much longer than anticipated from small-scale experiments. Washing the pasty powder was so troublesome that it could not

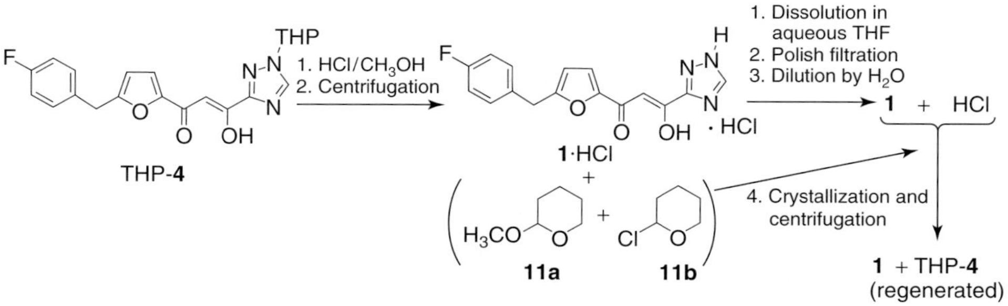

Scheme 11.6 THP deprotection, isolation of HCl salt, and regeneration of THP-4.

be washed sufficiently, and a trace of THP-4 (0.06% by HPLC) was observed in solid **1**·HCl. Subsequently, the **1**·HCl obtained was suspended in aqueous THF to liberate HCl and give a crystalline powder of **1**. Salt-free **1** was collected by filtration, however, it contained THP-4 (0.6% by HPLC), which increased 10 times from the 0.06% of the starting **1**·HCl. The cause of the contamination was the fact that the crystalline **1**·HCl contained significant amounts of two THP residues **11a** and **11b**, which were not detected by HPLC but identified later by NMR. The increased THP-4 was thought to be derived from the reverse reaction of **1** with THP-residue **11a**, which was present in **1**·HCl, under the strong acidic conditions rendered by released HCl (Scheme 11.6). This contamination made it necessary to rework the starting **1**·HCl, by reslurrying **1**·HCl in 2-propanol to remove **11**, collecting **1**·HCl by filtration, and subsequent washing. The reworked **1**·HCl eventually gave uncontaminated API **1** by a standard process of suspension in aqueous THF, which was collected by centrifugation.

Apart from the temporary reworking, a more substantial solution was required to overcome the difficulties encountered. We therefore examined the crystallization conditions of **1**·HCl and found the optimum conditions, which would give a favorable particle size of **1** HCl that would sediment quickly and allow collection by simple centrifugation and facile but thorough washing. The resulting precipitates of **1**·HCl were dissolved in aqueous THF and filtered, and the filtrate was subjected to crystallization of **1** by dilution with a large amount of water. Precipitates were collected by filtration, washed, and dried to give API of **1** in 83% yield [9]. The following campaign 2 was carried out by this improved method.

11.4.2.2 Quality Assurance and Productivity

For quality assurance of API, care must be taken in the preparation to avoid contamination with insoluble particles. They could come from a chemical process or from the environment in minute quantities. They are typically chemical impurities, dust, rust, fibers, charcoals, polymers, metals, insects, silica, and alumina. In order to reduce the risk of API being contaminated with insoluble particles, we routinely filter a solution of API or its precursor to remove potential insoluble particles. This filtration called *polish filtration* is usually carried out by an in-line cartridge filter or a Buchner with a porous (synthetic fiber, paper) filtration medium prior to the final crystallization [21].

Apart from the incomplete deprotection of THP-4, another drawback of the THP route is the low productivity of the final sequence of dissolution–polish filtration–crystallization to assure no contamination of API with insoluble particles. As API **1** and **1**·HCl are sparingly soluble in most solvents of class 3 (most recommended for low toxic potential) [22] commonly used for the final recrystallization, THF (class 2 solvents being of limited use) had to be used for dissolution-polish filtration of **1**·HCl before crystallization of **1** because **1**·HCl was relatively more soluble in aqueous THF. This forced us to filter an aqueous THF solution of **1**·HCl and dilute the filtrate by adding a large amount of water for crystallization of **1** in the final purification steps. As a result, volume efficiency (e.g., product weight (g)/total volume of the mixture (l)) of the crystallization step of **1** in the THP route was low (56 g l^{-1}), and this led to low productivity of the manufacturing of **1**, which eventually resulted in a high cost of API **1** due to the need for a large amount of solvents and large vessels for dissolution and crystallization. Although we recognized these drawbacks of the THP route, this method was employed for the scalable synthesis and several tens of kilograms of API was prepared at a pilot plant at Shionogi.

11.5
Process Development of S-1360 and Commercial Route by Methoxyisopropyl (MIP) Protection

Choice of a proper protective group was essential for process development in terms of the quality and productivity of **1** as shown in the preceding section, and we again surveyed several protective groups using a variety of alkyl vinyl ethers as protecting reagents. We tried five vinyl ethers as protecting reagents, that is, ethyl, propyl, butyl, isobutyl vinyl ethers, and 2-methoxypropene (Table 11.2). In order to find the most effective protective group, the entire sequence of reactions and purifications was repeated for the five protective groups starting from protection of triazole ester **3** all the way through isolation of **1**. We compared crystal forms, melting points,

Table 11.2 Protection **3** by vinyl esters, Claisen condensation, and deprotection.

Entry	Protecting reagents		Claisen condensation products		Deprotection
	Structure	Boiling points (°C)	Protect-4	Melting points (°C)	Residual protect-4
1	O-Et	33	R = H, R' = Et	86–87	Incomplete 2%
2	O-Pr	65	R = H, R' = Pr	67–68	Incomplete (not determined)
3	O-Bu	94	R = H, R' = Bu	Oil	Incomplete (not determined)
4	O-i-Bu	82–83	R = H, R' = i-Bu	70	Incomplete 16%
5	OCH$_3$ / CH$_3$	34–36	R = CH$_3$, R' = CH$_3$ (MIP-4)	111	Complete 0%

solubility, yields, and purities of the five final protected intermediates (protect-4), and subsequently, subjected them to deprotection and crystallization to **1** to check whether deprotection was complete and what the yields and purities of **1** were.

11.5.1
MIP Route

The methoxyisopropyl (MIP-, 1-methyl-1-methoxyethyl) group derived from 2-methoxypropene proved to be the best protective group in terms of complete deprotection under conventional acidic conditions (Table 11.2, entry 5). Other protective groups were deprotected incompletely and posed a potential risk of contaminating the API with the final intermediate protect-**4** similar to that observed with THP protection in the previous section. All the reactions and purifications were modified and optimized for the MIP route, and total synthesis was effected successfully and efficiently (Scheme 11.7).

Triazole ester **3** was protected by treatment with 2-methoxypropene in the presence of a catalytic amount of *p*-toluenesulfonic acid to give MIP-**3** quantitatively without a regioisomer. The following Claisen condensation of **2** and MIP-**3** was carried out with the addition of 28% NaOCH$_3$/CH$_3$OH, and the mixture was stirred at 60 °C for 3 h. The mixture was diluted with methanol, and then aqueous 14% acetic acid was added to neutralize the solution to pH 6. The mixture was diluted by water to complete crystallization of MIP-**4** and the resulting slurry was stirred at 5 °C for 1 h, centrifuged, and washed with aqueous methanol. The obtained wet MIP-**4** was subjected, without drying, to recrystallization from methanol and dried to obtain the final intermediate MIP-**4** in 84% yield.

MIP-**4** was dissolved in aqueous acetone and filtered to remove potential insoluble particles before deprotection. Deprotection was carried out by addition of aqueous 1.3% HCl to the filtered solution of MIP-**4** and heating it at 58 °C for 1 h. The

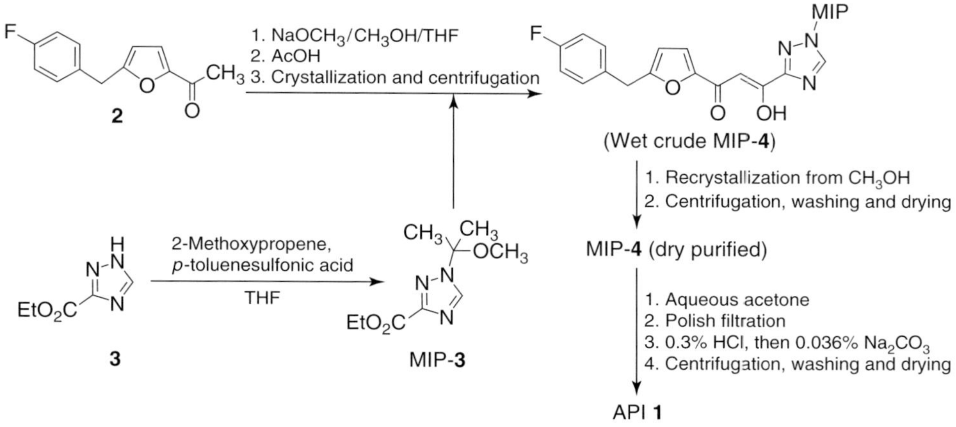

Scheme 11.7 MIP protection, Claisen condensation, and deprotection.

seed crystals of **1** and aqueous Na_2CO_3 were added to the mixture at 55 °C and the resulting slurry was stirred at 2 °C for 1 h, and then centrifuged. The collected crystalline **1** was washed with water and dried to give API of **1** in 98% yield [9]. Compared with the THP route, the MIP route was concise and straightforward, and it had an additional advantage in the purification of the final intermediate MIP-**4** and API **1**. The MIP route required no filtration of **1**·HCl, which was required in the THP route to remove the remaining THP-**4** and insoluble particles that might be present. In addition, the MIP route used acetone, a Class 3 solvent, in the final step. In practice, in order to alleviate potential contamination by insoluble particles, filtration of MIP-**4** was applied prior to deprotection, and then the subsequent deprotection of the MIP group was carried out under acidic conditions. The procedure of storing the filtrate, deprotection of MIP-**4**, and crystallization of **1** by adding water was done in one vessel installed in a clean room, and the following centrifugation, drying, and packaging of API **1** were done consecutively using the apparatuses in the same clean room. The MIP route has the minor disadvantage of using inflammable and relatively expensive 2-methoxypropene. However, we concluded that the MIP route has enough advantages over the THP route for it to be used for manufacturing API for phase 2 trials. Campaign 3 of API for Phase 2 clinical trial was carried out by the MIP route at Shionogi's subsidiary Nichia Pharmaceutical.

11.5.2
Further Improvement of Productivity

The volume efficiency of the deprotection-crystallization step in this campaign 3 of the MIP route was almost the same ($57\,g\,l^{-1}$) as that of the THP route ($56\,g\,l^{-1}$). The MIP route was finalized by scale-down experiments for the future manufacturing, and the results indicated that the volume efficiency increased to twice ($106\,g\,l^{-1}$) that of the THP route or the MIP route of the previous campaign 3. Suppose these conditions were employed for further commercial manufacturing, 230 kg of API **1** was assumed to be prepared in a 2000-l vessel for filtration, deprotection, and crystallization in one batch [9]. In this way, we established the commercial route of S-1360 while paying attention to speed, cost of development, yield and purity of API, productivity, robustness, safety, and the environment of the entire process.

11.6
Summary and Outlook

Process chemists utilize a number of strategies, tactics, and techniques for efficient and effective processes for development and manufacturing. Some of them have been demonstrated in the preceding sections including (i) short and convergent syntheses of **1** and **2**, (ii) one-pot synthesis of the final intermediates and protect **4** from the two starting materials **2** and **3**, (iii) telescoping of the intermediary wet final intermediate MIP-**4** to the subsequent recrystallization, and (iv) robust and

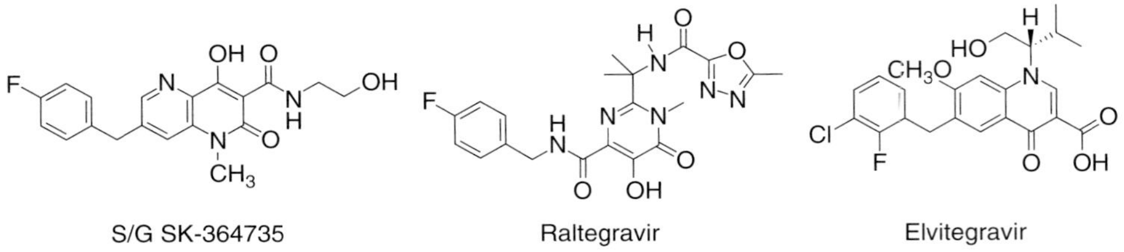

Figure 11.2 New integrase inhibitor.

high throughput deprotection–crystallization sequence in the final API-producing step. In addition, all the raw materials, reagents, and solvents are readily available and inexpensive, and every step is practicable in a multipurpose plant under mild conditions. The overall process had no serious issues with respect to the environment, health, and safety (EHS) and it was to be transferred to overseas plants for use there. In contrast to these merits, the kinds of troubles and faults that process chemists sometimes encounter at manufacturing sites occurred with the THP route.

Here is a sequel to the development of S-1360 by Shionogi. Codevelopment of S-1360 (1) was begun by the joint venture Shionogi–GlaxoSmithKline Pharmaceuticals, LLC in 2001, but development was terminated in 2003 owing to the rapid drug metabolism of S-1360 in humans [23] and its insufficient efficacy. Although S-1360 was not launched, R&D of integrase inhibitors was continued and some virological studies of clinical candidates including S-1360 and S/GSK-364735 were reported by coworkers at Shionogi and GlaxoSmithKline (Figure 11.2) [24]. The candidates S/GSK1349572 and S/GSK1265744 are once-a-day, unboosted integrase inhibitors with unprecedented antiviral activity in 10-day monotherapy and offer a superior resistance profile. S/GSK1349572 is currently in Phase 2b. Development of integrase inhibitors has been continued by several pharmaceutical companies and Merck's raltegravir was the first one to progress to phase 3 and was approved in 2007 [25]. JT/Gilead's elvitegravir, a novel quinolone β-keto acid, is under phase 3 clinical trials [26]. Recently cross-resistance of raltegravir and elvitegravir were reported *in vitro* and in clinic, and these findings underline the need for second-generation integrase inhibitors, avoiding cross-resistance with raltegravir and related compounds with improved potency, pharmacokinetics, and decreased toxicity [27].

More than 20 drugs are on the market for AIDS, however no drug is perfect. During the 25 years of research and development aiming at AIDS therapy, new drugs having new mechanisms of action have emerged. The fusion inhibitor, Roche/Trimeris's enfuvirtide [28], was approved in 2003, but it is a complex peptide, which is difficult to make and is administered by injection. Meanwhile, the first entry coreceptor CCR5 antagonist maraviroc from Pfizer was launched in 2007 [29]. Despite the number of new drugs approved and other new classes under development, the remaining pipeline is thin and no approvals are likely until 2010. In spite of many successful antiretroviral drugs, there is no cure or vaccine to prevent HIV infection. Considering emerging multidrug resistance, discovering treatments with existing drugs and developing new drugs are challenging themes for HIV chemotherapy [30].

Acknowledgments

We would like to express our sincere thanks to Drs H. Mikamiyama, K. Izumi, Messrs. T. Endoh, M. Kabaki, M. Uenaka, and all those who participated and contributed to the project.

References

1. Guidelines for the Use of Antiretroviral Agents in HIV-1-infected Adults and Adolescents December 1, 2009. Developed by the DHHS Panel on Antiretroviral Guidelines for Adults and Adolescents. http://www.aidsinfo.nih.gov/Guidelines/, November 3.
2. Pommier, Y., Johnson, A.A., and Marchland, C. (2005) *Nat. Rev. Drug Discovery*, **4**, 236.
3. Goldgur, Y., Craigie, R., Cohen, G.H., Fujiwara, T., Yoshinaga, T., Fujishita, T., Sugimoto, H., Endo, T., Murai, H., and Davies, D.R. (1999) *Proc. Natl. Acad. Sci. U.S.A.*, **96**, 13040.
4. Fujishita, T., Yoshinaga, T., and Sato, A. (2000) PCT International Applications. WO Patent 2000039086A.
5. Leese, P., Lippert, C., Russell, T., and Fujiwara, T. (2002) Abstracts of Papers, XIV International AIDS Conference, Barcelona, Spain; The Henry J. Kaiser Family Foundation, Washington, DC, Abstract TuPeB4436.
6. Wai, J.S., Egbertson, M.S., Payne, L.S., Fisher, T.E., Embrey, M.W., Tran, L.O., Melamed, J.Y., Langford, H.M., Guare, J.P. Jr., Zhuang, L., Grey, V.E., Vacca, J.P., Holloway, M.K., Naylor-Olsen, A.M., Hazuda, D.J., Felock, P.J., Wolfe, A.L., Stillmock, K.A., Schleif, W.A., Gabryelski, L.J., and Young, S.D. (2000) *J. Med. Chem.*, **43**, 4923.
7. Hazuda, D.J., Anthony, N.J., Gomez, R.P., Jolly, S.M., Wai, J.S., Zhuang, L., Fisher, T.E., Embrey, M., Guare, J.P. Jr., Egbertson, M.S., Vacca, J.P., Huff, J.R., Felock, P.J., Witmer, M.V., Stillmock, K.A., Danovich, R., Grobler, J., Miller, M.D., Espeseth, A.S., Jin, L., Chen, I.-W., Lin, J.H., Kassahun, K., Ellis, J.D., Wong, B.K., Xu, W., Pearson, P.G., Schleif, W.A., Cortese, R., Emini, E., Summa, V., Holloway, M.K., and Young, S.D. (2004) *Proc. Natl. Acad. Sci. U.S.A.*, **101**, 11233.
8. Smith, M.B. and March, J. (2001) *Advanced Organic Chemistry*, 5th edn, John Wiley & Sons, Inc., New York, p. 569.
9. Shimizu, S., Endo, T., Izumi, K., and Mikamiyama, H. (2007) *Org. Process Res. Dev.*, **11**, 1055.
10. Izumi, K., Kabaki, M., Uenaka, M., and Shimizu, S. (2007) *Org. Process Res. Dev.*, **11**, 1059.
11. Katritzky, A.R. and Taylor, T. (1990) *Adv. Heterocycl. Chem.*, **47**, 102.
12. Valenta, M. and Koubek, I. (1976) *Collect. Czech. Chem. Commun.*, **41**, 78.
13. Hayashi, E., Takahashi, Y., Itoh, H., and Yoneda, N. (1993) *Bull. Chem. Soc. Jpn.*, **66**, 3520.
14. Vemishetti, P., Leiby, R.W., Abushanab, E., and Panzica, R.P. (1988) *J. Heterocycl. Chem.*, **25**, 651.
15. Chiipen, G.I. and Grinshtein, V.Y. (1965) *Chem. Heterocycl. Compd.*, **1**, 420.
16. Zhang, P., Dong, Z.E., and Cleary, T.P. (2005) *Org. Process Res. Dev.*, **9**, 583.
17. Torriani, F.J., Rodriguez-Torres, M., Rockstroh, J.K., Lissen, E., Gonzalez-García, J., Lazzarin, A., Carosi, G., Sasadeusz, J., Katlama, C., Montaner, J., Sette, H. Jr., Passe, S., De Pamphilis, J., Duff, F., Schrenk, U.M., and Dieterich, D.T. (2004) *N. Engl. J. Med.*, **351**, 438.

18. Wuts, P.G.M. and Greene, T.W. (2006) *Greene's Protective Groups in Organic Synthesis*, 4th edn, John Wiley & Sons, Inc., Hoboken, pp. 25, 59, 77, 872.
19. Carey, J.S., Laffan, D., Thomson, C., and Williams, M.T. (2006) *Org. Biomol. Chem.*, **4**, 2337.
20. Atsumi, T., Kai, Y., and Katsube, J. (1977) Japan Kokai Tokkyo Koho, JP 52068182.
21. Anderson, N.G. (2000) *Practical Process Research & Development*, Academic Press, San Diego, p. 216.
22. ICH Guideline Q3C (R4) Impurities: Guidelines for Residual Solvents, February 2009. http://www.ich.org/cache/compo/363-272-1.html.
23. Rosenmond, M.J.C., St. John-Williams, L., Yamaguchi, T., Fujishita, T., and Walsh, J.S. (2004) *Chem. Biol. Interact*, **147**, 129.
24. Kobayashi, M., Nakahara, K., Seki, T., Miki, S., Kawauchi, S., Suyama, A., Wakasa-Morimoto, C., Kodama, M., Endoh, T., Oosugi, E., Matsushita, Y., Murai, H., Fujishita, T., Yoshinaga, T., Garvey, E., Foster, S., Underwood, M., Johns, B., Sato, A., and Fujiwara, T. (2008) *Antiviral Res.*, **89**, 213.
25. Summa, V., Petrocchi, A., Bonelli, F., Crescenzi, B., Donghi, M., Ferrara, M., Fiore, F., Gardelli, C., Paz, O.G., Hazuda, D.J., Jones, P., Kinzel, O., Laufer, R., Monteagudo, E., Muraglia, E., Nizi, E., Orvieto, F., Pace, P., Pescatore, G., Scarpelli, R., Stillmock, K., Witmer, M.V., and Rowley, M. (2008) *J. Med. Chem.*, **51**, 5843.
26. Sato, M., Motomura, T., Aramaki, H., Matsuda, T., Yamashita, M., Ito, Y., Kawakami, H., Matsuzaki, Y., Watanabe, W., Yamashita, K., Ikeda, S., Kodama, E., Matsuoka, M., and Shinkai, H. (2006) *J. Med. Chem.*, **49**, 1506.
27. Nakahara, K., Wakasa-Morimoto, C., Kobayashi, M., Miki, S., Noshi, T., Seki, T., Kanamori-Koyama, M., Kawauchi, S., Suyama, A., Fujishita, T., Yoshinaga, T., Garvey, E.P., Johns, B.A., Foster, S.A., Underwood, M.R., Sato, A., and Fujiwara, T. (2009) *Antiviral Res.*, **81**, 141.
28. Ketas, T.J., Frank, I., Klasse, P.J., Sullivan, B.M., Gardner, J.P., Spenlehauer, C., Nesin, M., Olson, W.C., Moore, J.P., and Pope, M. (2003) *J. Virol.*, **77**, 2762.
29. Åhman, J., Birch, M., Haycock-Lewandowski, S.J., Long, J., and Wilder, A. (2008) *Org. Process Res. Dev.*, **12**, 1104.
30. Thayer, A.M. (2008) *Chem. Eng. News*, **86** (33), 17, 25, 29.

12
An Efficient Synthesis of the Protein Kinase Cβ Inhibitor JTT-010

Takashi Inaba

12.1
Introduction

Patients with diabetes mellitus are at risk for developing microvascular complications such as neuropathy, nephropathy, and retinopathy, especially in the later stages of the disease. These complications seriously impair the quality of life for the patients suffering from long-term diabetes. Recently, protein kinase Cβ (PKCβ) activation has been implicated in the signal transduction pathway for the onset of these complications; hence, PKCβ selective inhibitors are expected to become therapeutic agents for these troubling conditions [1]. JTT-010, the staurosporine-related PKCβ inhibitor discovered in our laboratory, selectively inhibits PKCβ1 and β2 with IC_{50} values of 4.0 and 2.3 nM, respectively [2]. This compound is orally bioavailable and showed ameliorative effects in a variety of animal models having diabetic complications including retinopathy, neuropathy, and nephropathy. JTT-010 is structurally unique, possessing chiral aminomethyl-substituted fused pyrrolidine and anilino-indolylmaleimide components. These components define the biological features of JTT-010; the former is crucial for its favorable oral bioavailability while the latter functions as its pharmacophore. However, incorporation of these unique structural characteristics also entailed a synthetic challenge. The production methods of JTT-010 in the early discovery stages involved multiple steps including burdensome protection–deprotection and chromatographic purification procedures, which were unacceptable from the viewpoint of process chemistry. Hence, a new method that could be applied to the large-scale synthesis, when this compound was adopted as a candidate for further development, was in need. In this chapter, we describe an efficient synthesis of JTT-010 accomplished by a novel formal [3 + 2] cycloaddition of electron-deficient chiral cyclopropanes and indoles having a leaving group at the 2-position, followed by a three-component coupling reaction of the thus-obtained chirally substituted indole-fused pyrrolidine, maleimide, and aniline components [3]. This newly established route enabled us to provide sufficient amounts of JTT-010 for a variety of tests in the development stage. We also examined other synthetic methods during the course of this study, and for comparison we present some related reports from other laboratories.

Pharmaceutical Process Chemistry. Edited by Takayuki Shioiri, Kunisuke Izawa, and Toshiro Konoike
Copyright © 2011 WILEY-VCH Verlag GmbH & Co. KGaA, Weinheim
ISBN: 978-3-527-32650-1

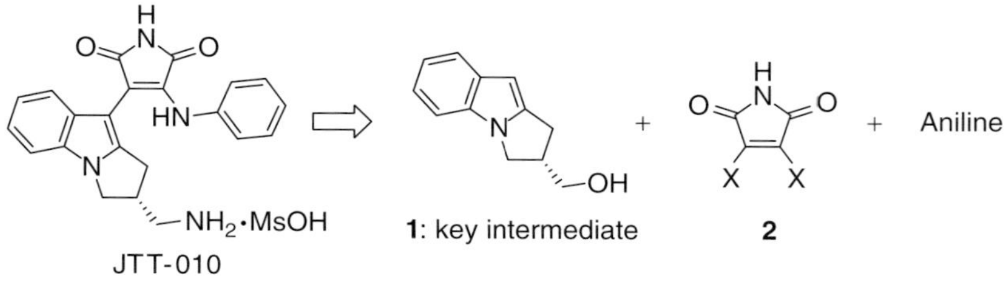

Scheme 12.1 Three components for the synthesis of JTT-010.

12.2
Synthetic Strategies

In our retrosynthetic analysis, JTT-010 was divided into three components including chirally substituted indole-fused pyrrolidine **1** (assigned as a key intermediate), maleimide **2**, and aniline because the assembly using these three components should be the most convergent and was the shortest conceivable route. In addition, the nucleophilic character of **1** at C(3) was seen as being potentially useful in reactions for its coupling with **2** (Scheme 12.1). This retrosynthetic prospect also conforms to the general principle that the most expensive component, which is **1** in this case, should be efficiently incorporated into the target molecule as late as possible in a synthetic sequence. In the earlier synthetic attempts for JTT-010, **1** represented the major part of production costs, and thus the significant improvements in the synthesis of key intermediate **1** was another indicator of the success of our approach. When this project was begun, no reaction applicable to the large-scale synthesis of **1** had been reported. The major issue to be addressed was the efficiency in the introduction of the chirality on the pyrrolidine ring. Hence, we began the search for a new efficient method that would give **1** in optically pure form.

12.3
Key Intermediate Synthesis

12.3.1
Optical Resolution

Optical resolution of racemic compounds by diastereomeric methods is a generally inefficient methodology to obtain chiral compounds since this method is usually accompanied by the waste of the undesired isomer and its yield never exceeds 50%. However, if the desired isomer can be conveniently separated by recrystallization and the undesired isomer in the filtrate can be isomerized to the initial isomeric ratio for recycling use, the efficiency of this methodology is improved. Moreover, if CIDT (crystallization-induced diastereomeric transformation) [4] or CIET (crystallization-induced enantiomeric transformation) [5] is possible, the efficiency of optical resolution may dramatically be improved. To accomplish this, the asymmetric center should be ready to be isomerized. In this context, an optical resolution was initially attempted on the carboxylic acid *rac*-**6** and its corresponding esters of chiral alcohols since its asymmetric center, α to the carboxyl group, was thought to be easily racemized by treatment with an appropriate base. We prepared *rac*-**6** from **4** via reduction of the carbonyl group with Raney Ni followed by saponification of the thus-obtained *rac*-**5** (Scheme 12.2). The Michael addition of **3** to ethyl acrylate followed by intramolecular condensation afforded **4** [6]. Despite various trials using diastereomeric salts of *rac*-**6** with chiral amines and its diastereomeric esters with chiral alcohols, we did not achieve the resolution, either by crystallization or even by chromatography. Eventually, the separation was chromatographically performed on the diastereomeric mixture of Boc-L-alanine ester of **7**, although the undesired isomer was not recyclable in this case [2a]. The Pd-catalyzed reduction of β-keto ester **4** gave the diastereomerically pure, but racemic, *cis*-hydroxyl ester **7**, which was then coupled with Boc-L-alanine to give a mixture comprising two diastereomers. The desired isomer **8** was chromatographically separated from the mixture in 22% yield. Hydrogenolysis of **8** in the presence of Pd catalyst provided optically pure **5**, which was finally converted to the key intermediate **1** by reduction with LiAlH$_4$. Although **8** was crystalline, this

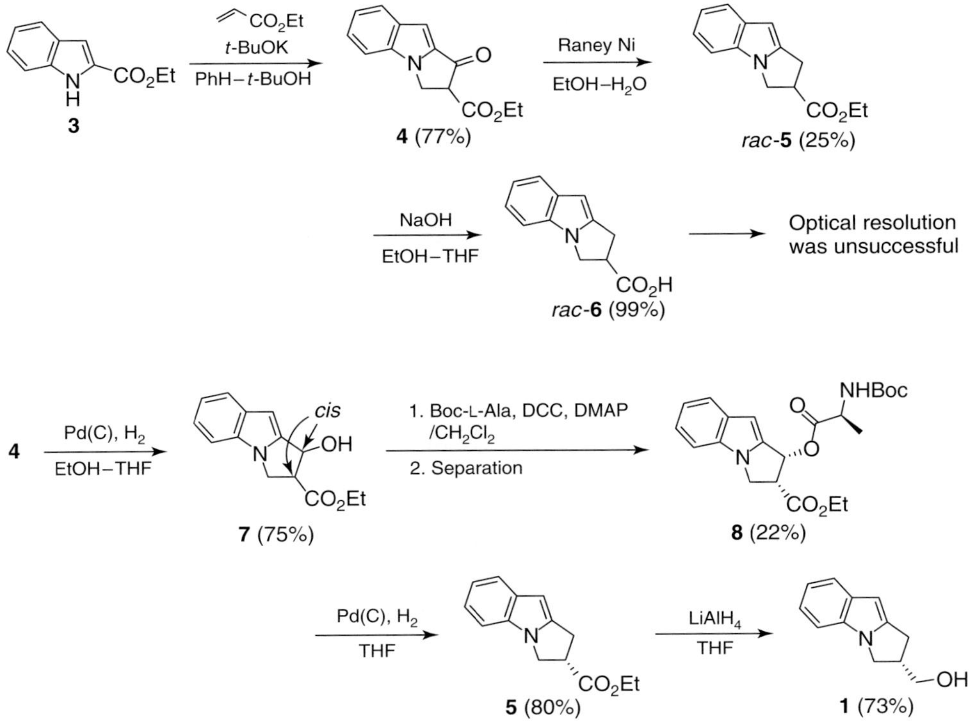

Scheme 12.2 Synthesis of the key intermediate **1** by optical resolution.

isomer could not be efficiently isolated by recrystallization from the diastereomeric mixture, and so chromatographic separation was required. Given the low total yield and inconvenience in separation, this method was impractical from the viewpoint of process chemistry and was employed only for the small-scale syntheses in the early discovery stage.

12.3.2
Enzymatic Chiral Induction

Enzymatic chiral induction is one of the most convenient options to obtain chiral compounds [7]. This methodology is most efficiently applied to prochiral symmetric substrates since high reaction conversion and high optical purity are both concurrently achievable by the enzymatic differentiation of two symmetric functional groups in the substrates. We applied lipase PS-promoted monoacetylation to the prochiral symmetric diol **11** [2a], a compound in the class of well-studied lipase substrates possessing the 2-substituted-1,3-propanediol substructure for enantioselective enzymatic desymmetrizations (Scheme 12.3) [8]. Compound **11** was obtained by Knoevenagel condensation of indole-2-carbaldehyde **9** with diethyl malonate followed by stepwise 1,4- and 1,2-reduction of the resulting unsaturated diester **10** with $NaBH_4$ and $LiAlH_4$, respectively. The lipase PS-promoted monoacetylation of diol **11** was carried out under acyl transfer conditions in vinyl acetate. The enzyme-catalyzed reaction proceeded successfully to give (R)-monoacetate **12** with 95% ee. However, this enantiomeric preference was inconvenient for the production of **14** that was required for the preparation of the key intermediate of JTT-010. If the enzymatic esterification had provided the antipode of **12**, the remaining hydroxyl group could have been directly converted to a leaving group for the subsequent cyclization into pyrrolidine. In reality, however, the newly acetylated hydroxyl group of **12** had to be converted into a leaving group via several steps prior to the cyclization into **14**. After protection of the remaining hydroxyl group of **12** with TBDPS (*t*-butyldiphenylsilyl), the enzymatically acetylated group was hydrolyzed

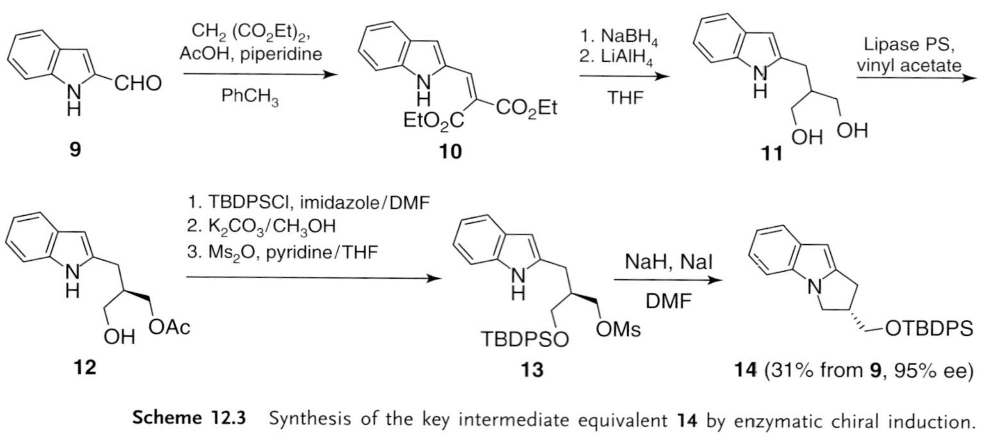

Scheme 12.3 Synthesis of the key intermediate equivalent **14** by enzymatic chiral induction.

to alcohol, which was then converted into a methanesulfonyloxy group to obtain **13**. Treatment of **13** with NaH in the presence of NaI gave the key intermediate equivalent **14** in 31% overall yield. Although the observed optical purity of 95% in the enzymatic reaction was acceptable and the chemical conversion was good enough to be employed, the number of the subsequent steps shown above was so large that a shorter route was needed. The large amount of hydrides required for the reduction of a double bond and two ester groups of **10** was another disadvantage of this enzymatic route, and therefore we did not pursue screening of other enzymes or modification of the substrate structure [8c] to obtain the enantiomer in a form ready to be cyclized.

12.3.3
C–H Bond Activation by a Chiral Catalyst

The transition metal–catalyzed activation and functionalization of otherwise unreactive carbon–hydrogen bonds allow great flexibility in chemical process designs. Bergman et al. applied their chiral variant of a rhodium-catalyzed intramolecular imine-chelation assisted C–H bond activation/olefin insertion reaction to the synthesis of the PKCβ inhibitor **19**, a derivative of JTT-010 selected from our patent publication [2c, 9]. They obtained the chirally substituted indole-fused pyrrolidine **17** in 90% ee by activating the C(2)–H bond of N-allylindole **15** in the presence of a rhodium catalyst bearing chiral ligand **16** (Scheme 12.4). The key intermediate equivalent **18** was obtained from **17** by deformylation using RhCl(dppp)$_2$. This route is very short and the methodology is sufficiently versatile to be applied to a variety

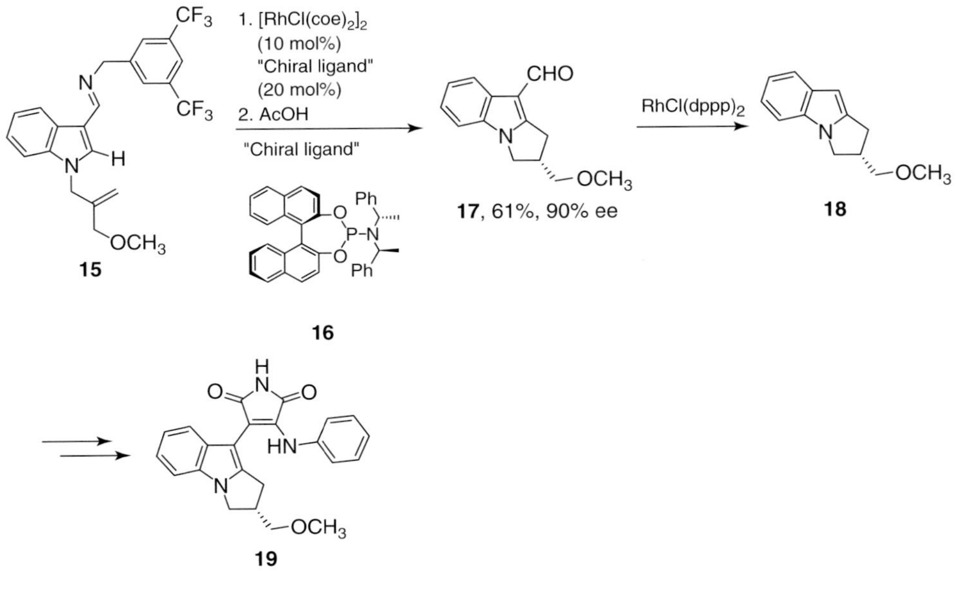

Scheme 12.4 Synthesis of the key intermediate equivalent **18** by a chiral catalyst.

of polycyclic aromatic compounds. The enantiomeric selectivity of this reaction, which achieved 90% ee of **17**, was also acceptable if the enantiomeric excess could be increased with some supplemental processes. However, the need for 20 mol% of the expensive chiral ligand to achieve this selectivity did not fulfill our requirements from the economical point of view. The moderate yield in the preparation of the substrate **15** was also a crucial issue that remained to be solved before adopting this route for future commercial-scale production of this key intermediate.

12.3.4
Formal [3 + 2] Cycloaddition Using Chiral Cyclopropane

One of the best established methods for the synthesis of substituted pyrrolidines is the [3 + 2] cycloaddition of activated cyclopropanes and a C=N double bond of aldimines [10]. Chirally substituted pyrrolidines are similarly available as long as chiral cyclopropanes are employed. These C=N double bonds could be interpreted to act as ylide equivalents in these reactions. Despite the number of reports on this reaction, to the best of our knowledge, this procedure had never been applied to the synthesis of heteroaromatic-1,2-fused pyrrolidines, probably due to the difficulty in generating an ylide form at the heteroaromatic N(1)–C(2)-position. It was envisioned that the N(1)–C(2) single bond of heteroaromatics **20**, in this case an indole, with a leaving group at the (C)2-position, could also act as an ylide equivalent in the stepwise formal [3 + 2] cycloaddition reaction shown in Scheme 12.5 [3]. The nitrogen anion of **20**, generated by appropriate base, was expected to open the electron-deficient cyclopropane **21** to afford the carbanion **22**, which was stabilized by the EWGs (electron-withdrawing groups). The carbanion was then anticipated to replace the leaving group on the indole (C)2-position to complete the formal [3 + 2] cycloaddition, thereby generating indole-fused pyrrolidine **23**. We considered that

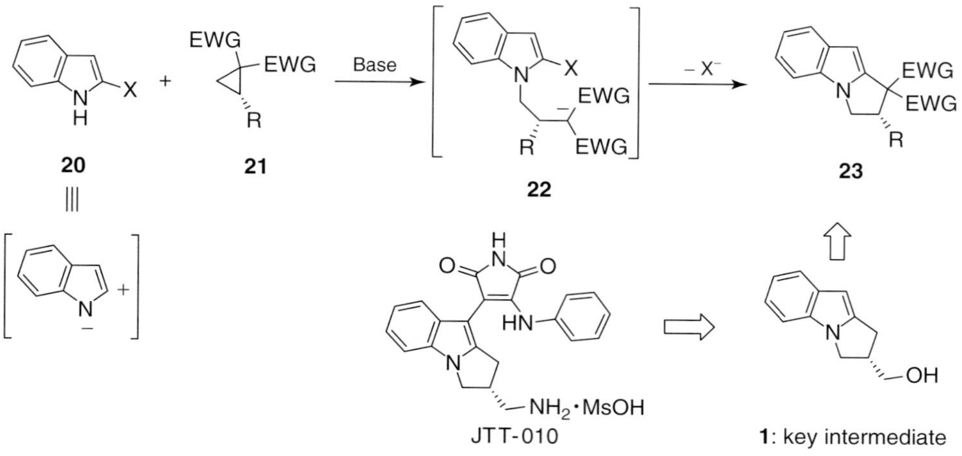

Scheme 12.5 Mechanistic concept for the synthesis of indole-fused pyrrolidines via formal [3 + 2] cycloaddition.

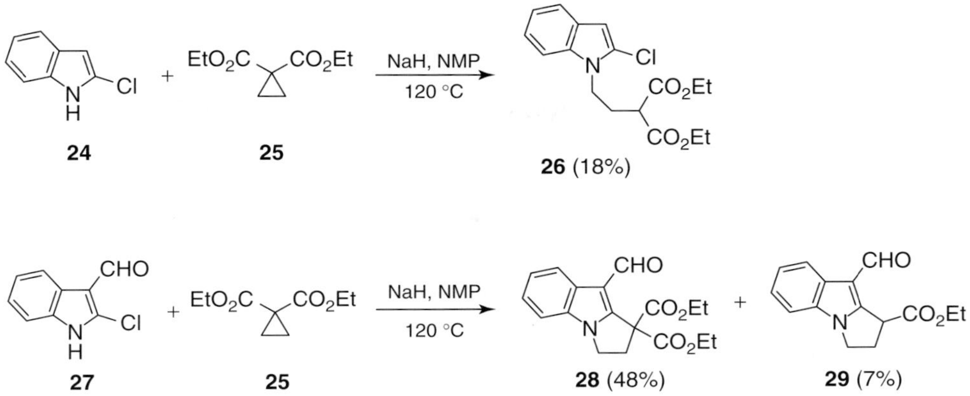

Scheme 12.6 Reaction of 2-chloroindoles with **25**.

the thus-obtained **23** could become a good precursor for the key intermediate **1** of JTT-010 as long as the cyclopropane is chiral and these two EWGs could be removed.

To assess the feasibility of this strategy, we initially examined the model reaction of indole **24** with cyclopropane **25**, which has no substituent other than two ester groups, in the presence of NaH (Scheme 12.6). However, the product was rather the acyclic compound **26** and not the desired pyrrolidine, which indicated that although the nitrogen anion generated from **24** had opened the cyclopropane ring of **25** the subsequent ring closure did not take place even at high temperatures of 120 °C. To facilitate the final ring closure step, the same reaction was conducted with **27**, in which a formyl group was introduced to the C(3)-position of the indole core so that the replacement could proceed smoothly via an addition–elimination pathway. This modification proceeded as expected to give the desired indole-fused pyrrolidine **28** in 48% yield together with partially decarboxylated product **29** (7%). This methodology was also applicable to other heteroaromatics as shown in Scheme 12.7. Under the

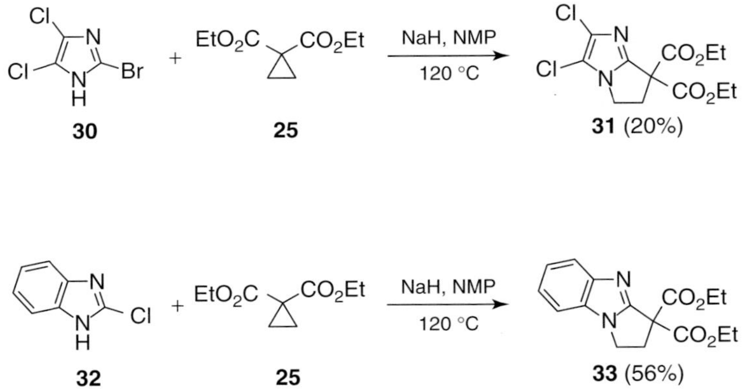

Scheme 12.7 Reaction of 2-haloimidazoles with **25**.

same reaction conditions, 2-haloimidazole **30** and benzimidazole **32** could be converted to the corresponding pyrrolidine-fused products **31** and **33** in 20 and 56% yield, respectively.

These findings were not only the first example of heteroaromatic-1,2-fused pyrrolidine synthesis by the formal [3 + 2] cycloaddition, but they also prompted us to apply this methodology to the chiral cyclopropane **34** possessing a benzyloxymethyl group that we anticipated would be converted into the aminomethyl group of JTT-010 (Scheme 12.8). Contrary to our expectations, however, no desired formal [3 + 2] cycloaddition product was obtained under the same conditions employed above and a decent amount of **34** was recovered. The resistance of **34** to the nucleophilic attack of the nitrogen anion of **27** may be due to the enhancement of steric hindrance and/or a change in the electronic state. In response to the lower reactivity observed in the substituted cyclopropane, we next focused on cyclopropane **35**, envisaging it to be more reactive because of its highly strained lactone-fused structure. We anticipated that the hydroxyl group, protected as a lactone form, would be converted to the aminomethyl group of JTT-010. In addition to this idea, using **35** had a significant advantage in its availability, that is, **35** was easily accessible from diethyl malonate and (R)-epichlorohydrin with >97% ee via a single step [11]. As expected, the reaction of **35** with **27** proceeded smoothly even with the action of a milder base such as K_2CO_3 and at lower temperatures of 85 °C in DMSO (dimethyl sulfoxide). However, we did not obtain the desired lactone-fused **38** but, rather, the unanticipated cyclopropane-fused **36** (67%) together with a small amount of decarboxylated product **37**. We speculated that the major product **36** was produced through **38** by the nucleophilic attack of Cl⁻ generated *in situ*, because **38** was isolated in a small amount when the reaction was quenched before completion, and treatment of isolated **38** with KCl prompted decarboxylation to give **36**. Although **36** was not the expected compound, we thought that it may be useful as a precursor or a substitute for the key intermediate, and several attempts

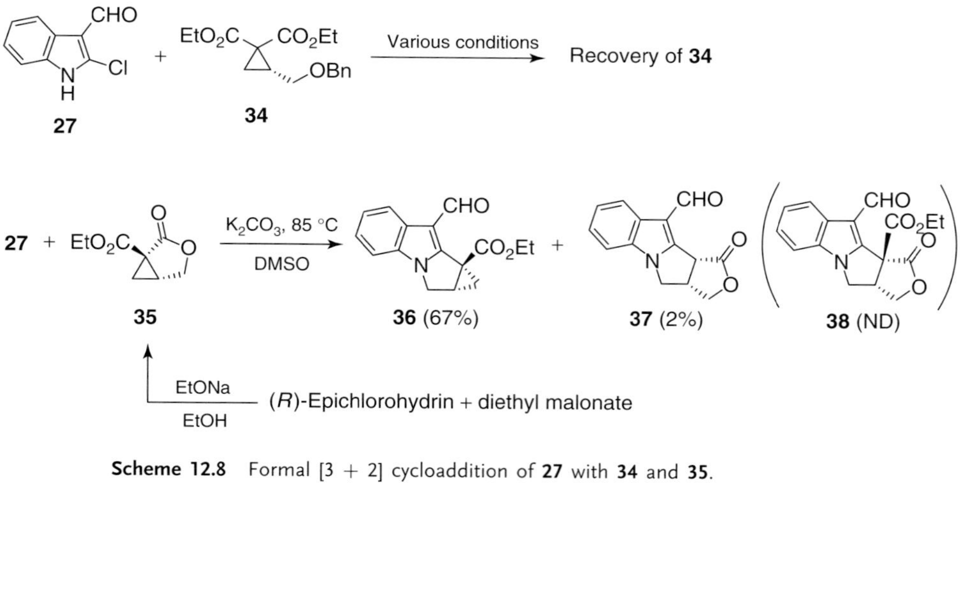

Scheme 12.8 Formal [3 + 2] cycloaddition of **27** with **34** and **35**.

Table 12.1 Reaction of **39** with **35**.

39a (X = Cl)
39b (X = OTs)

Entry	39 (X)	Solvent	Time (h)	Recovered 39 and product (%)			
				39	40	41	42
1	39a (Cl)	DMSO	12	0	77	5	10
2	39b (OTs)	DMSO	12	4	76	7	7
3	39b (OTs)	DMF	15	2	76	9	3

were made to open the cyclopropane of **36** with some appropriate nucleophiles. An attempt to introduce nitrogen atom directly using potassium phthalimide resulted in the formation of a complex mixture. Treatment of **36** with HBr in acetic acid quantitatively gave the corresponding γ-bromoester accompanied by opening of the cyclopropane ring, but the rapid recyclization to **36** took place under the basic conditions required for the subsequent replacement of the bromine atom with nitrogen nucleophiles. These results constrained us to find reaction conditions that selectively afforded a lactone-fused pyrrolidine such as **38**. Hence, we changed the EWG at C(3) and the leaving group at the C(2) indole, and found that employment of **39a** with an ester group at C(3) instead of its formyl analog **27** gave the desired lactone-fused pyrrolidine **40** in 77% yield with significant suppression of the formation of cyclopropane-fused pyrrolidine **42** (Table 12.1, entry 1). Thus, a small change in the EWGs at the indole C(3) resulted in significant change in the products. An attempt to avoid the formation of **42** by using tosylate **39b**, which generates the less nucleophilic TsO$^-$ rather than Cl$^-$, brought about similar results (entry 2), with 7% of **42** being detected in the reaction mixture before the completion of the reaction. When the same reaction was carried out in DMF (dimethylformamide), however, the formation of **42** was diminished while the amount of partially decarboxylated lactone-fused pyrrolidine **41** was slightly increased (entry 3). We eventually adopted these conditions in entry 3 as the final production method since we expected **41** to be converted to the key intermediate **1** under the same conditions that would convert **40** to **1**. Given this idea, contingent contamination of the product with **41** was thought to be acceptable. Furthermore, employing **39b** was advantageous because of its ready availability (see below). The desired compound **40**, formed in 76% under the conditions of entry 3, was then practically isolated in 68% yield as an optically pure form by simple precipitation from the reaction mixture upon the addition of water followed by trituration in CH_3OH. The thus-obtained crystalline **40** containing 1.6% of **41** was pure enough for the next step. Synthesis of the two starting materials for the production of **40** is also practical. The tosylate **39b**

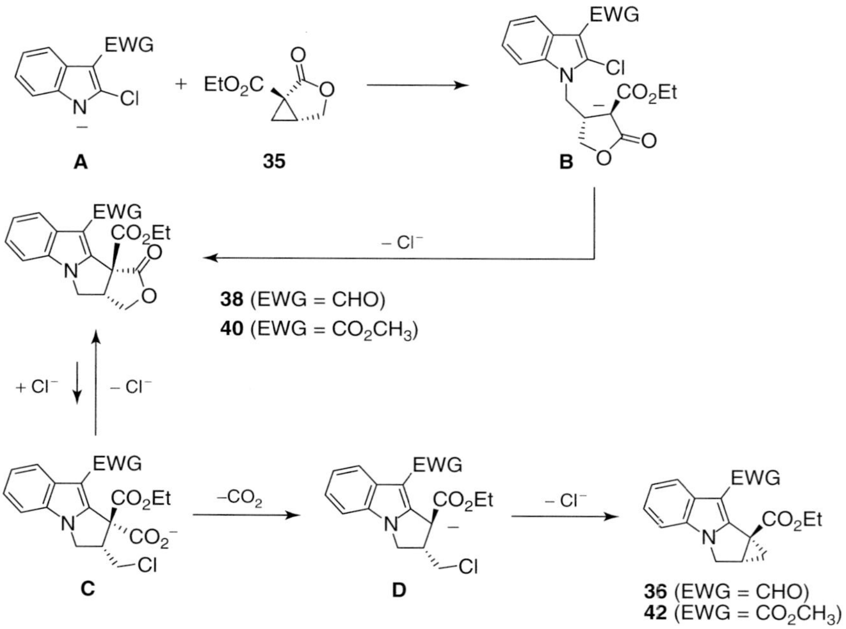

Scheme 12.9 Speculated reaction mechanism for the formation of **38/40** and **36/42**.

was conveniently prepared from oxindole in 57% yield after recrystallization. In a one-pot reaction, the anion generated from oxindole through the action of NaH was sequentially treated with dimethyl carbonate and 4-toluenesulfonyl chloride. Interestingly, its methanesulfonyl variant was not afforded by the same method. The melting point of cyclopropane **35**, prepared from (R)-epichlorohydrin and diethyl malonate, was too low for its practical purification by recrystallization. However, the use of its crude extract did not have an impact on the yield or purity of **40**.

A speculative mechanism for the reaction is shown in Scheme 12.9. The indole nitrogen anion **A** attacks the cyclopropane methylene carbon of **35** to afford the carbanion **B**, stabilized by the two ester groups, which then replaces the chlorine atom at indole C(2) to generate the desired lactone-fused indole-fused pyrrolidine **38/40**. The nucleophilic attack of Cl$^-$ on the lactone methylene carbon atom of **38/40** gives the carboxylate **C**, which is thought to be in equilibrium with **38/40**. We considered that when the EWG is a formyl group, decarboxylation of **C** is accelerated to afford highly stabilized carbanion **D**, which subsequently cyclizes to give cyclopropane-fused **36**. When the EWG is an ester group, its milder carbanion-stabilizing feature may render **C** less prone to undergo decarboxylation to **D**, thereby facilitating the retro reaction to regenerate desired **40**. Thus, a small change in the strength of the EWG can have a profound effect on the product profile.

For the conversion of **40** to the key intermediate **1**, the three carboxyl groups, including two esters and one lactone, must be removed. This triple decarboxylation

Table 12.2 Triple decarboxylation of **40**.

Iteration number	Conditions	Product on ratio (%)[a]	
		1	43
1	6 N NaOH (11.0 equiv.) then cHCl (11.5 equiv.)	64	36
2	6 N NaOH (2.5 equiv.) then cHCl (2.5 equiv.)	88	12
3	6 N NaOH (1.1 equiv.) then cHCl (1.1 equiv)	97	3
4	6 N NaOH (0.6 equiv.) then cHCl (0.6 equiv.)	99	1

[a]HPLC area% (214 nm).

should take place smoothly by taking advantage of the easy decarboxylation feature of indole-3-carboxylic acids and indole-2-acetic acids, as in fact occurred [12]. The simple three- to four-times one-pot iterative set of operations comprising alkaline hydrolysis with NaOH and subsequent neutralization with cHCl accomplished the triple decarboxylation to give the key intermediate **1** in quantitative yield (Table 12.2). This iterative operation was necessary for the completion of the reaction because regeneration of lactone **43** from the corresponding γ-hydroxycarboxylic acid was competitive with the decarboxylation to **1** during the neutralization step. As the number of repetitions increased, the remaining **43** was decreased and converted into **1**. During the operation, we were concerned to avoid explosive evolution of CO_2 gas as a result of the accumulation of the free carboxylic acids. Therefore, during the decarboxylation reaction, cHCl was added after alkaline hydrolysis while maintaining the reflux conditions of the reactions so that the concentration of the free carboxylic acids could be minimized. The gas evolution rate was carefully controlled by the dropping rate of acid addition. The extracted **1** from the reaction mixture was so pure that it did not need any further purification, but just azeotropic removal of residual EtOH with toluene for the next step. Compound **1** thus obtained was also optically pure.

This newly found indole-fused pyrrolidine synthesis, comprising the formal [3 + 2] cycloaddition of **35** and **39b** followed by the one-pot triple decarboxylation of the thus-obtained **40**, was adopted as the large-scale synthesis of the key intermediate **1**. With this method in hand, **1** had become possible to be prepared via three steps both from (R)-epichlorohydrin, one of the most inexpensive chiral C3 units, and from oxindole.

12.4
Replacement of the Hydroxyl Group of 1 with an Amino Group

By using the Gabriel synthesis – one of the most convenient methods for the preparation of amines, which consists of replacement of halides with potassium phthalimide followed by dephthalation using hydrazine – the hydroxyl group of the key intermediate **1** was replaced with an amino group. However, to apply this method to **1**, first the hydroxyl group had to be converted into a leaving group, in this case, as the mesylate. The challenge was how to bring these three steps together into a one-pot operation to shorten the route. The initial mesylation of **1** was carried out with methanesulfonyl chloride in DMA (dimethylacetamide) in the presence of NMM (N-methylmorpholine) to give **44** (Scheme 12.10). The choice of DMA as a solvent was based on its general usefulness in replacement reactions such as the subsequent replacement with potassium phthalimide and on its compatibility in the final dephthalating reaction using hydrazine. NMM was adopted as a base because its good solubility in water facilitated its efficient removal in the final extraction. After completion of the mesylation, potassium phthalimide was added to the reaction mixture to convert **44** to **45**, which was then dephthalated with hydrazine in the same reactor giving **46**. During the procedure, the reaction was diluted with toluene to maintain efficient stirring. The added toluene also acted as an extracting solvent for the product **46**. The thus-obtained **46** as a toluene solution was treated with Cbz–Cl affording **47** in 78% overall yield from **40** after recrystallization (Scheme 12.10). Consequently, **47** was obtained from **1** through four chemical reactions, three of which were conducted in a single flask.

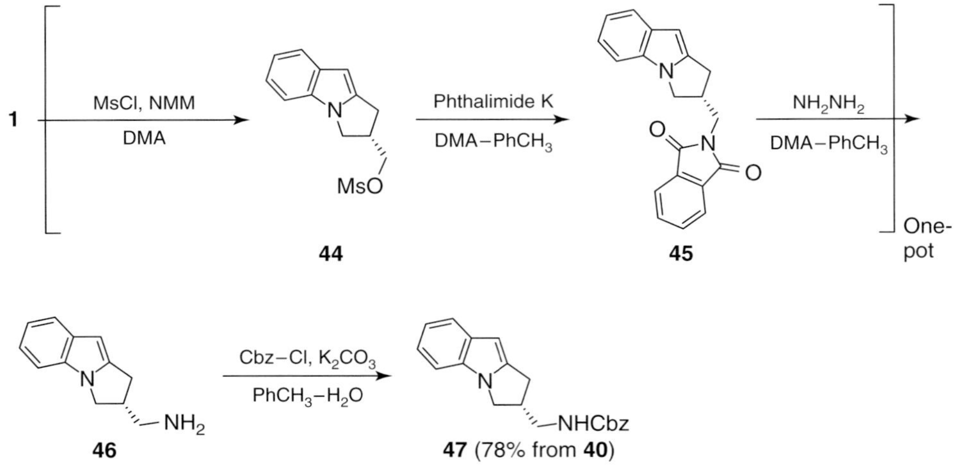

Scheme 12.10 Replacement of the hydroxyl group of **1** with an amino group.

12.5 Construction of JTT-010

12.5.1 Stepwise Maleimide Construction

Indolylmaleimides, as derivatives of staurosporine, have been synthesized following a procedure in which the maleimide parts were constructed stepwise on indole C(3) through multiple steps [13]. The maleimide part of JTT-010 was also built up via five linear steps modeled after these precedents as shown in Scheme 12.11 [2a]. Treatment of the key intermediate equivalent **14** with oxalyl chloride followed by quenching with ammonia gave α-ketoamide **48**, which was then sequentially reduced with NaBH$_4$ and Et$_3$SiH to afford acetamide **49**. Condensation of **49** with dimethyl oxalate in the presence of t-BuOK completed the construction of the maleimide to give **50**. Although **50** could be converted easily into JTT-010 through **51**, the linear multistep process of the maleimide construction involving two C–C bond formations and the stepwise reduction of the ketone to the methylene was onerous even in the gram-scale synthesis and therefore needed to be replaced with a more practical process.

12.5.2 Convergent Coupling Reaction to JTT-010

Direct coupling of the indole and maleimide components is the most convergent and shortest conceivable procedure for the synthesis of indolylmaleimides. In this context, there have been earlier reports on the coupling of indoles and dihalomaleimides. Faul *et al.* reported that reaction of dichloromaleimide **2a** with an excess amount of indole Grignard reagent **52** gave monoreplacement product **53** in good yield (Scheme 12.12) [14]. However, application of this method is

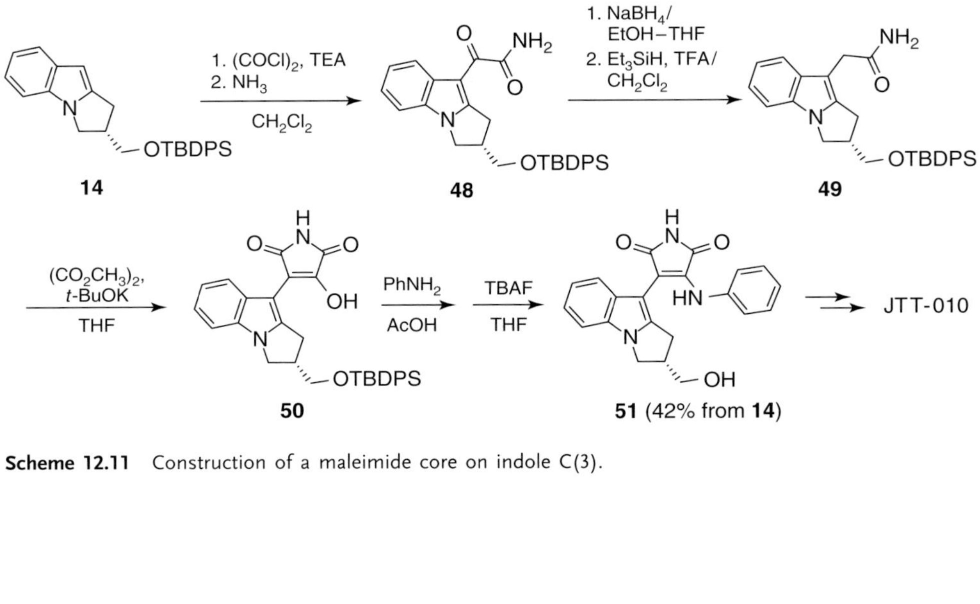

Scheme 12.11 Construction of a maleimide core on indole C(3).

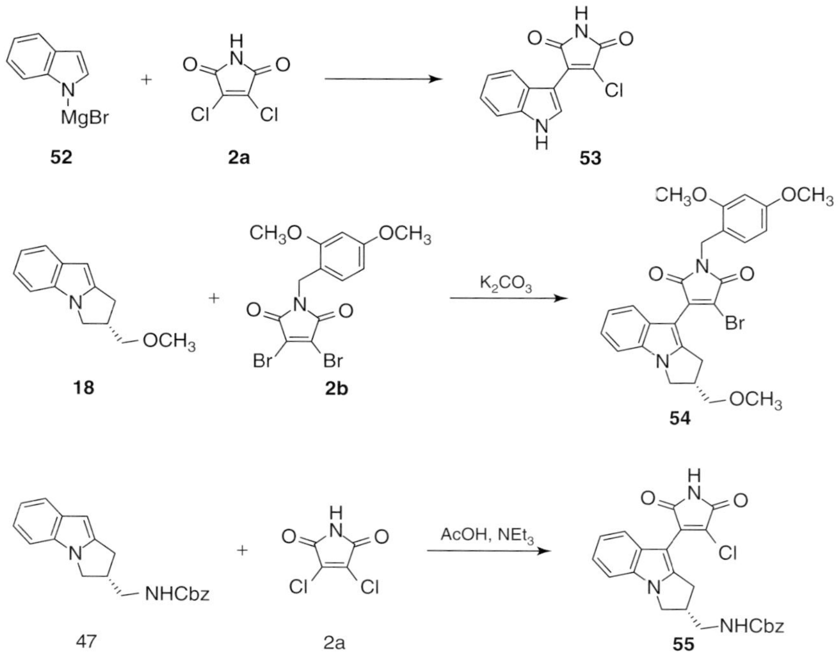

Scheme 12.12 Coupling reactions of indoles with dihalomaleimides.

limited to free *NH*-indoles. To overcome this limitation, Bergman *et al.* took advantage of using *N*-protected dibromomaleimide **2b** [9b]. In this study, C(3) at *N*-alkylated indole **18** successfully replaced one of the bromines in **2b** in the presence of K_2CO_3 to give **54**, which was converted into JTT-010 analog **19** [2c]. We also addressed this issue by attempting to find conditions that promote the replacement of one of two chlorines of *N*-unprotected dichloromaleimide **2a** with *N*-alkylated indole **47** to afford **55**, thereby shortening the total process of JTT-010 synthesis. We found that this reaction, which had not been previously reported for an *N*-unprotected maleimide, proceeded in acetic acid in the presence of triethylamine. The reaction also successfully afforded **55** in neutral solvents buffered with NaH_2PO_4. Maintaining the medium under weakly acidic conditions was thought to be important for the smooth progress of this reaction because basic conditions such as reactions in neutral solvents in the presence of triethylamine resulted in low yield. The final coupling reaction in the preparation of JTT-010 was the replacement of the remaining chlorine atom of **55** with aniline and this reaction occurred by simply heating **55** with an excess of aniline.[1] Given these results, we expected that these sequential coupling reactions of the three

---

1) Bergman *et al.* conducted the coupling reaction of **54** with aniline via a Buchwald reaction in the presence of $Pd(OAc)_2$ and (*R*)-BINAP. See [9b].

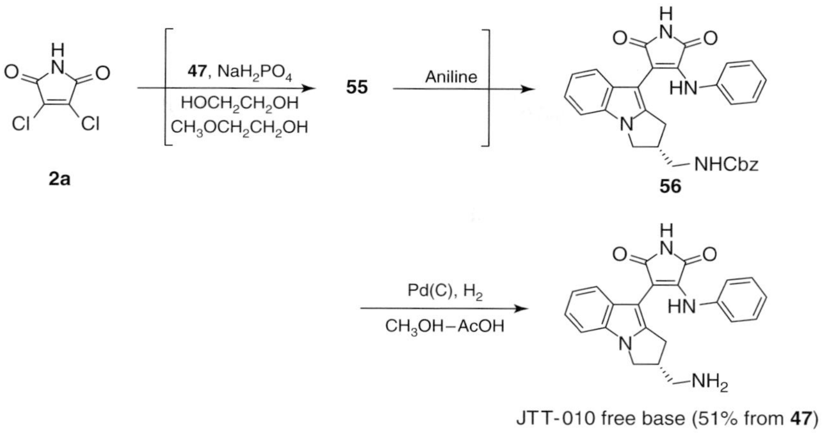

Scheme 12.13 One-pot coupling reaction of **2a**, **47**, and aniline.

components including **2a**, **47**, and aniline could be conducted in one-pot without isolation of the intermediate **55**. Consequently, we found that successive addition of **47** and aniline to the solution of **2a** in the mixed solvent of ethylene glycol and 2-methoxyethanol in the presence of NaH_2PO_4 successfully gave **56** (Scheme 12.13). This combined solvent system was selected by taking the convenience of the isolation procedure as well as isolation yield and purity of the product into account. This one-pot, three-component coupling reaction greatly contributed to shortening the total process of JTT-010 synthesis. The product **56** was precipitated from the reaction mixture by the sequential addition of acetic acid and water, and was sequentially purified by washing with diluted hydrochloric acid, charcoal treatment, and recrystallization. Finally, hydrogenolysis of the Cbz-protected **56** using Pd–carbon followed by recrystallization gave JTT-010 as the free base in 51% yield from **47**. Needless to say, the thus-obtained JTT-010 was optically pure.

12.6
Conclusion

The incorporation of the formal [3 + 2] cycloaddition reaction of **35** and **39b**, and the three-component coupling reaction of **2a**, **47**, and aniline dramatically shortened the route to JTT-010 synthesis. The optically pure key intermediate **1**, precursor of **47**, was synthesized from (R)-epichlorohydrin via three steps and **1** was converted to JTT-010 through seven chemical transformations, five of which were performed in two one-pot reactions. With this route in hand, the large-scale, efficient, and practical production of JTT-010 is possible. This route requires no chromatographic purification, expensive or hazardous reagents, or extremely high- or low-temperature reaction conditions. Finally, pyrrolidines fused with indole or imidazole are often found as a core structure of a variety of compounds of biological interest including PKC inhibitors identified by us and others [2, 15], 5HT$_{2C}$ receptor agonists [16], and CSBP ligands [17]. The formal [3 + 2] cycloaddition described herein is expected to provide rapid access to a variety of compounds of these classes.

References

1. (a) Inoguchi, T., Battan, R., Handler, E., Sportsman, J.R., Heath, W., and King, G.L. (1992) *Proc. Natl. Acad. Sci. U.S.A.*, **89**, 11059; (b) Inoguchi, T., Xia, P., Kunisaki, M., Higashi, S., Feener, E.P., and King, G.L. (1994) *Am. J. Physiol. Endocrinol. Metab.*, **267**, E369; (c) Ishii, H., Jirousek, M.R., Koya, D., Takagi, C., Xia, P., Clermont, A., Bursell, S.E., Kern, T.S., Ballas, L.M., Heath, W.F., Stramm, L.E., Feener, E.P., and King, G.L. (1996) *Science*, **272**, 728; (d) Nakamura, J., Kato, K., Hamada, Y., Nakayama, M., Chaya, S., Nakashima, E., Naruse, K., Kasuya, Y., Mizubayashi, R., Miwa, K., Yasuda, Y., Kamiya, H., Ienaga, K., Sakakibara, F., Koh, N., and Hotta, N. (1999) *Diabetes*, **48**, 2090.
2. (a) Tanaka, M., Sagawa, S., Hoshi, J., Shimoma, F., Yasue, K., Ubukata, M., Ikemoto, T., Hase, Y., Takahashi, M., Sasase, T., Ueda, N., Matsushita, M., and Inaba, T. (2006) *Bioorg. Med. Chem.*, **14**, 5781; (b) Sasase, T., Yamada, H., Sakoda, K., Imagawa, N., Abe, T., Ito, M., Sagawa, S., Tanaka, M., and Matsushita, M. (2005) *Diabetes Obes. Metab.*, **7**, 586; (c) Inaba, T., Tanaka, M., and Sakoda, K. (2000) International Patent Applications WO Patent 00/06564, February 10.
3. Tanaka, M., Ubukata, M., Matsuo, T., Yasue, K., Matsumoto, K., Kajimoto, Y., Ogo, T., and Inaba, T. (2007) *Org. Lett.*, **9**, 3331.
4. Brands, K.M.J. and Davies, A.J. (2006) *Chem. Rev.*, **106**, 2711.
5. Arai, K. (1986) *J. Synth. Org. Chem. Jpn.*, **44**, 486.
6. Remers, W.A. and Weiss, M.J. (1965) *J. Med. Chem.*, **8**, 700.
7. García-Urdiales, E., Alfonso, I., and Gotor, V. (2005) *Chem. Rev.*, **105**, 313.
8. For recent examples, see: (a) Yokomatsu, T., Takada, K., Yasumoto, A., Yuasa, Y., and Shibuya, S. (2002) *Heterocycles*, **56**, 545; (b) Takabe, K., Hashimoto, H., Sugimoto, H., Nomoto, M., and Yoda, H. (2004) *Tetrahedron: Asymmetry*, **15**, 909; (c) Takabe, K., Iida, Y., Hiyoshi, H., Ono, M., Hirose, Y., Fukui, Y., Yoda, H., and Mase, N. (2000) *Tetrahedron: Asymmetry*, **11**, 4825.
9. (a) Thalji, R.K., Ellman, J.A., and Bergman, R.G. (2004) *J. Am. Chem. Soc.*, **126**, 7192; (b) Wilson, R.M., Thalji, R.K., Bergman, R.G., and Ellman, J.A. (2006) *Org. Lett.*, **8**, 1745.
10. For recent examples, see: (a) Meyers, C. and Carreira, E.M. (2003) *Angew. Chem., Int. Ed.*, **42**, 694; (b) Saigo, K., Shimada, S., and Hasegawa, M. (1990) *Chem. Lett.*, 905; (c) Bertozzi, F., Gustafsson, M., and Olsson, R. (2002) *Org. Lett.*, **4**, 3147; (d) Yamago, S., Yanagawa, M., and Nakamura, E. (1999) *Chem. Lett.*, 879; (e) Oh, B.H., Nakamura, I., Saito, S., and Yamamoto, Y. (2001) *Tetrahedron Lett.*, **42**, 6203; (f) Taillier, C. and Lautens, M. (2007) *Org. Lett.*, **9**, 591; (g) Scott, M.E. and Lautens, M. (2008) *J. Org. Chem.*, **73**, 8154; (h) Jackson, S.K., Karadeolian, A., Driega, A.B., and Kerr, M.A. (2008) *J. Am. Chem. Soc.*, **130**, 4196.
11. Sekiyama, T., Hatsuya, S., Tanaka, Y., Uchiyama, M., Ono, N., Iwayama, S., Oikawa, M., Suzuki, K., Okunishi, M., and Tsuji, T. (1998) *J. Med. Chem.*, **41**, 1284.
12. (a) For decarboxylation of an indole-2-acetic acid derivative, see: Ho, B.T. and Walker, K.E. (1988) *Org. Synth.*, **6**, 965; (b) For decarboxylation of indole-3-carboxylic acid, see: Challis, B.C. and Rzepa, H.S. (1977) *J. Chem. Soc., Perkin Trans. 2*, 281.
13. (a) Neel, D.A., Jirousek, M.R., and McDonald, J.H. III (1998) *Bioorg. Med. Chem. Lett.*, **8**, 47; (b) Rooney, C.S., Randall, W.C., Streeter, K.B., Ziegler, C., Cragoe, E.J. Jr., Schwam, H., Michelson,

S.R., Williams, H.W.R., Eichler, E., Duggan, D.E., Ulm, E.H., and Noll, R.M. (1983) *J. Med. Chem.*, **26**, 700.

14. (a) Reaction of N-alkylated dibromomaleimide: Brenner, M., Rexhausen, H., Steffan, B., and Steglich, W. (1988) *Tetrahedron*, **44**, 2887; (b) Reaction of **2a**: Faul, M.M., Sullivan, K.A., and Winneroski, L.L. (1995) *Synthesis*, 1511.

15. Bit, R.A., Davis, P.D., Elliott, L.H., Harris, W., Hill, C.H., Keech, E., Kumar, H., Lawton, G., Maw, A., Nixon, J.S., Vesey, D.R., Wadsworth, J., and Wilkinson, S.E. (1993) *J. Med. Chem.*, **36**, 21.

16. (a) Bentley, J.M., Bickerdike, M.J., Hebeisen, P., Kennett, G.A., Lightowler, S., Mattei, P., Mizrahi, J., Morley, T.J., Plancher, J.-M., Richter, H., Roever, S., Taylor, S., and Vickers, S.P. (2002) International Patent Applications. WO Patent 02/051844 A1, July 4; (b) Adams, D.R., Bentley, J.M., Roffey, J.R.A., Hamlyn, R.J., Gaur, S., Duncton, M.A.J., Davidson, J.E.P., Bickerdike, M.J., Cliffe, I.A., and Mansell, H.L. (2000) International Patent Applications. WO Patent 00/12510, March 9; (c) Peters, R., Waldmeier, P., and Joncour, A. (2005) *Org. Process Res. Dev.*, **9**, 508.

17. Gallagher, T.F., Seibel, G.L., Kassis, S., Laydon, J.T., Blumenthal, M.J., Lee, J.C., Lee, D., Boehm, J.C., Fier-Thompson, S.M., Abt, J.W., Soreson, M.E., Smietana, J.M., Hall, R.F., Garigipati, R.S., Bender, P.E., Erhard, K.F., Krog, A.J., Hofmann, G.A., Sheldrake, P.L., McDonnell, P.C., Kumar, S., Young, P.R., and Adams, J.L. (1997) *Bioorg. Med. Chem.*, **5**, 49.

13
Process Development of Oral Carbapenem Tebipenem Pivoxil, TBPM-PI

Takao Abe and Masataka Kitamura

13.1
Introduction

Carbapenems are noted for their broad and potent antibacterial activities, resistance to β-lactamases, and are used as highly secure antibiotics as a last resort. Since imipenem was launched in 1987 by Merck Co., five more carbapenems – panipenem, meropenem, biapenem (BIPM), doripenem, and ertapenem (Figure 13.1) – have been in the market. All these compounds were developed for parenteral use, and the development of carbapenems for oral use has been expected for a long time.

Although many pharmaceutical companies attempted to develop oral carbapenems [1b], such as CS-834, GV-118819, and DZ-2640 (Figure 13.2), most efforts were suspended because of higher manufacturing costs for oral β-lactam antibiotics; chemical instability of carbapenem compounds in gastric, intestinal juice, and even in neutral solution; safety concerns; or unclear therapeutic positioning of the agents in the market.

Tebipenem pivoxil (TBPM-PI) [1, 2] was discovered by the research group of Lederle Japan in 1993, developed by Meiji Seika, and launched as the first oral carbapenem in 2009, having been appreciated for its ability. TBPM-PI was noticed by clinicians because it demonstrated a broad and strong antibacterial activity against various clinical isolates including major resistant pathogens of respiratory tract infections such as penicillin-resistant *Streptococcus pneumoniae* (PRSP) and β-lactamase-negative ampicillin-resistant (BLNAR) *Hemophilus influenzae*, excellent oral absorbability with pharmacokinetics profile like a parenteral drug, and its safety profile.

13.2
Discovery of TBPM-PI

The characteristic features of TBPM-PI are 1-(1,3-thiazolin-2-yl)azetidine-3-thio group at C2-position in the carbapenem skeleton and pivaloyloxymethyl (POM,

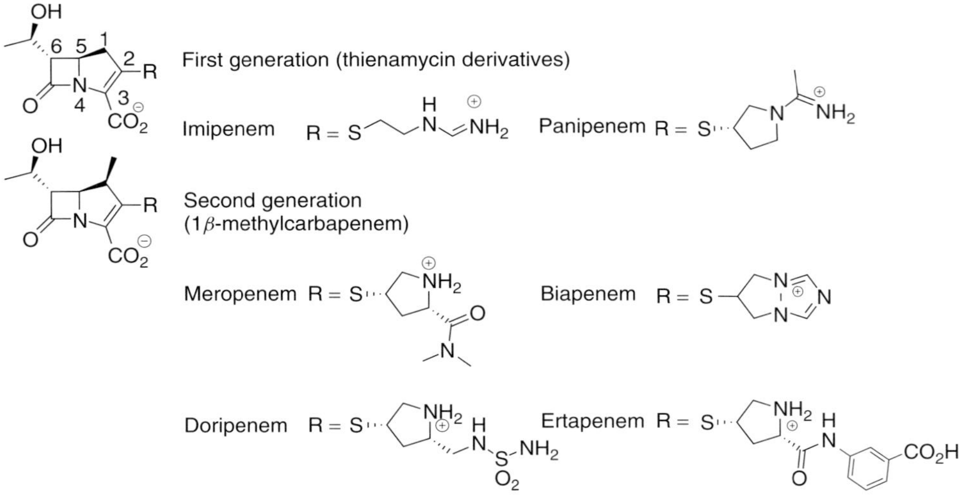

Figure 13.1 Parenteral carbapenem antibiotics.

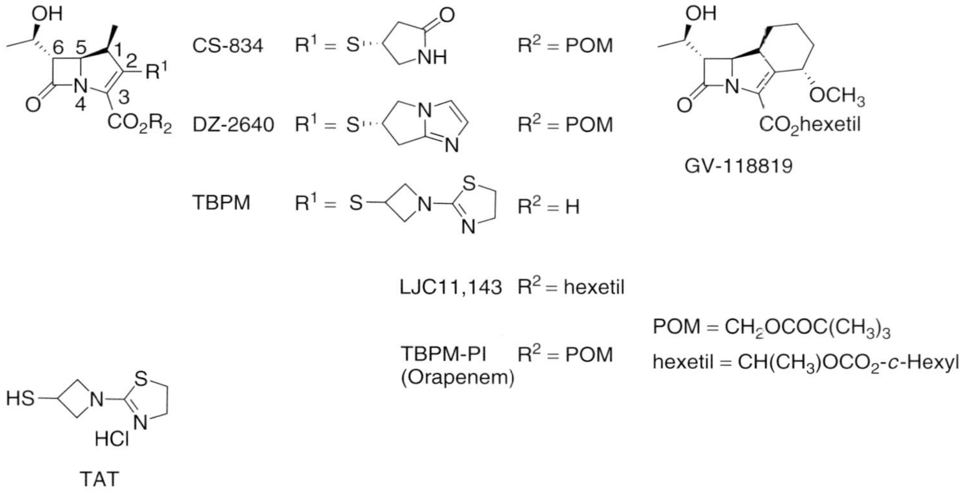

Figure 13.2 Oral 1β-methyl-carbapenem candidates.

POM = $CH_2OCOC(CH_3)_3$
hexetil = $CH(CH_3)OCO_2$-c-Hexyl

pivoxil) ester of C3 carboxylic acid as a prodrug moiety. Our strategy was to research an orally active parent molecule and apply prodrug approach if its oral absorbability is insufficient because prodrug approach is impossible to apply to an orally inactive compound. In the early stage of discovery, we found that a C2 substituent having S-Csp3-planar structure is important to show potent oral absorbability. We then designed the C2 substituent of TBPM considering planar azetidine ring conjugated with 1,3-thiazoline moiety, which would show moderate lipophilicity and positive charge to demonstrate potent oral absorbability and carbapenem's characteristic broad and strong antibacterial activity, improvement of safety concern for the simple azetidine group, and novelty in the patent situation. TBPM showed potent oral absorbability as a parent molecule among tested, sufficient chemical and biological stability, and broad and strong antibacterial activity, like parenteral carbapenems, against clinically isolated pathogens except for *Pseudomonas aeruginosa*.

Although the oral absorbability of TBPM itself was insufficient to use as an oral drug, we speculated that prodrug approach would be applicable to TBPM since the PK (pharmacokinetics) profile was sufficient. In fact, biological availability of TBPM-PI exceeded 40% when administered to rats. On the other hand, not having planar C2 substituent, improvement of oral absorbability was not observed by prodrug approach, indicating that the shape of the C2 substituent is important to show excellent oral absorbability [1a,b]. Absorption rate of TBPM-PI is very fast, so we thought that TBPM would be actively absorbed via an unknown transporter. After subsequent studies, it was revealed that TBPM was absorbed via an active transporter, human organic anion transporting polypeptide (OATP) family [3]. With regard to the selection of the prodrug moiety, α-branched carboxylic acid shows safety concerns about sedative effects and well-known pivalic acid reduces *in vivo* carnitine, so the prodrug moiety of TBPM was studied initially on cyclohexyloxycarboxyethyl (hexetil) ester having little influence of cyclohexanol on *in vivo* carnitine. However, the moiety was substituted with chemically stable POM ester because the chemical stability of the hexetil ester, LJC11,143, was insufficient [1] as an active pharmaceutical ingredient (API) and the influence of POM ester on *in vivo* carnitine was found to be tolerable.

The synthetic route is outlined in Scheme 13.1. TBPM-PI is synthesized by three processes: (i) introduction of 1-(1,3-thiazolin-2-yl) azetidine-3-thiol hydrochloride (TAT) onto *p*-nitrobenzyl (1*R*,5*R*,6*S*)-2-(diphenylphosphoryloxy)-6-[(*R*)-1-hydroxyethyl]-1-methylcarbapen-2-em-3-carboxylate (MAP), (ii) deprotection of C3 *p*-nitrobenzyl (PNB) ester of *p*-nitrobenzyl (1*R*,5*S*,6*S*)-6-[(*R*)-1-hydroxyethyl]-1-methyl-2-[1-(1,3-thiazolin-2-yl)azetidine-3-yl]thio-1-carbapen-2-em-3-carboxylate (L-188), and (iii) POM esterification of C3 carboxylic acid of TBPM. It was to our advantage that we had experienced establishing the process of parenteral BIPM and had some knowledge and experience of carbapenem chemistry. After the synthetic

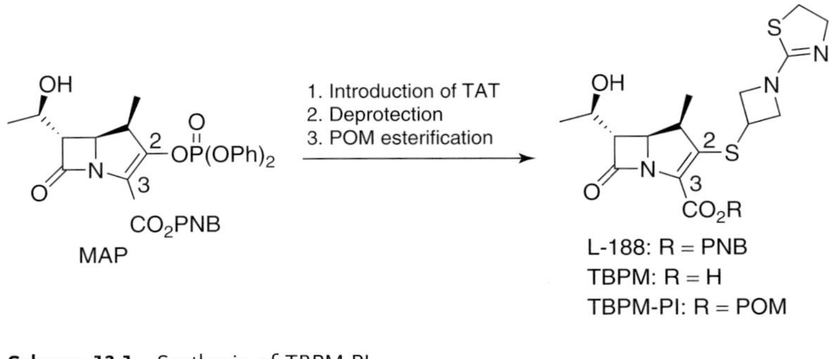

Scheme 13.1 Synthesis of TBPM-PI.

13.3
Synthetic Process of Side Chain on the C2-Position of TBPM, TAT

The synthetic challenges were (i) formation of azetidine ring, (ii) introduction of 1,3-thiazoline moiety, and (iii) introduction of mercapto group.

13.3.1
Original Synthetic Process of TAT Starting from Benzhydrylamine

The original synthetic method of TAT is shown in Scheme 13.2. TAT was synthesized by the introduction of 1,3-thiazoline moiety and subsequent SH group to 3-hydroxyazetidine (**3**) prepared from benzhydrylamine [1c, 2a, 4a]. Although **3** and its analogs could be versatile building blocks, the conventional procedures had not been fully studied because azetidine ring had been considered to be difficult to cyclize, and they could not be acceptable because of their low yield and production cost. Starting benzhydrylamine is too expensive and free form of N-benzhydryl-3-hydroxyazetidine (**2**) must be purified by column chromatography. Moreover, hydrochloric acid salt **3** must be isolated by drying up the concentrated residue of the aqueous reaction mixture for the next coupling reaction

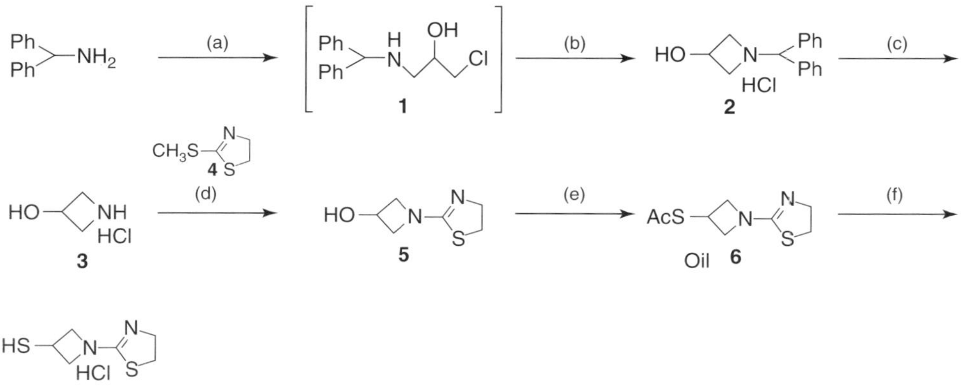

Scheme 13.2 Original synthetic method of TAT. Reagents and conditions: (a) eplchlorohydrin (1.0 equiv.), CH_3OH, rt 1 d, 89%; (b) DMSO, 50 °C 3 d, 54%; (c) H_2 (350 kPa), 10% Pd–C, H_2O–EtOH, rt 4 h, 94%; (d) **4** (1.0 equiv.), $NaOCH_3$ (0.9 equiv.), CH_3OH, reflux, 1 d, 82%; (e) Ph_3P (2.0 equi.), DEAD (2.0 equiv.), AcSH (2.0 equiv.), THF, 10 °C 1 h, rt ¯ h, 65%; (f) KOH (1.1 equiv.), IPA, 5 °C, 10 min, then HCl (2.4 equiv.)–CH_3OH, 95%.

with 2-methylthio-1,3-thiazoline (**4**), because the reaction proceeds only in weak acidic conditions and is sensitive to the amount of base (NaOCH$_3$ 0.9 mol equiv.). A coupled compound (**5**) was obtained as a solid form after workup. Following Mitsunobu reaction for the introduction of acetylthio group also needs anhydrous conditions, and chromatographic purification is required to obtain viscous oil of acetylthio derivative (**6**). Thus, these processes are not suitable for further scale-up.

13.3.2
Practical Synthetic Process of TAT from Benzylamine

We investigated the synthetic method for **3** in detail and established an efficient and facile synthetic process of **3** (Scheme 13.3). Starting benzhydrylamine was replaced with inexpensive benzylamine and the reaction condition was optimized. Into a 1 M aqueous solution of benzylamine was added epichlorohydrin in four portions over 3 h at 5 °C to give N-benzyl-3-chloro-2-hydroxypropylamine (**7**) as a solid form in 89% yield after collecting the precipitated solid from the reaction mixture [4d]. Isolation of **7** is necessary to remove the by-product **7a**.

Depending on the base and solvent combination, cyclization yields of **7** varied. The reaction proceeded in i-PrOH or t-BuOH in the presence of Et$_3$N (Table 13.1, entry 1, 2). Moreover, the yield of N-benzyl-3-hydroxyazetidine (**8**) dramatically increased to 90% in 0.5 M CH$_3$CN solution when KHCO$_3$ was used as a base (entry 9). We then examined the reaction in a pilot scale under the improved conditions, but the yield of **8** decreased to 73%. In searching the reason for the poor reproducibility, we found that the particle size of KHCO$_3$ affected the yield of the reaction (Table 13.2). Commercially available KHCO$_3$ was considered to be

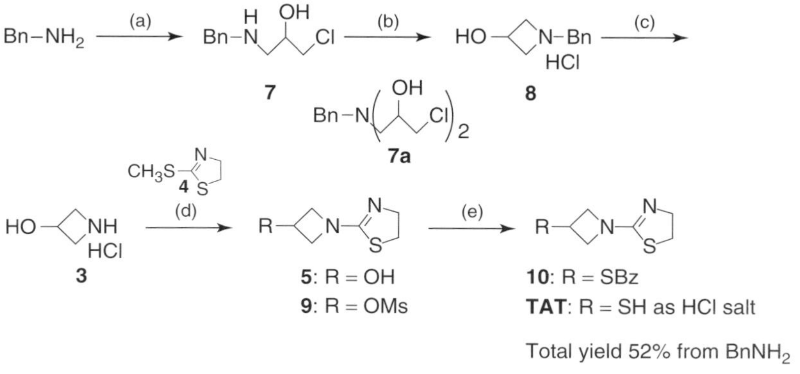

Scheme 13.3 Improved process of TAT. Reagents and conditions: (a) epichlorohydrin (0.95 equiv.), H$_2$O (1 M), 5 °C 3 h, rt 18 h, 89%; (b) NaHCO$_3$ (2.0 equiv.), CH$_3$CN (0.5 M), reflux, 6.5 h, then HCl (1.0 equiv.) –dioxane, 85%; (c) H$_2$ (500 kPa), 10% Pd–C, H$_2$O–EtOH, 50 °C 18 h, quant.; (d) (i) **4** (1.0 equiv.), KHCO$_3$ (0.7 equiv.), CH$_3$OH, reflux, 1 d, 95%; (ii) MsCl (1.1 equiv.), Et$_3$N (1.2 equiv.), DMAP (0.01 equiv.), THF, 5 °C, 0.5 h, 94%; (e) (i) KSBz (1.5 equiv.), tBuOAc, reflux, 5.5 h, 86%; (ii) KOH (1.1 equiv.), CH$_3$OH, 5 °C, 10 min, then HCl (2.4 equiv.) –CH$_3$OH, 90%.

Total yield 52% from BnNH$_2$

Table 13.1 Optimization of the cyclization conditions.

Bn—NH—CH(OH)—CH₂Cl (**7**) → Base, Solvents, reflux 6 h → HO—◁—N—Bn (**8**)

Entry	Solvent	Concentration (M)	Base (mol equiv.)	Yield (%)
1	i-PrOH	1.0	Et_3N (2)	60
2	t-BuOH	1.0	Et_3N (2)	70
3	i-PrOH	1.0	$KHCO_3$ (2)	40
4	t-BuOH	1.0	$KHCO_3$ (2)	57
5	CH_3OH	1.0	$KHCO_3$ (2)	NR
6	EtOH	1.0	$KHCO_3$ (2)	NR
7	CH_3CN	1.0	$KHCO_3$ (2)	58
8	t-BuOH	0.5	$KHCO_3$ (2)	85
9	CH_3CN	0.5	$KHCO_3$ (2)	90
10	CH_3CN	0.5	$NaHCO_3$ (2)	93

NR, no reaction.

Table 13.2 Effect of the particle size of inorganic base.

Bn—NH—CH(OH)—CH₂Cl (**7**) → Base (2), CH_3CN (0.5 M), reflux 6 h → HO—◁—N—Bn (**8**)

Entry	Base	Particle size (μm)	Distribution ratio (%)	HPLC yield (%)
1	$KHCO_3$	335–500	53.5	58
2	$KHCO_3$	250–355	14.8	54
3	$KHCO_3$	177–250	13.0	98
4	$KHCO_3$	150–177	3.3	97
5	$KHCO_3$	<150	15.4	97
6	$NaHCO_3$	<250	96.9	98

an inappropriate reagent because of its broad distribution of particle size, whereas $NaHCO_3$ showed a narrow distribution range in small particle size showing higher reaction yields. Therefore, the best conditions were refluxing with 2 M equiv. of commercially available $NaHCO_3$ powder for 6 h in a 0.5 M CH_3CN solution. Pure **8** was obtained as a solid HCl salt in an 85% yield after treating the concentrated mixture with HCl–dioxane. N-Benzyl group of **8** was deprotected by hydrogenolysis on Pd–C to give **3** in an overall yield of 77% from benzylamine.

In the coupling reaction with **4**, various bases and workup procedures were investigated. $NaOCH_3$ was substituted with $KHCO_3$ (0.7 mol equiv.) for easier

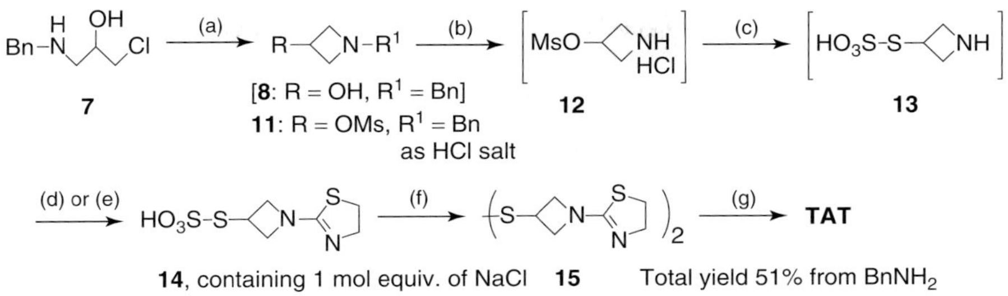

13.3 Synthetic Process of Side Chain on the C2-Position of TBPM, TAT

Scheme 13.4 Optimized process of TAT. Reagents and conditions: (a) (i) NaHCO$_3$ (2.0 equiv.), CH$_3$CN, reflux, 6.5 h; (ii) MsCl (1.1 equiv.), Et$_3$N (1.2 equiv.), CH$_3$CN, 5 °C, 7 h, then HCl–CH$_3$CN, 80% from 7 in 2 steps; (b) H$_2$ (500 kPa), 10% Pd–C, H$_2$O–CH$_3$OH, 40 °C, 24 h, 95%; (c) Na$_2$S$_2$O$_3$ (1.0 equiv.), H$_2$O–CH$_3$OH, 50 °C, 22 h; (d) 4 (1.1 equiv.), H$_2$O–CH$_3$OH, reflux, 17 h, 42% from 11 in 3 steps; (e) (i) Et$_3$N (1.1 equiv.), 2-chloroethyl isothiocyanate (1.1 equiv.), H$_2$O–CH$_3$OH, 5 °C, 0.5 h; (ii) NaOCH$_3$ (1.1 equiv.); (iii) 1.33 M HCl–CH$_3$OH (1.2 equiv.), 82% from 11 in 3 steps; (f) (i) cHCl (5.0 equiv.) 55 °C, 2 h, then KHCO$_3$ (8 equiv.), H$_2$O$_2$ (0.5 equiv.), 5 °C, 45 min, 93%; (g) Ph$_3$P (1.2 equiv.), H$_2$O (2.0 equiv.), 8 M HCl–CH$_3$OH (2.5 equiv.), CH$_3$CN, rt, 1.5 h, 93%.

handling. After completion of the reaction, the mixture was treated with excess KHCO$_3$ to remove remaining acid instead of extraction procedure. Mitsunobu reaction was changed to stepwise method via isolable mesylate (**9**) and benzoylthio derivative (**10**). All the intermediates (**7**, **8**, **3**, **5**, **9**, and **10**) were obtained as a solid form without chromatographic purification. After alkaline hydrolysis of **10**, TAT was isolated as HCl salt in a crystalline form. TAT is not a deliquescent solid, so it can be handled without special care. Thus, the first-generation process of TAT was developed via six isolable intermediates with the sequence of seven reactions from starting benzylamine in an overall yield of 52% [4d, 5].

13.3.3
Industrial Synthetic Process of TAT: Back to Classic Bunte's Salt

Next, expensive mercaptobenzoic acid was successfully substituted with inexpensive Na$_2$S$_2$O$_3$ to obtain Bunte's salt (**13**) from 3-mesyloxyazetidine hydrochloride (**12**) in aqueous CH$_3$OH (Scheme 13.4). Isolated **13** smoothly reacted with **4** in aqueous CH$_3$OH to produce a coupled compound (**14**), which can be converted to TAT by acidic hydrolysis [4e]. Unlike **3**, it was not necessary to add KHCO$_3$ during the coupling reaction of **13** with **4**. Compared to the first-generation process, isolation of **8** was skipped, S-function was introduced before coupling with **4** and reaction solvent was commoditized to aqueous CH$_3$OH. As the step reaction yields of **12**–**14** were quantitative by HPLC, we envisaged that the three sequential reactions could also proceed in a one-pot manner without isolating **12** and **13**. After removal of the catalysts of the hydrogenolysis of 3-mesyloxy-1-benzylazetidine hydrochloride (**11**), Na$_2$S$_2$O$_3$ was successively added to the resulting solution of **12** at 50 °C and then **4** under reflux. Unexpectedly, the overall yield of **14** in a one-pot manner was disappointingly low (42% from **11**). The reason for the low yield of **14** might be due to the harsh reaction conditions of **13** and **4** (reflux 17 h).

Consequently, **4** was replaced with 2-chloroethylisothiocyanate, which reacts under milder conditions [2d], successfully producing **14** from **11** in 82% yield without isolating **12** and **13**. 2-Chloroethylisothiocyanate would be less expensive than **4** as long as a thiophosgen production line is available, and this method would reduce environmental burden because of not producing odorous methanethiol as in **4**. Deprotection of **14** was carried out by acid hydrolysis with concentrated HCl to obtain TAT. However, disulfide (**15**) and sulfate salts of TAT were concomitantly produced in the acid hydrolysis and could not be separated. Therefore, crude TAT was once converted to disulfide **15** by treating with H_2O_2 in a similar way to the SH-substituent of BIPM [6b], and purified TAT was isolated as a crystalline form after reductive cleavage of the disulfide bond with Ph_3P [4g, 5]. As described above, we established the facile manufacturing process of TAT via four isolable intermediates (**7, 11, 14,** and **15**) with the sequence of eight reactions from starting benzylamine in an overall yield of 51% without using expensive materials and as a higher throughput process.

We also developed an alternative short synthetic process via 1-azabiclo[1.1.0] butane (**17**) (Scheme 13.5) [4a,b, 7], where TAT can be synthesized from allyl amine in five steps. However, we decided not to adopt the process because of the use of expensive n-BuLi, insufficient step yields, complicated isolation procedure due to production of considerable amount of by-products, and safety concern of **17**.

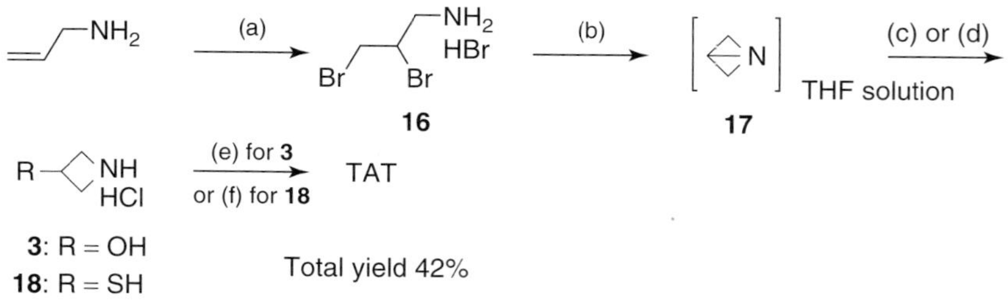

Scheme 13.5 Synthesis of TAT via azabicyclo[1.1.0]butane (**17**). Reagents and conditions: (a) Br_2 (2.1 equiv.), EtOH, ice-bath-rt 18 h, 95%; (b) (i) n-BuLi (3.1 equiv.), THF, −78 °C, 1 h; (ii) 50% KOH quench, dlstn; (iii) dried over K_2CO_3; (c) HCO_2H (1.2–16 equiv.), −40 °C, 10 h then 2.7 M HCl–CH_3OH (8 mol equiv.), rt 18 h, 61% in 2 steps; (d) (i) AcSH (3.3–16 equiv.), THF, −4 °C to rt, 18 h, 68% in 2 steps; (ii) 3 M HCl (5 equiv.), reflux, 1.5 h, quant; (e) (i) 4 (1.0 equiv.), $KHCO_3$ (0.7 equiv.), CH_3OH, reflux, 1 d, 92%; (ii) MsCl (1.1 equiv.), Et_3N (1.2 equiv.), DMAP (0.01 equiv.), THF, 5 °C, 0.5 h, 93%; (iii) KSBz (1.5 equiv.), tBuOAc, reflux, 5.5 h, 85%; (iv) KOH (1.1 equiv.), CH_3OH, 5 °C, 10 min, then HCl (2.4 equiv.) −CH_3OH, 95%; (f) 4 (0.8 equiv.), PH_3P (0.1 equiv.), H_2O–CH_3OH, reflux, 6 h, 65%.

13.4
Synthetic Process of TBPM-PI from 4-Nitrobenzyl (1R,5R,6S)-2-diphenylphosphoryloxy-6-[(R)-1-hydroxyethyl]-1-methyl-1-carbapen-2-em-3-carboxylate, MAP

Synthetic process of L-188 and TBPM–$4H_2O$ was developed based on that of BIPM [6a,b]. MAP was selected as a starting material considering easily crystallizable intermediates, commercial availability, and cost reduction effect by commoditization of the starting material with that of BIPM. Prodrug esterification reaction was studied by taking conversion yields, reproducibility, ease of isolation, solvent efficiency, and throughput into account.

13.4.1
Synthesis of PNB Ester of TBPM, L-188

MAP and TAT were suspended in cold CH_3CN, followed by dropwise addition of i-Pr_2EtN to initiate the reaction. The reaction proceeded smoothly in the suspension state (about 0.25 M of MAP) and L-188 was easily isolated by collecting precipitated solid after completion of the reaction in a similar way to the process of PNB ester of BIPM. The problem was the loss of L-188 to mother liquor over 10% and contamination of L-188 with 20–30% of poorly soluble HCl salt of L-188 (Table 13.3). Concentration of the mother liquor increased the proportion of the HCl salt of L-188 to L-188. The contaminated HCl salt of L-188 not only slowed down the next reaction and decreased the reaction yields, but also brought the pH of the solution into acidic region where TBPM decomposed (Table 13.4). Since the HCl salt of L-188 could not be reconverted to free form of L-188 by treatment with amine or aqueous $NaHCO_3$ in a suspended state, it was necessary to increase the volume of the reaction solvent or add another base for an efficient deprotection of L-188. Considering solvent efficiency and throughput of the next deprotection process, it was not acceptable to increase the solvent volume. We therefore investigated the

Table 13.3 Solubility of L-188.

Solvent	Solubility (% w/v)	
	L-188	L-188-HCl salt
DMF	>2	0.358
THF	1.03	0.0008
CH_3CN	0.52	0.015
n-BuOH	0.071	0.005
H_2O	0.004	ND

DMF, dimethylformamide; THF, tetrahydrofuran.
ND, not determined.

Table 13.4 Influence of HCl salts of L-188 on deprotection step.

$$\text{L-188} \xrightarrow[\text{solvent, rt, 1.5 h}]{\text{H}_2 \text{ (400 kPa), 10\% Pd/C (50\% wet, 50\% w/w)}} \text{TBPM-4H}_2\text{O}$$

Entry	Contents of L-188-HCl salt (%)	Solvent (v/w)	Additive (mol equiv.)	pH of aqueous Phase	HPLC yield (%)
1	0	A + B (20, 20)	–	6.1	90
2	20	A + B (20, 20)	–	–	73
3	50	A + B (20, 20)	–	–	66
4	100	A + B (20, 20)	–	3	60
5	100	A + B (40, 40)	–	–	70
6	100	A + B (100, 100)	–	–	92
7	100	A + B (20, 20)	NaHCO$_3$ (1)	6.6	88
8	0	A + C (20, 20)	NaHCO$_3$ (1)	8.3	91
9	100	A + C (20, 20)	NaHCO$_3$ (1)	–	88

A, n-BuOH; B, 0.05 M phosphate buffer (pH 6.5); C, H$_2$O.

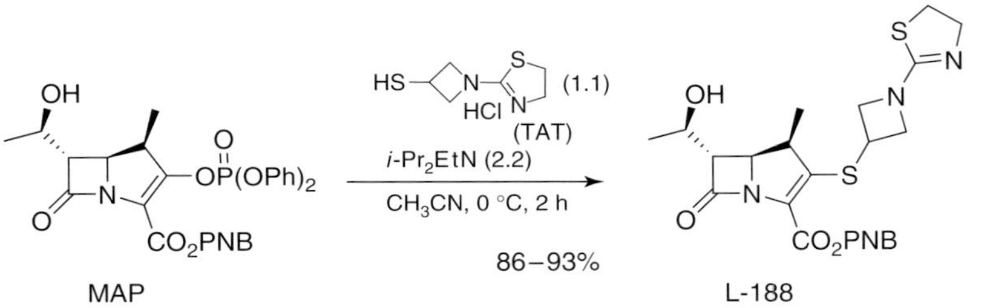

Scheme 13.6 Synthetic process of PNB ester, L-188.

workup procedure of L-188 in detail and found that it was most effective to add water as the poor solvent to the reaction mixture after completion of the reaction. Consequently, it not only increased the isolation yield up to 86–93% (Scheme 13.6), but also made the next deprotection process reproducible [2c]. Wet-formed L-188 was used without drying for the next step, showing higher throughput than dried L-188.

13.4.2
Synthesis of TBPM-4H$_2$O

First, deprotection of PNB ester of L-188 in the discovery stage was carried out by zinc powder reduction [2a, 6c] considering the inhibitory effect of 1,3-thiazoline moiety on Pd–C in hydrogenolysis. However, it did not seem to be reproducible in a pilot scale because zinc powder is too heavy to be suspended, so we performed catalytic hydrogenolysis instead, finding it better than the zinc reduction. A

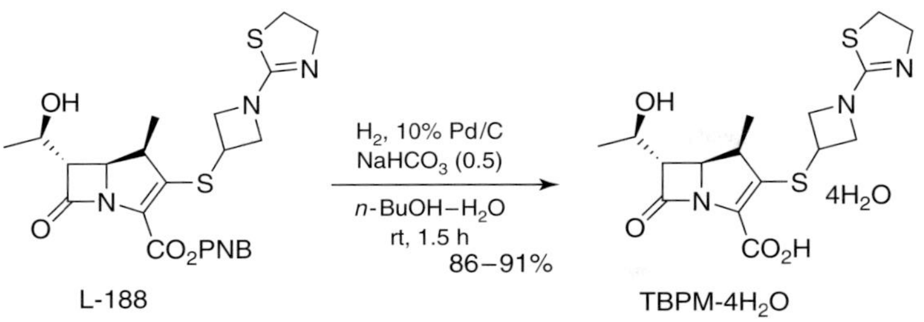

Scheme 13.7 Synthetic process of TBPM–4H$_2$O.

two phase solvent mixture of phosphate buffer and n-BuOH was employed for hydrogenolysis in order to maintain the pH of the reaction mixture around 6.5 where tebipenem remains stable, and also to remove p-toluidine derived from PNB group into n-BuOH layer. Solubility of L-188 is low (Table 13.3), so the reaction proceeded in five-component suspension system (L-188, Pd–C, n-BuOH, phosphate buffer, and H$_2$ gas). Pressure filtration apparatus is more efficient than vacuum filtration one to filter the catalyst from the aqueous reaction mixture. Although the reaction proceeded smoothly with a reaction yield of 90%, the isolation yield of TBPM was only 70% because of degradation and loss of TBPM in the aqueous solution due to desalting with resin chromatography for a long period, concentration of aqueous active fractions, and lyophilization. And, from these results, it is conceivable that carbapenems is, in general, unstable in the aqueous solution.

After the establishment of the process of L-188 and acquisition of the crystals of TBPM–4H$_2$O, we took advantage of the opportunity and replaced the phosphate buffer with water to perform a direct crystallization of TBPM–4H$_2$O from the reaction solution. The volume of water was set based on the solubility of TBPM–4H$_2$O (about 2% w/w to water) and the reaction was carried out under hydrogen atmosphere (400 kPa) considering reproducibility of the process at room temperature. As per the procedure, acetone was added as a poor solvent to the aqueous layer after the workup; consequently, the isolation yield of TBPM–4H$_2$O was increased to 86–91% (Scheme 13.7). Finally, the solvent volume was further reduced to one-third by increasing the solubility of TBPM by adding 0.5 mol equiv. of NaHCO$_3$ without affecting the yield and quality of TBPM–4H$_2$O. Single-crystal X-ray diffraction analysis of the crystalline materials of TBPM–4H$_2$O confirmed the planar structure of the azetidine ring of the C2 substituent of TBPM and four crystalline water molecules (Figure 13.3).

13.4.3
Prodrug Esterification: Synthesis of TBPM Hexetil, LJC11,143

Conventionally, prodrug esterification of oral β-lactam antibiotics was carried out by the reaction of their metal carboxylate salt with alkyl iodide such as hexetil iodide or POM iodide, so we first investigated the reaction with hexetil iodide to prepare prodrug LJC11,143. The amine base was mainly screened because deprotonation of

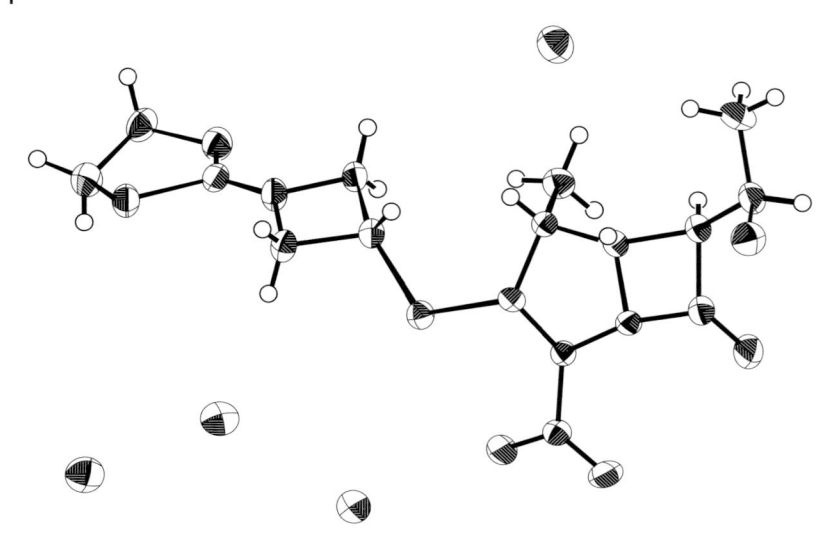

Figure 13.3 ORTEP (Oak Ridge Thermal Ellipsoid Plot) drawing of TBPM–4H$_2$O.

TBPM was difficult with inorganic base such as NaHCO$_3$. The reaction yield stayed around 80% and many impurities imposed a large burden on the purification step (Table 13.5, entry 1–3).

Since this is attributed to the reactivity of 1,3-thiazoline moiety of C2 substituent on the alkyl iodide, we studied other reaction conditions using less reactive hexetil chloride based on literature search [8]. The reaction proceeded in the presence of quaternary ammonium salts. Among the various kinds of quaternary ammonium salts tested, benzyltriethylammonium chloride (BnEt$_3$N$^+$Cl$^-$) was selected because of its lower hygroscopicity and good phase separation during workup, and further investigation was carried out [2c,d]. The reaction speed depends on the concentration of the quaternary ammonium salt and the reaction temperature, and it was found that the reaction was complete under saturated conditions (about 2 M BnEt$_3$N$^+$Cl$^-$), without forming any by-products even at room temperature. Interestingly, there appeared several reports on mild reaction conditions for prodrug formation using quaternary ammonium salts for oral carbapenem DZ-2640 [9], oral cephalosporin [10], and oral penem MEN11505 [11], and on similar acceleration effect of inorganic salts for oral carbapenem CS-834 [12].

In spite of the improvement of the reaction conditions, there still remained other issues regarding chemical instability of LJC11,143, especially in the presence of moisture, because the final material is an amorphous solid of a mixture of two diastereomers derived from the hexetil group. Although various approaches – separation of each isomer by crystallization, followed by mixing them and retransforming to amorphous solid, or study on various salts of LJC11,143 – were further investigated, eventually its development was suspended because of the

Table 13.5 Prodrug esterification reaction in the presence of quaternary ammonium salt.

TBPM–4H$_2$O + (structure: X-C(O)-O-C(O)-O-cyclohexyl, 2 mol equiv.) →(Amine, DMF) LJC11,143

Entry	X	Amine (mol equiv.)	Additive (mol equiv.)	DMF (M)	Temperature (°C)	Time (h)	HPLC yield (%)
1	I	i-Pr$_2$NEt (2)	–	0.22	−10	2	82
2	I	Et$_3$N (2)	–	0.22	−10	2	77
3	I	Py (2)	–	0.22	−10	2	30
4	Cl	i-Pr$_2$NEt (1.1)	–	0.22	65	1	12
5	Cl	i-Pr$_2$NEt (1.1)	n-Bu$_4$NBr (1)	0.22	65	1	55
6	Cl	i-Pr$_2$NEt (1.1)	n-Bu$_4$NBr (2)	0.22	65	2	85
7	Cl	i-Pr$_2$NEt (1.1)	n-Bu$_4$NBr (2)	0.22	65	2	62
8	Cl	i-Pr$_2$NEt (1.1)	NaI (1) n-Bu$_4$NCl (2)	0.22	65	2	84
9	Cl	i-Pr$_2$NEt (1.1)	Et$_4$NCl (2)	0.22	65	2	87
10	Cl	i-Pr$_2$NEt (1.75)	(CH$_3$)$_4$NCl (2)	0.22	65	2	63
11	Cl	i-Pr$_2$NEt (1.75)	BnEt$_3$NCl (2)	0.22	65	2	87
12	Cl	i-Pr$_2$NEt (1.75)	BnEt$_3$NCl (2)	0.22	45	6	92
13	Cl	i-Pr$_2$NEt (1.75)	BnEt$_3$NCl (2)	0.88	35	16	93

instability of API. We reinvestigated other prodrug esters to find out TBPM-PI possessing chemical stability and excellent oral absorbability, which is applicable to purification by crystallization. Consequently, we switched our development candidate from LJC11,143 to TBPM-PI (Table 13.6) [2].

13.4.4
Synthesis of TBPM-PI

The synthetic process of TBPM-PI was optimized to further mild conditions based on the reaction conditions of LJC11,143 (Scheme 13.8). The reaction proceeded quantitatively without forming impurities. Taking advantage of the basicity of TBPM-PI, the product was transferred into a cold aqueous–acid solution and washed to remove excess reagents. Further workup processes were optimized to reduce losses by considering its chemical properties such as stability in solution state, solubility, and partition ratio between solvents and aqueous phases. The material was dried *in vacuo* to give crystalline TBPM-PI in 93–98% yield from TBPM–4H$_2$O. It showed excellent chemical stability in storage conditions (40 °C, RH (Relative Humidity) 75%) and during wet granulation in formulation process. No crystal polymorphism was found.

Table 13.6 Stability of prodrug esters of TBPM.

Promoiety	Description	Rat BA (%) 20 mg kg^{-1}	Remaining rate (%) of prodrug storage conditions time (month)					
			40 °C				40 °C, 75% RH	
			0.5	1	2	3	0.5	1
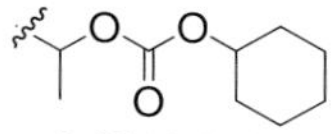 LJC11,143	Amorphous	36	99	96	94	94	32	0
	Crystals	36	100	100	98	98	80	47
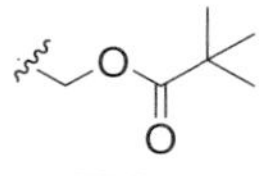 TBPM-PI	Amorphous	–	90	80	–	–	0	–
	Crystals	38	100	98	–	–	99	98
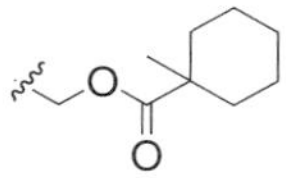	Amorphous	–	–	68	–	37	5	–
	Crystals	41	–	99	–	99	–	99
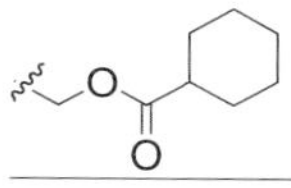	Amorphous	–	85	71	48	38	0	–
	Crystals	30	100	99	97	94	97	74

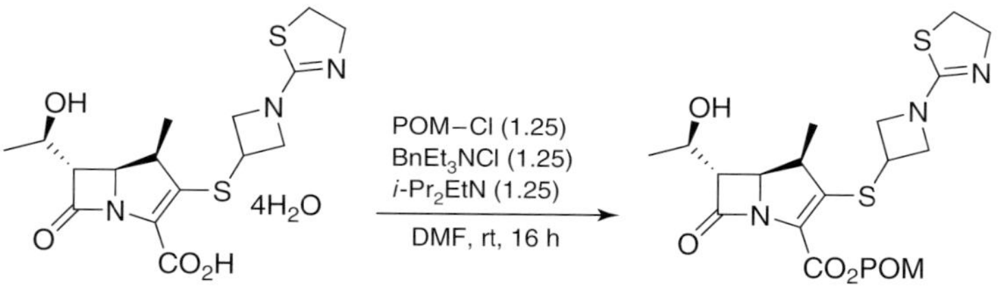

Scheme 13.8 Synthetic process of TBPM-PI.

13.5
Summary and Outlook

After establishing manufacturing process of TBPM-PI, it took a long time to design the protocol of the clinical tests and carry them out to evaluate the therapeutic value and profitability of the agent. TBPM-PI was launched as the first oral carbapenem, showing excellent performance as a last resort of the oral β-lactam antibiotics for community-acquired respiratory tract infection caused by resistant pathogens.

A practical synthetic process of TAT and TBPM-PI was established. TBPM-PI can be manufactured in 69–83% overall yield from the commercially available MAP. This process has been scaled up to over 200 kg of TBPM-PI. We established an efficient and facile synthetic process for versatile **8** and analogs of TAT and developed a mild and efficient prodrug esterification method using quaternary ammonium salts without producing impurities.

Acknowledgments

We thank Dr Yasushi Murai for helpful discussion and advice and Dr Takashi Watanabe for offering X-ray crystallographic structure of TBPM. We also thank all the process researchers in Meiji Seika Kaisha, Ltd., and the former synthetic researchers in Lederle Japan for their endeavors to establish the manufacturing process for TBPM-PI.

References

1. (a) Abe, T., Hayashi, K., Mihira, A., Sato, C., Tamai, S., Yamamoto, S., Hikida, M., Kumagai, T., and Kitamura, M. (1998) 38th Interscience Conference on Antimicrobial Agents and Chemotherapy, Abstract F-64; (b) Kumagai, T., Abe, T., and Hikida, M. (2002) *Curr. Med. Chem.-Anti-Infective Agents*, **1**, 1; (c) Isoda, T., Ushirogochi, H., Satoh, K., Takasaki, T., Yamamura, I., Sato, C., Mihira, A., Abe, T., Tamai, S., Yamamoto, S., Kumagai, T., and Nagao, Y. (2006) *J. Antibiot.*, **59**, 241.
2. (a) Abe, T., Isoda, T., Sato, C., Mihira, A., Tamai, S., and Kumagai, T. (1996) Japan Kokai Tokkyo Koho, Japanese Patent 8053453; (b) Abe, T., Isoda, T., Sato, C., Mihira, A., Tamai, S., and Kumagai, T. (1996) Japan Kokai Tokkyo Koho, Japanese Patent 8253482; (c) Abe, T., Isoda, T., and Shimada, O. (1997) Japan Kokai Tokkyo Koho, Japanese Patent 9110868; (d) Abe, T., Isoda, T., and Shimada, O. (1998) Japan Kokai Tokkyo Koho, Japanese Patent 10195076; (e) Hayashi, K., Sato, C., and Tamai, S. (2002) Japan Kokai Tokkyo Koho, Japanese Patent 2002012593; (f) Tamai, S. and Kumagai, T. (1996) Japan Kokai Tokkyo Koho, Japanese Patent 8253481; (g) Wang, Y., Bolos, J., and Serradell, N. (2006) *Drugs Fut.*, **31**, 676.
3. Kato, K., Kuraoka, E., Sekiguchi, T., Kikuchi, A., Iguchi, M., Suzuki, H., Shibasaki, S., Kurosawa, T., Shirasaka, Y. and Tamai, I. (2007) *Drug Metab. Rev.*, **39** (1), 74.
4. (a) Hayashi, K., Sato, C., and Tamai, S. (1997) Japan Kokai Tokkyo Koho, Japanese Patent 9136888; (b) Hayashi, K., Sato, C., and Tamai, S. (1997) Japan Kokai Tokkyo Koho, Japanese Patent 9077770; (c) Tamai, S., Shimizu, H., and Yamamura, I. (1997) Japan Kokai Tokkyo Koho, Japanese Patent 9278776; (d) Tamai, S. and Yamamura, I. (2000) Japan Kokai Tokkyo Koho, Japanese Patent 2000001474; (e) Isoda, T. and Yamamura, I. (2000) Japan Kokai Tokkyo Koho, Japanese Patent 2000355592; (f) Isoda, T. and Yamamura, I. (2002) Japan Kokai Tokkyo Koho, Japanese Patent 2002003490; (g) Isoda, T. and Yamamura, I. (2002) Japan Kokai Tokkyo Koho, Japanese Patent 2002003491.
5. Isoda, T., Yamamura, I., Tamai, S., Kumagai, T., and Nagao, Y. (2006) *Chem. Pharm. Bull.*, **54**, 1408.
6. (a) Kumagai, T., Tamai, S., Abe, T., Matsunaga, H., Hayashi, K., Kishi, I., Shiro, M., and Nagao, Y. (1998) *J. Org. Chem*, **63**, 8145; (b) Abe, T., Tamai, S., and Nagase Y. (1992) Japan Kokai Tokkyo Koho, Japanese Patent 4230286; (c) Kumagai, T., Abe, T., Fujimoto, Y., Hayashi, T., Inoue, Y., and Nagao, Y. (1993) *Heterocycles*, **36**, 1729.

7. Hayashi, K., Sato, C., Hiki, S., Kumagai, T., Tamai, S., Abe, T., and Nagao, Y. (1999) *Tetrahedron Lett.*, **40**, 3761.
8. (a) Perboni, A., Rossi, T., Gaviraghi, G., Ursini, A., and Targzia, G. (1992) International Patent Application WO 9203437; (b) Nilsaake, J. (1994) International Patent Application WO 9415904.
9. Nishi, T., Ishida, Y., and Ohtsuka, M. (1998) Japan Kokai Tokkyo Koho, Japanese Patent 10158199.
10. (a) Wang, H.P. and Lee, J.S. (1996) US Patent 5,498,787; (b) Lee, H.W., Kang, T.W., Kim, E.N., Shin, J., Cha, K.H., Cho, D.O., Choi, N.H., Kim, J.W., and Hong, C. II (1998) *Synth. Commun.*, **28**, 4345.
11. Arcamone, F.M., Altamura, M., Perrotta, E., Crea, A., Manzini, S., Poma, D., Salimbeni, A., Triolo, A., and Maggi, C.A. (2000) *J. Antibiot*, **53**, 1086.
12. Kobayashi, K., Shimizu, M., Tanabe, A., Nabatame, J., Fukuhara, H., and Takebayashi, T. (1998) Japan Kokai Tokkyo Koho, Japanese Patent 10130270.

14
Some Progress in Organic Synthesis of Pharmaceuticals in China
Delong Liu and Wanbin Zhang

14.1
Introduction

The Chinese history of pharmaceutical production could be traced back to the ancient Shennong Era. To some extent, the development of pharmaceutical production has witnessed the vicissitudes of long Chinese history. Chinese herbal medicine (CHM), a major aspect of traditional Chinese medicine (TCM), has been used in China to treat a variety of human diseases for thousands years. It consists of natural medicines and processed components made from herbal, animal, mineral, and some chemical and biological substances. A central concept of CHM is to promote and maintain health via restoring energy balance rather than to treat particular diseases.

As the story goes, the first herbalist was Shennong, a mythical personage, who is said to have tasted hundreds of herbs and imparted his knowledge of medicinal and poisonous plants to farmers. The first Chinese manual on pharmacology *Shennong Bencao Jing* was written in the first century C.E. and lists some 365 medicines, of which 252 are herbs. The most complete and comprehensive medical book in Chinese history is *Bencao Gangmu* that included descriptions of almost all the plants, animals, minerals, and other objects that were believed to have medicinal properties.

Some familiar CHMs are lingzhi, ginseng, wolfberry, dongquai, *Coptis chinensis*, and yuanhu (*Corydalis*) (Figure 14.1).

Although the mechanisms of CHM are not fully comprehended, CHM has been and is still being used by the world's population. It is estimated that more than 8000 kinds of herbal medicines are currently being produced in China, 600 of which are commonly used. These medicines not only meet the demands of domestic people but also are exported to 80 countries and regions.

This chapter attempts to provide the readers with a concise summary on some progress in organic synthesis of pharmaceuticals in China. First, China's large territory, geographical abundance, and climatic variety have given rise to different ecological environments and a great variety of herbal plants. In comparison to modern medicines, CHM have much lesser adverse reactions. Producing CHM

Pharmaceutical Process Chemistry. Edited by Takayuki Shioiri, Kunisuke Izawa, and Toshiro Konoike
Copyright © 2011 WILEY-VCH Verlag GmbH & Co. KGaA, Weinheim
ISBN: 978-3-527-32650-1

Lingzhi	Ginseng	Wolfberry
Dongquai	*Coptis chinensis*	Yanhusuo (*Corydalis*)

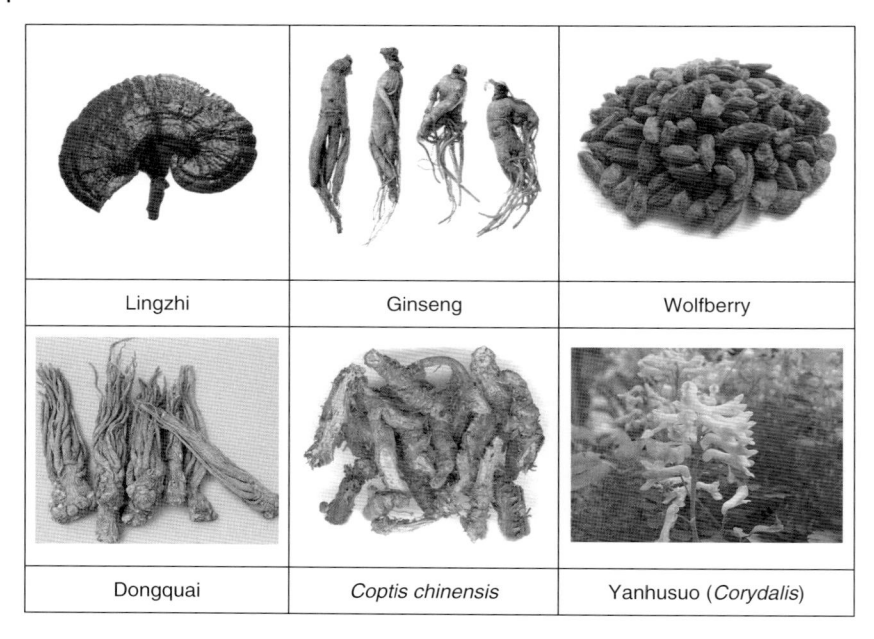

Figure 14.1 Examples of Chinese herbal medicine (CHM).

either by extraction or synthesis has been an important method in developing new drugs. The first section of this chapter focuses on the organic synthesis of some CHMs. Secondly, drugs with greater biological activity could be obtained through suitable structural modification of CHM. The following section summarizes some new pharmaceuticals derived from CHM through structural modification. Lastly, some organic synthesis of pharmaceuticals such as vitamin C and vitamin B7 (biotin) are summarized and presented in the last section.

14.2
Industrial Synthesis of Chinese Herbal Medicines

14.2.1
Industrial Synthesis of Berberine

Berberine (Figure 14.2, **1**), a natural alkaloid derived from plants, has a long history of use in Chinese and Ayurvedic medicines with antidiarrheal, antimicrobial, antioxidant, and hepatoprotective properties. It is present in the roots, rhizomes, and stem bark of various plants including *Hydrastis canadensis* (goldenseal), *C. chinensis* (coptis or goldenthread), *Berberis aquifolium* (Oregon grape), *Berberis vulgaris* (barberry), and *Berberis aristata* (tree turmeric). Berberine has a dibenzo[a,f]quinolizidine ring system structure. The quaternary salts with an aromatic C ring and a 2,3-methylenedioxy moiety are essential for its antibacterial activities [1].

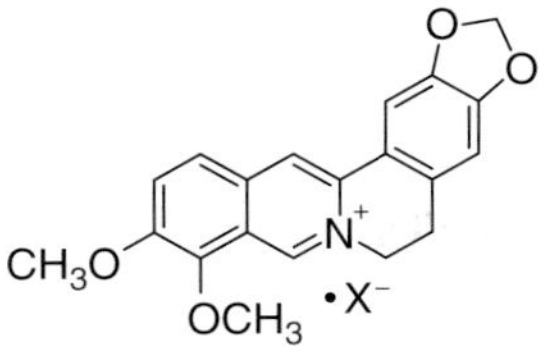

Figure 14.2 The molecular formula of berberine.

1
Berberine

Berberine extracts and decoctions have demonstrated significant antimicrobial activity against a variety of organisms including bacteria, viruses, fungi, protozoans, helminthes, and *Chlamydia* [2, 3]. It has been used in its purified form for over 20 years in the treatment of many ailments, most notably diarrheal diseases. Recent observations indicated that berberine may have beneficial metabolic effects in animal models of diabetes and hyperlipidemia [4].

Salt formation is known to influence a number of physicochemical properties of the drugs, such as solubility, dissolution rate, stability, and hygroscopicity. These properties, in turn, will affect the bioavailability and medicinal value of the corresponding drugs. Commonly used salt forms of berberine included berberine chloride, berberine sulfate, berberine bisulfate, and berberine tannate [5].

Current berberine is supplied by two methods: extraction and chemical synthesis. Originally, berberine was extracted from plants such as *Coptis*, *Phellodendron*, and *Radix berberidis*. This extraction method is not environment-friendly, and it affected the ecological balance severely since only 4–10% of berberine exists in *Coptis*. For example, an eight-year-old *Phellodendron* can only give 75 g of berberine. It is estimated that more than 53.7 million plants of *Phellodendron* need to be harvested every year to meet the annual demand of 500 tons of berberine in China. Thus, it is imperative to pursue alternative method from both environmental and economic perspectives. The best way to obtain berberine is via chemical synthesis.

The Northeast General Pharmaceutical Factory (NEGPF) developed a method to synthesize berberine chloride **18**, and higher purity (>99%) than plant extraction (97%) could be achieved. The synthesis started from phenol (**2**), and consists of 17 steps shown in Scheme 14.1. In 1981, annual production capacity reached 30 tons. However, this chemical synthetic method was stopped in 1994 due to the high cost, laggardly technology, and long synthetic route [6].

The search for a more efficient synthetic route focused on the key intermediate 2-(3,4-methylenedioxyphenyl)ethylamine (homopiperonylamine, **14**). In the original method indicated in Schemes 14.1 and 14.2, it took seven steps to prepare **14** from **7**. However, the alternative route described in Scheme 14.2 took only two steps ($S_N 2$ substitution and hydrogenation) [7].

Scheme 14.3 shows the method developed by Chen. The starting material **5** reacts with CH_2Cl_2 to produce cycloether **6**, which was then converted to the key intermediate **14** with the alternative strategy mentioned above. As shown in Scheme 14.3, the new method has the advantages of lower cost, higher purity, and increased performance.

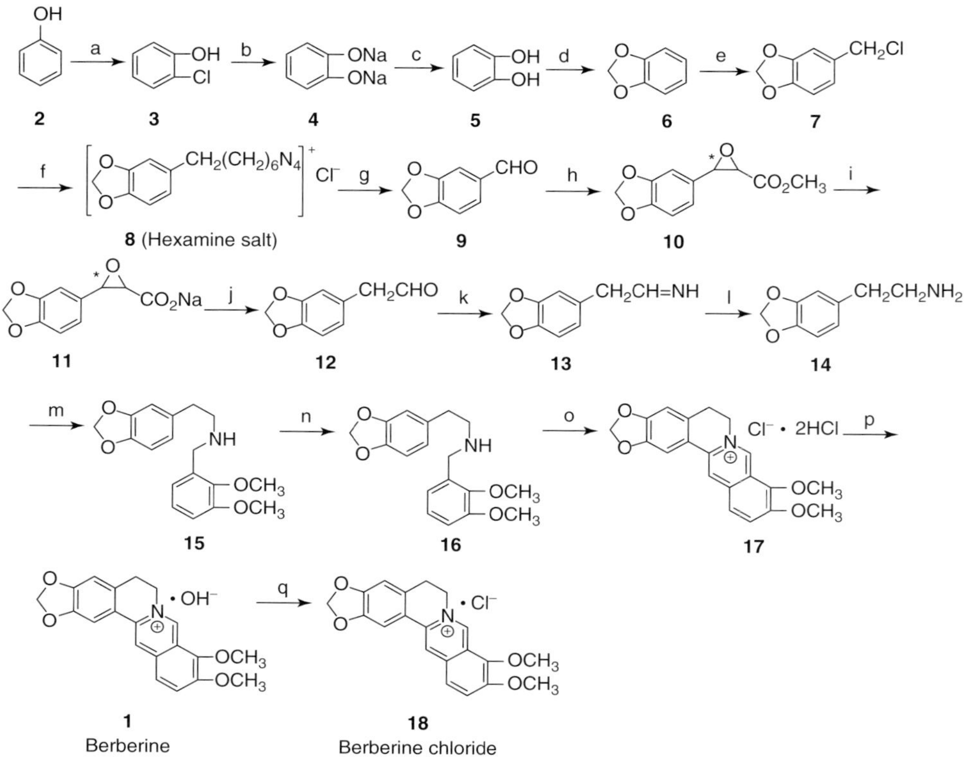

Conditions: (a) Cl$_2$/PhH, 24~28 °C; (b) NaOH/CuSO$_4$, 180~190 °C; (c) HCl, pH 3~3.5; (d) CH$_2$Cl$_2$/NaOH, DMSO, 110~115 °C; (e) 1,3-Trioxane, HCl, 25~30 °C; (f) Hexamethylenetetramine, CH$_3$CO$_2$H; (g) CH$_3$CO$_2$H/H$_2$O, 100~102 °C; (h) ClCH$_2$CO$_2$CH$_3$, CH$_3$ONa/CH$_3$OH, 17~20 °C; (i) NaOH/CH$_3$OH; (j) CH$_3$CO$_2$H, 45~45 °C; (k) NH$_3$/CH$_3$OH, 10~20 °C; (l) H$_2$/Ni, EtOH, 30~40 kg/cm^2; (m) 2,3-Dimethoxybenzaldehyde, H$_2$/Ni, EtOH, 40 kg/cm^3; (n) HCl/EtOH, pH 3~3.5; (o) Oxalaldehyde, CuSO$_4$, NaCl, Ac$_2$O, AcOH: (p) CaO/H$_2$O, pH 8~8.5: (q) HCl / H$_2$O, pH 1~2.

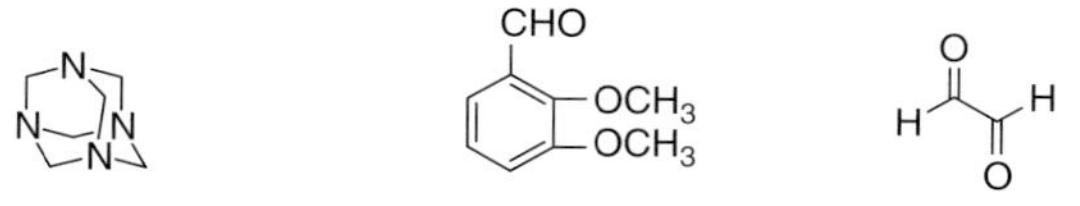

Hexamethylenetetramine   2,3-Dimethoxybenzaldehyde   Oxalaldehyde

Scheme 14.1 The primary synthetic route for berberine.

In 1999, large-scale industrial production of berberine became possible. A sales revenue of about RMB17 million was achieved in the same year and more than 10 tons output was achieved in 2006 [7c].

Chen of Changzhou Yabang Pharmaceuticals developed a novel synthetic route for berberine in 2007 [8]. Owing to the industrial production of homopiperonylamine, the company started from dimethoxylphenylmethanol via a six-steps reaction. Yield of >90% was achieved in every reaction step and the overall yield was up to 67%. The details are shown in Scheme 14.4.

Initial method for the intermediate

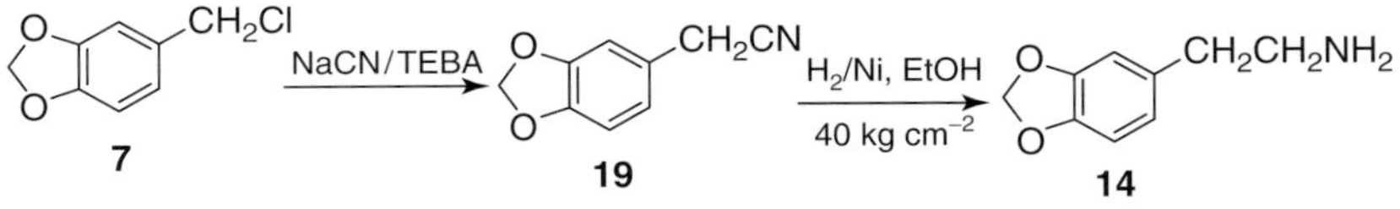

Scheme 14.2 The improved synthetic route for the intermediate of berberine.

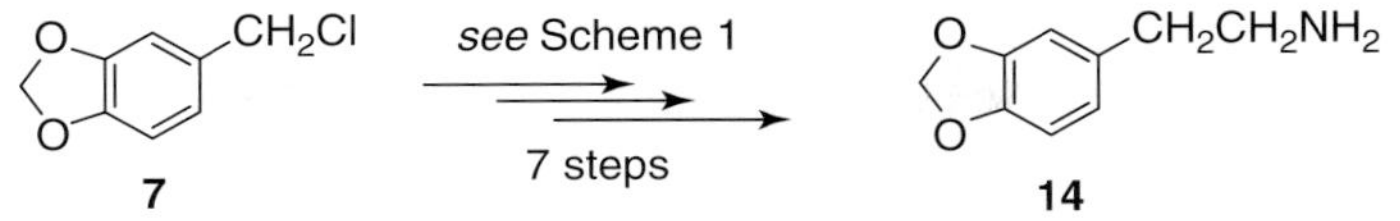

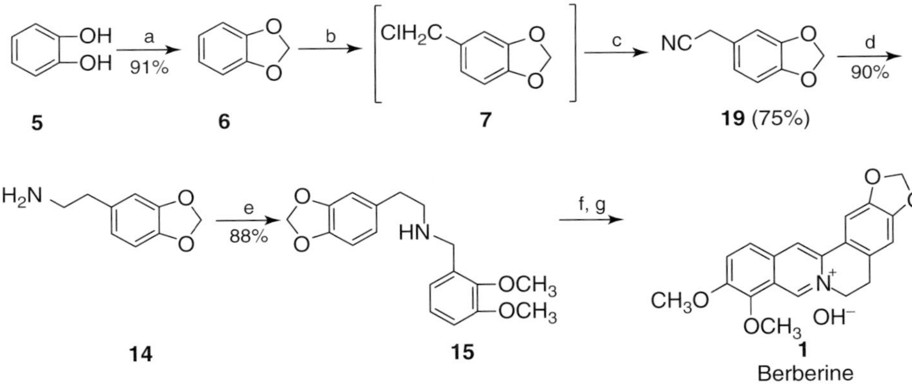

Improved method for the intermediate

Conditions: (a) CH$_2$Cl$_2$, NaOH, H$_2$O, Bu$_4$NBr, 90–100 °C; (b) 1,3,5-trioxane, HCl, PCl$_3$, 70–90 °C; (c) NaCN, NaOH, Bu$_4$NBr, H$_2$O; (d) NH$_3$, H$_2$(4.0 Mpa)/Ni, CH$_3$OH; (e) 2,3-dimethoxybenzaldehyde, H$_2$/Ni, CH$_3$OH; (f) oxalaldehyde, CuCl$_2$, NaCl, Ac$_2$O, AcOH; (g) NH$_3$, H$_2$O.

Scheme 14.3 The improved synthetic procedure for berberine.

The procedure was soon applied in large-scale industrial production of berberine in Changzhou Yabang Pharmaceutical. A sales income of about RMB20 million was achieved in 2007.

14.2.2
Industrial Synthesis of D,L-Tetrahydropalmatine (THP)

Rhizoma corydalis (yanhusuo or yuanhu), a CHM, was obtained from the dry stem of *Corydalis yanhusuo* W.T. Wang [9]. It was first recorded in *Ben Cao Shi Yi* and now widely cultivated in Zhejiang province. As a widely used medicinal herb to

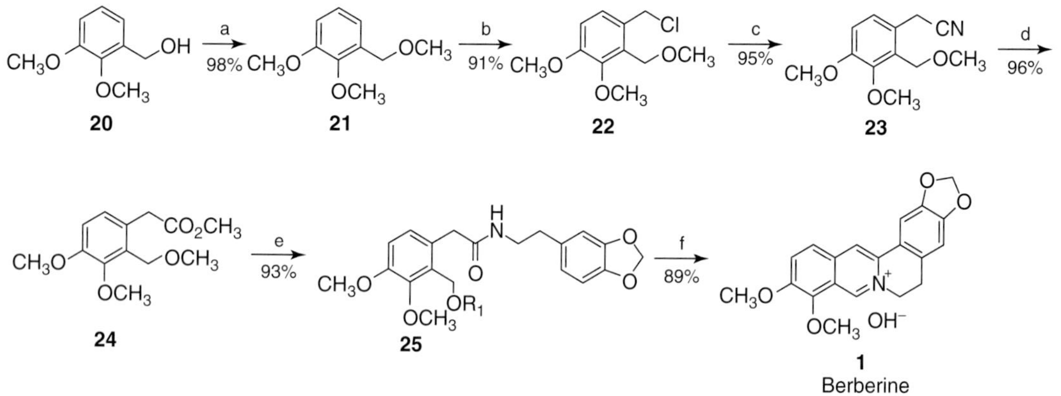

Scheme 14.4 The new synthetic procedure for berberine.

Conditions: (a) 30% NaOH (aqueous), $(CH_3)_2SO_4$, 80 °C; (b) $(CH_2O)_n$/HCl (aqueous), toluene, 30 °C; (c) NaCN, $(Bu)_4NBr$, Toluene, 100 °C; (d) CH_3OH, H_2SO_4 (catalyst), ruflux; (e) homopiperonylamine (14), 170 °C; (f) PCl_5, CH_2Cl_2, 40 °C.

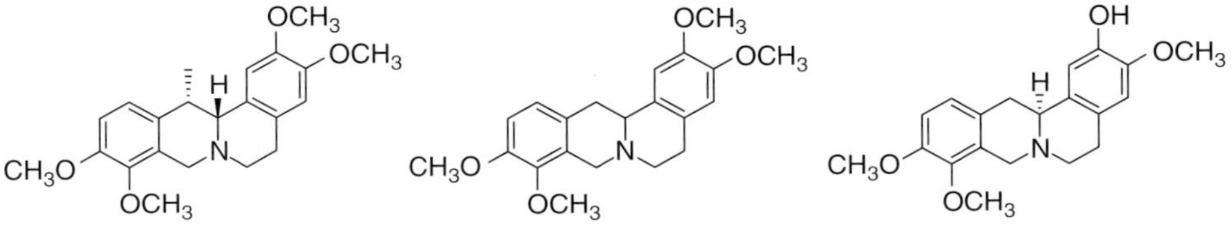

Figure 14.3 The molecular formula of some main components of yanhusuo (*Corydalis*).

treat spastic pain, abdominal pain, and other pains due to injuries, *R. corydalis* was systematically studied by Chou during 1928–1936, and was known to be rich in alkaloids, mainly D-corydaline, D,L-tetrahydropalmatine (THP, Figure 14.3, **26**) and corydalis L [10]. Of these, THP is the most active ingredient with hypnotic and anodynic activities [11]. It has been listed in the Chinese pharmacopoeia since 1977, as an analgesic with sedative and hypnotic effects. Pharmacological studies revealed that THP blocks dopamine receptors in the central nervous system, accounting for most of its analgesic activity [12].

THP, also known as *protoberberine*, exists in racemic mixtures and is characterized by a tetracyclic skeleton containing an isoquinoline core. Because of the restricted existence of *C. yanhusuo* W.T. Wang plus the low content of THP, the early preparation of THP was through extraction from the wild plant *Fibraurea recisa* Pierre as well as hydrogenation of palmatine. However, the extraction method is also not environment-friendly and less than 4.5% of palmatine exists in *Fibraurea tinctoria* Lour. The preparation of THP by this method is not advisable due to the limitation of the starting materials and thus, the exploration of alternative processes is imperative.

In the 1970s, researchers of Guangzhou Pharmaceutical Industrial Research Institute developed an innovative way for total synthesis of THP (Scheme 14.5). First, treatment of catechol (**5**) with dimethyl sulfate in aqueous NaOH (30%) was used to obtain 1,2-dimethoxybenzene in 90% yield. Then, 1,2-dimethoxybenzene reacted with acrylonitrile in the presence of AlCl$_3$, followed by partial hydrolysis and Hofmann rearrangement to obtain the key intermediate 1,2-dimethoxyphenylethanamine (**30**). After condensation, cyclization, reduction, and salification, THP was obtained with an overall yield of <6% (with respect to the original material **5**) [13].

This synthetic procedure had some limitations such as the harsh reaction conditions and poor overall yield. Cheng and coworkers developed a novel synthetic route to obtain THP via six steps (Scheme 14.6) [14]. Here, treatment of guaiacol (2-methoxyphenol, **35**) with dimethyl sulfate in aqueous NaOH (30%) was started to obtain **27** which was then converted to the key intermediate **30** via chloromethylation, substitution, and NH$_3$-reduction. After condensation, cyclization, and reduction, THP was obtained with a yield of over 14% (with respect to the original material **35**).

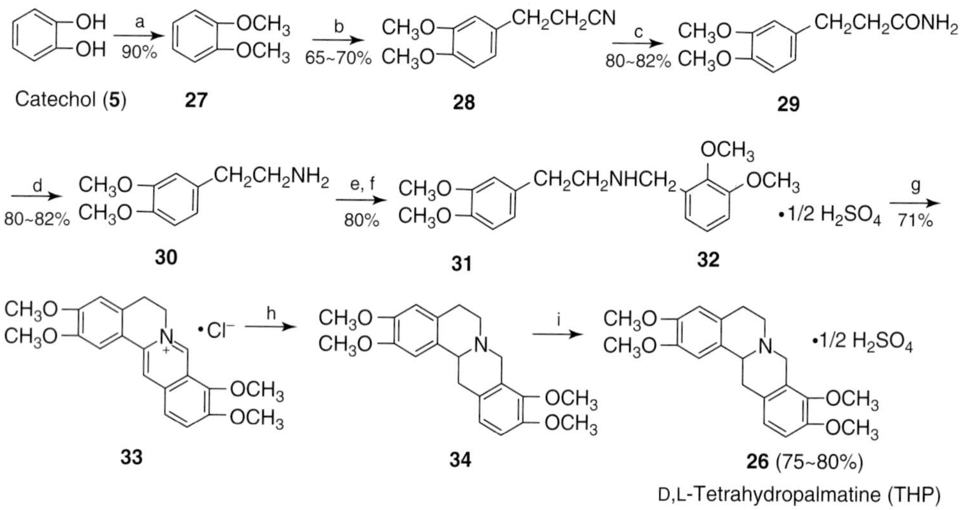

Scheme 14.5 The primary synthetic route to D,L-tetrahydropalmatine (THP).

Conditions: (a) $(CH_3)_2SO_4$, 30% NaOH (aqueous), H_2O; (b) $CH_2=CHCN$, $AlCl_3$; (c) H_2SO_4; (d) 10% NaOH, Cl_2; (e) 2,3-dimethoxy-benzaldehyde, 110 °C; (f) H_2, Ni; (g) succinaldehyde, AcOH, $CuSO_4$, NaCl; (h) H_2-Ni, NaOAc, EtOH; (i) H_3SO_4, H_2O.

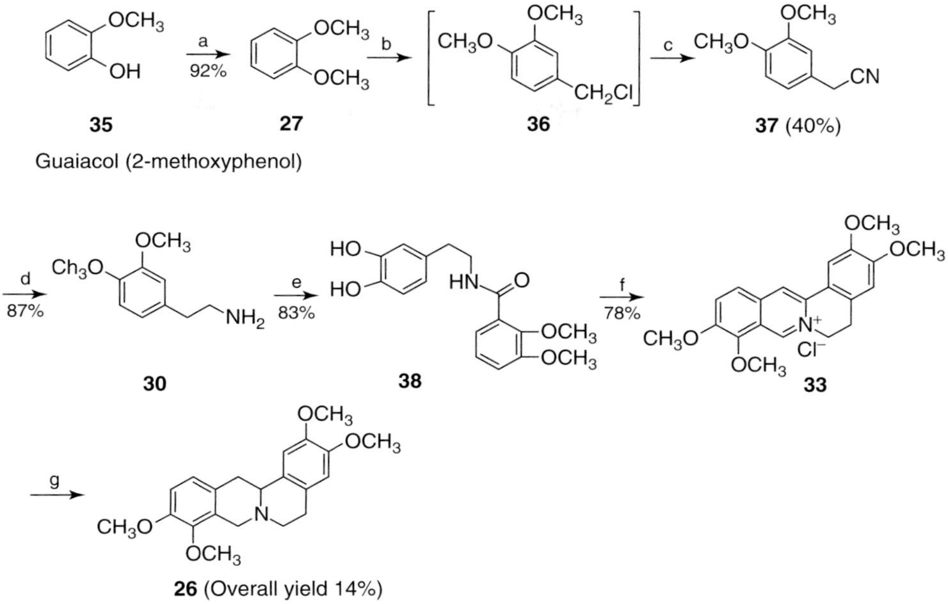

Scheme 14.6 The improved synthetic procedure for D,L-tetrahydropalmatine (THP).

Conditons: (a) $(CH_3)_2SO_4$, NaOH, H_2O; (b) HCHO, HCl, $POCl_3$, CH_3Cl; (c) NaCN, NaOH, Catalyst $PhCH_2NEt_3Cl$; (d) NH_3, Ni, 90 °C, 4 MPa, EtOH; (e) 2,3-Dimethoxybenzoic acid, H_2SO_4, Ni, EtOH; (f) Oxalaldehyde H_2SO_4, NaCl, AcOH; (g) H_2, AcONa, Ni, EtOH.

14.3
New Agents Derived from Chinese Herbal Medicines

14.3.1
Bifendate and Bicyclol

14.3.1.1 Bifendate

Hepatocellular carcinoma (HCC) accounts for 7.5 and 3.5% of all cancers in men and women, respectively [15]. Most cases of HCC in China usually occur in patients infected with viral hepatitis [16]. HCC is relatively insensitive to chemotherapy, radiotherapy, and other available treatment modalities such as radiofrequency ablation, ethanol injection, cryotherapy, and transarterial chemoembolization [17]. The incidence of hepatitis B in China is the highest in the world making HCC the second most common cancer in China [16].

Schisandra chinensis (Chinese wuweizi) has long been used in CHM for a variety of biological activities. Research shows that the fruit of *S. chinensis* is rich in dibenzocyclooctadiene derivatives, such as schizandrins A, B, C, schizandrol A, B, and schizandrers A, B [18]. It was found that the biaryl structures are essential for

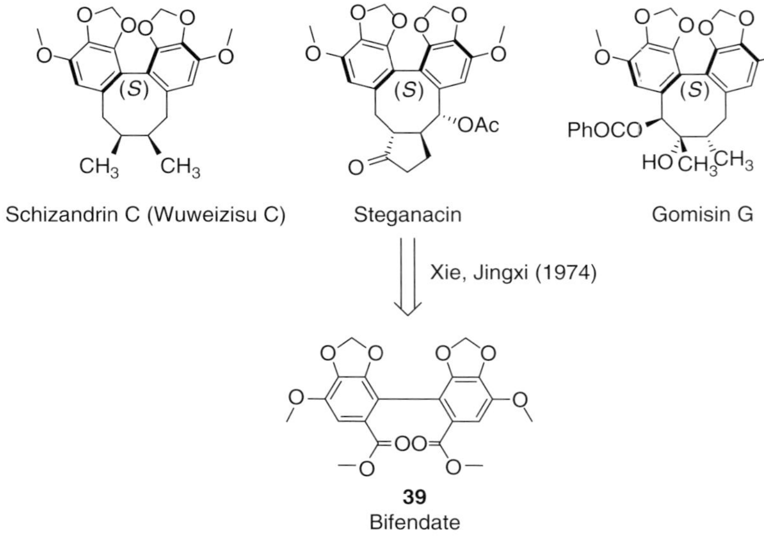

Figure 14.4 The discovery of bifendate.

various pharmacological activities, most notably antihepatotoxic activity against liver injury induced by CCl_4 [19]. Therefore, considerable attention had been focused on studying compounds with biaryl structures, such as schizandrin C, steganacin, and gomisin G (Figure 14.4). Many biaryl compounds containing dibenzocyclooctadiene structure and their analogs were developed exhibiting versatile biological activities and pharmacological activities.

Bifendate, (dimethyl 4,4′-dimethoxy-5,6,5′,6′-dimethenedioxy-biphenyl-2,2′-dicarboxylate, also named as α-DDB, Figure 14.4, **39**), a synthetic intermediate as well as an analog of schizandrin C and other schizandrins, were discovered by Xie *et al.* in the course of synthesis of schizandrin C. It showed almost the same activity as schizandrin C [20] and was also found to be protective against drug-induced liver injury [21]. In addition, bifendate may also have anti-HIV activity [22].

Now, bifendate, clinically used for the treatment of hepatitis with minimal side effects, is regarded as a positive control for exploring other hepatoprotective agents [23]. Moreover, bifendate treatment has been shown to attenuate hepatic steatosis in mice with hypercholesterolemia induced by cholesterol/bile salt or a high-fat diet [24].

The complete synthesis of bifendate is represented in Scheme 14.7 [20]. First, gallic acid (**40**) was treated with methanol in the presence of catalytic amounts of sulfuric acid to produce an esterification product (**41**), which was then selectively protected to obtain **42**. After cyclization with CH_2I_2 and then electrophilic bromination, **44** were obtained from **42** via **43** as intermediate with moderate yield. Finally, the target product bifendate was obtained via Cu-catalyzed coupling reaction in an overall yield of 22%.

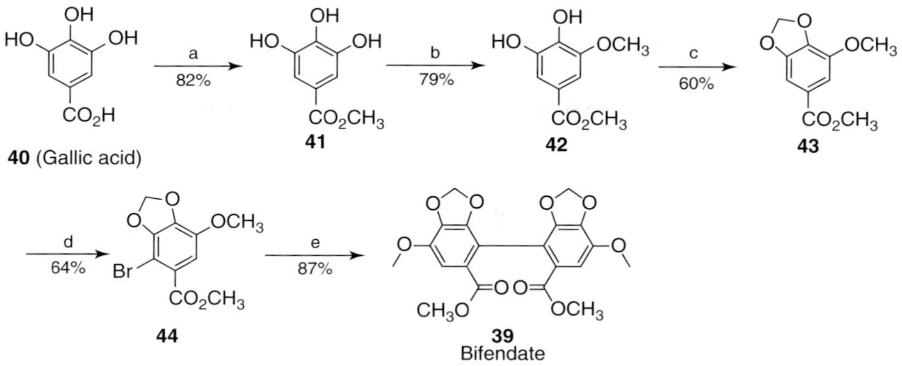

Conditions: (a) H₂SO₄, CH₃OH; (b) (CH₃)₂SO₄; (c) CH₂I₂, NaOH; (d) Br₂/HOAc; (e) Cu, DMF.

Scheme 14.7 The Xie's synthetic procedure for bifendate.

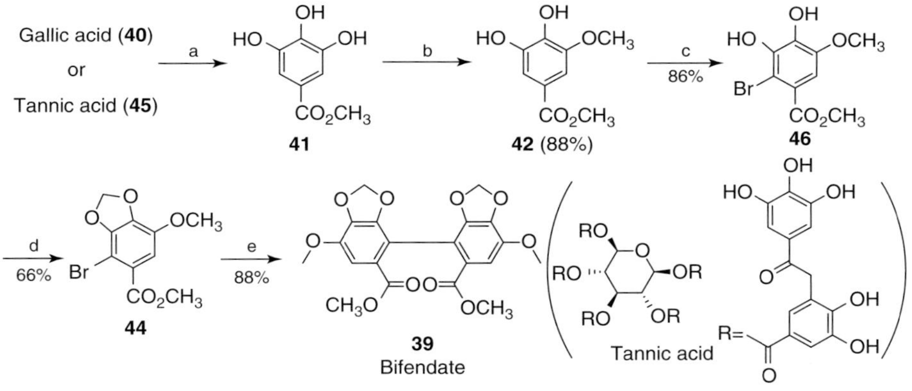

Conditions: (a) H₂SO₄, CH₃OH; (b) (CH₃)₂SO₄; (c) Br₂/HOAc; (d) CH₂I₂, NaOH; (e) Cu, DMF.

Scheme 14.8 The Zheng's synthetic procedure for bifendate.

The reaction route has some disadvantages, including (i) the reagents used were somewhat expensive or difficult to reuse; (ii) poor regioselectivity was observed in the bromination and **44** was obtained with the by-product of another substituted product [25].

Zheng developed a more feasible synthetic route to synthesize bifendate **39** (Scheme 14.8) [26]. The same procedure was adopted to obtain **42** with up to 88% overall yield using optimized conditions. Subsequently, electrophilic bromination was carried out followed by cyclization, to give **44**. The reversed route is more efficient as compared with the one mentioned above. This improved route has an improved overall yield of 31%.

It has been shown that bifendate holds axial chirality, resulting from the restricted rotation of the four *ortho*-substituents, which consequently resulted in two enantiomers. Each isomer showed different biological activity. Till date, there have been some reports on the asymmetric preparation of bifendate besides classical resolution of its racemic isomers [27].

14.3.1.2 Bicyclol

Hepatitis B virus (HBV) infection may cause cirrhosis, HCC, and death. More than 1 million deaths every year are caused by hepatitis B worldwide, and it is estimated that 350 million patients are chronically (lifelong) infected [28]. Hepatitis B is endemic in parts of Asia, the South-Pacific region, and sub-Saharan Africa. In China and sub-Saharan Africa, HCC associated with HBV is one of the leading causes of cancer in men [29].

Xie collaborated with other Chinese biochemists and medicinal chemists, and developed the drug bifendate 39 [20], which is now widely used for the treatment of hepatitis with minimal side effects. During bifendate development, however, there was no patent system in China. Liu and researchers from the Chinese Academy of Medical Sciences and Peking Union Medical College developed other drugs with IP protection to treat chronic hepatitis [30]. The second generation of drugs such as bicyclol (4,4′-dimethoxy-2,3,2′,3′-dimethylene-dioxy-6-hydroxymethyl-6′-carbonyl-biphenyl) has shown remarkable effectiveness for chronic hepatitis and a better safety profile.

Bicyclol (Figure 14.5, 47), a derivative of CHM with hepatoprotective property, has been reported to suppress hepatitis virus replication and may have a potential role in the chemoprevention of HCC. Clinical trials have revealed that bicyclol can improve liver function and liver histology, suppress the development and progression of hepatic fibrosis, and inhibit the replication of HBV and HCV in patients with chronic hepatitis [31]. In animal studies, bicyclol has been shown to have significant hepatoprotective and anti-fibrosis effect against liver injury induced by hepatotoxins, in rodents [32]. This agent has also been shown to have an antiviral effect in duck viral hepatitis [33].

Bicyclol is the first national first-class anti-hepatitis drug with independent intellectual property rights [34]. Clinical trials started in 1996, and the drug was approved only two years later. It was produced and marketed by Beijing Union Pharmaceutical Factory (BUPF) for the treatment of liver diseases such as chronic HBV, HCV infection, and alcoholic liver injury [31]. Presently, bicyclol has been patented in 16 countries all over the world for 20 years, enjoys a 12 year term of executive protection, and has been in the market since September 2001.

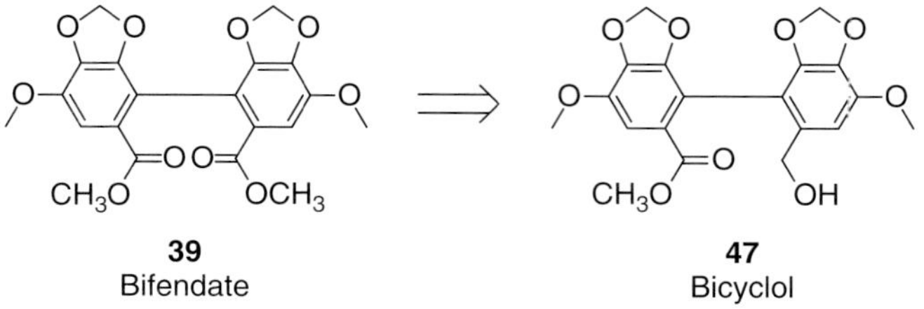

Figure 14.5 Finding of new anti-hepatitis drug bicyclol.

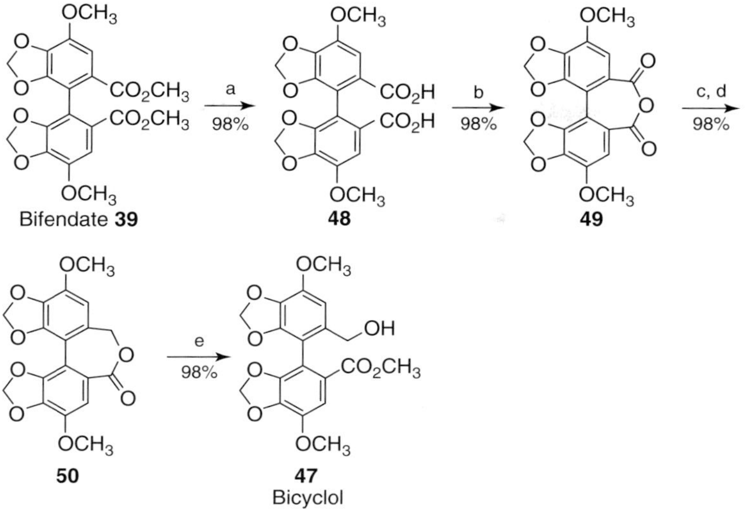

OCH₃ Bifendate **39** → (a, 98%) → **48** → (b, 98%) → **49** → (c, d, 98%) → **50** → (e, 98%) → **47** Bicyclol

Conditions: (a) NaOH, CH₃OH-H₂O, Concentrated HCl; (b) Ac₂O, reflux, 4 h; (c) NaBH₄, THF; (d) p-TsOH, PhH, reflux, 3 h; (e) anhydrous NaOAc, CH₃OH, reflux, 7 h.

Scheme 14.9 The synthesis of bicyclol.

The synthesis of bicyclol is described in EP0353358 [35]. In the reaction Scheme 14.9, bifendate (**39**) was used as starting material and hydrolyzed using sodium hydroxide to give the dicarboxylic acid (**48**). Then, acetic anhydride was heated together to give an anhydride (**49**) which was then reduced with sodium borohydride, followed by heating with p-toluenesulfonic acid to obtained lactone **50**. Finally, bicyclol was obtained via the reaction of lactone **50** with methanol in the presence of anhydrous NaOAc. The overall yield was 39%.

Presently, the sale of bicyclol in BUPF has exceeded RMB100 million in three years and it is exported to Ukraine, Korea, Indonesia, and other countries.

14.3.2
Qinghaosu

14.3.2.1 Synthesis

Three billion people are at risk of getting malaria. The causative agents of malaria are species from the *Plasmodium* genus that are transmitted from human to human via *Anopheles* mosquitoes. An estimated 247 million cases of malaria occur every year, resulting in about 1 million deaths, mostly in children aged less than five years.

Quinine has been used historically to prevent/treat malaria in South America. A number of substituted quinines were synthesized in the last century. Chloroquine, (7-chloro-4-(4′-diethylamino-1′-methylbutylamino) quinoline, SN-7618) (Figure 14.6, **51**), was discovered by a German chemist, Hans Andersag, in 1934.

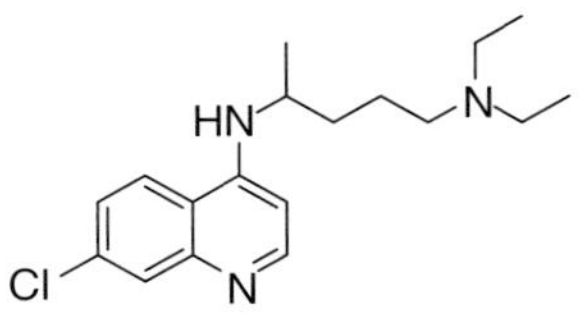

51
Chloroquine

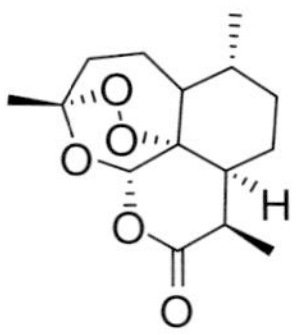

52
Qinghaosu/artemisinin

Figure 14.6 The structure of chloroquine and qinghaosu/artemisinin.

He named his compound resochin. Through a series of lapses and confusion brought about during the war, chloroquine was finally recognized and established as an effective and safe antimalarial drug in 1946 by British and US scientists. In fact, chloroquine has become one of the most widespread means of chemotherapy against malaria in a number of regions of the world (predominantly in Southeast Asia).

Upon extended exposure, *Plasmodium* may develop resistance to chloroquine [36]. The emergence and spread of resistance has created a need for novel antimalarial agents. Newer antimalarials have been discovered, but are either expensive or have undesirable side effects. Moreover, after a variable length of time, the parasites, especially the *falciparum* species, have started showing resistance to these drugs [37].

Artemisia annua L., also known as *qinghao*, has been used to treat fever for many centuries in China. The effective constituent qinghaosu (artemisinin, Figure 14.6, **52**) was isolated by Chinese investigators in 1972 and shown to be sesquiterpene lactone [38]. Artemisinin has significant activity against chloroquine-resistant malaria, and a fast onset of action and remarkably few side effects [39].

In 1983, total synthesis of artemisinin was completed by Swiss chemists Schmid and Hofheinz using (−)-isopulegol (**53**) as starting material via a 10-step reaction. The overall yield was 18% (Scheme 14.10) [40].

Almost at the same time, Chinese chemist Chou and coworkers reported the semisynthesis of artemisinin. They used artemisinic acid (**60**), an extract from artemisinin, as starting material via nine steps with an overall yield of 5.6% [41]. The details are shown in Scheme 14.11.

In order to achieve the total synthesis of artemisinin, the synthesis to **61** was also carried out by Chou's group [42]. They started with R-(+)-citronellal (**66**) via a 12-step reaction (Scheme 14.12). The overall yield was 6%.

Synthetic processing of artemisinin has many limitations, including a long synthetic route, low yield, and high cost which encumbers the industrial preparation of artemisinin [43]. As of now, artemisinin is still obtained through extraction [44].

In 2008, the production of artemisinin in China reached 150–200 tons, 90% of which was exported. Currently, three companies dominate the market: Holley Pharmaceuticals, Shanghai Fosun Pharmaceuticals, and Guangzhou Pharmaceuticals Ltd.

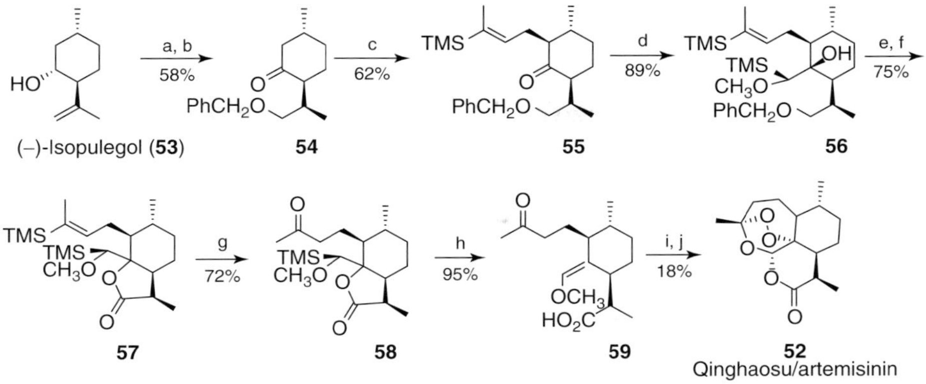

14.3 New Agents Derived from Chinese Herbal Medicines

Conditions: (a) ClCH$_2$OCH$_3$, PhN(CH$_3$)$_2$, CH$_2$Cl$_2$, rt; B$_2$H$_6$, THF, 0 °C; alkaline H$_2$O$_2$; (b) PhCH$_2$Br, KH, 4:1 THF, DMF, 0 °C; PCC, CH$_2$Cl$_2$, rt; (c) LDA, THF, TMS (CH$_3$)C=CHCH$_2$I,; (d) TMS(Li)C (OCH$_3$)H, THF, −78 °C; (e) Li-NH$_3$(l); (f) PCC, CH$_2$Cl$_2$; (g) m-CPBA, THF, TFA; (h) n-Bu$_4$NF, THF; (i) O$_2$, CH$_3$OH, hv, -78 °C; (j) HCO$_2$H, CH$_2$Cl$_2$.

Scheme 14.10 The total synthesis of artemisinin by Schmid and Hofheinz.

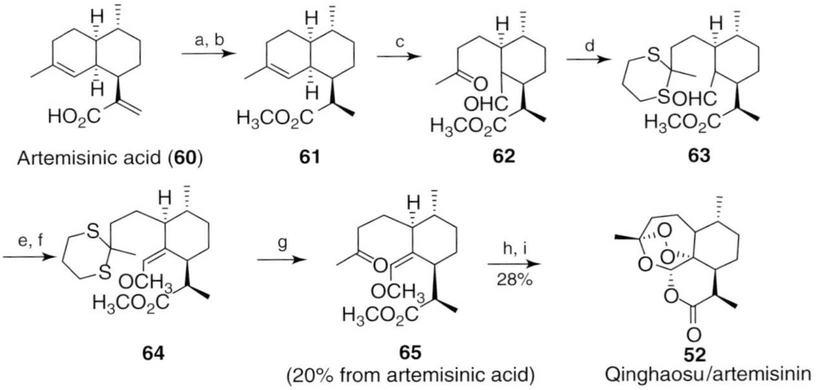

Conditions: (a) CH$_2$N$_2$, Et$_2$O; (b) NaBH$_4$, NiCl$_2$·6H$_2$O, CH$_3$OH; (c) O$_3$, CH$_2$Cl$_2$, CH$_3$OH, (CH$_3$)$_2$S; (d) HS(CH$_2$)$_3$SH, BF$_3$·Et$_2$O, CH$_2$Cl$_2$; (e) HC(OCH$_3$)$_3$, p-TsOH; (f) Xylene, heating; (g) HgCl$_2$, CaCO$_3$, 80% CH$_3$CN/H$_2$O; (h) O$_2$, CH$_3$OH, Rose Bengal, hv; (i) 70% HClO$_4$, THF, H$_2$O.

Scheme 14.11 The synthesis of artemisinin by Zhou.

14.3.2.2 Fixed-Dose Riamet/Coartem

Artemisinin is poorly soluble in water or oil. The unusually large dose is also a significant obstacle. As a result, novel drugs with better potency and improved physical and bioavailability properties are urgently needed. The cyclic hemiacetal dihydroartemisinin [39b], a reductive product of artemisinin, has shown higher antimalarial activity than its parent compound [39b]. A number of ethers and esters of dihydroartemisinin with improved biological activity, stability, and bioavailability have been produced [45].

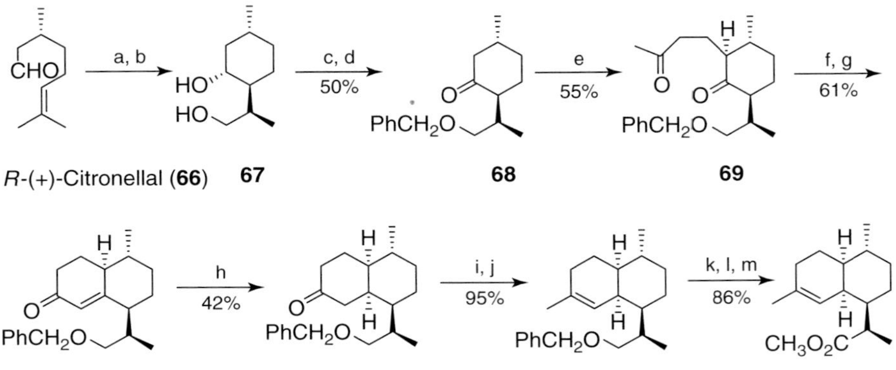

R-(+)-Citronellal (**66**) **67** **68** **69**

70 **71** **72** **61**

Conditions: (a) $ZnBr_2$; (b) B_2H_6, H_2O_2, NaOH; (c) $PhCH_2Cl$, NaH, DMF, 10~15 °C; (d) Oxidation; (e) LDA/THF (−78 °C), $CH_2{=}C((CH_3)_3Si)COCH_3$; (f) $Ba(OH)_2 \cdot 8H_2O$, EtOH, rt; (g). 2.5% $(COOH)_2$/EtOH; (h) $NaBH_4$-Py.; (i) CH_3MgI; (j) *p*-TsOH; (k) Na-NH_3 (l) Jone's Oxidation; (m) CH_2N_2.

Scheme 14.12 The synthesis of artemisinic acid by Zhou.

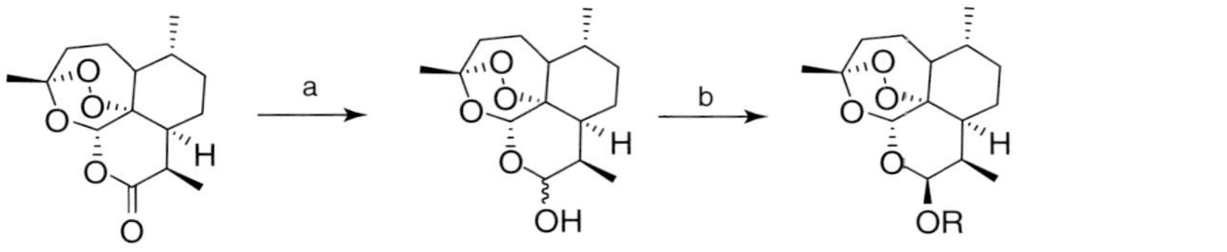

52 Qinghaosu

73 Dihydroartemisinin

Artemether (**74a**, 59.5%): R=-Me;
Arteether (**74b**, 64.15%): R=-Et.

Conditions: (a) $NaBH_4$; (b) $BF_3 \cdot Et_2O$, ROH.

Scheme 14.13 The synthesis of dihydroartemisinin and its derivatives.

Artemisinin (**52**) can be reduced to dihydroartemisinin (**73**), with approximately 10 times efficacy than that of the parent compound *in vitro*. The semisynthesis of the dihydroartemisinin was first achieved by Chinese chemist Chou and coworkers [46]. Li and co-workers went on the synthesis of a number of ether derivatives of dihydroartemisinin such as artemether **74a**, arteether **74b**, and other types of derivatives while searching for more effective and soluble drugs (Scheme 14.13) [47].

Many dihydroartemisinin derivatives were developed by Li and Brossi [47, 48], as shown in Figure 14.7 (**72–76**). One such compound – artemether [49] (Figure 14.7) – has 5–14 times higher antimalarial activities as compared to artemisinin. Artemether also has the advantage of having better lipid solubility, high stability, and high efficiency. It was also observed that artemether has a short drug effect interval and high recrudescence rate.

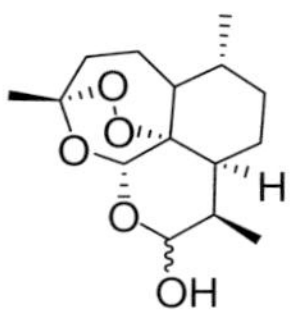

72 Dihydroartemisinin (OH)

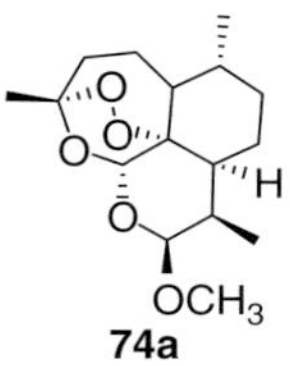

74a Artemether (OCH$_3$)

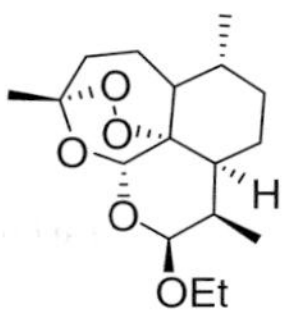

74b Arteether (OEt)

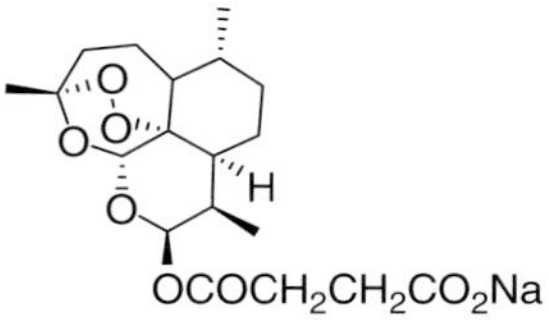

75 Sodium artesunate (OCOCH$_2$CH$_2$CO$_2$Na)

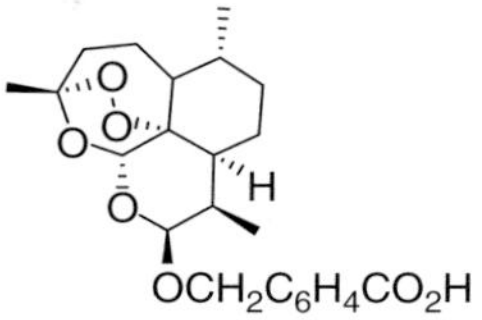

76 Artelinic acid (OCH$_2$C$_6$H$_4$CO$_2$H)

Figure 14.7 A series of derivatives of artemisinin.

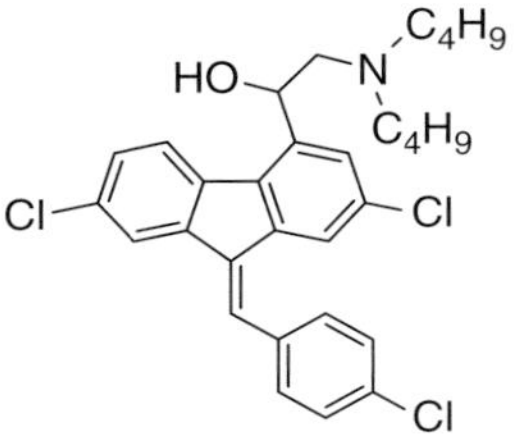

Figure 14.8 The structure of benflumetol/lumefantrine.

77 Benflumetol/lumefantrine

From more than 300 newly synthesized compounds [50], Deng and coworkers from the Institute of Microbiology and Epidemiology, Academy of Military Medical Sciences in China, identified a promising compound: benflumetol/lumefantrine (Figure 14.8, **77**). The drug has a low recrudescence rate (<5%) and high efficiency (>95%) in eliminating *Plasmodium*. Disadvantage includes slow onset of action and development of resistance.

The synthesis of the benflumetol was shown in Scheme 14.14 [51]. Here, the 9*H*-fluorene (**78**) was treated with FeCl$_3$ in AcOH to give the dihalogenated 2,7-dichloro-9*H*-fluorene (**79**), which was then followed by Friedel–Crafts reaction and reduction with NaBH$_4$ to deliver epoxidation product 4-oxiranyl-2,7-dichloro-9*H*-fluorene (**81**). Finally, benflumetol was obtained via ring-opening reaction and condensation with **82** as intermediate. The overall yield was up to 19%.

It was obvious that both artemether and benflumetol hold their advantages and shortcomings while having complementary characters. What results can we expect if we use the drug combination of artemether with benflumetol? Zhou mixed artemether and benflumetol together and carried out a number of

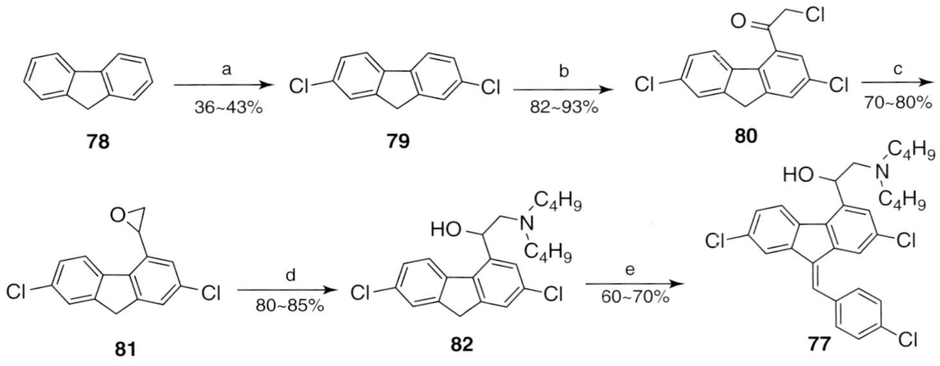

Conditions: (a) FeCl$_3$/CH$_3$CO$_2$H, Cl$_2$; (b) ClCH$_2$COCl, AlCl$_3$/CH$_2$ClCH$_2$Cl; (c) KBH$_4$, EtOH; (d) NH(C$_4$H$_9$)$_2$, EtOH; (e) p-ClC$_6$H$_4$-CHO, CaCl$_2$, NaOH, EtOH.

Scheme 14.14 The five step synthesis of benflumetol/lumefantrine14.

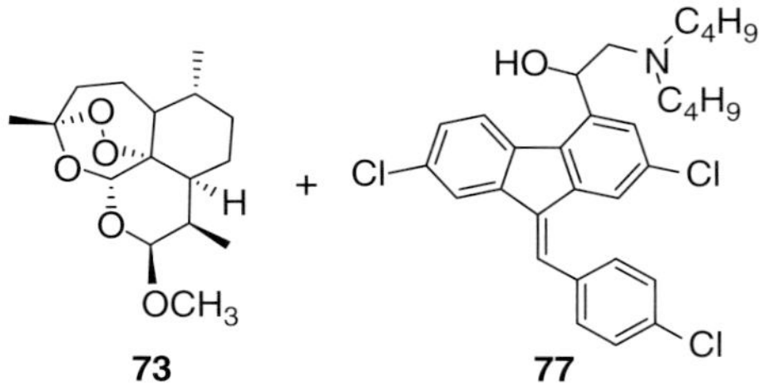

Figure 14.9 The fixed-dose Riamet/Coartem.

experiments to examine the efficiency of different combinations. A high efficient fixed-dose came out with the trade name "Riamet/Coartem" (Figure 14.9). This new fixed-dose presents the merits of the two components while avoiding their individual limitations. Meanwhile, the recrudescence to drug was largely lowered [52].

Coartem facilitates compliance and supports optimal clinical efficiency. It was codeveloped by Novartis in collaboration with Chinese partners, and it is currently registered in 79 countries world wide. Coartem has been extensively studied in multicenter clinical trials involving more than 3000 patients. Under an agreement with the World Health Organization (WHO), Novartis has supplied Coartem on a nonprofit basis to health authorities in malaria-endemic developing nations since 2001. Production of coartem, the leading artemisinin-based combination therapy, has increased from 100 000 treatments in 2002 to up to 65 million in 2006.

14.4
Process Chemistry for L-Ascorbic Acid and Biotin

14.4.1
Two Steps Fermentation Method for the Preparation of L-Ascorbic Acid (Vitamin C)

Vitamin C, also known as L-*ascorbic acid* (Figure 14.10, **83**), is a water-soluble vitamin. It was first isolated in 1928 by Nobel Prize winner Szent-Gÿorgyi [53], a Hungarian biochemist. The structure was determined by Hirst *et al.* five years later [54]. Vitamin C is needed for the formation of collagen to hold the cells together and for healthy teeth, gums, and cardiovascular systems; it improves iron absorption and resistance to infection [55].

Vitamin C can be synthesized by higher plants, and nearly all higher animals, as well as a number of yeasts [56]. However, humans and a few other mammals do not have the ability to make their own vitamin C. It is found in many fresh vegetables and fruits, such as lemon, cabbage, pineapple, strawberry, and most citrus fruits. The following table shows the relative abundance in different raw plant sources (Table 14.1) [57].

The global demand for vitamin C is estimated to be at 110–120 thousand tons in 2008. Approximately 50% is used in the food industry where it is added during processing to prevent pigment discoloration and enzymatic browning, prevent loss of flavor and aroma, protect or enhance nutrient content, and extend shelf-life [56]. Thirty percent is used in the pharmaceutical industry for either medical applications

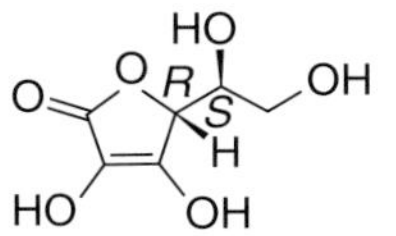

Figure 14.10 The molecular formula of L-ascorbic acid (vitamin C).

L-Ascorbic acid (vitamin C)

Table 14.1 The relative abundance of vitamin C in different raw plant sources.

Entry	Fruit	Amount (mg/100 g)	Entry	Fruit	Amount (mg/100 g)
1	Apricot	10	11	Loquat	1
2	Banana	9	12	Lychee	72
3	Grapefruit	34	13	Mango	28
4	Lemon juice	46	14	Orange	53
5	Lime juice	29	15	Peach	7
6	Longan	84	16	Pear	4
7	Tomato	19	17	Pineapple	15
8	Watermelon	10	18	Strawberry	57
9	Spinach	30	19	Potato	20
10	Carrot	9	20	Cucumber	3

or vitamin supplements. The remainder is used in animal feeds (13%), the cosmetic industry (5%), and other industrial processes (2%) [58].

China is the largest vitamin C exporting country in the world, producing 100 thousand tons per year. Industrial synthesis of vitamin C is accomplished using two major methods. One method starts from D-glucose via the seven steps of the Reichstein process established in 1933 and is still currently employed by Roche, BASF, and Takeda [59]. The process involves six chemical steps and one fermentation step (Scheme 14.15). First, D-glucose (84) is converted into D-sorbitol (85) via Ni-catalyzed hydrogenation under high H_2 pressure. Then, the fermentation step is applied for the oxidation of 85 to L-sorbose (86) with *Sorbose bakteriums*. L-Sorbose (86) is transformed into di-acetone-ketogulonic acid (DAKS, 88) in a two-stage chemical process. The first step involves a reaction with acetone. This produces di-acetone sorbose (87), which is then oxidized using chlorine and sodium hydroxide to produce DAKS. The key intermediate 2-keto-L-gluconic acid (2-KLG, 89) is obtained via hydrolysis of 88. To obtain the lactone vitamin C 83, the acid-catalyzed intramolecular esterification 89 is carried out with a strong acid such as sulfuric acid or hydrochloric acid. At the moment, with the acquired improvements, the practical yield of the Reichstein method is estimated to be around 50%.

The currently used method is highly energy consuming and requires high temperatures and/or pressures for many steps. In addition, most of the chemical transformations involve considerable quantities of organic and inorganic solvents and reagents such as acetone, sulfuric acid, and sodium hydroxide. Although some of the compounds can be recycled, stringent environmental control is required, resulting in significant waste disposal costs. These and other economic and environmental disadvantages prompted synthetic chemists to pursue microbial biotransformations in the manufacturing of L-ascorbic acid [60]. Significant achievements were made in this field: among them, Chinese Yin's two-step fermentation

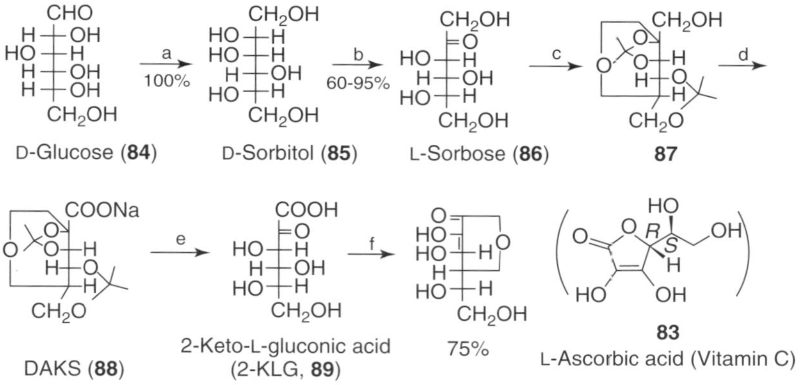

Conditons: (a) Ni/H_2(80–125 atm); (b) Sorbose-bakteriums/[O]; (c) CH_3COCH_3, H_2SO_4, SO_3; (d) NaOCl, $NiSO_4$; (e) H_3O; (f) Concentrated HCl.

Scheme 14.15 The Reichstein process for the synthesis of vitamin C.

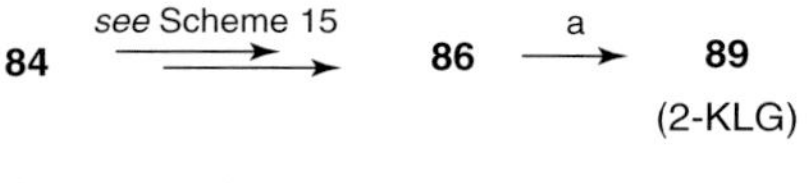

84 $\xrightarrow{\text{see Scheme 15}}$ 86 $\xrightarrow{a}$ 89
(2-KLG)

Conditions: (a) *Pseudomonas striata*/[O]

Scheme 14.16 Yin's two-step fermentation method for the synthesis 2-KLG.

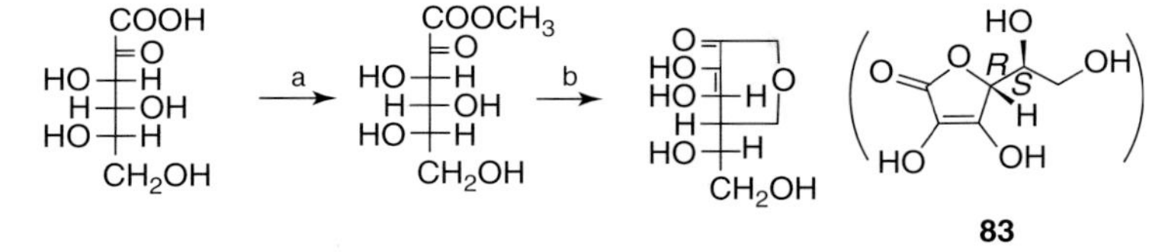

2-Keto-L-gluconic acid (2-KLG, **89**) → Methyl 2-keto-L-gluconate (2-KLGM, **90**) → Overall yield 92.6% → L-Ascorbic acid (vitamin C) **83**

Conditions: (a) CH_3OH, H_2SO_4; (b) $NaHCO_3$, H_2SO_4.

Scheme 14.17 The transformation of 2-KLG into vitamin C via the basic method.

process has been proved to be the most practical and has been adopted quickly in industrial-scale production in China as well as other countries [61].

As shown in the Scheme 14.16, the first two steps are similar to those of the Reichstein process. The most ingenious step was the direct biosynthetic oxidation of L-sorbose to 2-KLG (**89**), performed by *Pseudomonas striata* and *Gluconobacter oxydans*. This one-step fermentation process replaced the four chemical steps in the Reichstein process with mild reaction conditions, high yield (about 80%), and low environmental impact.

Generally, the transformation of 2-KLG into L-ascorbic acid could be conducted via two processes, acid transformation and basic transformation (Scheme 14.17). Compared with acid method used in the Reichstein process, basic transformation here has higher yield and lower corrosion to reaction equipment.

In summary, the Reichstein process in vitamin C production has come to the end of its reign. A two-stage fermentation technology for the production of 2-KLG represents a cleaner and more efficient method with overall production cost savings of about a third as compared to the Reichstein process. In the two-stage fermentation process, a second fermentation step replaces the chemical reactions used to produce DAKS (**88**) in the Reichstein method. Compared to the Reichstein process, the two-stage fermentation process also used less toxic solvents and reagents.

Many patents for the production of vitamin C by different routes have been published. There have been two recent foci of research and development work [60, 62]. The first focuses on fermentation to transform glucose into 2-KLG, either directly or via sorbitol, sorbose, or diketogluconic acid. These methods include sorbital pathway, L-idonic acid pathway, L-gulonic acid pathway, 2-keto-D-gluconic acid pathway, and 2,5-diketo-D-gluconic acid pathway. The second area of research is direct conversion of glucose into vitamin C by fermentation using mutant microalgae. That is the so-called 2-keto-L-gluconic acid pathway.

Table 14.2 The main manufacturers of vitamin C in the world.

Entry	Company	Output/tons (× 1000)[a]	Country
1	Shijiazhuang Pharmaceutical Group Co., Ltd	∼30.0	China
2	Northeast Pharmaceutical Group Co., Ltd	∼25.0	China
3	Jiangsu Jiangshan Pharmaceutical Co., Ltd	∼20.0	China
4	Northchina Pharmaceutical Group Co., Ltd	∼20.0	China
5	Shandong Luwei Pharmaceutical Co., Ltd	∼20.0	China

[a]The data were gained from the Internet.

Table 14.2 shows the output of major vitamin C manufacturers of the world. The top five account for more than 90% of the production. China is by far the largest vitamin C manufacturer and the largest exporter in the world.

14.4.2
Total Synthesis of Biotin (Vitamin B7)

Biotin (also known as *vitamin B7 or H*, Figure 14.11, **91**) is a water-soluble B-complex vitamin. Biotin was discovered in 1901, and isolated from egg yolk in 1936 [63]. The molecular formula and configuration of biotin were identified in 1941 and 1942, respectively by du Vigneaud and coworkers [64].

Biotin is a cofactor in the metabolism of fatty acids and leucine, and plays a key role in the growth and maintenance of hair, nails, and bone. Hair loss and brittle nails have been correlated with biotin deficiency, and these symptoms are often alleviated when optimal biotin levels are achieved. The human body can make its own biotin. Biotin is also found in many foods, including oatmeal, vegetables, peanuts, mushrooms, egg yolks, rice, nuts, spinach, potatoes, poultry, and beef.

The first total synthesis of biotin was achieved by Harris in 1943 [65] and a new synthetic route was reported by Baker four years later [66]. However, neither approach could be applied in industrial production. In 1947, Goldberg and Sternbach, assignors to Hoffmann–La Roche Inc., completed total synthesis with fumaric acid (**92**) as the starting material, and 15 step reactions. This milestone work enabled Roche to become the first biotin producer in the world (Scheme 14.18) [67].

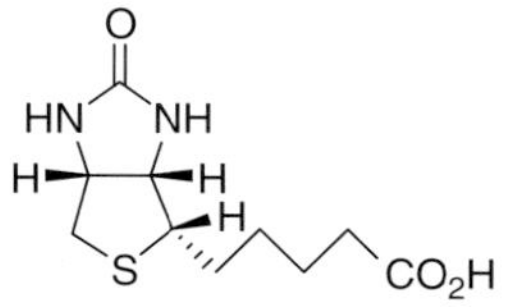

91
Biotin (vitamin B7)

Figure 14.11 The molecular formula of biotin (vitamin B7).

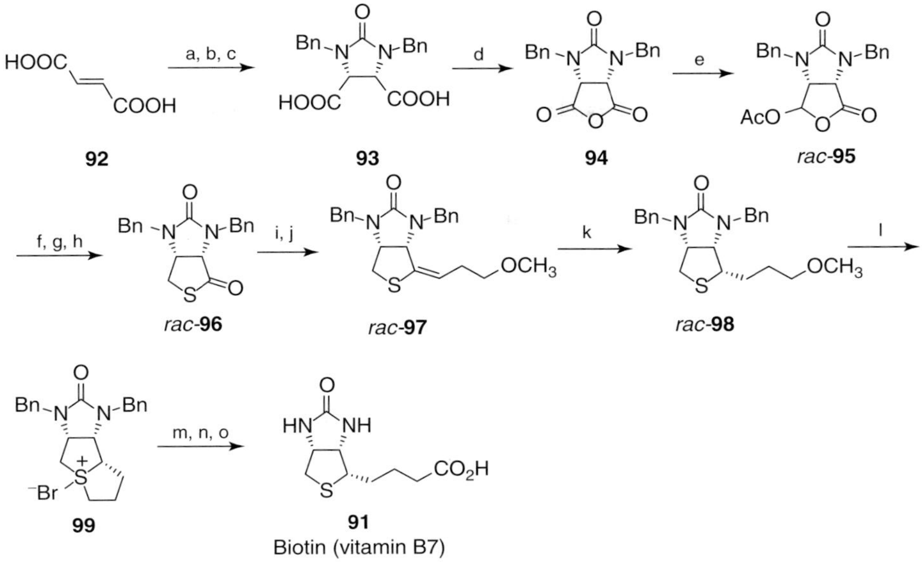

Scheme 14.18 The original Goldberg–Sternbach approach for the synthesis of biotin (vitamin B7).

Conditions: (a) Br$_2$; (b) PhCH$_2$NH$_2$, EtOH; (c) COCl$_2$, KOH; (d) Ac$_2$O; (e) Zn, Ac$_2$O, HOAc; (f) H$_2$S, HCl; (g) KHS, EtOH; (h) Zn, HOAc; (i) ClMg(CH$_2$)$_3$OCH$_3$; (j) HOAc; (k) H$_2$,PdO; (l) HBr; (m) Silver d-camphorsulfonate, followed by fractional crystallization; (n) NaCH(COOEt)$_2$; (o) 48% HBr.

Many researchers have attempted to improve the so-called Goldberg–Sternbach route [68]. In 1970, Gerecke, Zimmerman, and Aschwanden from Hoffmann–La Roche (Basel and Paris) reported an important modification to the original Goldberg–Sternbach scheme allowing for the efficient production of the key thiolactone **96** in the required enantiomeric starting from *meso*-acid **93** or *meso*-anhydride **94** [69]. Crucial to this new development was the finding that lactone **101** could be converted in very high yields (>90%) into the corresponding thiolactone **96** with retention of its configuration (Scheme 14.19).

Another improvement involved the formation of hemiester **100a** from *meso*-anhydride **94** (Scheme 14.19). The anhydride **94** is converted to a racemic mixture of hemiesters *rac*-**100a** after treatment with cyclohexanol, and followed by chiral resolution with suitable resolving agents to give the optically pure (+)-**100a** [70]. The improved synthetic route had the advantages of shorter steps (14 steps instead of 15 steps) as well as higher yield.

While the Goldberg–Sternbach approach has been thoroughly optimized for many years, it has some fundamental drawbacks: (i) it is still a multistep synthesis (14 steps) with the use of toxic reagents and intermediates and (ii) it involves theoretically impractical diastereomeric or enzymatic resolution of racemic materials. These obstacles prompted synthetic chemists to develop more efficient ways to

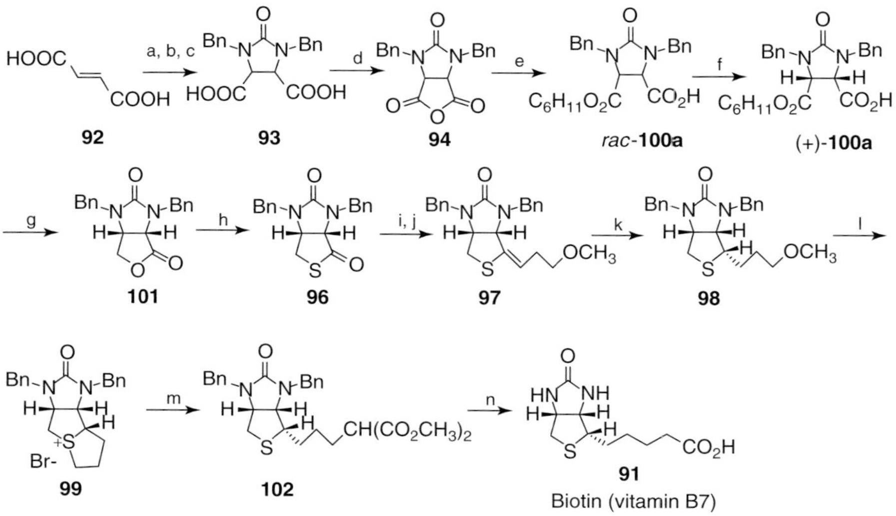

Scheme 14.19 The Improved Goldberg–Sternbach approach for the synthesis of biotin (vitamin B7).

Conditions: (a) Br_2; (b) $PhCH_2NH_2$, EtOH; (c) $COCl_2$, KOH; (d) Ac_2O, AcOH; (e) $C_6H_{11}OH$; (f) (+)-Ephedrine; optical resolution; (g) $LiBH_4$; (h) AcSK; (i) $CH_3O(CH_2)_3MgCl$; (j) H_3O; (k) H_2–Ni; (l) HBr; (m) $CH_2(CO_2CH_3)_2$; (n) 48% HBr.

obtain biotin [68]. In my view, Chen's 10-step synthetic method [71], as shown in Scheme 14.20, represents a major breakthrough.

The first improvement is the synthesis of hemiester (+)-**100b** during the preparation of the key intermediate lactone **101**. In Chen's 10-step synthetic method, **94** was directly converted into enantiomer-rich (+)-**100b** in 95% yield via asymmetric desymmetrization using chiral amine reagents as asymmetric catalyst. Then (+)-**100b** was reduced by KBH_4 followed by cyclization to give lactone **101** in high yield (90%) with excellent enantioselectivity (98.5% ee, Scheme 14.21, Eq. (1)) [7b,g]. However, two steps, including the chiral resolution process, were adopted in the above improved Goldberg–Sternbach approach (Eq. (2)). Further, 50% of the hemiester was discarded during the chiral resolution process.

The second improvement is the introduction of the carboxylbutyl side chain to the intermediate thiolactone. By the reaction of thiolactone **96** with an alkyldimagnesium dibromide reagent, derived from 1,4-dibromobutane, with subsequent treatment of carbon dioxide, gives the hydroxy acid **103**. Without purification, **103** was then subject to catalytic hydrogenation under 40 atm of hydrogen using 5% Pd/C in the presence of anhydrous $ZnCl_2$ in toluene at 120 °C providing N,N-dibenzylbiotin **104** stereoselectively in 90% yield (Scheme 14.22, Eq. (3)). However, up to seven step reactions were required in the Goldberg–Sternbach method (Eq. (4)).

The efficiency and practicality of Chen's method were subsequently applied in large-scale industrial production. Zhejiang Medicine Co., Ltd., took the lead

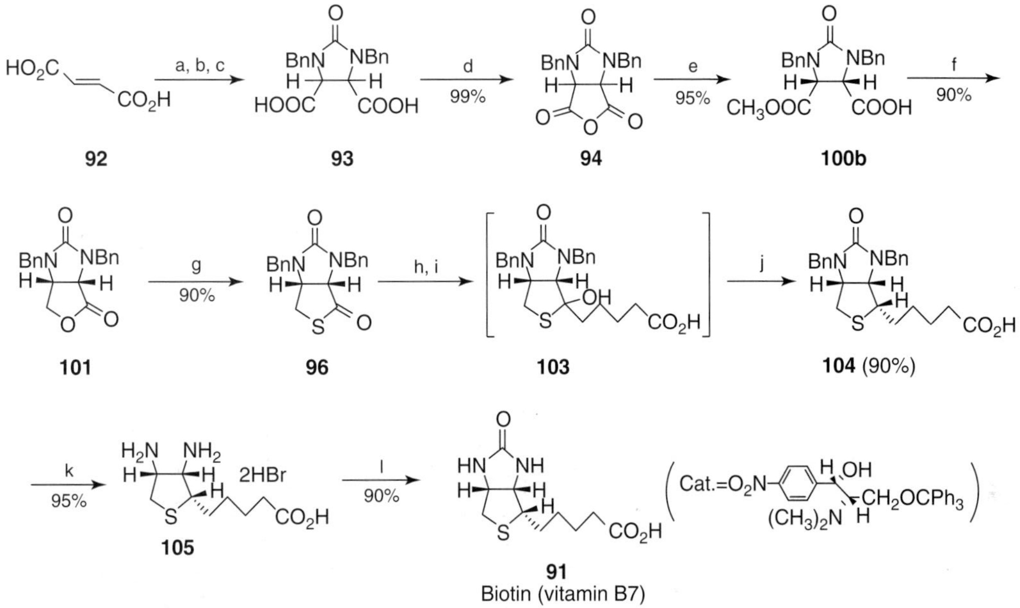

Scheme 14.20 The Fener Chen's 10-step synthetic method for biotin (vitamin B7).

Conditions: (a) Br$_2$; (b) PhCH$_2$NH$_2$, EtOH; (C) COCl$_2$, KOH; (d) Ac$_2$O, Toluene 130 °C, 5 h; (e) CH$_3$OH, (1S,2S)-2-(dimethylamineo)-1-(4-nitrophenyl)-3-(trityloxy)propan-1-ol (Cat.) THF, -15~-10 °C, 36 h; (f) KBH$_4$, CaCl$_2$, Et$_2$O. rt, 18 h; (g) EtSCS$_2$K/DMF, 125 °C, 5 h; (h) Br(CH$_2$)$_4$Br, Mg; (i) CO$_2$, MTBE; (j) 5% Pd-C/H$_2$ (75–80 Kg/cm^{-2}), ZnCl$_2$, Toluene, 120 °C, 3 h; (k) 45% HBr/Xylene, 130 °C, 45 h; (l) Cl$_3$COCOOCCl$_3$, 10% NaOH, 1,4-Dioxane, rt, 10 h.

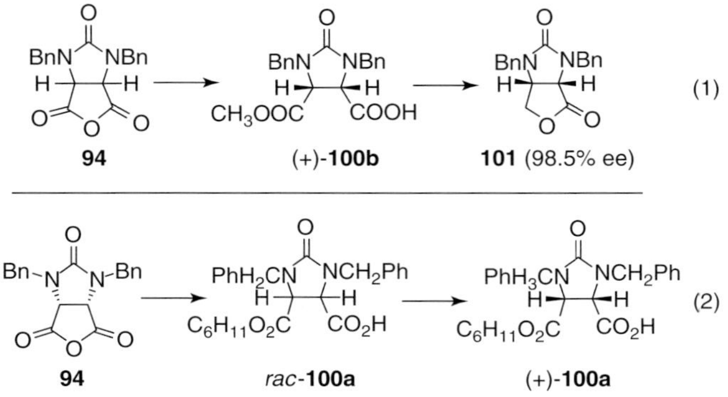

Scheme 14.21 The first improvement of Fener Chen's synthetic method.

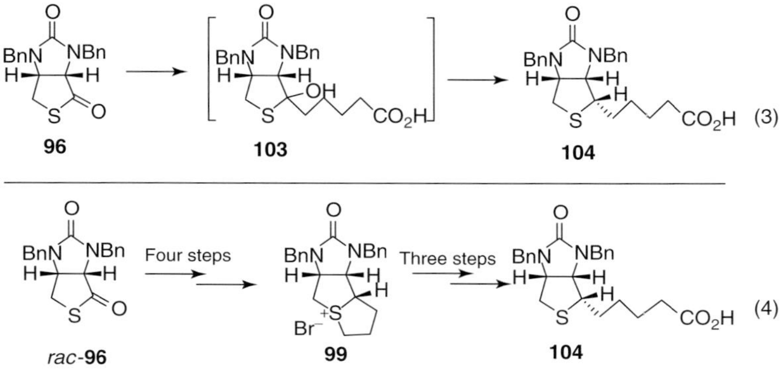

Scheme 14.22 Another improvement of Fener Chen's synthetic method 14.

in producing biotin by utilizing Chen's method. As of now, three companies in China account for about 80% overall biotin production of the world. They are Zhejiang Medicine, Zhejiang Shengda Pharmaceutical Co., Ltd, and Zhejiang NHU Company Ltd (NHU).

14.5
Conclusion and Perspectives

In 2008, the output of the Chinese pharmaceutical industry was RMB866.68 billion, an increase of 25.7% compared with former years. The output value of China's pharmaceutical industry is expected to top RMB1 trillion in 2009, ranking seventh globally, according to a recent report released by the Chinese Pharmaceutical Enterprises Association. China is forecasted to become the world's fifth-largest pharmaceutical market by 2010 and the world's largest pharmaceutical market by 2050.

However, the industry is still small-scale with a scattered geographical layout, duplicated production processes, and outdated manufacturing technology and management structure. The Chinese pharmaceutical industry also has a lower market concentration and weak international trading competitiveness, coupled with lack of patented pharmaceuticals developed in-house.

We remain confident about the prospects for China's pharmaceutical industry despite the challenging global and domestic environment. With the development of science and technology, China has made great progress in natural science, especially in pharmaceutical chemistry. More and more new drugs have been developed and China will keep pace with developed countries such as North American and Europe. On the basis of the predominance of CHM, great progress will be continued to be made in the organic synthesis of CHMs such as the total synthesis of qinghaosu/artemisinin. Meanwhile, various new pharmaceuticals derived from CHMs will also be developed in the nondistant future. Furthermore, these developments will give impetus to process chemistry of Chinese pharmaceuticals.

Abbreviations

Bicyclol	4,4'-dimethoxy 2,3,2',3'-dimethylene dioxy 6-hydroxymethyl 6'-carbonyl biphenyl
BUPF	Beijing Union Pharmaceutical Factory
CHB	chronic viral hepatitis B
CHC	chronic viral hepatitis C
CHM	Chinese herbal medicine
Cy	cyclohexyl
DDB	dimethyl-4,4'-dimethoxy-5,6,5',6'-dimethenedioxy-biphenyl-2,2'-dicarboxylate
DMF	dimethylformamide
DAKS	di-acetone-ketogulonic acid
GDP	gross domestic product
HBV	hepatitis B virus
HCC	hepatocellular carcinoma
2-KLG	2-keto-L-gluconic acid
Ms	methylsulfonyl
MOM	methoxymethyl
MTBE	methyl t-butyl ether
NEGPF	northeast general pharmaceutical factory
NEPG	the core of northeast pharmaceutical group corporation
SFDA	state food and drug administration
THP	D,L-tetrahydropalmatine
WTO	world trade organization

References

1. (a) Iwasa, K., Kam imauchi, M., and Ueki, M. (1996) *Eur. J. Med. Chem.*, **31**, 469; (b) Iwasa, K. and Kam imauchi, M. (1997) *Planta. Med.*, **63**, 196; (c) Iwasa, K., Nanba, H., and Lee, D.-U. (1998) *Planta. Med.*, **64**, 748.
2. Anonymous (2000) *Berberine Altern. Med. Rev.*, **5**, 175.

3. Verma, R.L. (1933) *Ind. Med. Gaz.*, **68**, 122.
4. Birdsall, T.C. and Kelly, G.S. (1997) *Altern. Med. Rev.*, **2**, 94.
5. Miyazaki, S., Oshiba, M., and Nadai, T. (1981) *Chem. Pharm. Bull.*, **29**, 883, The literatures cited.
6. Yang, X. and Zhang, C. (2006) *Chin. J. Med. Guide.*, **10**, 107.
7. (a) Lin, W. (1981) *Contem. Chem. Ind.*, 16; (b) Chen, S., Lin, W., Zhang, J., Liu, J., Liu, S., Zhang, L., and Song, J. (2001) CN Patent 1, 312,250.
8. Chen, Z.X., Tao, F., Xia, Z.J., and Ma, S.M. (2008) CN Patent 1, 012,450,64.
9. Pharmacopoeia Commission of RPC (2005) *Chinese Pharmacopoeia, Part I*, Chemical Industry Press, Beijing, p. 94.
10. (a) Chou, T.Q. (1928) *Chin. J. Physiol.*, **2**, 201; (b) Chou, T.Q. (1929) *Chinese J. Physiol.*, **3**, 301; (c) Chou, T.Q. (1933) *Chin. J. Physiol.*, **7**, 35; (d) Chou, T.Q. (1934) *Chin. J. Physiol.*, **8**, 155; (e) Chou, T.Q. (1936) *Chin. J. Physiol.*, **10**, 507.
11. (a) Hong, Z., Fan, G., Chai, Y., Yin, X., and Wu, Y. (2005) *Chirality*, **17**, 293; (b) Hong, Z.Y., Fan, G.R., Chai, Y.F., Yin, X.P., and Wu, Y.T. (2005) *J. Chromatogr. B*, **826**, 108; (c) Zhai, Z.D., Shi, Y.P., Wu, X.M., and Luo, X.P. (2006) *Anal. Bioanal. Chem.*, **384**, 939.
12. Jin, G.Z. (1987) *Acta Pharm. Sinica*, **22**, 472.
13. Guangzhou Pharmaceutical Industrial Research Institute (1979) *Pharm. Ind.*, **1**, 1.
14. Cheng, B., Ru, R., Sun, K., Xie, F., Wang, S., Li, J., Wei, E., Miao, C., and Huang, W. (1993) CN Patent 1, 068,113.
15. Ferlay, J.B.F. and Pisani, P.G. (2001) *Cancer Incidence, Mortality and Prevalence Worldwide*, IARC Press, Lyon, 2001.
16. Tang, Z.Y. (2000) *J. Gastroenterol Hepatol.*, **15**, 1.
17. (a) Fong, Y., Sun, R.L., Jarnagin, W., and Blumgart, L.H. (1999) *Ann. Surg.*, **229**, 790; (b) El-Serag, H.B. (2004) *Gastroenterol*, **127**, 27; (c) Decaens, T., Roudot, T.F., Bresson, H.S., Meyer, C., Gugenheim, J., Durand, F., Bernard, P.H., Boillot, O., Boudjema, K., Calmus, Y., Hardwigsen, J., Ducerf, C., Pageaux, G.P., Dharancy, S., Chazouilleres, O., Dhumeaux, D., Cherqui, D., and Duvoux, C. (2005) *Liver Transpl.*, **11**, 767; (d) Wands, J.R. (2004) *N. Engl. J. Med.*, **351**, 1567; (e) Lodato, F., Festi, G., Mazzella, D., Azzaroli, F., Colecchia, A., and Roda, E. (2006) *World J. Gastroenterol.*, **12**, 7239.
18. Liu, G.T. (1989) *Chin. Med. J.*, **102**, 740.
19. Gottlieb, O.R. (1977) *New Products and Plant Drug with Pharmacological, Biological or Therapeutical Activity*, Springer-Verlag, Berlin, Herdelberg, p. 227.
20. (a) Xie, J.X., Zhou, J., Zhang, C.Z., Yang, J.H., and Chen, J.X. (1981) *Kexue Tongbao*, **28**, 430; (b) Xie, J.X., Zhou, J., Zhang, C.Z., Yang, J.H., Jin, H.Q., and Chen, J.X. (1982) *Yaoxue Xuebao*, **17**, 23; (c) Xie, J.X., Zhou, J., and Zhang, C.Z. (1983) *Sci. Sin., Ser. B*, **26**, 1291; (d) Guo, R.Y., Chang, J.B., Chen, R.F., Xie, J.X., and Yan, D.Y. (2001) *Chin. Chem. Lett.*, **12**, 491.
21. (a) Xie, J.X., Zhou, J., Zhang, C.Z., Yang, J.H., and Chen, J.X. (1981) *Acta Pharmacol. Sin.*, **16**, 306; (b) Ko, R.K.M. and Mak, D.H.F. (2004) in *Modern Herbal Medicine: Molecular Basis in Health and Disease Management* (eds L. Packer, B. Halliwell, and C.N. Ong), Marcel Dekker, New York, p. 289.
22. Lee, K.H., Kashiwada, Y., Xie, L., Cosentino, L.M., Manak, M., Xie, J.X., Cheng, Y.C., and Kilkulskie, R. (1997) US Patent 5, 612,341.
23. (a) Cui, S., Wang, M., and Fan, G. (2002) *Chin. Med. J.*, **82**, 538; (b) Guan, L.P., Nan, J.X., Jin, X.J., Jin, Q.H., Kwak, K.C., Chai, K.Y., and Quan, Z.S. (2005) *Arch. Pharm. Res.*, **28**, 81.
24. Pan, S.Y., Yang, R., Dong, H., Yu, Z.L., and Ko, K.M. (2006) *Eur. J. Pharmacol.*, **552**, 170.
25. Zhan, S.L. and Zhang, C.Z. (1992) *Chin. Chem. Lett.*, **3**, 29.
26. Zheng, D. (1993) CN Patent 1, 073,438.
27. (a) Zhang, C.Z. (1987) *Kexue Tongbao*, **32**, 72; (b) Yeng, F.W., Chiu, J.Y., and Wang, C.L.J. (1996) US Patent 5, 504,221.
28. Purcell, R.H. (1993) *Gastroenterology*, **104**, 955.
29. Lok, A.S.F. and McMahon, B.J. (2001) *Hepatology*, **34**, 1225.

30. (a) Liu, G.T. (2001) *Chin. J. New Drugs*, **10**, 325; (b) Yao, G., Ji, Y., Wang, Q., Zhou, X., Xu, D., Chen, X., and Zhang, Q. (2002) *Chin. J. New Drugs Clin. Rem.*, **21**, 457.
31. (a) Wu, T., Xie, L., Liu, G.J., Hao, B., and Harrison, R.A. (2006) *Cochrane Database Syst. Rev.*; **4**, (b) Yang, X., Zhuo, Q., Wu, T., and Liu, G.J. (2007) *Cochrane Database Syst. Rev.* **2**,
32. (a) Wang, H. and Li, Y. (2006) *Eur. J. Pharmacol.*, **534**, 194; (b) Liu, G.T., Li, Y., Wei, H.L., Zhang, H., Xu, J.Y., and Yu, L.H. (2005) *Liver Int.*, **25**, 872; (c) Hu, Q.W. and Liu, G.T. (2006) *Life Sci.*, **79**, 606.
33. Liu, G.T. (2009) *Med. Chem.*, **5**, 29.
34. Liu, G.T., Zhang, C.Z., and Li, Y. (1999) *Natl. Med. J. China*, **14**, 51.
35. Gu, S.-J. (2004) EP Patent 0, 353,358.
36. Coatney, G.R. (1963) *Am. J. Trop. Med. Hyg.*, **12**, 121.
37. (a) Olliaro, P.L. and Yuthavong, Y. (1999) *Pharmacol. Ther.*, **81**, 91; (b) Frédérich, M., Dogné, J.-M., Angenot, L., and De Mol, P. (2002) *Curr. Med. Chem.*, **9**, 1435; (c) Rosenthal, P.J. (2003) *J. Exp. Biol.*, **206**, 3735; (c) Gelb, M.H. (2007) *Curr. Opin. Chem. Biol.*, **11**, 440.
38. Liu, J.-M., Ni, M.-Y., Fan, Y.-F., Tu, Y.-Y., Wu, Z.-H., Wu, Y.-L., and Chou, W.-S. (1979) *Acta Chim. Sin.*, **37**, 129.
39. (a) Klayman, D.L. (1985) *Science (Washington, D.C.)*, **228**, 1049; (b) Luo, X.D. and Shen, C.C. (1987) *Med. Rec. Rev.*, **7**, 29.
40. Schmid, G. and Hofheina, W. (1983) *J. Am. Chem. Soc.*, **105**, 624.
41. Xiu, X.-X., Zhu, J., Huang, D.-Z., and Chou, W.-S. (1983) *Acta Chim. Sin.*, **41**, 574.
42. (a) Xiu, X.-X., Zhu, J., Huang, D.-Z., and Chou, W.-S. (1984) *Acta Chim. Sin.*, **42**, 940; (b) Xiu, X.-X., Zhu, J., Huang, D.-Z., and Chou, W.-S. (1986) *Tetrahedron*, **42**, 819.
43. (a) Avery, M.A., Jennings-White, C., and Chong, W.K.M. (1987) *Tetrahedron Lett.*, **28**, 4629; (b) Ravindra-nathan, T., Kumar, M.A., Menon, R.B., and Hiremath, S.V. (1990) *Tetrahedron Lett.*, **31**, 755; (c) Liu, H.-J., Yeh, W.-L., and Chew, S.Y. (1993) *Tetrahedron Lett.*, **34**, 4435.
44. (a) Zhang, Y.Z. (1981) *Yaoxue Tongbao*, **16**, 197; (b) Luo, X.D. and Shen, C.C. (1987) *Med. Res. Rev.*, **7**, 29.
45. China Cooperative Research Group on Qinghaosu and its Derivative Antimalarials (1982) *J. Trad. Chin. Med.*, **2**, 9.
46. Liu, J.-M., Ni, M.-Y., Fan, J.-F., Tu, Y.-Y., Wu, Z.-H., Wu, Y.-L., and Chou, W.-S. (1979) *Acta Chim. Sin.*, **37**, 129.
47. (a) Li, Y., Yu, P.-U., Chen, Y.-X., Li, L.-Q., Gai, Y.-Z., Wang, D.-S., and Zhang, Y.-P. (1979) *Yaoxue Tongbao*, **24**, 667; (b) Li, Y., Yu, P.-U., Chen, Y.-X., Li, L.-Q., Gai, Y.-Z., Wang, D.-S., and Zhang, Y.-P. (1981) *Acta Pharm. Sin.*, **16**, 429.
48. Brossi, A., Venugopalan, B., Domingueg, G.L., Yeh, H.J.C., Flippend-Anderson, J.L., Buchs, P., Luo, X.D., Milhous, W., and Peters, W. (1988) *J. Med. Chem.*, **31**, 646.
49. (a) Yang, Y.-H., Li, Y., Shi, Y.-L., Yang, J.-D., and Wu, B.-A. (1995) *Bioorg. Med. Chem. Lett.*, **5**, 1791; (b) Posner, G.H., Ploypradith, P., Parker, M.H., O'Dowd, H., Woo, S.-H., Northrop, J., Krasavin, M., Dolan, P., Kensler, T.W., Xie, S., and Shapiro, T.A. (1999) *J. Med. Chem.*, **42**, 4275; (c) Avery, M.A., Alvim-Gaston, M., Vroman, J.A., Wu, B., Ager, A., Peters, W., Robinson, B.L., and Charman, W. (2002) *J. Med. Chem.*, **45**, 4321; (d) Singh, C. and Tiwari, P. (2002) *Tetrahedron Lett.*, **43**, 7235; (e) Kim, B.J. and Sasaki, T. (2004) *J. Org. Chem.*, **69**, 3242; (f) Boehm, M., Fuenfschilling, P.C., Krieger, M., Kuesters, E., and Struber, F. (2007) *Org. Process Res. Dev.*, **11**, 336.
50. Deng, R.X., Yu, L.B., Zhang, H.B., Geng, R.L., Ye, K.L., and Zhang, D.F. (1981) *Acta Pharma. Sin.*, **12**, 920.
51. Deng, R.X., Zhong, L.X., and Zhao, D.C. (1990) CN Patent 1, 042,535A.
52. Beutler, U., Fuenfschilling, P.C., and Steinkemper, A. (2007) *Org. Process Res. Dev.*, **11**, 341.
53. Szent-Györgyi, A. (1928) *Biochem. J.*, **22**, 1387.
54. (a) Herbert, R.W., Hirst, E.L., Percival, E.G.V., Reynolds, R.J.W., and Smith, F. (1933) *J. Chem. Soc.*, 1270. (b) Hirst, E.L.E., Percival, G.V., and Smith, F. (1933) *Nature*, **131** 617.

55. Banwell, M.G., Blakey, S., Harfoot, G., and LongmoreB, R.W. (1999) *Aust. J. Chem.*, **52**, 137.
56. (a) Heick, H.M.C., Gra, G.L.A., and Humpers, J.E.C. (1969) *Can. J. Microbiol.*, **18**, 597; (b) Asard, H., May, J.M., and Smirnoff, N. (2004) *Vitamin C. Functions and Biochemistry in Animals and Plants Bios.*, Scientific Publishers, London.
57. (a) Gould, S., Tressler, D.K., and King, C.G. (1936) *Food Res.*, **1**, 427; (b) Schatzlein, C. and Fox-Timmling, E. (1940) *Z. Unters. Lebensm.*, **79**, 157.
58. Marz, U. (2002) World Markets for Citric, Ascorbic, Isoascorbic Acids: Highlighting Antioxidants in Food. Report No. FOD031A, Business Communications Company Inc.
59. (a) Reichstein, T., Grussner, A., and Oppenauer, R. (1933) *Helv. Chim. Acta*, **16**, 1019; (b) Reichstein, T. and Grussner, A. (1934) *Helv. Chim. Acta*, **17**, 311; (c) Reichstein, T. (1935) GB Patent 4, 251,98; (d) Reichstein, T. (1935) GB Patent 4, 288,14; (e) Reichstein, T. (1935) GB Patent 427286.
60. (a) Zhang, L., Wang, Z., Xia, Y., Kai, G., Chen, W., and Tang, K. (2007) *Crit. Rev. Biotechnol.*, **27**, 173; (b) Hancock, R.D. (2009) *Recent Pat. Food, Nu. Agric.*, **1**, 39.
61. Yin, G., Tao, Z., Yu, L., Wang, D., Tan, J., Yan, Z., Nin, W., Wang, C., Wang, S., Jiang, H., Zhang, X., Feng, X., Zhao, Q., and Wei, W. (1980) *Acta Microbiol. Sin.*, **20**, 246.
62. (a) Hancock, R.D. and Viola, R. (2002) *Trends Biotechnol.*, **20**, 299; (b) Sauer, M., Branduardi, P., Valli, M., and Porro, D. (2004) *Appl. Environ. Microbiol.*, **70**, 6086.
63. Kogl, F., Tonnis, B., and Hoppe-Seyl, Z. (1936) *Physiol. Chem.*, **242**, 43.
64. (a) du Vigneaud, V., Hofmann, K., Melville, D.B., and Rachele, J.R. (1941) *J. Biol. Chem.*, **140**, 763; (b) du Vigneaud, V., Melville, D.B., Fokkers, K., Wolf, D.E., Mozingo, R., Keresztesy, J.C., and Harris, S.A. (1942) *J. Biol. Chem.*, **146**, 475; (c) Melville, D.B., Moyer, A.W., Hofmann, K., and du Vigneaud, V. (1942) *J. Biol. Chem.*, **146**, 487.
65. Harris, J.A., Wolf, D.E., and Fokkers, R. (1943) *Science*, **97**, 447.
66. Baker, B.R., Querry, M.V., Bernstein, S., Safir, S.R., and Subbarow, Y. (1947) *J. Org. Chem.*, **12**, 167.
67. (a) Goldberg, M.W. and Sternbach, L.H. (1949) US Patent 2, 489,232; (b) Goldberg, M.W. and Sternbach, L.H. (1949) US Patent 2, 489,235; (c) Goldberg, M.W. and Sternbach, L.H. (1949) US Patent 2, 489,238.
68. For a review, see: DeClercq, P.J. (1997) *Chem. Rev.*, **97**, 1755.
69. Gerecke, M., Zimmerman, J.-P., and Aschwanden, W. (1970) *Helv. Chim. Acta*, **53**, 991.
70. Reiff, F., Mueller, H.R., Hedinger, A., Herold, T., Casutt, M., Schweickert, N., Stoehr, J.S., and Kuhn, W. (1986) EP Patent 0, 173,185.
71. For patents: (a) Chen, F.E., Jia, H.Q., and Yan, M.G. (2002) CN Patent 1, 356,320; (b) Chen, F.E., Peng, X.H., and Sun, G.F. (2002) CN Patent 1, 358,725; (c) Chen, F.E. (2002) CN Patent 1, 374,312; (d) Chen, F.E. (2003) CN Patent 1, 434,039; (e) Chen, F.E. (2003) CN Patent 1, 443,766; (f) Chen, F.E. (2003) CN 1, 445,229; (g) Chen, F.E. (2004) CN Patent 1, 473,832; Selected Papers: (a) Chen, F.E., Fu, H., Meng, G., Luo, Y.F., and Yan, M.G. (2002) *Chem. J. Chin. Univ.*, **23**, 1060; (b) Chen, F.E., Yuan, J.L., Dai, H.F., Kuang, Y.Y., and Chu, Y. (2003) *Synthesis*, **14**, 2155; (c) Chen, F.E., Dai, H.F., Kuang, Y.Y., and Jia, H.Q. (2003) *Tetrahedron: Asymmetry*, **14**, 3667.

15
The Use of Continuous Processing to Make AZD 4407 Intermediates
Andrew S. Wells

15.1
Green Chemistry and the Drive for Sustainability

Good process chemistry and engineering has always been about making manufacturing processes more efficient, reducing the cost of goods, and reducing waste, alongside minimizing hazard and risks to operators and the environment. In the synthesis of pharmaceuticals and chemical products, this has been highlighted by Sheldon whose famous E factor (kilogram products per total kilogram material used) has been adopted as a means of quantifying increases in environmental performance [1, 2]. As we move forward in the twenty-first century with ever-increasing population, escalating demand, and decreasing finite resources, striving to maximize materials efficiency, removing materials of concern (ozone depletors, ecotoxic chemicals, etc.), and minimizing energy consumption in our manufacturing processes is ever more urgent. We can be guided by the 12 principles of green chemistry published by Anastas [3], refined by Winterton to include engineering [4], and combined and extended by Poliakoff [5]. The 12 principles are summarized below [3].

1) Prevent generation of waste.
2) Use safer reagents/solvents.
3) Design less hazardous materials
4) Use catalysis.
5) Use sustainable feedstocks.
6) Avoid derivatization.
7) Maximize atom economy.
8) Use safer solvents and reagents.
9) Minimize potential for accidents.
10) Use real-time analysis.
11) Increase energy efficiency.
12) Design more benign products.

Pharmaceutical Process Chemistry. Edited by Takayuki Shioiri, Kunisuke Izawa, and Toshiro Konoike
Copyright © 2011 WILEY-VCH Verlag GmbH & Co. KGaA, Weinheim
ISBN: 978-3-527-32650-1

Of course, design of a synthetic sequence or route; adoption of new, more efficient chemical technologies; and rational choice of solvent, solvent reduction, and recycling all go toward producing greener synthesis. But what role does technology, such as reaction engineering, have in helping us design and develop greener and more sustainable manufacturing processes?

15.2
Advantages of Chemistry in Continuous-Flow Reactors

In the fine chemicals and pharmaceutical industry, processes have traditionally been scaled up from lab to pilot plant, from pilot plant to manufacturing in stirred-tank reactors. There are of course, a number of reasons for this mode of operation, but an increasing number of reports are seen of switching from batch to continuous-flow chemistry. The estimate of the number of organic synthetic transformations that could be switched from batch to flow operation varies between 40 and 70% and the number that would show real benefits, from 15 to 40%. So which chemistries have big drivers to switch from batch to flow chemistry? Obvious areas are reactions using energetic and unstable materials. Bristol-Myers Squibb reported use of various flow chemistries to generate and use hydroperoxides in reactions using O_2 or H_2O_2 [6], and highly exothermic 4-hydroxy 2,2,6,6-Tetramethylpiperidinyloxy (TEMPO)-catalyzed oxidations have been reported to work well in flow mode where selectivity is also improved [7]. Nitration is usually highly exothermic and prone to thermal runaway. DSM has used a microchannel plate reactor for making nitrate esters like naproxcinod safely on a very large scale [8]. The high rates of heat removal in microreactors make them ideal for exothermic reactions, generating thermosensitive intermediates or products. The consequences of running large-scale reactions with highly energetic materials in an "all in" batch mode were tragically demonstrated by the recent tragic incident in Jacksonville, Florida [9].

Minimizing inventory of certain materials is also an advantage. The *in situ* generation and use of highly reactive and toxic intermediates like methylsulfonyl cyanide in a flow process has been developed by Lonza for the manufacture of intermediates for HIV protease inhibitors [10].

The scale of operation of continuous-flow processes is also expanding. This now ranges from the small-scale synthesis of milligrams to grams of target molecules and natural products [11] to complete multistage flow process to make active pharmaceutical ingredients, active pharmaceutical ingredient (API) [12, 13], and intermediates [14, 15] on the multiton scale. Taken to its logical conclusion, flow synthesis has been fully integrated with continuous product separation and purification, solvent recovery, and recycling and waste treatment [14, 16]. Also expanding at a rapid pace is the range of reactor types that can be employed for screening and scale-up/out-of-flow reactions. These include lab-on-a-chip microfluidics [17], microchannel reactors [8], high-speed tube reactors [7], static mixer reactors [15], and continuously stirred-tank reactors [13].

15.3
Introduction to AZD 4407

Approximately 300 000 000 people suffer from respiratory inflammatory disorders such as asthma and chronic obstructive pulmonary disease (COPD) that disable, lower the quality of life, and often result in premature death [18]. Unfortunately, this number is growing due to a number of factors such as pollution and consumption of tobacco products. One potential treatment pathway for such conditions is to interrupt the arachidonic acid pathway and prevent the formation of biologically active leukotrienes by inhibiting the enzyme 5-lipooxygenase [19]. AZD 4407 was in development by Zeneca and AstraZeneca as a 5-LO (lipoxygenase) inhibitor until termination of the project in the mid-2000s [20]. The structure is shown in Figure 15.1. The molecule consists of an N-methyl oxindole linked at the 5-position to a thiophene that in turn, is attached to a complex pyranol with two stereo centers. The API is the (2S,4R) enantiomer [20].

Over the lifetime of AZD 4407 as a development project, a number of synthetic routes were discovered and demonstrated in the laboratory and in the pilot plant. A number of these involved the generation of reactive organolithium and organomagnesium species and subsequent reaction with electrophiles. This chapter details the conversion of one of these reactions from batch to a continuous-flow process using two linked microchannel reactors.

15.4
Comparison of the Synthetic Routes Used to Prepare AZD 4407

During the lifetime of this development project, a number of synthetic routes to AZD 4407 were identified and exemplified. These are outlined in the following section. The active isomer of AZD 4407 is (2S,4R). The 2(S) stereocenter was supplied via (S)-2-methyl tetrahydropyran-4-one. After construction of the 3-thiophene-tetrahydropyran carbon–carbon bond, mixtures of (2S,4R) and (2S,4S) isomers are formed. The desired stereoisomer was obtained by acid-catalyzed equilibration to give ~98% (2S,4R). This was normally carried out on the final API molecule, but could be done with earlier intermediates [21, 22].

The first route developed for AZD 4407 is shown in Scheme 15.1 [21]. This was used for initial supplies of API but had several major drawbacks for large-scale production. The synthesis started from 2,3,5-tribromothiophene and involved two very low temperature metallation steps run at −100 °C to minimize migration of halogens around the thiophene ring. Thiol emissions and a low-yielding final step

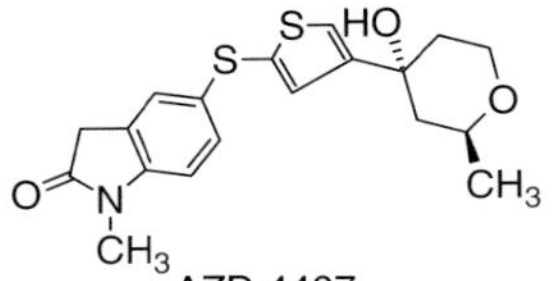

Figure 15.1 Structure of AZD 4407, (5-[4-((2S,4R)-4-hydroxy-2-methyl-tetrahydro-pyran-4-yl)-thiophen-2-ylsulfanyl]-1-methyl-1,3-dihydro-indol-2-one).

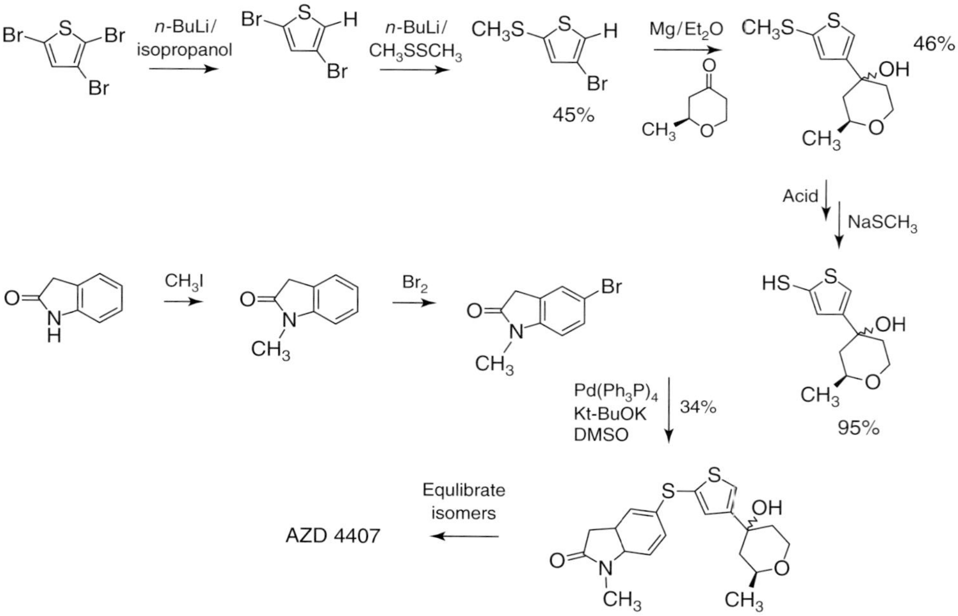

Scheme 15.1 Synthesis of AZD 4407: coupling at final stage.

were also unfavorable for further scale-up. The crude product from the Pd-catalyzed coupling reaction always had to be further purified by column chromatography to obtain API of acceptable quality. The coupling reaction also employed a high loading (10 mol%) of Pd(Ph$_3$P)$_4$[21].

The second route to AZD 4407 employed the much cheaper and more readily available 3-bromothiophene as a starting material (Scheme 15.2) [21, 22]. It was deemed desirable to use a less basic organometallic intermediate to couple with the pyranone to minimize yield loss via deprotonation α to the ketone function. This was achieved by generating the 3-lithio derivative and *in situ* conversion to a titanate species followed by reaction with the pyranone. The resulting *t*-alcohol was protected as a trimethylsilyl ether via reaction with trimethylsilyl chloride (TMSCl) and imidazole. A bulky group was required to ensure that subsequent metallation occurred selectively at the 5-position rather than the 2-position. The approach was then to generate the 5-lithiothiophene and couple this with a suitable electrophile to introduce the thiol moiety. The use of dimethyl disulfide gave the 5-thiomethyl ether. Conveniently, after a solvent change to *N*-methylpyrrolidinone, the generated bi-product, CH$_3$SLi, demethylated the thioether to give the free thio compound [21].

Pd- or Ni-catalyzed coupling with *N*-methyl-5-bromooxindole gave AZD 4407 as a mixture of isomers. This new route improved the overall yield of API from ∼7 to ∼22%. However, as with the first route, this step was plagued by low yields (30–40%) in the final step, and the need for chromatography to access material of acceptable purity.

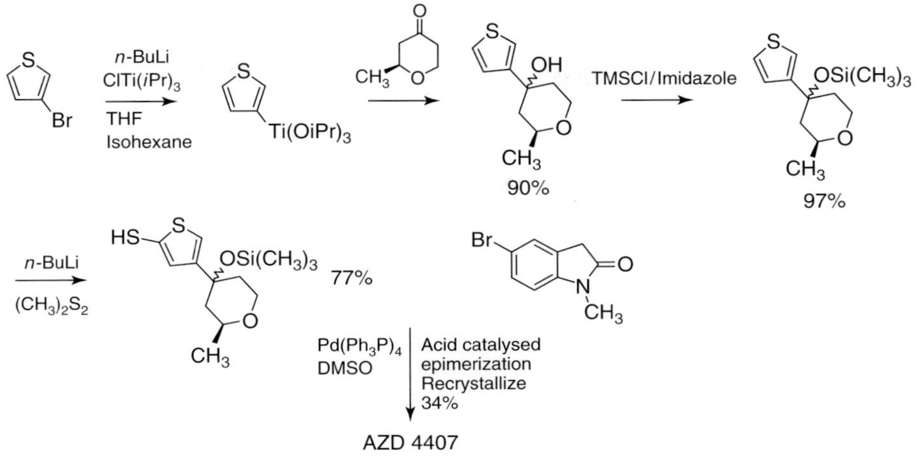

Scheme 15.2 Second route to AZD 4407.

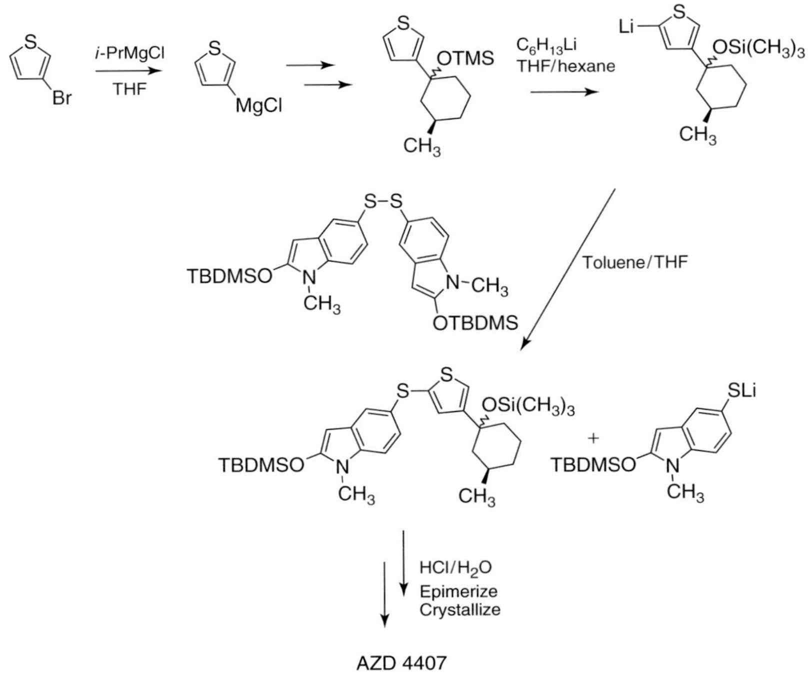

Scheme 15.3 Disulfide route to AZD 4407.

A major goal for the next synthetic iteration was to remove the complex and low-yielding coupling reaction with a more efficient transformation. 3-Bromothiophene was converted to the Grignard reagent with i-PrMgCl to avoid formation of the titanate intermediate [22]. Following reaction with the (S) pyranone, the t-alcohol was then O-silylated as given in Scheme 15.3. The new synthetic strategy to avoid the Pd-catalyzed cross-coupling reaction was to execute a regioselective lithiation of the thiophene followed by carbon–sulfur bond formation with an electrophilic sulfur species that would directly introduce the oxindole group (Scheme 15.3). Initially, it was decided to use the symmetrical disulfide shown in Scheme 15.3. Of course, the use of this disulfide to introduce the oxindole moiety is straightforward, but very atom-inefficient. After the linking reaction and quenching, the AZD 4407 isomer mixture could be extracted into an organic solvent with the lithium salt of the 5-thiooxindole cleanly portioning into the aqueous phase, so in principle, the redundant sulfide portion could be recovered and readily converted back to disulfide via oxidation. However, during the development process, more efficient sulfur-leaving groups were identified; see Scheme 15.8 [22].

15.5
Conversion of Batch to a Flow Process

While the thiophene was the more expensive synthon in this sequence, the reaction yield was limited by undercharging disulfide (0.85 equiv.) since excess disulfide left after reaction could cause problems with the efficient isolation of the AZD 4407 API. After lithiation of the thiophene and quenching, the solution yield of AZD 4407 seemed to plateau ~65% and despite extensive investigations, this was never really understood. There were concerns regarding the stability of the lithiated thiophene intermediate and by-products produced if the temperature of the lithiation exceeded −20 °C.

A program of work was undertaken to see if there would be a benefit in converting the lithiation and linking reaction with the disulfide from a batch to a flow reaction. It was envisaged that the following may be achieved:

1) use of higher temperatures for the lithiation reaction;
2) dispense with building an inventory of unstable lithiated intermediates;
3) increase optimum consumption of thiophene;
4) combine lithiation and reaction with disulfide;
5) obtain a higher yield of AZD 4407.

15.5.1
Preparation of Synthons Used in the Microreactor Study

15.5.1.1 Disulfide Synthesis
The disulfide electrophile was prepared from N-methyl oxindole as the starting material by regioselective chlorosulfonation, followed by reduction with hydrogen

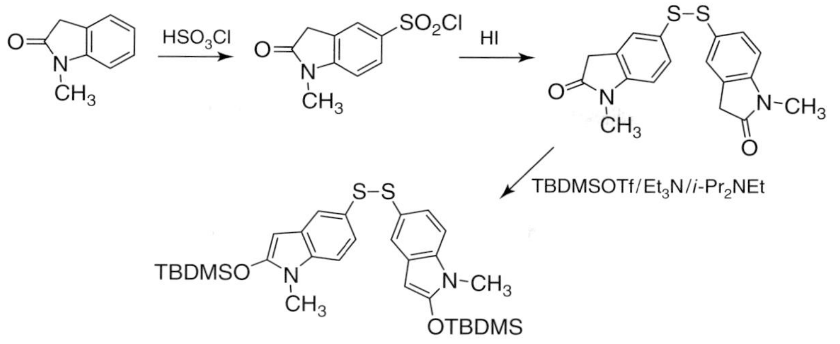

Scheme 15.4 Synthesis of oxindole disulfide.

iodide to give the disulfide [22]. Initial investigations utilizing the unprotected disulfide as an electrophilic coupling partner were disappointing due to quenching of any basic organometallic intermediates by the acidic methylene protons of the oxindole [21, 22]. This issue was circumvented by conversion to the bis-silylated indole (Scheme 15.4). The protecting group, *t*-butyldimethyl silyl (TBDMS), also rendered the highly insoluble parent disulfide much more soluble in solvents traditionally used for the generation and use of organometallic intermediates such as tetrahydrofuran (THF) and toluene [22]. This proved to be vital when converting reactions of the disulfide from batch to continuous-flow processes.

15.5.1.2 (S)-2-Methyl Tetrahydropyran-4-one

The 2-(S)-methyltetrahydro-4H-pyran-4-one required for AZD 4407 was prepared via resolution of the *cis*-racemate derived from the Prins reaction between acetaldehyde and butene-1-ol [23]. After separation of the *trans*-isomers, resolution with vinyl butyrate and *Candida antarctica* B lipase (Novozym 435) (Scheme 15.5) [23] gave the (4-*R*), (2-*S*) butyrate in >95 % enantiomeric excess (ee). Hydrolysis and

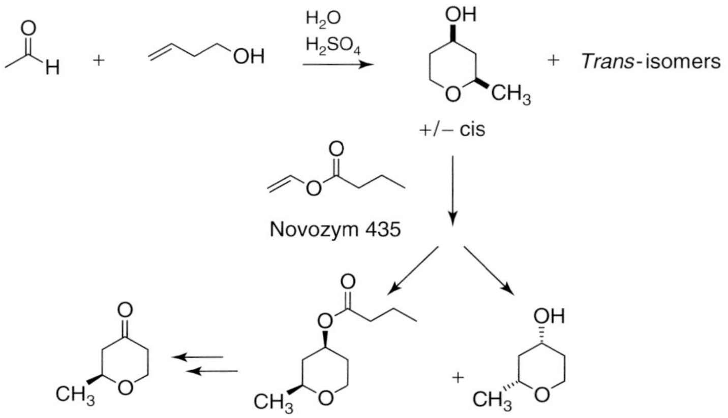

Scheme 15.5 Preparation of (S)-pyranone via biocatalytic resolution.

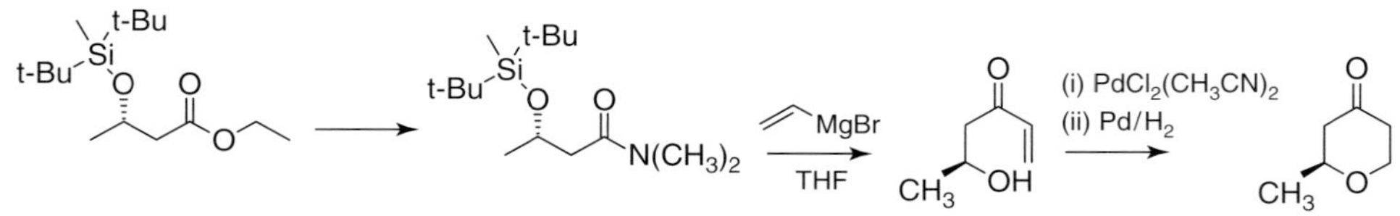

Scheme 15.6 Synthesis of pyranone from (S)-ethyl hydroxybutyrate.

subsequent oxidation of the 4-hydroxy function with Jones reagent furbished the required (S)-pyranone.

While the above process provided multikilogram supplies of pyranone in high chiral purity, overall it was not very efficient and toward the end of the project, alternative routes were identified that moved away from resolution and could deliver (S)-pyranone in a more mass-efficient manner from (S)-ethyl hydroxylbutyrate as outlined in Scheme 15.6 [24].

15.5.1.3 Thiophene Synthon
The thiophene was prepared as outlined in Scheme 15.3 via metallation of 3-bromothiophene, reaction with (S)-pyranone, and subsequent silyl protection.

15.5.2
n-Hexyl Lithium Versus n-Butyl Lithium

Throughout this work, n-hexyl lithium was used as a metallation reagent rather than the more commonly employed n-butyl lithium. This choice was not based on a greater efficiency of lithiation, but for safety and operability on scale. The generation of n-butane, especially in cooled reactions, can lead to the sudden release of highly flammable n-butane gas when the reaction is warmed up or quenched [25]. Depending on the type and efficiency of abatement technologies available, this may lead to breach of local permits and safety hazards. For the conversion of a batch to a flow process, the potential generation of gas bubbles in a nonpressurized reactor cell can lead to poor mixing and variable flow rates and hence, irreproducible results.

The efficiency of the conversion of thiophene to the 2-lithio derivative was measured by quenching aliquots from reaction mixtures with excess TMSCl and quantitation of the bis-trimethylsilyl compound by gas–liquid chromatography (GLC; Scheme 15.7).

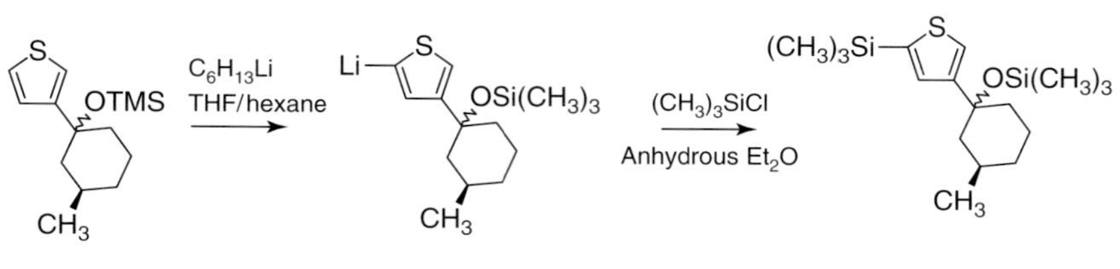

Scheme 15.7 Efficiency of the lithiation reaction.

15.5 Conversion of Batch to a Flow Process

Efficiency of the reaction between lithiated thiophene and disulfide was measured by quenching reactions into aqueous acid, stirring at ambient temperature to allow hydrolysis of the TMS- and TBDMS-protecting groups followed by extraction and quantitation by high-pressure liquid chromatography (HPLC).

15.5.3
Batch Reactions: Lithiation Reaction and Disulfide Linking

In batch mode, a solution of thiophene in THF–toluene is cooled to $-25\,°C$ and 1.15 equiv. of n-hexyl lithium is added slowly, keeping the temperature at $-25\,°C$. The mixture was then stirred for 30 min and conversion to the 2-lithio thiophene measured by quenching an aliquot with TMSCl. Typically on a 10-g scale, conversion was >98%. On larger scales, this could decrease slightly to 90% but could be increased to >95% by careful titration of extra n-hexyl lithium. The charge of n-hexyl lithium needed to be carefully controlled to prevent dilithiation and subsequent by-product formation [22]. A solution of the disulfide in toluene was added slowly to the lithiated thiophene (0.85 equiv. based on lithiated thiophene), and the reaction was then stirred at $-25\,°C$ for 30 min. and then quenched by the addition of aqueous sodium chloride. The quench mixture was then acidified and stirred for several hours to complete the hydrolysis of all the silicon-protecting groups. The lithiation reaction was very exothermic (adiabatic temperature rise of $55\,°C$) and the reaction with the disulfide mildly exothermic.

15.5.4
Microreactor Description

The flow reactor chosen to study the lithiation and linking reaction was the CYTOS™ reactor [26] developed by Cellular Process Chemistry, Frankfurt [26].

The CYTOS™ reactor system consists of a microchannel reactor of 2 ml volume followed by a number of residence units (Residos™) to complete the reaction. The reactor consists of a large number of submillimeter channels giving very fast and efficient mixing. The number of residence units can be altered to accommodate varying reaction times. Rotary piston pumps feed the reagent solutions with a capacity of $0.04-90\,\text{ml min}^{-1}$. The reactor and residence units are arranged as parallel plates with the reactor at the bottom. These are linked to a PC and Huber units to maintain set temperature in the microreactor and residence units [26]. The reagent flow delivered by the pumps can be monitored by decrease in volume or weight or reagent feed vessels (Figure 15.1). A number of groups have demonstrated the use of the CYTOS™ microchannel flow reactor system in a wide variety of transformations such as the nitration of aromatics and heteroaromatics (pyrazoles, pyridine N-oxides [27], Buchwald–Hartwig aminations of 4-bromotoluene [28], carbamate synthesis [29], high temperature Newman–Kuart rearrangements [29], and ring-expansion reactions using diazoesters [29].

Prior to starting runs, the rotary piston pumps on the flow rig were calibrated and the reactor flushed with dry solvent to remove air and water and to bring the reactor to constant temperature. The reagent feeds were maintained under an argon atmosphere. After starting charging, the system was left to attain the steady state before sampling the product stream.

Thiophene was used as a solution in THF–toluene (~2:1 v/v). Disulfide was used initially as a solution in toluene, but several instances of pump blockage were traced to disulfide prematurely crystallizing in the pumps. This premature crystallization of the disulfide was not generally observed in batch reactions, and the cause of this phenomenon in the flow reactor is unclear; possibly, cavitation in the pump heads initiated crystallization. This was solved by the addition of 20% by volume THF which kept the disulfide in solution. n-Hexyl lithium was commercially supplied as a 2.3–2.4 M solution in hexane. During the development of the flow process, the actual concentration of n-hexyl lithium was determined by titration with diphenyl acetic acid immediately prior to use.

The eventual goal of the project was to link both, the lithiation reaction and reaction with the disulfide in a continuous-flow reactor train. Initially, both reactions were examined independently in the CYTOS™ reactor.

15.5.5
Lithiation Reaction in Flow Mode

Since the metallation reaction was very fast, the reactor and one residence unit were employed. A 0.66 M solution of thiophene in THF–toluene was reacted with a 2.3 M solution of n-hexyl lithium in n-hexane. To attain the required stoichiometry of 1.15 equiv. n-hexyl lithium, the flow rate was 3:1. The residence time was 7.5 min. The product stream was quenched into diethyl ether/excess TMSCl and the extent of lithiation determined. Variation of the reactor temperature showed little effect on efficiency of conversion (Table 15.1).

Owing to the highly efficient heat transfer of the microchannel reactor, the metallation reaction could be run for short time periods at higher temperature, without concern for the stability of the unstable lithiated thiophene intermediate. Of course, making the lithiated thiophene in the microreactor and then accumulating batches for later consumption did lead to issues with falling yields and irreproducible results due to decomposition of the stored lithiated intermediate.

Table 15.1 Lithiation efficiency versus temperature.

Sample	Temperature (°C)	Conversion to lithiothiophene (%)
1	−20	93
2	−20	97
3	−5	95
4	−5	98
5	7	96
6	7	96

Table 15.2 Results of continuous disulfide reaction.

Sample	Temperature (°C)	HPLC area (%) of disulfide	HPLC area (%) of AZD 4407	Solution yield of AZD 4407 (%)
1	−5	36	49	77
2	−5	41	46	65
3	3	58	30	48
4	3	61	28	41
5	13	62	25	35
6	13	66	21	30

15.5.6
Coupling Reaction with Disulfide in Flow Mode

Initial attempts to use disulfide as a 35 wt% solution in toluene were hampered by pump blockages, so this solution was diluted with 20% THF which held the disulfide in solution. This solution was used in reactions, with a solution of the lithiated thiophene generated in a separate run and stored. Since the reaction is less exothermic and slower than the lithiation step, three residence cells were used. The concentrations of the feedstocks were adjusted to obtain a stoichiometry of 0.85 equiv. disulfide with a flow ratio of 1 : 1. The product stream was subject to an acidic aqueous quench and analyzed by HPLC (Table 15.2).

It can be seen from Table 15.2 that at lower temperatures, the reaction is reasonably successful but conversion of disulfide and yield of AZD 4407 drops rapidly with increasing temperature. Although the feed solution of lithiated thiophene was kept cold, it was felt that this drop in yield was most likely due to the instability of this intermediate, especially in the non–temperature controlled sections of the reactor (probably the pump heads), before the feed stream entered the thermostatted microreactor. This can be a common problem with flow reactors when transfer or mixing of unstable reagents or highly exothermic reactions occurs outside a thermostatted region of the reactor and/or the temperature in regions of initial mixing cannot be accurately monitored and controlled. It was also observed that repeating runs with aged solutions of lithiated thiophene consistently gave lower conversions of disulfide and lower yields of AZD 4407.

15.5.7
Linking the Lithiation and Reaction with Disulfide in Flow Mode

Owing to the instability of the lithiated thiophene, the sensible option was deemed to be linking both reactions in two flow reactors and not building up an inventory of lithiated thiophene. This was achieved by linking two CYTOS™ reactors in series with insulated transfer lines to minimize any temperature increase between the two reactors. The initial reactor comprised one reactor and one residence unit

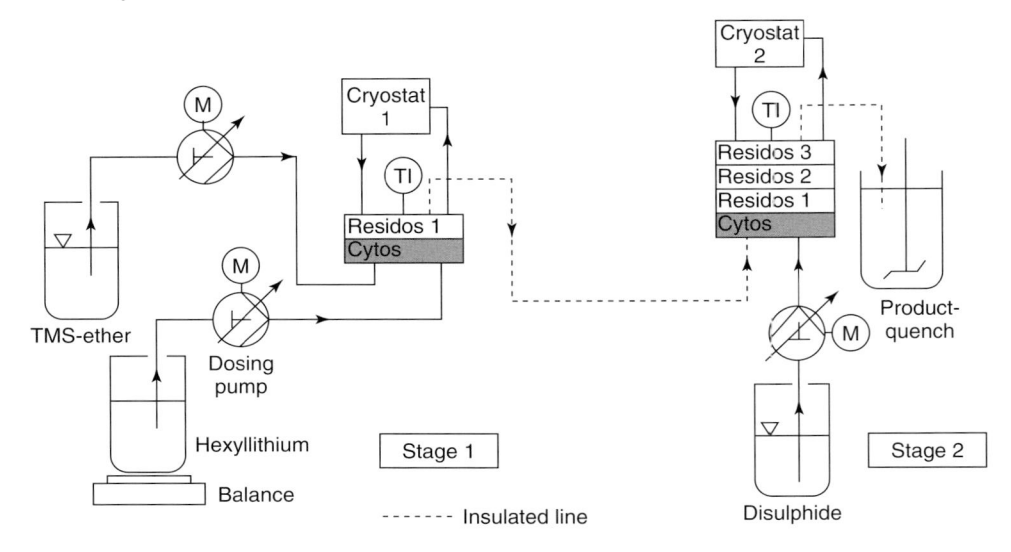

Figure 15.2 Flow reactor configuration for linked lithiation and reaction with disulfide followed by aqueous quench. Courtesy of CPC Microsystems.

(15 ml total) for lithiation and a second unit of one reactor and three residence units (45 ml) for reaction with the disulfide (Figure 15.2).

Solutions of thiophene (0.61 M), n-hexyl lithium (2.1 M), and disulfide (0.39 M) gave a ratio of flow rates of $3:1:4$ ($1.5:0.5:2\,\text{ml}\,\text{min}^{-1}$) to maintain the desired stoichiometry. After reaching steady state, the temperature was varied and samples quenched for HPLC analysis (Table 15.3) and HPLC peak-area ratio measurement (Table 15.4).

The results show good conversions with yields higher than that seen in batch reactions and at higher operating temperatures. The yield increased at the higher temperatures examined, indicating that residence time was not sufficient for complete reaction at lower temperatures. The isomer ratio of AZD 4407 (2S,4R) to (2S,4S) produced in the flow reactor was $\sim 2.4:1$ and did not seem to vary greatly

Table 15.3 Yield of AZD 4407 versus temperature in flow reactor.

Sample	Temperature (°C)	Solution yield of AZD 4407 based on disulfide (%)	Solution yield of AZD 4407 based on TMS ether (%)
1	−20	83	71
2	−20	81	69
3	−10	88	75
4	−10	87	74
5	0	87	74
6	0	87	74

Table 15.4 Reaction profile of AZD 4407 versus temperature in flow reactor.

Sample	Temperature (°C)	HPLC area (%) of disulfide	HPLC area (%) of AZD 4407
1	−20	19	64
2	−20	19	59
3	−10	10	67
4	−10	11	64
5	0	7	66
6	0	9	65

with temperature or from the typical ~2.5 : 1 ratio that was typically produced in batch reactions.

The level of residual disulfide of up to 10% found in the flow reactor was greater than that seen in small-scale batch reactions at ~1%, although this level of disulfide did increase to comparable levels in higher-scale batch reactions. As described earlier, in batch, the linking reaction reached maximum conversion after the consumption of ~0.85 equiv. of disulfide, and proved recalcitrant to any efforts to increase the conversion and hence, yield of AZD 4407. A limited investigation was undertaken to further explore this phenomenon in the flow reactor. Variation of the ratio of disulfide from 0.75 to 0.95 did not measurably effect the solution yield of AZD 4407, dropping slightly at the higher ratios. At the end of the reaction, there was always disulfide present, but exact quantitation was hampered by losses of disulfide over time. This material appeared to be somewhat unstable in air and/or light in the quench mixture (Figure 15.3), so for accurate quantitation,

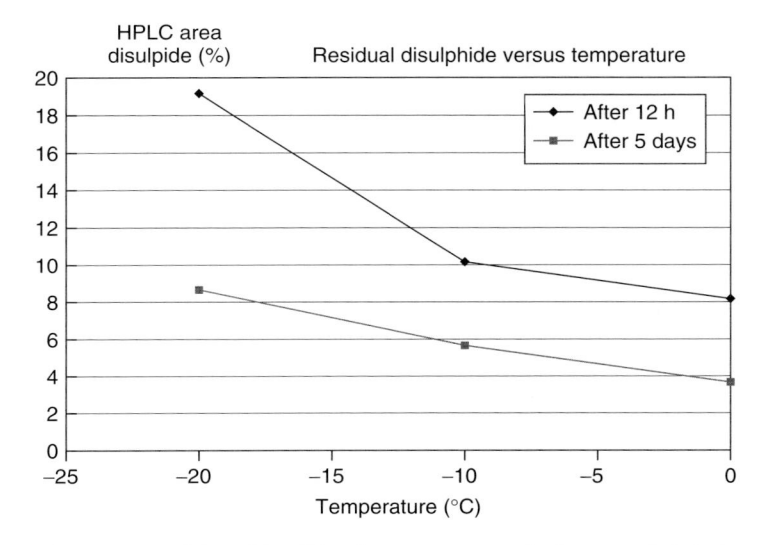

Figure 15.3 Stability of disulfide after aqueous quench. Courtesy of CPC Microsystems.

measurements had to be taken soon after quench and the storage of samples avoided. AZD 4407 was determined to be quite stable in the quench medium.

15.6
Conclusions

Overall, the microreactor study proved to be a success. While it did not meet every criteria set at the onset of the project, the following were achieved. (i) Increase in solution yield of 15–20% of the API: toward the end of a multistage process, this increase in efficiency would produce a considerable decrease in the E factor for the overall process [1]; (ii) the exothermic lithiation reaction could be run at a temperature that was ~20 °C higher; (iii) linking the metallation and reaction with disulfide meant that accumulation and storage of the unstable lithiated thiophene would not be required.

The results are summarized in Table 15.5.

Although never getting to production scale, the linked-flow reaction as a means of producing AZD 4407, did promise improvements in the areas of environmental and operational efficiency when compared to batch operation. Kralisch and Kreisel studied the production of m-anisaldehyde from m-bromoanisole (lithiation with n-butyl lithium followed by quenching with N,N-dimethylformamide) in a series of CYTOS™ microreactors compared with traditional batch production in a 400 l reactor [30]. The use of microreactors produced an increase of 22% in yield and thus, a much-improved E factor. The use of energy was also reduced, the flow reactor working at 0 °C as compared to −80 °C for the 400 l batch reactor. Across most indicators, the flow process was deemed to have a positive life cycle impact as compared to batch chemistry, especially in reduced energy requirements, mainly from not having to use large amounts of liquid N_2. DSM examined the causes of scale-up failure or deviation in batch reactions. The top three were reported as being due to mass transfer (mixing), extended processing times, and heat exchange (temperature control) [31], all factors which can be reduced or eliminated by the use of flow chemistry in microreactors.

Shortly after the successful demonstration of the lithiation and reaction with disulfide in the flow reactor, an even more efficient route to AZD 4407 was discovered based on the selective metallation of 2,4-dibromothiophene and reaction

Table 15.5 Comparison of reaction modes.

Reaction type	Scale	Temperature (°C)	Yield 4407 (%)	Accumulation of lithiothiophen
Batch	Multikilogram	−25	65	Yes
Flow	Hundreds of grams	0	87	No

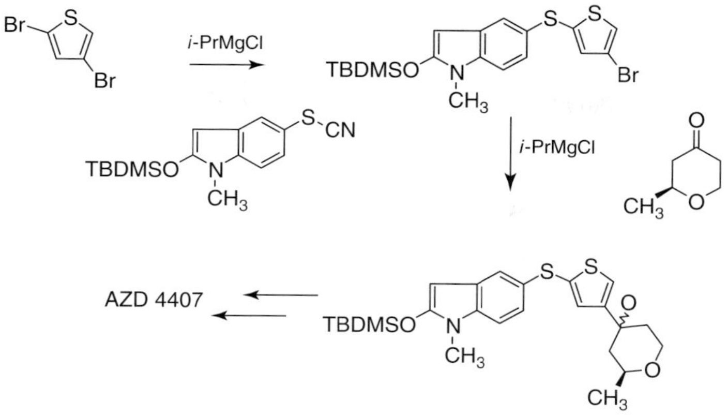

Scheme 15.8 Final route to AZD 4407.

with a much more atom-efficient sulfur electrophile. This route is shown in Scheme 15.8 [22].

Of course, there would have been the possibility of running the chemistry depicted in Scheme 15.8 in continuous-flow mode, but shortly after the discovery of this route, the AZD 4407 development project was terminated by AstraZeneca.

Acknowledgments

The author would like to acknowledge the assistance of scientists from Cellular Process Chemistry, Dr Michael Hohmann, Dr Axel Kleemann, Dr Wolfgang Stirner, Shahriyar Taghavi-Moghadam, and Dr Thomas Schwalbe, who ran the flow chemistry project at CPC and the AZD 4407 team, Marie-Lyne Alcaraz, Stephanie Atkinson, Philip Cornwall, Alison C. Foster, Duncan M. Gill, Lesley A. Humphries, Philip S. Keegan, Richard Kemp, Eric Merifield, Robert A. Nixon, Allison J. Noble, Darren O'Beirne, Zakariya M. Patel, Jacob Perkins, Paul Rowan, Paul Sadler, John T. Singleton, James Tornos, Andrew J. Watts, Ian A. Woodland, Jonathan Hutton, Andrew D. Jones, Stanley A. Lee, David M. G. Martin, Brian R. Meyrick, Ian Patel, Rachel F. Peardon, and Lyn Powell.

References

1. (a) Sheldon, R.A. (1992) *Chem. Ind. (London)*, **23**, 903; (b) Sheldon, R.A. (2007) *Green Chem.*, **9**, 1273; (c) Sheldon, R.A. (2008) *Chem. Commun.*, **29**, 3352.
2. Dunn, P., Williams, M., and Wells, A. (eds) (2010) *Green Chemistry in the Pharmaceutical Industry*, Wiley-VCH Verlag GmbH, Weinheim (in press).
3. (a) Anastas, P.T. and Zimmerman, J.B. (2006) in *Sustainability Science and Engineering Defining Principles* (ed. M.A. Abrahams), Elsevier, p. 11; (b) Anastas, P. and Warner, J.C. (1998) *Green Chemistry: Theory and Practice*, Oxford University Press, Oxford.
4. Winterton, N. (2001) *Green Chem.*, **3**, G73.

5. (a) Tang, S.L.Y., Smith, R.L., and Poliakoff, M. (2005) *Green Chem.*, **7**, 761; (b) Tang, S.L.Y., Borne, R., Smith, R.L., and Poliakoff, M. (2008) *Green Chem.*, **10**, 268.
6. (a) LaPorte, T., Hamedi, M., DePue, J.S., Shen, L., Watson, D., and Hsieh, D. (2008) *Org. Process Res. Dev.*, **12**, 956; (b) Crispino, G.A., Hamedi, M., LaPorte, T.L., Thornton, J.E., Pesti, J.A., Xu, Z., Lobben, P.C., Leahy, D.K., Muslehiddinoglu, J., Lai, C., Spangler, L.A., and Discordia, R.P. (2007) US Patent 0249610.
7. Fritz-Langhals, E. (2005) *Org. Process Res. Dev.*, **9**, 577.
8. (a) Braune, S., Poechlauer, P., Reintjens, R., Steinhofer, S., Winter, M., Lobet, O., Guidat, R., Woehl, P., and Guermeur, C. (2009) *Chim. Oggi.*, **27** (1), 26; (b) Braune, S., Steinhofer, S., Poechlauer, P., Reintjens, R., Wilhelmus, E., Linssen, N., Theodora, W., and Thathagar, M. (2009) WO Patent 2009080755.
9. Bresland, J.S., Visscher, G. Wark, W., and Wright, W., U.S. Chemical Safety and Hazard investigation board report 'T2 Laboratories runaway reaction' report http://www.csb.gov/assets/document/T2_Final_Copy_9_17_09.pdf Published 2009 (accessed July 2010).
10. Rouhi, A.M. (2003) C&E News, *Simplifying Syntheses is always a Key Goal* p. 81.
11. (a) Baxendale, R., Hayward, J.J., Lanners, S., Ley, S.V., and Smith, C.D. (2008) Heterogeneous reactions, in *Microreactors in Organic Synthesis and Catalysis* (ed. T. Wirth); Wiley-VCH Verlag GmbH, Weinheim. 84–122. (b) Tanaka, K. and Fukase, K. (2009) *Org. Process Res. Dev.*, **13**, 983.
12. (a) Letendre, L.J., McGhee, W.D., Snoddy, C., Klemm, G., and Gaud, H.T. (2003) PCT International Applications, WO Patent 2003099794; (b) Hawkins, J.M., Guiness, S., Jackson, R.P., and am Ende, D.J. (2008) 15th International Process Development Conference, Annapolis, Maryland, May 21, 2008.
13. Hawkins, J.M. (2010) in *Green Chemistry in the Pharmaceutical Industry* (eds P. Dunn, M. Williams, and A. Wells), Wiley-VCH Verlag GmbH, Weinheim. 221.
14. Proctor, L.D. and Warr A.J. (2003) WO Patent 2003097581.
15. Proctor, L.D. and Warr, A.J. (2002) *Org. Process Res. Dev.*, **6**, 884.
16. Proctor, L.D. (2010) in *Green Chemistry in the Pharmaceutical Industry* (eds P. Dunn, M. Williams, and A. Wells), Wiley-VCH Verlag GmbH, Weinheim. 221.
17. Watts, P. and Haswell, S.J. (2005) *Chem. Soc. Rev.*, **34**, 235.
18. Atkins, P.J., Woodcock, A., Blinova, O., Khan, J., Stechert, R., Wright, P., and Yizhong, Y., Medical Aerosols, IPCC/TEAP Special Report: Safeguarding the Ozone Layer and the Global Climate System, Chapter 8. http://www.ipcc.ch/pdf/special-reports/sroc/sroc08.pdf (accessed July 2010).
19. Crawley, G.C., Briggs, M.T., Dowell, R.I., Edwards, P.N., Hamilton, P.M., Kingston, J.F., Oldham, K., Waterson, D., and Whalley, D.P. (1993) *J. Med. Chem.*, **36**, 295.
20. Bird, T., Ple, P., Crawley, G., and Large, M. (1994) EP Patent 94400190.
21. Hutton, J., Jones, A.D., Lee, S.A., Martin, D.M.G., Meyrick, B.R., Patel, I., Peardon, R.F., and Powell, L. (1997) *Org. Process Res. Dev.*, **1**, 61.
22. Alcaraz, M.-L., Atkinson, S., Cornwall, P., Foster, A.C., Gill, D.M., Humphries, L.A., Keegan, P.S., Kemp, R., Merifield, E., Nixon, R.A., Noble, A.J., O'Beirne, D., Patel, Z.M., Perkins, J., Rowan, P., Sadler, P., Singleton, J.T., Tornos, J., Watts, A.J., and Woodland, I.A. (2005) *Org. Process Res. Dev.*, **9**, 555.
23. Holt, R.A., Rigby, S.R., and Waterson, D. (1997) WO Patent 97/19185.
24. (a) Tornos, J. and Atkinson, S. (2003) WO Patent 03/051862; (b) Crawley, G.C. and Briggs, M.T. (1995) *J. Org. Chem.*, **60**, 4264; (c) Anderson, K.R., Atkinson, S.L.G., Fujiwara, T., Giles, M.E., Matsumoto, T., Merifield, E., Singleton, J.T., Saito, T., Sotoguchi, T., Tornos, J.A., and Way, E.L. (2009) *Org. Process Res. Dev.*, ASAP article. **14**, 58.
25. (a) Baenziger, M., Mak, C.-P., Muehle, H., Nobs, F., Prikoszovich, W., Reber,

J.-L., and Sunay, U. (1997) *Org. Process Res. Dev.*, **1**, 395; (b) Manley, P.W., Acemoglu, M., Marterer, W., and Pachinger, W. (2003) *Org. Process Res. Dev.*, **7**, 436.

26. Taghavi-Moghadam, S., Kleemann, A., and Golbig, K. (2001) *Org. Process Res. Dev.*, **5**, 652.

27. Panke, G., Schwalbe, T., Stirner, W., Taghavi-Moghadam, S., and Wille, G. (2003) *Synthesis*, **18**, 2827.

28. Mauger, C., Buisine, O., Caravieilhes, S., and Mignani, G. (2005) *J. Organometal. Chem.*, **690**, 3627.

29. Zhang, X., Stefanick, S., and Villani, F.J. (2004) *Org. Process Res. Dev.*, **8**, 455.

30. Kralisch, D. and Kreisel, G. (2007) *Chem. Eng. Sci.*, **62**, 1094.

31. Hulshof, B. (2004) 10th International Conference on Organic process Research and Development, Scientific Update, 2004, Vancouver.

16
Sustainable Processes Based on Enzymes Enabling 100% Yield and 100% ee Concepts

Oliver May

16.1
Introduction

A rapidly growing fraction of our society is interested in the ecological footprint of the products they consume and use. The emphasis on sustainability, which includes ecological, economical, and societal aspects, has shifted from an image claim of the past to a real differentiator today. Starting in 2005, the ACS Green Chemistry Institute® (ACS GCI) and global pharmaceutical corporations and recently also service providers have formed the ACS GCI Pharmaceutical Roundtable to encourage innovation while catalyzing the integration of green chemistry and green engineering in the pharmaceutical industry. The activities of the Roundtable reflect the joint belief that the pursuit of green chemistry and green engineering is imperative for a sustainable business and world environment. In the context of green chemistry and sustainability in general, biocatalysis is one of the areas which is considered as important [1–3]. The reason for this is the intrinsic properties of enzymes which are often extremely stereo-, enantio-, and regioselective and highly efficient under mild reaction conditions. Furthermore, enzymes can be prepared from renewable resources and are fully biodegradable.

Indeed, biocatalysis is receiving an increasing attention not only because it can very well address the above described needs for sustainable manufacturing solutions with respect to cost, quality, and ecologic footprint but also because of the groundbreaking technological breakthroughs in enabling technologies that happened during the last 10 years. For example, our ability to clone DNA from environmental samples (metagenome) [4], breakthroughs in enzyme engineering by evolutionary design [5], exponentially reduced cost, and increased speed of DNA sequencing [6] and synthesis [7] led to an exponential increasing number of available enzymes and an accelerated development speed as well as reduced development costs.

This chapter focuses on enzymatic concepts enabling the synthesis of chiral molecules in up to a theoretical 100% yield and >99% ee. Such concepts are very much preferred, from an economic as well as ecologic perspective, over resolution concepts which generate either 50% waste or requiring costly and waste generating

Pharmaceutical Process Chemistry. Edited by Takayuki Shioiri, Kunisuke Izawa, and Toshiro Konoike
Copyright © 2011 WILEY-VCH Verlag GmbH & Co. KGaA, Weinheim
ISBN: 978-3-527-32650-1

recycling steps. The discussed concepts are divided into (i) asymmetric synthesis, (ii) desymmetrization, (iii) deracemization, and (iv) dynamic kinetic resolution (DKR) concepts. Asymmetric introduction of oxygen by oxygenases which also enables 100% yield and 100% ee concepts is, however, beyond the scope of this chapter. The interested reader is referred to some excellent reviews that have been published elsewhere [8–10].

The aim of this chapter is not to be comprehensive but to provide an overview on some general concepts, discuss recent progress, and highlight industrially relevant examples of processes that are operated on a significant scale to demonstrate the practical use of the described approaches.

16.2
Asymmetric Synthesis

Asymmetric synthesis is a traditional term used for stereoselective synthesis of chiral molecules [11]. In this paragraph a more narrow definition of asymmetric synthesis is used, which includes only reactions where a chiral center is newly formed, for example, by C–C coupling reactions or reactions on flat prochiral centers such as carbonyl compounds or C=C bonds.

16.2.1
C–C Bond Formations

Lyases catalyze the formation and breakage of various bonds, of which enzymes forming C–C (EC 4.1.X.X), C–O (EC 4.2.X.X), and C–N (EC 4.3.X.X) are industrially most relevant.

One of the earliest reports of industrial asymmetric enzymatic synthesis is the application of a lyase for production of ephedrine which was introduced already in the 1920s [12, 13]. The process is based on a pyruvate decarboxylase from *Saccharomyces cerevisiae* catalyzing the stereoselective condensation of benzaldehyde and fermentatively produced acetaldehyde to yield (R)-1-hydroxy-1-phenylpropanone as shown in Scheme 16.1.

More than 80 years later, this process is still applied at BASF [14], which clearly demonstrates the efficiency of enzymatic C–C coupling reactions.

Aldolases and their natural functions have been extensively studied between end of the 1960s and beginning of the 1970s. Reactions catalyzed by these enzymes are shown in Table 16.1.

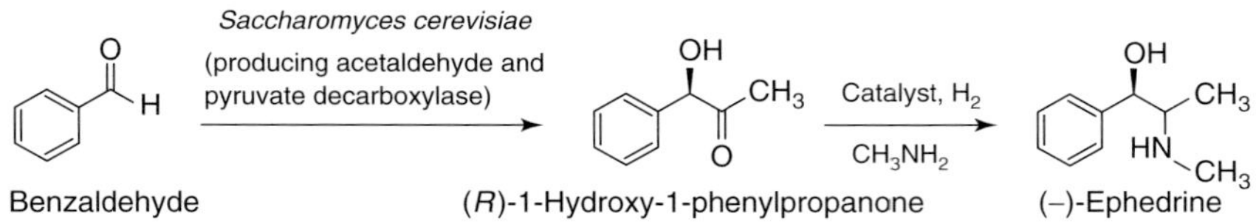

Scheme 16.1 Pyruvate decarboxylase–catalyzed reaction used in the synthesis of ephedrine.

Table 16.1 Reactions catalyzed by aldolases.

Group	Donor (nucleophile)	Acceptor (electrophile)	Product
dihydroxyaceton phosphate-dependent aldolases, for example, fructose-1,6-bisphospate aldolase	HO–C(=O)–CH₂–OP	R–CHO	R–CH(OH)–CH(OH)–C(=O)–CH₂–OP
Pyruvate-dependent aldolases, for example, sialic acid aldolase	CH₃–C(=O)–CO₂H and CH₂=C(OP)–CO₂H	R–CHO	R–CH(OH)–CH(OH)–C(=O)–CO₂H
2-Deoxyribose-5-phosphate aldolase	CH₃–CHO	R–CHO	R–CH(OH)–CH₂–C(=O)–H
Glycine-dependent aldolase, for example, threonine aldolase	H₂N–CH₂–CO₂H	R–CHO	R–CH(OH)–CH(NH₂)–CO₂H

First patents about their applications in organic synthesis appear in the 1990s [15–17] and first ton scale applications were reported in 1997 for the production of neuraminic acid [18]. The latter compound is used as an intermediate for zanamivir, an antiviral drug.

Another remarkable example of a green process based on enzymatic C–C coupling of very cheap and simple precursor molecules is the production of statin intermediates according to Scheme 16.2.

This process was pioneered by Wong *et al.* who reported for the first time in 1994 an asymmetric tandem aldol reaction using 2-deoxyribose-5-phosphate aldolase (DERA) and some years later its application for production of a statin intermediate [19, 20]. Cheap starting compounds (acetaldehyde and chloroacetaldehyde), high isolated yields and productivities (>30 g (l h)$^{-1}$), high optical purities (>99.5% ee, >99.5% de), use of only one single enzyme, and no need for hydride donors or chiral auxiliaries are the characteristic of this process which makes it economically very attractive [21]. It is perhaps one of the greenest routes for production of statin intermediates, implemented on industrial scale [22]. An important breakthrough to achieve economically viable productivities and lower enzyme costs was the development of more stable and active enzyme variants obtained by directed evolution [23].

Asymmetric HCN addition to aldehydes is catalyzed by hydroxynitrile lyases (EC 4.1.2.X). Excellent isolated yields (>90%), high productivities (>250 g (l d)$^{-1}$), and very high stereoselectivities (>99% ee) make hydroxynitrile lyase–catalyzed

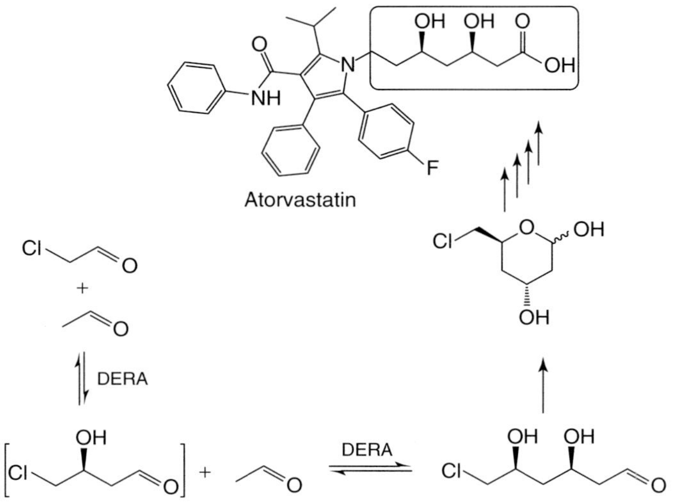

Scheme 16.2 DERA-catalyzed tandem aldol reaction used for the production of statin intermediates (example shown for atorvastatin).

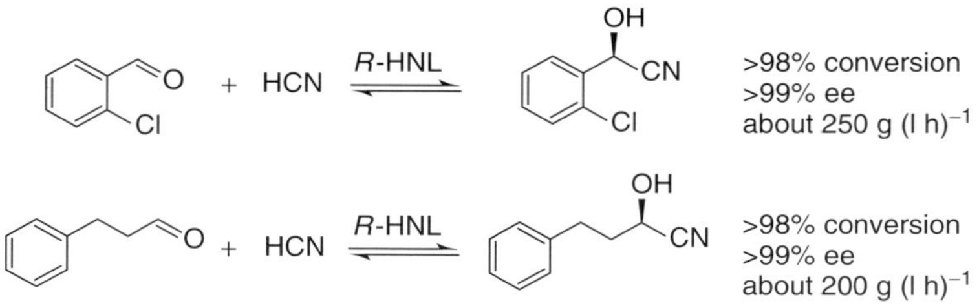

Scheme 16.3 Examples of HNL-catalyzed synthesis of pharma intermediates.

HCN additions more cost effective and sustainable than many other alternative routes toward optically pure cyanohydrins [24]. The impressive process economics were enabled by breakthroughs in recombinant production of acid-stable (R)- and (S)-selective enzymes and design of efficient catalysts with improved activities and enantioselectivities, as well as significant progress in reaction engineering using biphasic systems [25, 26]. Scheme 16.3 shows some examples of pharma intermediates that have been implemented at multitons scale at DSM for production of (R)-2-mandelonitrile and (R)-2-hydroxy-4-phenylbutyronitrile. These compounds are used as intermediates for different cardiovascular drugs that have been launched.

One of the first industrial applications of ammonia lyases was reported by Yamada et al. for the production of L-phenylalanine by the addition of ammonia to cinnamic acid, the reversal of the physiological reaction of the enzyme [27]. More recently, DSM has replaced a seven-step chemical process for production of a cyclic

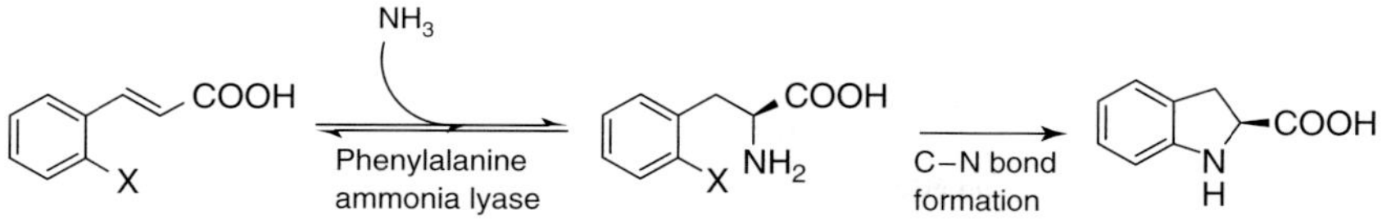

Scheme 16.4 Phenylalanine ammonia lyase–catalyzed production of a phenylalanine derivative used in the synthesis of (S)-2,3-dihydro-1H-indole-2-carboxylic acid.

amino acid derivative with a two-step chemoenzymatic process based on ammonia lyases. The process root is depicted in Scheme 16.4.

For the desired high conversions (>90% yield), high ammonia concentration (>10% wt) and basic conditions (pH >10) were required. Furthermore, substrate inhibition had to be overcome in order to reach industrial relevant productivities and product titers. In collaboration with Verenium (former Diversa), screening of an enzyme collection consisting of more than 20 different ammonia lyases revealed one very robust enzyme that could be operated under the required process conditions that are quite extreme and incompatible for many other enzymes described in the literature. The result also shows the importance of having access to a unique biodiversity to establish efficient industrial enzymatic processes.

A new concept has been recently reported by Janssen et al. for the application of a phenylalanine aminomutase from *Taxus chinensis* for highly enantioselective ammonia addition to (E)-cinnamic acid derivatives for the production of α- and β-amino acids [28]. For practical applications, enzymes with higher activity as well as a practical strategy to avoid the epimerization of the α- and β-amino group during the ammonia addition still need to be developed for implementing this elegant concept on industrial scale.

16.2.2
Production of Chiral Alcohols by Enzymatic Reduction of Ketones

Enzymatic reductions based on recombinantly produced enzymes and cheap recombinant whole-cell biocatalysts (see also Chapter 9) are now widely accepted as a highly competitive technology for the production of chiral alcohols. The competitiveness is actually confirmed by the growing number of processes applied at industrial scale for production of a number of pharma intermediates and other fine chemicals [29, 30]. Although the Nobel Prize winning technology of metal-catalyzed asymmetric hydrogenation of ketones is and will remain a highly competitive technology for the production of chiral alcohols [31], an increasing number of examples show that alcohol dehydrogenases (ADHs) can outperform chemical alternatives based on the advantages in (i) high regio-, stereo-, and enantioselectivity, (ii) application of mild aqueous reaction conditions, and (iii) lower catalyst cost. An important breakthrough for all enzymatic reduction technologies was achieved by the development of cost efficient cofactor regeneration systems [32]. In 2002,

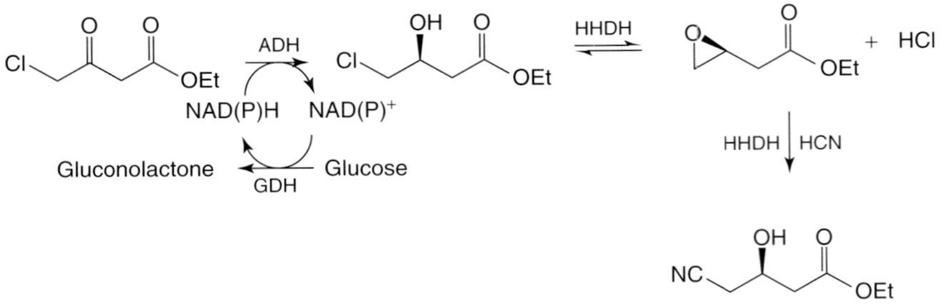

Scheme 16.5 Enzymatic process for the production of the intermediate of atorvastatin using an alcohol dehydrogenase (ADH) coupled with a halohydrine dehalogenase (HHDH) and a glucose dehydrogenase for cofactor regeneration.

the very prestigious German Future Prize was awarded to Kula and Pohl for the development of the formate dehydrogenase system [33].

In 2006, several years after some pioneering work of the Shimizu group [34], the Presidential Green Chemistry Challenge Award from the United States Environmental Protection Agency (USEPA) was awarded to the development of a green process for a statin intermediate. As shown in Scheme 16.5, this process is based on an enzymatic reduction using an ADH coupled with a cyanation using a halohydrin dehalogenase [35]. Reported advantages are the substitution of hydrogen bromide with a renewable and biodegradable hydride donor (glucose) and less waste production especially due to the cleaner enzymatic cyanidation step which also eliminates the need for high-vacuum fractional distillation used in the classical chemical process [36].

More recently, several other preparative-scale (>1 kg) processes were reported, for example, for the production of key intermediates for Montelukast [37], carbapenem antibiotics [38], chemokine receptor inhibitor [39], talsaclidine, and revatropate [40], as well as for an neurokinin-1 receptor antagonist [41]. Some of those pharma intermediates and important process characteristics confirming the efficiency of the enzymatic reduction of ketones are shown in Table 16.2.

16.2.3
Reduction of Activated C=C Bonds

One of the first industrial useful application of C=C bond reducing enzymes was reported by Prof. Shimizu *et al.* who developed in collaboration with Roche Vitamins (now DSM) an enzymatic process for actinol [42, 43]. This process is based on a coupled reduction using the old yellow enzyme from *Candida macedoniensis* and a levodione reductase from *Corynebacterium aquaticum* as shown in Scheme 16.6. Using a recombinant *Escherichia coli* coexpressing the old yellow enzyme and a glucose dehydrogenase from *Bacillus megaterium*, impressive productivities of almost $100\,g\,l^{-1}$ within 6 h with molar yields of 98.7% and an enantiomeric excess of >90% ee were reported for the production of the actinol intermediate

Table 16.2 Recent examples of pharma intermediates produced at >1 kg scale using alcohol dehydrogenases.

Intermediate product	Drug	Product concentration (g l^{-1})	Reaction time (h)	Conversion (%)	% ee	References
	Statin intermediate (e.g., Lipitor)	208	12	96	>99 (S)	[34]
	Montelukast (Singulair)	100	n.d.	99	>99 (S)	[37]
	Chemokine receptor inhibitor drug candidate	100	10 h	96–98	>99 (R)	[39]
	Talsaclidine and revatropate	100	21	98.6	>99.9 (R)	[40]
	NK-1 receptor antagonists drug candidates	100	<24	99	99.5 (S)	[41]

Scheme 16.6 Two-step process for conversion of ketoisophorone into actinol using, in the first step, an old yellow enzyme to produce (6R)-levodione followed by another reduction using an alcohol dehydrogenase.

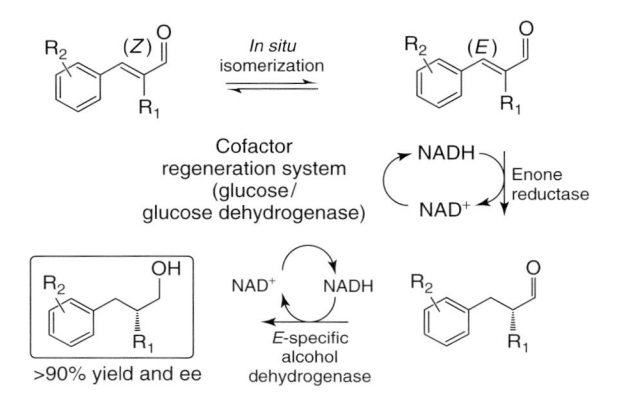

Scheme 16.7 One-step double reduction process based on in situ E/Z isomerization and a coupled enzyme system consisting of an enone reductase, alcohol dehydrogenase, and a glucose dehydrogenase for NADH regeneration.

(6R)-levodione [44, 45]. Since then, a large number of new enzymes were discovered, which led to an increased industrial interest in this reduction technology [46, 47]. For a recent review on C–C double bond reduction see, for example, Faber et al. [48].

Recently, an elegant one step-synthesis was developed by us using an enone reductase in combination with an ADH for the production of a pharma intermediate, the process root of which is shown in Scheme 16.7. An important feature of this process and prerequisite for reaching high yields are an efficient in situ isomerization of the E/Z isomers of the α,β-unsaturated compound as well as an ADH with very high selectivity toward the reduced intermediate. This process delivered the target compound in >90% yield and enantiomeric excess from the E/Z mixture of the α,β-unsaturated substrate (Oliver May, Personal communication (2009)).

16.2.4
Reductive Amination of α-Keto-Acids for the Production of α-Amino Acids

Amino acid dehydrogenases (AADHs; EC 1.4.1.X) are found in a large number of organisms, both eukaryotes and prokaryotes. Their natural function is to catalyze the oxidative deamination of amino acids, which is a first step in the degradation of amino acids for use as carbon and nitrogen source as well as for energy. Many synthetic applications rely on their property to catalyze the reductive amination of α-keto acids to α-amino acids, with the concomitant oxidation of the cofactor NAD(P)H.

Of the large number of AADHs identified in nature, only a few are industrially relevant for the synthesis of enantiomerically pure α-H-α-amino acids [49]. Besides alanine dehydrogenases (AlaDHs, EC 1.4.1.1) and glutamate dehydrogenases (GluDHs, EC 1.4.1.2–4), these are particularly phenylalanine dehydrogenases (PheDHs, EC 1.4.1.20) and even more leucine dehydrogenases (LeuDHs, EC 1.4.1.9). Because of the highly stereoselective hydride transfer catalyzed by the enzyme all of the above mentioned dehydrogenases catalyze the reductive conversion of α-keto acids leading to the corresponding L-amino acids ((S)-configuration) with very high optical purity. Owing to thermodynamic reasons, the equilibrium of the AADH-catalyzed reaction is usually far on the side of the amino acid. The Keq for the leucine/ketoleucine (2-oxo-4-methyl-pentanoic acid) reaction, for instance, equals 9×10^{12} M^{-2} [50], whereas for the phenylalanine/phenylpyruvate reaction a calculated Keq of 2.2×10^{13} M^{-2} has been reported [51]. The conversion yield of AADH-catalyzed amination reaction is therefore almost quantitative. High yields, very high enantioselectivity, and the broad spectrum of available enzymes make enzymatic reductive amination a very attractive tool for production of α-amino acids [52].

A large number of synthetic applications have been reported with LeuDHs from several *Bacilli* and from *Thermoactinomyces intermedius* [53]. LeuDHs from *Bacillus stearothermophilus*, *Bacillus cereus*, and *Bacillus sphaericus* have remarkably similar substrate spectra, and accept α-keto acids with hydrophobic, aliphatic, branched, and unbranched carbon chains from four up to and including six C atoms in a straight chain. Furthermore, they can convert some alicyclic α-keto acids, but do not have any activity for aromatic substrates, for example, L-phenylalanine [51]. The specific activities of these LeuDHs for the reference substrate 2-oxo-4-methyl-pentanoic acid (ketoleucine) vary significantly from rather low for synthetic applications (3.3 U mg^{-1}, *B. sphaericus*) to very good (120 U mg^{-1}, *B. stearothermophilus*) [52].

The fact that aromatic amino acids are poor substrates for LeuDHs can be addressed by PheDHs, for example, from *Brevibacterium* sp. [54], *Rhodococcus* sp. [55], and *T. intermedius* [56]. Studies showed that the substrate specificity of these PheDHs is very broad, not only accepting the similar type of substrates as the LeuDHs but also α-keto acids with an aromatic side chain. Consequently, the substrate range of LeuDHs and PheDHs are complementary. Recently, synthesis of D-amino acids via enzymatic reductive amination has also been reported based on an engineered 2,6-diaminopimelic acid dehydrogenase of the enzyme from *Corynebacterium glutamicum* [57]. The substrate scope of this enzyme is still limited, making it a target for future improvement by enzyme engineering approaches.

A number of different process concepts have been explored to efficiently operate the reductive amination process. Kula, Wandrey, and coworkers developed a process with isolated enzymes in an enzyme-membrane reactor (EMR), whereas others use batch or fed-batch processes in standard stirred tank reactors employing isolated enzymes or whole-cell processes (see also Chapter 10).

At Bristol-Myers Squibb, for instance, a recombinant *Pichia pastoris* coexpressing a *T. intermedius* PheDH gene and the endogenous FDH was applied for the production of (S)-2-amino-5-(1,3-dioxolan-2-yl)-pentanoic acid (allysine ethylene acetal), a key intermediate of the former drug candidate omapatrilat, with 97% yield and >98% ee [58].

Galkin *et al.* constructed amongst others two different recombinant *E. coli* strains both expressing the FDH from *Mycobacterium vaccae* in combination with either the LeuDH or the PheDH from *T. intermedius*. The whole-cell biocatalysts produced L-leucine, L-valine, L-norvaline and L-methionine (combination LeuDH–FDH), and L-phenylalanine, as well as L-tyrosine (combination PheDH–FDH) in high molar yields (>88%) and excellent optical purities (>99.9% ee) [59].

The former Degussa Company (now Evonik Industries) also reported on the construction of a whole-cell biocatalyst for the asymmetric reductive amination of α-keto acids as a cost-attractive alternative to their current route, which is based on the application of (expensive) isolated and purified enzymes [60]. Application of this new whole-cell biocatalyst for the synthesis of L-*tert*-leucine showed that this reaction proceeded even without the addition of external cofactor. However, substrate concentrations of 0.5 M and higher led to incomplete conversion, which was in line with the earlier observations by Galkin *et al.* [59]. This problem could be solved by the addition of a low amount of additional cofactor (1–10 µM) or, more attractively, by fed-batch operation. Continuous addition of 1.25 M solution of the substrate trimethylpyruvic acid over a period of 12 h (equaling a final substrate concentration of 1 M) resulted in an overall conversion of >95% after 24 h without the need for an external cofactor. The enantiomeric excess of the L-*tert*-leucine obtained after workup (yield 84%) was >99%. The same whole-cell biocatalyst has also been successfully applied for the synthesis of L-neopentylglycine with >95% conversion and >99% ee at substrate concentrations of up to 88 g l^{-1}. At this high concentration, the amino acid formed precipitates from the reaction mixture. Nevertheless, the resting cells stayed intact and the reaction proceeded without addition of external cofactor [61].

Recently, BMS reported another industrial application in the production of (S)-3-hydroxyadamantylglycine, a key intermediate for the synthesis of saxagliptin [62]. In this process, a PheDH mutant from *T. intermedius* in combination with a formate dehydrogenase from *P. pastoris* was used. Almost, a quantitative conversion of the keto acid could be achieved at 30 kg scale within 49 h producing exclusively the *S*-enantiomer.

16.2.5
Transamination Reactions

Transaminases (EC 2.6.1.X) are industrially applied for the production of optically pure amines and amino acids starting from the corresponding ketones and keto acids, respectively [63, 64]. Amino acid transaminases strictly require a carboxylic group at the α-position of the substrate to be converted, whereas ω-transaminases (EC 2.6.1.18, also called *β-alanine pyruvate transaminase*) accept a much broader spectrum of ketones and amines. Therefore, production of optically pure amines and β-amino acids make use of ω-transaminases, whereas α-amino acids are synthesized using amino acid transaminases. These enzymes can be used in a resolution mode, but more interestingly also in a theoretical 100% yield asymmetric synthesis concepts.

Both transaminases catalyze the amine transfer from a given amine donor via their cofactor pyridoxal-5-phosphate (PLP). In contrast to reductive amination via dehydrogenases (Section 16.2.4), the equilibrium of the amination reaction starting from a ketone is, in many cases, unfavorable, even at a large excess of the amine donor [65]. For obtaining high yields, methods to shift the equilibrium toward the desired amine product are required. These methods are usually based on pulling the equilibrium by removal of the formed ketone derived from the amine donor. Depending on the used amine donor, the reaction can be pulled, for example, by biphasic systems or by coupled enzymatic methods. For industrial production of α-amino acids, a large number of different approaches for shifting the equilibrium have been developed as reviewed by Fotheringham *et al.* [66, 67].

For the synthesis of chiral amines using ω-transaminases, alanine is an often used amine donor. Lactate dehydrogenases [65, 68], pyruvate decarboxylases [69], and AlaDHs [70] have been shown to be effective for the removal of pyruvate that is formed from alanine. The latter approach when coupled to a cofactor regeneration system as shown in Scheme 16.8 is conceptually very elegant. Theoretically, it requires only the target ketone, ammonia, catalytic quantities of alanine, and glucose for cofactor regeneration. Despite the high yields reported for most of the tested amines (>90%), the requirement of high excess of alanine (fivefold) and low

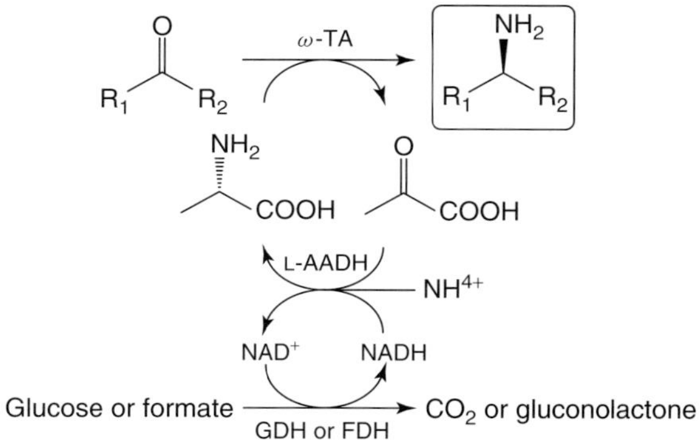

Scheme 16.8 Synthesis of chiral amines from ketones, ammonia, and catalytic amounts of alanine using a coupled ω-transaminase, L-alanine dehydrogenase, and a cofactor regeneration system (based on glucose or formate dehydrogenase).

substrate/product concentrations of 50 mM are not yet satisfactory for industrial applications. However, if these hurdles can be overcome, this concept might become another very powerful enzymatic synthesis method for chiral primary amines.

A quite general concept for driving the equilibrium of transaminase-catalyzed reactions was recently developed by Fotheringham and Oswald using peroxide to degrade pyruvate into acetic acid during the enzyme reaction, which significantly increased the reaction yields [71].

Some of the commercialized processes of ω-transaminases use isopropylamine as amine donor. The advantage of this system is an easy removal of acetone by evaporation. One of the earliest reports and a landmark example of the industrial application of ω-transaminases is the production of (S)-methoxyisopropylamine, a herbicide intermediate developed by Celgene in the early 1990s [72–74]. Using an engineered transaminase, a highly efficient process was developed which allowed for a 97% conversion of 180 g l^{-1} methoxyacetone into the optically pure amine (>99% ee) within 7 h. Other tons scale processes using this approach are reported for the production of 7-methoxy-2-aminotetralin and D-amphetamine [75].

16.3
Enzymatic Desymmetrization

The desymmetrization of achiral symmetrical compounds can create chiral products at a theoretical 100% yield as shown in Table 16.3. The basis for this is the unique property of enzymes to discriminate chemically identical reactive groups. These groups can be either located at the same (Scheme 16.9a) or different carbon atoms (Scheme 16.9b).

A lot of very efficient syntheses are reported based on enzymatic desymmetrization processes, which were reviewed by Gotor et al. in 2005 [86]. Classical examples are conversions of diesters by hydrolases such as pig liver esterase (PLE). Some examples are shown in Table 16.3, which indicate the versatility of the desymmetrization method.

For more than 100 years, only crude enzyme extract from pig liver (called *pig liver esterase*) was available as biocatalyst, which contains a mix of different esterase isoforms having different selectivities. Given the fact that the isoform composition can vary from batch to batch leading to irreproducible results and the need for a nonanimal substitute for pharma applications, Bornscheuer et al. cloned and successfully expressed for the first time various isoforms in P. pastoris and E. coli [87–91]. Recently, Schwab et al. developed together with DSM a highly efficient expression system for the production of various isoforms of PLE [92–94]. On the basis of this highly efficient system, different isoforms of PLE are now available for which a first tons scale application for the production of a pharma intermediate was reported (Oliver May, personal communication (2009)).

In addition to many enzymatic ester hydrolysis or acetylation reactions [95], also redox [96–98], hydratation, and hydrolysis reactions of dinitriles [99] have

Table 16.3 Examples of products obtained by desymmetrization of diesters by pig liver esterase.

Product	Conversion yield (%)	Enantiomeric excess (% ee)	References
4-F-C6H4-CH(CO2CH3)-CH2-CO2H	89	95	[76]
cyclopropane-CO2CH3/CO2H	88	97	[77]
norbornadiene-CO2nPr/CO2H	99	71	[78]
cyclohexane with CH3, OAc, CH3, OH	100	>99.5	[79]
Ph-P(=O)(CO2CH3)(CO2H)	92	72	[80]
decalin-OH/OAc	60–70	98	[81]
hydantoin: BnN-CO2H/BnN-CO2CH3	90	91	[82]
diazine-cyclohexane-OH/OAc	92	90	[83]

(*continued overleaf*)

Table 16.3 (continued)

Product	Conversion yield (%)	Enantiomeric excess (% ee)	References
(S)-configured cyclohexane with OH, O-sec-butyl, CO₂Et, HO, CO₂H substituents	98	96–98	[84]
H₃C, CO₂H, CO₂CH₃, phenyl-substituted stereocenter	90	81	[85]

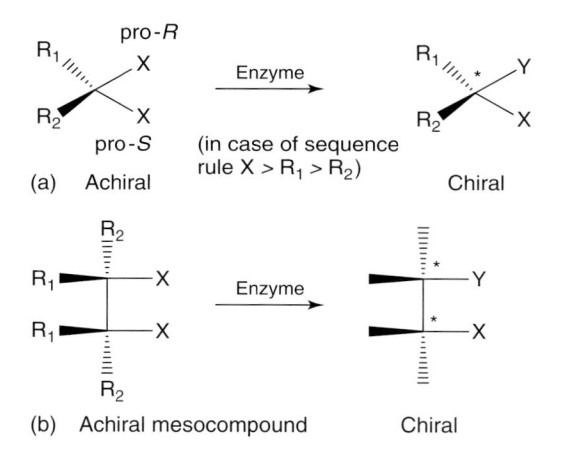

Scheme 16.9 Enzyme-catalyzed desymmetrization of achiral compounds into chiral products at a theoretical 100% yield.

(a) Achiral tetrahedral center with R₁, R₂, and two X groups (pro-R and pro-S, in case of sequence rule X > R₁ > R₂) → Chiral product with R₁, R₂, Y, X.

(b) Achiral mesocompound with R₁, R₂, X groups → Chiral product.

been used in desymmetrization processes. For example, an elegant synthesis for the production of the statin intermediate (R)-4-cyano-3-hydroxybutyrate has been reported by Dow in collaboration with Verenium (former Diversa). This process is based on the desymmetrization of 3-hydroxyglutaronitrile using a nitrilase, which quantitatively converts 330 g l^{-1} of the substrate in 16 h into the desired optically pure (99% ee) product [100].

A recent example of an industrial enzymatic desymmetrization process operated at kilogram scale was reported by BMS for the production of (1S,2R)-2-(methoxycarbonyl)cyclohex-4-ene-1-carboxylic acid, an active pharmaceutical ingredient intermediate for a potential chemokine receptor modulator [101]. Similar compounds have been synthesized before using PLE, which, however, delivered the opposite enantiomer [102, 103]. By screening of a hydrolase collection, several commercially available enzymes were identified with opposite selectivity as PLE. The best performing enzyme (Novozyme 435, *Candida antartica* Lipase B) was used to convert approximately $100\,\mathrm{g\,l^{-1}}$ of the diester within approximately 24 h into the desired product with a total yield of >98% and an optical purity of >99.9% ee.

16.4
Enzymatic Deracemization

Racemic compounds can be converted into single enantiomers at 100% theoretical yield by a so-called deracemization process [104]. This involves a transformation of racemic mixtures (or one of the enantiomers) into nonchiral compounds, which are subsequently converted enantioselectively to one single enantiomer as shown in Scheme 16.10.

Deracemization processes often involve redox cycles based on dehydrogenases or oxidases for which many examples are reported for the production of amines, amino acids, and alcohols [105–107]. However, epoxide hydrolases, sulfhydrolases, and very recently ω-transaminases have also been used [70, 108, 109]. The later application of ω-transaminases developed by Kroutil *et al.* is based on the reaction shown in Scheme 16.11. High conversions can be reached by running the reaction in two consecutive steps employing first a resolution of a racemic mixture of amines with one *R*- or *S*-selective ω-transaminase followed by heat inactivation and a subsequent stereoselective amination with another ω-transaminase having an opposite enantioselectivity. In contrast to the use of amine oxidases for which at the moment no efficient *R*-selective enzymes are available, the transaminase approach offers access to both enantiomers of a broad range of amines for which examples are shown in Scheme 16.12. However, a drawback might be the requirement of four enzymes compared to one (plus one chemo catalyst) for the amine oxidase approach developed by Turner *et al.* [110, 111]. The latter can also be used for the production

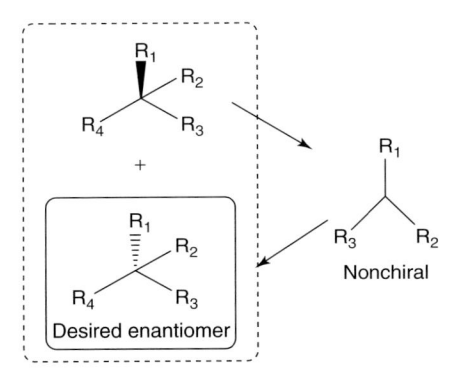

Scheme 16.10 Deracemization of a racemic mixture into one single desired enantiomer by stereoinversion.

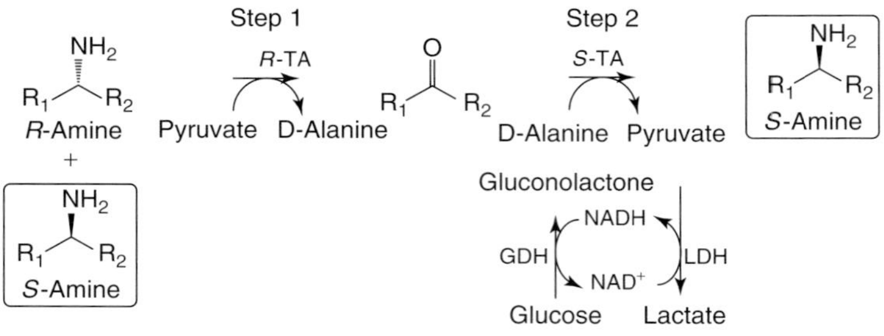

Scheme 16.11 Reaction scheme for deracemization of racemic amines using ω-transaminases of opposite enantioselectivities. In the first step, an R-selective ω-transaminase (R-TA) is converting the R-amine and pyruvate into the corresponding ketone, respectively, D-alanine. After heat inactivation of the R-TA, the ketone is converted by an S-selective ω-transaminase (S-TA) into the corresponding S-amine. The equilibrium is pulled by converting pyruvate into lactate using a lactate dehydrogenase (LDH)- and a glucose dehydrogenase-based cofactor regeneration system.

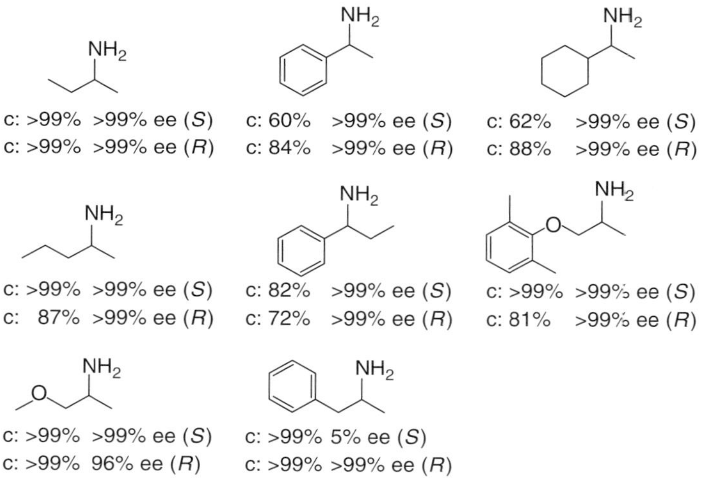

Scheme 16.12 Examples of amines obtained at a certain conversion yield (c) and optical purity (% ee) by deracemization of racemic amines using ω-transaminase of opposite enantioselectivities according to the reaction shown in Scheme 16.11.

of secondary and tertiary amines which are not accessible via transaminases [112, 113].

A recent example of a deracemization process operated at kilogram scale was reported by Patel et al. for the production of (R)-2-amino-3-(7-methyl-1H-indazol-5-yl)propanoic acid [114]. This drug intermediate was prepared in 68% yield with >99% ee from the racemic amino acid using a L-amino acid deaminase from *Proteus mirabilis* in combination with a D-transaminase using D-alanine as amine donor.

16.5
Dynamic Kinetic Resolution

DKR involves the combination of an enantioselective transformation with an *in situ* racemization process as shown in Scheme 16.13. This allows, in principle, for a quantitative conversion of a racemic mixture into an optically pure compound without the need to isolate and recycle intermediates or the nonreactive enantiomer, a limitation of classical kinetic resolutions. Given the obvious attractiveness of DKR, this concept is very actively explored and a large number of applications have been reported, which were recently reviewed by Pellissier *et al.*, Bäckvall *et al.*, and Abdul Halim *et al.* [115–117].

For chemoenzymatic DKRs, a mild racemization method compatible with the enantioselective enzymatic reaction is the key. Such racemization methods have been developed, for example, based on racemases [118–120], base, and metal catalysts [121], or on the intrinsic property of some compounds to racemize spontaneously [122, 123].

A large number of different enzymes have been applied. Proteases and lipases are often used for the production of chiral alcohols, acids, esters, or amines [124–128]. Typical examples of pharmaceutically relevant compounds that have been synthesized by DKR using lipases and trioctylamine or sodium hydroxide as base catalyst are (S)-naproxen, (S)-fenoprofen, and (S)-ibuprofen [129–131]. Not surprisingly, lipases and proteases are very often used because of their typically high stability. This is required, as nonenzymatic racemization conditions are usually rather demanding for enzymes.

For the production of L- and D-amino acids, hydantoinases [132], acylases [133], and amidases [134] are applied. These are often coupled to the corresponding racemases, which sometimes requires compromises as optimal reaction conditions can be different for the racemases and hydrolases.

Hydroxyacids are compounds that are accessible, for example, by conversions of cyanohydrins using nitrilases, whereas halohydrin dehalogenases have been used for DKR of epihalohydrins [135–137]. Furthermore, the use of ADHs, Baeyer–Villiger monooxygenases, and very recently also the use of ω-transaminases has been reported [138–140].

Despite a very large number of scientific literatures, the numbers of large-scale DKR processes is still rather small. A well-known example of an enzymatic DKR process operated on tons scale by various companies is the hydantoinase process for the production of D-*p*-OH-phenylglycine and other D- and L-amino acids based

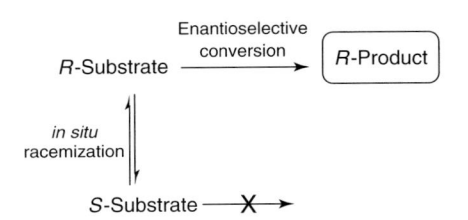

Scheme 16.13 Concept of dynamic kinetic resolution (DKR).

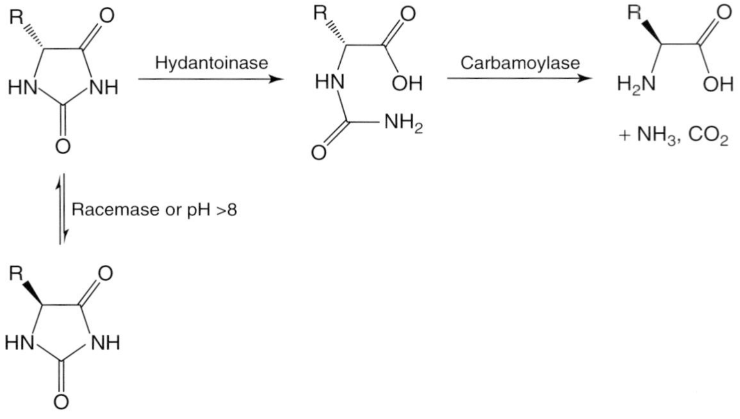

Scheme 16.14 Reaction scheme of the hydantoinase process.

on the reaction root shown in Scheme 16.14 [141]. The advantages of this process are easily accessible substrates from cheap aldehydes as well as extremely high optical purities (typically >99%) of the produced amino acids. The latter is secured by the consecutive reaction of two enantioselective enzymes (hydantoinase and carbamoylase). Another example of an industrial DKR process is the nitrilase process for the production of (R)-mandelic acid operated at BASF [17].

16.6
Summary and Outlook

Biocatalysis has come a long way from being considered as a "diluted chemistry" with little industrial value in the past to a well-established and for some transformations even a "first choice technology" of today. The technological progress has been tremendous during the last 10 years that enabled faster access to a broader spectrum of enzymes offering very efficient synthetic concepts. Such concepts that allow for the production of optically pure compounds at 100% yield have been the focus of this chapter. The intrinsic efficiency of these concepts holds great promise in chemistry in general and brings urgently required medicines developed in the pharmaceutical industry to the patient with minimum impact on the environment at affordable costs and therefore in a sustainable manner.

However, the full value of enzymes will only be captured if we manage to integrate biocatalysis as a standard organic chemistry tool in our route scouting and process R&D activities. There is still room for significant improvement in this integration of chemistry and biotechnology. Furthermore, despite all the progress made since the first enzyme designed for laundry applications 25 years ago [142], the technology still needs further improvement. And, it will be further improved by increasingly efficient enzyme design methods. These will allow us to further drive down enzyme costs faster, with lower development cost and risk. Furthermore, new enzymes will be discovered and designed which will expand the range of reactions that can be efficiently catalyzed by enzymes. It is conceivable, and first promising results are reported, that enzymes can be created *de novo* to catalyze novel reactions not found in nature [143]. For sure, it will still take some years until the first *de novo* designed enzyme will be implemented on an industrial scale. But until then, there are already plenty of opportunities to use the currently existing enzymes for developing more sustainable processes for manufacturing increasingly complex drug molecules with 100% yield and 100% ee, the concepts of which have been reviewed in this chapter.

References

1. Constable, D.J.C., Dunn, P.J., Hayler, J.D., Humphrey, G.R., Leazer, J.L. Jr., Linderman, R.J., Lorenz, K., Manley, J., Pearlman, B.A., Wells, A., Zaks, A., and Zhang, T.Y. (2007) *Green Chem.*, **9**, 411.
2. Challenger, S., Dudin, L., DaSilva, J., Dunn, P., Govaerts, T., Hayler, J., Hinkley, B., Houpis, Y., Hunter, T., Jellet, L., Leazer, J.L. Jr., Lorenz, K., Mathew, S., Rammeloo, T., Sudini, R., Wan, Z., Welch, C., Wells, A., Vance, J., Xie, C., and Zhang, F. (2008) *Org. Process Res. Dev.*, **12**, 807.
3. Andrews, I., Cui, J., DaSilva, J., Dudin, L., Dunn, P., Hayler, J., Hinkley, B., Hughes, D., Kaptein, B., Kolis, S., Lorenz, K., Mathew, S., Rammeloo, T., Wang, L., Wells, A., White, T., Xie, C., and Zhang, F. (2009) *Org. Process Res. Dev.*, **13**, 397.
4. Green, B.D. and Keller, M. (2006) *Curr. Opin. Biotechnol.*, **17**, 236.
5. Tracewell, C.A. and Arnold, F.H. (2009) *Curr. Opin. Chem. Biol.*, **13**, 3.
6. Du, L. and Egholm, M. (eds) (2008) in *Next Generation Genome Sequencing* (ed. M. Janitz), Wiley-VCH Verlag GmbH & Co. KGaA, Weinheim, pp. 43–56.
7. Van den Brulle, J., Fischer, M., Langmann, T., Horn, G., Waldmann, T., Arnold, S., Fuhrmann, M., Schatz, O., O'Connell, T., O'Connell, D., Auckenthaler, A., and Schwer, H. (2008) *BioTechniques*, **45**, 340.
8. Urlacher, V.B. and Schmid, R.D. (2006) *Curr. Opin. Chem. Biol.*, **10**, 156.
9. Julsing, M.K., Cornelissen, S., Buehler, B., and Schmid, A. (2008) *Curr. Opin. Chem. Biol.*, **12**, 177.
10. Leak, D.J., Sheldon, R.A., Woodley, J.M., and Adlercreutz, P. (2009) *Biocatal. Biotransform.*, **27**, 1.
11. IUPAC (1997) in *Compendium of Chemical Terminology*, 2nd edn (the "Gold Book"). Compiled by (eds A. D. McNaught and A. Wilkinson), Blackwell Scientific Publications, Oxford.
12. Neuberg, C. and Hirsch, J. (1921) *Biochem. Z.*, **115**, 282.
13. Neuberg, C. and Ohle, H. (1922) *Biochem. Z.*, **128**, 610.
14. Breuer, M., Ditrich, K., Habicher, T., Hauer, B., Kesseler, M., Stuermer, R., and Zelinski, T. (2004) *Angew. Chem. Int. Ed.*, **43**, 788.
15. Brockamp, H.P., Kula, M.R., and Goetz, F. (1991) DE Patent 3, 940,431.
16. Fessner, W.D. (1992) DE Patent 4, 111,971.
17. Wong, C.-H. (1992) US Patent 5, 162,513.
18. Muhmoudian, M., Noble, D., Drake, C.S., Middleton, R.F., Montgomery, D.S., Piercey, J.E., Ramlakhan, D., Todd, M., and Dawson, M.J. (1997) *Enzyme Microb. Technol.*, **20**, 393.
19. Gijsen, H.J.M. and Wong, C.-H. (1994) *J. Am. Chem. Soc.*, **116**, 8422.
20. Machajewski, T.D., Wong, C.-H., and Lerner, R.A. (2000) *Angew. Chem., Int. Ed.*, **39**, 1352.
21. Greenberg, W.A., Varvak, A., Hanson, S.R., Wong, K., Huang, H., Chen, P., and Burk, M.J. (2004) *Proc. Natl. Acad. Sci. U.S.A.*, **101**, 5788.

22. Thayer, A.M. (2006) *Chem. Eng. News*, **84**, 26.
23. Jennewein, S., Schuermann, M., Wolberg, M., Hilker, I., Luiten, R., Wubbolts, M., and Mink, D. (2006) *Biotechnol. J.*, **1**, 537.
24. Purkarthofer, T., Skranc, W., Schuster, C., and Griengl, H. (2007) *Appl. Microbiol. Biotech.*, **76**, 309.
25. Glieder, A., Weis, R., Skranc, W., Poechlauer, P., Dreveny, I., Majer, Sandra, S., Wubbolts, M., Schwab, H., and Gruber, K. (2003) *Angew. Chem. Int. Ed.*, **42**, 4815.
26. Weis, R., Gaisberger, R., Skranc, W., Gruber, K., and Glieder, A. (2005) *Angew. Chem. Int. Ed.*, **44**, 4700.
27. Yamada, S., Nabe, K., Izuo, N., Nakamichi, K., and Chibata, I. (1981) *Appl. Environ. Microbiol.*, **42**, 773.
28. Wu, B., Szymanski, W., Wietzes, P., de Wildeman, S., Poelarends, G.J., Feringa, B.L., and Janssen, D.B. (2009) *Chembiochem*, **10**, 338.
29. Buchholz, S. and Groeger, H. (eds) (2007) in *Biocatalysis in the Pharmaceutical and Biotechnology Industries* (ed. R.N. Patel), CRC Press LLC, Boca Raton, pp. 757–790.
30. De Wildeman, S.M.A., Sonke, T., Schoemaker, H.E., and May, O. (2007) *Acc. Chem. Res.*, **40**, 1260.
31. Knowles, W.S., Noyori, R., and Sharpless, K.B. (2002) *Angew. Chem. Int. Ed.*, **41**, 1998.
32. Wichmann, R. and Vasic-Racki, D. (2005) *Adv. Biochem. Eng.:Biotechnol.*, **92**, 225.
33. Kula, M.R. and Pohl, M. (2002) Deutscher Zukunftspreis see http://www.deutscherzukunftspreis.de/archiv/02.htm 2009.
34. Kizaki, N., Yasohara, Y., Hasegawa, J., Wada, M., Kataoka, M., and Shimizu, S. (2001) *Appl. Microbiol. Biotechnol.*, **55**, 590.
35. Davis, S.C., Grate, J.H., Gray, D.R., Gruber, J.M., Huisman, G.W., Ma, S.K., Newman, L.M., Sheldon, R.A., and Wang, L.A. (2004) WO Patent 2, 004,015,132.
36. For the 2006 Greener Reaction Conditions Award of Codexis see http://www.epa.gov/greenchemistry/pubs/pgcc/winners/grca06.html 2009.
37. Rozzell, D. and Liang, J. (2008) *Speciality Chem. Mag.*, **28**, 36.
38. Campaipano, O., Mundorff, E., Borup, B., and Voladri, R. (2009) WO Patent 2, 009,046,153, Codexis, Inc., USA.
39. Kosjek, B., Nti-Gyabaah, J., Telari, K., Dunne, L., and Moore, J.C. (2008) *Org. Process Res. Dev.*, **12**, 584.
40. Uzura, A., Nomoto, F., Sakoda, A., Nishimoto, Y., Kataoka, M., and Shimizu, S. (2009) *Appl. Microbiol. Biotechnol.*, **83**, 617.
41. Pollard, D., Truppo, M., Pollard, J., Chen, C.-Y., and Moore, J. (2006) *Tetrahedron: Asymmetry*, **17**, 554.
42. Kataoka, M., Kotaka, A., Hasegawa, A., Wada, M., Yoshizumi, A., Nakamori, S., and Shimizu, S. (2002) *Biosci. Biotechnol. Biochem.*, **66**, 2651.
43. Shimizu, S. and Wada, M. (2003) WO Patent 2, 030,70959.
44. Kataoka, M., Kotaka, A., Thiwthong, R., Wada, M., Nakamori, S., and Shimizu, S. (2004) *J. Biotechnol.*, **114**, 1.
45. Wada, M., Yoshizumi, A., Noda, Y., Kataoka, M., Shimizu, S., Takagi, H., and Nakamori, S. (2003) *Appl. Environ. Microbiol.*, **69**, 933.
46. Kosjek, B., Fleitz, F.J., Dormer, P.G., Kuethe, J.T., and Devine, P.N. (2008) *Tetrahedron: Asymmetry*, **19**, 1403.
47. Hall, M., Stueckler, C., Hauer, B., Stuermer, R., Friedrich, T., Breuer, M., Kroutil, W., and Faber, K. (2008) *Eur. J. Org. Chem.*, **9**, 1511.
48. Stuermer, R., Hauer, B., Hall, M., and Faber, K. (2007) *Curr. Opin. Chem. Biol.*, **11**, 203.
49. Ohshima, T. and Soda, K. (eds) (2000) in *Stereoselective Biocatalysis* (ed. R.N. Patel), Marcel Dekker, Inc., New York, pp. 877–902.
50. Sanwal, B.D. and Zink, M.W. (1961) *Arch. Biochem. Biophys.*, **94**, 430.
51. Brunhuber, N.M.W., Thoden, J.B., Blanchard, J.S., and Vanhooke, J.L. (2000) *Biochemistry*, **39**, 9174.
52. Bommarius, A.S. (ed.) (2002) in *Enzyme Catalysis in Organic Synthesis*, 2nd edn, vol. 3 (eds K. Drauz and

H. Waldmann), Wiley-VCH Verlag GmbH, Weinheim, pp. 1047–1063.
53. Ohshima, T., Nishida, N., Bakthavatsalam, S., Kataoka, K., Takada, H., Yoshimura, T., Esaki, N., and Soda, K. (1994) *Eur. J. Biochem.*, **222**, 305.
54. Hummel, W., Weiss, N., and Kula, M.R. (1984) *Arch. Microbiol.*, **137**, 47.
55. Hummel, W., Schuette, H., Schmidt, E., Wandrey, C., and Kula, M.R. (1987) *Appl. Microbiol. Biotechnol.*, **26**, 409.
56. Ohshima, T., Takada, H., Yoshimura, T., Esaki, N., and Soda, K. (1991) *J. Bacteriol.*, **173**, 3943.
57. Vedha-Peters, K., Gunawardana, M., Rozzell, J.D., and Novick, S.J. (2006) *J. Am. Chem. Soc.*, **128**, 10923.
58. Hanson, R.L., Howell, J.M., LaPorte, T.L., Donovan, M.J., Cazzulino, D.L., Zannella, V., Montana, M.A., Nanduri, V.B., Schwarz, S.R., Eiring, R.F., Durand, S.C., Wasylyk, J.M., Parker, W.L., Liu, M.S., Okuniewicz, F.J., Chen, B.-C., Harris, J.C., Natalie, K.J., Ramig, K., Swaminathan, S., Rosso, V.W., Pack, S.K., Lotz, B.T., Bernot, P.J., Rusowicz, A., Lust, D.A., Tse, K.S., Venit, J.J., Szarka, L.J., and Patel, R.N. (2000) *Enzyme Microb. Technol.*, **26**, 348.
59. Galkin, A., Kulakova, L., Tishkov, V., Esaki, N., and Soda, K. (1995) *Appl. Microbiol. Biotechnol.*, **44**, 479.
60. Menzel, A., Werner, H., Altenbuchner, J., and Groeger, H. (2004) *Eng. Life Sci.*, **4**, 573.
61. Groeger, H., May, O., Werner, H., Menzel, A., and Altenbuchner, J. (2006) *Org. Process Res. Dev.*, **10**, 666.
62. Hanson, R.L., Goldberg, S.L., Brzozowski, D.B., Tully, T.P., Cazzulino, D., Parker, W.L., Lyngberg, O.K., Vu, T.C., Wong, M.K., and Patel, R.N. (2007) *Adv. Synth. Catal.*, **349**, 1369.
63. Crump, S.P. and Rozzell, J.D. (eds) (1992) in *Biocatalytic Production of Amino Acids Derivatives* (eds J.D. Rozzell and F. Wagner), John Wiley & Sons, Inc., New York, pp. 43–58.
64. Hoehne, M. and Bornscheuer, U.T. (2009) *ChemCatChem*, **1**, 42.
65. Shin, J.-S. and Kim, B.-G. (1999) *Biotechnol. Bioeng.*, **65**, 206.
66. Taylor, P.P., Pantaleone, D.P., Senkpeil, R.F., and Fotheringham, I.G. (1998) *Trends Biotechnol.*, **16**, 412.
67. Li, T., Kootstra, A.B., and Fotheringham, I.G. (2002) *Org. Process Res. Dev.*, **6**, 533.
68. Truppo, M.D., Turner, N.J., and Rozzell, J.D. (2009) *Chem. Commun.*, 2127.
69. Hoehne, M., Kuehl, S., Robins, K., and Bornscheuer, U.T. (2008) *ChemBioChem*, **9**, 363.
70. Koszelewski, D., Clay, D., Rozzell, D., and Kroutil, W. (2009) *Eur. J. Org. Chem.*, **14**, 2289.
71. Fotheringham, I. and Oswald, N. (2008) US Patent 2, 008,213,845.
72. Matcham, G. (1992) *Speciality Chem. Mag.*, **12**, 178, 180.
73. Matcham, G.W. and Stg Bowen, A.R. (1996) *Chim. Oggi*, **14**, 20.
74. Matcham, G., Bhatia, M., Lang, W., Lewis, C., Nelson, R., Wang, A., and Wu, W. (1999) *Chimia*, **53**, 584.
75. Scarlato, G. (2009) *Specialty Chem. Mag.*, **6**, 56.
76. Yu, M.S., Lantos, I., Peng, Z.-Q., Yu, J., and Cacchio, T. (2000) *Tetrahedron Lett.*, **41**, 5647.
77. Sabbioni, G. and Jones, J.B. (1987) *J. Org. Chem.*, **52**, 4565.
78. Kashima, Y., Liu, J., Takenami, S., and Niwayama, S. (2002) *Tetrahedron: Asymmetry*, **13**, 953.
79. Bohm, C., Austin, W.F., and Trauner, D. (2003) *Tetrahedron: Asymmetry*, **14**, 71.
80. Kielbasinski, P., Zurawinski, R., Albrycht, M., and Mikolajczyk, M. (2003) *Tetrahedron Asymmetry*, **14**, 3379.
81. Chenevert, R., Courchesne, G., and Jacques, F. (2004) *Tetrahedron: Asymmetry*, **15**, 3587.
82. Chen, F.-E., Chen, X.-X., Dai, H.-F., Kuang, Y.-Y., Xie, B., and Zhao, J.-F. (2005) *Adv. Synth. Catal.*, **347**, 549.
83. Chenevert, R. and Jacques, F. (2006) *Tetrahedron: Asymmetry*, **17**, 1017.
84. Zutter, U., Iding, H., Spurr, P., and Wirz, B. (2008) *J. Org. Chem.*, **73**, 4895.

85. Toone, E.J. and Jones, J.B. (1991) *Tetrahedron: Asymmetry*, **2**, 1041.
86. Garcia-Urdiales, E., Alfonso, I., and Gotor, I.V. (2005) *Chem. Rev.*, **105**, 313.
87. Musidlowska, A., Lange, S., and Bornscheuer, U.T. (2001) *Angew. Chem. Int. Ed.*, **40**, 2851.
88. Lange, S., Musidlowska, A., Schmidt-Dannert, C., Schmitt, J., and Bornscheuer, U.T. (2001) *ChemBioChem*, **2**, 576.
89. Musidlowska-Persson, A. and Bornscheuer, U.T. (2003) *Protein Eng.*, **16**, 1139.
90. Boettcher, D., Bruesehaber, E., Doderer, K., and Bornscheuer, U.T. (2007) *Appl. Microbiol. Biotechnol.*, **73**, 1282.
91. Hummel, A., Bruesehaber, E., Boettcher, D., Trauthwein, H., Doderer, K., and Bornscheuer, U.T. (2007) *Angew. Chem., Int. Ed.*, **46**, 8492.
92. Steinbauer, G., Stanek, M., Pojarliev, P., Skranc, W., Schwab, H., Wubbolts, M., Kierkels, J., Pichler, H., Hermann, M., and Zenzmaier, C. (2007) WO Patent 2, 007,073,847.
93. Hermann, M., Kietzmann, M.U., Ivancic, M., Zenzmaier, C., Luiten, R.G.M., Skranc, W., Wubbolts, M., Winkler, M., Birner-Gruenberger, R., Pichler, H., and Schwab, H. (2008) *J. Biotechnol.*, **133**, 301.
94. Kietzmann, M., Schwab, H., Pichler, H., Ivancic, M., May, O., and Luiten, R.G.M. (2009) WO Patent 2, 009,004,093.
95. Morgan, B., Dodds, D.R., Zaks, A., Andrews, D.R., and Klesse, R. (1997) *J. Org. Chem.*, **62**, 7736.
96. Snajdrova, R., Braun, I., Bach, T., Mereiter, K., and Mihovilovic, M.D. (2007) *J. Org. Chem.*, **72**, 9597.
97. Rial, D.V., Bianchi, D.A., Kapitanova, P., Lengar, A., van Beilen, J.B., and Mihovilovic, M.D. (2008) *Eur. J. Org. Chem.*, **7**, 1203.
98. Fujii, M., Takeuchi, M., Akita, H., and Nakamura, K. (2009) *Tetrahedron Lett.*, **50**, 4941.
99. Kielbasinski, P., Rachwalski, M., Kwiatkowska, M., Mikolajczyk, M., Wieczorek, W.M., Szyrej, M., Sieron, L., and Rutjes, F.P.J.T. (2007) *Tetrahedron: Asymmetry*, **18**, 2108.
100. Bergeron, S., Chaplin, D.A., Edwards, J.H., Ellis, B.S.W., Hill, C.L., Holt-Tiffin, K., Knight, J.R., Mahoney, T., Osborne, A.P., and Ruecroft, G. (2006) *Org. Process Res. Dev.*, **10**, 661.
101. Goswami, A. and Kissick, T.P. (2009) *Org. Process Res. Dev.*, **13**, 483.
102. Schneider, M., Engel, N., Hoenicke, P., Heinemann, G., and Goerisch, H. (1984) *Angew. Chem.*, **96**, 55.
103. Laumen, K., Reimerdes, E.H., Schneider, M., and Goerisch, H. (1985) *Tetrahedron Lett.*, **26**, 407.
104. Steinreiber, J., Faber, K., and Griengl, H. (2008) *Chem.--A Eur. J.*, **14**, 8060.
105. Azerad, R. and Buisson, D. (2000) *Curr. Opin. Biotechnol.*, **11**, 565.
106. Turner, N.J. (2004) *Curr. Opin. Chem. Biol.*, **8**, 114.
107. Turner, N.J. (ed.) (2008) in *Asymmetric Organic Synthesis with Enzymes* (eds V. Gotor, I. Alfonso, and E. Garcia-Urdiales), Wiley-VCH Verlag GmbH & Co. KGaA, Weinheim, pp. 115-131.
108. Simeo, Y. and Faber, K. (2006) *Tetrahedron: Asymmetry*, **17**, 402.
109. Gadler, P. and Faber, K. (2007) *Trends Biotechnol.*, **25**, 83.
110. Alexeeva, M., Enright, A., Dawson, M.J., Mahmoudian, M., and Turner, N.J. (2002) *Angew. Chem. Int. Ed.*, **41**, 3177.
111. Carr, R., Alexeeva, M., Enright, A., Eve, T.S.C., Dawson. M.J., and Turner, N.J. (2003) *Angew. Chem. Int. Ed.*, **42**, 4807.
112. Carr, R., Alexeeva, M., Dawson, M.J., Gotor-Fernandez, V., Humphrey, C.E., and Turner, N.J. (2005) *ChemBioChem*, **6**, 637.
113. Dunsmore, C.J., Carr, R., Fleming, T., and Turner, N.J. (2006) *J. Am. Chem. Soc.*, **128**, 2224.
114. Hanson, R.L., Davis, B.L., Goldberg, S.L., Johnston, R.M., Parker, W.L., Tully, T.P., Montana, M.A., and Patel, R.N. (2008) *Org. Process Res. Dev.*, **12**, 1119.

115. Pellissier, H. (2008) *Tetrahedron*, **64**, 1563.
116. Martin-Matute, B. and Backvall, J.-E. (2008) *Asymmetric Organic Synthesis with Enzymes*, Wiley-VCH, pp. 89–113.
117. Kamaruddin, A.H., Uzir, M.H., Aboul-Enein, H.Y., and Abdul Halim, H.N. (2009) *Chirality*, **21**, 449.
118. Asano, Y. and Yamaguchi, S. (2005) *J. Am. Chem. Soc.*, **127**, 7696.
119. Boesten, W.H.J., Raemakers-Franken, P.C., Sonke, T., Euverink, G.J.W., and Grijpstra, P. (2003) WO Patent 2, 003,106,691.
120. Schnell, B., Faber, K., and Kroutil, W. (2003) *Adv. Synth. Catal.*, **345**, 653.
121. Pamies, O. and Backvall, J.-E. (2004) *Trends Biotechnol.*, **22**, 130.
122. Kakuchi, T. and Kaga, H. (2003) *Tetrahedron: Asymmetry*, **14**, 1581.
123. Paal, T.A., Forro, E., Liljeblad, A., Kanerva, L.T., and Fueloep, F. (2007) *Tetrahedron: Asymmetry*, **18**, 1428.
124. Kim, M.-J., Chung, Y.I., Choi, Y.K., Lee, H.K., Kim, D., and Park, J. (2003) *J. Am. Chem. Soc.*, **125**, 11494.
125. Boren, L., Martin-Matute, B., Xu, Y., Cordova, A., and Baeckvall, J.-E. (2006) *Chem. A Eur. J.*, **12**, 225.
126. Brand, S., Jones, M.F., and Rayner, C.M. (1995) *Tetrahedron Lett.*, **36**, 8493.
127. Reetz, M.T. and Schimossek, K. (1996) *Chimia*, **50**, 668.
128. Pamies, O. and Backvall, J.-E. (2002) *Adv. Synth. Catal.*, **344**, 947.
129. Lu, C.-H., Cheng, Y.-C., and Tsai, S.-W. (2002) *Biotechnol. Bioeng.*, **79**, 200.
130. Chen, C.-Y., Cheng, Y.-C., and Tsai, S.W. (2002) *J. Chem. Tech. Biotechnol.*, **77**, 699.
131. Fazlena, H., Kamaruddin, A.H., and Zulkali, M.M.D. (2006) *Bioprocess Biosyst. Eng.*, **28**, 227.
132. Syldatk, C., Cotoras, D., Moeller, A., and Wagner, F. (1986) *Biotech-Forum*, **3**, 9.
133. May, O., Verseck, S., Bommarius, A., and Drauz, K. (2002) *Org. Process Res. Dev.*, **6**, 452.
134. Yamaguchi, S., Komeda, H., and Asano, Y. (2007) *Appl. Environ. Microbiol.*, **73**, 5370.
135. Yamamoto, K., Oishi, K., Fujimatsu, I., and Komatsu, K. (1991) *Appl. Environ. Microbiol.*, **57**, 3028.
136. Lutje Spelberg, J.H., Tang, L., Kellogg, R.M., Richard, M., and Janssen, D.B. (2004) *Tetrahedron: Asymmetry*, **15**, 1095.
137. Haak, R.M., Berthiol, F., Jerphagnon, T., Gayet, A.J.A., Tarabiono, C., Postema, C.P., Ritleng, V., Pfeffer, M., Janssen, D.B., Minnaard, A.J., Feringa, B.L., and de Vries, J.G. (2008) *J. Am. Chem. Soc.*, **130**, 13508.
138. Luedeke, S., Richter, M., and Mueller, M. (2009) *Adv. Synth. Catal.*, **351**, 253.
139. Berezina, N., Alphand, V., and Furstoss, R. (2002) *Tetrahedron: Asymmetry*, **13**, 1953.
140. Koszelewski, D., Clay, D., Faber, K., and Kroutil, W. (2009) *J. Mol. Catal. B: Enzym.*, **60**, 191.
141. Pietzsch, M. and Syldatk, C. (2002) in *Enzyme Catalysis in Organic Synthesis*, 2nd edn, (eds K. Drauz and H. Waldmann), Wiley-VCH Verlag GmbH, Weinheim, pp. 2761–2799.
142. Estell, D.A., Graycar, T.P., and Wells, J.A. (1985) *J. Biolog. Chem.*, **260**, 6518.
143. Roethlisberger, D., Khersonsky, O., Wollacott, A.M., Jiang, L., DeChancie, J., Betker, J., Gallaher, J.L., Althoff, E.A., Zanghellini, A., Dym, O., Albeck, S., Houk, K.N., Tawfik, D.S., and Baker, D. (2008) *Nature (London, UK)*, **453**, 190.

17
Development of a Novel Synthetic Method for RNA Oligomers
Tadaaki Ohgi and Junichi Yano

17.1
Introduction

Since the discovery of ribonucleic acid interference (RNAi), a specific gene-silencing mechanism mediated by small RNA molecules, in 1998 [1], there has been an explosive growth in RNA research that has contributed to the discovery of new mechanisms of cellular defense and the control of gene expression [2, 3]. This RNA renaissance has focused on the analysis of new classes of noncoding RNA, and it has had a major impact not only in the life sciences and basic medical research but also in the applied fields of therapeutic medicine and drug discovery. In particular, there are great hopes for the development of RNA medicines based on RNAi, including small interfering ribonucleic acid (siRNA) [4] as well as microribonucleic acid (miRNA) [5] and its precursor pre-miRNA. Other RNA medicine candidates such as ribozymes [6] and aptamers [7], which though not based on RNAi recognize biological molecules with high specificity, are also under development.

The accelerating pace of RNA research has led to an increased demand for synthetic RNA, and there is now a greater-than-ever need for a cost-effective method of synthesizing highly pure RNA in high yield. Chemical rather than enzymatic synthesis of RNA is desirable because it avoids the errors and inefficiencies associated with *in vitro* transcription [8]. Chemical synthesis is also desirable for the ease of incorporation of modified nucleosides, which cannot be incorporated by enzymatic methods. The incorporation of modified nucleosides is particularly important for research on RNA medicines, to improve their nuclease resistance, biodistribution, and biological activity, as well as to increase the thermodynamic stability of a duplex, avoid unfavorable off-target effects, and suppress unwanted immunostimulatory effects [9]. A chemical synthetic method that can be easily scaled up is especially desirable for research on the development and production of RNA medicines.

At present, oligonucleotides of more than 20 residues are synthesized by solid-phase methods. In particular, phosphodiester oligonucleotides are synthesized by the phosphoramidite method, which is the best method from the viewpoint of ease of preparation of the activated monomer as well as the coupling yield obtained. The principle of the phosphoramidite method was established for the

Pharmaceutical Process Chemistry. Edited by Takayuki Shioiri, Kunisuke Izawa, and Toshiro Konoike
Copyright © 2011 WILEY-VCH Verlag GmbH & Co. KGaA, Weinheim
ISBN: 978-3-527-32650-1

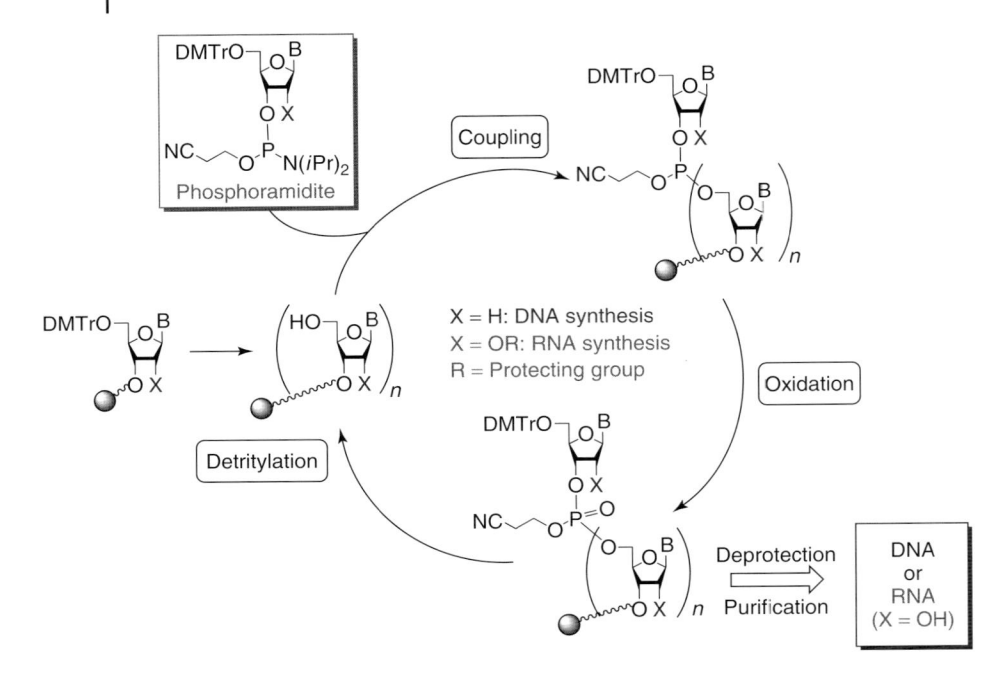

Scheme 17.1 Solid-phase synthesis of DNA/RNA. Because the starting material is CPG with the first nucleoside attached, n−1 cycles are required to synthesize an n-mer.

synthesis of DNA almost 30 years ago in the laboratory of Caruthers [10]. A general synthetic scheme for the solid-phase synthesis of DNA or RNA by the phosphoramidite method is shown in Scheme 17.1. The synthetic cycle consists of three major stages, coupling, oxidation, and detritylation. Because of the necessity of protecting the 2′-hydroxyl group of the ribose moiety during the synthetic cycle, RNA is more difficult to synthesize than DNA, and it is fair to say that the success of an RNA synthetic method depends almost entirely on the selection of a suitable 2′-hydroxyl protecting group. For example, to obtain a high coupling yield in the elongation of the RNA oligomer in solid-phase synthesis, a 2′-hydroxyl protecting group with low steric bulk should be selected, because bulky 2′-hydroxyl protecting groups interfere with the elongation reaction by creating steric hindrance at the vicinal 3′ phosphorus center. The 2′-hydroxyl protecting group must also be stable throughout the synthetic cycle, and after synthesis it must be completely removable under conditions under which the RNA oligomer is stable [11].

For the 5′-hydroxyl function, the most widely used protecting group is the 4,4′-dimethoxytrityl (DMTr) group. With DMTr as the 5′-hydroxyl protecting group, the 2′-hydroxyl function is usually protected with a fluoride-cleavable silyl group [12], a photocleavable group [13], or an acid-cleavable acetal [14, 15]. t-Butyldimethylsilyl (TBDMS) [12] has long been a popular choice of 2′-hydroxyl protecting group, and the protected phosphoramidites (often termed simply *amidites*) are commercially available. Nevertheless, primarily because of the large steric bulk of TBDMS and its relative difficulty of removal after the synthesis of the oligomer, RNA synthesis with TBDMS as the 2′-hydroxyl protecting group leaves something to be desired from the standpoint of the yield and purity of the final product [16–19]. In particular, the synthesis of RNA oligomers longer than about 50 nucleotides by this method is extremely difficult. To solve these problems, several new 2′-hydroxyl protecting groups with lower steric bulk than TBDMS, such as bis(2-acetoxyethoxy)methyl (ACE) [20] and triisopropylsilyloxymethyl (TOM) [21], have been developed (Figure 17.1). Solid-phase synthesis with ACE instead of TBDMS as the 2′-hydroxyl protecting group gives a considerably improved coupling yield, but the 5′-silyl-protected ACE amidites have a complicated chemical structure [11] and so are laborious to prepare. Furthermore, because a fluoride anion is needed to remove the 5′-silyl protecting group at each elongation step, the synthesizer must be specially adapted because of the incompatibility of glass materials with fluoride anion. In addition, while a much-improved coupling yield is obtained with TOM as the 2′-hydroxyl protecting group, allowing the synthesis of RNA oligomers of up to 84 nucleotides, capillary gel electrophoresis of the 84mer reveals appreciable amounts of residual impurities resulting from incompletely suppressed side reactions [21]. It is also difficult to monitor the deprotection reaction by HPLC because of the hydrophobic nature of the TOM group. ACE and TOM, therefore, though they represent considerably improved 2′-hydroxyl protecting groups, are not completely satisfactory.

Pfleiderer and coworkers [15] investigated the use of acetal groups with electron-withdrawing substituents, such as 2-cyanoethoxyethyl (CEE), as 2′-hydroxyl protecting groups (Figure 17.1). However, such protecting groups were found to

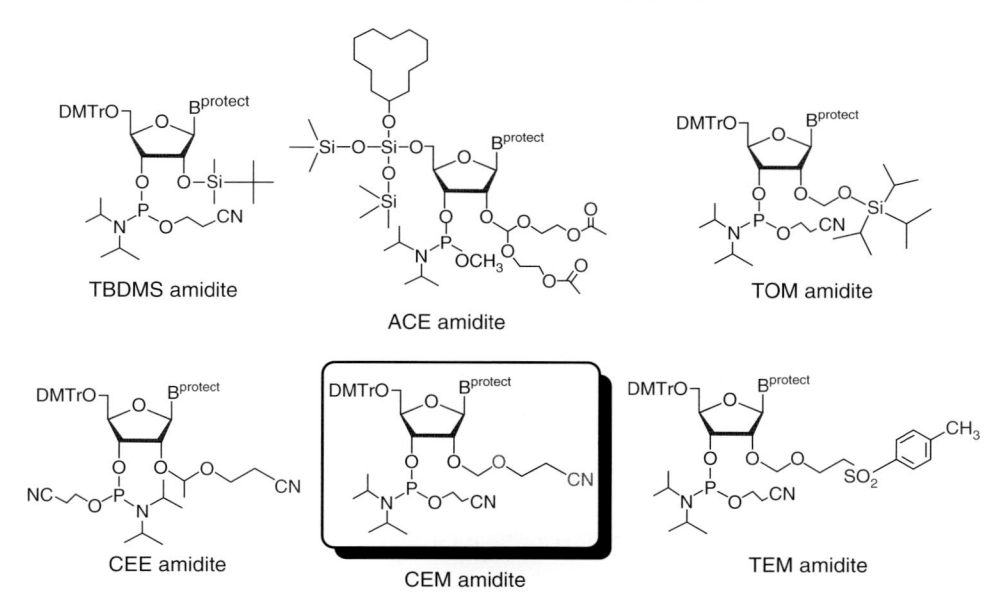

Figure 17.1 Phosphoramidites for RNA synthesis.

be too acid stable to be removed under normal deprotection conditions, and their development as acid-cleavable 2′-hydroxyl protecting groups was abandoned. These workers observed unwanted loss of the CEE group in polar aprotic solvents due to fluoride anion, but this loss could be prevented by adding acetic acid. Subsequently, Umemoto and Wada [22] took advantage of this selective fluoride cleavability of CEE in aprotic solvents to demonstrate the usefulness of CEE as a 2′-hydroxyl protecting group in the synthesis of dinucleotides.

With these considerations in mind, we developed a novel solid-phase RNA synthetic method based on the use of 2-cyanoethoxymethyl (CEM) as the 2′-hydroxyl protecting group [23–25]. The CEM group has the following advantages:

1) Ease of preparation of cyanoethyl methylthiomethyl ether (CEM-SCH$_3$), a novel alkylating agent for introducing CEM at the 2′ position, from moderately priced starting materials.
2) Ease of introduction of the protecting group at the 2′ position, facilitating the large-scale production of nucleotide monomer block.
3) Small overall size of the protecting group, with correspondingly low steric hindrance at the elongation reaction site and hence a much improved coupling yield in oligomer synthesis.
4) A minimum of substituents on the carbon attached to the 2′ oxygen (ribose–2′–O–CH$_2$–O–R), avoiding the generation of an asymmetric center during the attachment of the protecting group.
5) Ease of analysis of the CEM-protected RNA by HPLC for monitoring the deprotection steps carried out by ammonia treatment.
6) Rapid and complete postsynthetic removal of the protecting group by fluoride anion (for oligomers of <25 nucleotides, removal is complete after 1 h at room temperature, compared to 8 h at room temperature for the TBDMS group).
7) Compatibility of the deprotection procedure with standard unmodified DNA synthesizer equipment.

After our discovery of CEM as a new 2′-hydroxyl protecting group, Chattopadhyaya and his coworkers in Sweden [26] reported the use of 2-(4-tolylsulfonyl)ethoxymethyl (TEM) as a new 2′-hydroxyl protecting group based on CEM (Figure 17.1). While TEM is slightly more stable than CEM under basic conditions, it is somewhat more difficult to remove completely after synthesis, so it may be difficult to apply it to the synthesis of very long RNA oligomers.

We have proposed a mechanism for the removal of the CEM group by fluoride anion derived from tetrabutylammonium fluoride (TBAF) in polar aprotic solvents (Scheme 17.2). (The general properties of fluoride anion derived from TBAF as a base in organic synthesis are described in a review by Clark [27]). First, the fluoride anion abstracts a proton from the carbon adjacent to the cyano group with the release of acrylonitrile by a β-elimination process to form a transient cation-stabilized alkoxide intermediate (Scheme 17.2b). Then, on neutralization with buffer, intermediate **b** rapidly and spontaneously releases formaldehyde to give the final deprotected product. We assume that a cation-stabilized alkoxide intermediate forms because we observe no cleavage of the phosphodiester linkage,

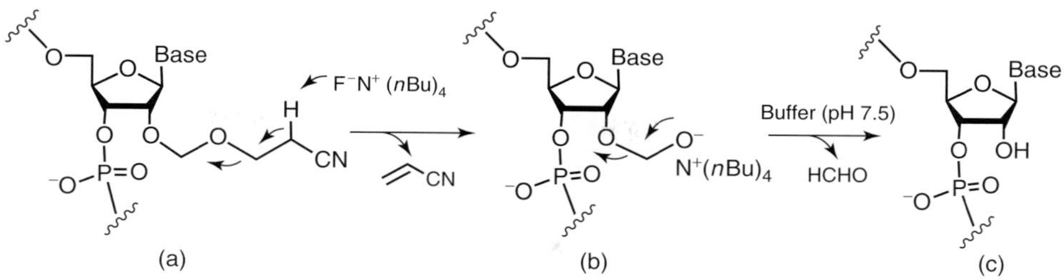

Scheme 17.2 Proposed mechanism for the removal of the CEM group.

which could occur through attack of ribose-2′-O⁻ on the neighboring phosphorus center. In protic solvents with TBAF (and in protic or aprotic solvents with triethylamine trihydrofluoride) as the source of fluoride anion, this reaction does not proceed. This is because under these conditions the fluoride anion is already protonated, so that it cannot serve as a nucleophile in β-elimination. The acrylonitrile produced in the conversion of **a** to **b** can react with amino groups of A, C, and G (i.e., all nucleobases except U) to form cyanoethylated side-products, but this can be prevented by adding nitromethane or n-propylamine/bis(2-mercaptoethyl) ether as an acrylonitrile scavenger. Furthermore, the formaldehyde produced in the conversion of **b** to **c** is captured by the neutralization buffer (Tris-HCl, pH 7.5). If any of the above side reactions *do* occur, their products can easily be detected by mass spectrometry.

The CEM group is one of the smallest effective 2′-hydroxyl protecting groups developed to date, and the significantly improved coupling yield obtained with it allows its application to the synthesis of RNA oligomers longer than 50 nucleotides. We now describe the synthesis of the CEM amidites and give examples of their use in the solid-phase synthesis of RNA oligomers of up to 110 nucleotides.

17.2 Synthesis of CEM Amidites

The four kinds of CEM amidites corresponding to the four natural bases of RNA are synthesized as shown in Scheme 17.3. For introduction of the CEM group at the 2′ position of **1a–d**, a novel alkylating agent, CEM-SCH$_3$, is prepared by treating a solution of 2-cyanoethanol in dimethylsulfoxide (DMSO) with acetic anhydride and acetic acid [24] (Scheme 17.4). Alkylation of **1a–c** with CEM-SCH$_3$ proceeds in 80–90% yield to give **2a–c** (Scheme 17.3, step (i)). Because the yield of the adenine derivative **2d** was unsatisfactory in step (i), an alternative two-step route via intermediate **6** was devised to give **2d** in good yield (Scheme 17.3, steps (ii) and (iii), 79 and 90% yield, respectively). Owing to the mild conditions used in the conversion of **6** to **2d** (step (iii)), even very bulky acid-labile tertiary alcohols can be introduced at the 2′ position instead of cyanoethanol. Therefore, compound **6** *is a valuable intermediate for the general preparation of 2′-O-alkoxymethyl ribose derivatives.* As it happens, diols **3a–d** can be easily purified by crystallization from

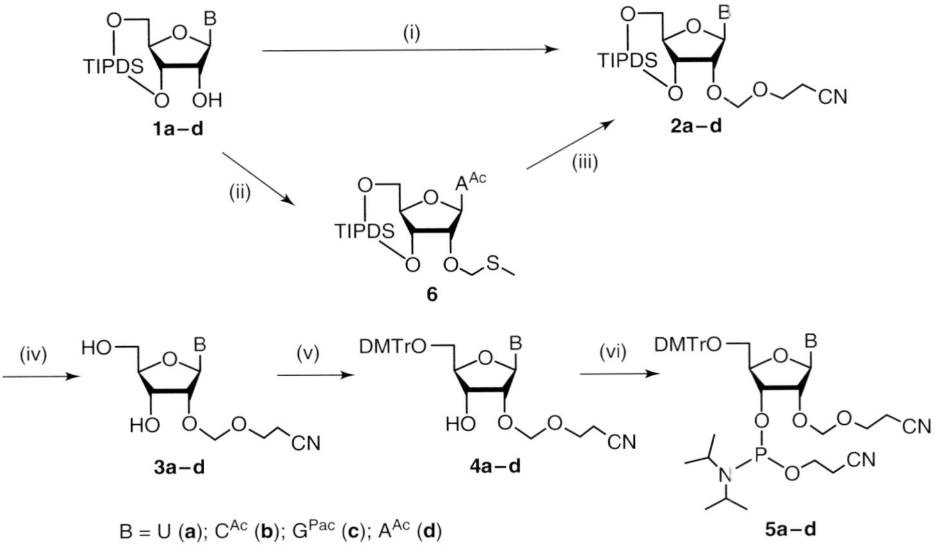

Scheme 17.3 Synthesis of CEM amidites. (i) 2-Cyanoethyl methylthiomethyl ether, molecular sieves 4A (MS 4A), N-iodosuccinimide (NIS), CF$_3$SO$_3$H, THF, −45 °C; (ii) DMSO, acetic anhydride, acetic acid; (iii) 3-hydroxypropionitrile, MS 4A, NIS, CF$_3$SO$_3$H, THF, −45 °C; (iv) NH$_4$F, CH$_3$OH, 50 °C or TEA·3HF (triethylamine trihydrofluoride), THF, 45 °C; (v) 4,4′-dimethoxytrityl chloride, THF, pyridine, MS 4A, room temperature; and (vi) diisopropylammonium tetrazolide, bis(N,N-diisopropylamino)cyanoethyl phosphite, CH$_3$CN, 40 °C.

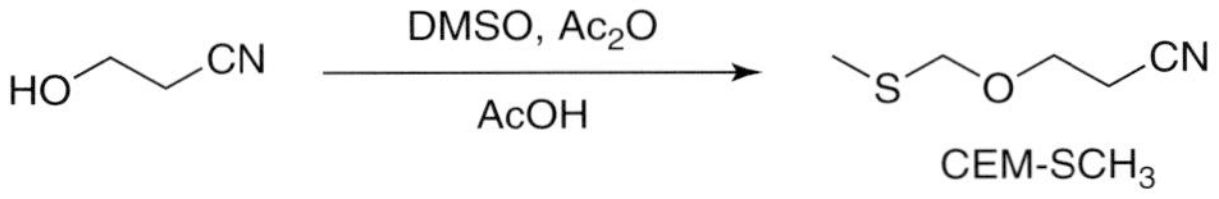

Scheme 17.4 Preparation of 2-cyanoethyl methylthiomethyl ether (CEM-SCH$_3$).

the reaction mixture either directly during the progress of the reaction (**3b** and **d**) or after workup (**3a** and **c**), so that the need for tedious, time-consuming, and costly column chromatography is eliminated. Crystallization is probably assisted by the fact that stereoisomers are not generated during the introduction of the CEM group at the 2′ position (unlike, for example, the CEE group). Purification by crystallization is especially suited to the large-scale preparation of amidites. Finally, compounds **4a–d** are phosphitylated to give the CEM amidites **5a–d**.

17.3
Synthesis of RNA Oligomers from CEM Amidites

For oligomer synthesis with CEM amidites [23, 24], controlled-pore glass (CPG) is used as the solid support and 5-ethyl-1H-tetrazole as the activating agent with a coupling time of 150 s. Unlike CPG, polystyrene resins swell during the synthetic cycle, so that although they can be used for the synthesis of shorter oligomers, in our experience they are unsuitable for the synthesis of longer oligomers.

First, we synthesized U_{40} on a micromolar scale. We chose U_{40} for the initial synthesis because uridine is the only nucleoside for which base-protecting groups are not needed, allowing us to evaluate the coupling efficiency more easily. After synthesis, the crude oligomer was cleaved from the CPG resin with a mixture of concentrated ammonia and EtOH at 40 °C for 4 h, conditions that also serve to remove the protecting groups of the phosphodiester linkage and nucleobases except uridine. We initially tried to remove the phosphate- and nucleobase-protecting groups with CH_3NH_2 in $EtOH/H_2O$. However, under those conditions, we observed substantial loss of the CEM group with resulting chain cleavage. For oligomers of up to 50 nucleosides, CPG resin with the standard pore size of 500 Å can be used. However, for the synthesis of oligomers longer than this, resins with larger pore sizes, such as 1000 or 2000 Å, give better results despite their lower nucleoside loading. This is because the synthetic reactions occur mainly in the pores of the resin, and larger pores are needed to accommodate longer oligomers. After removal of the CEM protecting group with 1 M TBAF in tetrahydrofuran (THF), U_{40} was isolated in 65% yield. Since an average yield of 99% over the 39 coupling steps corresponds to a calculated yield of 67%, the actual average coupling yield obtained is close to 99%, comparable to that obtained in DNA synthesis. HPLC profiles of the fully deprotected crude U_{40} prepared by the CEM method (Figure 17.2a) and DNA of the same chain length (dT_{40}; Figure 17.2b) illustrate the relatively high purity of the product even before purification. In the synthesis of RNA oligomers incorporating nucleobases other than U, the acrylonitrile formed on removal of the CEM group can react with nucleobase amino groups. To prevent

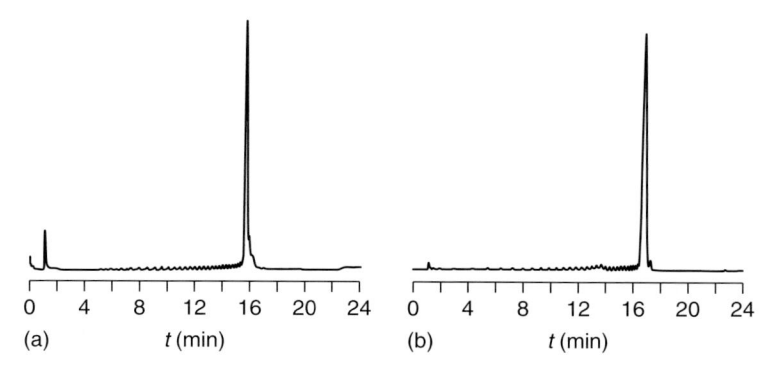

Figure 17.2 HPLC of RNA and DNA homooligomers. (a) Unpurified fully deprotected U_{40}, DNAPac PA-100 anion-exchange column (4.6 × 250 mm; Dionex); buffer A, 10% CH_3CN, 25 mM Tris-HCl, pH 8.0; buffer B, 10% CH_3CN, 25 mM Tris-HCl, pH 8.0, containing 700 mM $NaClO_4$; gradient from 5 to 50% buffer B in 20 min; flow rate, 1.5 ml/min; 50 °C. (b) Unpurified fully deprotected dT_{40} (DNA). Buffer A, 25 mM Tris-HCl, pH 8.0; buffer B, 25 mM Tris-HCl, pH 8.0, containing 700 mM $NaClO_4$; gradient from 10 to 30% buffer B in 20 min; flow rate, 1 ml/min; 40 °C. UV detection was at 260 nm.

this, 2% nitromethane is included in the 1 M TBAF solution as an acrylonitrile scavenger. Alternatively, 10% n-propylamine and 1% bis(2-mercaptoethyl) ether can be added as an acrylonitrile scavenger. When a 20mer incorporating all four bases, 5'-CUUACGCUGAGUACUUCGAU-3', was synthesized from CEM amidites, an isolated yield of 58% was obtained, compared with about 20% when the same 20mer was synthesized from TBDMS amidites (data not shown).

Synthesis of the 55mer 5'-UGAACACAAAUCACAGAAUCGUCGUAUGCAGU GAAAACUCUCUUCAAUUCUUUAdT-3' from CEM amidites gave an isolated yield of 15% and a final purity of >95% by HPLC [23]. To check for side reactions, we completely digested the 55mer to nucleosides with nuclease P1 and alkaline phosphatase. HPLC analysis of the digest showed no modified nucleosides, but only the expected constituent nucleosides, including the single 3'-terminal dT residue used to prime the solid support. In particular, the absence of enzyme-resistant 2', 5'-linked dimers shows that no 3' → 2' migration of the phosphodiester linkage had occurred (Figure 17.3). (This rearrangement can occur through attack of the 2'-hydroxyl group on the neighboring phosphodiester linkage.) This analysis demonstrates the high quality of RNA oligomers synthesized from CEM amidites.

As an example of the synthesis of an even longer RNA oligomer from CEM amidites, we synthesized the 110mer precursor of miR-196a, an miRNA that regulates the human homeobox (*Hox*) genes, which are master regulators of embryonic morphogenesis (Figure 17.4) [25, 28, 29]. As far as we know, this 110mer is the longest RNA oligomer that has been entirely chemically synthesized. The synthesis was done on a standard unmodified commercial DNA/RNA synthesizer, and the synthetic conditions are shown in Table 17.1.

The protocol differs from that used for shorter oligomers mainly in the capping conditions. For oligomers of up to at least 82 nucleotides (data not shown), N-methylimidazole (NMI) can be used as the base catalyst in the capping reagent as described by Eadie and Davidson [30]. However, in the synthesis of a 110mer,

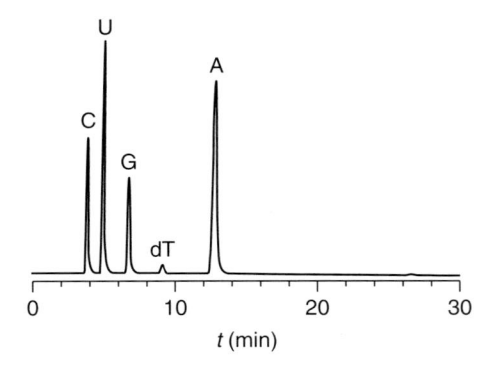

Figure 17.3 HPLC analysis of 55mer RNA after enzymatic digestion. Develosil ODS-UG-5 reverse-phase column (4.6 × 250 mm); buffer, 5% CH_3OH, 5 mM $(nBu)_4NHSO_4$, 50 mM phosphate, pH 7.5; flow rate, 1 ml/min; 35 °C. UV detection was at 260 nm.

Figure 17.4 Nucleotide sequence of 110mer precursor of miR-196a.

Table 17.1 Synthetic conditions for 110mer RNA.

Step	Operation	Reagent	Time (s)
1	Deblocking	4% CCl_3COOH in CH_2Cl_2	60
2	Coupling	0.075 M amidite in CH_3CN + 0.25 M 5-benzylmercapto-1H-tetrazole in CH_3CN	150
3	Capping	0.1 M Pac_2O in THF + 6.5% 2-DMAP, 2% NMI and 10% 2,6-lutidine in THF	170
4	Oxidation	0.1 M I_2 in THF/H_2O/pyridine (7/1/2)	5
5	Capping	0.1 M Pac_2O in THF + 6.5% 2-DMAP, 2% NMI and 10% 2,6-lutidine in THF	30

the capping reaction does not proceed to completion under these conditions. To further promote the capping reaction, we carried it out under more basic conditions by using 2-(dimethylamino)pyridine (2-DMAP) together with NMI in the capping reagent. There is a side reaction involving the formation of 2,6-diaminopurine (2,6-DAP) from guanine base [30]. This side reaction was almost completely suppressed by the modified protocol even though the synthetic cycle was repeated more than 100 times. *It is important to use 2-DMAP instead of the isomeric 4-DMAP, which is usually used.* This is because in 2-DMAP, but not 4-DMAP, attack of the pyridine ring nitrogen at the 6-position of guanine is sterically hindered by the dimethylamino group, so that the formation of 2,6-DAP is practically prevented. Also, we used phenoxyacetic anhydride (Pac_2O) in the capping reagent instead of the Ac_2O generally used. This is because, when we tried using Ac_2O in the capping reagent, removal of the phosphate- and nucleobase-protecting groups by ammonia treatment under our conditions (28% NH_4OH/EtOH at 35 °C for 24 h for the 110mer) resulted in incomplete deprotection of acetylguanine nucleobase only (i.e., incomplete conversion of acetylguanine to guanine) in the oligomer, as described by Chaix et al. [31]. When Pac_2O was used instead, however, no residual protected guanine nucleobase was detected.

Note that cleavage of the oligomer from the resin and removal of the protecting groups are carried out in two steps. In Step 1, the resin is treated with concentrated ammonia in ethanol at 40 °C for several hours to remove the oligomer from the resin and, at the same time, to remove the phosphate- and nucleobase-protecting groups from the oligomer, leaving only the CEM protecting group. It is important to carry out Step 1 under mild conditions so that CEM is not removed at this stage. If the CEM is removed at this stage, chain cleavage occurs. Then, in Step 2, CEM is removed with TBAF in THF or DMSO containing 2% nitromethane for acrylonitrile capture.

The 110mer was synthesized on a 0.8 µmol scale [25]. Capillary gel electrophoresis carried out after deprotection but before purification (Figure 17.5) showed that even in the crude reaction mixture the desired product was already fairly pure. After a simple two-step purification, a highly pure final product was obtained, as shown by analytical reverse-phase HPLC (Figure 17.6a), ion-exchange HPLC (Figure 17.6b), capillary gel electrophoresis (Figure 17.6c), and polyacrylamide gel electrophoresis (Figure 17.6d). The absence of by-products (except for a tiny trace of 2,6-DAP) was confirmed by analytical reverse-phase HPLC of the purified final product after complete enzymatic digestion to nucleosides (data not shown). A good overall yield of 5.5% (1.35 mg) was obtained. Further analysis of the 110mer was carried out by matrix-assisted laser desorption/ionization time-of-flight (MALDI-TOF) mass spectrometry after partial digestion by MazF, an endonuclease that cleaves on the 5′ side of ACA, and complete digestion by RNase T1, an endonuclease that cleaves specifically on the 3′ side of G. Partial digestion by MazF produced all three expected fragments with the masses predicted from the sequence of the 110mer (Figure 17.7). The parent ion was also detected, and its mass is consistent with a length of 110 nucleotides. Digestion by RNase T1 first yields the 2′,3′-cyclic phosphate, which is then hydrolyzed to the 2′-phosphate or the 3′-phosphate (Scheme 17.5), giving rise to pairs of fragments differing by 18 Da. For instance, fragments L and L′ (AACUCG; sequence 13) in Figure 17.8 and Table 17.2 gave observed molecular weights of 1939.5 Da (fragment L; calculated 1939.2) and 1921.5 Da (fragment L′; calculated 1921.2). The Na^+ and K^+ salts of both fragments were also observed as smaller peaks, and many of the very small peaks can be similarly assigned to specific fragments. Except for monomers, dimers, and one trimer, which were hard to distinguish among the noise of the spectrum, all other expected fragments and

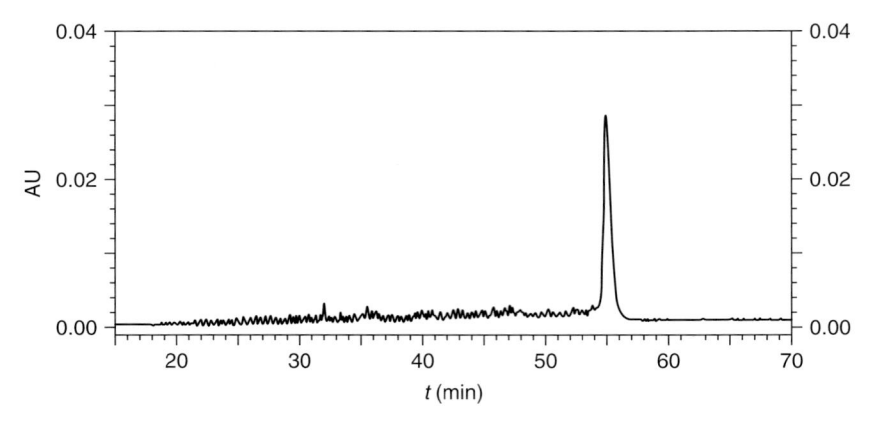

Figure 17.5 Capillary gel electrophoresis of crude 110mer after removal of all protecting groups. UV detection was at 254 nm.

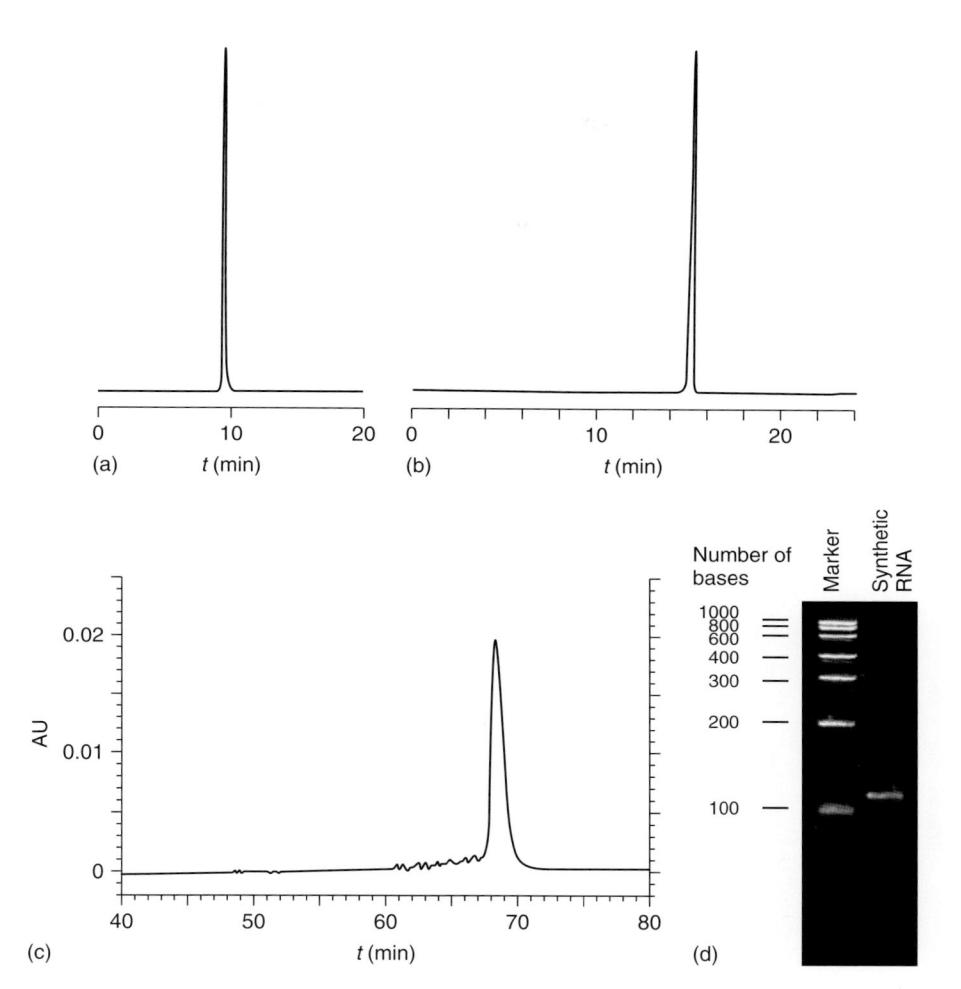

Figure 17.6 HPLC and electrophoresis of purified 110mer RNA. (a) A PLRP-S 300 Å reverse-phase column was operated at a flow rate of 1 ml/min and maintained at a temperature of 80 °C. The solvent system was buffer A (5% acetonitrile, 50 mM triethylammonium acetate (TEAA), pH 7.0) and buffer B (90% acetonitrile, 50 mM TEAA, pH 7.0), and the RNA was eluted with a linear gradient from 0 to 50% buffer B in 20 min. UV detection was at 260 nm. (b) A DNAPac PA-100 anion-exchange column was operated at a flow rate of 1.5 ml/min and maintained at a temperature of 70 °C. The solvent system was buffer C (10% acetonitrile, 25 mM Tris-HCl, pH 8.0) and buffer D (10% acetonitrile containing 700 mM NaClO$_4$, 25 mM Tris-HCl, pH 8.0), and the RNA was eluted with a linear gradient from 5 to 50% buffer D in 20 min. UV detection was at 260 nm. (c) Capillary gel electrophoresis. UV detection was at 254 nm. (d) Polyacrylamide gel electrophoresis. The synthetic RNA was analyzed on a 5% polyacrylamide gel and stained with the cyanine dye SYBR Green II.

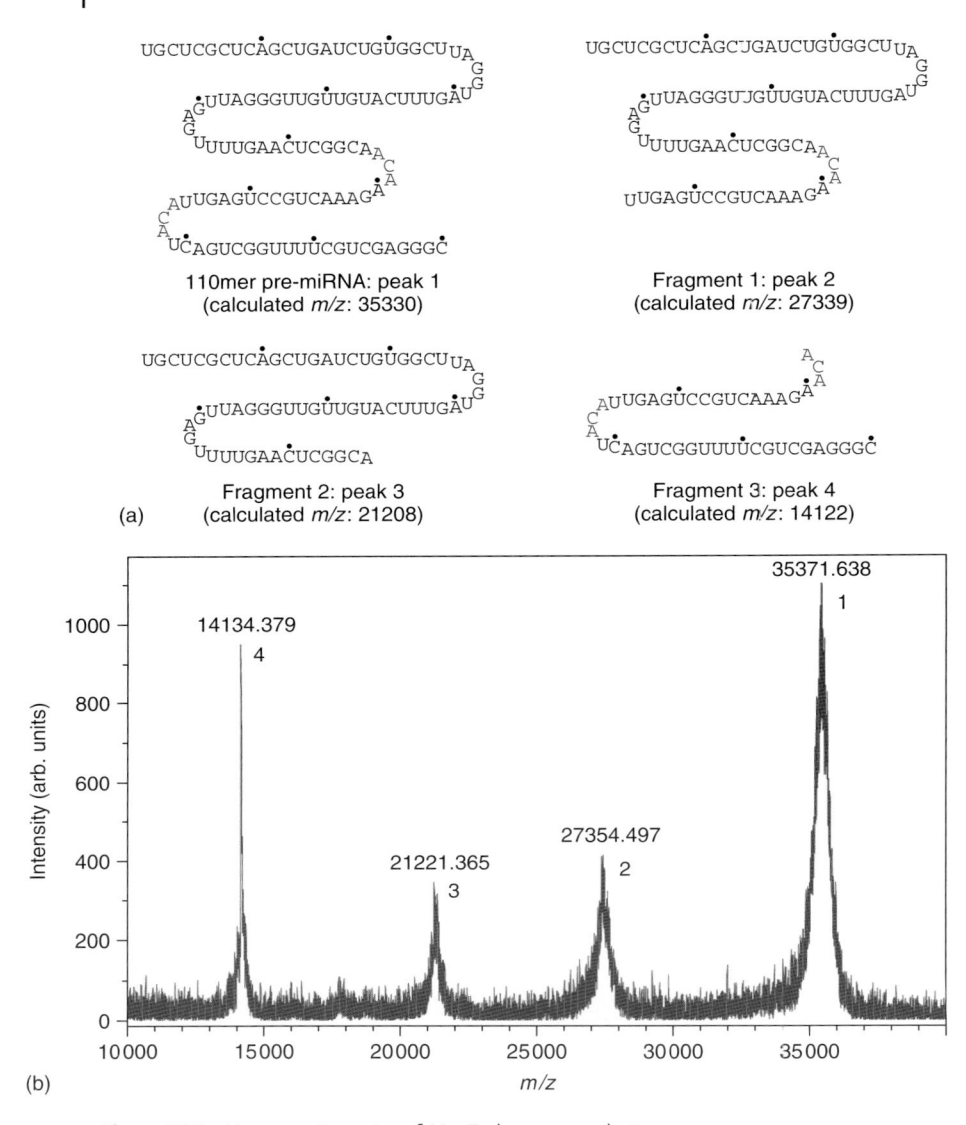

Figure 17.7 Mass spectrometry of MazF cleavage products of 110mer RNA. (a) Sequence of 110mer and the predicted cleavage products. (b) MALDI-TOF mass spectrum of MazF cleavage products.

most fragment pairs (2′,3′-cyclic phosphate and 2′- or 3′-phosphate) were similarly identified. The mass spectrometric analysis of the enzyme digestion products thus provides evidence of the correct length and sequence of the synthetic 110mer.

We next tested the biological activity of the 110mer pre-miR-196a-2 with a luciferase reporter assay system in which a 55 bp fragment of the homeobox gene b-8

17.3 Synthesis of RNA Oligomers from CEM Amidites

Scheme 17.5 Mechanism of cleavage of RNA by RNase T1.

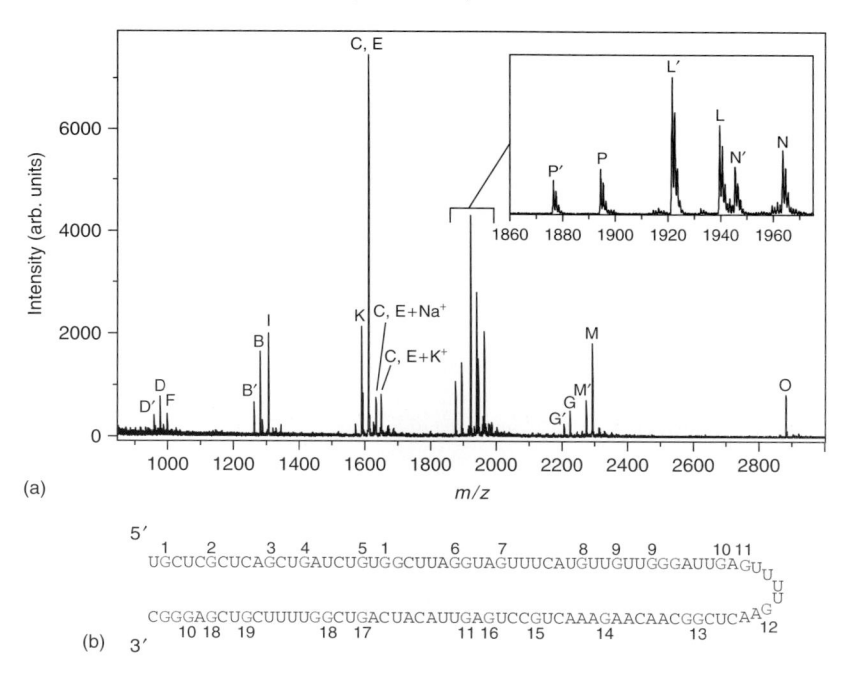

Figure 17.8 Mass spectrometry of RNase T1 cleavage products of 110mer RNA. (a) MALDI-TOF mass spectrum. The inset shows an expanded view of the mass range 1860–1980 Da. (b) Sequences of cleavage products. The numbers shown represent the entry numbers of the fragments listed in Table 17.2. Cleavage occurs on the 3′ side of G, and the numbers are placed at the G residues where cleavage occurs.

Table 17.2 Sequences and masses of fragments produced by RNase T1 digestion.

Entry number[a]	Sequence (5'→3')	2'- or 3'-phosphate			2',3'-Cyclic phosphate		
		Peak	Calculated mass (Da)	Observed mass (Da)	Peak	Calculated mass (Da)	Observed mass (Da)
1	UG	A	670.4	NI[b]	A'	652.4	NI.
2	CUCG	B	1280.8	1280.9	B'	1262.8	1262.9
3	CUCAG	C	1610.0	1610.2	C'	1592.0	ND[c]
4	CUG	D	975.6	975.6	D'	957.6	957.6
5	AUCUG	E	1611.0	1611.2	E'	1593.0	1593.2
6	CUUAG	E	1611.0	1611.2	E'	1593.0	1593.2
7	UAG	F	999.6	999.7	F'	981.6	ND
8	UUUCAUG	G	2223.3	2223.6	G'	2205.3	2205.6
9	UUG	H	976.6	NI	H'	958.6	NI
10	AUUG	I	1305.8	1306.0	I'	1287.8	1287.9
11	AG	J	693.5	NI	J'	675.5	NI
12	UUUUG	K	1588.9	1589.1	K'	1570.9	ND
13	AACUCG	L	1939.2	1939.5	L'	1921.2	1921.5
14	CAACAAG	M	2291.4	2291.8	M'	2273.4	2273.7
15	AAACUG	N	1963.2	1963.5	N'	1945.2	1945.5
16	CCUG	B	1280.8	1280.9	B'	1262.8	1262.9
17	UUACAUCAG	O	2880.7	2881.1	O'	2862.7	ND
18	UCG	D	975.6	975.6	D'	957.6	957.6
19	UUUUCG	P	1894.1	1894.4	P'	1876.1	1876.4

[a]The entry numbers correspond to the numbered fragments shown in Figure 17.8.
[b]NI, not identified.
[c]ND, not detected.

(HoxB8) containing the miR-196a-2 target site was inserted downstream from the luciferase gene in a reporter plasmid. miR-196a is thought to negatively regulate HoxB8 by targeting the HoxB8 3'-untranslated region (UTR) [28]. In the luciferase reporter assay system, a reporter plasmid containing the luciferase gene and the HoxB8 3'-UTR was transfected into human embryonic kidney–derived G3T-hi cells. In this assay system, a gene-silencing effect is seen as a reduction in the expression of luciferase by the cells. Our synthetic 110mer had a gene-silencing effect that was almost the same as that of the mature synthetic double-stranded 22mer miR-196a-2 used as a positive control (Figure 17.9a). This result provides evidence that the synthetic pre-miR-196a-2 had been successfully processed in the cell and that the mature miRNA derived from it functioned as a silencing miRNA. Next, the luciferase assay system was modified by inserting the 55 bp HoxB8 3'-UTR fragment into the plasmid in the reverse direction, providing a predicted target for the complementary strand of miR-196a-2. In this modified assay system, the mature double-stranded 22mer miR-196a-2 had about the same gene-silencing activity as in the original

110mer pre-miR-196a:

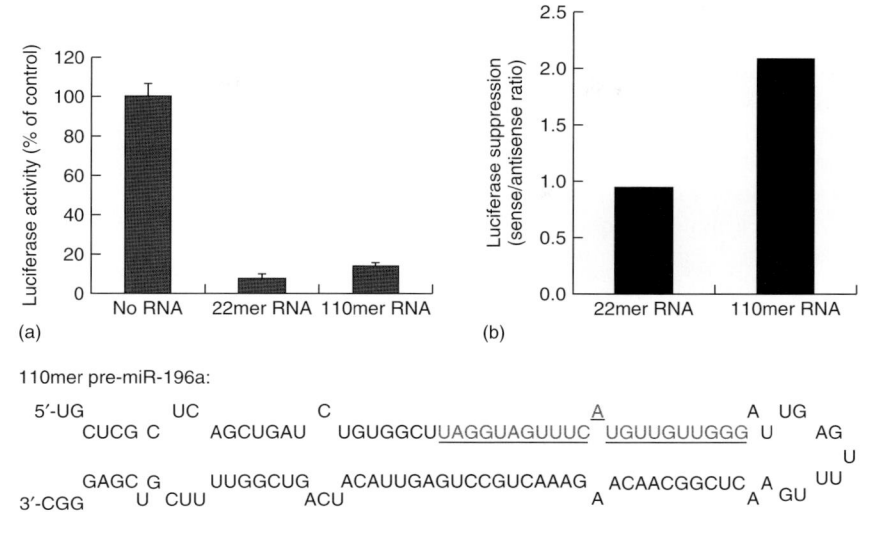

```
5'-UG       UC           C                         A            A  UG
   CUCG C     AGCUGAU    UGUGGCUUAGGUAGUUUC UGUUGUUGGG U      AG
                                                                   U
     GAGC G    UUGGCUG   ACAUUGAGUCCGUCAAAG   ACAACGGCUC A     UU
3'-CGG   U CUU         ACU                 A          A  GU
```

Synthetic 22mer miR-196a duplex (positive control):

5'-CAACAACAUGAAACUACUUAAG-3'
3'-GGGUUGUUGUACUUUGAUGGAU-5'
(c)   (same as underlined sequence above)

Figure 17.9 Gene silencing by chemically synthesized 110mer RNA. (a) Effect of 110mer RNA on expression of luciferase target gene. G3T-hi cells were transfected with pHOXB-Luc reporter plasmid and effector RNA (100 nM) and reporter assays were performed 48 h after transfection. Each value is the average of three independent experiments, and the error bars indicate the standard deviations. (b) Comparison of the ratio of sense to antisense luciferase suppression for 110mer RNA. G3T-hi cells were transfected with pHOXB-Luc or pHOXB-Luc-antisense reporter plasmid and effector RNA (30 nM) and reporter assays were performed 48 h after transfection. The results are presented as the ratio of the average percent suppression of luciferase containing sense target to the average percent suppression of luciferase containing antisense target by 110mer RNA (pre-miRNA) or 22mer mature miRNA. (c) Sequences of the effector RNAs used. The underlined parts of the 110mer pre-miR-196a-2 and the synthetic duplex 22mer miR-196a represent the sequence of the mature miRNA-196a.

assay (i.e., the ratio of the sense to the antisense suppression activity was about 1.0), whereas the 110mer precursor had weaker silencing activity (Figure 17.9b). In the modified assay, therefore, the 110mer pre-miRNA (Figure 17.9c) showed target-strand selectivity, whereas the 22mer mature miRNA did not. It may be that, in the medical application of miRNA, pre-miRNA will prove more useful than mature double-stranded 22mer miRNA by displaying target-strand selectivity in the cell, thereby allowing the avoidance of off-target effects of the complementary strand [32, 33]. As described above, we have succeeded in synthesizing a 110mer RNA oligomer, which we believe to be the first example of a wholly chemically synthesized RNA molecule longer than 100 nucleotides, and we have provided evidence of its purity, physicochemical identity, and biological activity.

RNA synthesis from CEM amidites, when compared with conventional RNA synthesis from TBDMS amidites, has been shown to proceed in high yield and to give a final product of extremely high purity. The synthesis of long RNA oligomers by the CEM method is comparable in ease to DNA synthesis and permits the synthesis of RNA molecules longer than 50 nucleotides, which are not easily synthesized from other kinds of amidites. CEM amidites are easily prepared and RNA synthesis from them can be achieved with commercially available unmodified DNA/RNA synthesizer equipment. These and other advantages of the CEM method make it a useful tool not only for biological research but also for the development and manufacture of RNA medicines.

The CEM method has recently been applied to the synthesis of diastereomeric diadenosine boranophosphates [34]. The CEM group has also been used to protect the 3′-O-position in the so-called "sequencing-by-synthesis" method of DNA sequencing [35]. We believe that the CEM method of RNA synthesis has the potential to become a standard synthetic method with application to the small- and large-scale synthesis of both short and long RNA oligomers.

In summary, we have developed a novel solid-phase method for the synthesis of RNA oligomers with CEM as the 2′-hydroxyl protecting group. The new method allows the synthesis of oligoribonucleotides with an efficiency and final purity comparable to that obtained in DNA synthesis. To verify the potential of the method, we synthesized a very long RNA oligomer, a 110mer with the sequence of a precursor-miRNA candidate, and obtained the final product in good yield. We confirmed the physicochemical identity and biological activity of the 110mer. To the best of our knowledge, this is the longest chemically synthesized RNA oligomer reported to date. This chapter describes the development of CEM as a 2′-hydroxyl protecting group for RNA synthesis and compares it with other 2′-hydroxyl protecting groups. The advantages of CEM include its ease of introduction at the 2′ position, low steric hindrance leading to a high coupling yield, ease of removal under mild conditions, and application to the synthesis of very long oligomers.

Acknowledgments

The research described in this article was supported in part by grants from the New Energy and Industrial Technology Development Organization (NEDO) of Japan for its Functional RNA Project. We thank Dr. G. E. Smyth, Discovery Research Laboratories, Nippon Shinyaku Co., for helpful discussions and suggestions concerning the manuscript.

References

1. Fire, A., Xu, S., Montgomery, M.K., Kostas, S.A., Driver, S.E., and Mello, C.C. (1998) *Nature*, **391**, 806.

2. The FANTOM Consortium and RIKEN Genome Exploration Research Group and Genome Science Group (2005) *Science*, **309**, 1559.

3. RIKEN Genome Exploration Research Group and Genome Science Group and the FANTOM Consortium (2005) *Science*, **309**, 1564.
4. Elbashir, S.M., Harborth, J., Lendeckel, W., Yalcin, A., Weber, K., and Tuschl, T. (2001) *Nature*, **411**, 494.
5. Hammond, S.M. (2006) *Trends Mol. Med.*, **12**, 99.
6. Plehn-Dujowich, D. and Altman, S. (1998) *Proc. Natl. Acad. Sci. U.S.A.*, **95**, 7327.
7. Ng, E.W., Shima, D.T., Calias, P., Cunningham, E.T. Jr., Guyer, D.R., and Adamis, A.P. (2006) *Nature Rev. Drug Discov.*, **5**, 123.
8. Helm, M., Brulé, H., Giegé, R., and Florentz, C. (1999) *RNA*, **5**, 618.
9. Manoharan, M. and Rajeev, K.G. (2008) in *Antisense Drug Technology: Principles, Strategies, and Applications*, 2nd edn (ed. S.T. Crooke), CRC Press, Boca Raton, p. 437.
10. Beaucage, S.L. and Caruthers, M.H. (1981) *Tetrahedron Lett.*, **22**, 1859.
11. Reese, C.B. (2002) *Tetrahedron*, **58**, 8893.
12. Usman, N., Ogilvie, K.K., Jiang, M.-Y., and Cedergren, R.J. (1987) *J. Am. Chem. Soc.*, **109**, 7845.
13. Schwartz, M.E., Breaker, R.R., Asteriadis, G.T., de Bear, J.S., and Gough, G.R. (1992) *Bioorg. Med. Chem. Lett.*, **2**, 1019.
14. Yamakage, S., Sakatsune, O., Furuyama, E., and Takaku, H. (1989) *Tetrahedron Lett.*, **30**, 6361.
15. Matysiak, S., Fitznar, H.P., Schnell, R., and Pfleiderer, W. (1998) *Helv. Chim. Acta*, **81**, 1545.
16. Gasparutto, D., Molko, D., and Téoule, R. (1990) *Nucleosides Nucleotides Nucleic Acids*, **9**, 1087.
17. Reese, C.B. (2001) *Current Protocols in Nucleic Acid Chemistry*, Chapter 2:Unit 2.2. John Wiley & Sons.
18. Gough, G.R., Miller, T.J., and Mantick, N.A. (1996) *Tetrahedron Lett.*, **37**, 981.
19. Welz, R. and Müller, S. (2002) *Tetrahedron Lett.*, **43**, 795.
20. Scaringe, S.A., Wincott, F.E., and Caruthers, M.H. (1998) *J. Am. Chem. Soc.*, **120**, 11820.
21. Pitsch, S., Weiss, P.A., Jenny, L., Stutz, A., and Wu, X. (2001) *Helv. Chim. Acta*, **84**, 3773.
22. Umemoto, T. and Wada, T. (2004) *Tetrahedron Lett.*, **45**, 9529.
23. Ohgi, T., Masutomi, Y., Ishiyama, K., Kitagawa, H., Shiba, Y., and Yano, J. (2005) *Org. Lett.*, **7**, 3477.
24. Ohgi, T., Kitagawa, H., and Yano, J. (2008) *Current Protocols in Nucleic Acid Chemistry*, Chapter 2:Unit 2.15. John Wiley & Sons.
25. Shiba, Y., Masuda, H., Watanabe, N., Ego, T., Takagaki, K., Ishiyama, K., Ohgi, T., and Yano, J. (2007) *Nucleic Acids Res.*, **35**, 3287.
26. Zhou, C., Honcharenko, D., and Chattopadhyaya, J. (2007) *Org. Biomol. Chem.*, **5**, 333.
27. Clark, J.H. (1980) *Chem. Rev.*, **80**, 429.
28. Griffiths-Jones, S., Grocock, R.J., van Dongen, S., Bateman, A., and Enright, A.J. (2006) *Nucleic Acids Res.*, **34** (Database issue), D140.
29. Yekta, S., Shih, I.H., and Bartel, D.P. (2004) *Science*, **304**, 594.
30. Eadie, J.S. and Davidson, D.S. (1987) *Nucleic Acids Res.*, **15**, 8333.
31. Chaix, C., Molko, D., and Téoule, R. (1989) *Tetrahedron Lett.*, **30**, 71.
32. Matranga, C., Tomari, Y., Shin, C., Bartel, D.P., and Zamore, P.D. (2005) *Cell*, **123**, 607.
33. Rand, T.A., Petersen, S., Du, F., and Wang, X. (2005) *Cell*, **123**, 621.
34. Enya, Y., Nagata, S., Masutomi, Y., Kitagawa, H., Takagaki, K., Oka, N., Wada, T., Ohgi, T., and Yano, J. (2008) *Bioorg. Med. Chem.*, **16**, 9154.
35. Földesi, A., Keller, A., Stura, A., Zigmantas, S., Kwiatkowski, M., Knapp, D., and Engels, J.W. (2007) *Nucleosides Nucleotides Nucleic Acids*, **26**, 271.

18
Process Research with Explosive Reactions
Hiromu Kawakubo

18.1
Introduction

Under certain conditions in explosive chemical process development, the introduction of nitro group ($-NO_2$), diazo group ($-N_2$), or azido group ($-N_3$) is required. Over the past 70 years, we have manufactured several hundred tons of explosive compounds, including nitroglycerin and pentaerythritol tetranitrate (Figure 18.1), and have established procedures for their synthesis and evaluation.

In this report, we introduce our method for assessing the risk of explosive compounds and present the results of a study on nitroacetic acid ethyl ester and nitrobenzene derivatives as an example of cost-efficient explosive compound process development that is suitable for industrialization.

18.2
Safety Evaluation of an Explosive Chemical Process

Under certain conditions, an explosive compound may ignite and explode during the course of a chemical process. In such a case, the susceptibility to ignition or explosion (sensitivity), the amount of energy generated by the ignition or explosion, and the rate of ignition or explosion (power) may vary depending on the physicochemical energy applied to the explosive compound.

In a chemical process that involves a compound with potential energy for explosion, there are risk factors specific to each chemical process, including temperature, pressure, atmosphere, shock, and friction. The improper control of the operating conditions during the manufacture of an explosive compound could lead to an explosion, which may threaten human safety and seriously damage the environment. Our method for assessing the risk of explosion in a chemical process is shown in Figure 18.2.

If we consider the process up to explosion, the method we use to assess the risk of explosion in a chemical process and preventive measures we take based on this assessment are as follows:

Pharmaceutical Process Chemistry. Edited by Takayuki Shioiri, Kunisuke Izawa, and Toshiro Konoike
Copyright © 2011 WILEY-VCH Verlag GmbH & Co. KGaA, Weinheim
ISBN: 978-3-527-32650-1

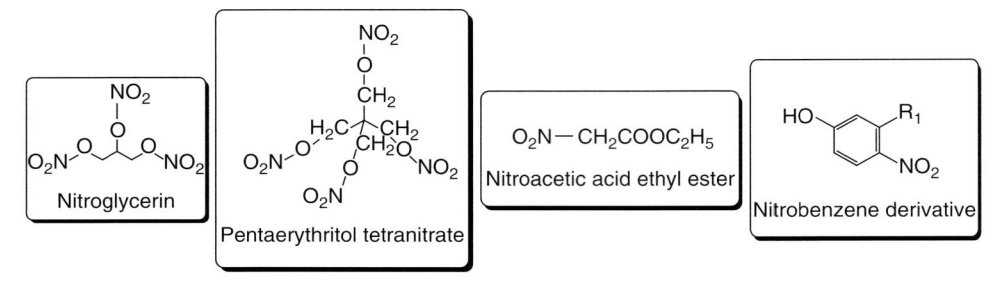

Figure 18.1 Examples of explosive compounds for risk assessment.

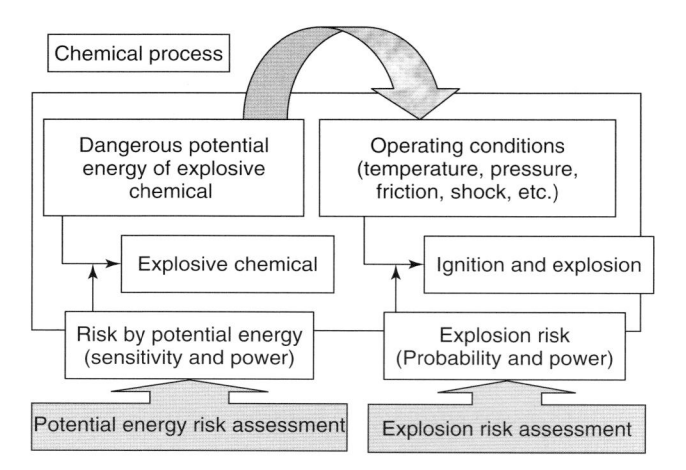

Figure 18.2 Explosion risk assessment during a chemical process.

1) Understand the risk factors and operating conditions in a chemical process: measure the explosive sensitivity and establish safe conditions.
2) Potential energy risk assessment of an explosive compound: determine the sensitivity to explosion and the explosive power of the potential energy.
3) Explosion risk assessment in a chemical process: assess the risk of explosion (explosion probability and power) of the process obtained from 1 and 2 above.
4) Disaster-prevention measures for a chemical process: examine the risk factors and operating conditions of the explosive compound and the process, and preventive measures.
5) Assessment of environmental impact of a chemical process: assess the environmental impact (explosion probability and power), considering preventive measures for a chemical process.

18.3
Standard Procedures for Risk Assessment

Before a specific explosive compound is synthesized, a risk assessment is carried out, as shown in Figure 18.3. The data obtained are used to confirm safety in chemical process development.

18.3.1
CHETAH (Chemical Thermodynamic and Energy Release Evaluation Program) Calculation

Chemical Thermodynamic and Energy Release Evaluation Program (CHETAH), a program developed by the E-27.07 committee of American Society for Testing and Materials (ASTM), is typically used to evaluate the thermodynamics (heat generated, entropy, specific heat, and heat of combustion) and the energy released by chemical compounds.

CHETAH calculations are carried out for all chemical compounds that are considered to be present in a reaction system at the time of process development to predict risks.

The advantages of such a thermochemical calculation are as follows: no test sample is required and the potential risk can be evaluated solely from the chemical formula of the compound used in the chemical reaction. This ability to predict the risk of potential energy of the process (e.g., heat of reaction and heat of decomposition in a reaction system and a generation system) is an effective tool in the early stages of research, particularly for medicine, agricultural compounds, dyes, paints, and pigments.

18.3.2
DSC (Differential Scanning Calorimetry)

Differential scanning calorimetry (DSC) is a thermoanalytical technique (as shown in Figure 18.4) that can be used to simultaneously measure the susceptibility

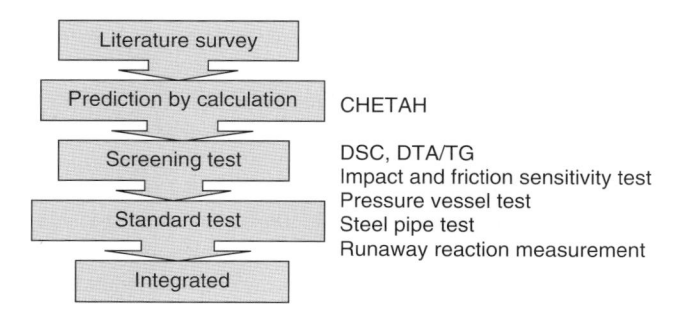

Literature survey
↓
Prediction by calculation — CHETAH
↓
Screening test — DSC, DTA/TG
Impact and friction sensitivity test
Pressure vessel test
↓
Standard test — Steel pipe test
Runaway reaction measurement
↓
Integrated

Figure 18.3 Procedures for risk assessment.

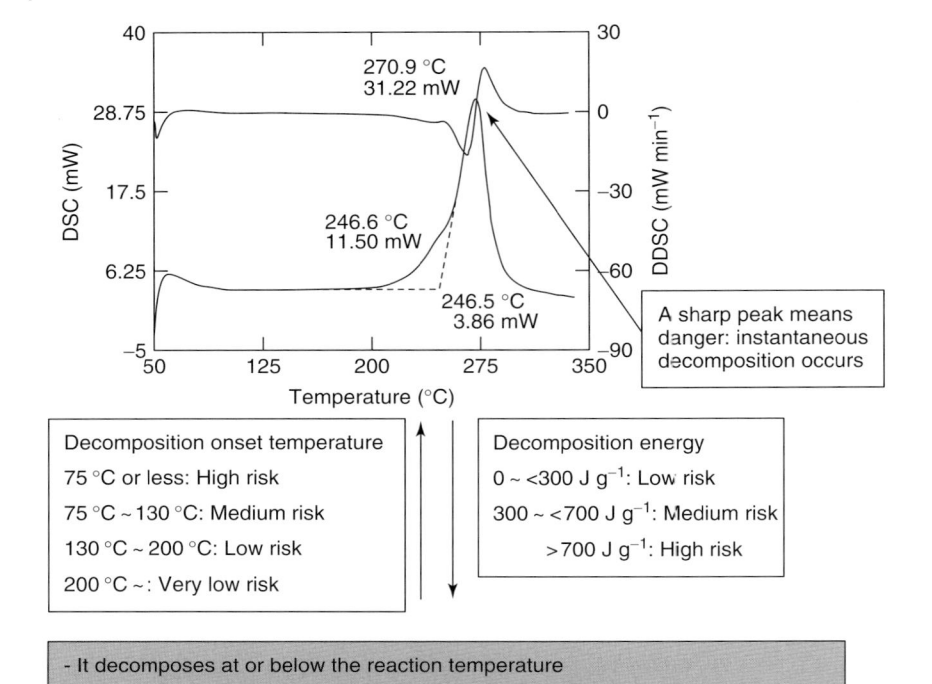

18 Process Research with Explosive Reactions

Figure 18.4 DSC chart.

DSC chart showing:
- 270.9 °C, 31.22 mW
- 246.6 °C, 11.50 mW
- 246.5 °C, 3.86 mW
- A sharp peak means danger: instantaneous decomposition occurs

Decomposition onset temperature
- 75 °C or less: High risk
- 75 °C ~ 130 °C: Medium risk
- 130 °C ~ 200 °C: Low risk
- 200 °C ~ : Very low risk

Decomposition energy
- 0 ~ <300 J g^{-1}: Low risk
- 300 ~ <700 J g^{-1}: Medium risk
- >700 J g^{-1}: High risk

- It decomposes at or below the reaction temperature
- Decomposition energy : ≥300 J g^{-1} for self-reactive substances, according to the UN Recommendation

to heat generation, such as pyrolysis, the runaway reaction (the temperature at which decomposition starts), and the amount of heat generated (decomposition energy). Therefore, DSC is often used as an initial screening method to examine the risk of pyrolysis and chemical reactions. It has been adopted as one of the tools for evaluating class 5 hazardous materials of the Fire Service Law (self-reactive substances) [1]. Since DSC requires only a small amount of sample (a few milligrams), there is little risk in handling and expensive samples, such as newly developed compounds, can be measured.

The UN Recommendation [2] defines a self-reactive substance as one with a decomposition energy ≥ 300 J g^{-1}, while we use a risk ranking which also considers the decomposition temperature and energy. A chemical process for which the reaction temperature is higher than the decomposition onset temperature is judged to be high risk. To avoid such risks, we usually recommend that the reaction temperature should be controlled to a value, that is, 100 °C [3] lower than the decomposition onset temperature, as evaluated by DSC.

Figure 18.4 shows an example of risk assessment for a class 5 hazardous material by DSC. In this method, the abscissa is the logarithm of 0.8 times the decomposition onset temperature minus 25 °C in the case of benzoyl peroxide (BPO) and 0.7 times

Table 18.1 Decomposition temperatures and energies generated for samples A and B.

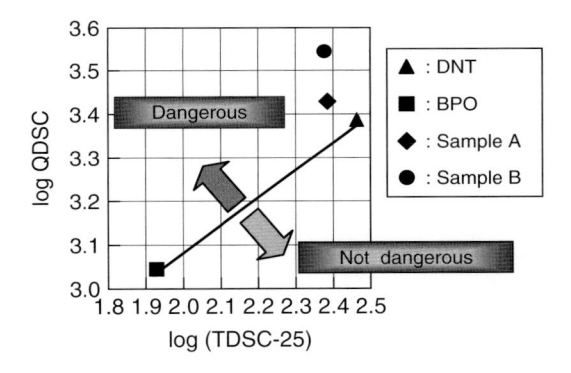

Measurement number	Sample A		Sample B	
	Decomposition onset temperature (°C)	Energy generated (J g^{-1})	Decomposition onset temperature (°C)	Energy generated (J g^{-1})
1	268.0	2864	259.0	3453
2	267.0	2714	259.0	3524
3	268.0	2580	259.0	3737
4	267.0	2583	260.0	3146
5	268.0	2779	259.0	3791
Average	267.6	2704	259.2	3530

Figure 18.5 Identification of class 5 hazardous material by DSC.

the decomposition onset temperature minus 25 °C in the case of 3,5-dinitrotoluene (DNT), and the ordinate is the logarithm of the decomposition energy generated. A line is drawn between the point for BPO and that for DNT: the area above the line is judged to be dangerous and that below the line is judged to not be dangerous.

As a specific example, the measured decomposition onset temperatures and energies generated for samples A and B are shown in Table 18.1. When plotted in Figure 18.5, the points representing the two compounds lie in the dangerous area and, therefore, they are judged to be class 5 hazardous materials.

18.3.3
DTA (Differential Thermal Analysis) and TG (Thermogravimetry)

The thermal stability and weight changes of chemical compounds can be measured using Differential Thermal Analysis (DTA) and Thermogravimetry (TG). As shown in Figure 18.6, the results of DTA and TG indicate that this compound decomposed pyrolytically, since weight reduction and heat generation occurred at the same

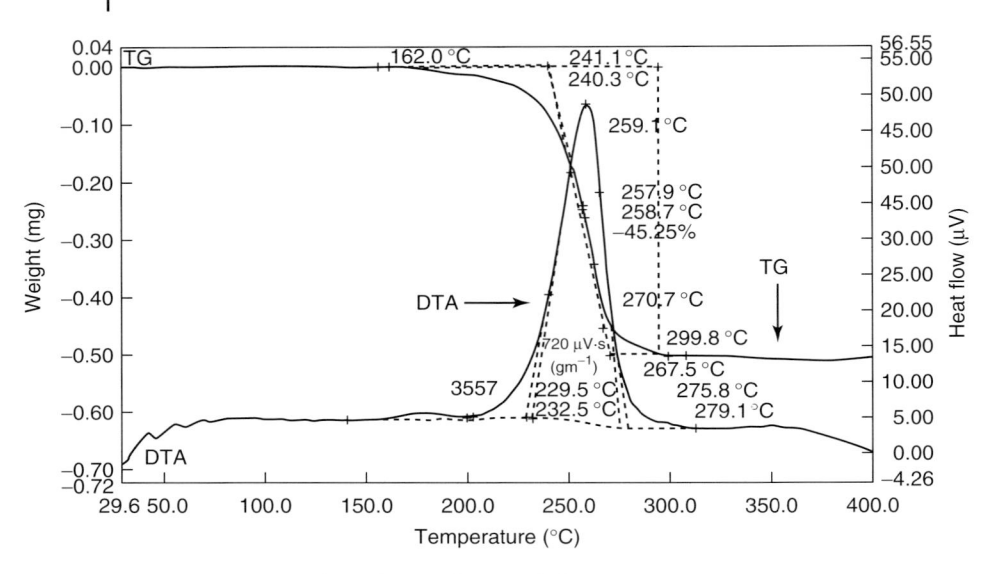

Figure 18.6 DTA and TG chart. Judgment of phase transition, decomposition, evaporation, etc. DTA: heat generation; TG: decrease ... pyrolysis (including explosion). DTA: endothermic reaction; TG: decrease ... evaporation. DTA: endothermic reaction; TG: no change ... melting, phase transition.

temperature. The example compound on Figure 18.6 is judged to undergo changes, including phase transition and evaporation.

18.3.4
Impact Sensitivity Test

The possibility of decomposition and ignition due to shock and handling problems can be evaluated by the impact sensitivity test. Sensitivity to shock is measured quantitatively and sensitivity to impact is evaluated by the 1/6 explosion point shown in Table 18.2 according to the Japanese Industrial Standard (JIS) method [4]. The tester is shown in Figure 18.7.

18.3.5
Friction Sensitivity Test

The possibility of decomposition and ignition due to friction and handling problems can be evaluated by the friction sensitivity test. Sensitivity to shock is measured quantitatively and sensitivity to friction is evaluated by the 1/6 explosion point shown in Table 18.2 according to the JIS method [5] using the Bundesanstalt für Materialforschung und prüfung (BAM) friction sensitivity tester shown in

Table 18.2 1/6 explosion point by impact sensitivity test and friction sensitivity test.

Grade	Dropping height (cm) on 1/6 explosion point	N on 1/6 explosion point
1	<5	<9.8
2	>5 to <10	>9.8 to <19.6
3	>10 to <15	>19.6 to <39.2
4	>15 to <20	>39.2 to <78.5
5	>20 to <30	>78.5 to <156.9
6	>30 to <40	>156.9 to <353.0
7	>40 to <50	>353.0
8	>50	–

Figure 18.7 Impact sensitivity tester.

Figure 18.8. The results obtained from the impact sensitivity and friction sensitivity tests are shown in Table 18.3. In both tests, compounds of grade 3 or higher are judged to be dangerous.

18.3.6
Pressure Vessel Test

The explosive power required for the decomposition of explosive compounds can be evaluated by the explosion of an orifice board (pore diameter: 9 or 1 mm) in the pressure vessel test. The principle that underlies this test is based on the judgment test for class 5 hazardous materials of the Fire Service Law. A schematic of the

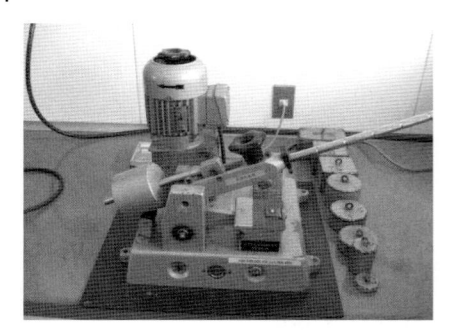

Figure 18.8 BAM friction sensitivity tester.

Table 18.3 Results of impact sensitivity and friction sensitivity tests.

Impact sensitivity test		Friction sensitivity test	
Lead azide, TNT[a]	Grade 8	Lead azide	Grade 1
Dynamite	Grade 4–5	Penthrite[b]	Grade 3
Blasting powder	Grade 5	TNT	Grade 5
Penthrite	Grade 4	Blasting powder	Grade 7

[a] 2,4,6-Trinitrotoluene.
[b] Pentaerythritol tetranitrate.

test container is shown in Figure 18.9. The judgment test for class 5 hazardous materials consists of heating an aluminum cylinder containing 5 g of sample to 400 °C at a temperature ramp rate of 40 °C min^{-1}. The specific criteria for judging a compound as a category I, II, or III class 5 hazardous material are listed in Table 18.4.

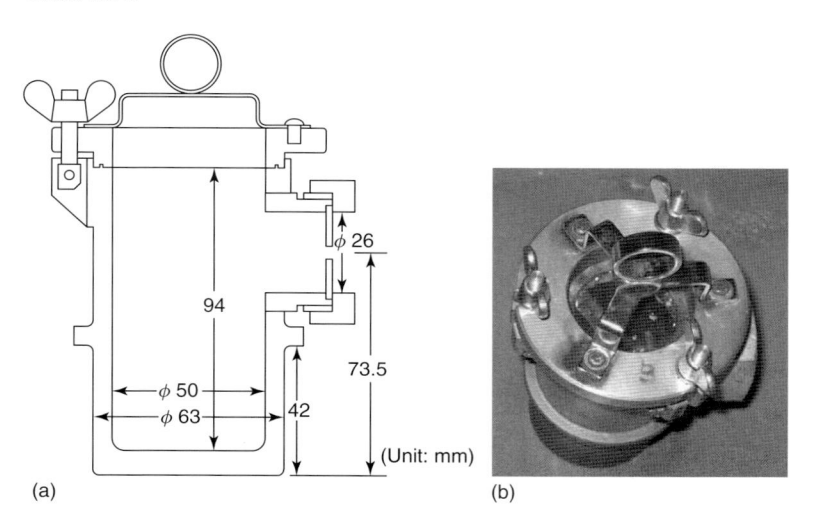

Figure 18.9 Pressure vessel tester.

Table 18.4 Judgment criteria for category I, II, or III of class 5 hazardous materials.

Thermal analysis test results from DSC	Pressure vessel test		
	Category I	Category II	Category III
Compound with risk	I	II	II
Compound with no risk	I	II	–

Judgment criterion for category I : ≥ 5 explosions in 10 tests with an orifice board with a pore diameter of 9 mm.
Judgment criterion for category II : ≥ 5 explosions in 10 tests with an orifice board with a pore diameter of 1 mm.
Judgment criterion for category III : ≤ 4 explosions in 10 tests with an orifice board with a pore diameter of 1 mm.

18.3.7
Steel Pipe Test

The steel pipe test [6] measures the susceptibility to shock and the magnitude of the explosion of an explosive compound. The test is performed with a 22 mm steel pipe and a No.6 percussion cap [7]. A schematic is shown in Figure 18.10. The classification of risk according to the steel pipe test is shown in Table 18.5.

18.4
Safety Evaluation of Nitroacetic Acid Ethyl Ester

We developed the process of anxiolytic AP521 [8] in seven steps, using benzo[b]thiophene (**1**) as a starting material, as shown in Scheme 18.1.

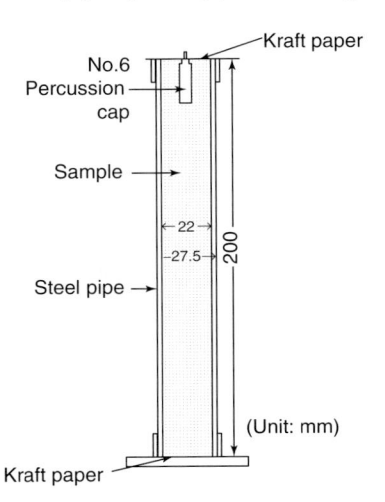

Figure 18.10 Steel pipe tester.

Table 18.5 Assessment of risk by the steel pipe test.

State of steel pipe	Risk assessment
The steel pipe is blown to pieces	Very high risk
About half of the steel pipe is demolished	High risk
Crack near the percussion cap or expansion bigger than the blank	Medium risk
Deformation similar to that with the blank	Low risk

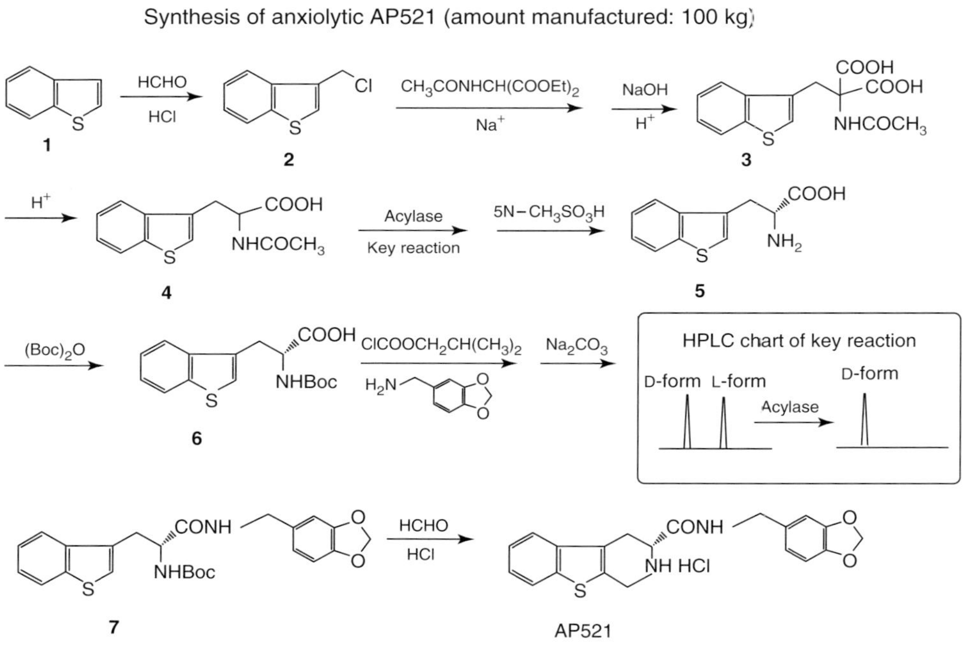

Synthesis of anxiolytic AP521 (amount manufactured: 100 kg)

Scheme 18.1 Synthesis of the anxiolytic AP521.

2-Acetylamino malonic acid diethyl ester [9] was used in the reaction step from compound **2** to compound **3**. However, as shown in Scheme 18.2, nitroacetic acid ethyl ester is more reactive than 2-acetylamino malonic acid diethyl ester for the synthesis of nonnatural amino acid (**5**) since the reaction with nitroacetic acid ethyl ester can progress with a weak base such as triethylamine, while a strong base is required with 2-acetylamino malonic acid diethyl ester. This is clearly one of the most attractive methods for obtaining compound **5** in high yields. Therefore, the safety of nitroacetic acid ethyl ester was evaluated.

Scheme 18.2 Comparison of the reactivities of nitroacetic acid ethyl ester and 2-acetylamino-malonic acid diethyl ester.

18.4.1
Various Safety Evaluations of Nitroacetic Acid Ethyl Ester

As shown in Figure 18.11, the maximum decomposition enthalpy ΔH_{max} and the heat of combustion ΔH_c of nitroacetic acid ethyl ester were obtained by CHETAH calculations, and a plot of $\Delta H_{max} : |\Delta H_c - \Delta H_{max}|$ was created for evaluation.

The risk assessment of nitroacetic acid ethyl ester was performed using DSC, the impact sensitivity test, the friction sensitivity test, and a heat of combustion

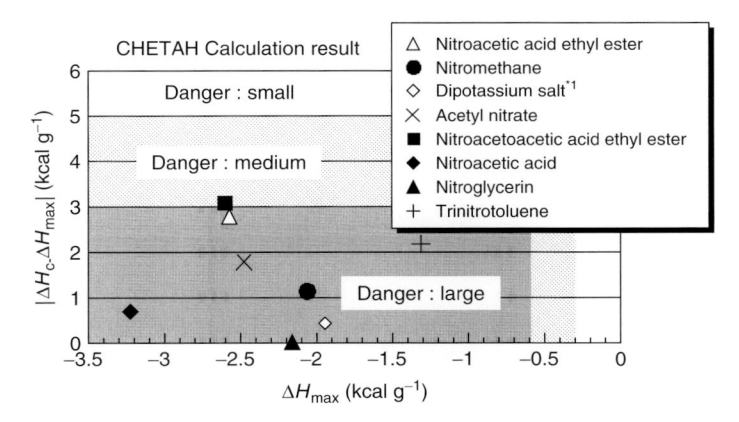

[*1] Dipotassium salt is $KO_2N=CHCO_2K$ in Figure 18.14

Figure 18.11 Results of CHETAH calculations for nitroacetic acid ethyl ester and its derivatives.

Table 18.6 Results of risk assessment for nitroacetic acid ethyl ester.

CHETAH calculation	High risk
DSC analysis	Decomposition onset temperature: 270.9 °C
	Decomposition energy: 1900 J g^{-1}
Impact sensitivity	Grade 8
Friction sensitivity	Grade 4
Heat of combustion	15.6 kJ g^{-1}

measurement, in addition to the CHETAH calculation. The results are listed in Table 18.6.

Nitroacetic acid ethyl ester is predicted to be a high-risk compound based on both CHETAH calculation and DSC analysis (decomposition energy: 1900 J g^{-1}). The impact sensitivity and friction sensitivity tests indicated that the risks of ignition and explosion were somewhat high due to friction. Although the heat of combustion was low compared to the value derived from the CHETAH calculation (22.4 kJ g^{-1}), it was confirmed to be a high-risk compound even if the CHETAH calculation result is corrected using the above measurements. On the basis of the above results, the issues that should be considered when manufacturing nitroacetic acid ethyl ester are indicated in Table 18.7. Since nitroacetic acid ethyl ester shows both, high decomposition energy and friction sensitivity, we examined the risk of this manufacturing method as follows.

18.4.2
Synthesis of Nitroacetic Acid Ethyl Ester and Its Risk Assessment

As shown in Scheme 18.3, the nitromethane method and the acetyl nitrate method are well-known methods for the synthesis of nitroacetic acid ethyl ester.

Table 18.7 Considerations in the manufacture of nitroacetic acid ethyl ester.

Extreme care is needed in a manufacturing method that uses nitromethane [10] as a starting material. Such a method is a high-risk method if it involves a step where nitromethane is handled at temperatures ≥ 100 °C
The high risk predicted by CHETAH calculations means that nitroacetic acid ethyl ester has an extremely high potential risk. Extreme care should be taken in handling this compound in large volumes
Since its decomposition temperature is 246.5 °C, nitroacetic acid ethyl ester is considered to be relatively stable. (It does not ignite during measurement of the amount of heat generated on explosion in a nitrogen atmosphere.) However, three measurements of the heat of combustion using different amounts of samples yielded almost the same values for the heat of combustion. This means that it completely combusts once it ignites
As a grade 4 compound, it is highly sensitive to friction. Extreme care should be taken in handling

1. Nitromethane method

$$2CH_3NO_2 + 2KOH \longrightarrow KO_2N=CHCO_2K + 2H_2O + NH_3$$
$$KO_2N=CHCO_2K + H_2SO_4 \xrightarrow{C_2H_5OH} O_2NCH_2CO_2C_2H_5 + H_2O + K_2SO_4$$

2. Acetyl nitrate method

$$CH_3COCH_2CO_2C_2H_5 + (CH_3CO)_2O + HNO_3 + H_2SO_4 \text{ (Catalyst)}$$
$$\longrightarrow O_2NCH_2CO_2C_2H_5 + 2CH_3CO_2H$$

Scheme 18.3 Nitromethane method and acetyl nitrate method.

First, the steel pipe test was performed for nitromethane, the starting material in the nitromethane method. The steel pipe partly exploded in one of three trials in the quiescent state (22 °C) and completely exploded in the high-temperature state (70–80 °C).

When the steel pipe did not explode, it swelled and when it exploded, it was blown to pieces. Figure 18.12 shows the results of the quiescent test (top: complete explosion, bottom: no explosion) and Figure 18.13 shows the results of the high-temperature test (photo of complete explosion).

The pressure vessel test was performed for the intermediate $KO_2N=CHCO_2K$. As shown in Figure 18.14, in the nine tests performed with an orifice board with a pore diameter of 9 mm, a large explosion was noted five times and a small explosion was noted twice. As a result, $KO_2N=CHCO_2K$ was judged to be a self-reactive substance equivalent to a class 5 category I hazardous material.

Figure 18.12 Steel pipe test for nitromethane in the quiescent state.

Figure 18.13 Steel pipe test for nitromethane in the high-temperature state.

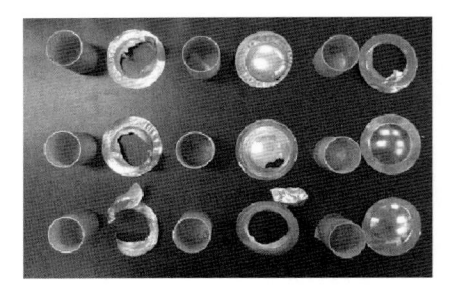

Figure 18.14 Pressure vessel test for $KO_2N=CHCO_2K$.

A runaway reaction measurement (accelerating rate calorimetry (ARC)) was performed for $KO_2N=CHCO_2K$. As shown in Figure 18.15, the results of ARC measurement indicated that the temperature at which heat generation by $KO_2N=CHCO_2K$ started was 100.5 °C and the calculated values showed that the time to runaway was very short: 1 min. As a result, $KO_2N=CHCO_2K$ was judged to be a high-risk material.

On the basis of the above results, the risks associated with $KO_2N=CHCO_2K$ were as follows:

1) It is a self-reactive substance equivalent to a class 5 category I hazardous material.
2) It easily ignites.
3) There is variability among synthesized lots, with some exploding at 100 °C. In such a case, the time to runaway is very short, suggesting that ignition and explosion may occur immediately after self-heating begins.
4) The temperature at which self-heating begins is 100 °C or less, and it is considered to be an unstable material that easily changes with time. On the basis of the above results, special care should be taken in handling $KO_2N=CHCO_2K$ in the manufacture of nitroacetic acid ethyl ester.

On the basis of the results of the above risk assessment, we decided not to use the nitromethane method, since both nitromethane (starting material) and $KO_2N=CHCO_2K$ (intermediate) are high-risk materials. Instead, we established a safe acetyl nitrate manufacturing process ($CH_3COCH_2CO_2C_2H_5$, $(CH_3CO)_2O$, HNO_3, and H_2SO_4) which does not require materials with explosion risks. With this method, we manufactured one lot (10 kg) of nitroacetic acid ethyl ester.

18.5
Development of an Efficient Method for the Synthesis of Nitrobenzene Derivatives

It is extremely difficult to achieve the regioselective nitration of an aromatic ring even if a variety of reaction conditions are used. Moreover, separation by silica gel column chromatography is not suitable for large-scale manufacture. In the nitration reaction of phenol derivatives shown in Scheme 18.4, two by-products are generated.

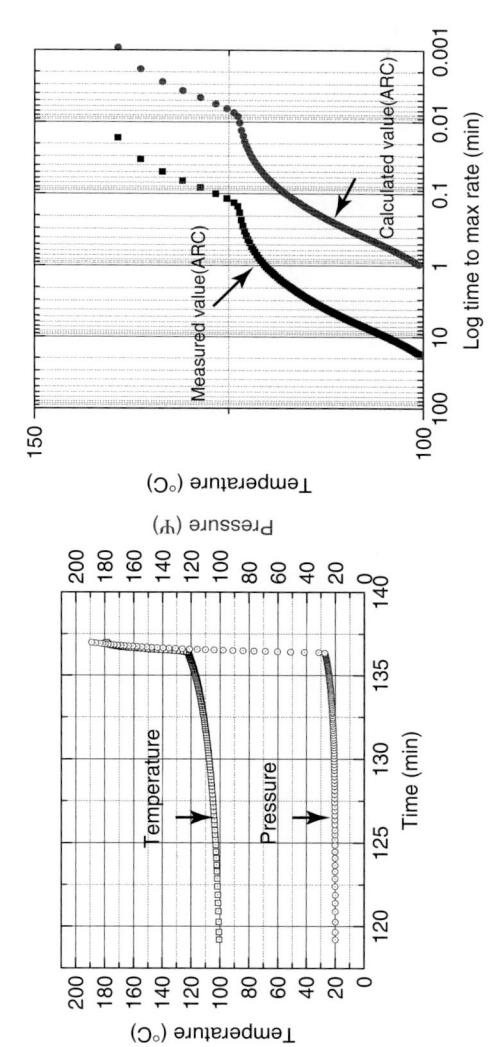

Figure 18.15 Runaway reaction measurement (ARC) for $KO_2N=CHCO_2K$. Generation of heat: beginning temperature −100.5 °C (left); time to maximum rate − 1 min (calculation value, right).

Scheme 18.4 Nitration reaction states of phenol derivatives.

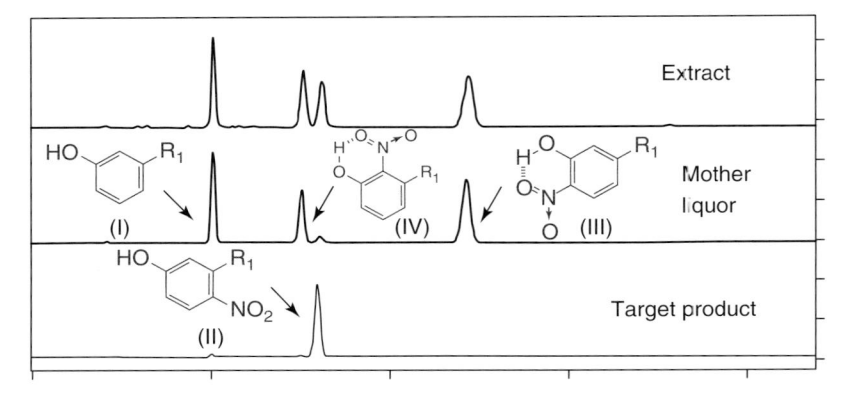

Figure 18.16 HPLC chart for each process in the nitration reaction of phenol derivatives.

Upon the formation of a hydrogen bond between the hydroxyl group and nitro group, the hydrophobicity of the by-products (**III**) and (**IV**) is increased, and therefore they are highly soluble in toluene. In contrast, the target product (**II**) does not form hydrogen bonds and exhibits low solubility in toluene. The target product (**II**) was selectively obtained in 46% yield and 97.5% purity by recrystallization from toluene. The HPLC charts of the extract, the mother liquor, and the target product are shown in Figure 18.16.

Table 18.8 and Figure 18.17 show the results of CHETAH calculations for the mononitro and dinitro derivatives generated in this reaction process, along with the trinitro derivative that is expected to be generated.

Table 18.8 Results of CHETAH calculations for nitrophenol derivatives.

	Mononitro	Dinitro	Trinitro
ΔH_{max} (kcal g^{-1})	−0.879	−1.08	−1.214
$|\Delta H_c - \Delta H_{max}|$ (kcal g^{-1})	4.024	2.703	1.821
Judgment	Medium risk	High risk	High risk

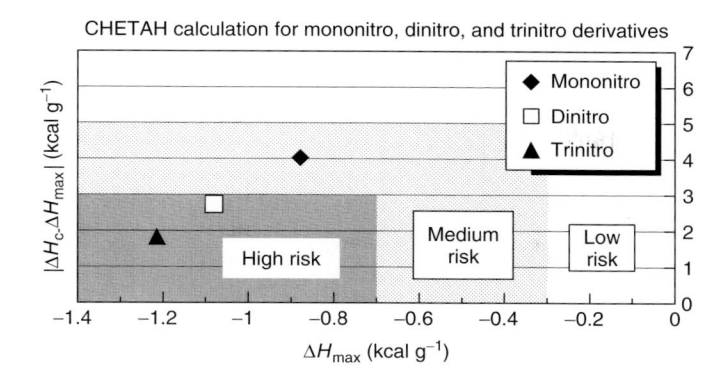

18.5 Development of an Efficient Method for the Synthesis of Nitrobenzene Derivatives

Figure 18.17 Results of CHETAH calculations for nitrophenol derivatives.

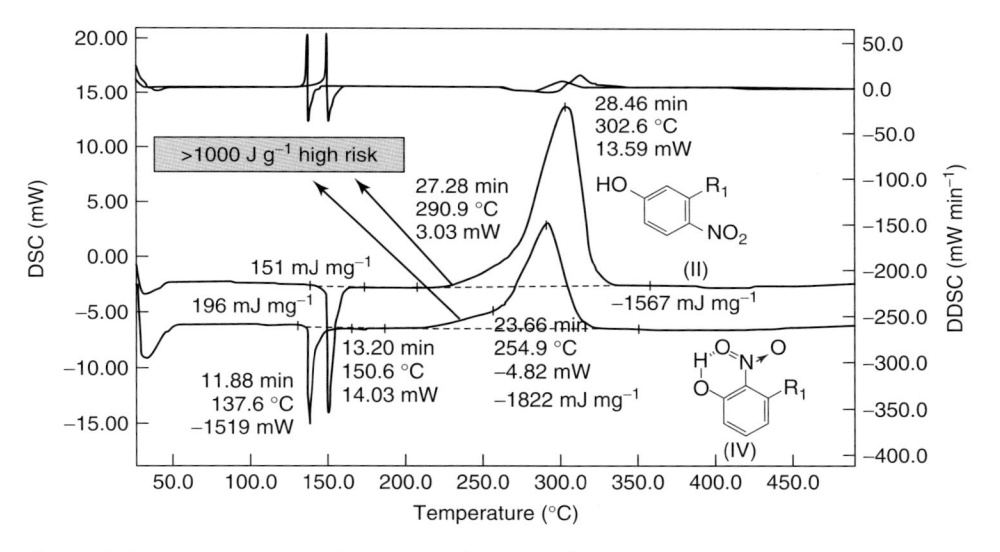

Figure 18.18 DSC measurement for target product (II) and by-product (IV).

The mononitro derivative is in the medium-risk region, and the dinitro and trinitro derivatives fall into the high-risk region. Thus, there is a correlation between the results of DSC measurement and the CHETAH calculation for the target product (II) and by-product (IV), as shown in Figure 18.18.

Although high-risk dinitro and trinitro derivatives were not generated in this process, the target product (II) and by-product (IV) were judged to be high risk by DSC.

In the manufacture of nitrobenzene derivatives, various safety assessments must be performed. Manufacturing safety is a critical issue and it is essential that safety assessments be performed for all compounds that may be generated by the process prior to starting any manufacturing process, even though such assessments are time-consuming.

References

1. Japanese Ministry of Internal Affairs and Communications (1948) Shobo Hou (Fire Service Law) of Japan http://law.e-gov.go.jp/htmldata/S23/S23HO186.html.
2. United Nations (2007) GHS *Globally Harmonized System of Classification and Labelling of Chemicals*, 2nd revised edn, Part 2, Chapter 2.8. http://www.unece.org/trans/danger/publi/ghs/ghs_rev02/English/02e_part2.pdf.
3. Tamura S. 10 (ed.) (2000) Kagaku Porosesu Anzenn Hanndobukku, Kagakubusshitu Orobi Kagakupurosesu No Sougou Anzennsei Hyoka, Asakurashotenn, pp. 136–138.
4. JIS (2003a) JIS K 4810 5-8. *Testing Methods of Explosives*, Japanese Industrial Standard.
5. JIS (2003b) JIS K 4810 8-11. *Testing Methods of Explosives*, Japanese Industrial Standard.
6. Shadanhoujin Kayaku Gakkai (1996) Kayaku Gakkai Kikaku(IV)(Kandosikenn Houhou). ES-32 (1) pp. 83–85.
7. JIS (2003) JIS K 4810 11. *Testing Methods of Explosives*, Japanese Industrial Standard.
8. Kawakubo, H., Takagi, S., Yamaura, Y., Katoh, S., Ishimoto, Y., Nagatani, T., Mochizuki, D., Kamata, T., and Sasaki, Y. (1993) *J. Med. Chem.*, **36**, 3526.
9. (a) Little, D.A. and Weisblat, D.I. (1947) *J. Am. Chem. Soc.*, **69**, 2218; (b) Zen, S. and Kaji, E. (1970) *Bull. Chem. Soc. Jpn.*, **43**, 2277.
10. Steinkopf, W. (1909) *Ber.*, **42**, 2030; (1923), *Ann.*, **434**, 21.

19
Scientific Strategy for Optical Resolution by Salt Crystallization: New Methodologies for Controlling Crystal Shape, Crystallization, and Chirality of Diastereomeric Salt

Rumiko Sakurai and Kenichi Sakai

19.1
Introduction

Since the discovery of the basic concept of optical resolution via diastereomeric salt crystal formation by Pasteur [1], various racemic compounds have been resolved and valuable enantiomerically pure compounds have been supplied to pharmaceutical, agrochemical, and other functional materials industries. In particular, it is estimated that more than half of the enantiopure drugs in the current pharmaceutical market are produced by the diastereomeric salt formation method using an enantiopure resolving agent because of its ease of operation and wide applicability [2]. Since FDA (Food and Drug Administration) announced in the early 1990s that undesired enantiomers are recognized as an impurity and should be removed, various enantiomerically pure pharmaceuticals have been intensively investigated and the method has been utilized as one of the major technologies for their production. Currently, chiral key intermediates for important chiral drugs such as indinavir (antiviral) [3], sertraline (antidepressant) [4], orlistat (antiobsessional) [5], and duloxetine (antidepressant) [6] are widely known to be efficiently produced by this method on an industrial scale. Despite its obvious importance in the pharmaceutical industry, other than some reinforcement and improvement made by industrial chemists, not many scientists have been interested in and have paid attention to the resolution mechanism of the method. In fact, until today, a universal methodology to find or optimize resolution conditions has not yet been proposed except for improvement of the original technical method of Pasteur.

In the last decade, some new technologies useful for optical resolution via diastereomeric salt crystallization have advanced with some scientific discoveries by the Sakai and Sakurai research groups, among others. They are (i) *control of crystal shape* to give easily separable salt crystals with higher chiral purity, (ii) *control of crystallization* by the concept of space filler, and (iii) *control of chirality* based on dielectrically controlled resolution (DCR) phenomena. These new methodologies are based on the control of the molecular recognition mechanism during the resolution reaction. This review focuses on the concepts of these three methodologies.

Pharmaceutical Process Chemistry. Edited by Takayuki Shioiri, Kunisuke Izawa, and Toshiro Konoike
Copyright © 2011 WILEY-VCH Verlag GmbH & Co. KGaA, Weinheim
ISBN: 978-3-527-32650-1

19.2
Control of Crystal Shape: Crystal Habit Modification

In typical resolution research, finding a suitable resolving agent for the target racemate is the first concern. In this case, alcohols, water, or their mixture, are selected as standard solvents for the resolution. The crystal shape of the resulted salt is rarely of concern in most cases. Practically, however, even if resolution conditions optimized in the lab are applied on an industrial scale, unexpected scale-up issues are often observed: solid–liquid separability of the salt from the mother liquor is a typical example that arises from the shape of the salt crystal. The unsuccessful separation remarkably affects chiral purity of the salt as a result of contamination of the separated salt crystals with the residual mother liquor containing the undesired enantiomer.

During process development of the resolution of (RS)-1-phenylethylamine (PEA) with enantiopure mandelic acid (MA), we encountered such a low-salt separability issue resulting in unexpectedly lower chiral purity as compared with that of lab data. This scale-up issue has been solved by controlling the crystal shape of the salt to be easily separable with a chiral additive derived from one of the components of the resolution system without any deterioration in industrial operation. In this section, causes and solution measures of the crystal shape issue in the production site are described based on an example of the resolution of (RS)-PEA with (R)-MA.

19.2.1
Significance of Crystal Shape in Industrial-Scale Production

Solid–liquid separation is one of the most influential steps in industrial-scale production. Purity of the salt separated is often affected by the residual mother liquor resulting in lower chemical and chiral purities. This phenomenon is apt to occur at a larger scale. In the case of the salt being a thin plate, in general, solid–liquid separation by a centrifuge becomes very difficult, because the crystals become arranged perpendicularly to the centrifugal force, resulting in contamination with a considerable amount of the mother liquor (Figure 19.1). The optical purity of the obtained diastereomeric salt therefore becomes lower than that expected from the preceding experiment in the lab. In such a case, to change the crystal shape into an easily separable one is one of the solutions for the issue of separability. For example, as shown in Figure 19.1, short and thick crystals will separate more easily from the mother liquor. To obtain such thick crystals, crystal habit modification is required.

Lahav et al. have extensively studied the habit modification of amino acids by resembling a coexisting additive structure [7] and found that the absolute configuration of the target amino acid can be determined by the crystal habit modification with the additive [8]. In conclusion, they proposed a two-step mechanism involving "binding" and "inhibition" for this habit modification [7c]. It is widely known that PEA can be resolved with enantiopure MA via diastereomeric salt formation in industrial-scale production [9]. However, the optical purity of the PEA obtained was

19.2 Control of Crystal Shape: Crystal Habit Modification | 383

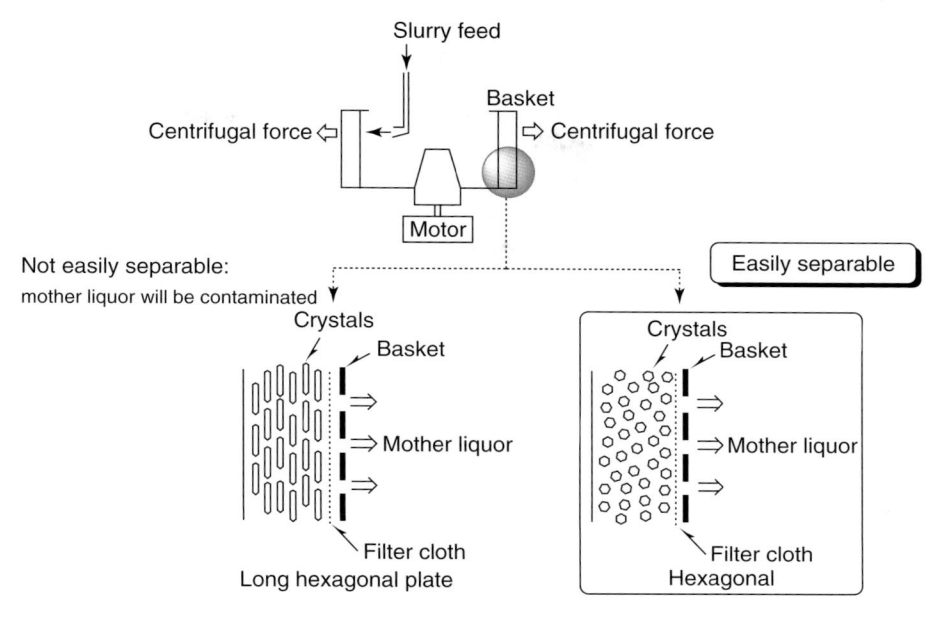

Figure 19.1 Crystal shapes and arrangements in a centrifuge.

Scheme 19.1 Resolution of (RS)-PEA with (R)-MA.

not stable (97–99% ee), whereas more than 99% ee has been steadily obtained in the laboratory (Scheme 19.1).

In order to clarify this unexpected fact, manufacturing conditions were carefully reviewed. As a result, it was found that there were two crystal shapes of the salt: thin, long, and hexagonal, and thick and hexagonal. The former shape caused difficulties in separating the mother liquor containing an undesired enantiomer (Figure 19.2). Moreover, it was found that thin plates crystallized when fresh (newly produced) racemic PEA was used, whereas thick crystals crystallized when racemized PEA was mainly used. This observation indicated that some impurities contained in the racemized PEA clearly affected the crystal shape of the salt.

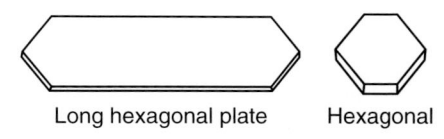

Figure 19.2 Crystal shapes of the diastereomeric salt composed of (R)-PEA and (R)-MA.

19.2.2
Effective Additive for Controlling Crystal Shape

In order to find a key substance (effective impurity) to change crystal shape, the impurity profile in the resolution process was investigated. As a result, it was found that major impurities were a Schiff base and *sec*-amine derived from the racemization step using an alkaline metal catalyst (Figure 19.3).

These impurities were isolated from racemized PEA and added individually into the resolution reaction. As a result, it was found that only *sec*-amine (bis-PEA) gives thick hexagonal salt crystals. However, *sec*-amine has three stereoisomers, (R,R), (R,S), and (S,S), based on two chiral centers. These three stereoisomers were synthesized and added into the resolution reaction. Test results are summarized in Table 19.1.

As shown in Table 19.1, it was found that the stereochemistry of *sec*-amine (bis-PEA) dramatically affected the morphological change of the salt crystals. Moreover, it was also found that the presence of a coexisting salt is also a key to influencing the change in crystal shape. These results indicate that *sec*-amine (bis-PEA), having at least one of the same absolute configurations of PEA in the diastereomeric salt, effectively gives thick salt crystals. If the target less-soluble salt is (R)-PEA:(R)-MA, the effective concentrations of (R,R)- and (R,S)-bis-PEA are 0.007 and 0.29 mol%, respectively, to (RS)-PEA in the resolution reaction. None of the other compounds were effective. These results clearly revealed that crystal habit modification with a chiral additive is extremely structure- and stereo-specific to the target substrate in the salt.

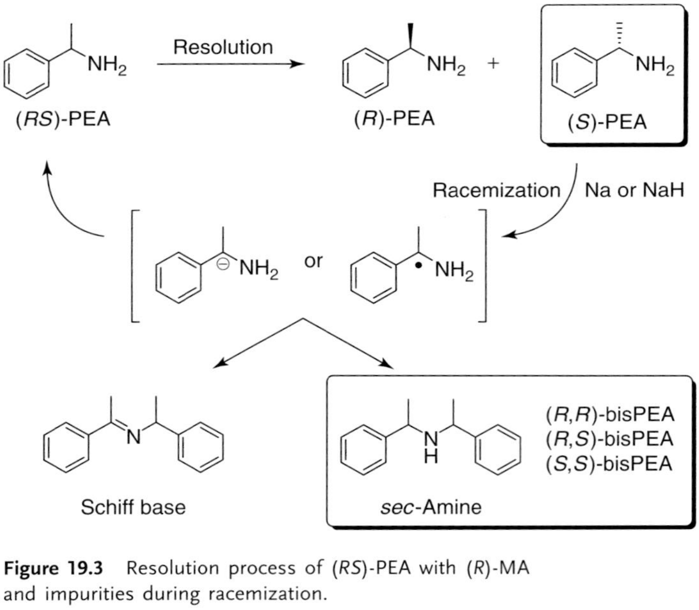

Figure 19.3 Resolution process of (RS)-PEA with (R)-MA and impurities during racemization.

19.2 Control of Crystal Shape: Crystal Habit Modification

Table 19.1 Effective additive for morphological change of (R)-PEA·(R)-MA salt.

Entry	Additive	Concentration of additive, mol% to less-soluble salt	Coexistence salt	Morphological change
1	(R,R)-bis-PEA·HCl	0.007	PEA·HCl	A
2	(R,S)-bis-PEA·HCl	0.29	PEA·HCl	A
3	(S,S)-bis-PEA·HCl	1.00	PEA·HCl	U
4	(R,R)-bis-PEA·HCl	1.00	None	U
5	(R,S)-bis-PEA·HCl	1.00	None	U
6	(S,S)-bis-PEA·HCl	1.00	None	U
7	(R,R)-bis-PEA·HCl	0.05	BnNH$_2$·HCl	A
8	(R,R)-bis-PEA·HCl	0.05	EtNH$_2$·HCl	A
9	(R,R)-bis-PEA·HCl	0.05	NH$_4$Cl	A
10	(R,R)-bis-PEA·HCl	0.05	NaCl	A
11	(R,R)-bis-PEA·HCl	0.05	AcONH$_4$	A
12	(R)-N-BnPEA·HCl	2.41	PEA·HCl	U
13	Et$_2$NH·HCl	0.50	PEA·HCl	U

A and U means the formation of affected hexagonal crystal and unaffected long hexagonal crystals, respectively.

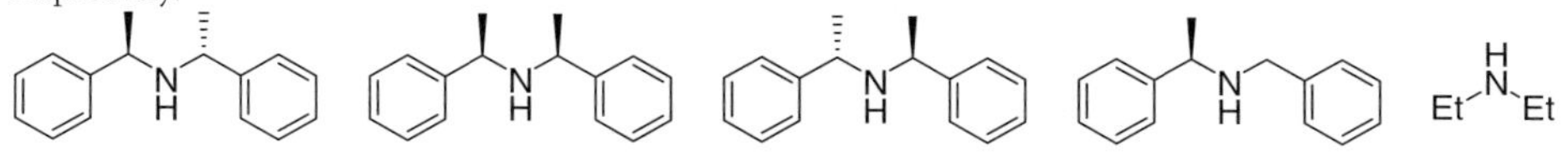

(R,R)-bis-PEA   (R,S)-bis-PEA   (S,S)-bis-PEA   (R)-N-Bn-PEA   Et$_2$NH

19.2.3
Mechanism of Crystal Habit Modification

The mechanism of crystal shape modification was investigated by observing crystal shape and by X-ray crystallographic analysis. In order to determine the direction of crystal growth and inhibition, the angles of the affected and unaffected crystals were microscopically measured. Model shapes are shown in Figure 19.4. The obtuse angle (α) and the acute angle (β) in both affected and unaffected crystals were measured by a protractor using 20 crystals shown in micrographs. The α and β angles observed in both crystals were perfectly identical ($\alpha = 128°$ and $\beta = 103°$). This means that the longest axis of the unaffected crystal was shortened by

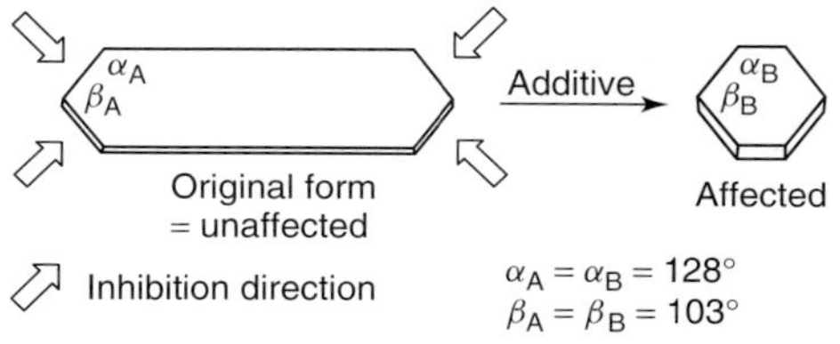

$\alpha_A = \alpha_B = 128°$
$\beta_A = \beta_B = 103°$

Figure 19.4 Crystal morphologies of affected and unaffected salt crystals.

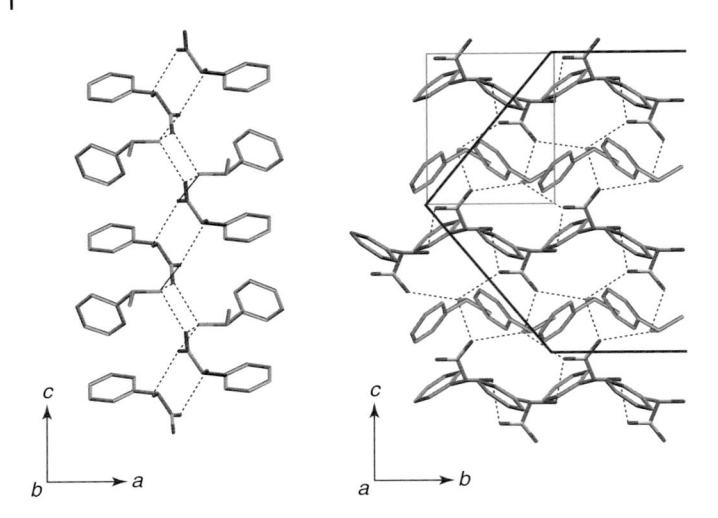

Figure 19.5 Crystal structure of (R)-PEA:(R)-MA diastereomeric salt.

inhibition of crystal growth. These crystal angles were identical to the angles calculated by crystal parameters obtained in the X-ray crystallographic analysis [10].

The crystal structure of the diastereomeric salt (R)-PEA:(R)-MA obtained in the resolution of (RS)-PEA with (R)-MA from the resolution solvent (water) was determined by X-ray crystallographic analysis (Figure 19.5).

As shown in Figure 19.5, hydrogen bonding networks were observed only on the plane composed of crystallographic b–c axes. No hydrogen bonding was observed along the a-axis: this means that the thickness of the crystal is along the a-axis. PEA and MA molecules are individually lined along the b-axis, making single-molecule layers by means of hydrogen bonds. MA molecules bind to each other by O–H···O hydrogen bonds to make rigid layers along the b-axis. The PEA molecule seems to be sandwiched by N–H···O and O–H···N hydrogen bonds between two layers of MA along crystallographic +c and −c sides and making a 2_1 column. Namely, the PEA molecule is unequivocally recognized by a cavity made by these MA layers. Because enantiomerically pure (R)-PEA (>99% ee) is exclusively recognized with (R)-MA during resolution, the cavity composed of (R)-MA layers must play a key role in recognizing (R)-PEA.

A crystal lattice is shown in Figure 19.5. Crystal angles were calculated from the crystal data obtained. Angles α and β were 129° and 101°, respectively, and they were perfectly identical to those observed by the direct observation of the salt crystals with a protractor. These results clearly suggest that the direction of crystal growth is along the b-axis. In other words, it is certain that the cavity composed of MA layers along the b-axis exclusively recognizes the oncoming (R)-PEA molecule. Attached (R)-PEA binds to two molecules of (R)-MA located on the +c- and −c-axis by two hydrogen bonds. Primary amine (R)-PEA attached to the crystal surface can bind another MA molecule with its residual hydrogen bonding hand. The crystal grows successively in this way.

Detailed crystal habit modification mechanisms of the phenomenon [11] and its application to other compounds [12] are reported in the literature.

19.3
Control of Crystallization: Concept of Space Filler

In general, finding an optimum resolving agent for the target racemate intended to be resolved is the first objective of the research. However, tedious trial and error experiments have been unavoidable because there was no theory to find one such agent thus far. Therefore, researchers have mainly been paying attention to the establishment of a quick methodology to find a better resolving agent. Borghese *et al.* proposed an efficient fast-screening methodology to find a resolving agent on the basis of eutectic composition and solubility of the salt [13]. Dyer *et al.* also proposed a new methodology based on thermal analysis of diastereomeric salts [14]. However, these methodologies need a number of experiments to collect basic physicochemical data prior to application to the resolution experiment. Kellogg *et al.* offered a new quick methodology named *Dutch resolution* using a structure-related family of resolving agents in one resolution experiment [15]. This method gives a remarkably quick result in a survey of the suitable resolving agent. However, there is unavoidable drawback: namely, that the target stereoisomer is not always crystallized with a single resolving agent because in some cases, a multicomponent salt is favorably crystallized. Saigo *et al.* proposed three key functions, relative molecular lengths (MLs), CH-π, and van der Waals interactions between resolving agent and racemic substrate for the molecular recognition during resolution, and designed resolving agents fit for the given target racemic molecule [16]. However, those tailored resolving agents are expensive and usually very costly for industrial-scale production. Regrettably, at present, a number of economical and readily available resolving agents are still limited on the market. Therefore, we should first consider how to aptly use readily available resolving agents.

The concept of space filler is a new methodology to quickly find a suitable resolving agent for the given target substrate [17]. The concept is based on realizing the closest molecular packing in a salt crystal lattice, the most essential factor for obtaining successful resolution. In order to realize close packing, the critical key is molecular size. First, we determined how to evaluate the size of molecules.

19.3.1
Evaluation of Molecular Size

Molecular sizes are evaluated as an ML shown in Figure 19.6. ML is determined by counting a number of heavy atoms from the α-atom to the far side of the molecule along a bond connection. For example, molecular length of mandelic acid (ML$_{MA}$) (Figure 19.6b) is determined to be "5." A molecule with a *p*-substituent is continuously counted from the phenyl group end; in the case of *p*-methoxy-1-phenyethylamine, ML$_{pCH_3OPEA}$ is determined to be "7."

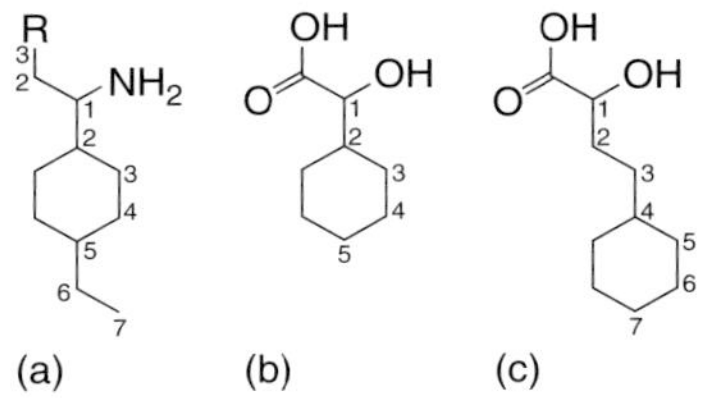

Figure 19.6 Molecular lengths of basic skeletons of amine and acids.

On the optical resolution via diastereomeric salt crystal formation, amine and acid molecules recognize each other by various interactions based on their molecular structures and functional groups. During salt crystallization, these molecules arrange in order and pack as tightly as possible while forming the closest packing mode by using hydrogen bonds, CH-π, and van der Waals interactions in the crystal lattice. In the salt crystal, functional groups of amine and acid molecules face each other to form a hydrophilic layer along the 2_1 column constructed with hydrogen-bond networks as observed in the literature [16, 17]. These observations led to the following concept.

19.3.2
Concept of Space Filler

In order to demonstrate the role of relative ML between racemic substrate and resolving agent during salt crystallization in optical resolution, typical molecular arrangements of the less-soluble diastereomeric salt crystals are schematically depicted in Figure 19.7a on the basis of various crystal structures of the diastereomeric salts found in the literature [16, 18, 19].

If MLs of base and acid are the same or similar (Figure 19.7a), the outer surface of the column becomes smooth and flat (planar surface). In such a case, the salt crystal is usually precipitated with high chiral purity (Figure 19.7a), while molecules are packed tightly and the closest packing is realized. On the other hand, if MLs of base and acid molecules differ (Figure 19.7b), the outer surface of the column becomes uneven (nonplanar surface). Moreover, even if the outer surface is flat, in a case in which an empty space is shown in the hydrophilic layer (Figure 19.7c), low chiral purity is also expected. In both cases, the closest packing will not be realized due to empty spaces caused by differences in ML. The concept of space filler is applied to such cases. The concept is useful for making crystals with closest packing by compensating an empty space with another molecule with hydrogen bonding ability (Figure 19.7d). Specifically, it is a protic molecule such as water, methanol, and so on. In the following section, this application is presented with an example.

19.3.3
Resolution of MMT

Optical resolution of enantiopure 3-(methylamino)-1-(2-thienyl)propan-1-ol (MMT), a key intermediate for duloxetine (LY-248686), (S)-(+)-N-methyl-3-(1-naphthyloxy)-3

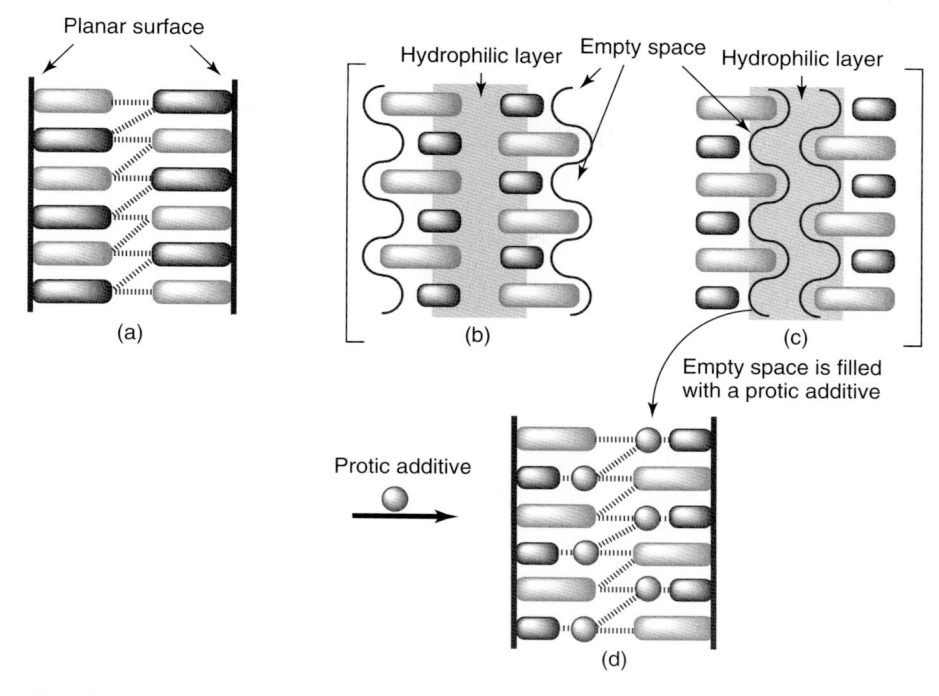

Figure 19.7 Typical molecular arrangement in the diastereomeric salt crystal.

-(2-thienyl)propylamine, was investigated. MMT could not be resolved with (S)-MA and gave no crystals at all, although its analog 3-(dimethylamino)-1-(2-thienyl)propan-1-ol (DMT) could be resolved with the same resolving agent. The concept of space filler was applied to this case.

MLs of MMT and MA are 6 and 5, respectively (Figure 19.8). According to the concept of space filler, an appropriate protic molecule to fill the space arising from the difference in MLs, was determined to be a water molecule having one heavy atom, based on the calculation of MMT(6)−MA(5). Thus, resolution solvents were optimized using water and water-containing solvents. As a result, we successfully found that 2-butanol containing 2 equiv. of water gave a monohydrated salt with the highest resolution efficiency [20].

Figure 19.8 Resolution of (RS)-MMT with (S)-MA.

19.3.4
Crystal Structures of the Salts

In order to elucidate the role of water molecules during salt formation in optical resolution, the crystal structure of (S)-MMT:(S)-MA salt was investigated [20b].

In the crystals of the ordinary less-soluble diastereomeric salts consisting of MA and primary 1-arylalkylamines, such as PEA, 1-(2-methylphenyl)ethylamine, and 1-(3-methoxyphenyl)ethylamine, hydrogen bonds between the NH (amine) and the O (carboxylate) are essential to form a fundamental unit, and generally exist to form a 2_1 column in the hydrophilic layer [16a,b, 21]. Moreover, α-OH(acid)$\cdots$O (carboxylate) hydrogen bonds between MA molecules give rigidly structured chiral MA layers for recognizing amine molecules. In sharp contrast, the crystal structures of (S)-MMT:(S)-MA:H$_2$O shown in Figure 19.9 are unique compared with ordinary salt crystals. In the (S)-MMT:(S)-MA:H$_2$O crystal, water molecules participate in crystal formation as a connector between MA molecules to form a hydrogen-bonded chiral MA layer along the b-axis, and MMT molecules are sandwiched by the MA layers from both + and − sides along the a-axis, while functional groups of the molecules face each other to form a hydrophilic layer based on the 2_1 column. As described above, the most important key for obtaining salt crystals with higher chiral purity is to realize the planar outer surface of the column. The difference in MLs between MMT and MA was determined to be 1. Therefore, this resolution system was evaluated to be unfavorable on the basis of the concept of space filler. In fact, no crystal was obtained from a water-free solvent, only from water-containing alcoholic solvents. This indicates that water is essential to crystallize the (S)-MMT:(S)-MA:H$_2$O salt; namely, an empty space, which arose from the relative ML, is filled with a water molecule.

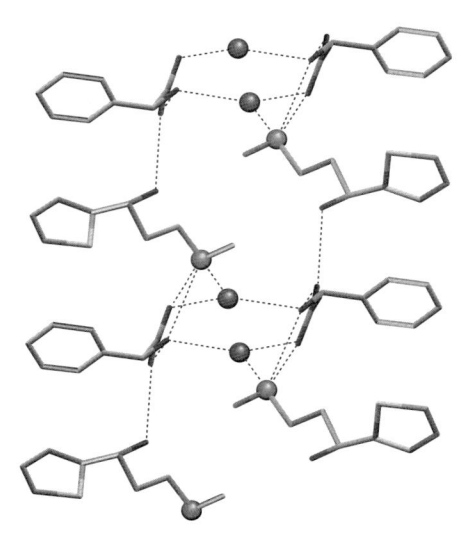

Figure 19.9 Crystal structure of (S)-MMT:(S)-MA:H$_2$O.

19.4
Control of Chirality: Dielectrically Controlled Optical Resolution (DCR)

As described in former sections, diastereomeric salt formation is one of the most frequently used resolution methods in industrial production of enantiopure compounds. Despite this, regrettably, there is weakness in chiral recognition between acid and base molecules. Namely, it is generally assumed that chiral recognition between acid and base molecules is in a chiral one-to-one relationship. Therefore, it was believed that both enantiomers of the resolving agent are needed for obtaining both enantiomers of target molecules. It seems that there is no way to change this one-to-one situation. Recently, however, we have accidentally discovered quite an unusual phenomenon due to which we should reconsider common sense on chiral discrimination. Specific chiral molecules (i.e., resolving agents) can not only recognize both enantiomers individually to be deposited as the less-soluble salt from the different solvents but can also control the diastereomeric excess (% de) of the salt when the dielectric constant (ε) of the solvent used is adjusted. This phenomenon was termed *dielectrically controlled resolution*. In this section, the DCR observed in optical resolution of racemic amines with enantiopure acids as resolving agents is presented. A strategic resolution process based on DCR is also described.

19.4.1
DCR

Resolution experiments were performed under overhead-stirring without seeding in order to avoid preferential crystallization with specific seed crystals. An equimolar resolving agent was used to the target racemate and solvent weight was determined by solubility of the solid substances at 50 °C in each resolution experiment.

19.4.1.1 DCR in Resolution of (RS)-ACL with (S)-TPA
DCR was found during the resolution process development of (RS)-α-amino-ε-caprolactam (ACL) with N-tosyl-(S)-phenylalanine (TPA) as the resolving agent (Scheme 19.2). Surprisingly, it was found that the chirality of the target substrate deposited as the less-soluble diastereomeric salt varied depending on the solvent

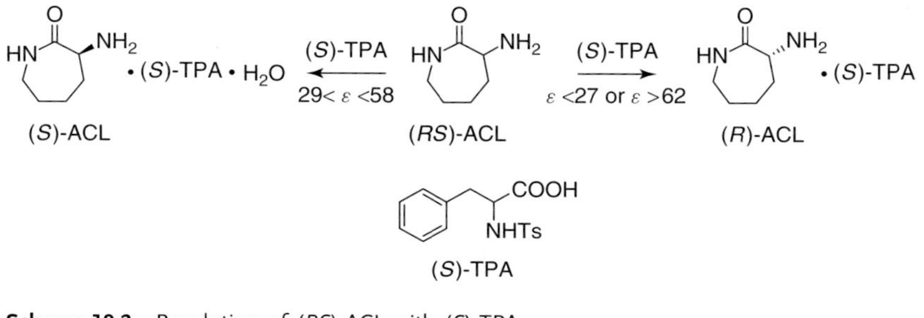

Scheme 19.2 Resolution of (RS)-ACL with (S)-TPA.

Table 19.2 Resolution of (RS)-ACL with (S)-TPA in alcohol–water solvents[a].

Entry	Solvent[b]	Solvent weight (vs (RS)-ACL) (w/w)	Yield[c] (%)	Diastereomeric excess (% de)	Resolution efficiency (E) (%)	Absolute configuration
1	CH_3OH	10	30	93	56	S
2	60% CH_3OH	11	9	95	17	S
3	45% CH_3OH	8	48	3	3	S
4	35% CH_3OH	6	16	13	4	R
5	10% CH_3OH	19	37	35	26	R
6	EtOH	32	68	7	10	R
7	90% EtOH	15	60	10	12	S
8	81% EtOH	12	24	99	48	S
9	2-PrOH	50	64	32	41	R
10	89% 2-PrOH	11	59	29	34	R
11	Water	18	30	28	17	R

[a] Resolving agent (S)-TPA/(RS)-ACL = 1.0 (molar ratio).
[b] Mixed solvent are indicated with alcohol contents in weight.
[c] Yield is calculated based on (RS)-ACL.

used (Table 19.2). These data indicated that only CH_3OH appeared to be effective for obtaining the (S)-stereoisomer, whereas the (R)-stereoisomer was obtained from water, EtOH, or 2-PrOH, although their enantiomeric purities were lower.

In order to rationalize this unusual phenomenon, various physicochemical properties of solvents were examined. As a result, it was found that change in chirality and enantiomeric purity of the salt obtained was closely related with the dielectric constant (ε) of the solvent used. Resolution experiments were extended

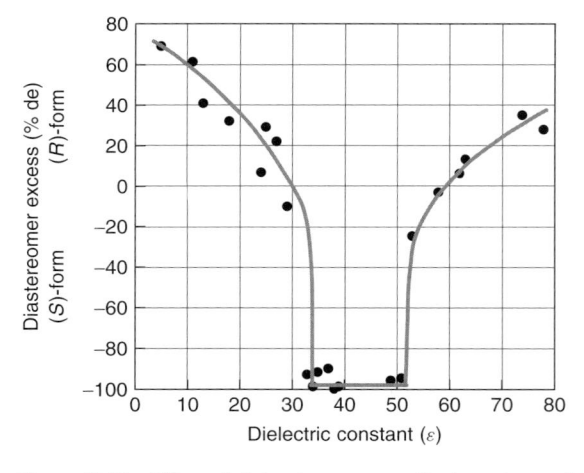

Figure 19.10 Effect of dielectric constant of solvent in resolution of (RS)-ACL with (S)-TPA.

to cover the full range of the ε value (5–78). Experimental results are summarized in Figure 19.10 [22].

As can be seen in Figure 19.10, the (S)-salt, mainly containing (S)-ACL, was deposited as the less-soluble diastereomeric salt from the solvents with a relatively medium range of dielectric constant ($29 < \varepsilon < 58$) such as 45–100% CH_3OH, 70–90% EtOH, DMSO, DMF (dimethylformamide), and so on. On the other hand, the (R)-salt, mainly containing (R)-ACL, was deposited as a less-soluble diastereomeric salt from solvents with an outer ε range for (S)-ACL; $\varepsilon < 27$ such as EtOH, 85–100% 2-PrOH, EDC (ethylene dichloride) or $\varepsilon > 62$ such as 30% EtOH, 10–35% CH_3OH, or water. From X-ray crystallographic analyses of those crystals, it was found that they were (S)-ACL:(S)-TPA:H_2O and (R)-ACL:(S)-TPA, respectively. That is, the water molecule played a key role in changing the molecular recognition system.

Strategic Resolution Utilizing DCR A practical continuous resolution process was devised with a solvent switch method based on DCR (Figure 19.11). CH_3OH was the best solvent for the first resolution to obtain the (S)-ACL:(S)-TPA:H_2O salt ((S)-salt), although 89% 2-PrOH was suitable for the second resolution to obtain the (R)-ACL:(S)-TPA salt ((R)-salt; 60%, 29% de, E 35%). After the (S)-salt (30%, 93% de, E 56%) was collected by filtration in the first resolution from CH_3OH, an equimolar mixture of (R)-enriched ACL (40% de) and (S)-TPA was recovered as a condensate by evaporating the mother liquor of the first resolution, and 89% 2-PrOH was added. The mixture was heated to dissolve the condensate and gradually cooled to crystallize the (R)-salt. The (R)-salt was obtained in extremely high resolution efficiency (41%, 93% de, E 75%) as compared with the single

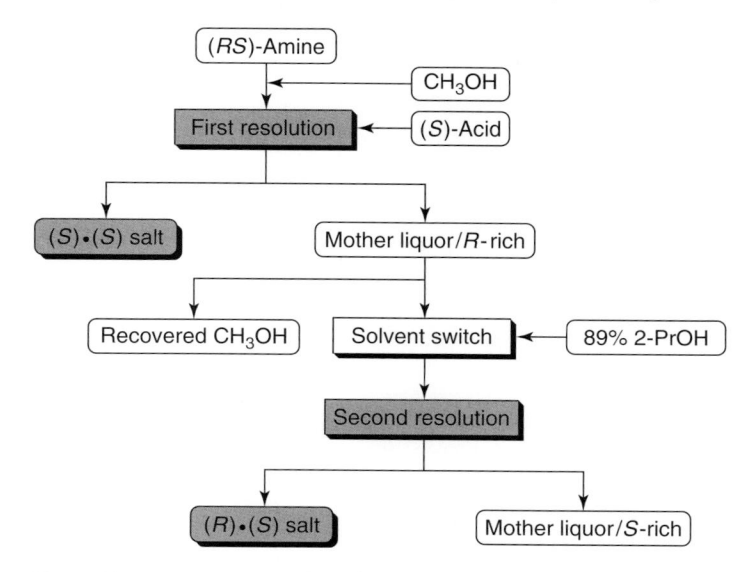

Figure 19.11 Strategic optical resolution procedure.

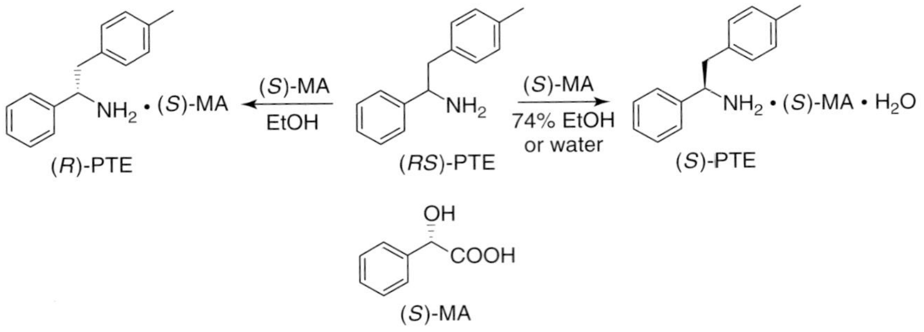

Scheme 19.3 Resolution of (RS)-PTE with (S)-MA.

resolution result (29% de, E 35%). The resolution processes were repeated and fairly reproducible results were obtained (first resolution: 29–31%, 91–93% de, E 54–56%; second resolution: 41–42%, 91–94% de, E 75–77%).

19.4.1.2 DCR in Resolution of (RS)-PTE with (S)-MA

DCR was observed in the resolution of (RS)-1-phenyl-2-(4-methylphenyl)ethylamine (PTE) with (S)-MA as the resolving agent (Scheme 19.3). At first, we examined six sorts of alcohols from C1 to C4 (dielectric constant $\varepsilon = 16-33$) and water ($\varepsilon = 78$) as the resolution solvent while an equimolar resolving agent (S)-MA was used to the racemate (RS)-PTE. Resolution results are summarized in Table 19.3 [23].

As can be seen from Table 19.3, chirality of the less-soluble diastereomeric salt crystallized was variable depending on the sort of solvent. Namely, the (R)-PTE:(S)-MA salt was crystallized with relatively higher diastereomeric excess (de %) and resolution efficiency (E) from nonbranched alcohols such as 1-butanol, 1-propanol, and ethanol (Table 19.3, entries 2, 4, and 5) except for methanol which is a very basic nonbranched alcohol. The (RS)-PTE:(S)-MA salt was crystallized from branched alcohols such as 2-butanol and 2-propanol, and from methanol (Table 19.3, entries 1, 3, and 6). On the other hand, to our surprise,

Table 19.3 Resolution of (RS)-PTE with (S)-MA in alcohols and water.

Entry	Solvent	ε	Solvent weight vs (RS)-PTE (w/w)	De (%)	R/S	Yield	E
1	2-BuOH	16	47	1	RS	55	4
2	1-BuOH	17	18	92	R	25	46
3	2-PrOH	18	49	1	RS	54	1
4	1-PrOH	22	21	23	R	23	11
5	EtOH	24	12	96	R	16	31
6	CH$_3$OH	33	4	0	RS	27	0
7	Water	78	228	98	S	24	47

(S)-MA/(RS)-PTE = 1.0 molar ratio.

the (S)-PTE:(S)-MA:H$_2$O salt crystallized only from water, although the weight of water was impractical for industrial-scale production (Table 19.3, entry 7). The resolution results obtained from various alcohols obviously reveal that the molecular recognition between PTE and MA molecules was affected by the solvent molecular structures. Although the function of the solvent molecular structure is not yet clear, a crowded solvent association state structured by the branched alcohols may spoil the chiral discrimination ability of the (S)-MA molecule to PTE molecules. In other words, the solvent association state structured by the nonbranched alcohol with less steric hindrance may enhance the discrimination ability of (S)-MA. The result from methanol may be exceptional because of its specific physicochemical properties when compared with those of other alcohols. Moreover, the fact that oppositely chiral (S)-PTE was crystallized only from water indicates that the water molecule plays a distinctive role in the chiral discrimination of (S)-MA to (S)-PTE, as observed in the ACL–TPA resolution system. These facts unequivocally suggest that chirality and chiral purity of PTE could be controlled by using a mixture solvent of alcohols and water; it can be expected that chirality and chiral purity of the target substrate PTE in the salt will be inclined to the (S)-enantiomer side as water content increases. In particular, ethanol and its mixture with water are expected to be successful for obtaining both enantiomers of PTE only by controlling water contents, since ethanol and water gave (R)-PTE and (S)-PTE respectively, with relatively higher chiral purities.

Next, we examined various mixture solvents of alcohol and water as a resolution solvent ($\varepsilon = 16$–78). Water was added to each alcohol as long as a homogeneous clear solution formed at ambient temperature. In order to efficiently display and

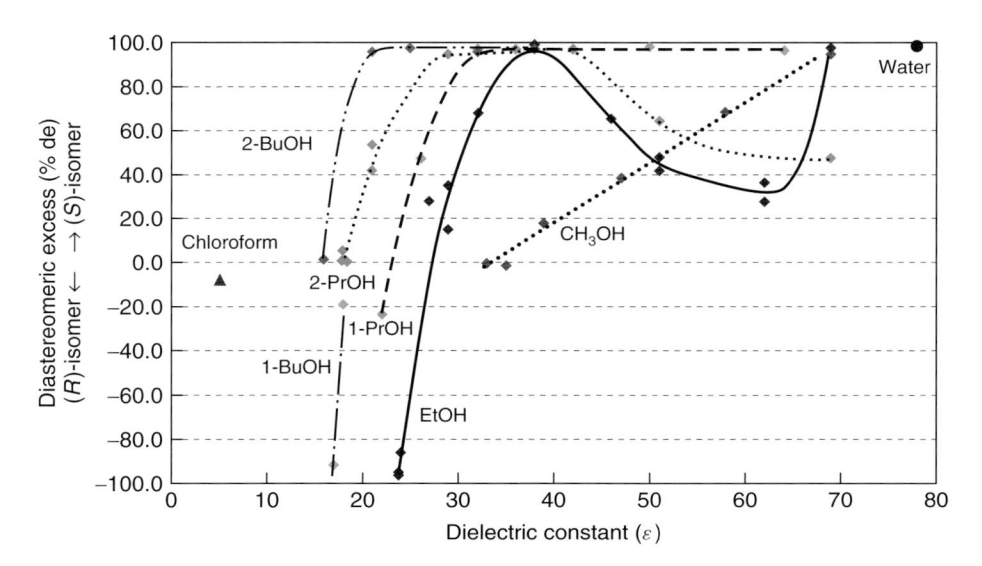

Figure 19.12 Effect of dielectric constant of the solvent in resolution of (RS)-PTE with (S)-MA.

compare the resolution results obtained from different solvents, the trends of the chirality changes are depicted by the order of the solvent dielectric constant ε in Figure 19.12. Concentrations of 1-butanol and 2-butanol were limited and the maximum concentrations were 99 and 65%, respectively, whereas other alcohols were completely miscible with water at any concentration.

As shown in Figure 19.12, all alcoholic solvents, except for methanol, exhibited the same tendency in a range of about $\varepsilon < 40$ and their chiral purities were rapidly inclined to be (S)-enriched from (R)-enriched or (RS)-form according to the increase in water content; namely, to increase the solvent ε value. On the other hand, chiral purities of the salts obtained from aqueous ethanol with $\varepsilon > 40$ were decrease to be 36% de by $\varepsilon = 62$ and retrieved at $\varepsilon = 69$ to afford enantiopure (S)-PTE (98% de). Although the reason for this phenomenon is not yet clear, it suggests that the presence of a water molecule is not a single factor for chiral discrimination. Namely, it proves that the reaction environment controlled by the solvent dielectric constant plays a very distinctive role in utilizing the water molecule for crystallizing the less-soluble diastereomeric salt (S)-PTE:(S)-MA:H_2O during the resolution reaction. From a practical viewpoint, it is noteworthy that chirality of the salt has been drastically turned from enantiopure (R)-PTE to enantiopure (S)-PTE when water was added to the nonbranched alcohol, ethanol; (R)-PTE (96% de) and (S)-PTE (99% de) were crystallized from ethanol ($\varepsilon = 24$) and 74% ethanol ($\varepsilon = 38$), respectively. On the other hand, when water was added to the branched alcohols such as 2-butanol and 2-propanol, chiral purities of the salts crystallized were improved to give the enantiopure (S)-form from the racemic form at $22 < \varepsilon < 38$ and $29 < \varepsilon < 42$, respectively. These results suggest that we should experiment with changing the reaction environment along the solvent dielectric constant even if a racemic salt was obtained from a pure solvent in the first trial of the resolution experiment. Interestingly, a completely different trend was observed when methanol was used; chiral purity was linearly improved from the (RS)- to the (S)-form as water content increased, and maximum chiral purity of (S)-PTE (95% de) was obtained at 20% CH_3OH ($\varepsilon = 69$). This phenomenon may be related to particular physicochemical properties of methanol.

Strategic Resolution Utilizing DCR Phenomenon In order to develop the optimum resolution conditions of (RS)-PTE with (S)-MA suitable for industrial-scale production, namely, to produce both (R)-PTE and (S)-PTE with high efficiency, ethanol ($\varepsilon = 24$) and 74% ethanol ($\varepsilon = 38$) were selected as resolution solvents while considering a simple combination solvent from a practical viewpoint. Since the resolution efficiency of (R)-PTE from ethanol was relatively lower, the first resolution was designed to produce (S)-PTE from 74% ethanol. Accordingly, the second resolution was designed to produce (R)-PTE by using ethanol from mother liquor of the first resolution containing (R)-enriched PTE. As a result, (S)-PTE:(S)-MA:H_2O was obtained with efficiency identical (33%, 96% de, E 64%) to the single resolution result (34%, 99% de, E 67%). The mother liquor of the first resolution was evaporated to dryness to recover the mixture of (R)-enriched PTE (49% de) and equimolar (S)-MA, followed by switching the solvent to ethanol for the second resolution. As intended, the resolution efficiency of the (R)-PTE:(S)-MA salt was extremely improved (23%, 98% de, E 46%) as compared with the result observed in the single resolution (16%, 96% de, E 31%).

Other examples of DCR are reported in the literature [24].

19.5
Conclusion and Prospect

Optical resolution via crystallization is considered to be old-fashioned. However, it is still an attractive method not only for chemists in industry but also for chemists in the lab who wish to readily obtain enantiomerically pure compounds. In this review, crystal habit modification with a chiral additive, the concept of space filler for realizing salt crystallization, and chirality control based on the phenomenon of DCR were introduced. These three methods are useful not only for finding the optimum resolution conditions but also for comprehending the molecular recognition mechanism. At the same time, chemists should pay attention to the solvent effects in addition to the usual concerns of chiral discrimination, such as molecular structures of resolving agents and racemic substrates and their hydrogen bonding abilities. In other words, if we could illuminate the mechanisms of solvent effect and chiral discrimination among molecules participating in the resolution reaction, we will be able to quickly settle on much more reliable strategies for a cost-effective resolution process.

References

1. (a) Pasteur, L.M. (1848) *Compt. Rend.*, **26**, 535; (b) Pasteur, L.M. (1848) *Ann. Chim. Phys.*, **243**, 442.
2. Rouhi, A.M. (2003) *Chem. Eng. News*, May **5**, 46.
3. Indinavir: (a) Merck Index, (2001) 13ed, No. 4970; (b) Murakami, H., Tobiyama, T., and Sakai, K. (1997) JP Kokai Patent 1997-48762; (c) Murakami, H., Satoh, S., Tobiyama, T., Sakai, K., and Nohira, H. (1996) EP Patent 710,652; (d) Murakami, H., Satoh, S., Tobiyama, T., Sakai, K., and Nohira, H. (1996) US Patent 5,792,869; *Chem. Abstr.*, **125**, 87209.
4. Sertraline: (a) Merck Index, (2001) 13ed, No. 8541; (b) Quallich, G.J. (2000) US Patent 6593496; *Chem. Abstr.*, **134**, 41980; (c) Mendelovici, M., Nidam, T., Pilarsky, G., and Gershon, N. (2001) US Patent 6552227; *Chem. Abstr.*, **135**, 242019; (d) Taber, P.G., Pfisterer, D.M., and Colberg, J.C. (2004) *Org. Proc. Res. Dev.*, **8**, 385.
5. Orlistat: (a) Merck Index, (2001) 13ed, No. 6935; (b) Karpf, M. and Zutter, U. (1991) EP Patent 443449; *Chem. Abstr.*, **115**, 255980.
6. Duloxetine: (a) Merck Index, (2001) 13ed, No. 3498; (b) Berglund, R.A. (1994) EP Patent 650965; (1995) US 5362886, JP 1995-188065; *Chem. Abstr.*, **122**, 132965; (c) Sakai, K., Sakurai, R., Yuzawa, A., Kobayashi, Y., and Saigo, K. (2003) *Tetrahedron: Asymmetry*, **14**, 1631.
7. (a) Addadi, L., van Mil, J., and Lahav, M. (1981) *J. Am. Chem. Soc.*, **103**, 1249; (b) Addadi, L., Gati, E., and Lahav, M. (1981) *J. Am. Chem. Soc.*, **103**, 1251; (c) Addadi, L., Berkovitch-Yellin, Z., Domb, N., Gati, E., Lahav, M., and Leiserowitz, L. (1982) *Nature*, **296**, 21.

8. Berkovitch-Yellin, Z., Addadi, L., Idelson, M., Leiserowitz, L., and Lahav, M. (1982) *Nature*, **296**, 27.
9. (a) Sakai, K., Murakami, H., Saigo, K., and Nohira, H. (1994) JP Appl. 1994-1757; *Chem. Abstr.*, **120**, 298229; (b) Murakami, H., Sakai, K., and Tobiyama, T. (2000) JP Appl. 2000-297066; *Chem. Abstr.*, **133**, 296269.
10. (a) Orthorhombic, $P2_12_12_1$, $Z = 4$, $a = 25.581(7)$, $b = 6.867(4)$, $c = 8.348(2)$, $V = 1474.4(9)$, $R = 0.048$, $Rw = 0.051$; Analytical results were identical to data by Brianso (crystal obtained from ether-ethanol solvent system): Brianso, M.-C. (1979) *Acta. Crystallogr.*, **B35**, 2751; (b) Polymorphological data were also reported: $P2_1$, $Z = 2$, Hashimoto, K., Sumida, Y., Terada, S., and Okamura, K. (1993) *J. Mass. Spectrometry. Soc. Jpn. (Shitsuryo Bunseki)*, **41**, 87.
11. Sakai, K. (1999) *J. Org. Synth. Chem. Jpn*, **57**, 458.
12. Sakai, K., Yoshida, S., Hashimoto, Y., Kinbara, K., Saigo, K., and Nohira, H. (1998) *Enantiomer*, **3**, 23.
13. Borghese, A., Libert, V., Zhang, T., and Charles, A.A. (2004) *Org. Proc. Res. Dev.*, **8**, 532.
14. Dyer, U.C., Henderson, D.A., and Mitchell, M.B. (1999) *Org. Proc. Res. Dev.*, **3**, 161.
15. (a) Kellogg, R.M., Nieuwenhuijzen, J.W., Pouwer, K., Vries, T.R., Broxterman, Q.B., Grimbergen, R.G.P., Kaptein, G., La Crois, R.M., de Wever, E., Zwaagstra, K., and van der Laan, A.C. (2003) *Synthesis*, **10**, 1626; (b) Nieuwenhuijzen, J.W., Grimbergen, R.G.P., Koopman, C., Kellogg, R.M., Vries, T., Pouwer, K., van Echten, E., Kaptein, B., Hulshof, L.A., and Broxterman, Q.B. (2002) *Angew. Chem. Int. Ed.*, **41**, 4281; (c) Vries, T., Wynberg, H., van Echten, E., Koek, J., ten Hoeve, W., Kellogg, R.M., Broxterman, Q.B., Minnaard, A., Kaptein, B., van der Slues, S., Hulshof, L.A., and Kooistra, J. (1998) *Angew. Chem. Int. Ed.*, **37**, 2349.
16. (a) Kinbara, K., Sakai, K., Hashimoto, Y., Nohira, H., and Saigo, K. (1996) *Tetrahedron: Asymmetry*, **7**, 1539; (b) Kinbara, K., Sakai, K., Hashimoto, Y., Nohira, H., and Chem, J. (1996) *Soc. Perkin Trans.*, **2**, 2615; (c) Kinbara, K., Hashimoto, Y., Sukegawa, M., Nohira, H., and Saigo, K. (1996) *J. Am. Chem. Soc.*, **118**, 3441; (d) Kinbara, K., Harada, Y., and Saigo, K. (1998) *Tetrahedron: Asymmetry*, **9**, 2219; (e) Kinbara, K. and Saigo, K. (2003) *Topics Stereochem.*, **23**, 207, references there in.
17. (a) Sakai, K. (2004) *CSJ Chem. Chem. Ind. (Kagaku To Kogyo)*, **5**, 507; (b) Sakai, K., Saigo, K., Murakami, H., and Nohira, H. (1993) Symposium on Chiral Compounds, Tokyo, October 22; (c) Sakai, K. (1994) Molecular Recognition Mechanism on the Optical Resolution via Crystallization of Chiral Amines, PhD Thesis, Saitama University (Japan).
18. Kinbara, K., Sakai, K., Hashimoto, Y., Nohira, H., and Saigo, K. (1996) *Tetrahedron: Asymmetry*, **7**, 1539.
19. (a) Sakai, K., Maekawa, Y., Saigo, K., Sukegawa, M., Murakami, H., and Nohira, H. (1992) *Bull. Chem. Soc. Jpn.*, **65**, 1747; (b) Sakai, K., Yoshida, Y., Hishimoto, Y., Kinbara, K., Saigo, K., and Nohira, H. (1998) *Enantiomer*, **3**, 23.
20. (a) Sakai, K., Sakurai, R., Yuzawa, A., and Hatahira, K. (2002) JP Appl. 2002-289068; (b) Sakai, K., Sakurai, R., Yuzawa, A., Kobayashi, Y., and Saigo, K. (2003) *Tetrahedron: Asymmetry*, **14**, 1631; (c) Sakai, K. (2003) Symposium Molecular Chirality, Shizuoka, Japan, IL-8, October 19; (d) Sakai, K. (2004) Japan Process Chemistry Winter Symposium, Tokyo, December 3; (e) Sakurai, R., Yuzawa, A., Murakami, H., Kobayashi, Y., Saigo, K., and Sakai, K. (2005) Japan Process Chemistry, Summer Symposium, Tokyo, Japan, July 28, p. 45; (f) Sakai, K., Sakurai, R., Yuzawa, A., and Hatahira, K. (2006) USP Appl. 2006/0063943.
21. Sakai, K., Hashimoto, Y., Kinbara, K., Saigo, K., Murakami, H., and Nohira, H. (1993) *Bull. Chem. Soc. Jpn.*, **66**, 3414.
22. (a) Sakai, K. (2003) Symposium Molecular Chirality, Shizuoka, Japan, IL-8, October 19; (b) Sakurai, R.,

Sakai, K., Yuzawa, A., and Hirayama, N. (2003) Symposium Molecular Chirality, Shizuoka, Japan, PA-11, October 19; (c) Sakai, K., Sakurai, R., Yuzawa, A., and Hirayama, N. (2003) *Tetrahedron: Asymmetry*, **14**, 3713;   (d) Sakai, K., Sakurai, R., Yuzawa, A., and Hatahira, K. (2003) JP Appl. 2003-338118;   (e) Sakai, K., Sakurai, R., and Hirayama, N. (2004) *Tetrahedron: Asymmetry*, **15**, 1073.

23. Sakai, K., Sakurai, R., Nohira, H., Tanaka, R., and Hirayama, N. (2004) *Tetrahedron: Asymmetry*, **15**, 3495.

24. (a) Sakurai, R., Yuzawa, A., Sakai, K., and Hirayama, N. (2006) *Cryst. Growth Design*, **6**, 1606;   (b) Sakai, K., Yokoyama, M., Sakurai, R., and Hirayama, N. (2006) *Tetrahedron: Asymmetry*, **17**, 1541;   (c) Sakai, K., Sakurai, R., and Hirayama, N. (2006) *Tetrahedron: Asymmetry*, **17**, 1812.

20
Development of New Drug and Crystal Polymorphs

Mitsuhisa Yamano

20.1
Introduction

Crystal polymorph study has been indispensable and of major importance in programs for the development of new drugs. This field continuously expands and many excellent books on polymorphic phenomena, including those describing recent significant progress, have been published [1]. However, it is not easy to adapt these recent findings concerning polymorphs to practical drug development because the study of polymorphic phenomena is based on multiple disciplines, such as synthetic organic chemistry, analytical chemistry, physical chemistry, computational chemistry, chemical engineering, crystallography, and formulation sciences.

Recent progress in high-throughput technology for biological assays in drug discovery has produced important hit compounds. Some of these hit compounds have low solubility in water. Others are being developed into more complex structures with many functional groups, which have the possibility of forming a large number of hydrogen bonds and also exhibit flexible conformations; both of these effects may cause polymorphic phenomena. Therefore, physicochemical research on drug substances including polymorph study has become important in drug discovery. In particular, polymorph screening as part of physicochemical studies has made remarkable progress.

Meanwhile, to scale-up the manufacturing of drug substances in process research, key issues include the setting up of process parameters of crystallization for drug substances and/or synthetic intermediates to establish an optimal synthetic route. In a large number of crystallization processes, polymorph phenomena are important and, especially at the final step of drug substance manufacturing, careful polymorph control is required to obtain the requisite form of crystals. In this chapter, a perspective for process chemists to tackle polymorphic phenomena is provided and some case studies are described.

20.2
Scope of Crystal Polymorphs

Crystal polymorphs are strictly defined as different crystal arrangements of the same chemical composition [1]. Crystals of drug substances often have more than one component. Some of them are ionized while others include a solvent and/or water. Recently, a number of trials to obtain cocrystals of drug substances have been carried out [1]. Furthermore, a variety of combinations of these multicomponent systems can be considered (Figure 20.1). Crystal polymorphs can exist in each case [2].

Solvate crystals emerge in a number of crystallization processes during the production of drug substances, depending upon crystallization conditions. Solvate crystals have been referred to as *pseudopolymorphs* [3] for a long time. Pseudopolymorphs should be treated similar to polymorphs in thermodynamic consideration. If solvate crystals are obtained in the crystallization of drug substances, the residual solvent should be removed from the crystals in successive drying processes. Since solvent removal processes often accompany a change in crystal form, careful control of crystal form in the drying process is also required.

20.3
Late-Appearing Polymorphs

In crystallization processes, the most thermodynamically stable form rarely emerges in a supersaturated solution spontaneously. Instead, a metastable form first appears and is sequentially transformed into a more stable form. This is called *Ostwald's stage rule* [4].

The appearance of the new form often occurs too late and years pass before the metastable form appears, which is called a *late-appearing polymorph* (Figure 20.2).

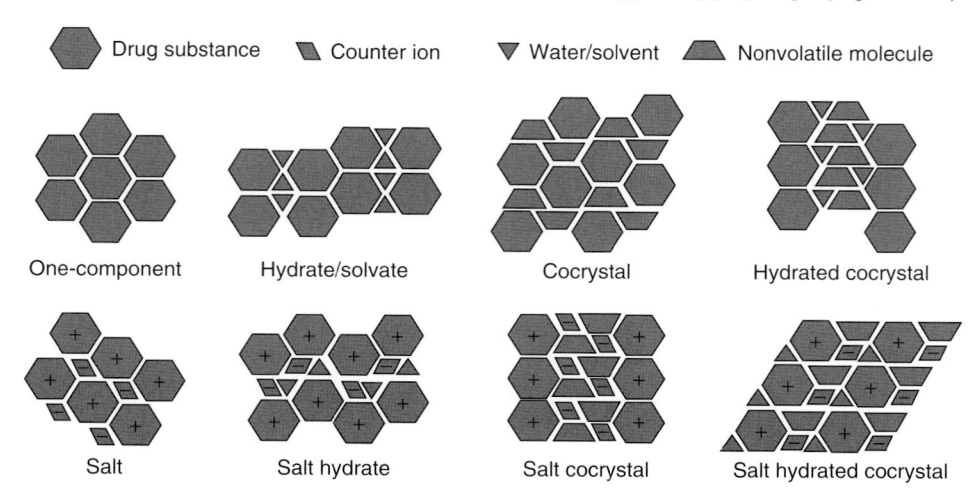

Figure 20.1 Multicomponent crystals [2].

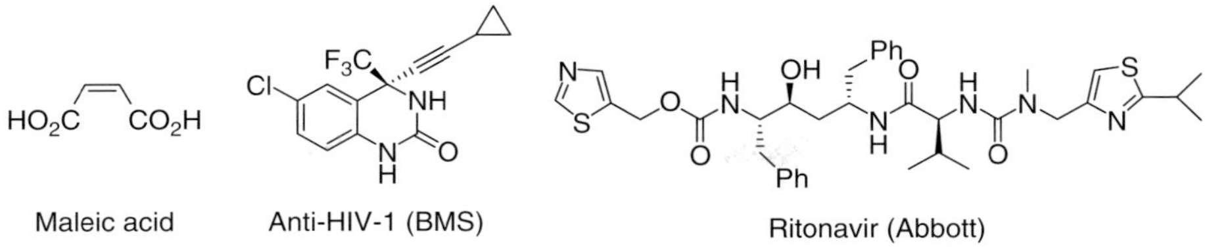

Figure 20.2 Examples of late-appearing polymorphs.

An example of this is maleic acid, a well-known reagent for the salt formation of drugs, of which the annual production is more than 1000 tons. It had no identified polymorph for 124 years, nevertheless, recently a new stable form appeared [5d]. For drug substances, new forms often appear during development. Researchers at BMS reported a case study of an HIV-1 inhibitor that suddenly exhibited a new stable form at low temperature in a batch subjected to a 30th drying process [5c]. Furthermore, a change of formulation was needed in the case of ritonavir, which suddenly exhibited a new stable form II after its launch [5a,b].

20.4
Late-Appearing Polymorphs as a Process Research Issue

The possibility of late-appearing polymorphs is not limited to drug substances, but should also be considered for each crystalline intermediate in the drug synthetic processes. Therefore, reconsideration of the crystallization condition with the appearance of a new form is needed in process development. As with similar phenomena, synthetic organic chemists often encounter the difficulty of reproduction in previously simple reactions, in which a lower solubility of a new crystal form of the substrate is exhibited. In extreme cases, the reaction has to be discarded. Synthetic organic chemists, who frequently handle new compounds, have provided a number of examples in which it became difficult to efficiently purify a compound by column chromatography because of poor solubility in eluent with first crystallization of the compound.

A large number of manufacturing processes for chiral drugs have an optical resolution step using a diastereomer salt method. An example [6] of this is that the late-appearing stable crystal form of an undesired diastereomer salt with lower solubility caused poor resolution, although it previously had excellent resolution efficiency (Figure 20.3). The chiral resolution process for racemate of a synthetic intermediate (TQA,1-[6-chloro-1,2,3,4-tetrahydroquinolin-3-yl]-*N*,*N*-dimethylmethanamine) of a chiral drug candidate has been established using tosyl-D-valine (TDV) as a resolving reagent. The desired optical isomer ((*S*)-TQA) could be prepared by this chiral resolution step. Initially, the diastereomer salt for the desired optical isomer (*S*)-TQA·TDV had been obtained with an optical purity of 89.4% de. The diastereomer salt crystals obtained by the chiral resolution had the same X-ray diffraction (XRD) pattern as optically pure diastereomer salt crystals with 100% de. The XRD pattern of the diastereomer salt crystals obtained by chiral

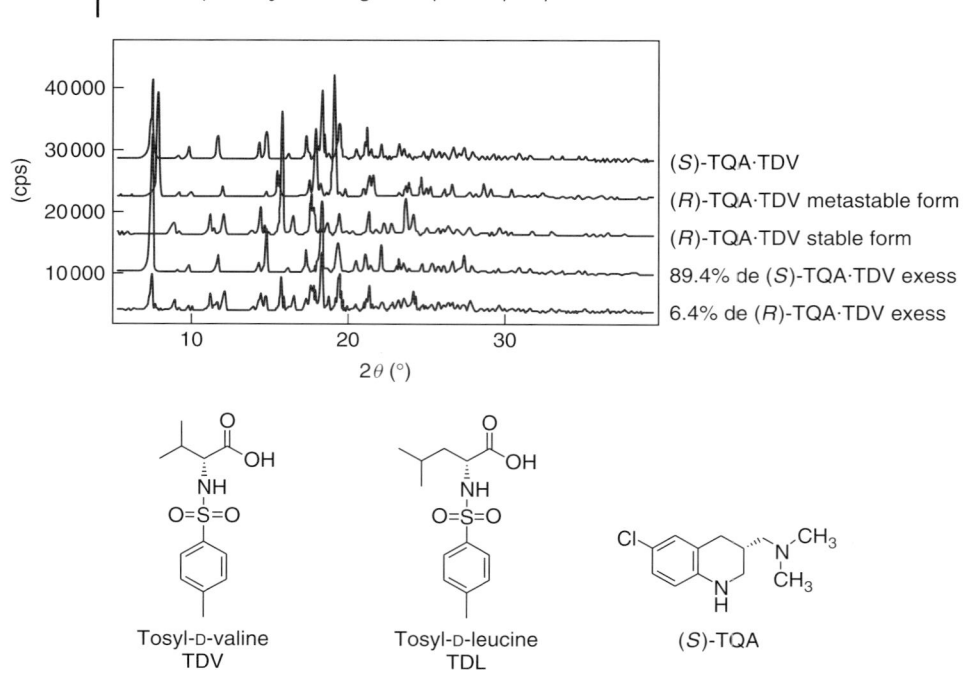

Figure 20.3 X-ray powder diffraction patterns for late-appearing new polymorphs in the optical resolution of TQA by the diastereomer salt method [6].

resolution implies no contamination of the undesired stereoisomer (R)-TQA·TDV, which has higher solubility. However, during scale-up production, optical purity of some batches gradually decreased and finally nearly racemic crystals with 6.4% de of undesired (R)-TQA·TDV were obtained. Examination of this incident proved that a new stable crystal form of undesired (R)-TQA·TDV with a lower solubility appeared in the scaled-up production and contaminated crystallization of chiral resolution with undesired diastereomer salt crystal of (R)-TQA·TDV; the prior form of (R)-TQA·TDV was found to be metastable by XRD analysis. Therefore, the chiral resolution process using TDV was discarded and an alternative resolution process using tosyl-D-leucine (TDL) with one extra carbon in the chain was adopted instead of TDV, although this new process had a less efficient resolution.

20.5
Drug Substance Form Selection

Solid-state properties of drug substances have become one of the key issues in the selection of drug candidates because they greatly affect bioavailability, formulation property, shelf life of drug substances, and drugs. Therefore, the drug substance form selection is now carried out before the nomination of a drug candidate.

If the drug candidate has poor solid-state properties with an acid/base functional group, salt screening would initially be carried out, because salt selection affects solubility greater than polymorphs. Successively, polymorph screening is carried out to select drug substance form. Recently, polymorph screening at this stage has become as thorough as possible, within the limits of time and cost, to avoid late-appearing polymorphs. The issue of which form should be selected is difficult to resolve. Ideally, the best crystal form should be comprehensively decided by consideration of all the solid-state properties of solubility, dissolution rate, hygroscopy, melting point, and handling property. However, the most stable form tends to be selected as the drug substance form to avoid late-appearing polymorphs. This tendency confers the disadvantage of poor solubility because the most stable form tends to be the least soluble. The selection of the metastable form is rare and limited to cases in which the most stable form has (i) too low solubility, (ii) improper dissolution rate for quick-relief formulations, (iii) manufacturing difficulty, (iv) IP issues, and (v) chemical instability due to topochemical factors [1b]. In the case of manufacturing difficulty of the stable form, it would be important for process research to suggest that an alternative form should be selected for the drug substance form.

20.6
Polymorph Screening

In addition to polymorph screening to select the drug substance form, consecutive complete screening is required to seek latent metastable forms, because information on a metastable form, including solvate, is important for setting parameters of crystallization, as well as for formulation, regulatory issues, and IP issues. The polymorph screenings, both for drug substance form selection and for seeking the metastable form, require numerous crystallization experiments. High-throughput apparatus that has recently been greatly improved is useful for this aim.

The work flow in the polymorph screening is roughly divided into the crystallization experiment and the process of analyzing the obtained crystals. Unlike routine screening, the crystallization experiment is required to have as wide a diversity as possible to comprehensively examine the polymorphs of the given compound. For this diversity, it is important to plan well-designed experiments, including the selection of a crystallization method, such as cooling, antisolvent addition, evaporation, and setting of parameters such as temperature, concentration, and kind of solvent [7]. The best strategy for the planned experiments have been examined by many pharmaceutical companies, although recent high-throughput apparatus can carry out approximately 10 000 kinds of experiments.

However, to simply apply the high-throughput method to crystallization screening is not the ultimate solution for the issue of polymorphs [7a]. It is not only difficult to encompass whole metastable forms, but failure to find the most stable forms can also occur. In some recent development programs, new stable forms emerged in phase II or phase III studies [8]. In many cases, the samples for polymorph screening are supplied as single batch products. Many of these samples supplied at the preformulation stage in an early development program have a lower purity than the late-stage samples for clinical studies. It is known that coexisting substances in supersaturated solution often inhibit nucleation [9] and, in some instances, impurities may cause a late-appearing polymorph. Therefore, in polymorph screening, the most stable form may not be found owing to the use of a sample contaminated by impurities. Basically, samples for polymorph screening should have a high purity equal to the analytical standard. Ideally, for evaluation of the effect of impurities, polymorph screening using samples that include potential impurities is suggested as additional screening [8].

20.7
Thermodynamically Stable Polymorphs

20.7.1
One-Component System

Viewed from a thermodynamic perspective, the most stable crystal form has the lowest free energy, and all other crystal forms are metastable forms. Whenever new crystal forms emerge, reconsideration of thermodynamic equilibrium is required. A polymorph of a drug substance is usually treated as a one-component system thermodynamically, without any other component like humidity or solvent considered, with pressure assumed to be constant and only the temperature being considered as a variable parameter. The polymorph in a one-component system is classified as enantiotropism, in which the stable form is reversed with the temperature region, and monotropism, in which stable form is the same at all temperatures (Figure 20.4) [1]. The crystal form with a higher melting point is the stable polymorph for monotropism or the high-temperature stable form for enantiotropism. These considerations are chiefly on the basis of the result of differential scanning calorimetry (DSC) analysis. For enantiotropism, the stable form at room temperature is often selected as a drug substance form.

20.7.2
Multicomponent System

Meanwhile, from the perspective of process research on the crystallization process, the stable crystal form from the result of DSC analysis is only for the one-component system, which is applicable to melt crystallization. In process research on drug manufacturing, which mainly treats solution crystallization, the thermodynamic system is more complex because it is a multicomponent system with at least more than two components, including the solvent as another component. In solution crystallization, the crystal form having the lowest solubility is the most stable form under constant conditions of solvent kind, solution concentration, and temperature. Therefore, the most stable form in solution crystallization can be

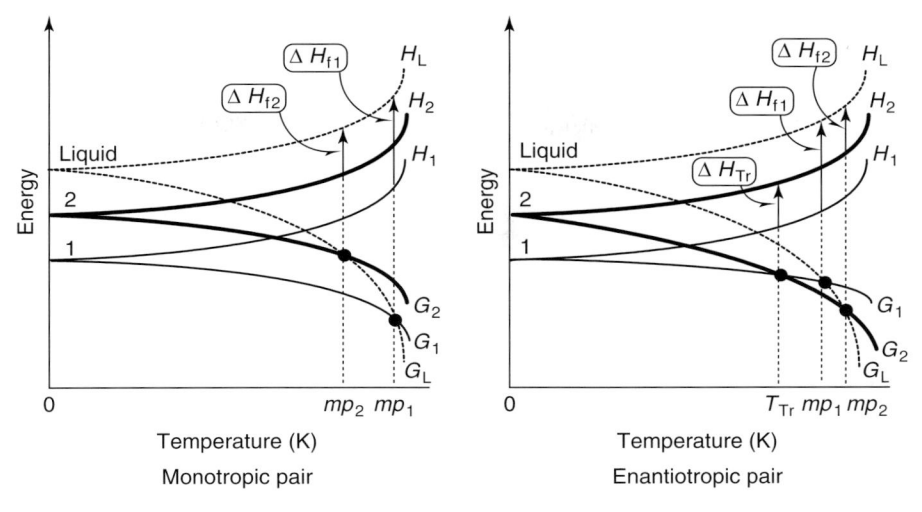

Figure 20.4 Energy temperature diagram for the one-component system. G is free energy, H is enthalpy, T is temperature, mp is melting point, subscripts 1 and 2 and L refer to the polymorph 1, polymorph 2, and liquid phase, respectively, and subscripts f, and Tr refer to fusion and transition point, respectively.

different depending on the kind of solvent. It may also be different from the most stable form in a one-component system. Furthermore, solution crystallization has the possibility of resulting in solvate crystals, which should be included in the considered thermodynamic system. Although the most stable form in a one-component system tends to be selected for the drug substance form, the most stable form in the crystallization process, including solvate crystals, should be selected for robust manufacturing.

Furthermore, in crystallization by the antisolvent method, solvent composition should be considered thermodynamically because antisolvent crystallization is carried out in a mixed solvent system. For example, in the case of TAK-029, which has antithrombotic efficacy [10], the crystallization of TAK-029 was performed in a mixed solvent system of H_2O–ethanol by the antisolvent method where H_2O was the antisolvent. The anhydrous crystal form has a lower solubility in the region of high ethanol concentration, while the hydrate crystal form has a lower solubility in the region of high H_2O concentration (Figure 20.5) [6]. Each stable crystal form was easily selectively obtained in each region. In fact, the anhydrous crystal form is the stable form in the region of higher ethanol concentration, while the hydrate crystal form is the stable form in the region of higher H_2O concentration. Although only the trihydrate crystal form had previously existed as the hydrate form, a tetrahydrate crystal form with a lower solubility suddenly emerged as a late-appearing polymorph. The tetrahydrate crystal form became the stable form that had a lower solubility in the entire region where previously trihydrate had existed stably. In this way, the region where the trihydrate form had existed stably

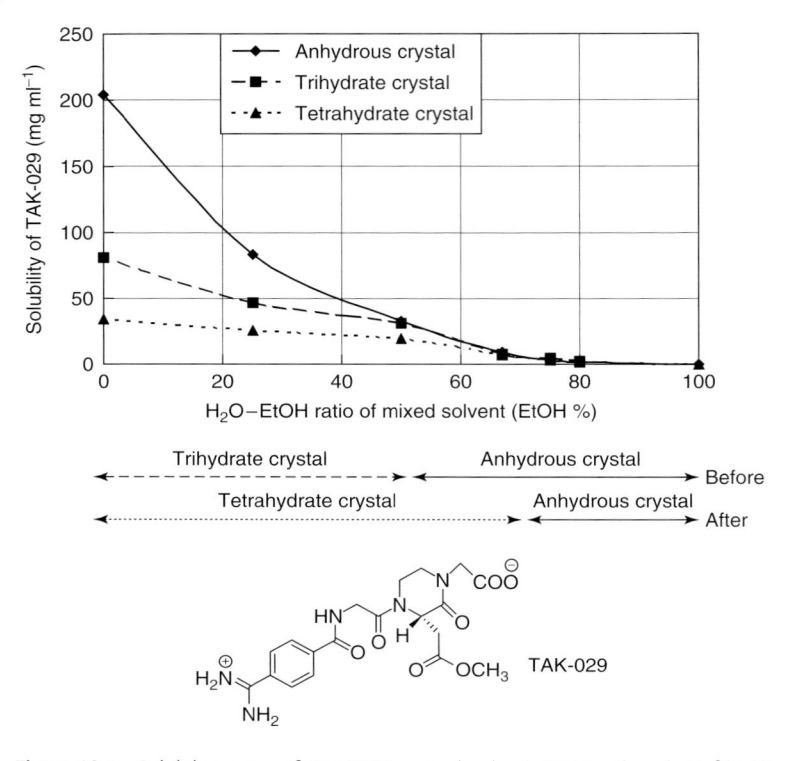

Figure 20.5 Solubility curve of TAK-029 in mixed solvent (H_2O–ethanol, 25 °C) [6].

had vanished. However, the region where the anhydrate crystal form stably existed remained at a higher ethanol concentration, although the region became slightly narrower. Therefore, it has been possible to produce the anhydrate crystal form as a drug substance without any interruption because of the continued existence of the stable region.

In contrast to the case of the anhydrate form of TAK-029 for which the thermodynamically stable region remained with the late-appearing polymorph, the stable region for the previous form completely vanished in the well-known example of ritonavir developed by Abbott as an HIV protease inhibitor [5a,b]. Form II of ritonavir with lower solubility as a late-appearing polymorph suddenly emerged two years after its launch. This caused an interruption in the manufacturing of form I as the drug substance form, which had previously existed stably. Six years passed after the first synthesis of ritonavir until form II emerged. The crystallization of ritonavir has been carried out with a mixed solvent of ethyl acetate–heptane. The stable region for form I completely vanished in almost the entire composition of the mixed solvent in which form II has lower solubility, as shown in the solubility curve (Figure 20.6) [5a,b]. In the case of ritonavir, form I became the metastable form, and the most stable form II has lower solubility in the entire region of mixed solvent of ethyl acetate–heptane.

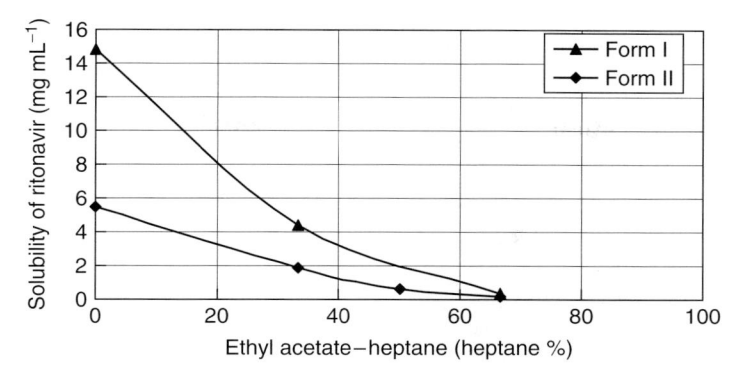

Figure 20.6 Solubility curve of ritonavir in mixed solvent (ethyl acetate–heptane, 25 °C) [5a].

In the practical issue of thermodynamic consideration of the crystal form, the chemical stability of drug substances should also be considered. Many drug substances have poor thermal stability at high temperature and this can prevent existence of the high-temperature stable form. The zwitterion crystal form of SCE-2787, which is a cephem antibiotic [11], had previously existed only as the trihydrate α form. Four years after the initial acquisition of the α form, the heptahydrate γ form of SCE-2787 zwitterion, which has an unusually low solubility of one-tenth of that of the trihydrate form, suddenly appeared as a late-appearing polymorph [6]. The solubility of the γ form in water was measured and compared with the solubility of the α form, which had previously been measured. The γ form had a lower solubility in the entire temperature region where solubility was measured (Figure 20.7) [6]. However, the result of a van't Hoff plot suggested that the solubility of the γ form had a much larger temperature dependence than that of the α form, and that the α form had a lower solubility at above 95 °C. However, cephem compounds like SCE-2787 are extremely chemically unstable in water solution at high temperature near 100 °C, and would decompose before crystallization. Therefore, the α form has no stable region, and it has actually been recognized as a metastable form.

20.8
Polymorph Control

In crystallization processes, thermodynamically stable forms are not always obtained, but the obtained product can include a metastable form and/or the reproducibility of the obtained crystal form is lost. These cases are usually called *concomitant polymorphs* [12]. For a concomitant polymorph, the desired crystal form should be obtained by the control of crystallization parameters to control the rate of crystallization. The establishment of the crystallization process to selectively obtain the desired crystal form would be a highly probable result of process research [13].

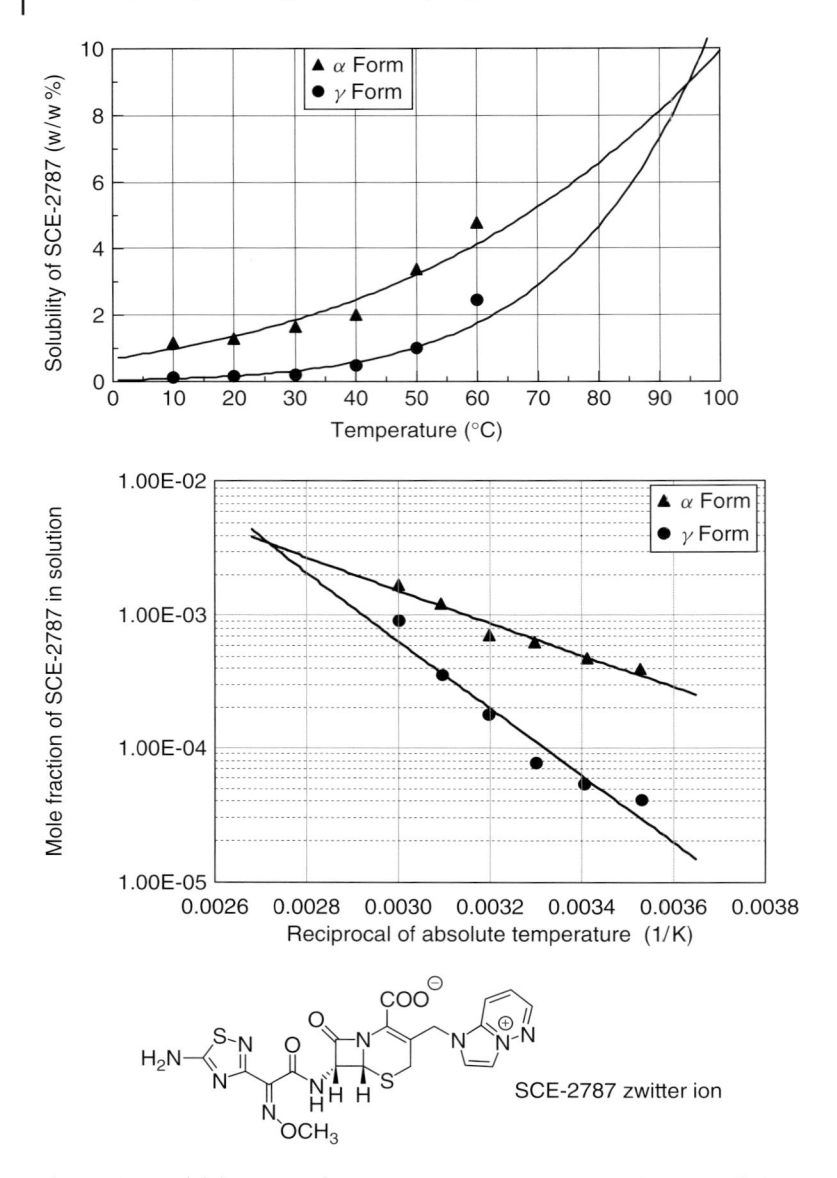

Figure 20.7 Solubility curve of SCE-2787 zwitter ion in water and van't Hoff Plot [6].

It is relatively easy to obtain a desired crystal form that has retained the stable region, such as TAK-029 anhydrate. However, the loss of crystal form control often causes difficulty in selectively obtaining the desired crystal form. In extreme cases, it may cause disappearing polymorphs, examples of which have been documented [14]. The reason why crystal form control becomes difficult in some cases will now be examined.

20.8.1
Nucleation and Seeding

The obtained crystal form is determined by the rate of crystallization, which is divided into nucleation and consecutive crystal growth [15]. Nucleation (Figure 20.8) in particular, would affect crystal form to a large degree, and secondary nucleation would be dominant in industrial crystallization. Therefore the seeding is carried out by adding the crystals of desired form to control the secondary nucleation under the condition of inhibited primary nucleation, and this seeding procedure is called *intentional seeding* [14]. In contrast, crystallization sometimes occurs unintentionally by contamination of the supersaturated solution with a very small amount of crystalline nuclei in an ambient environment; this phenomenon is called *unintentional seeding*. Unintentional seeding can compete with intentional seeding. If unintentional seeding becomes dominant, it would be difficult to control the crystal form. The previous metastable crystal form is not always obtained because of a substantial effect of unintentional seeding, and this phenomenon obviously results in a disappearing polymorph. The fact that the late-appearing polymorph spreads like an infection via human movement can be explained by unintentional seeding. The metastable form of SCE-2787 zwitterion crystal α form has never been obtained owing to the emergence of the most stable form, the γ form, as a late-appearing polymorph.

20.8.2
Ostwald's Stage Rule

Primary nucleation occurs in the first crystallization of newly synthesized compound or in late-appearing polymorphs. Therefore, it would be critical that primary nucleation is efficiently generated in a limited number of experiments in polymorph screening. According to Ostwald's stage rule, the metastable form would occur more easily in a supersaturated solution because it has a lower energy barrier than the stable form in classical nucleation theory (Figure 20.9). The derivation of the energy barrier is that the surface free energy (surface term) between the crystal nucleus and the solution is superior to the driving force (volume term) for

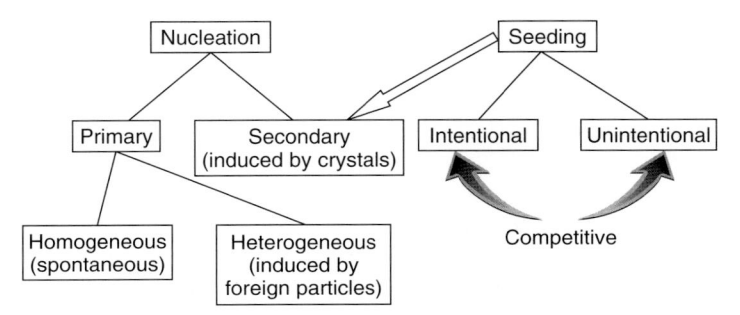

Figure 20.8 Nucleation in crystallization.

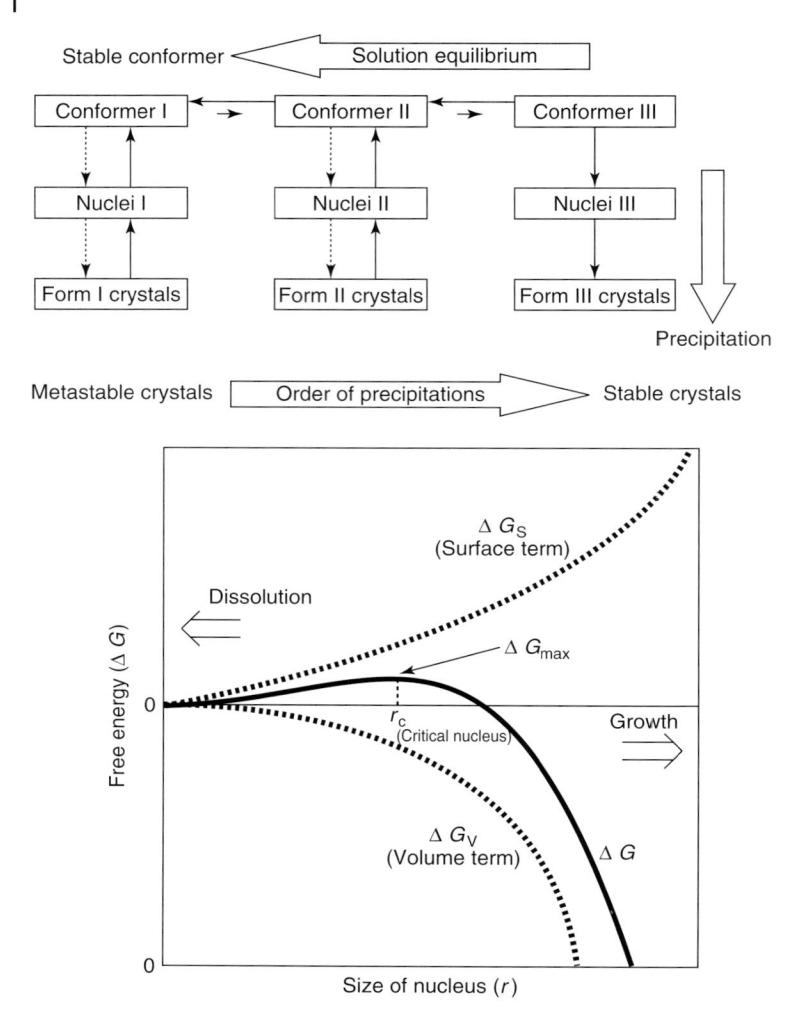

Figure 20.9 Ostwald's stage rule in solution crystallization and plot of Gibbs free energy in the classic nucleation theory.

crystallization by supersaturation in ultramicro-scale. The crystal size where the energy barrier has maximum value is called the *critical nucleus size* (r_c). The crystal particles larger than the critical nucleus grow rapidly and the smaller ones dissolve as clusters.

For polymorphic phenomena, it is important that the transformation process occurs among crystal polymorphs, in addition to nucleation and crystal growth by usual crystallization. Many polymorphs of organic compounds are conformational polymorphs [16] and have different conformers in each crystal form because of large conformational freedom of the compounds. Since the surface free energy of the crystal form, which consists of similar conformation molecules with thermodynamically stable conformers in supersaturated solution, is the lowest, the metastable form tends to crystallize first with primary nucleation. Afterwards, the passage of sufficient time would lead to the previous metastable form being replaced by more stable forms with lower solubility, with the process of dissolution and crystallization being repeated for each form (Figure 20.9).

This analysis was applied to the example of ritonavir. The result of X-ray crystal structure analysis showed that metastable crystal form I had a *trans* isomer of an amide bond while stable form II had a *cis* isomer of an amide bond (Figure 20.10) [5b]. Therefore, form I would have a thermodynamically stable conformation, which is likely to be similar to the stable conformer in the supersaturated solution. As such, form I would be expected to have a lower surface free energy. This would be because of the first appearance of form I as a metastable form obeying Ostwald's stage rule. Despite appearance of stable form II, metastable form I is still initially crystallized in mixed solvent of ethyl acetate–heptane without seeding and gradually changes into stable form II. For the ritonavir case, it would be relatively easy to control the polymorph by seeding, and a selective crystallization method for form I has already been established.

In contrast to the polymorph of ritonavir, polymorph control by seeding has proved difficult for the pseudopolymorph of SCE-2787 zwitterion. In this case, after emergence of the stable γ form, only the stable γ form can be obtained as the SCE-2787 zwitterion crystal form, and the α form became a disappearing

Ritonavir form I
(*trans* amide bond)

Ritonavir form II
(*cis* amide bond)

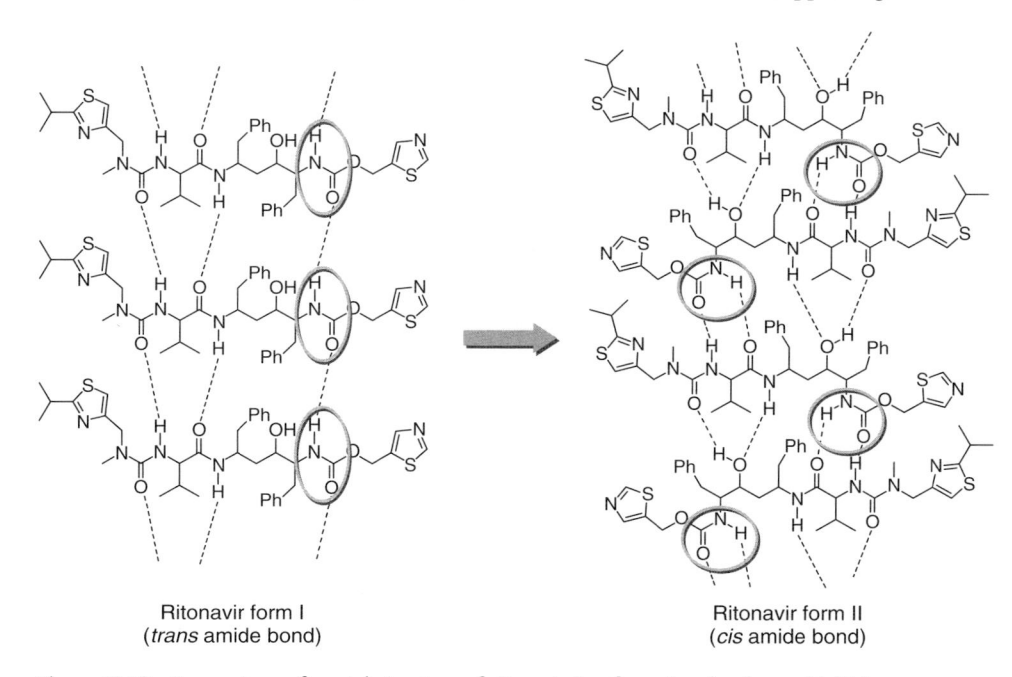

Figure 20.10 Comparison of crystal structure of ritonavir (conformational polymorph) [5b].

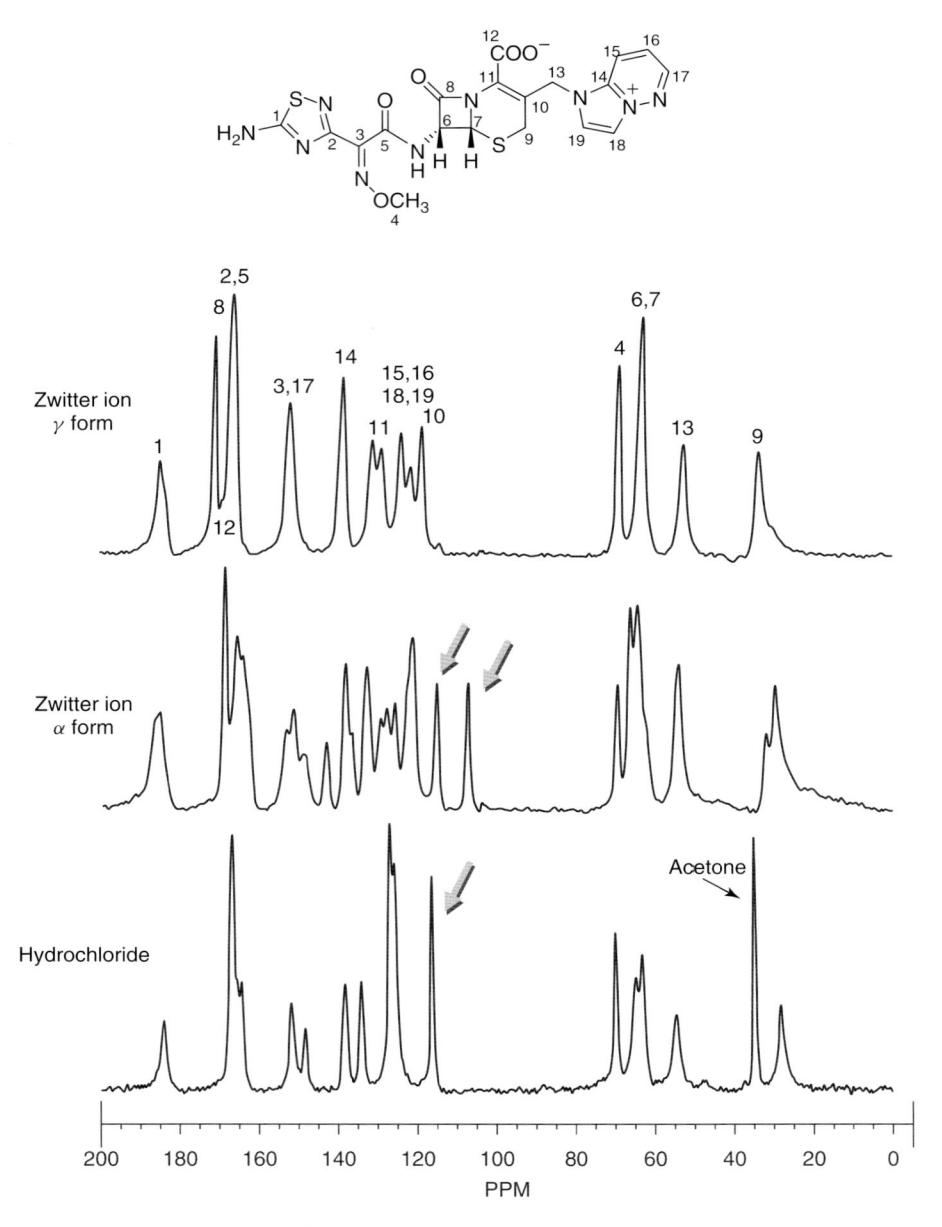

Figure 20.11 Solid-state ^{13}C NMR spectra of SCE-2787 crystals. Numbers in spectrum of γ form indicate signals assigned by solution ^{13}C NMR spectra. Arrowed signals in spectra have large difference from solution ^{13}C NMR spectra in chemical shift [6].

polymorph. The comparison of solid-state ^{13}C NMR spectra for various crystal forms of SCE-2787 showed that the γ form had quite a different spectrum from the metastable α form and SCE-2787 hydrochloride salt crystals. The signals indicated by arrows denote large chemical shift differences (Figure 20.11). These large chemical shift differences implied that these crystal forms exhibit large differences in molecular conformation within each crystal. This estimation was partly supported by the results of X-ray crystal structure analysis (Figure 20.12), which showed large conformational differences between the γ form and hydrochloride salt. However, the γ form had chemical shifts similar to those of the ^{13}C NMR solution spectra for SCE-2787 zwitterion aqueous solution and for SCE-2787 hydrochloride salt solution. Therefore, the thermodynamically stable conformers in zwitterion aqueous solution and hydrochloride salt solution would have quite different conformations than the molecules in the metastable α form crystal and in the hydrochloride salt crystal; however, surprisingly, these conformers in the SCE-2787 solutions were likely to have a similar conformation to the stable γ form before the appearance of the γ form. This implies that the stable γ form would have lower surface free energy than the metastable α form. This assumption is opposite to that made in the ritonavir case.

Stable γ form crystal structure has highly networked hydrogen bonds with as many as seven crystal waters (Figure 20.13). This crystal structural feature provides many sites on the crystal surface where hydrogen bonds can easily form and may also cause lower surface free energy.

γ form of SCE-2787 zwitter ion   SCE-2787 hydrochloride

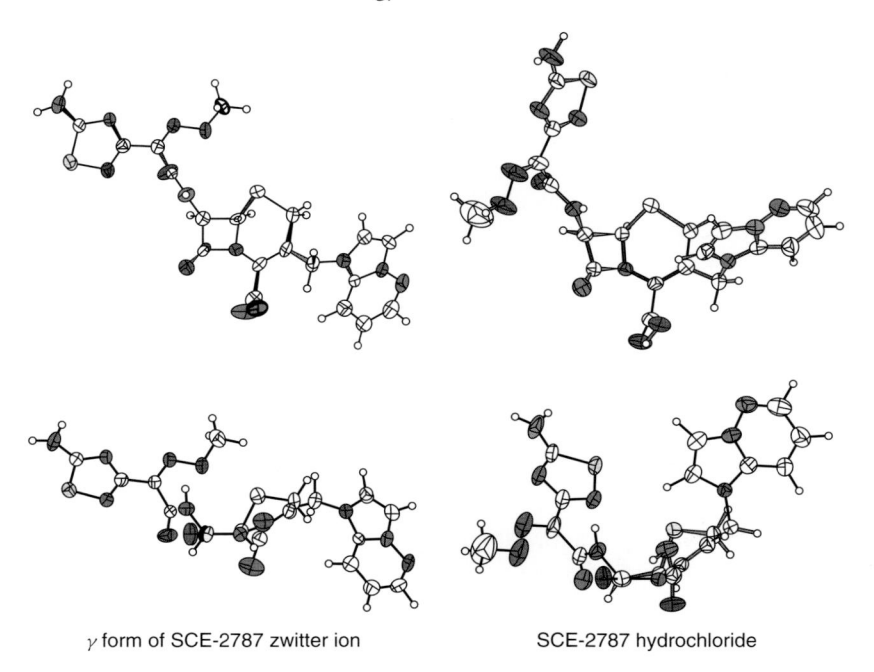

Figure 20.12 Conformation of SCE-2787 molecule in crystal.

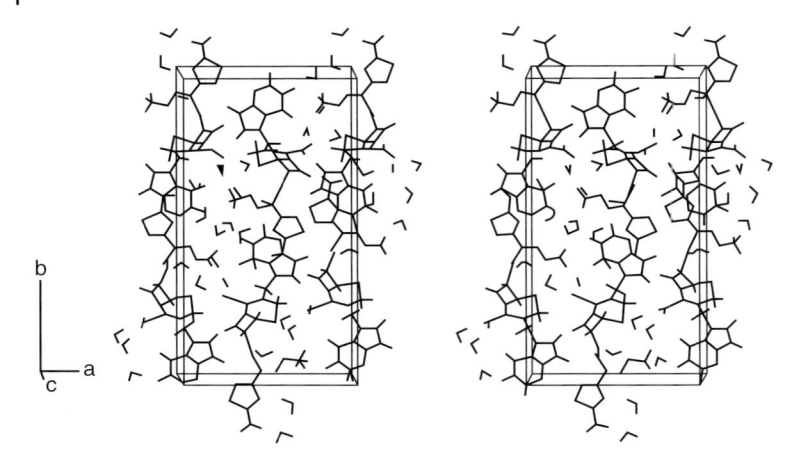

Figure 20.13 Stereoview of crystal structure of SCE-2787 zwitter ion γ form [6].

20.8.3
Unintentional Seeding

As mentioned above, although the γ form of SCE-2787 zwitterion is the most stable form; its the surface free energy is likely to be expected to be extremely low. This would cause an extremely small critical nucleus of the γ form of SCE-2787 zwitterion. If the stable crystal form with low solubility has an extremely small critical nucleus, small and trace amounts of crystals, which usually never have seeding effects, can have the effect of unintentional seeding. As a result, the previous stable form never reappears and becomes a disappearing polymorph with difficulty of control of polymorph. However, the most stable form would usually have larger critical nuclei with higher surface energy, and the effect of unintentional seeding is supposed to be negligible. In addition, the difficulty of detecting the effect of unintentional seeding may lead to confusion of this phenomenon with primary nucleation.

20.9
Primary Nucleation

During the last decade, various novel methods of crystallization [17] such as tailor-made additives, self-assembled monolayers, polymer templating, mechanical grinding, supercritical fluids crystallization, microporous membranes crystallization, capillary crystallization, nanoscopic confinement, contact line crystallization, sonocrystallization, and nonphotochemical laser-induced nucleation, have been developed in the search for diversity of crystal form. These methods have brought diverse conditions of primary nucleation. Among various methods, the timescale of crystallization is a key issue in primary nucleation, including traditional crystallization experiments (Figure 20.14) [17]. As a general principle, slow operation

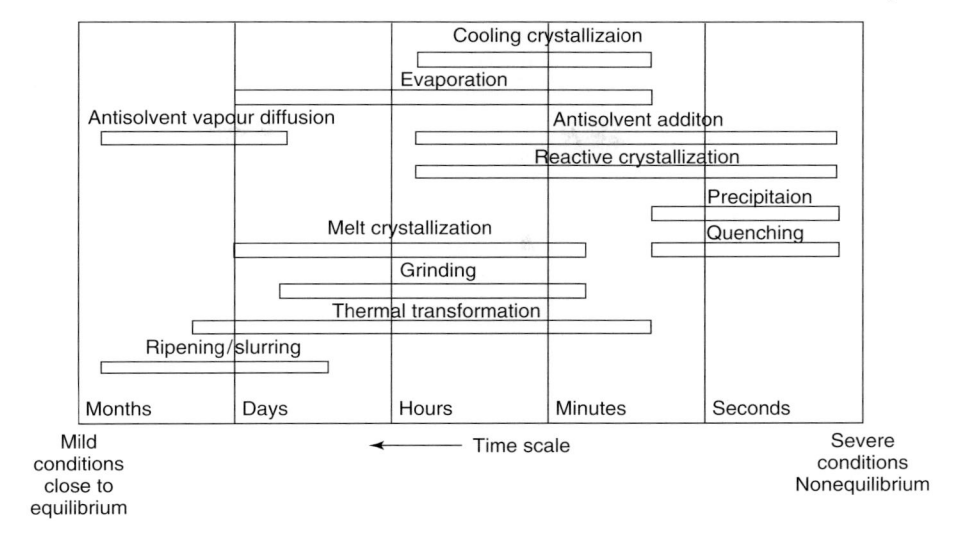

Figure 20.14 Various timescales depending on crystallization methods [1c, 17].

under mild conditions tends to produce a stable form. In searches for the most stable form, a large amount of time may be required.

A late-appearing polymorph would be caused by a large delay in primary nucleation of the most stable crystal form, or by a long waiting time until primary nucleation. For example, until recently, the most stable crystal form of maleic acid had not exhibited primary nucleation for 124 years. In the case of the SCE-2787 zwitterion crystal form, although the most stable γ form would be expected to exhibit easy primary nucleation owing to small critical nuclei, it actually required as many as four years for primary nucleation. Therefore, it may be unreasonable that the classical nucleation theory used so far is applied to the rate process for all primary nucleation. Alternative rate processes for the rate-determining step of primary nucleation should be considered, replacing classical nucleation theory related to nucleus size. Therefore, for a crystal with a complex structure involving a hydrogen bond network due to crystal waters, such as the γ form of SCE-2787 zwitterion crystal, the self-organization process [18] related to crystal structure at the molecular level (Figure 20.13) can become the rate-determining step for nucleation, and this process can delay primary nucleation.

It would be difficult to obtain the most stable form perfectly by the present experimental technology; however, a computational method of predicting all polymorphs of a given compound facilitates the achievement of this goal [19].

20.10
Summary

For process research of polymorphs in solution crystallization, the thermodynamic equilibrium (solubility curve) of a multicomponent system should be carefully considered. In this case, the stable form may be different from the most stable form in a one-component system. For exploring new polymorphs, it would be desirable for the most stable polymorph to be crystallized at an earlier stage of development to avoid a late-appearing polymorph. Application of high-throughput polymorph screening to this strategy has become helpful. However, some compounds would have a long waiting time until primary nucleation. Therefore, it would be difficult to produce all possible polymorphs comprehensively by automated screening. The long waiting time for primary nucleation cannot be explained by classical nucleation theory. Another rate process should be examined, such as a self-organization process. A computational method of predicting all polymorphs has become complementary to the goal of obtaining all the polymorphs of a given compound.

Acknowledgments

This chapter is based on the outstanding contributions of many scientists at Takeda Pharmaceutical Company. The author hereby expresses sincere gratitude to their dedication in the development of (S)-TQA, TAK-029, and SCE-2787.

References

1. (a) Bernstein, J. (2002) *Polymorphism in Molecular Crystals*, Oxford University Press, New York; (b) Hilfiker., R. (ed.) (2006) *Polymorphism in the Pharmaceutical Industry*, Wiley-VCH Verlag GmbH, Weinheim; (c) Brittain., H. G. (ed.) (2009) *Polymorphism in Pharmaceutical Solids*, 2th edn, Informa Healthcare USA, Inc., New York.
2. Childs, S. (2007) Proceedings of The 7th International Conference and Exhibition of Polymorphism and Crystallisation, November 29–30, Clearwater Beach.
3. Nangia, A. (2006) *Cryst. Growth Des.*, **6**, 2.
4. Ostwald, W.F. (1897) *Z. Phys. Chem.*, **22**, 289.
5. (a) Chemburkar, S.R., Bauer, J., Deming, K., Spiwek, H., Patel, K., Morris, J., Henry, R., Spanton, S., Dziki, W., Porter, W., Quick, J., Bauer, P., Donaubauer, J., Narayanan, B.A., Soldani, M., Riley, D., and McFarland, K. (2000) *Org. Process Res. Dev.*, **4**, 413; (b) Bauer, J., Spanton, S., Henry, R., Quick, J., Dziki, W., Porter, W., and Morris, J. (2001) *Pharm. Res.*, **18**, 859; (c) Desikan, S., Parsons, R.L. Jr., Davis, W.P., Ward, J.E., Marshall, W.J., and Toma, P.H. (2005) *Org. Process Res. Dev.*, **9**, 933; (d) Day, G.M., Trask, A.V., Motherwell, W.D.S., and Jones, W. (2006) *Chem. Commun.*, 54.
6. Yamano, M. (2007) *J. Synth. Org. Chem. Jpn.*, **65**, 907.
7. (a) Hilfiker, R., Berghausen, J., Blatter, F., Burkhard, A., De Paul, S.M., Freiermuth, B., Geoffroy, A., Hofmeier, U., Marcolli, C., Siebenhaar, B., Szelagiewicz. M., Vit, A., and von Raumer, M. (2003) *J. Therm. Anal. Calorim.*, **73**, 429; (b) Morissette, S.L., Almarsson, O., Peterson, M.L., Remenar, J.F., Read, M.J., Lemmo, A.V., Ellis, S., Cima, M.J., and Gardner, C.R. (2004) *Adv. Drug Deliv. Rev.*, **56**, 275.
8. Laird, T. (2004) *Org. Process Res. Dev.*, **8**, 301.
9. (a) Blagden, N., Davey, R.J., Rowe, R., and Roberts, R. (1998) *Int. J. Pharm.*, **172**, 169; (b) Rubin-Preminger, J.M. and Bernstein, J. (2005) *Cryst. Growth Des.*, **5**, 1343; (c) Kwon, O.-P., Kwon, S.-J., Jazbinsek, M., Choubey, A., Losio,

P.A., Gramlich, V., and Günter, P. (2006) *Cryst. Growth Des.*, **6**, 2327; (d) Gu, C.-H., Chatterjee, K., Young, V., Jr., and Grant, D.J.W. (2002) *J. Cryst. Growth*, **235**, 471.

10. (a) Sugihara, H., Fukushi, H., Miyawaki, T., Imai, Y., Terashita, Z., Kawamura, M., Fujisawa, Y., and Kita, S. (1998) *J. Med. Chem.*, **41**, 489; (b) Kitamura, S., Fukushi, H., Miyawaki, T., Kawamura, M., Terashita, Z., Sugihara, H., and Naka, T. (2001) *Chem. Pharm. Bull.*, **49**, 258.

11. Miyake, A., Yoshimura, Y., Yamaoka, M., Nishimura, T., Hashimoto, N., and Imada, A. (1992) *J. Antibiot.*, **45**, 709.

12. Bernstein, J., Davey, R.J., and Henck, J.-O. (1999) *Angew. Chem. Int. Ed.*, **38**, 3440.

13. (a) Hulliger, J. (1994) *Angew. Chem., Int. Ed.*, **33**, 143; (b) Beckmann, W., Nickisch, K., and Budde, U. (1998) *Org. Process Res. Dev.*, **2**, 298; (c) Beckmann, W. (2000) *Org. Process Res. Dev.*, **4**, 372; (d) Beckmann, W., Otto, W., and Budde, U. (2001) *Org. Process Res. Dev.*, **5**, 387.

14. (a) Dunitz, J.D. and Bernstein, J. (1995) *Acc. Chem. Res.*, **28**, 193; (b) Henck, J.-O., Bernstein, J., Ellern, A., and Boese, R. (2001) *J. Am. Chem. Soc.*, **123**, 1834.

15. Mullin, J.W. (2001) *Crystallization*, 4th edn, Elsevier and Butterworth-Heinemann, Oxford, p. 181.

16. (a) Hagler, A.T. and Bernstein, J. (1978) *J. Am. Chem. Soc.*, **100**, 6349; (b) Nather, C., Nagel, N., Bock, H., Seitz, W., and Havlas, Z. (1996) *Acta Crystallogr. Sect. B.*, **52**, 697; (c) Reed, S.M., Weakley, T.J.R., and Hutchison, J.E. (2000) *Cryst. Eng.*, **3**, 85; (d) Yu, L., Reutzel-Edens, S.M., and Mitchell, C.A. (2000) *Org. Process Res. Dev.*, **4**, 396.

17. Llinas, A. and Goodman, J.M. (2008) *Drug Discov. Today*, **13**, 198.

18. (a) Gavezzotti, A. and Filippini, G. (1998) *Chem. Commun.*, 287; (b) Davey, R.J., Blagden, N., Righini, S., Alison, H., Quayle, M.J., and Fuller, S. (2001) *Cryst. Growth Des.*, **1**, 59; (c) Davey, R.J., Allen, K., Blagden, N., Cross, W.I., Lieberman, H.F., Quayle, M.J., Righini, S., Seton, L., and Tiddy, G.J.T. (2002) *Cryst. Eng. Commun.*, **47**, 257.

19. Price, S.L. (2009) *Acc. Chem. Res.*, **42**, 117.

21
Development of LIPOzymes Based on Biomembrane Process Chemistry

Hiroshi Umakoshi, Toshinori Shimanouchi, and Ryoichi Kuboi

21.1
Introduction

Our intent is to develop a novel bio/chemical process based on the construction of biomembrane mimics. In conventional bio/chemical processes, multistaged reactors along with a series of unit operations are often used for the reaction and isolation of the target compound. Although process improvements are consistently being made, these efforts are still being carried out by trial and error to increase the efficiency of the product recovery. Therefore, we never reach the "ecological" goal of the process design to effectively utilize resources and energy to obtain a target compound.

A bioprocess to produce a useful enzyme with a host cell (i.e., *Escherichia coli*) may involve many stages such as cell cultivation, cell disruption, solid–liquid separation, extraction, several types of chromatography, precipitation or crystallization, and drying. In general, most engineers are apt to set operational conditions at the minimum stress. Since the operational conditions are different from the equilibrium conditions of each reaction/separation system, the target molecules (i.e., enzymes) are inevitably exposed to stress conditions during the period of bioprocessing, leading to denaturation by destruction of the enzyme structure. The situation is similar in the chemical process to produce pharmaceutical compounds, where extraction, crystallization, chromatography, or membrane separation is often used for separation of the target at the end of each reaction step. One cannot escape such a situation, because almost all the unit operations require their own respective optimal conditions governed by the thermodynamic equilibrium of the system.

Improvement of the conventional process has been undertaken by considering the effective use of "stress" to activate the potential functions of the biomaterials themselves. In the late 1990s, the concept of a "stress-mediated bioprocess" was presented, in which the conventionally problematic stress is positively utilized for the induction and control of the latent (potential) functions of the biological cells and target molecules [1]. There are some examples of bioprocesses to produce a target protein in an integrated process [2–4]. It has recently been recognized that the key to effective utilization of the "stress–response function" of biological

Pharmaceutical Process Chemistry. Edited by Takayuki Shioiri, Kunisuke Izawa, and Toshiro Konoike
Copyright © 2011 WILEY-VCH Verlag GmbH & Co. KGaA, Weinheim
ISBN: 978-3-527-32650-1

cells and biomaterials is their successful integration into a "biomembrane," which consists of a self-assembly of lipids and other elemental biomaterials, followed by possible induction of latent (potential) or synergistic functions under the specific stress conditions [5–7]. A process design to make the best of the latent (potential) functions of the "biomembrane" is a major direction of "biomembrane process chemistry (BMPC)" based on a new concept. The combination/integration of separation (recognition) and catalytic functions could be achieved as the final goal by selecting a biomimetic membrane (liposome) as the core material [8–10].

This chapter describes the latest results regarding LIPOzymes (*lipo*some + en*zyme*) and their related techniques, followed by the introduction of the basic concept of BMPC as well as the characteristic features of biomimetic membranes (liposomes).

21.2
From "Process Chemistry" to "Biomembrane Process Chemistry"

In recent years, a new type of chemistry, "process chemistry [11]," which involves developing good manufacturing methods for the target materials (i.e., pharmaceuticals), has attracted much attention among (chemical) engineers and scientists (chemists). In process chemistry, the safety, quality, delivery, cost, environment (SQDCE) is important for the design and development of the pharmaceutical process (Figure 21.1) [12]. The conventional process chemists seem to primarily focus on the design of a novel synthesis route to minimize the number of reaction (separation) steps without a loss of product quality, to minimize the formation of harmful or valueless by-products and the use of the harmful solvent, and to minimize the economical costs in the design of the bio/chemical process [13]. Various kinds of effective catalysts have also been designed to meet the expectations of the process engineers [14]. There have been some reports that the catalytic reactions are combined with separation techniques (i.e., membrane separation, solvent extraction, etc.) for the purpose of removing products and/or by-products and the supply of the substrate in the system [15]. However, the present strategies have not provided us with new innovations. This is because almost all the current strategies are still based on trial-and-error improvements of the previous process.

From the perspective of chemical engineers, separation technology (such as distillation, adsorption, chromatography, extraction, crystallization, and drying) has also played an important role as a unit operation in achieving the above purposes in process chemistry. The separation steps are indispensable, because the intermediate chemicals must be purified before being introduced into the next reactor. Separation technology has been classified by Giddings, taking into consideration the type of phase equilibrium of the system (chemical potential) and the type of transport of the materials [16], as schematically shown in Figure 21.2. Almost all of the separation techniques can herewith be classified based on the above concept. As such, if different separation techniques can be selected, there

21.2 From "Process Chemistry" to "Biomembrane Process Chemistry" | 423

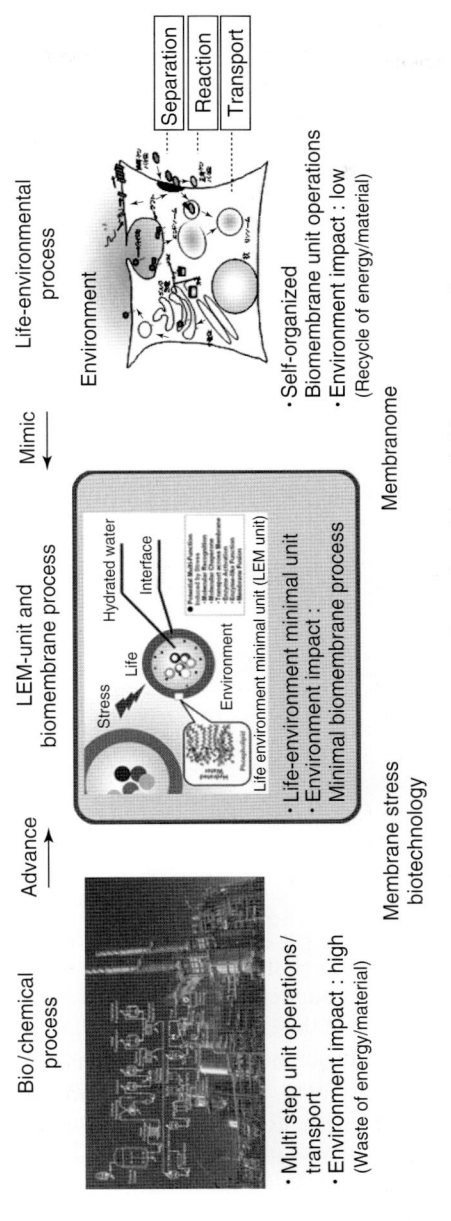

Figure 21.1 Biomembrane process in between artificial (bio/chemical) and natural (life-environmental) processes.

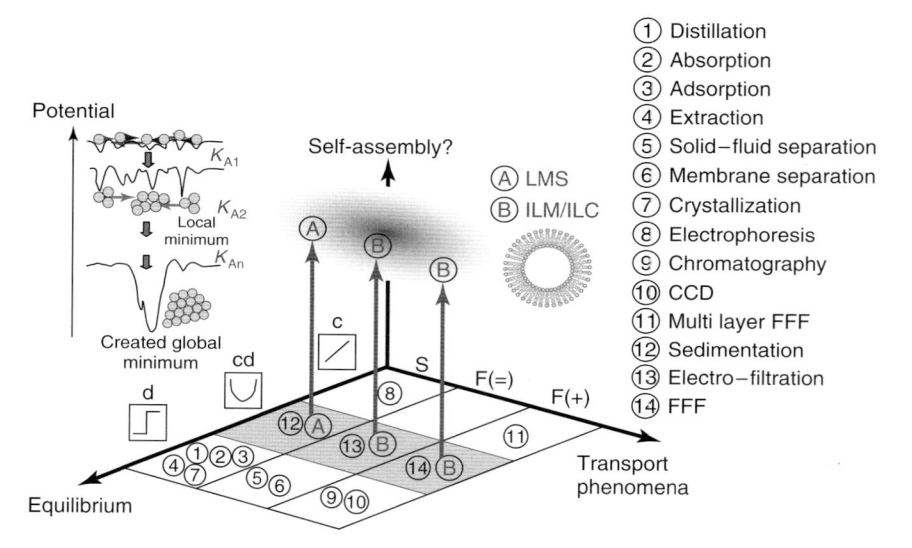

Figure 21.2 Classification of conventional separation technology and a new separation based on self-assembly.

could be different equilibriums throughout the entire process and one could employ multistep unit operations to obtain the target material with excess exposure of the target to "stress" conditions. A fusion of concepts from both "process chemistry" and "separation technology" becomes important when attempting to achieve a SQDCE in the design and development of an ecological process, for example, by employing the structure and function of a "self-assembly" system, which lets us design the entropy-driven process in addition to the enthalpy-driven conventional one (Figure 21.2).

What is an ecological (environmentally friendly) process on earth for achieving a SQDCE-based concept? One possible answer is a "biological cell" (Figure 21.1). A biological cell is a minimal unit of "life," where complexed and parallel biochemical reactions, molecular transport, and selective separation are cooperatively achieved in a well-organized manner through the interactions among various active interfaces (biomembranes). A new process chemistry, life-environment friendly biomembrane-based process chemistry, could be established if one could understand the highly organized strategy of the biological system and could utilize it for the process design. A "biomembrane" is a key interface for achieving integration of the production and separation of target materials. "Membranomics" can herewith be defined as a systematic methodology for research regarding information obtained from the structures and functions of biomimetic membranes and can be gradually recognized as an important research methodology for investigating the potential functions of biomembranes and applying them to bioprocess design [17]. A biomimetic membrane "liposome" possesses several benefits with regard to the recognition of (bio)molecules, as it can recognize them with (i) electrostatic interactions, (ii) hydrophobic interactions, and (iii) a stabilization effect of hydrogen bonds in the hydrophobic lipid environment [18]. Molecular recognition with high selectivity can be achieved by the simple liposome membrane on its surface through the combination of (i–iii) through its stress–response dynamics [19, 20]. Novel biomembrane-based catalysis can be designed and developed by integrating the above recognition sites and a simple catalytic center on the liposome surface. Some examples of the above have previously been reported as LIPOzyme [8–10, 21–29]. A process design utilizing LIPOzyme chemistry and the recognition (separation) functions of model biomembranes (liposomes) is the basic strategy in "BMPC" (Figure 21.1). In other words, the core of BMPC is just the conversion of technology based on the equilibrium theorem at a macroscopic level into a novel technology for utilizing a potential structure consisting of a set of metastable states at a nanoscopic level (Figure 21.2).

In the following, examples of LIPOzyme design are shown by employing the "antioxidative" LIPOzyme as an example of BMPC applications, followed by introduction of the recognition (separation) function of liposomes.

21.3
Recognition (Separation) Function of Liposomes

Separation technology is essential for the design and development of bio/chemical processes. The scientific part of the separation technology has previously been introduced by Giddings, where the separation technologies can be classified by considering "chemical potential (phase equilibrium)" and "transport phenomena" (Figure 21.2) [16]. For example, in the extraction process, the target molecule can be partitioned, depending on the physicochemical potentials between the organic and aqueous phases, and separation of the target material can be achieved by mechanical recovery of the organic phase. In another case, bioseparation can often be achieved by utilizing the antigen–antibody reaction, where the solid support modifying the antibody can be easily recovered together with target materials recognized by the antibody. As described above, the separation (recognition) of the target materials can be conventionally achieved based on a thermodynamic theorem. The potential structure can be described as static and the most stable state for the temperature and composition of the system (Figure 21.3). The basic characteristics of the separation techniques have been applied in recovery processes for the target materials and in the development of separation/analysis techniques in aqueous two-phase systems. A conventional separation/purification process (Figure 21.1) often becomes multistage and consumes energy, because it relies on a separation technique based on equilibrium theory.

In contrast, a liposome composed of designed lipids and biomolecules is a nonequilibrium system connected in a complicated manner with the phase equilibrium [30, 31]. Liposomes are self-aggregates of amphiphiles (i.e., phospholipids, fatty acids, cholesterols, and detergents) with closed bilayer membranes, which harbor a hydrophobic nano-thin layer (about 5 nm), and they are known as *model biomembranes* (or *biomimetic membranes*). There have been many reports regarding the phase-transition behaviors of liposomes as a function of various operational factors (i.e., lipid composition, temperature, and pH). In the case of the static condition, interactions between the liposomes and other solutes can be roughly related to (i) electrostatic interactions, (ii) hydrophobic interactions, and (iii) stabilization effects of hydrogen bonds in the hydrophobic lipid environment [18]; in other words, the physicochemical properties of liposomes can be governed by the temperature and lipid composition, and the system stability under the normal (static) conditions [30, 31].

However, the liposome characteristics cannot simply be explained under stress conditions (i.e., under variations of temperature or pH and in the presence of hydrophobic peptides, metal ions, etc.), and the liposome could reveal potential (latent) functions to recognize the biomaterials [22–24]. Liposomes under such stress conditions often have many metastable states (a local minimum) as part of their stress–response dynamics [32, 33]. The metastable state shifts to the most stable structure (a global minimum) in response to changes in the external environmental conditions through the exchange of materials and energy on the liposome surface. In other words, the liposome system has potential structures consisting of sets of several dynamic metastable states (Figure 21.3), and these structures are progressively shifted into in an ordered-state (for example, an active center) on the lipid membrane surface. In practice, there are some examples of the selective recognition of partly damaged proteins (carbonic anhydrase, lysozyme, etc.) under the stress condition (i.e., heating, acidification, and reductive condition) [29] and oxidized fragment of superoxide dismutase (SOD) [8]. This result explains why selective recognition and separation of the material in the cell (Figure 21.1) is easily achieved through the dynamic function of the liposome itself. The notable characteristics of the liposome can be regarded as the change in the potential structure in response to the external environment to express potential functions such as recognition and catalytic functions. There are some examples of the potential functions appearing only in a variable environment [34, 35]. Although the details of such liposome-based recognition remain unclear due to their complexity, a new tool for evaluating various characteristics of the membrane surface as a

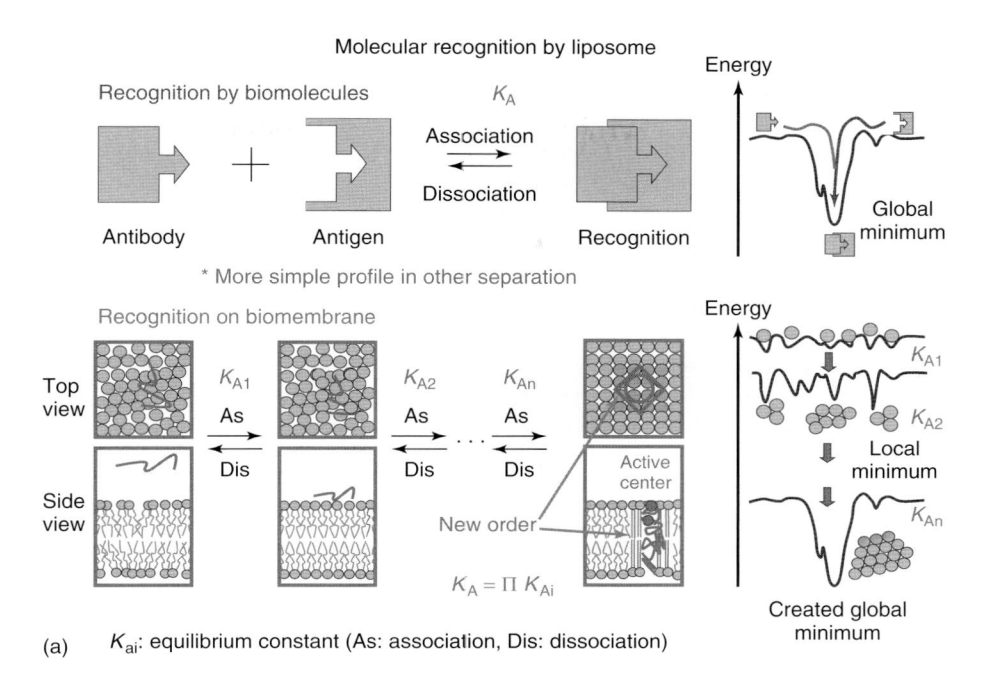

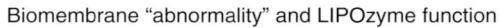

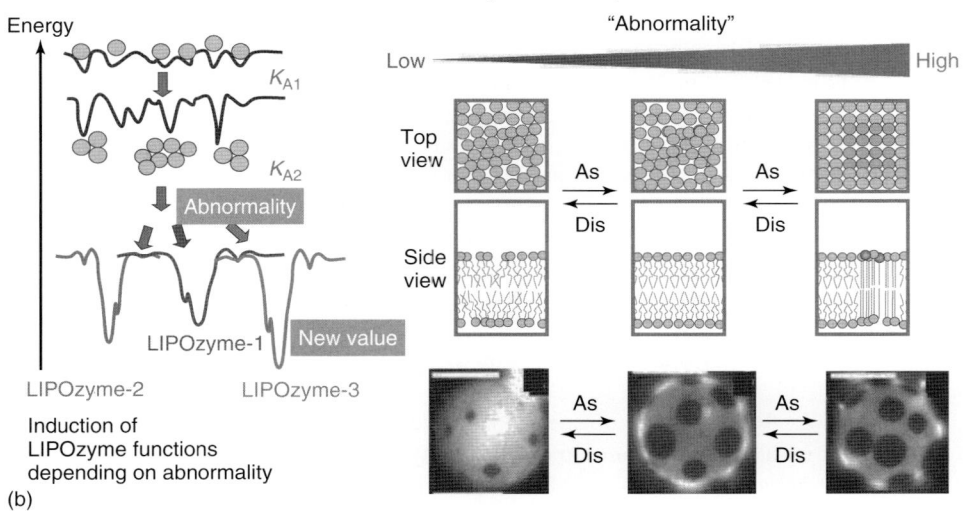

Figure 21.3 Basic concept of recognition (separation) function of biomimetic membrane (liposome). (a) Comparison of recognition functions of antigen–antibody and liposome-stressors (peptides). (b) Basic concept of LIPOzyme preparation based on membrane abnormality.

rather simple parameter is now being developed as a "membrane chip" for the "liposome design" [36–38].

It is the core of the membranome and the basic strategy for process design utilizing LIPOzyme to contribute to our understanding of the potential functions of biomembrane that is dynamically expressed under stress conditions from the Build-Up approach and for utilizing the potential functions of the biomembrane itself for the design and development of materials and processes from the Break-Down approach [8, 9].

21.4
LIPOzyme: Liposome with Enzyme-Like Activity?

The liposome or self-assembly (i.e., micelle) has previously been used as a platform for a catalytic center, because it can provide a novel hydrophobic environment on the membrane surface [39, 40]. In BMPC, to achieve high recognition of the target material and its conversion, it is important to develop a catalytic process utilizing the "stress–response functions" of the biomembrane as an example of the essential characteristics of self-assembly. Enzyme-like activity (LIPOzyme) can be induced through the recognition (separation) of target elements on the liposome surface and also through the creation of a minimal active center through membrane stress–response dynamics [8, 23–27]. There are two kinds of strategies for designing LIPOzymes such as (i) "Break-Down" type LIPOzyme and (ii) "Build-Up" type LIPOzyme (Figure 21.4). The former LIPOzyme type (type (i)) can be prepared by displaying the specific peptide fragment, broken down under strong stress conditions, to induce its potential functions on the liposome surface [8, 23–25]. The latter (type (ii)) can be prepared through self-assembly of the minimal molecular elements (i.e., functional ligands) on the liposome surface [9, 28, 29]. Although the above two strategies differ from each other, the common approach is to highly utilize the stress–response dynamics of the liposome membrane itself. In the following, some examples of the above two types of LIPOzymes are introduced by selecting the antioxidative enzyme as a target.

21.4.1
Break-Down Type LIPOzyme

Cu,Zn-Superoxide dismutase (Cu,Zn-SOD) is a metallo-enzyme that catalyzes the chemical reaction from superoxide to hydrogen peroxide and is known as an *antioxidative enzyme* (Figure 21.5a). In some cases, SOD is known to be inactivated in the presence of excess product (hydrogen peroxide) [41]. It has been reported that liposomes can assist with SOD activity under such a strong oxidative stress condition [21]. The peptide fragment of H_2O_2-treated Cu,Zn-SOD has recently been shown to be reactivated with liposomes prepared by using zwitterionic phospholipid (1-palmitoyl-2-oleoyl-sn-glycero-3-phosphocholine; POPC) [22–24]. The H_2O_2-treated SOD, which loses its activity at different incubation times, was

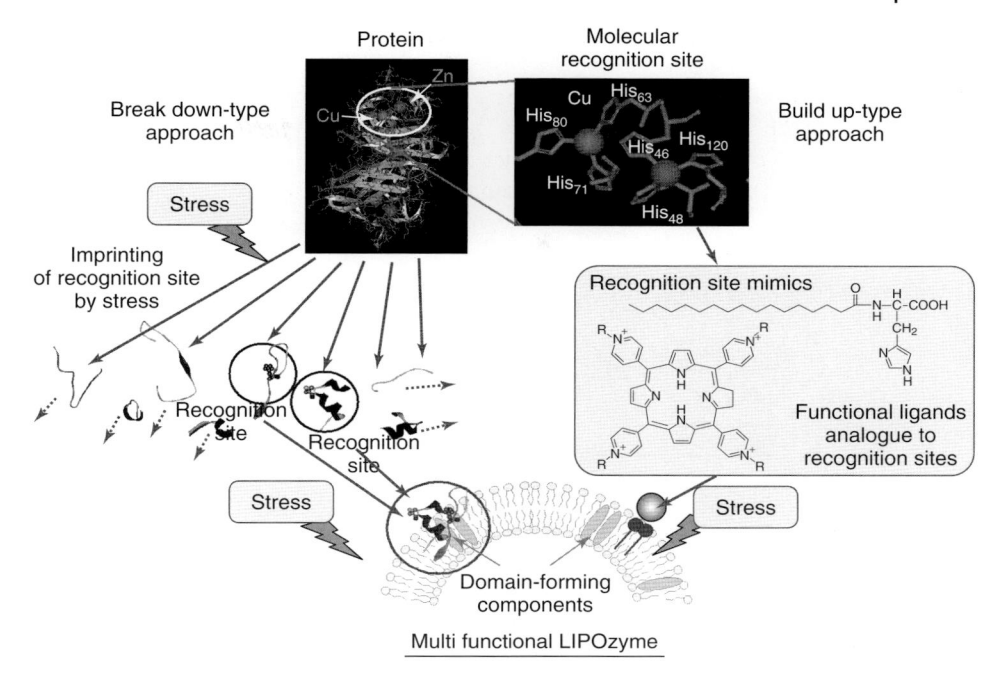

Figure 21.4 Strategies of LIPOzyme preparation.

found to be dramatically reactivated by adding only POPC liposomes [22]. The ultrafiltration and reverse phase-HPLC (RP-HPLC) analyses of H_2O_2-treated SOD coincubated with liposomes show that some specific peptide fragments (4.1 and 8.2 kDa) of the oxidized and fragmented SOD can interact with POPC liposomes [22]. The effects of several kinds of liposomes on the recruited activity of oxidized and fragmented SOD were furthermore investigated [24], where the addition of zwitterionic liposomes with high membrane fluidity or that with positive charge was found to increase the SOD-like activity of fragmented SOD, although the negatively charged liposome had no effect on its activity. The SOD-like activity was shown to be strongly dependent on the adsorbed amount of peptides on the liposome surface. The characteristics of the specific peptide analyzed by LC/MS (data not shown), together with the adsorption behaviors of the peptide fragment on various kinds of liposomes, imply that the liposome-recruited activity of the fragmented SOD was strongly related to liposome recognition of the SOD fragment by a combination of electrostatic and hydrophobic interactions, and by hydrogen bonding between the peptide and liposome membrane. Surprisingly, the SOD fragment also induced the activity of another type of enzyme (catalase) to decompose the hydrogen peroxide by selecting the appropriate liposome surface and, also, by applying the appropriate oxidative stress condition [8, 22, 23]. These results show that liposomes can recruit (recognize) the potentially active fragment of the SOD among various fragments and can create an enzyme (SOD)-like active center on its surface, resulting in

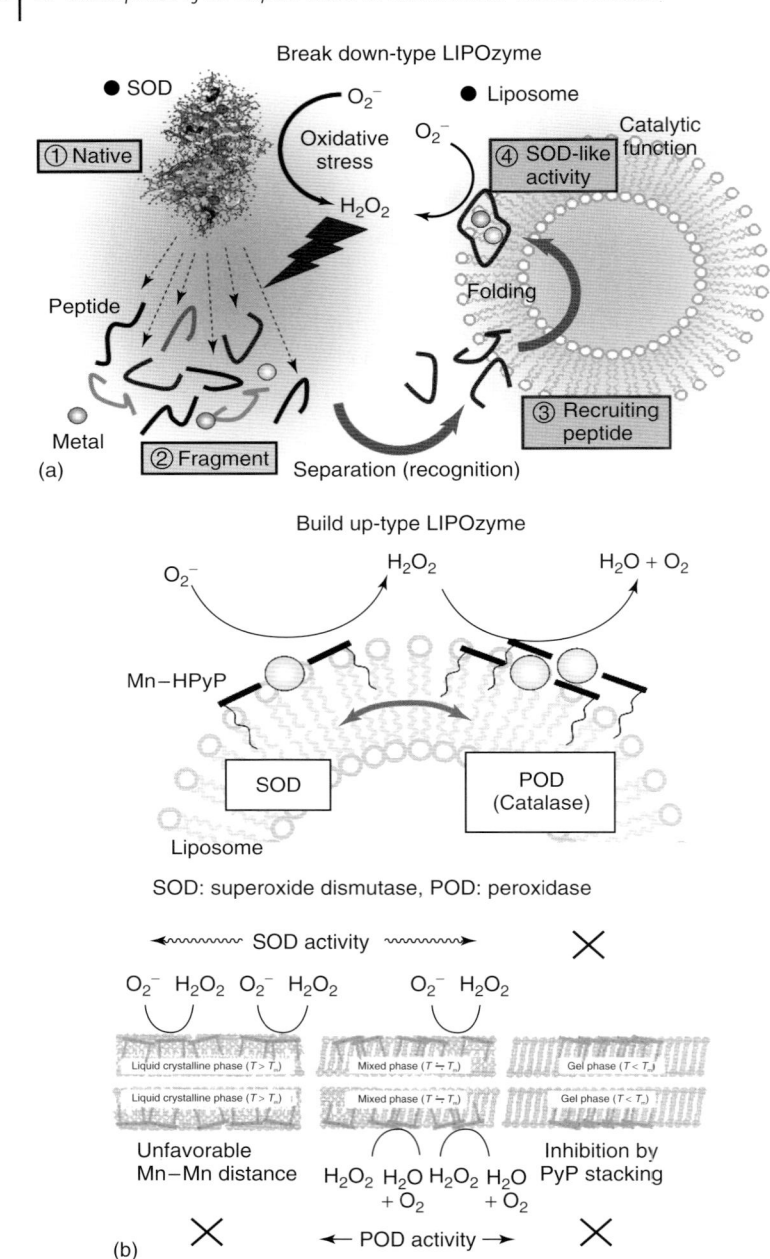

Figure 21.5 Examples of LIPOzymes. (a) Break-down type LIPOzyme. (b) Build-up type LIPOzyme.

one-pot conversion of the superoxide to oxygen and water by way of hydrogen peroxide on a liposome surface.

21.4.2
Build-Up Type LIPOzyme

The above SOD can also be prepared through another approach. In previous research, there have been several attempts to carry out ligand modification of the liposome surface (i.e., vitamin B_{12}–modified liposome) to achieve liposome-harboring catalytic activity [39] (Figure 21.5b). However, it seems that their main purpose has been modification of the hydrophobic environment of the catalytic core (i.e., porphyrin), resulting in lower activity. An enzyme is known to induce its catalytic activity through modulation of its structure, suggesting that one should consider dynamic modulation of the structure or assembly of the ligands. An antioxidative liposome catalysis that mimics both SOD and peroxidase (POD) activities has recently been developed using the liposomes modified with lipophilic Mn–(5,10,15,20-tetrakis[1-hexadecylpyridium-4-yl]-$21H,23H$-porphyrin) (Mn–HPyP) [8, 27] or Dodecanoyl-His (Dodec-His) [28, 42]. In the former case, the SOD- and POD-like activities of the Mn–HPyP-modified liposome were first investigated by varying the type of phospholipid. Higher SOD-like activity was obtained in the case of liposomes prepared by using 1,2-dipalmitoyl-sn-glycero-3-phosphocholine (DPPC) and 1,2-dimyrystoyl-sn-glycero-3-phosphocholine (DMPC), in which the ligands were well dispersed on the membrane in the liquid crystalline phase. The POD-like activity was maximal in the case of DMPC liposome, in which the Mn–HPyP complex was appropriately clustered on the membrane in the gel phase. On the basis of the above results, the co-induction of SOD and POD activities to eliminate the superoxide and also hydrogen peroxide as a one-pot reaction was finally performed by using the Mn–HPyP-modified DMPC liposome, resulting in increased efficiency of the elimination of both superoxide and hydrogen peroxide.

Two important strategies for preparing the LIPOzyme have thus been introduced. The appropriate activity level of SOD- and POD-LIPOzyme has also been obtained in both Break-Down and Build-Up approaches, although it has not reached the activity level of an enzyme (Figure 21.6). The merits of LIPOzyme can herewith be described as follows: (i) higher tolerance against variable conditions, (ii) display of multiple activity points on the same liposome surface, and (iii) controllable activity via environmental stress. Actually, two types of enzyme-like activities (i.e., SOD and catalase (CAT)), prepared via both "Break-Down" and "Build-Up" approaches, can be induced on the liposome surface, and the dual-activity LIPOzyme can be used as a "one-pot catalyst" to convert the superoxide to oxygen and water by way of hydrogen peroxide. In recent years, the concept of LIPOzyme has been developed further, and there have been several reports of other types of LIPOzymes (Table 21.1). Chitosanase, which hydrolyzes chitosan to give oligochitosan, can also be displayed on the surface without processing of the signal peptide (SP) (SP-Chitosanase LIPOzyme) [43]. A subunit

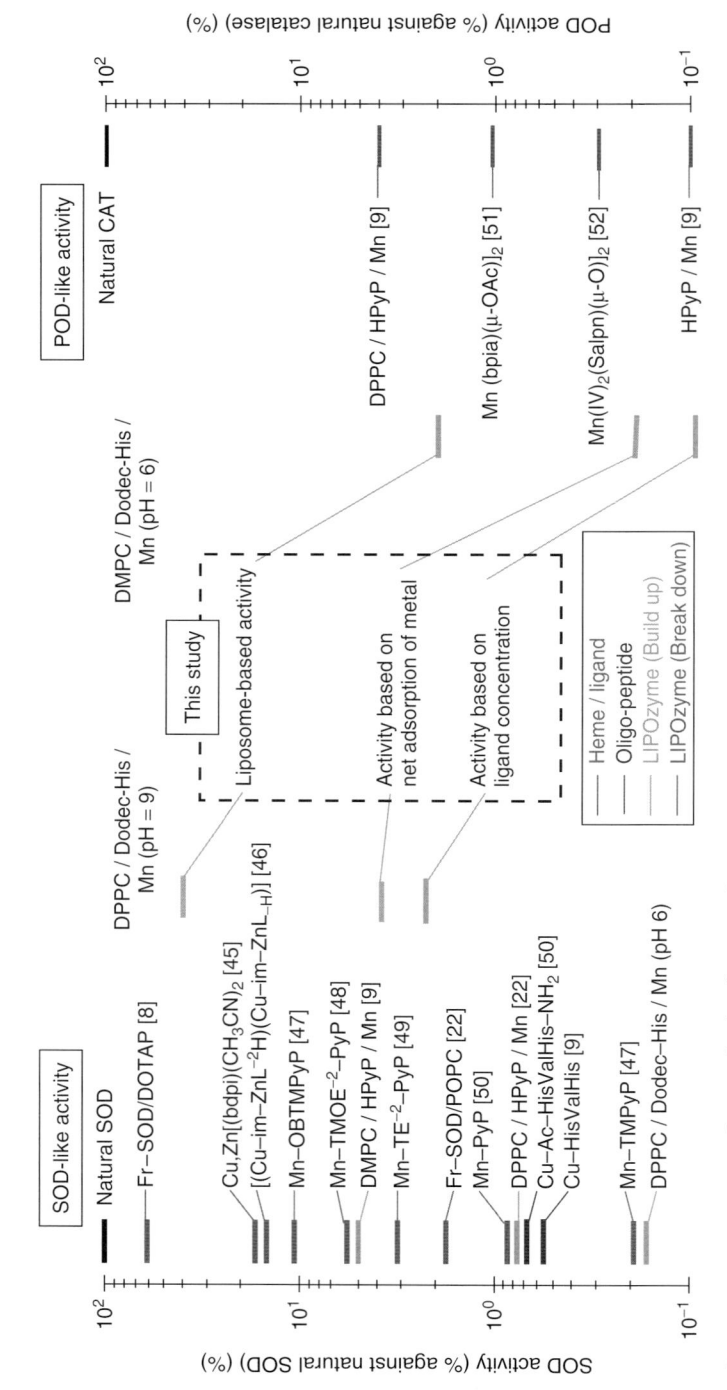

Figure 21.6 Comparison of activity levels of SOD- and POD-LIPOzyme.

of the multimeric enzyme alcohol dehydrogenase (ADH) can also be displayed on the liposome surface to give its original activity (ADH LIPOzyme) to oxidize the alcohol molecules with a short chain (i.e., methanol, ethanol, and butanol) (H. Umakoshi et al. unpublished results). The amyloid β-peptide induces oxidase-like activity (i.e., cholesterol to oxysterol and dopamine to dopaquinone) on the surface of the liposome [44]. The above results show that the high value-added LIPOzyme can be designed by suitable design of the liposome surface via stress conditions.

21.5
Biomembrane Interference

Biomembrane interference phenomena have also been suggested for their apparent LIPOzyme functions (Figure 21.7) [10, 54–56]. It is commonly known that a central dogma in molecular biology assumes a series of elemental steps to complete the gene expression initiated by the gene code of DNA. Recently, the RNA interference phenomena have been reported to be a powerful tool for gene silencing, where specific gene expression can be controlled via the complex formation of target mRNA and introduced oligonucleotides [57]. The above basic stance on gene expression has been employed in most conventional approaches to gene regulation. It has recently been shown that the "biomembrane" can also interfere with gene expression through its interaction with various kinds of biological molecules (i.e., DNA, RNA, and protein) [10], where the elemental steps of gene expression such as (i) transcription, (ii) translation, and (iii) folding can be affected by the addition of the biomimetic membrane (liposome). A systematic investigation of the role of liposomes in gene expression has been performed in an *in vitro* gene expression system (cell-free translation system) by selecting the green fluorescent protein (GFP) as a reporter protein. The GFP expression was performed in the above system in the presence of various liposomes with various surface properties. The upregulation of the GFP expression has been observed in the case of (i) zwitterionic liposome (i.e., POPC) [10, 56] and (ii) cholesterol-modified liposome (i.e., POPC/Cholesterol (Ch)) [10], while its downregulation has been observed in (iii) negatively charged liposome (i.e., POPC/1-palmitoyl-2-oleoyl-sn-glycero-3-phosphatidylglycerol (POPG)) [54, 56], and (iv) positively charged liposome (i.e., 1,2-dioleoyl-3-trimethyl-ammonium propane (DOTAP)) [55, 56]. The role of liposome in the elemental steps of gene expression was further analyzed by measuring the mRNA production (for transcription step) and net GFP production (for translation step), together with the above GFP fluorescence (for folding step). The downregulation of GFP expression has been shown to be dependent on the type of liposome surface, with negatively charged liposomes possibly interfering with the final folding step and positively charged liposomes interfering with the transcription and translation steps because of the recognition of biomolecules (such as DNA, RNAs, and polypeptides) by the liposome itself [10]. As a result, it has been shown that the recognition of biomolecules can be achieved through the combination of electrostatic interactions, hydrophobic interactions,

Table 21.1 Characteristics of various LIPOzymes.

LIPOzyme type	Study on type of designed liposome				Resources	Catalytic function		Catalytic level (relative value against naturals)	Application	References
	Neutral	Negative charge	Positive charge	Others		Description	Reaction			
(Liposome itself) Molecular chaperon	Yes	Yes	Yes	Fatty acids Cholesterol	None	Conformational change of partly denatured protein (carbonic anhydrase, lysozyme, transglutaminase, ribonuclease, etc.)	Protein (denatured) → Protein (intermediate) → Protein (native)	50–100% (against molecular chaperone)	Protein refolding	[5]
Catalase-like function (LIPOzyme)	Yes	No	No	Effect of membrane fluidity	None	Decomposition of hydrogen peroxide	$H_2O_2 \to H_2O + 1/2\, O_2$	0.005% (against catalase)	Catalase-like function	[53]
SOD/catalase LIPOzyme	Yes	Yes	Yes	Cholesterol Oxidized lipid	Oxidized SOD fragment and Cu/Zn	SOD-like function (dismutation of superoxide) and catalase-like function (Decomposition of hydrogen peroxide)	(SD) $2O_2^- + 2H^+ \to H_2O_2 + O_2$ (Catalase) $H_2O_2 \to H_2O + 1/2\, O_2$	100% (against SOD) 4–5% (against Catalase)	Antioxidative drug Metal removal (chelating) drug	[8, 22, 23]

LIPOzyme			Substrate	Function	Reaction	Activity	Application	Reference
SOD/peroxidase LIPOzyme	Yes	Yes	No	Cholesterol	(1) Hexadecanoyl-Mn-porphyrin (2) Dodecanoyl Histidine with Cu/Zn or Mn	SOD-like function (dismutation of superoxide) and peroxidase-like function (oxidation of methanol with hydrogen peroxide)	(SD) $2O_2^- + 2H^+ \rightarrow H_2O_2 + O_2$ (Peroxidase) $H_2O_2 \rightarrow H_2O + 1/2O_3$ $H_2O_2 + CH_3OH \rightarrow 2H_2O + HCHO$	2–10% (against SOD) 1–2% (against peroxidase)	Antioxidative drug Metal removal (chelating) drug	[9, 24, 42]
ADH LIPOzyme	Yes	Yes	Yes	Cholesterol	Monomer unit of alcohol dehydrogenase (tetrameric protein) with NAD$^+$	Oxidation (dehydrogenation) of alcohol and reduction of NAD+	$CH_3CH_2OH + NAD^+ \rightarrow CH_3CHO + NADH + H^+$	4% (against ADH)	Alcohol oxidation Alcohol production from aldehyde	H. Umakoshi et al. unpublished results
SP-chitosanase LIPOzyme	Yes	Yes	Yes	Direct preparation under the controlled interaction between cells and liposome	Signal peptide (SP)-harboring chitosanase without SP-processing	Hydrolysis of chitosan	Glycol chitosan $\rightarrow$ oligochitosan + glucosamine	150% (against natural chitosanase)	Production of oligochitosan and glucosamine	[43]
Translgutaminase LIPOzyme	Yes	No	No	–	Oxidized TG fragment by hydrogen peroxide	Transamidation reaction	Hydroxamate formation from benzyloxycarbonyl-L-glutamyl-glycine and hydroxylamine	5–10% (against natural transglutaminase)	Extracellular matrix promotion Biological glue of cells	Unpublished data

(continued overleaf)

Table 21.1 (continued)

LIPOzyme type	Study on type of designed liposome				Resources	Catalytic function		Catalytic level (relative value against naturals)	Application	References
	Neutral	Negative charge	Positive charge	Others		Description	Reaction			
Cholesterol oxidase LIPOzyme	Yes	No	No	Cholesterol at different level	Amyloid β-peptide with Cu	Oxidation of cholesterol	Cholesterol → oxidized cholesterol (hydroxyl- or keto-form)	—	Modulation of cholesterol level in Alzheimer disease	[25, 26]
Dopamine oxidase LIPOzyme	Yes	No	No	Membrane fluidity	Amyloid β-peptide with Cu	Oxidation of dopamine	Dopamine → oxidized dopamine (dopaquinone)	—	Fibril solubilization by oxidized product in Alzheimer Disease	[44]
Protein expression	Yes	Yes	Yes	DOTAP	Minimal elements for protein expression (in vitro protein synthesis system)	Up- and downregulation of final gene product	Catalytic reactions at all the elemental steps (transcription, translation, and folding)	0–170% (against in vitro control)	Protein production system	[10, 54–56]

21.5 Biomembrane Interference

Basic concept of "biomembrane interference"

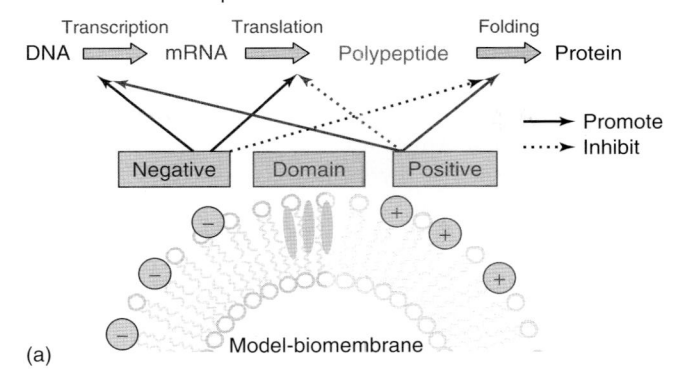

(a)

Recognition of biomolecules by "biomembrane"

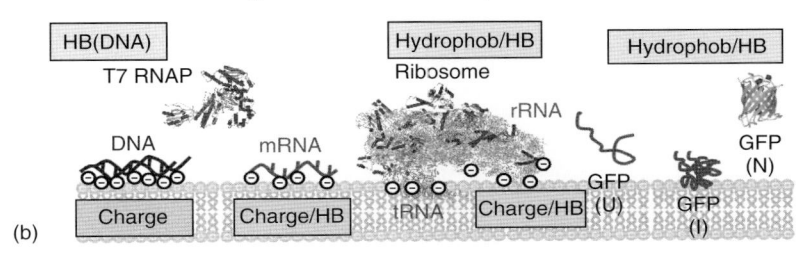

(b)

Spectra of "biomembrane interference"

	Neutral	Microdomain	Negative charge	Positive charge	
	POPC, PMPC, DPPC, etc	POPC/Ch etc	POPC/POPG	POPC/SA	DOTAP
DNA ↓ Transcription ↓ RNA	+	++	++	++	--
RNA ↓ Translation ↓ Polypeptide	+	++	~0	--	--
Polypeptide ↓ Folding ↓ Protein	+	+	–	+/–	+/–

(c)

Figure 21.7 Biomembrane interference of gene expression. (a) Basic Concept of "Biomembrane Interference". (b) Recognition of biomolecules by "Biomembrane". (c) Spectra of "Biomembrane Interference".

and hydrogen bond stability on the liposome surface and can affect the elemental steps of gene expression [10, 54–56]. It has also been found that on–off control of the above gene interference effect by the biomembrane can also be achieved by control of the environmental conditions. At present, the above phenomena have been examined only with regard to the role of the model biomembrane in the gene expression process. However, it has been gradually revealed that the above interference phenomena can be regulated by recognition of the RNAs on the liposome surface [8] (K. Suga *et al.* unpublished results). It is expected that protein synthesis and folding could also be performed on the same liposome surface by recruiting the minimal elements onto the liposome surface through its recognition function.

Although it is only a technology of the potent, it is possible to design a LIPOzyme capable of functioning as a one-pot-type catalyst and to use it as a material for integrated reaction/separation (material production). A new catalyst that can control a chemical reaction with minimum energy input is expected to be designed by utilizing the recognition (separation) function of the biomembrane (liposome) as a core material that can be flexibly controlled.

21.6
Summary

The significance of the basic concept of BMPC has thus been introduced by introducing the examples of LIPOzymes that are being used as a core material in BMPC. Such liposome-based materials (LIPOzyme) could be utilized in the practical bioprocess design by developing a useful tool that will enable us to simplify the practical operations. Various tools for separation and analytical use have been developed by using the liposome as a recognition (separation) element (Figure 21.8): (i) immobilized liposome chromatography (ILC) and immobilized liposome membrane (ILM) (module) [58, 59], (ii) immobilized liposome sensor (ILS) [60, 61], and (iii) dielectric dispersion analytical (DDA) method) [62, 63]. The recognition (separation) function of the liposome can thus be analyzed as a combination of nonspecific interactions such as electrostatic interactions, hydrophobic interactions, and hydrogen bond stability under normal static conditions and, also, rather specific interactions driven by a combination of the above interactions through the stress–response dynamics of the membrane (Figure 21.3) [10, 18, 24]. In addition, another tool that can be used to analyze the stability of the metal–ligand complex on the membrane surface, such as (v) metal affinity immobilized liposome chromatography (MA-ILC), has been developed [18]. Progress has also been made with (vi) membrane chip development [36–38], which can make various liposome arrays. The separation and analysis technique utilizing the liposome as a core material offers a fundamental design factor to guide molecular recognition of the liposome itself. The new bio/chemical process of integrated reaction/separation (material production) is expected to be developed by using (i) LIPOzyme as functional materials [8–10], (ii) membrane chip as a monitoring tool [36–38, 64], and (iii) ILM-Module as a separator tool [65, 66]. The designed liposome surface can also achieve optical resolution of

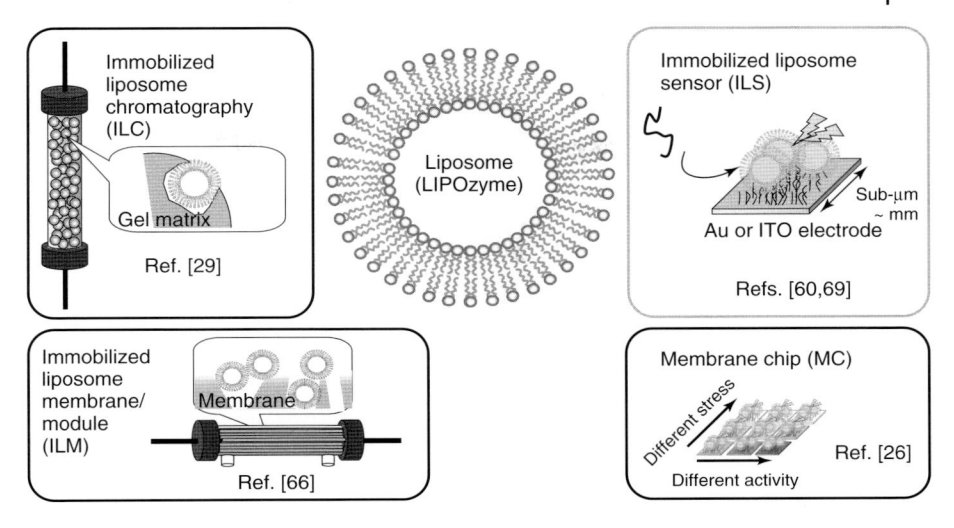

Figure 21.8 Liposome(LIPOzyme)-immobilized support for industrial uses.

chiral compounds (i.e., amino acids, dipeptides (D/L-Trp-D/L-Trp), and bilirubin) [67, 68], implying that LIPOzyme with an optical resolution function can easily be developed. A new type of LIPOzyme that can be used by the pharmaceutical industry is expected to be developed with the above tools.

Acknowledgments

The fundamental concept of this study was supported by the research group of "Membrane Stress Biotechnology" (http://www.membranome.jp). The authors appreciate the contributions to this research by Drs L.Q. Tuan, H.T. Bui, K.X. Ngo, H. Ishii, and H.T. Vu, and Mrs K. Morimoto and K. Suga. It was partially supported by a Grant-in-Aid for Scientific Research (nos. 19656203, 19656220, 20360350, 20760539, 21246121) from the Ministry of Education, Culture, Sports, Science, and Technology of Japan, a grant from Sigma Multidisciplinary Research Laboratory "Membranomics", a grant from the Global COE program Bio-Environment Chemistry of Japan Society for the Promotion of Science (JSPS), and a grant from Japan Science and Technology (JST) Agency.

References

1. Kuboi, R. and Umakoshi, H. (1998) *Adv. Biosep. Eng. (Chem. Eng. Symp. Ser.)*, **65**, 208.
2. Kuboi, R., Maruki, T., Tanaka, H., and Komasawa, I. (1994) *J. Ferment. Bioeng.*, **78**, 431.
3. Kuboi, R., Umakoshi, H., and Komasawa, I. (1995) *Biotechnol. Prog.*, **11**, 202.
4. Umakoshi, H., Yano, K., Kuboi, R., and Komasawa, I. (1996) *Biotechnol. Prog.*, **12**, 51.

5. Kuboi, R., Yoshimoto, M., Walde, P., and Luisi, P.L. (1997) *Biotechnol. Prog.*, **13**, 828.
6. Umakoshi, H., Yoshimoto, M., Shimanouchi, T., Kuboi, R., and Komasawa, I. (1998) *Biotechnol. Prog.*, **14**, 218.
7. Yoshimoto, M., Walde, P., Umakoshi, H., and Kuboi, R. (1999) *Biotechnol. Progr.*, **15**, 689.
8. Tuan, L.Q., Umakoshi, H., Shimanouchi, T., and Kuboi, R. (2008) *Langmuir*, **24**, 350.
9. Umakoshi, H., Morimoto, K., Ohama, Y., Nagami, H., Shimanouchi, T., and Kuboi, R. (2008) *Langmuir*, **24**, 4451.
10. Bui, H.T., Umakoshi, H., Nishida, M., Shimanouchi, T., and Kuboi, R. (2008) *Langmuir*, **24**, 10537.
11. Busca, G. (2009) *Ind. Eng. Chem. Res.*, **48**, 6486.
12. Zhang, T.Y. (2006) *Chem. Rev.*, **106**, 2583.
13. Butters, M., Catterick, D., Craig, A., Curzons, A., Dale, D., Gillmore, A., Green, S.P., Marziano, I., Sherlock, J.-P., and White, W. (2006) *Chem. Rev.*, **106**, 3002.
14. Astruc, D., Lu, F., and Aranzaes, J.R. (2005) *Angew. Chem.*, **44**, 7852.
15. Ozdemir, S.S., Buonomenna, M.G., and Drioli, E. (2006) *Appl. Catal. A*, **307**, 167.
16. Giddings, J.C. (1991) *Unified Separation Science*, Chapter 7, Wiley-Interscience, p. 141.
17. http://www.membranome.jp.
18. Kuboi, R. and Umakoshi, H. (2006) *Solv. Extr. Res. Dev., Jpn.*, **13**, 9.
19. Sakai, N. and Matile, S. (2004) *Chem. Biodiv.*, **218**, 28.
20. Vigh, L., Horvath, I., Maresca, B., and Harwood, J.L. (2007) *Trends Biochem. Sci.*, **32**, 367.
21. Nagami, H., Yoshimoto, N., Umakoshi, H., Shimanouchi, T., and Kuboi, R. (2005) *J. Biosci. Bioeng.*, **99**, 423.
22. Tuan, L.Q., Umakoshi, H., Shimanouchi, T., and Kuboi, R. (2008) *Membrane*, **33**, 173.
23. Tuan, L.Q., Umakoshi, H., Shimanouchi, T., and Kuboi, R. (2009) *Enzyme Microb. Tech.*, **44**, 101.
24. Umakoshi, H., Tuan, L.Q., Shimanouchi, T., and Kuboi, R. (2009) *Biochem. Eng. J.*, **46**, 313.
25. Yoshimoto, N., Tasaki, M., Shimanouchi, T., Umakoshi, H., and Kuboi, R. (2005) *J. Biosci. Bioeng.*, **100**, 455.
26. Shimanouchi, T., Tasaki, M., Huong, V.T., Ishii, H., Yoshimoto, N., Umakoshi, H., and Kuboi, R. (2010) *J. Biosci. Bioeng.*, **109**, 145.
27. Nagami, H., Umakoshi, H., Shimanouchi, T., and Kuboi, R. (2004) *Biochem. Eng. J.*, **21**, 221.
28. Umakoshi, H., Tuan, L.Q., Morimoto, K., Ohama, Y., Shimanouchi, T., and Kuboi, R. (2008) *Membrane*, **33**, 180.
29. Kuboi, R., Yoshimoto, M., Shimanouchi, T., and Umakoshi, H. (2000) *J. Chromatogr. B*, **743**, 93.
30. Veatch, S.L. and Keller, S.L. (2003) *Biophys. J.*, **85**, 3074.
31. de Almeida, R.F.M., Fedorov, A., and Prieto, M. (2003) *Biophys. J.*, **85**, 2406.
32. Talmon, Y., Evans, D.F., and Ninham, B.W. (1983) *Science*, **221**, 1047.
33. Menger, F.M. and Angelova, M.I. (1998) *Acc. Chem. Res.*, **31**, 789.
34. Tsong, T.Y. (1991) *Biophys. J.*, **60**, 297.
35. Saitoh, A., Takiguchi, K., Tanaka, Y., and Hotani, H. (1998) *Proc. Natl. Acad. Sci. U.S.A.*, **95**, 1026.
36. Shimanouchi, T., Ishii, H., Yoshimoto, N., Umakoshi, H., and Kuboi, R. (2009) *Colloid Surf. B*, **73**, 156.
37. Ishii, H., Shimanouchi, T., Umakoshi, H., and Kuboi, R. (2009) *J. Biosci. Bioeng.*, **108**, 425.
38. Shimanouchi, T., Oyama, E., Ishii, H., Umakoshi, H., and Kuboi, R. (2009) *Membrane*, **34**, 342.
39. Murakami, Y. (1996) *Trends Biotechnol.*, **10**, 170.
40. Murakami, Y., Kikuchi, J., Hisaeda, Y., and Hayashida, O. (1996) *Chem. Rev.*, **96**, 721.
41. Salo, D.C., Pacifici, R.E., Lin, S.W., Giulivi, C., and Davies, K.J.A. (1990) *J. Biol. Chem.*, **265**, 11919.
42. Umakoshi, H., Morimoto, K., Yasuda, N., Shimanouchi, T., and Kuboi, R. (2010) *J. Biotechnol.*, **147**, 59.

43. Ngo, K.X., Umakoshi, H., Shimanouchi, T., and Kuboi, R. (2010) *J. Biotechnol.*, **146**, 105.
44. Vu, H.T., Shimanouchi, T., Yagi, H., Umakoshi, H., Goto, Y., and Kuboi, R. (2010) *J. Biosci. Bioeng*, **109**, 629.
45. Ohtsu, H., Shimazaki, Y., Yamauchi, O., Mori, W., and Itoh, S. (2000) *J. Am. Chem. Soc.*, **122**, 5733–5741.
46. Li, D.-F., Yang, D., Yu, J.H., Huang, J., Li, Y.Z., and Tang, W.X. (2003) *Inorg. Chem.*, **42**, 6071–6080.
47. Batinić-Haberle, I., Liochev, S.I., Spasojević, I., and Fridovich, I. (1997) *Arch. Biochem. Biophys.*, **343**, 225–233.
48. Batinić-Haberle, I., Spasojevic, I., Stevens, R.D., Hambright, P., Neta, P., Okado-Matsumoto, A., and Fridovich, I. (2004) *Dalton Trans.*, 1696–1702.
49. Batinić-Haberle, I., Spasojevic, I., and Fridovich, I. (2004) *Free Rad. Biol. Med.*, **37**, 367–374.
50. Boka, B., Myari, A., Sovago, I., and Hadjiliadis, N. (2004) *J. Inorg. Biochem.*, **98**, 113–122.
51. Triller, M.U., Hsieh, W.-Y., Pecoraro, V.L., Rompel, A., and Krebs, B. (2002) *Inorg. Chem.*, **41**, 5544–5554.
52. Larson, E.J. and Pecoraro, V.L. (1991) *J. Am. Chem. Soc.*, **113**, 3810–3818.
53. Yoshimoto, M., Miyazaki, Y., Umemoto, A., Walde, P., Kuboi, R., and Nakao, K. (2007) *Langmuir*, **23**, 9416.
54. Bui, H.T., Umakoshi, H., Suga, K., Nishida, M., Shimanouchi, T., and Kuboi, R. (2009) *Biochem. Eng. J.*, **46**, 154.
55. Bui, H.T., Umakoshi, H., Suga, K., Tanabe, T., Shimanouchi, T., and Kuboi, R. (2009) *Membrane*, **34**, 146.
56. Umakoshi, H., Suga, K., Bui, H.T., Nishida, M., Shimanouchi, T., and Kuboi, R. (2009) *J. Biosci. Bioeng.*, **108**, 450.
57. Fire, A., Xu, S., Montgomery, M.K., Kostas, S.A., Driver, S.E., and Mello, C.C. (1998) *Nature*, **391**, 806.
58. Yoshimoto, N., Yoshimoto, M., Yasuhara, K., Shimanouchi, T., Umakoshi, H., and Kuboi, R. (2006) *Biochem. Eng. J.*, **29**, 174.
59. Yoshimoto, M., Kuboi, R., Yang, Q., and Miyake, J. (1998) *J. Chromatogr. B*, **712**, 59.
60. Jung, H.S., Ishii, H., Shimanouchi, T., Umakoshi, H., and Kuboi, R. (2007) *Membrane*, **32**, 294.
61. Morita, S., Nukui, M., and Kuboi, R. (2006) *J. Coll. Poly. Sci.*, **298**, 672.
62. Morita, S., Shimanouchi, T., Sasaki, M., Umakoshi, H., and Kuboi, R. (2003) *J. Biosci. Bioeng.*, **95**, 252.
63. Noda, M., Shimanouchi, T., Suzuki, H., Okuyama, M., and Kuboi, R. (2006) *IEEE Microw. Theory Technol.*, **54**, 1983.
64. Vu, H.T., Shimanouchi, T., Ishii, H., Umakoshi, H., and Kuboi, R. (2009) *J. Coll. Interf. Sci.*, **336**, 902.
65. Sugaya, H., Umakoshi, H., Tohtake, Y., Oyama, E., Shimanouchi, T., and Kuboi, R. (2009) *Membrane*, **33**, 272.
66. Sugaya, H., Umakoshi, H., Fadzil, K.B.M.A., Tuan, L.Q., and Kuboi, R. (2010) *Desalin. Water Treat.*, **17**, 281.
67. Bombelli, C., Borocci, S., Lupi, F., Mancini, G., Mannina, L., Segre, A.L., and Viel, S. (2004) *J. Am. Chem. Soc.*, **126**, 13354.
68. Borocci, S., Ceccacci, F., Cruciani, O., Mancini, G., and Sorrentic, A. (2009) *Synlett*, **7**, 1023.
69. Jung, H.S. *et al.* (2003) *SERDJ*, **10**, 123.

22
Matching Chemistry with Chemical Engineering for Optimum Design and Performance of Pharmaceutical Processing

Amit V. Mahulkar, Parag R. Gogate, and Aniruddha B. Pandit

22.1
Concept of Molecule to Money

The pharmaceutical industry over the period has mastered the art of converting "molecules to money" and is continuously striving toward achieving excellence in it. This chapter aims to provide guidelines for taking a step for further excelling in this art.

In today's pharmaceutical market scenario, it is always a race against time to introduce new molecules and processes to manufacture them into the market. In order to make sure that the new molecule captures its maximum share in the market, the molecule and/or the market has to be the first of its kind. This priority is claimed by filing drug master file (DMF) for the specific molecule. Thus, the chemists in pharmaceutical industries are eager to file DMF without bothering about the yields, selectivity, impurities, and utility consumption so that the priority/claim remains. As the DMF is converted to a full-scale operation, several engineering aspects emerge that not only aim at manufacturing the molecule but also aim at manufacturing it at a minimal cost while following the rules and regulations for its purity, safety, health, and environmental aspects.

The actual manufacturing is essentially handled by the engineers; whereas the first part of DMF is handled by the chemist. The chemists are not rigorous and quantitative about the material and energy balance calculations, which the chemical engineers do regularly. When the chemical engineers start to look into DMF, they start questioning various unit operations or the equipments used to implement these unit operations. What happens many times is that the equipments that are used in DMF are not amenable to scale-up. As a result, cost of the drug can increase, thus making it uneconomical to manufacture or making it an unattractive choice in the market.

In this chapter, we discuss the normal procedure followed by the pharmaceutical industry from DMF to full-scale plant; at the same time, we point out the difficulties that could make a process uneconomical. Further, we show that chemists with some basic knowledge of chemical engineering and involvement of chemical engineers

Pharmaceutical Process Chemistry. Edited by Takayuki Shioiri, Kunisuke Izawa, and Toshiro Konoike
Copyright © 2011 WILEY-VCH Verlag GmbH & Co. KGaA, Weinheim
ISBN: 978-3-527-32650-1

in the synthesis of a drug molecule in labs can make the process of bringing a new molecule into the market quick and efficient.

22.2
Steps Involved in Bringing Molecule to Market

A normal procedure followed in bringing a new drug into the market is as follows:

1) Synthesis step (there is difference in the word synthesis and manufacturing)
2) Kilo lab
3) Pilot plant
4) Full-scale plant.

22.2.1
Synthesis in Lab

- Developing basic chemistry of the reaction (A + B → C)
- Finding reaction kinetics (stoichiometry and material balance)
- Optimizing the reaction parameters such as temperature, pressure, limiting reactant, and reaction kinetics for the desired reaction step
- Preparation of standard operating procedure (SOP) in line with DMF
- Standardization of the analytical procedure for the product
- Documentation of analytical procedure for the different products/by-products/intermediates.

The first step, that is, synthesis step in gram/milligram lab, involves the development of a *basic reaction route* to synthesize a molecule. This step involves developing a reaction mechanism to produce a particular "molecule" without concern for yield, conversion, purity, or selectivity. The aim at this stage is to find the feasibility of a desired chemical reaction, without its commercial feasibility. This is often done to beat the already established patent or to file a patent for preparing the new molecule. After deciding the synthetic route to the molecule, the *reaction kinetics* is developed by carrying out lab-scale experiments in stoichiometric proportions. This step also establishes analytical procedures for the products, by-products, and intermediates. Next, the reaction parameters such as temperature, limiting reactant, and reaction kinetics for the desired reaction steps are *optimized*. This gives an SOP for producing the molecule at lab scale.

In most of the cases, the process can be made hassle free by taking right decisions on the lab scale itself; for example, proper selection of solvent or a reaction medium for carrying out the reaction or separation by avoiding unnecessary steps or by opting for better reaction conditions can reduce several equipments required for large-scale operation. However, in most cases no consideration, to any engineering principles (which is where the real money is made), is given at this stage and the SOP thus generated essentially remains unoptimized and true to the DMF with respect to the final production plant. This, essentially premature SOP, is further taken to kilo-lab scale.

22.2.2
Kilo Lab

- Group involves chemists and chemical engineers
- Finding the rate-controlling steps
- Stoichiometry balance for various steps
- Utilities required for the reactions (cooling water, brine, vacuum, etc.)
- Selection of material of construction (MOC) of the reactor and other equipments
- Generating thermodynamic and kinetic data (exothermic/endothermic data of reaction, various rate constants)
- Preparation of good manufacturing practice (GMP)
- Validation of batches for pilot plant.

In kilo lab, chemical engineers are also involved in the study of scalability of the process. The synthetic route, or SOP obtained from gram lab, is not a single-step process but a collection of multiple steps, broadly involving the synthesis of the intermediates and the final product and downstream processing for its separation, purification, and/or recycle of catalysts, solvents, and intermediates. The time required for each of these steps is not the same; one of the steps that limits the throughput of the whole process is called the *rate-limiting step*. Engineers realize that the procedure can be scaled up by selecting only one or two process steps, which are rate-controlling steps, and so their job essentially is to identify the rate-controlling steps. Such a "rate-limiting step" could be the synthesis of an intermediate due to its very slow rate or a reaction with low equilibrium conversion, or a time and labor consuming separation step, for example, separation of compounds with similar physicochemical properties or polarity. Then, the scalability of the manufacturing process and sizing of the equipment are done based on this rate-limiting process.

Further, the *material balance* should be satisfied and the reactions should be carried out on the basis of the *stoichiometric proportions*. Special attention should be paid to purity and quantity of used chemicals, reagents, or treatment done. The engineering issues, which ultimately will allow you to make money, need to be involved at the kilo-lab stage. The quantities of raw material required to produce the given quantity of the product in trials, test formulations, and so on, should be estimated using material balance and costing should be done. This is not possible without engineering aspects.

Material of construction (MOC) is usually not a thing of concern for the chemist as most of the equipments are usually glass lined or in glass in gram lab and kilo lab. But when it comes to industrial manufacturing and if we do not have information on MOC, then the only choice that is left is glass-lined reactor. Glass-lined equipments have severe limitations of design and operation in terms of speed of operation, pressure, temperature, alkalinity, and so on.

In kilo lab, emphasis should also be on studying the *thermodynamics* of the reactions, which is further useful for the selection and designing of equipments and also for estimating the load on utilities like cooling water, brine, heating media, vacuum, and so on. It is recommended that some experiments should be done using differential scanning calorimeter (DSC) to find out the heat effects at various stages of production. Then, *GMP* is prepared, after which the batches are validated for pilot-plant study. Here, all the operational parameters such as agitation, mixing, blending, rate of addition of reactants, and suspension of solid/catalyst particles, and other unit operations like crystallization, filtration, and drying involved in the process are detailed out.

The objectives of the kilo-lab stage are to characterize the rate-limiting steps in either synthetic chemistry or process engineering equipments. A unit operation that is not rate limiting on a lab scale may become a bottleneck at a scaled-up level at a kilo stage, and subsequently become a real constraint at a pilot scale of operation.

An operation, such as filtration and washing, can be completed in only a few minutes on a lab scale, without showing any detrimental effect on the product remaining in contact with the mother liquor. However, the time required may be in several minutes or even few hours at a kilo-lab level. Such a problem is sometimes encountered and can only be understood at a kilo-lab stage. Hence, the experiments at kilo level, in addition to the scale limitation of the chosen chemistry, should also target the following answer:

1) Choice of MOC and feasibility of obtaining larger scale equipment in the selected MOC.
2) The choice of technique for performing a unit operation through the chosen equipment and its effect on the time of operation. Unit operations, where the state of operation can have a significant effect on the operational time, are (i) mixing, (ii) filtration, and (iii) drying.
3) Finalization of a pilot-plant configuration, having sufficient flexibility to optimize the eventually scaled-up configuration.

22.2.3
Pilot Plant

- Group involves the chemical engineers
- Targeting the rate-controlling step for optimization
- Selection of rate-controlling step and setting up of optimum operating conditions
- Estimation of quantity of utilities required for the reaction
- Preparation of DMF (this is the proper stage for preparation of DMF)
- Validation of batches for the full scale-up operation
- Treatment of waste for safe discharge.

Pilot-plant operations are generally handled by chemical engineers in western countries, and no chemist gets involved unless specified. This step again involves identification of the rate-controlling step or an operation and setting up of the optimum conditions for the chosen unit operation (such as type of mixer, crystallizer, etc.) and the quantity of utilities required for the reaction (cooling water, brine, vacuum, etc.). This is done by taking material and energy balance over the chosen equipments. Here, optimization of a chosen unit operation is done on the basis of economy of process and economy of scale. Pilot plant ensures that the targeted molecule can be obtained on a larger scale and aims at finalizing the unit operations and related equipments. Once the chemists have proposed an SOP based on standardized chemistry and incomplete understanding of the engineering issues, the production is handed over to chemical engineers, and chemists are not engaged in manufacturing. Now the chemical engineers start experiencing the difficulty in implementing the steps in SOP developed on a gram-lab scale and a kilo-lab scale. At this point, it may happen that the DMF has to be completely rewritten for overcoming the difficulties in scaling up for a cost-efficient production.

There is a tremendous resistance to go through this revision of DMF as it involves considerable time and expenses. This again results in unoptimized operation and compromises, which restricts further developments in the process efficiencies. The standardization of the effluent/waste discharge to the control standards needs to be frozen at this stage. With the new molecule being developed, it is also necessary to develop a novel, efficient treatment process. Failing which, the permission to operate itself can be questioned.

22.2.4
Full-Scale Plant

- Chemical engineers and plant personnel are involved
- Final recoveries are important
- Economic optimization of the process is done
- Solvent recovery units (essential on a plant scale and not on kilo or pilot scale)
- Utilities supply management
- Man power management.

Full-scale plant is concerned with yields, conversion, purity, and selectivity, and it involves many unit operations at a higher scale. This involves choosing appropriate equipment to carry out various unit operations optimally. Aspects such as solvent recovery are considered, because the reactions are usually carried out in solvents. Thus, the cost of solvent recovery and recycle should be considered, and this could occupy a significant portion of cost of an active pharmaceutical ingredient (API). Similarly, cost of utility is also monitored and controlled.

From the above discussion, it is obvious that correction or improvements are required to the process at pilot-plant level, but to a limited extent in order to preserve the validity of DMF, because the priority claim is lost if DMF is altered too much. Hence while filing DMF, if some chemical engineering principles are followed, then a DMF stating almost similar production process as that of full-scale plant can be filed. This capability needs to be developed for

1) designing a process of a full-scale plant based on synthetic route in lab scale itself;
2) process optimization/intensification of the above process;
3) use of dynamic/process simulator coupled with DSC or similar apparatus.

Let us discuss the first two points in detail to get some insight into achieving these. Third point is out of scope of this chapter.

22.3
Interrelation in Each Step and Concept of Unit Operations

It should be noted that every step at the time of synthesis in lab scale grows to a unit operation or a unit process at full-scale operation. Thus, a little know-how about the *unit operations* and *equipments* involved therein will make scientists in gram lab aware of problems that could arise at the full-scale plant. So, if a chemical engineer gets involved during the synthesis step itself, then the chemist also starts thinking in terms of material balance, energy balance, unit operations, and appropriate equipments to be used for final production. Thus, the DMF based on engineering principles could be filed right at the beginning.

This concept is explained with a classical example of manufacture of penicillin (Figure 22.1). Penicillin is manufactured by fermenting *Penicillium* fungi. The fermentation broth is then separated out in a series of downstream processes including filtration, extraction, crystallization, and drying to get penicillin. Fermentation is the rate-limiting step in this process, and the downstream unit operations can be made to operate in either batch or continuous manner. Since usually continuous processes have higher efficiency, the downstream processing equipments should be designed such that the cycle time of downstream processing exactly matches the batch fermentation time. This ensures that all the downstream process equipments are functional at all the times and operate at optimal level. This could be achieved by having multiple fermentation units, which operate in such a sequence that the downstream processing operations and equipments are made continuous.

One of the typical unit operations is mixing (with/without chemical reaction); usually in a pharmaceutical industry, the products are precipitated out, or an antisolvent is added to precipitate the product, which necessitates the mixing operation. Other unit operations that follow precipitation or crystallization are filtration and centrifugation (solid–liquid) operations, essentially followed by drying. The sequence of these operations will keep on changing depending on the types of synthetic steps that are involved in the manufacture of a particular API. Interestingly, the sequence of the steps will also decide the type of equipment that is to be used. Hence, no generalization rule can tell that a particular type of equipment can be used for filtration, drying, or crystallization. The selection of the type of equipment depends on where the unit operation is located in the entire sequence of synthetic steps.

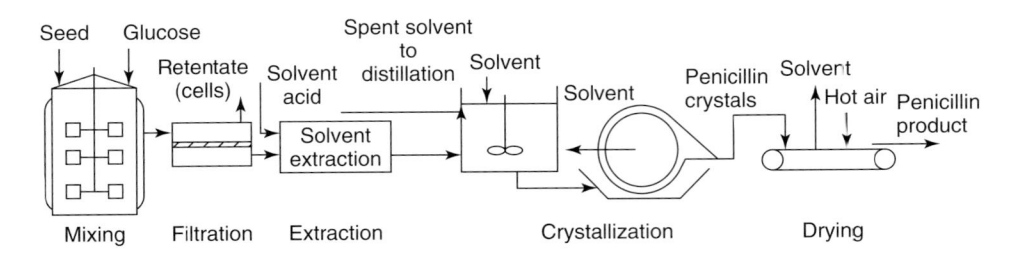

Figure 22.1 Penicillin manufacture: process flow.

22.4 Unit Operations

22.4.1 Mixing

22.4.1.1 Utility of Mixing

Mixing may involve the mixing of liquids, gases, or solids in any possible combination of two or more constituents. Mixing serves important purposes like blending, uniform heating/cooling, solid suspension, solid dissolution, homogenization, gas dispersion, and so on, to promote mass and heat exchange between the phases, to increase contact area between the phases, and to initiate or assist the chemical reaction between different phases or reduction of particle size. When the mixing is accompanied by chemical reaction between the phases, it becomes essential to maintain uniformity of concentration of reactant to control the quality of the final product in terms of its purity and yield. Depending on the application, an appropriate type of mixing device should be chosen from a wide range of mixing devices available. Inappropriate selection will not only result in increased operational cost but can also result in material degradation, excessive by-product formation, nonuniformity in temperature/concentration, and so on.

22.4.1.2 Types of Equipments

A variety of mixing equipments are available that can essentially be differentiated by the way they bring about mixing/homogeneity among the phases. Some equipments produce high bulk flow while others produce more of shear and turbulence. Mixing equipments that produce more of turbulence are chosen when disintegration of the dispersed phase (bubbles, drops, particles, lumps, etc.) and creation of new interfacial area are needed, similar to those for solids that have a tendency to form lumps, gas dispersion, liquid–liquid mixing, and so on, while mixing equipments generating high bulk flow are chosen when creating an interface is easier (low interfacial tension system) and only homogeneity in temperature and/or concentration is needed, for example, suspending non-agglomerating solid particle in liquid, blending of miscible liquids.

A variety of process functions are carried out in vessels stirred by rotating impellers. A standard agitated vessel is shown in Figure 22.2, and it consists of a cylindrical tank, baffles, cooling coil or cooling jacket, and impeller mounted on a shaft. Baffles are provided to avoid vortex formation. Vortex formation in an agitated tank simply keeps the liquid rotating with no significant vertical/axial mixing and hence needs to be avoided. Gas sparger can be provided for gas dispersion. Gas sparger is simply a plate with multiple holes, circular rings with holes, porous ceramic surface, or an open pipe.

Impeller is the most important component of mixing equipment, and it determines the type of flow that will be generated in the vessel. On the basis of the characteristics of the flow generated by the impellers, they can be categorized as

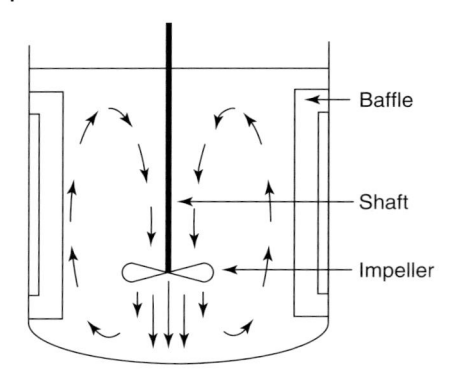

Figure 22.2 Flow pattern produced by an axial flow impeller.

- Axial flow impellers
- Radial flow impellers
- Mixed flow impellers.

Axial Flow Impellers These include all the impellers in which the blade makes an angle of less than 90° with the plane of rotation. Propellers, pitched blade turbines, and hydrofoils are examples of axial flow impellers. The direction of rotation is usually chosen to force the liquid downward (Figure 22.2). The flow currents leaving the impeller continue until deflected by the floor of the vessel. These are good for suspension and dispersing (non-agglomerate) solids and maintaining uniform suspension. These are preferred when lower fluid shear is expected, especially in fermentation or other biological processes where intense shear would destroy the microbes.

Radial Flow Impellers Straight blade turbine or a disk turbine is a typical example of radial flow impellers. Turbine impellers push the liquid radially outward and in tangential direction with no axial/vertical motion (Figure 22.3). The flow currents travel away from the turbine to the sidewalls and then go either upward or downward aided by the baffles. The turbines rotate with a tip speed of $1-5$ m s^{-1} and produce a lot of turbulence and lesser bulk flow. These are particularly useful for dispersing gas and suspending solids that tend to agglomerate. The gas is sparged beneath

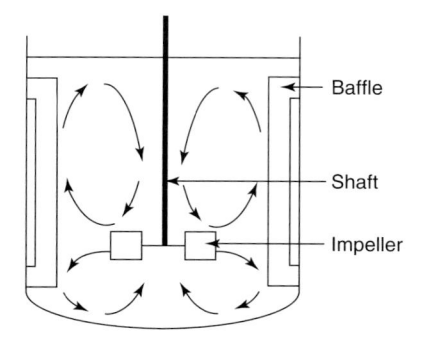

Figure 22.3 Flow pattern produced by a radial flow impeller.

the turbine. The gas bubbles/solid lumps get disrupted because of the turbulence and shear, and are dispersed over the entire cross section of the tank.

Mixed Flow Impellers Mixed flow impellers generate both the bulk flow and shear, and are especially useful if the same equipment is to be used for a variety of jobs. In a pitched blade turbine, by changing the pitch (angle between the impeller blade and the plane of rotation), the axial/radial components of the bulk flow and turbulence can be manipulated. They are multipurpose impellers, which are preferred in many processes.

For liquids with viscosity more than 5000 cP, *helix impeller* or *anchor impeller* is used. The diameter of the helix or anchor impeller is almost equal to the internal diameter of the vessel. These impellers do not produce much of a bulk flow, but they keep on scrapping the surface and remove the material from near the wall and maintain a good heat transfer near the wall. The impeller shape ensures the absence of dead zones in the vessel.

Gas Inducing Impellers *Gas inducing impellers* are used to introduce gas into the liquid without a sparger (Figure 22.4). Gas inducing impellers can be operated in dead end mode for the gas phase; for example, in hydrogenation reaction, the hydrogen gas needs to be completely absorbed/reacted in the reactor. In such cases, a gas inducing impeller can be designed that sucks the hydrogen from the head space of the mixer and reintroduces at the bottom of the vessel. At the same time, stirring is also carried out by the impeller blades [1].

22.4.1.3 Selection Criteria

The mixing equipments should be selected such that they produce only the required quantity of either bulk flow or turbulence or both. Producing more of turbulence or bulk flow will not only lead to wastage of energy but also spoil the quality of the final product. Two most important factors that need to be considered while selecting a mixing device are power consumption and mixing time. For further details on the type of mixing equipments and selection criteria, the readers are directed to [2, 3].

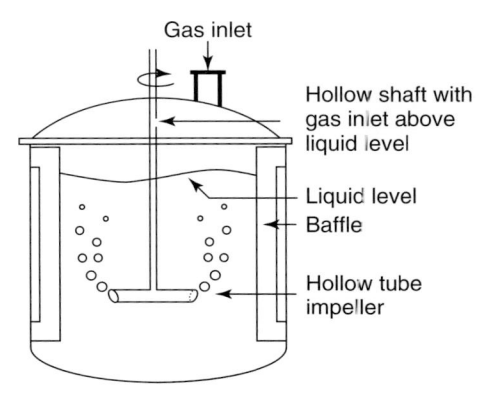

Figure 22.4 Gas inducing impeller.

22.4.2
Crystallization

Crystallization is essentially a purification process, because the word "crystal" in itself means pure. A crystal of proper shape is achieved only when it is pure. Hence the *crystallization process* is usually defined as a purification step, rather than a separation step. In principle, crystallization is a solidification step within a homogenous liquid phase. It involves transfer of mass from concentrated solution to a nuclei or a growing crystal. Similar to other phase-change processes, crystallization also involves simultaneous mass and heat transfers during the phase-change phenomena.

22.4.2.1 Utility of Crystallization
A well formed crystal is almost pure. Thus, the shape of a crystal is itself an indicator of purity. Crystalline substance is easy to handle, pack, store, and is high priced. When the crystals are to be sent to the end user, they need to be strong, non-aggregated, uniform in size, and noncaking when packed/stored. When the crystals are required to be processed further, reasonable and narrow size distribution is also required.

Crystallization is brought about by saturating the solution, which can be done by the following:

- *Cooling* the solution if solubility of solute drastically decreases with temperature.
- Concentrating the solution by *evaporation*, if solubility is almost independent of temperature.
- Adding an *antisolvent*, to the solution, which reacts with the primary solvent to form a third solvent in which the solute is insoluble.
- *Precipitation*: a secondary solute is added to the solution, which reacts with the primary solute to form a third insoluble solute, which is then precipitated.

22.4.2.2 Types of Equipments
A crystallizer can be operated in a batchwise or continuous manner. But mainly the design of a crystallizer is based on the method employed for supersaturating the mother liquor. Since the degree of supersaturation also determines the crystal size and size distribution, in all the designs primary focus is on maintaining the required and uniform concentration and temperature. A few main types of crystallizers, categorized on the basis of saturation method, are discussed below.

Cooling Crystallizer

Agitated Tank Crystallizer A simple crystallizer is an agitated tank with cooling arrangement. It simply consists of a stirred tank either jacketed or with coil for cooling and exit for crystals at bottom (Figure 22.5a). Cooling jackets are usually preferred to coils for cooling, because the latter often become encrusted with crystals. The inner surfaces of the crystallizer should be smooth and flat to

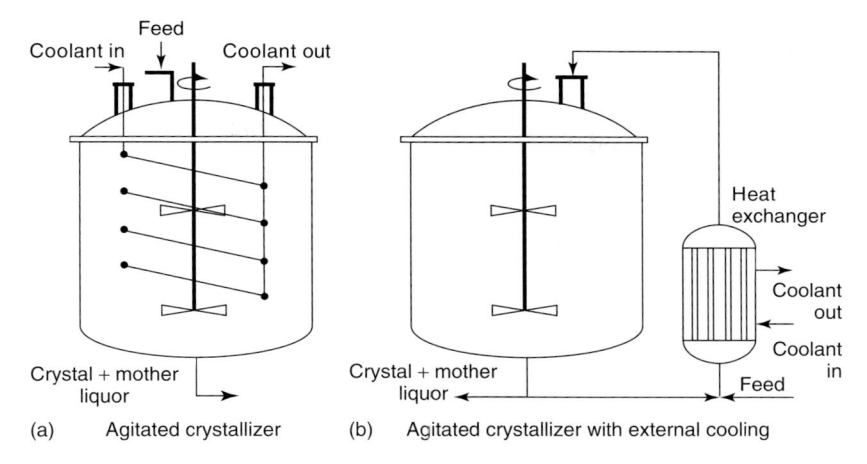

Figure 22.5 Agitated tank cooling crystallizers.

minimize encrustation. An agitator, located in the lower region of a draft tube, circulates the crystal slurry through the growth zone of the crystallizer. External circulation, as shown in Figure 22.5(b), allows good mixing inside the unit and promotes high rates of heat transfer between the liquor and the coolant, and an internal agitator may be installed in the crystallization tank if required. Since the liquor velocity in the tubes is high and even low temperature difference is usually adequate, the encrustation on heat transfer surfaces is reduced considerably. *Oslo-fluidized bed crystallizer* consists of external cooling arrangement and a bed of crystals, which is being fluidized by the upward flow of the saturated liquid in the annular region surrounding the downcomer draft tube. This is also called *ideal classified bed crystallizer*, as it separates crystals on the basis of their size.

Evaporative Crystallizer These types of crystallizers are widely used when solubility is a weak function of temperature. The majority of continuously operated evaporative crystallizers are of three basic types: forced circulation, fluidized-bed, and draft-tube agitated units. Forced-circulation crystallizer is essentially an evaporator with external heating arrangement. The vapors are separated in the evaporator, and the magma, containing crystals and mother liquor, is removed as a product, which is used for filtration and crystal separation. In *vacuum crystallizers*, supersaturation is achieved by adiabatic evaporative cooling, and these have higher energy efficiency as compared to those operating at atmospheric pressure. These crystallizers also offer an added advantage that they can be used for heat-sensitive materials.

22.4.2.3 Selection Criteria
The temperature–solubility relationship for the solute and solvent is of prime importance in the selection of a crystallizer. For solutions that yield appreciable amounts of crystals on cooling, a simple cooling unit is appropriate. An evaporating crystallizer would be used for solutions that change little in composition on cooling; otherwise, the antisolvent method is also used in certain cases. The shape, size,

and size distribution of the product are also important factors. For large uniform crystals, a controlled suspension unit fitted with suitable traps for fines, permitting the discharge of a partially classified product, would be suitable. This simplifies filtration, washing, and drying operations, and even screening of the final product may not be necessary.

Simple cooling-crystallizers are relatively inexpensive as compared to vacuum or evaporative crystallizers. Heavy crystal slurries can be handled in cooling units without liquor circulation, though cooling surfaces can get coated with crystals, thus reducing the heat transfer efficiency. Vacuum crystallizers with no cooling surfaces do not have this disadvantage, but they cannot be used when the liquor has a high boiling-point elevation. In terms of space, both vacuum and evaporating units usually require a considerable head and floor space.

Once a particular class of unit has been decided upon, the choice of a specific unit depends on the initial and operating costs, the space available, the type and size of the product, the characteristics of the feed liquor, and the need for corrosion-resistant MOC, and so on. Particular attention must be paid to liquor mixing zones, since the circulation loop includes many regions where flow streams of different temperature and composition mix. These are all points at which temporary high supersaturation may occur causing heavy nucleation, and hence encrustation, poor performance, and operating instabilities.

In crystallization since the crystal size and size distribution are of critical importance, it becomes critical to control the rate at which saturation is brought about. Sonocrystallization is one such technique that provides benefits of achieving control over the crystal size and polymorphs. Sonocrystallization is the application of high-intensity (100 W l^{-1}) low-frequency (20–60 kHz) ultrasound to promote and control crystallization. The main effect of ultrasound is to promote nucleation via cavitation. Recognizing the benefits that ultrasound can bring to crystallization; many researchers endeavored to study sonocrystallization at the laboratory and industrial scales [4–7].

22.4.3
Filtration and Centrifugation

These two unit operations are applied for solid–liquid and liquid–liquid separations. In filtration, separation is based on the size of the solid particle, while in centrifugation, separation is based on difference in the density. As a result of which in addition to solid–liquid separation, centrifugation can be used for liquid–liquid separation (decantation) also. Filtration is driven by gravity or fluid pressure while centrifugation is based on centrifugal force.

22.4.3.1 Utility of Filtration

Filtration is an operation whereby a solid is separated from a liquid or gas by means of a porous medium that retains the solid and allows the fluid to pass through. The valuable stream from the filter may be a stream of solids, like that in crystallization, or a liquid stream for removing the suspended solids, or, in some cases, even both

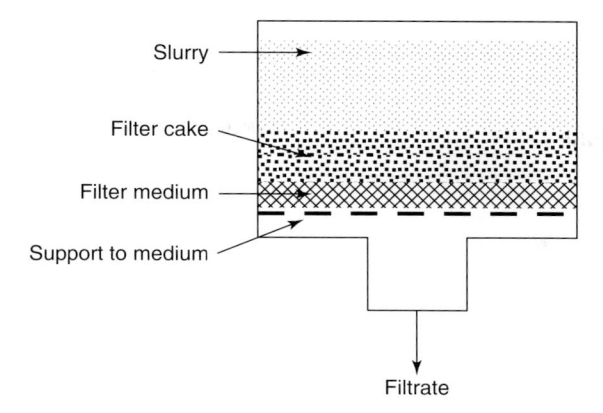

Figure 22.6 Schematic of basic filtration operation.

the streams. Usually, in a pharmaceutical industry, it is solids that are separated from the liquid. In rare cases, it is neither, as is the case when waste solids must be separated from the waste liquid prior to disposal. In the pharmaceutical industry, this operation can be referred to as *downstream processing*, wherein the product, either in liquid or in solid form, needs to be purified, and filtration is usually the second unit operation after crystallization. A basic filtration unit, shown in Figure 22.6, usually consists of the following:

1) Filter medium: it does basic filtration; it can be made of woven fabric, and can be a screen made of metal, plastic, or any substance, which is essentially strong enough to withstand the flow-induced stresses and is inert to filtrate and retentate.
2) Support to the filtration medium that provides strength to the filter medium to withstand pressure difference, solid loading, and so on. Similar to filter medium, it can be metallic, polymeric, or of any material which is strong enough to withstand stresses and is inert to the exposed chemicals.
3) Filter cake that is formed by the deposition of solids on the medium; filter cake also does filtration to a significant extent, after which the filtration needs to be stopped and cake be removed.

22.4.3.2 Types of Equipments

All filtrations are separation processes based on the size of solids, but owing to wide difference in properties of solids and solid concentration, several types of filtration equipments are available. Here, we discuss some of the most common filtration devices used in pharmaceutical industry. Based on the rate, filtration is broadly categorized as constant-rate filtration and constant-pressure filtration. In constant-rate filtration, the pressure drop is varied from a minimum at the start of the filtration to a maximum at the end of filtration, to maintain the rate of filtration throughout the run, whereas in constant-pressure filtration, the pressure drop over the filter is held constant and rate of filtration decreases over the run time.

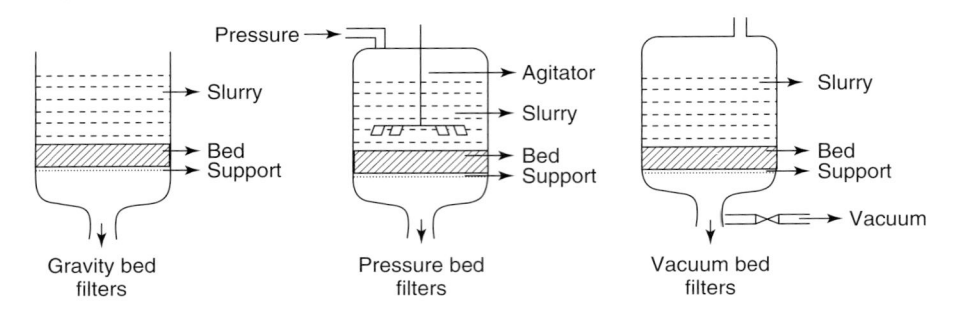

Figure 22.7 Types of filtration based on driving force.

Filters are also classified as those operating with a pressure above the atmospheric pressure, before the medium, termed as *"pressurized filters,"* and those having a subatmospheric pressure after the medium, termed as *"vacuum filters"* (Figure 22.7). Pressure filtration can be carried out by using either a pump or it may even be gravity driven. Gravity filters are restricted to very coarse screens for separating coarse solids like crystals. Industrially, gravity filters are restricted to draining liquids from very coarse crystals. A much faster rate of filtration can be achieved in vacuum filtration, but it is still limited to a maximum pressure differential of 1 atm. Pressurized filtration units are most common, which can give higher rate of filtration even with fine solids which form slimy cakes over the medium.

Mode of operation (batch or continuous) of filtration is usually decided by the mode of operation of other unit operations in the overall process. If crystallization is carried out in a continuous manner, then the following filtration is also carried out in the continuous mode. Inherently, all the filtration is essentially batch operation, because the filter medium needs to be cleaned of the solid cake deposited on it. However, there are filtration units in which one part of the filter is cleaned while the other part is used for filtration; thus, the overall filtration never stops.

Here, we discuss some of the most frequently used filters in the pharmaceutical industry. *Rotary vacuum drum filter*, shown in Figure 22.8, consists of a sheet metal drum of 1–15 ft diameter and 1–20 ft long. A filter medium such as canvas covers the curved face of the drum, which is partially submerged in the liquid. The drum has slotted face and turns at 0.1–2 rpm. Under the slotted surface of

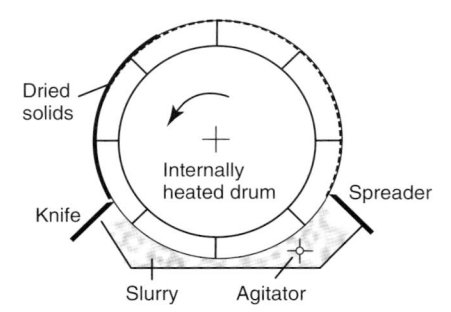

Figure 22.8 Drum dryers for slurries.

472 | *23 The Integration of Safety, Health, and Environmental Considerations into Process Development*

dosed for chronic periods as demanded by global drug regulators. This delivery is critical because changes in the impurity profile beyond this point could have significant impact on these toxicity studies. For example, if additional impurities are introduced, these could potentially invalidate the ongoing toxicity studies or trigger the need for additional bridging toxicity studies to be undertaken. The ability of the development team to fundamentally impact the design of the route is limited beyond this point of development, especially for the later synthetic stages within a route. Therefore, it is critical that the key SHE thinking be applied to the route before this point.

The fourth phase of development is Process Optimization and Understanding which focuses on the finer details of the process such as identifying optimum stoichiometry and concentrations of the process. At this stage of development, the ability to fundamentally change the SHE impact of a process is more limited. The SHE focus switches to developing control for any SHE issues that remain, such as identifying suitable abatement or containment strategies for any emissions that may result from the process. This is the render harmless phase which ends in the detailed process being frozen.

The final stage of process development, often called *technology transfer*, involves the transfer of knowledge of the process from the development team to the site of final manufacture that will be used to provide commercial API. The timing of this process can vary. However, it is important to ensure the SHE knowledge of the process is fully transferred.

Figure 23.2 illustrates some of the key triggers used within the AstraZeneca SHE Triggers model. It is not the complete model but covers a broad description of each phase that will be elaborated later in this chapter.

| Early delivery | Route evaluation | Process design | Optimisation and understanding | Tech transfer |

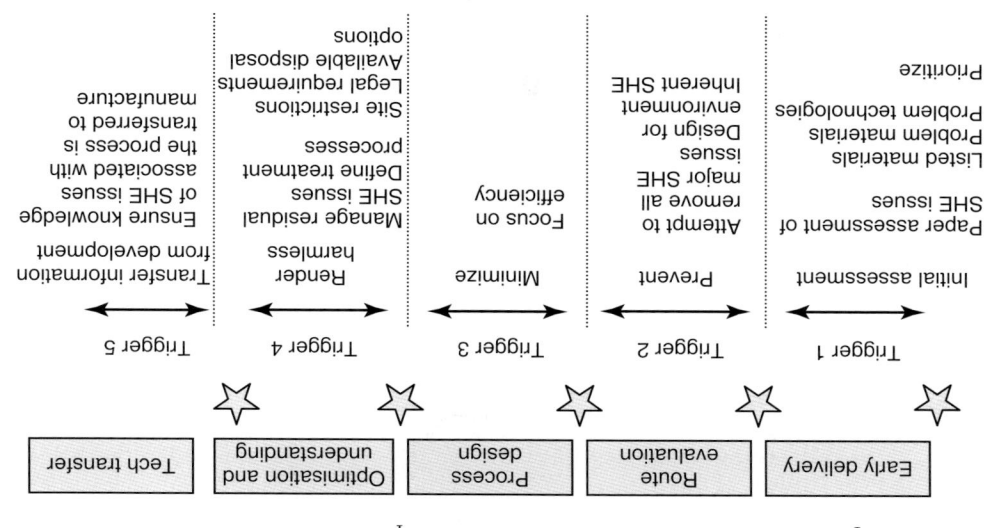

Trigger 1 — Initial assessment
- Paper assessment of SHE issues
- Listed materials
- Problem materials
- Problem technologies
- Prioritize

Trigger 2 — Prevent
- Attempt to remove all major SHE issues
- Design for environment Inherent SHE

Trigger 3 — Minimize
- Focus on efficiency

Trigger 4 — Render harmless
- Manage residual SHE issues
- Define treatment processes
- Site restrictions
- Legal requirements
- Available disposal options

Trigger 5 — Transfer information from development
- Ensure knowledge of SHE issues associated with the process is transferred to manufacture

Figure 23.2 The AstraZeneca SHE triggers model.

23.1 Introduction

These phases are

- early delivery,
- synthetic route evaluation,
- process design,
- process optimization and understanding,
- technology transfer.

When a molecule has been selected for development following a period of Lead Optimization, the first delivery of the process development team is often a few kilograms (typically 2–5 kg) to supply material for preclinical safety studies and Phase I clinical supplies. The priority during this phase is ensuring the synthesis used is fit for purpose and can deliver in time to meet the needs of the wider development program. The main aim during this phase is speed of delivery. It is common that the synthetic route used is not the final long-term route, although this is not exclusively the case. The focus from a SHE perspective is ensuring that the process used for this manufacture manages the risks acceptably, given the scale and speed of manufacture.

The next phase of development is associated with selecting the optimum long-term synthetic route for the particular API being developed. This is the point at which the development team needs to balance all the factors involved in long-term API supply, perhaps being capable of producing hundreds of tons per year at peak-year sales of the medicine. This phase will finish when the raw materials and intermediates used in the synthesis are fixed for the long-term. This is the period when the most impact can be made by the development team on the SHE performance of the process by preventing SHE issues. For example, if the number of stages used to make a particular API can be halved from 10 to 5, the reduction in material usage can be huge and therefore significantly reduce the environmental burden of the project. Similarly, if a route can be selected that avoids the isolation of dusty genotoxic intermediates, the need for costly containment options of personal protective equipment (PPE) can be significantly reduced. Often, this period of development will result in the first pilot plant campaign of between 10 and 50 kg which may supply Phase II clinical studies and later safety studies. It will also supply material for formulation development.

The third phase of development is Process Design which focuses on further refinement of the process such as selection of solvents and reagents used within the process. Significant improvements can still be made in the SHE performance of the process at this stage by minimizing SHE issues. Often the solvents used within a process will account for a significant proportion of the total material used within a synthesis. Therefore, selection of more benign solvents from a SHE perspective can be key. Similarly, selection of an optimum catalyst, which increases the yield of a particular stage, will reduce the mass intensity of the process. This period of development can culminate in production of a pilot plant campaign of 50–300 kg which will typically be used in Phase III clinical supplies and the long-term toxicity studies. These toxicity studies provide data to ensure that the drug is safe when

current global Environmental and Safety Legislation (Figure 23.1). This hierarchy works on the assumption that prevention of SHE-related issues is optimum. If an issue cannot be prevented, it should be minimized. The final step in the process after consideration and implementation has been given to prevention and minimization, as well as to render any remaining issues harmless.

For example, removal of an environmentally harmful solvent for a more benign solvent prevents the SHE issues associated with the use of this material. If that solvent cannot be substituted because of particular demands of the chemistry, then its use should be minimized, perhaps by increasing the concentration of the process. Finally, any impact of the use of that solvent should be rendered harmless by techniques such as appropriate abatement technology, to ensure, for example, that any emissions to the environment are adequately controlled.

Drug development is a highly regulated process. Synthetic route and process changes become more and more difficult to implement as clinical development progresses. With the constant drive to develop new medicines ever more quickly, the concept of SHE triggers maximizes the development of inherently sustainable processes. By triggering the relevant activities at the appropriate stages of development, the approach also ensures that SHE considerations remain off the critical path during development, again facilitating an efficient process to get new medicines to market.

During the early stages of the development process, the rate of attrition of novel medicines is very high, with many molecules being dropped from further development because of toxicity or lack of efficacy. Therefore, balancing the amount of effort invested early in the project versus the risk of attrition is always a challenge. The SHE Triggers model tries to achieve an optimum balance of highlighting the key issues early in the development process without a huge investment in SHE assessment, followed by more thorough assessment later in development, when the likelihood of success is higher.

In order to discuss the SHE Triggers model, it is first necessary to define the phases of development as used within the model. For the purposes of simple explanation of the model, the development process will be split into discrete periods. Clearly, the individual strategy used for a specific API will vary depending on the wider development program for the molecule.

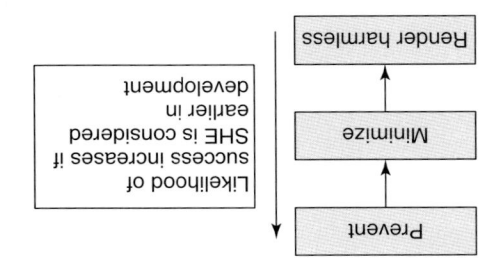

Figure 23.1 The prevent/minimize/render harmless hierarchy.

23
The Integration of Safety, Health, and Environmental Considerations into Process Development

Wesley White, Vyv Coombe, and Jonathan Moseley

23.1
Introduction

Development of a synthetic route for long-term production of an active pharmaceutical ingredient (API) requires the consideration of many factors to achieve the optimum solution. These factors include, amongst others, achieving delivery of a safe and consistent product to the patients; developing processes that can be manufactured safely while minimizing the environmental impact of the process; and achieving an acceptable total manufacturing cost which makes the product commercially viable. The development team needs to balance all of these factors, some of which are in conflict with each other.

This chapter describes one approach to ensuring that the safety, health, and environment (SHE) considerations of process design are fully integrated in the synthetic route development process. This approach is the AstraZeneca SHE Triggers Model. This model is included within the overall framework of Council Directive 96/61/EC, the European Union Environmental Legislation, as an example of "Best Available Techniques." The model is included as part of the Reference Document on Best Available Techniques for the Manufacture of Organic Fine Chemicals [1].

23.1.1
The SHE Triggers Model

To achieve the successful integration of SHE aspects into the development process, the synthetic design process must include SHE thinking throughout the process, from the day it enters development at the preclinical phase, through to the product launch phase; it cannot be achieved as an afterthought. The SHE Triggers model is one approach to achieving this.

The model simply defines the SHE activities that need to be considered during the various stages of development. The concept of triggers is used in that a trigger starts a course of action. It is built on the principle of the hierarchy of prevent, minimize, and render harmless which provides the foundation of much of the

Pharmaceutical Process Chemistry. Edited by Takayuki Shioiri, Kunisuke Izawa, and Toshiro Konoike
Copyright © 2011 WILEY-VCH Verlag GmbH & Co. KGaA, Weinheim
ISBN: 978-3-527-32650-1

References

9. Perry, R.H. (1997) in *Perry's Chemical Engineers Handbook*, 7th edn (eds R.H. Perry, D.W. Green, and J.O. Maloney), McGraw-Hill, p. 18.
10. Mullin, J.W. (2001) *Crystallization*, 4th edn, Butterworth-Heinemann, Oxford.
11. Ambler, C.M. (1978) in *Encyclopedia of Chemical Processing and Design*, vol. 7 (ed. J.J. Mc Ketta), Marcel Dekker, New York.
12. Lavanchy, A.C. and Keith, F.W. (1979) in *Encyclopedia of Chemical Technology*, vol. 4 (eds R.E. Kirk and D.F. Othmer), Wiley-Interscience, p. 710, Centrifugal separation.
13. Richardson, J.F., Harker, J.H. and Backhurst, J.R. (2002) *Coulson and Richardson's Chemical Engineering*, 5th edn, Butterworth-Heinemann, Oxford, vol. 2, p. 475.
14. Mujumdar, A.S. (2008) in *Guide to Industrial Drying: Principles, Equipment and New Developments* (ed. A.S. Mujumdar), Three S. Colors, India, p. 23.
15. Kudra, T., Mujumdar, A.S. and Mujumdar, A.S. (2007) in *Handbook of Industrial Drying*, 3rd edn, CRC Press, Taylor & Francis Group, Florida, p. 453.
16. Van't Land, C.M. (1991) *Industrial Drying Equipment: Selection and Application*, Marcel Dekker, New York.
17. Rajan, R. and Pandit, A.B. (2001) *Ultrasonics*, 39, 235.
18. Avvaru, B., Patil, M.N., Gogate, P.R. and Pandit, A.B. (2006) *Ultrasonics*, 44, 146.
19. Mujumdar, A.S. and Mujumdar, A.S. (2007) in *Handbook of Industrial Drying*, 3rd edn, CRC Press, Taylor & Francis Group, Florida.
20. Mujumdar, A.S. and Devahastin, S. (2000) in *Mujumdar's Practical Guide to Industrial Drying: Principles, Equipment and New Developments* (ed. S. Devahastin), Exergex Corporation, Montreal, p. 1.
21. Mullin, J.W. (1958) in *Chemical Engineering Practice*, vol. 6 (eds H.W. Cremer and T. Davies), Butterworths, p. 528.
22. Garside, J. (1985) *Chem. Eng. Sci.*, 40, 3.
23. Bennett, R.C., Fiedelman, H., and Randolph, A.D. (1973) *Chem. Eng. Prog.*, 69, 86.
24. Richardson, J.F., Harker, J.H. and Backhurst, J.R. (2002) *Coulson and Richardson's Chemical Engineering*, 5th edn, Butterworth-Heinemann, Oxford, vol. 2, p. 867.

operations under given circumstances. This would involve developing the model and validating it with operating plant data, and then it becomes very useful for quick optimization.

Once such process flow diagram is prepared on the basis of synthetic route developed in gram lab, and process optimization is achieved; then the DMF should be prepared on the basis of chemical engineering principles and is likely to be carried till full-scale plant without any hassle. Although kilo lab and pilot plant trials are essential for optimized final full-scale plant installations, making the chemist aware of how the things will happen on a full-scale plant and increasing the involvement of chemical engineers at synthesis lab will give correct estimated process flow steps, in the form of optimized SOPs and pre-DMF right at the synthetic step. Further, such optimized SOPs are easily scalable in kilo lab and pilot level studies.

22.7 Summary

To sustain the cut-throat competition, it is essential for a pharmaceutical enterprise that they maintain a priority claim for the molecule that is being launched. Thus, in an utter hurry, DMF is filed on the basis of information obtained from gram-lab data and some (rarely) pilot-scale data. Usually, such DMF is not very optimized, which later leads to either following the DMF being converted to an unoptimized full-scale plant or refiling the DMF.

In order to avoid being in such a situation, it is suggested that chemical engineering principles be incorporated for developing a DMF. Thus, a two-step method, involving developing process flow diagram and then optimizing the process, is suggested to obtain a DMF satisfying the date as well as the final configuration of the optimized plant. This can be done by making the chemist aware of the various unit operations in a full-scale plant. A chemist with the knowledge of such unit operations can make a better choice of synthetic routes. This will also avoid complications during subsequent scale-up. Further process optimization can be carried out on the basis of mathematical modeling and other data from the pilot scale.

References

1. Joshi, J.B. and Sharma, M.M. (1977) *Can. J. Chem. Eng.*, **55**, 683.
2. Pandit, A.B. and Joshi, J.B. (1983) *Chem. Eng. Sci.*, **38**, 1189.
3. Joshi, J.B., Pandit, A.B. and Sharma, M.M. (1982) *Chem. Eng. Sci.*, **37**, 813.
4. Raisimba, B., Biscans, B., Delmas, H., and Jenck, J. (1997) *Kona*, **17**, 38.
5. Thompson, L.H. and Doraiswamy, L.K. (2000) *Chem. Eng. Sci.*, **55**, 3085.
6. Kim, S., Wei, C., and Kiang, S. (2003) *Org. Process Res. Dev.*, **7**, 997.
7. Li, H., Wang, J., Bao, Y., Guo, Z., and Zhang, M. (2003) *J. Cryst. Growth*, **247**, 192.
8. Richardson, J.F., Harker, J.H. and Backhurst, J.R. (2002) *Coulson and Richardson's Chemical Engineering*, 5th edn. vol. 2, Butterworth-Heinemann, Oxford, p. 372.

22.6 Optimization and Intensification of Unit Operations

intensification of the selected unit operations. This can be carried out in following steps:

22.6.1
Step I: Establishing Material and Energy Balance across All the Equipments

This involves taking balance over each unit operation. Performing this will let us estimate the rates of mass and energy flow over the whole process. Once flow rates of each line and section are obtained, the equipments, which can handle the given load, need to be chosen. For example, if the rate of production of slurry from the crystallizer is seen to be too high, then appropriate filtration unit and subsequent dryer should be designed. This will give the time cycle and capacities of all the equipments, and the constraints on the overall process.

22.6.2
Step II: Preliminary Evaluation of All the Major Equipments

After complete process flow is prepared with the production loads on various units and equipments, a detailed evaluation of performance (efficiency calculations) of major equipments should be done. This would involve cross-checking of the given load and designed capacity. This can be done by comparing the performance of the same equipment under conditions similar to those obtained from the previous data. In case, if no experimental data are available, evaluation can be based on mathematical simulations of the equipment.

22.6.3
Step III: Analysis of All the Operations and Equipments at Relatively Intricate Levels

About 40–50 experiments should be performed in the laboratory on 200 ml to 1 l scale with an objective to determine the rate-controlling step with respect to operating conditions such as temperature, pressure, power consumption, catalyst, and so on, and also to determine the thermodynamic data and physical properties. For example, in the case of catalytic hydrogenation, overall operation may be controlled by mass transfer or heat transfer of catalyst suspension depending on the operating conditions and substrate to be hydrogenated. Similar analysis needs to be done for the cost of various unit operations per unit of the final product obtained.

22.6.4
Step IV: Mathematical Modeling of Individual Equipment and Subsequently the Entire Plant

Rates of transport processes are important for the estimation of the performance of various equipments, but experimental determination is most difficult due to the complexity of fluid mechanics. Thus, there is a need for mathematical modeling and simulation to predict the performance of the equipment used for specific

to be taken to preserve its shape and structure and a tray dryer should be used in such instances, and if the production rate is high, then a belt conveyor dryer is an economical option. For heat-sensitive materials, either spray dryers or flash dryers are a good option. It is even possible to carry out the drying operation partially in one dryer, and the remaining moisture can be removed in another dryer. However, it should be noted that drying being a phase-change phenomenon is very much energy intensive and should be chosen only when mechanical dewatering becomes uneconomical. For further details on the types of dryers and selection, the readers are referred to Mujumdar [19], Mujumdar and Devahastin [20].

22.5
Scale-up Problems

Design of any unit operation is based on kinetics of that process measured on laboratory- or pilot-scale units, and sometimes on both in critical cases. The main challenge in scaling up of equipments is in understanding the effect of changing hydrodynamics from lab scale to plant scale on various transformations involved in the unit operation. For example, crystallization consists of steps like nucleation and crystal growth. So, it is essential to characterize the particle–fluid hydrodynamics at various scales, that is, kilo-lab scale, pilot scale, and full scale and assess its effects on the kinetics of nucleation and crystal growth. In fluidized-bed crystallizers, for example, the crystal suspension velocity must be evaluated as a parameter that is related to crystal size, size distribution, and shape, as well as to bed voidage and other system properties such as density differences between the particles and the liquid, and viscosity of the solution.

In agitated vessels, the "just-suspended" agitator speed, that is, the minimum rotational speed necessary to keep all crystals/solids in suspension, must be determined, since not only do all the crystals/solids have to be kept in suspension, but also the development of "dead spaces" in the vessel must be avoided, and at the same time crystals/solid breakage needs to be avoided because of very high agitation speeds. Fluid and crystal/solid properties, together with vessel and agitator geometries, are important in establishing these values [21, 22]. Agitated vessel crystallizers are often successfully scaled up on the crude basis of either constant power input per unit volume or constant agitator tip speed, although Bennett et al. [23] have suggested that, in draft tube agitated vessels, the quantity (tip speed)^2/(vessel volume/volumetric circulation rate) should be kept constant [24]. For scaling up of a dryer, drying kinetics of the solids should be obtained beforehand on lab scale and kilo-lab scale, and then it should be integrated with the hydrodynamics prevailing in the full-scale dryer.

22.6
Optimization and Intensification of Unit Operations

After developing a process for a full-scale operation on the basis of the synthetic route developed on lab scale, one needs to carry out optimization and process

which dissolves easily in water and organic solvents and shows better bioavailability upon pharmaceutical formulation. The chief advantage of spray dryers is very short residence time, which permits the drying of heat-sensitive materials.

Rotary Filter Dryer *Rotary filter dryer* is a combination of filter and dryer, and is most commonly used in the pharmaceutical industry for its advantages like automatic operation, minimal dismantling, closed operation to avoid contamination, and minimal solvent loss. In rotary filter dryer, it is very easy to manipulate the filtration time and drying time, and hence precise control over the quality of dried material can be obtained. It is described in detail earlier in Section 22.4.3.2.

Double Cone Dryer *Double cone dryer* (Figure 22.12) is a good option for free-flowing solids. These dryers can be operated under vacuum, with indirect heating for drying solids at low temperature in an air-free environment to avoid oxidation and for solvent recovery. The conical shape facilitates faster loading and unloading of dryer even in the stationary position. The vapor outlet in these dryers is stationary and passes through the hollow central shaft. The outlet is protected by the dust filter.

22.4.5.3 Selection Criteria

Dryer selection is mainly done on the basis of the type of feed, mode of heating, mode of operation, production rate needed, and the final form of the desired product. Selection of dryers based on feed type is proposed by Mujumdar [14]. Tray dryers can handle almost any type of feed except liquids and slurries, but at low production rates, while liquid feeds can be handled best in drum dryers, spray dryers, and fluidized-bed dryers. Drum dryers and spray dryers can be scaled up for any production rates. On the other hand, drum dryers are not very suitable options for hot feed or with liquids having high vapor pressure, which can be handled easily in spray dryers. When a dried product is fragile and crystalline, special care needs

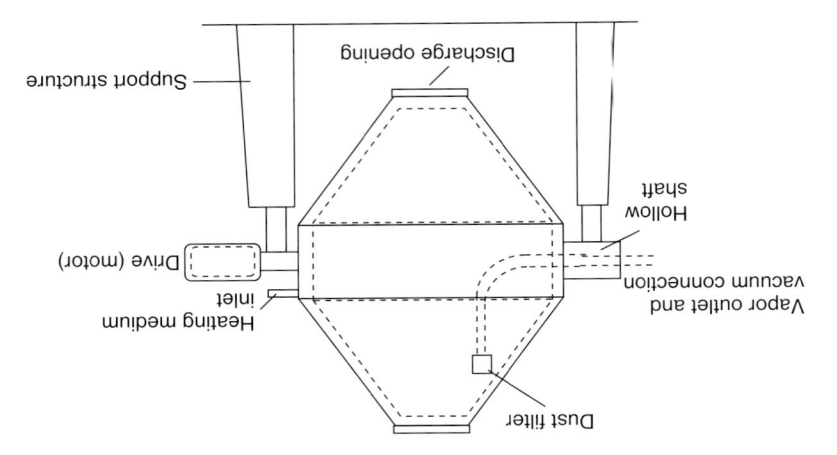

Figure 22.12 Double cone dryer.

462 22 Matching Chemistry with Chemical Engineering for Pharmaceutical Processing

etc.) to the wet solid by conduction mostly through the metal wall. Since no gas is present on the wet solid side, some driving force like vacuum or gentle gas flow is needed to remove the evaporated moisture. Heat transfer surfaces may range in temperature from −40 °C (as in freeze drying) to about 300 °C. Vacuum operation can allow the recovery of solvents by direct condensation. Dust recovery is obviously simpler due to low volumes of the carrier gas so that such dryers are especially suited for drying of toxic, dusty products, which must not be entrained in gases. Heat may also be supplied by radiation (using electric or natural gas-fired radiators) or volumetrically by placing the wet solid in dielectric fields in the microwave or radio frequency range. Radiant dryers have found important applications, especially when product contamination is to be avoided and yet higher heat transfer rate is to be maintained. Here, we discuss some dryers that are typically used in the pharmaceutical industry.

Spray Dryer In a spray dryer, slurry is sprayed in the form of fine drops in a hot gas stream (Figure 22.11). The liquid rapidly vaporizes from the drop, leaving behind the solid mass. The flow of gas may be cocurrent or countercurrent with respect to the solids. The size of the particles and hence the drying characteristics are mainly determined by the atomizer used in the spray dryer. Pressure nozzles, twin fluid atomizer, spinning disk, and spinning cup atomizer are frequently used for atomization in spray dryers. Ultrasonic atomization has added advantages like uniform drop sizes (hence uniform drying), low nozzle exit velocity (hence reduced dryer volume), ability to handle slurries with high solid content, ability to handle fluctuating flow rate of slurry and yet produce consistent drop size, and no need for excess pressure energy [17, 18]. Obtained solid is usually amorphous, and a spray dryer is intentionally employed to convert a crystalline solid to an amorphous solid.

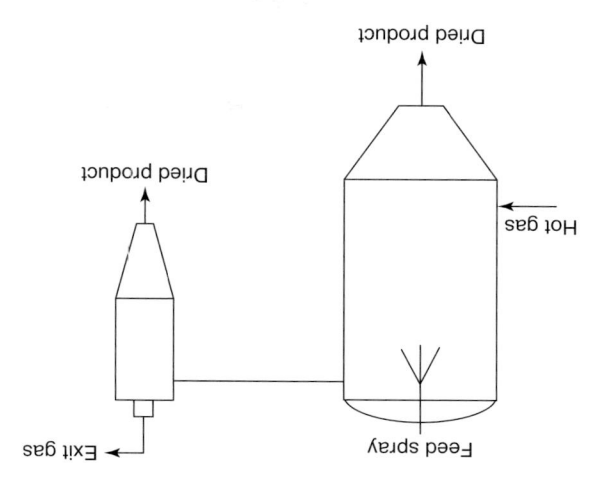

Figure 22.11 Spray dryer.

22.4.5.2 Types of Equipments

There are various ways to classify the dryers available for various applications [14–16]. Table 22.1 gives some of the common criteria used to classify drying equipments.

In *direct contact dryer*, the material to be dried comes in direct contact with the drying medium. These are also known as *convective dryers*. Air is the most common drying medium for manufacturing API. The drying medium not only supplies heat to the material to be dried directly but also carries away the evaporated vapors. Drying gas temperatures may range from 50 to over 100 °C depending on the stability of the material to be dried. Dehumidified gas may be needed when drying highly heat-sensitive or hygroscopic materials to carry the drying at lower temperature. An inert gas such as nitrogen may be used to avoid oxidation or contamination of API, or when an organic solvent is to be removed for environmental concerns and for recycling. In *indirect dryer*, heat is supplied to the drying material without direct contact with the heat transfer medium, that is, heat is transferred from the heat transfer medium (steam, hot gas, thermal fluids,

Table 22.1 Criteria for classification of dryers and dryer types.

Criterion	Types
Mode of operation	Batch
	Continuous
Heat input type	Convection, conduction, radiation, electromagnetic fields, combination of heat transfer modes
	Intermittent or continuous
	Adiabatic or nonadiabatic
State of material in the dryer	Stationary
	Moving, agitated, dispersed
Operating pressure	Vacuum
	Atmospheric
Drying medium	Air
	Superheated steam
	Flue gases
Drying temperature	Below boiling point
	Above boiling point
	Below freezing point
Relative motion between drying medium and solids	Cocurrent
	Countercurrent
	Mixed flow
Number of stages	Single
	Multiple
Residence time	Short (<1 min)
	Medium (1–60 min)
	Long (>60 min)
Contact type	Direct contact
	Indirect contact

major modifications. The feed cone is provided to smoothly take the slurry to the perforated walls to retain solids. The liquid discharges through the filter, and a cake is formed on the walls. The reciprocating plate then pushes the cake by few inches toward the lip of the basket and in return stroke it opens up a space on the wall for the new feed to enter. By the time the cake reaches the other end, it is drained of the liquid and it then falls in a solid collector or a chute. Such reciprocating pusher arrangement does not break the solids/crystals.

22.4.4.3 Selection Criteria

Depending on the load (output rate) and capacity of the centrifuge, it can be either a batch centrifuge, a semicontinuous centrifuge, or a continuous (both phases) centrifuge. In batch operation, the slurry is loaded to the centrifuge and the perforated basket is rotated till the required quantity of liquid is removed from the retained solids. Batch centrifuges are very uncommon and are used to handle very small volumes of slurry. Pusher-type centrifuge is selected when continuous operation is needed in both phases. If solids are to be removed from slurry, then the centrifuge is essentially perforated type, whereas for liquid–liquid separation (decantation) the basket should be non-perforated. Valve-discharge-type centrifuge can be used for separating solids and two liquids in a single stage. For cases when the separation is difficult (very low difference in density of phases), a disk-type centrifuge can be used. For further details on the types and selection of centrifuges, readers are directed to Mullin [10]; Ambler [11]; Lavanchy and Keith [12]; and Richardson et al. [13].

22.4.5 Drying

22.4.5.1 Utility of Drying

Drying is usually the last operation in the manufacturing process where the solvent is to be removed from the solids. Drying is mainly carried out as a means of preservation/storage, to reduce the cost of transportation, and to achieve the desired product quality. Drying competes with distillation as the most energy-intensive unit operation. Mechanical dewatering (filtration, sedimentation, centrifugation, etc.) is much cheaper (up to 100 times cheaper than drying). Thus, drying is usually preceded by some mechanical dewatering methods to reduce the energy consumption in the drying step. The major cost for dryer is in its operation rather than its initial investment cost, and hence it is the major factor in selection of a dryer. Drying is a complex heat and mass transfer operation, which most of the times results in changes in the product quality (morphology). Physical changes that may occur include shrinkage, puffing, balling, crystallization or glass transition, and so on. In some cases, undesirable chemical or biochemical reactions may occur, leading to the change in color, texture, odor, or other properties of the solid product.

22.4 Unit Operations

centrifuge, which controls the thickness of the cake. The excess solids or crystals are taken out of the centrifuge, while the filtrate is continuously removed from the outer cylinder. There can be considerable breakage or degradation of the crystals by the unloader knife.

Continuous Centrifuge Continuous centrifuge is useful for separating the coarse particle/crystals without the need to stop the centrifuge for discharging the solids (Figure 22.10). It is much similar to a basket type of centrifuge except for two

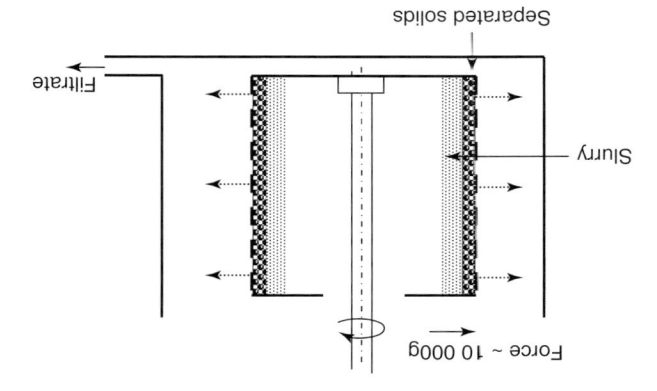

Figure 22.9 Top suspended centrifuge (semi-continuous).

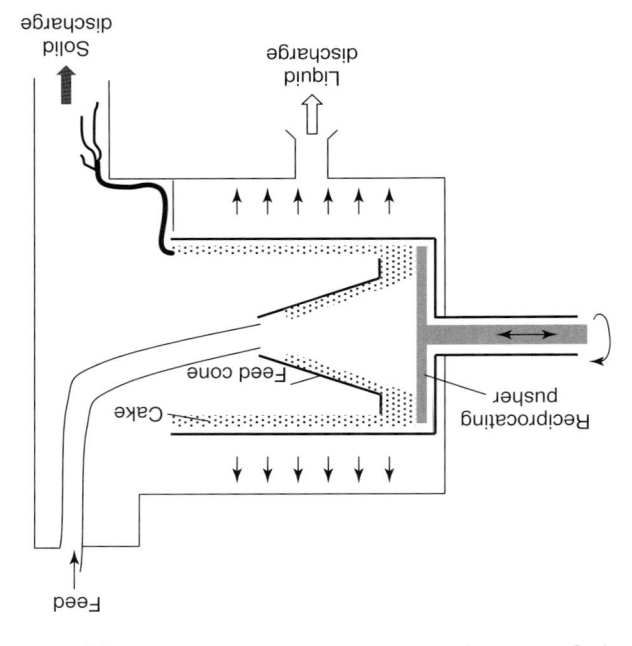

Figure 22.10 Continuous reciprocating centrifuge.

22.4.4 Centrifugation

22.4.4.1 Utility of Centrifugation

When the phases are to be separated on the basis of density or size, then some kind of force needs to be applied. Gravitational force is usually used for liquid–liquid separation (decantation) and solid–liquid separation without filter media (settling) by leaving the suspension to stand still for some time, which separates the phases. However, this is a time-consuming process, and a stronger force like centrifugal force can be applied for settling, decantation, and filtration operations. In both the cases, it replaces the weak force of gravity, resulting in rapid separation and solid cake containing less liquid. This force causes the solids to settle through a layer of liquid. Centrifugation is a sort of pressure filtration, discussed earlier, where pressure energy is replaced with centrifugal force to separate solids by the flow of the filtrate through a bed of porous solids held inside a perforated rotating container. A centrifuge consists of a rotating basket with perforated sidewalls. The slurry to be separated is fed to the rotating basket. Rotary motion of the basket produces centrifugal force, which pushes the liquid out and solids are retained on the filter medium. If centrifugation is used for liquid–liquid separation, then the sidewall of the basket is not perforated. Instead the basket holds all the liquid that is fed. The heavy liquid moves preferentially toward the wall and the liquid mixture separates into two layers, which can then be removed individually using adjustable weir. When the separated material must be dried by thermal means, considerable savings may result by the use of centrifuge. Centrifuges can also be used for separating liquids, breaking or concentrating emulsions.

22.4.4.2 Types of Equipments

The main types of centrifugal equipments are tubular bowl, disk with nozzle, disk with intermittent discharge, basket-type centrifuge (perforated basket centrifuge, Non-perforated basket centrifuge), continuous centrifuge, decanter centrifuge, and so on. Most commonly used suspended basket centrifuge is described here.

Suspended Centrifuge A common type of batch centrifuge in industrial processing is the top suspended centrifuge. The basket is held at the bottom with free end swinging shaft (Figure 22.9). The basket spins at 600–1800 rpm. The feed is spread near the sidewall. Liquid drains through the filter medium into the casing and leaves out through a discharge pipe. The solid cake of varying thickness can be obtained inside the basket by varying the feeding time. After obtaining the required thickness of the cake, feeding is stopped and the cake is washed and dried by spinning the basket further. Top suspended centrifuges are extensively used for separating crystals in short cycles of 2–3 min per load. Another type of batch centrifuge is driven from the bottom with the drive motor, basket, and all casing suspended from vertical legs mounted on the base plate. In bottom driven centrifuges, an additional advantage of necessary vibrational movement is obtained. In a semi-continuous centrifuge, a knife is held against the sidewall of the

the main drum, there is a smaller drum with solid curved face. The annular space between the drums is divided into compartments by radial partitions. In each compartment, an arrangement is provided to operate it individually under vacuum. As the compartment dips into the filtering solution, vacuum is applied to it. The filtrate flows through the canvas and enters the compartment, and then through the internal tubes it enters the collection tanks. As the compartment leaves the filtration section, vacuum is applied through the separate section, sucking wash liquid and air through the cake. The wash liquid is collected in a separate tank. After sufficient wash liquid is sucked through the cake, the air makes the cake dry. Further, the washed and dried cake is removed from the drum using doctor's knife. A little air is blown under the drum, which blows the cloth on the drum and cracks the cake. After the cake is dislodged, it is again taken to the filtration zone to repeat the cycle. The operation of each compartment is batchwise, but since each operation is carried out by one or the other compartment, the overall process is continuous.

Operation of the vacuum filter becomes difficult when the solids are very fine or the vapor pressure of the liquid is very high. In such cases the design of continuous rotary filter is adapted to operate under positive pressure up to 15 atm. For this, the entire dryer is enclosed in the pressurized chamber. High pressure drives the liquid through the filter medium into the annular sections inside the drum. However, the mechanical problems of discharging the solids from filters, their high cost, complex design, and their small size limit their application to special problems. Rotary vacuum filter is essentially a combination of filter and dryer. This filter-cum-dryer is commonly used, because it has a number of merits, such as containment of API and solvent vapor, and is an operator- and environment-friendly apparatus. Filter dryer is a combination of filter and dryer, and it is the best and a favorite apparatus due to its ability to prevent contamination with foreign particles.

22.4.3.3 Selection Criteria for Filtration

Filtration equipment should be chosen on the basis of the physicochemical properties (density, viscosity, and vapor pressure) of the filtrate. These parameters determine the pressure that needs to the applied to obtain the required rate of filtration. High-vapor-pressure liquid tends to vaporize and boil under vacuum operation, and consequently pressurized filtration should be chosen for these. Nature of solids, for example, size, shape, size distribution, and packing characteristics, is also equally important, similar to that of liquid properties, as it also determines the characteristics of the cake that will be formed. For coarse solids, usually gravity filters are used, but for the fine solids or for solids with wide size range that form a compact cake, pressurized filters need to be used. Since filtration is inherently a batch operation, the parameters like solid loading in the feed and necessity and quantity of wash liquid also decide the cycle time. When the cycle time is too small, it is better to go for filters that need nil or minimum dismantling, like the rotary drum filter. Other important parameters include operating temperature and the required filtration rate. More information on the types of filtration equipments and selection can be found elsewhere [8, 9].

23.1.2
Introduction to AZD4619

To illustrate the SHE Triggers model in the rest of this chapter, the real example of AZD4619 will be used. Development of AZD4619 incorporated the SHE Triggers approach and provides a good exemplification of how the consideration of SHE issues can aid the process development team in their decision-making process.

AZD4619 was a selective Peroxisome Proliferater-activated Recopter (PPAR)α agonist in development as a potential monotherapy in patients with combined dyslipidemias; the compound is no longer in development. It is a propionic acid derivative and a single enantiomer (Figure 23.3). Peak-year demand for the API was expected to reach hundreds of tons. The predicted manufacturing volume is a key aspect of the SHE thinking that should be applied during process development. Clearly, a process that is expected to deliver hundreds of tons of material has a larger potential SHE impact than a process which will deliver only tens of kilograms at peak demand, such as many specialist oncology compounds.

The first delivery of the molecule was a few kilograms to provide material for preclinical studies. Scheme 23.1 shows the synthetic route used during the mid phase of development.

Clearly, this chapter cannot focus on the whole of the development of this route. However, examination of the synthetic scheme quickly identifies areas for potential improvement from a SHE perspective.

From an environmental perspective, the fact that the desired product is a single enantiomer is of immediate interest. The initial manufacture used chiral chromatography to separate the enantiomers with disposal of the undesired enantiomer. While this is a highly effective way to deliver small quantities of material, for the long-term, a more resource-efficient option would be desirable. A number of synthetic routes can be found in the literature when searching for stereoselective synthesis of suitable intermediates toward AZD4619. The prospect of developing a stereoselective method for AZD4619 was very much dependent on the lability of the intermediates to racemization. It is well known that carboxylic esters with a sulfide in the α-position can be epimerized under basic to neutral conditions. An alternative approach would be to use an enzyme-catalyzed kinetic dynamic resolution.

From a process safety perspective, the route uses a diazotization to form the *rac*-chloro acid intermediate. Diazotization reactions are potentially hazardous because some diazonium salts, when dry, are capable of decomposing explosively and may be friction-, shock-, and heat-sensitive [2]. Clearly, this is something that would need careful examination during process development.

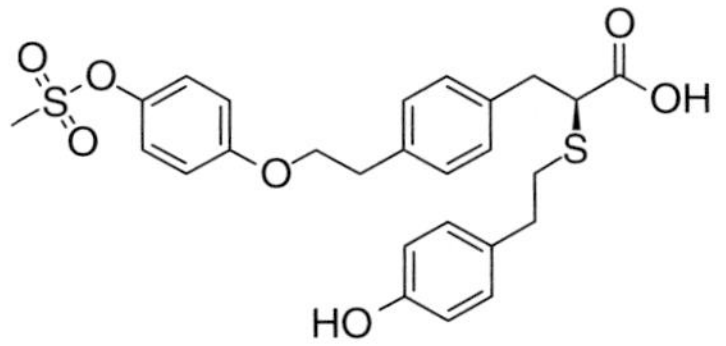

Figure 23.3 AZD4619.

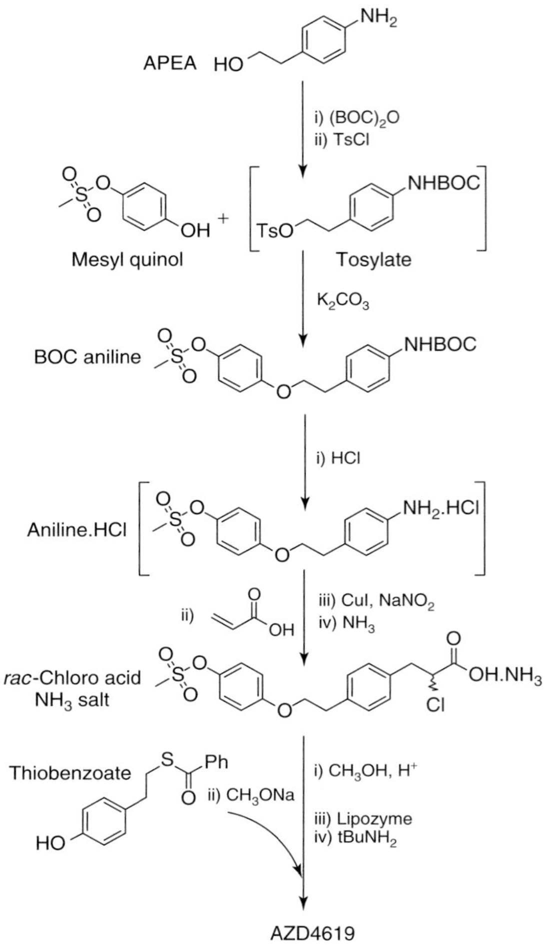

Scheme 23.1 Synthetic scheme for AZD4619.

From a health perspective, little would be known about the intermediates. But an immediate concern would be isolation of the tosylate intermediate. Tosyl groups are well known for their potential genotoxicity and therefore, isolation of this intermediate during long-term manufacture would be best avoided to reduce the likelihood of exposure of the process operators during product manufacture.

23.2
Process Safety

The major process safety hazards associated with the manufacture of pharmaceutical products include explosion, fire, runaway reaction, and uncontrolled gas evolution. These can be divided into what are generally referred to as *chemical and operational hazards*. Chemical hazards can be described as the hazards that arise from uncontrolled chemical reactions with operational hazards arising from the flammability of the individual materials.

The safety triggers ensure that, before a process is operated, the process safety hazards associated with it have been identified and their potential impact fully considered. The aim is to obtain sufficient data in order that the risk be assessed adequately. The depth of a chemical or operational hazard assessment should reflect the complexity of the reaction, the size of the risks involved and the scale of operation.

23.2.1
Assessment of Chemical Reaction Hazards

Chemical reaction hazards arise from one of the following three areas:

- thermal stability,
- heat of reaction,
- gas evolution.

It is important that information relating to the above is sufficient to allow safe operation of the process.

23.2.2
Assessment of Operational Hazards

Operational hazards arise due to the flammable nature of the materials used or produced in the process under the proposed operating conditions. The key question that needs to be answered is, are the process materials potentially flammable under the proposed operating conditions?

23.2.3
Basis of Safety

The assessments carried out as part of the SHE Triggers model allow the chemical and operational hazards associated with the process to be determined. With this information, it is possible to define an appropriate Basis of Safety. The Basis of Safety is the actual principle by which the process can be operated safely. The precautions needed to achieve it are also given.

Essentially there are two types of bases of safety, that is, preventive and protective. Preventive bases of safety operate by ensuring that the hazard does not occur. With operational hazards, these are avoidance of potential ignition sources and avoidance of flammable atmospheres. With chemical hazards, the preventive basis of safety is termed *process control*; that is, the process is closely controlled so that a hazard cannot occur. The protective bases of safety operate on the principle that it is not possible to ensure that the hazard will never happen and so in the event that it does, the protective measures employed will ensure that the consequences are acceptable. With operational hazards, the protective measures are explosion suppression, explosion venting, and explosion containment. With chemical hazards the protective measures are reaction inhibition, emergency venting, containment, and drown out.

23.2.4
Process Safety Triggers

23.2.4.1 Early Delivery
During initial laboratory work, experiments are generally carried out on less than 2 l scale. At this stage, safety is based on operation within a fume cupboard, the relatively small scale of operation, and the absence of any explosives. This last point is very important because the decomposition of even a fairly small quantity of a detonating explosive could negate the safety effects of the fume cupboard. Ensuring that standard laboratory working procedures are implemented protects against operational hazards at this scale.

At this stage the development team consult the process safety experts to ensure safety considerations are integrated into the early process development.

Requests to the process safety experts usually relate to a specific aspect of the process, that is, they are not full assessments. The work is often to quantify potentially serious issues within a process that may, if not identified at an early stage, result in significant wasted effort. Examples include the assessment of the thermal stabilities of individual compounds or reaction mixtures. This is often associated with "all-in" processes, that is, where all the materials are charged to the reactor at room temperatures and the mixture is then heated up followed by determination of whether a material is a potential explosive or not.

Operational hazards are addressed principally in a generic operational hazard assessment. This assessment addresses issues such as the identification of ignition sources and how to control them. The assessment also looks at the standard operating procedures carried out in the facility, for example charging and discharging from vessels, and gives precautions that enable these procedures to be carried out safely. It is important to realize that the above generic operational hazard assessment will not cover all possible scenarios. Consequently, before a process is operated within the facility, an assessment is made by the development team to determine whether an additional more specific/detailed assessment is required.

23.2.4.2 Later Process Safety Triggers
As the development team enters the route evaluation phase, the discussions continue with the process safety experts, following a similar approach to above. It is important that the process safety experts remain an extension of the development team. In this way, the development can be influenced to ensure that process safety thinking is inherent throughout the process. A major change during this stage of development, associated with larger-scale manufactures, is needed. Typically, these later-stage manufacturing campaigns involve reactors of 400 l and above. At

this scale, a full assessment of the chemical reaction hazards associated with the manufacture is carried out. The assessment will look at the thermal stability of both, individual materials and mixtures. The heats of reaction for the chemical transformations will be determined, as will the quantity and rate of any gas evolved. This will involve a number of experimental techniques including small-scale screening tests, isothermal calorimetry, and adiabatic calorimetry.

Operational hazards are again addressed initially with a generic assessment but this is often supplemented by an additional, more specific report if the process operates outside the generic basis of safety for the manufacturing plant.

23.2.4.3 Technology Transfer

At the point where the process is transferred to the strategic long-term site of manufacture, the chemical and hazard assessments completed throughout development will be thoroughly reviewed by the development team and the process safety experts. Any assessments of relevance will be transferred to the long-term supply site.

23.2.5
Application of Process Safety to AZD4619

At the end of the synthetic route evaluation stage and prior to the first pilot plant manufacture, a full chemical hazard assessment was completed for all the stages of the process. A brief summary of the results of the chemical hazard assessment for AZD4619 *rac*-chloro acid stage is summarized here. Note: process outline below is for an earlier process than shown in Scheme 23.1.

Process outline:

1) Aniline salt, copper (I) iodide, and acetone are charged to the reactor.
2) Water, acrylic acid, and hydrochloric acid are then charged.
3) Batch is cooled to 3 °C.
4) Aqueous solution of sodium nitrite is added evenly over 10 h, maintaining batch temperature at 3 °C.
5) It is stirred for a further 3 h at 3 °C.
6) It is warmed to 22 °C over 10 h.
7) An aqueous solution of urea is added to destroy excess nitrous acid.
8) The batch is worked up.

23.2.5.1 General Hazards of Aqueous Diazotization and the Basis of Safety

- Diazonium salt solutions/slurries gas slowly even at low temperatures. Therefore, transfer lines should not be sealed with diazonium salt solution contained in them, nor should pressure buildup between a pump and a closed valve be possible.
- Some diazonium salts when dry are capable of decomposing explosively and may be friction-, shock-, and heat sensitive [2]. Therefore, the diazonium salt solution

should not be allowed to dry out. After transfer of the diazonium salt solution, the reactor, transfer lines, and so on should be washed thoroughly.
- The potential for generation of nitrous/NOx fume exists in diazotization reactions. Within the chemical industry there is a history of incidents, for example, fires in vents, which have been attributed to amine/NOx interaction. The inadvertent contact between NOx fume and dry amines should be avoided.
- Sodium nitrite can form explosive mixtures with amines, sulfamates, and ammonium salts. Inadvertent mixing of sodium nitrite with other chemicals should therefore be avoided.

23.2.5.2 Additions of Acetone, Water, Hydrochloric Acid, and Acrylic Acid

The heat and gas generation associated with the additions of acetone, water, hydrochloric acid, and acrylic acid were determined using an isothermal reaction calorimeter fitted with gas measurement capability. Only the addition of hydrochloric acid generated heat and this was not sufficient to generate any hazard. None of the additions generated any gas.

Thermal stability testing, using differential scanning calorimetry of the reaction mixtures after each addition confirmed that there would be no hazard at the proposed operating temperatures.

23.2.5.3 Addition of Sodium Nitrite

- **Heat of reaction**: The addition of sodium nitrite was shown to be exothermic. Uncontrolled addition of the sodium nitrite would result in the batch temperature being raised to a level at which the diazotization reaction mixture would decompose, generating additional heat and gas. It was however shown that the rate of heat generation for a 10 h addition period could be safely dissipated by the reactor cooling system. In the event of cooling failure, the addition would be stopped.
- **Gas evolution**: Experimental work showed that gas was generated during the addition of sodium nitrite solution, a total volume and maximum rate being measured. It was ensured that the vent/scrubbing system was sufficient to handle these levels of gas generation.
- **Thermal stability**: Screening and adiabatic thermal stability tests were carried out on the reaction mixture. It was shown that if no coupling took place, that is batch temperature was too low and then cooling was lost, then the diazonium salt would self-heat. It was however shown that the decomposition reaction would not generate enough heat to result in a hazard, provided the increased rate of gas generation could be safely handled.

Additional testing was carried out on the reaction mixture generated in the absence of copper iodide. This maloperation was evaluated because it would result in a mixture where no coupling had taken place and consequently all of the diazonium salt would be available to decompose. In this case, it was shown that subsequent decomposition would generate enough heat to present a hazard. It was therefore stressed that adherence to the process was necessary.

In completing the chemical hazard assessment detailed above, the design team were able to ensure that the process could be operated safely.

The overall conclusion of the process safety assessment above confirmed that the process could be operated safely, providing the necessary precautions were taken.

23.3 Health

23.3.1 Introduction

When developing a long-term process for the manufacture of an API, the process design team must consider the impact of the materials used within the synthesis on the health of the process operators who will run the process. The demand for API may mean that the process being used will be run almost continuously, isolating many batches of each stage every year. The size of these batches could be many hundreds of kilos of isolated weight, giving significant potential for exposure. Therefore, using processes that avoid highly hazardous materials is preferable but not always practical. Many of the intermediates used in the synthetic route are, by design, highly reactive and this may result in some form of toxicity. If intermediates used are identified as particularly hazardous to health, it may be possible to minimize the potential for exposure by avoiding isolation of dusty solids and telescoping particular stages. Finally, once the options for prevention and minimization have been explored, the risk of exposure should be "rendered harmless" by the use of suitable containment technology or other control options. It should be recognized that containment technology is costly and can make the charging and final isolation of materials to and from the process very difficult. Therefore, the development team can make a significant long-term impact by reducing the use of highly hazardous materials.

This section focuses on the health triggers of the long-term process and not the health risk assessment used within the laboratory to ensure the protection of the development team. While this is very important and has some overlap with long-term health considerations, it is not covered in this chapter.

23.3.2 Health Triggers

23.3.2.1 Early Delivery

During this phase of development, the focus is on speed of delivery. The manufacture of the early phases will take place in a large-scale laboratory using vessels in the range of 20–100 l. In AstraZeneca, these vessels are located within fully draughted fume cupboards. Therefore, the principal method of risk management is minimization of exposure. In the absence of substance-specific data a precautionary approach is used assuming all custom raw materials, intermediates, and APIs in the synthesis present a high hazard with respect to health effects. Other materials used in the synthesis, such as reagents and solvents, will use a combination of internal expert assessment, suppliers' information, and publically available data to assess the health hazard for each material.

Even though the control is based on minimization of exposure, all the custom raw materials and intermediates will be assessed using *in silico* assessment for two key endpoints; genotoxicity and skin sensitization. This *in silico* assessment is based on structure–activity relationships (SARs). SARs attempt to predict a molecule's biological activity based on its chemical structure. The costs and time associated with SAR evaluation are relatively low. In the area of health risk assessment for the workplace, skin sensitization and genotoxity are two very important endpoints of interest. This, combined with a reasonable level of prediction for these particular endpoints, make *in silico* assessment an important part of AstraZeneca's SHE Triggers approach. It should be recognized that SAR assessment is not a definitive assessment of a molecule's hazard potential and more robust studies will often be used to confirm the results at later stages of development. At this stage, however, the use of SAR data effectively balances the need for key information with the high attrition rate in early development.

The results from the *in silico* assessment are used by the development team in the assessment of occupational health risks during early development. The basic assumption is that all materials without any significant toxicity information present a high potential hazard. Therefore, the likelihood of any modifications of containment or control measures as a result of the SAR assessment is low.

23.3.2.2 Synthetic Route Evaluation

During this stage of development, the results of the SAR evaluation are extremely useful to the process development team. Having obtained the information in the step above about potential health hazards of key raw materials and intermediates in the route, the development team is then able to consider these factors in alternate routes. If required, further *in silico* assessment can be completed on any alternate synthetic routes under consideration. This allows comparison of the routes from a health risk perspective. For example, if one route avoids the use of potentially genotoxic intermediates, then this may be an important factor in the route selection criteria leading to route freeze.

23.3.2.3 Process Design

Following route freeze in the stage above, the focus of the development team shifts from the prevent phase to minimize phase. The scale of manufacture also continues to increase during this stage. Until now, the data generation has been based on the use of SAR data. During process design, further information is obtained to aid the health risk assessment.

SAR genotoxicity data are evaluated by genetic toxicology experts and a decision is made as to whether further testing is required. If the decision is made to conduct further testing, an *in vitro* test such as the Ames test will be used. The Ames test is a biological assay to assess the mutagenic potential of a chemical compound. The result of an Ames test will supersede a SAR evaluation since the predictivity of the Ames test is considered to be higher. Other genotoxicity tests may sometime be selected based on the advice from the genetic toxicology experts.

During this stage of development, additional information may be gained for the custom intermediates used in the process. A set of four additional tests is commonly used for isolated intermediates, for which the probability of exposure is higher. The effects tested are short-term general toxicity, skin and eye corrosivity, and skin sensitization. This test package is intended to give information about some important toxicological effects of concern in the workplace. It is also in line with the tests required by some regulatory agencies for low-volume chemicals. Intermediates at later stages in the process route may have properties similar to the final API. For these substances, useful hazard information can be obtained through read-across from the API, for which a significant quantity of data will have been generated in support of the clinical program. Suitable assays can also be used to assess if there is any pharmacological effect. These test results improve the understanding of the specific hazards related to each material and will be used for transport and user classification/labeling, and to refine the risk assessment and control measures. At all times during the health assessment, alternatives to the use of animal testing will be utilized wherever possible and the principle of replacement, reduction, and refinement of animal testing will be applied by AstraZeneca.

Following the results of the above testing, the design team can employ various approaches to the minimization of any health hazards remaining within the synthetic route. If, for example, the selected route still involves the use of genotoxic intermediates, the risk could be minimized by avoiding isolating the intermediate of concern. Alternatively, the form in which the intermediate is isolated can be modified to reduce the exposure potential. For example, the material could be isolated as a wet paste, as opposed to a dry dusty solid.

23.3.2.4 Process Optimization and Understanding

During this later stage of development, the focus will be on ensuring that the manufacturing asset will be able to adequately control the toxicological risks remaining within the process by the use of appropriate containment technology and other control measures.

Additional testing may be triggered at this stage. Often, the individual regulations within the country where manufacture occurs may require specific testing to be undertaken. This could be for a number of reasons such as the need to transport the material across borders, or as the volumes increase above certain trigger levels. The process design team must remain aware of these needs and ensure any testing required is completed in line with national regulations.

23.3.2.5 Technology Transfer

As with process safety, it is vital that all the knowledge gained by the team during the development process is transferred to the site of long-term manufacture.

Information regarding the toxicological data of a particular intermediate will often be transferred using internationally recognized formats such as a Safety Data Sheet.

23.3.3
Application of Health Triggers to AZD4619

The use of SAR assessment, followed by appropriate toxicity testing, was used on AZD4619. The following discussion will focus on just three of the intermediates in the synthetic route. In the SAR assessment, the tosylate, BOC aniline, and the aniline TFA all had positive alerts associated with genotoxicity. This information was noted by the development team. As development progressed, the compounds were Ames-tested. Both the tosylate and the aniline TFA gave positive results in the Ames test while BOC aniline gave a negative result.

During the route design phase a number of alternate route options were explored, however the synthetic scheme above was deemed to be the best overall route for the manufacture of this molecule. Therefore, prevention of the use of the genotoxic intermediates was not an option when considered with all the other competing factors in the route design.

The design team used this information during the Process Design phase to minimize the risk associated with the use of these materials. The process was developed to avoid the isolation of both, the tosylate and the aniline TFA intermediates. Therefore, the likelihood of exposure to these hazardous compounds was very much reduced. This is a good example of how the process development team can minimize health effects associated with the route.

23.4
Environment

23.4.1
Introduction

Of the three issues in SHE, perhaps the least considered, traditionally, is that of Environment. This attitude has changed. Ever more stringent legislation coupled with an increasing public concern over unrecognized or untested environmental issues in the past have contributed to a step change in the way that chemical processes are developed, at least in the pharmaceutical industry. This was one of the main drivers that saw the introduction of the SHE Triggers model in AstraZeneca and why it remains important for the future.

23.4.2
Environment Triggers

23.4.2.1 Early Delivery and Synthetic Route Evaluation
Traditionally, little thought was given to environmental issues at this stage. One explanation for this could be based on the fact that the route used at this stage

may not be the long-term route. However, the long-term route is often based, in part, on the initial synthetic route used for early supplies. While the rate of both compound attrition and process change at this stage are high, a rapid prioritization of environmental issues achieves two things. First, it focuses the development team on where the main issues associated with the current synthetic process are. Secondly, it improves the development team's overall understanding of environmental issues, which may be of wider use in subsequent development programs.

The initial phase of the environmental assessment process is a desk-based exercise starting simply with the synthetic route. A computerized search is carried out to gather all the environmental information available for all of the materials used in the synthesis. Each of the materials is then allocated a "consequence" score from 1 (lowest environmental impact) to 4 (highest) with "unknown" environmental impact materials (i.e., no data available) being allocated a score of 3*. The rules for allocation of scores are set out in Table 23.1.

There then follows a requirement to make a judgment of the "likelihood" of a release of the material causing an environmental issue, that is, will it be high, medium, or low. This is a relatively subjective judgment. The way that this is approached in the SHE Triggers process is to calculate the kg substance/kg API × predicted annual tonnage. If this is >1000 then it would suggest that there may be a high risk, >100 a medium risk, and <100 a low risk. Other factors that may need to be taken into account when making this judgment include reactivity in the process; that is, is it consumed fully; volatility; and containment/treatment options (if these are feasible then it may reduce the "risk" considerably). This is only an indicator of the potential risks and is for guidance on prioritizing development

Table 23.1 AstraZeneca environmental hazard rating.

Consequence score (environmental hazard ranking)	Definition
4	Very high environmental impact materials: Volatile Organic Compound (VOC) directive R-45 substances, and so on. Water framework directive priority hazardous substances, Registration, Evaluation Authorisation and Restriction of Chemicals (REACH) restricted substances (Annex 17), and other high toxicity hazard materials
3*	No information available. Assumed high environmental hazard
3	High environmental impact materials: VOC directive R-40 substances, and so on. Water framework directive priority substances, REACH substances of very high concern
2	Medium environmental impact materials: effectively VOC class A
1	Low environmental impact materials: effectively VOC class B

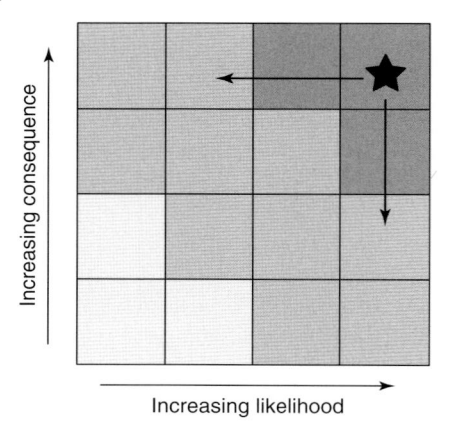

Figure 23.4 Risk matrix.

activities; that is, potential for replacing (eliminating) the high-impact materials primarily, if at all feasible.

The calculations are all performed automatically by spreadsheet after the development team have entered the required stage "metrics." These are simply the stage yield and the number of kilos of water, solvent, and "other" raw materials that are entered at each stage. All of this information is imported directly from electronic notebooks making transfer simple, quick, and less prone to errors. The spreadsheet then multiplies the "consequence" score by the "likelihood" score and allocates a combined "risk" score for each material in each stage which is also plotted in the form of a risk matrix (Figure 23.4).

The risk matrix is then used to prioritize those materials within a given route that present the highest environmental risks.

Later environmental assessments are simply iterations of this process, so that the preliminary environmental assessment may take place a number of times as process development continues and materials are added or removed from the route.

23.4.2.2 Process Design

Following route freeze, the possibility of eliminating environmentally harmful materials can be reduced. It may still be possible to remove compounds of concern, such as solvents. A further "Environmental Assessment at Route Freeze" is performed. This assessment concentrates on minimization of the environmentally hazardous materials that remain. It also clearly focuses on the actions that development chemists and engineers must take in order to maximize the chances of success.

23.4.2.3 Process Optimization and Understanding

Following process freeze, there is little opportunity for the removal or minimization of environmentally hazardous materials and the environmental assessment that is carried out at this stage concentrates on rendering harmless any environmental issues that remain. At this stage, detailed mass balances and desk-based estimations of waste steam compositions will be known and the responsibility of the development team at this stage is to ensure correct treatment and disposal of these various streams. Of particular concern is the fate of any API in the waste streams. APIs are often biologically active and may exquisitely target particular biological receptors. If these receptors are found in the environment as well as in the patient, an unintended biological effect may occur in the environment if the API is present at high enough concentrations. For this reason, a number of tools have been developed by the AstraZeneca SHE community to help chemists and engineers make environmental decisions.

To deal with the API issue, an "API Destruction Technology Selection" tool has been developed which enables the composition of the various waste streams to be entered and then uses expert-based "rules" to assess which of the various treatment technologies programmed into the tool can be combined to best treat the waste stream.

Other environmental tools have also been developed. One of the most commonly used is the "Solvent Selection Guide." Solvent selection is a key part of process design. There are a number of methodologies for environmentally benign solvent selection. The AstraZeneca approach is again included as part of the Reference Document on Best Available Techniques for the Manufacture of Organic Fine Chemicals [3].

In addition to the Solvent Selection Guide, other selection guides are available such as the Acid and Base Selection Guide and the Alkylating Agent Selection Guide which again allow environmentally informed choices of these reagents to be made quickly and easily.

The other main tool at the chemist's disposal is the "Substance Avoidance Database." The aim of the database is to provide access to information on substances appearing on regulatory lists indicating that the chemical is banned, severely restricted or has environmental emission limits. The database now holds information on well over 7000 substances.

23.4.2.4 Technology Transfer

Again, it is key that all the knowledge gained by the team during the development process is transferred to the site of long-term manufacture. During the Process Optimization and Understanding phase, the development team will work in partnership with the long-term manufacturing site to ensure that rendition of harmless solutions that are developed by the development team are consistent with the capabilities of the manufacturing site.

23.4.3
Application of Environmental Triggers to AZD4619

For AZD4619, the initial evaluation of the process clearly focused on achieving the most efficient synthetic route to producing the single desired enantiomer.

Significant chemistry effort was devoted to this goal. This evaluation included investigation of asymmetric approaches. Following evaluation of a number of different options, the synthetic approach using the Meerwein chemistry was selected as optimum from a chemistry perspective. Other synthetic approaches were viable but involved a greater number of stages and were therefore discounted.

Having selected this synthetic approach, achieving the optimum method of resolution became the major focus. Both classical and enzymatic resolution approaches were proven to deliver acceptable performance. Having selected a resolution approach, the focus of further development work was on recycling the undesired enantiomer. This work was ongoing at the time the project was removed from development.

The environmental assessment also identified the major issues with waste streams at the diazotization stage and the copper-based Meerwein chemistry. Copper is highly ecotoxic (as is iodide) and its removal was a high priority. In the event, although a great deal of work was carried out to achieve this, removal of the copper iodide proved impossible (at least prior to project termination) illustrating that the removal of *all* environmentally hazardous materials from process chemistry must remain an aspiration. At the time the project was removed from development, options to develop the most appropriate method of rendering the copper waste stream harmless were underway.

23.5
The Use of Risk Assessment

Often, the data gathered for a particular route selection or process design selection can be conflicting. The use of formal decision-making processes can aid these complex decisions to balance the multiple needs within the process. Even within the SHE considerations, the information can be difficult to assess in a balanced way.

One tool that can be used to aid process development to assess the relative risks associated with different SHE issues is a risk matrix. This is a similar approach as used in the Environmental Triggers described above. The risk matrix simply plots the likelihood of particular hazard occurring against the consequence of that hazard. It provides a graphical representation which quickly identifies the most important issues that the team should focus on. For AZD4619 this approach was used on the diazotization stage to allow the team to understand the comparative risks associated with the process.

The process involves a number of stages:

- define the scope and boundaries of the risk assessment;
- identify potential hazards;
- identify consequence if the hazard is realized;
- assess if consequence is tolerable;
- if not, identify potential for risk reduction.

The generic diagram (Figure 23.4) shows how the tool can be used to aid understanding.

The vertical axis represents the consequence of a particular hazard; the horizontal axis measures the likelihood of a hazard being realized. Risk is a function of likelihood and consequence. Therefore, the dark gray zone indicates the area of high risk, where both likelihood and consequence are high. The unshaded zone indicates the area of low risk. The lightly shaded zone is the area where the design team need to apply their knowledge to decide if a particular risk is tolerable. For example, if the exposure to a genotoxic compound is considered high in consequence, then the use of that intermediate in the synthesis will always be located somewhere on the top row. Improving the containment used while charging or discharging the compound will reduce the likelihood of exposure. This moves the risk position from right to left into the more acceptable area of the graph. Changing the synthetic route to remove the genotoxic impurity altogether would reduce the consequence from exposure for that equivalent step. This moves the risk position downward into the more acceptable region of the graph.

In the case of AZD4619 diazotization stage, the risks were identified and plotted on a similar diagram. This allowed the design team to assess which hazards were of most concern using a common and quantifiable framework. While it may seem likely that the chemical reaction hazards associated with the use of diazotization chemistry presented the highest risk, it was actually the use of the genotoxic intermediate, aniline TFA, during this stage that was the highest cause for concern. This tool can also be invaluable in discussions with wider stakeholders in the development community to demonstrate the risks associated with a particular process.

23.6 Conclusion

As stated in the introduction, development of a synthetic route for long-term production of an API requires the consideration of many factors to achieve the optimum solution. The likelihood of success of achieving this optimum solution increases, if the key factors are considered throughout the development of the synthetic process. This must be balanced against the likelihood of attrition of the wider project as preclinical and clinical activities progress.

AZD4619 was removed from development during the process design phase. By using the SHE Triggers, the process design team had ensured that the long-term SHE factors were fully considered in identifying a suitable route for long-term manufacture. From a safety perspective, the diazotization chemistry may not be a first choice; however, the chemical hazard assessment demonstrated how the process could be manufactured safely by employing the appropriate basis of safety. From a health perspective, the process employed the use of genotoxic intermediates. However, the risk associated with their use was minimized by avoiding the isolation of these intermediates. From an environmental perspective, Meerwein chemistry employed the use of copper which ends up in an aqueous effluent stream; the process design team had identified this issue and was already addressing how this waste stream could be rendered harmless. A viable asymmetric route had not been proven at the time the project was removed from development. Use of a classical or enzymatic resolution was under consideration, together with exploration of options for recycling the undesired enantiomer.

The route discussed is not perfect from a SHE perspective, but achieves a reasonable compromise when considered with all the other demands from the process.

Acknowledgments

We thank Marie Haag-Grönlund and Fredrik Waern (AZ, Sodertalje) for helpful discussions on the health section, and Steve Hallam (AZ Macclesfield) for contributions to the safety section.

References

1. European Commission (2006) Integration of EHS considerations into process development. *Reference Document on Best Available Techniques for Manufacture of Organic Fine Chemicals*, August 2006, section 4.1.2, pp. 92–93. Can be found under *http://ftp.jrc.es/eippcb/doc/ofc_bref_0806.pdf*.
2. Urben, P.G. (ed.) (1999) *Bretherick's Handbook of Reactive Chemical Hazards*, 6th edn, vol. 2, Butterworth Heinemann, Oxford, pp. 96–97, 101.
3. European Commission (2006) Example for a solvent selection guide. *Reference Document on Best Available Techniques for Manufacture of Organic Fine Chemicals*, August 2006, section 4.1.3, pp. 94–97. Can be found under *http://ftp.jrc.es/eippcb/doc/ofc_bref_0806.pdf*.

Index

a

accelerating rate calorimetry (ARC) 212, 376
acetaldehyde 322
acetanilide 114
3-acetoacetyl-2-oxazolidinone 55
acetophenone 132ff
acetyl nitrate method 375
acetyl xylan esterase 167
N-acetylamino acid 183
2-acetylamino malonic acid diethyl ester 372f
acid anhydride 113
acid phosphatase 173
acid–base combination chemistry
– design of dynamic salt catalyst 39ff
actinol 328
active pharmaceutical ingredient (API) 2ff, 222, 259, 304, 334, 427, 447, 471
– Destruction Technology Selection 485
– productivity 232
– purification 231
– quality assurance 232
N-acylamino acid 171
acylase 337
– L-acylase 184
– process 184
1,2-addition reaction 114ff
– Cu-catalyzed 120
– Rh-catalyzed enantioselective 116
1,4-addition reaction 114ff
– Rh-catalyzed enantioselective 116
– Pd-catalyzed enantioselective 118
agarose beads
– protein immobilization 131
agitated tank crystallizer 452
alanine 331
alanine dehydrogenase (AlaDH) 329ff

β-alanine pyruvate transaminase 331
β-alanine-poly(lysine)dendrimer 131
alcalase 168
– amidation of peptide C-terminal ester group 169
– CLEA 168f
alcohol
– chiral 198f, 325
– (S)-alcohol 201
alcohol dehydrogenase (ADH) 195ff, 325ff, 433
– ADH-LIPOzyme 433
– biocatalyst 195
– (R)-ADH 197ff
– (S)-ADH 195ff
– whole-cell biocatalyst 197ff
aldehyde 176
– allylation 121
– β-aryl- and β-alkyl-substituted α,β-unsaturated 61
– reductive amination 127ff
aldehyde collidine 6
aldol reaction
– enantioselective 121
aldolase 323
aliphatic benzyl ether derivative 80
alkene
– chemoselective hydrogenation 89
– Pd-PEI as a partial hydrogenation catalyst 93
alkenyl[2-(hydroxymethyl)phenyl]dimethylsilane 104ff
alkenyl(trimethyl)silane 103
alkenyldimethylphenylsilane 113
alkyl aryl sulfide 68ff
– Pt-mediated asymmetric oxidation 72
alkyl halide 113

alkyl[2-(hydroxyprop-2-yl)phenyl]
  diisopropylsilane 106
alkylamine borane 128
alkyltriorganosilicon reagent 105
alkyne
– chemoselective hydrogenation 89
– partial hydrogenation 93ff
– Pd-PEI 93ff
allyl(trimethoxy)silane 121
allylation 104ff
– Ag-catalyzed enantioselective 121
– aldehyde 121
– ketone 121
allylic alcohol 67
– metal-catalyzed asymmetric epoxidation 67
N-allylindole 243
allysine ethylene acetal 330
aluminum
– cis-β-η_2-hydroperoxoaluminum species 71
aluminum–salalen complex
– chiral 70
amidase 337
– CLEA 170
amidation
– peptide C-terminal ester group 169
amidocarbonylation 183
amine 122ff
– synthesis of primary amine 135
amine borane 127ff
– aromatic 129
– direct reductive amination 127ff
– type 128
amino acid 168
– α-amino acid 325ff
– β-amino acid 325
– D-amino acid 183ff, 337
– esterification 168
– L-amino acid 183ff, 337
amino acid dehydrogenase (AADH) 328
– biocatalyst 191ff
6-amino penicillanic acid (6-APA) 177
(RS)-α-amino-ε-caprolactam (ACL) 391ff
– (R)-ACL 393
(S)-2-amino-5-(1,3-dioxolan-2-yl)-pentanoic acid 330
(R)-2-amino-3-(7-methyl-1H-indazol-5-yl) propanoic acid 336
3-amino-1,2,4-triazole-5-carboxylic acid 228
aminoacylase 171
– L-aminoacylase 178, 183
D-2-aminobutyric acid 185

7-aminocephalosporanic acid (7-ACA) 167
9-aminoketone
– safe synthetic method 211
ammonium acetate 136
ampicillin 170
amrubicin 207ff
– bulk production 210
– stereoselective introduction of 7-hydroxy group 213
amrubicin hydrochloride 207ff
– drying method 217
– moisture adsorption 217
– polymorphism study 215
– stability 216
– stability in various water content 217
amrubicinol 207
anchor impeller 451
aniline 253
anilinium arenesulfonate 43
anilinium pentafluorobenzenesulfonate 48
anthracycline anticancer drug 207
anti-hepatitis drug 284
antioxidative enzyme 428
antisolvent 452
AP521
– anxiolytic 371f
API, see active pharmaceutical ingredient
aqueous diazotization 477
artemether 288f
artemisinic acid 287f
artemisinin 286ff
– synthesis 287
aryl halide 122
aryl sulfonate 112
aryl (trifluoro)silane 117
aryl[2-(hydroxymethyl)phenyl] dimethylsilane 104ff
aryl(trialkoxy)silane 117
aryl(triethoxy)silane 118
aryl(trifluoro)silane 111ff
aryl(trimethoxy)silane 111
α-arylalkyl ketone 110
3-arylbut-1-ene 108
α-arylcarboxylate 113
arylmethyl carbonate 106
L-ascorbic acid 291ff
– process chemistry 291
– Reichstein process 292ff
aspartic acid β-ester
– α-protected 168
Aspergillus niger feruloyl esterase 167
asymmetric epoxidation 60ff
– allylic alcohol 67
– metal-catalyzed 67

– unfunctionalized olefin 62
asymmetric Mannich-type catalyst 50
asymmetric oxidation 59ff
– hydrogen peroxide 59ff
– Pt-mediated 72
– sulfide 67f
– V-catalyzed 68
asymmetric sulfoxidation 69
asymmetric synthesis 322
atorvastatin 326
axial flow impeller 450
1-azabiclo[1.1.0]butane 264
AZD3409 130
AZD4407 305ff
– synthetic route 305ff
AZD4407 intermediate 303ff
– continuous processing 303ff
AZD4619 473
– environmental triggers 486
– health 482
– process safety 477
– synthetic scheme 474

b

Bacillus licheniformis alkaline protease 168
Baeyer–Villiger monooxygenase 337
basic reaction route 444
basis of safety 475ff
benflumetol 289f
benflumetol/lumefantrine 289f
– synthesis 290
benzaldehyde 175, 322
benzhydrylamine 260
benzimidazole 246
benzo[*b*]thiophene 371
benzoyl peroxide (BPO) 366
benzyl ether derivative
– aliphatic 80
benzylamine 261
benzylfuryl methyl ketone 225ff
benzylic alcohol
– tertiary 166
3-benzyloxy-1-phenyl-1-propene 78
berberine 274ff
– industrial synthesis 274
– primary synthetic route 276
– synthetic route 277f
1,1′-bi(2-naphthol) (BINOL) 50
biapenem (BIPM) 257
biclycol (4,4′-dimethoxy-2,3,2′,3′-
 dimethylene-dioxy-6-hydroxymethyl-6′-
 carbonyl-biphenyl) 284
– synthesis 285

bifendate (dimethyl 4,4′-dimethoxy-5,6,5′,6′-
 dimethenedioxy-biphenyl-2,2′-
 dicarboxylate, α-DDB) 281ff
– synthetic procedure 283
1,1′-binaphthyl-2,2′-disulfonic acid
 (BINSA) 50
– dynamic complexation 51
– (*S*)-BINSA 52
(*R*)-BINOL 51
biocatalyst
– amino acid manufacture 183ff
– industrial 159ff
– whole-cell 183ff
biomembrane interference 433
biomembrane process chemistry
 (BMPC) 421ff
biomimetic membrane 426
bio(mimetic)membrane interference
– gene expression 437
biotin 294ff
– Goldberg–Sternbach approach 295f
– synthetic method 297
bis(2-acetoxyethoxy)methyl (ACE) 347
N-bis(trimethylsilyl)methyl-substituted urea
 107
bleomycin 63
N-Boc-phenylaldimine 52
boron-based reagent 115
α-bromocarboxylate 113
Bunte's salt 263
Burkholderia cepacia lipase 166
– CLEA 166
n-butyl lithium 310
tert-butyl methyl sulfide 72
tert-butylamine borane (TBAB) 128ff
t-butyldimethyl silyl (*O*-TBDMS, TBDMS)
 protective group 83, 309, 346

c

C–C bond coupling reaction
– Suzuki–Miyaura 131
C–C bond formation 145, 154, 322
– silicon-based 101ff
– transition metal catalysis 101ff
C–C bond forming lyase 174
C=C bond 326
– activated 326
– reduction 326
C–H arylation reaction
– silicon-based 114
C–H bond
– activation by a chiral catalyst 243
– Ir-catalyzed silylation 122

C–O bond formation   146
Calsed®   207
Candida antarctica lipase A (CaLA)   166
– CLEA   166
Candida antarctica B lipase (CaLB)   165, 309, 335
– CLEA   165
Candida rugosa lipase   165
– CLEA   165
D-carbamoylamino acid   187
D-carbamoylase   185ff
carbamoylsilane   110ff
carbonyl addition reaction   114ff
– Cu-catalyzed   119
carboxylic acid
– bioisostere   222
– derivative   110
– membrane   155
carbapenem   257
– oral   257ff
catalase   173, 429ff
catalyst poison   77ff
catalyst support   90
N-Cbz protective group   94
N-Cbz-arylaldimine   53
N-Cbz-phenylaldimine   51
CEM amidite   349ff
– RNA synthesis   350ff
centrifugation   454ff
– equipment   458
– selection criteria   460
– utility   458
cephalexin   170
cephalosporin C   167
(*R*)-CH$_3$OBIPHEP-ligand   12
chemical engineering
– pharmaceutical processing   443ff
chemical hazard   475
chemical potential   425
chemical reaction hazards
– assessment   475
chemoselective heterogeneous catalyst   77, 90
chemoselective hydrogenation   77ff
– alkene   89
– alkyne   89
– Pd catalyst   77ff
chemoselective hydrogenation catalyst
– silk-fibroin-supported   90
chemoselective inhibition
– hydrogenolysis for *O*-benzyl protective group   77

CHETAH (chemical thermodynamic and energy release evaluation program) calculation   365ff
– nitrophenol derivative   378
Chinese herbal medicine (CHM)   273ff
– industrial synthesis   274
chirality
– diastereomeric salt   381ff
– dielectrically controlled optical resolution (DCR)   391ff
chitosanase   431
(7-chloro-4-(4′-diethylamino-1′-methylbutylamino)quinoline   285
5-chloro-2-picolinic acid   6
chlorofluorocarbon (CFC)   149
2-chloroindole   245
chloroperoxidase
– heme-dependent (CPO)   173
D-3-(4′-chlorophenyl)alanine   184
(*R*)-1-(4′-chlorophenyl)ethanol   200
(*S*)-1-(4′-chlorophenyl)ethanol   200
chloroquine   285
1-[6-chloro-1,2,3,4-tetrahydroquinolin-3-yl]-*N*,*N*-dimethylmethaneamine(TQA)   403f
cholesterol (ch)   426ff
α-chymotrypsin CLEA   178
(*E*)-cinnamic acid derivative   325
cinnamyl alcohol   203
5CITEP   222
R-(+)-citronellal   286
Claisen condensation
– S-1360   229
clinical candidate   2
Coartem   287ff
cofactor regeneration   193
combi-CLEA   164ff
– cascade process   175
continuous centrifuge   459
continuous-flow reactor   304
controlled-pore glass (CPG)   350
convective dryer   461
cooling   452
cooling crystallizer   452
copper   119f
– carbonyl addition reaction using organosilicon reagent   119
– Cu(I)   103
D-corydaline   279
corydalis L   279
coupling reaction   252, 313
critical nucleus   412, 416
cross-coupling reaction   102ff
cross-linked enzyme (CLE)   159ff
– industrial biocatalyst   159ff

cross-linked enzyme aggregate (CLEA®) 160ff
– acetyl xylan esterase   167
– alcalase   168f
– amidase   170
– amidation of peptide C-terminal ester group   169
– *Aspergillus niger* feruloyl esterase   167
– *Burkholderia cepacia* lipase   166
– *Candida antarctica* lipase A (CaLA)   166
– *Candida antarctica* B lipase (CaLB)   165
– *Candida rugosa* lipase   165
– esterase   165
– hydrolase   164
– microchannel reactor   177
– NHase-CLEA   174
– penicillin G amidase   170
– protease   168
– reactor design   176
– recyclable   172
– *Thermomyces lanuginosa* lipase   165
cross-linked enzyme crystal (CLEC)   160
cross-linking agent   160
crotylsilanolate   108
crystal habit modification   382
– mechanism   385
crystal polymorph   401ff
– scope   402
crystal shape
– controlling   381ff
– effective additive   384
– industrial-scale production   382
crystal structure
– salt   390
crystallization   342, 381ff
– control   381ff
– equipment   452
– multicomponent system   406
– nucleation   411
– one-component system   406
– selection criteria   453
– utility   452
crystallization-induced diastereomeric transformation (CIDT)   240
crystallization-induced enantiomeric transformation (CIET)   240
Curtius degradation   32f
2-cyanoethoxyethyl (CEE)   347
2-cyanoethoxymethyl (CEM)   348f
– amidite   349ff
cyanoethylmethylthiomethyl ether (CEM-SCH₃)   348ff
R-cyanohydrin   174
cyclic 1,3-diketone   54

[3 + 2] cycloaddition
– chiral cyclopropane   244
cyclohexen-2-one   115
cyclohexylaniline   140
cyclopropane   244
CYTOS™ reactor   311f

d

daunorubicin   207f
decantation   454
decarboxylation   249
dehydrative condensation catalyst   40f
dehydrative cyclization   43ff
dehydrative cyclocondensation catalyst   43
deoxy-D-ribose phosphate aldolase (2-deoxyribose-5-phosphate aldolase, DERA)   175, 323
deprotection   34, 55, 85, 107, 185, 209, 225ff, 259, 346
– methoxyisopropyl (MIP)   234
– S-1360   225ff
deracemization
– enzymatic   335
desymmetrization   332f
– enzymatic   332
di-acetone-ketogulonic acid (DAKS)   292
(aR,R)-di-μ-oxo titanium–salen complex   64
diadenosine boranophosphate   360
dialkyl sulfide   69ff
2,6-diaminopurine (2,6-DAP)   353
N,N-diarylamine   45
N,N-diarylammonium arenesulfonate catalyst   43
N,N-diarylammonium pentafluorobenzenesulfonate   43ff
N,N-diarylammonium sulfonate   45
N,N-diarylammonium tosylate   45
diarylmethane   105
diazotization
– aqueous   477
dichloroethyl ether   150
2,5-dichloropyridine   6
dicyclohexylamine   140
dielectric dispersion analytical (DDA) method   438
dielectrically controlled optical resolution (DCR)   391ff
– phenomenon   396
– strategic resolution   393ff
Diels–Alder reaction   29f
1,3-diene   115
diethyl malonate   246
differential scanning calorimetry (DSC)   365

differential thermal analysis (DTA)   367
dihalomaleimide   252
(S)-2,3-dihydro-1H-indole-2-carboxylic acid   325
dihydroartemisinin   287
– derivative   288
1,3-diketo acid (DKA)   221
1,3-diketone   56
– cyclic   54
N,N-dimesitylammonium cation   48
N,N-dimesitylammonium pentafluorobenzenesulfonate   45
N,N-dimesitylammonium tosylate   48
dimethoxycamphorsulfonylimine   72
4,4′-dimethoxytrityl (DMTr) group   346
dimethyl carbonate   248
dimethylamine borane (DMAB)   128
2-(dimethylamino)pyridine (2-DMAP)   353
3-(dimethylamino)-1-(2-thienyl)propan-1-ol (DMT)   389
4,6-dimethylnonan-3,5,7-trione   43
1,2-dimyrystoyl-sn-glycero-3-phosphocholine (DMPC)   431
3,5-dinitrotoluene (DNT)   367
1,2-dioleoyl-3-trimethyl-ammonium propane (DOTAP)   432ff
1,2-dipalmitoyl-sn-glycero-3-phosphocholine (DPPC)   431
diperoxo(oxo)molybdenum(IV) complex   64
N,N-diphenylammonium triflate   41
N-(2,6-diphenylphenyl)-N-mesitylamine   46ff
diphenylphosphoryl azide (DPPA)   30
1,3-diphenylpropane   86
1,3-diphenylpropane-1,3-dione   53
2,6-diphenylpyridine   52
diphenylsulfide   86
direct contact dryer   461
direct reductive amination   127ff
– amine borane   127ff
2,2′-disubstituted 1,1′-binaphthyl   50
disulfide linking   311
disulfide synthesis   308
1,3-dithiane   69ff
1,3-dithiolane   69
dodecanoyl-His (Dodec-His)   431
double cone dryer   463
downstream processing   455
doxorubicin   207f
drug development   401ff
– process   2
drug master file (DMF)   443ff, 466
drug substance form selection   404
drying   460
– equipment   461
– selection criteria   463
– utility   460
duloxetine   381ff
Dutch resolution   387
dynamic ammonium salt
– sulfonic acid   40
dynamic complexation
– BINSA   51
dynamic kinetic resolution (DKR)   337
dynamic salt catalyst   39ff
– acid–base combination chemistry   39ff

e

early delivery   476ff
elvitegravir   236
enantioselective epoxidation   60
energy balance   465
environment   482f
environment, health, and safety (EHS)   236
enzymatic chiral induction   242
enzyme
– synthetic   60
ephedrine   322
(R)-epichlorohydrin   246ff
epihalohydrin   337
epirubicin   207f
epoxidation   59ff
– asymmetric   60ff
equipment
– evaluation   465
ester condensation   42
esterification   267
– amino acid   168
– catalyst   41
(S)-ethyl hydroxybutyrate   310
ethyl 1,2,4-triazole-3-carboxylate   222
5-ethyl-2-methylpyridine   129ff
5-ethyl-2-methylpyridine borane (PEMB)   131ff
5-ethyl-1H-tetrazole   350
evaporation   452
evaporative crystallizer   453
Exfluor-Lagow method   148f
explosion risk assessment   364
explosive chemical process development   363ff
– safety evaluation   363f

f

(S)-fenoprofen   337
ferulate (4-hydroxy-3-methoxycinnamate) ester group   167

feruloyl esterase   167
filtration   454
– equipment   455
– selection criteria   457
– utility   454
fluorination
– application   149
– direct   145ff
– electrochemical (ECF)   146
– liquid-phase   145ff
– thermodynamics   147
– vapor-phase   147
fluorine   145ff
(R)-1-(4′-fluorophenyl)ethanol   200
5-(4-fluorophenyl)-2-furyl methyl ketone   222
fluoropolymer   156
flow reactor configuration   314
formal [3 + 2] cycloaddition
– chiral cyclopropane   244
formate dehydrogenase (FDH)   186ff, 330
– LeuDH/FDH biocatalyst   191
friction sensitivity test   368
Friedel–Crafts alkylation
– anhydrous $ZnCl_2$ in dichloromethane   226
– aqueous $ZnCl_2$ 226
full-scale plant   447
fumaric acid   294

g
galactose dialdehyde   161
galactose oxidase   172
β-galactosidase   172
gas evolution   478
gas inducing impeller   451
gluconolactone   193
glucose   193
glucose dehydrogenase (GDH)   193ff
glucose oxidase   172
glutamate dehydrogenase (GluDH)   329
glutamic acid γ-ester
– α-protected   168
glutaraldehyde   160
N-glutaryl-L-phenyl alanine   178
GMP (good manufacturing practice)   445f
green chemistry   303
guaiacol   279

h
halohydrine dehalogenase (HHDH)   326
2-haloimidazole   245f

health   479
heat of reaction   478
helix impeller   451
Hemophilus influenzae
– β-lactamase-negative ampicillin-resistant (BLNAR)   257
3,5-heptanedione   53
heterogeneous chemoselective hydrogenation catalyst   81ff
hexafluoropropylene oxide (HFPO)   146
– synthesis of PPVE   146
n-hexyl lithium   310
α-hexyl-β-hydroxy-δ-benzyloxy acid   8, 10
α-hexyl-β-hydroxy-δ-lactone   12
2-hexyl-methyl-acetoacetate   10
HIV integrase inhibitor S-1360   221ff
HIV protease inhibitor   13
Hofmann degradation   32
homeobox gene b-8 (*HoxB8*)   357
homopiperonylamine   275f
hydantoin racemase   185ff
hydantoinase   337
– D-hydantoinase   185ff
– L-hydantoinase biocatalyst   189
– process   338
– whole-cell biocatalyst process   185ff
hydrogen peroxide   59ff
hydrogenation
– chemoselective   77ff
– partial   93
hydrogenolysis
– chemoselective inhibition for O-benzyl protective group   77
hydrolase
– CLEA®   164
cis-$\beta\eta_2$-hydroperoxoaluminum species   71
hydrosilane   122f
hydrosilylation of alkyne   113, 123
– Ru-catalyzed   121
hydroxo-hydroperoxo-titanium species   70
(R)-2-hydroxy-4-phenylbutyronitrile   324
(R)-1-hydroxy-1-phenylpropanone   322
hydroxyacid   337
(S)-3-hydroxyadamantylglycine   330
3-hydroxybenzylhydantoin   190
S-α-hydroxycarboxylic acid amide   176
2′-hydroxyl protecting group   346ff
hydroxynitrile lyase   323
hydroxynitrile lyase from *Linus usitatissimum* (*Lu*HNL)   174
L-*m*-hydroxyphenylalanine (L-*meta*-tyrosine, L-*m*-Tyr)   190

i

(S)-ibuprofen   337
ideal classified bed crystallizer   453
imidazole   253
immobilized liposome chromatography (ILC)   438
immobilized liposome membrane (ILM)   438
immobilized liposome sensor (ILS)   438
impact sensitivity test   368
indinavir   381
indirect dryer   461
indole   252f
influenza neuraminidase inhibitor   16
Invirase™   5ff
N-(p-iodobenzyl)cinchona alkaloid salt   60
iodosylbenzene   62
iridium
– silylation of C–H bond   122
iron
– Fe(III)porphyrin complex   62
– FeCl$_3$-(N-tosyl-1,2-diphenylthylenediamine)   63
– iron–salan complex   72
(–)-isopulegol   286

j

jet fluorination   147
JTT-010   239ff
– construction   251
– convergent coupling reaction   251
– key intermediate synthesis   240
– maleimide construction   251
– optical resolution   240
– replacement of the hydroxyl group with an amino group   250
– synthetic strategy   240

k

ketal formation catalyst   215
α-keto-acid
– reductive amination   328
keto–enol tautomerism
– Pd-catalyzed   94
2-keto-L-gluconic acid (2-KLG)   292f
2-keto-L-gluconic acid pathway   293
ketoisophorone   328
ketone
– α,β-unsaturated   60, 113
– Ag-catalyzed enantioselective allylation using allyl(trimethoxy)silane   121
– enzymatic reduction   325
– reductive amination   132
kilo lab   445

l

L-188   259ff
– synthetic process   266
laccase   172
β-lactamase-negative ampicillin-resistant (BLNAR) Hemophilus influenzae   257
(S)-lactic acid piperidineamide   64
β-lactone   8, 10
LaMar fluorination   148
laurinic aldehyde   10
lazabemide™   5ff
LJC   11, 143, 267ff
Lentikat   162
D-leucine   185
L-tert-leucine (L-Tle)   185ff, 330
leucine dehydrogenase (LeuDH)   185ff, 329f
– LeuDH/FDH biocatalyst   191
(S)-t-leucinol   68
levansucrase   172
(6R)-levodione   328
lignocellulose   167
lipase
– CLEA   165
lipase inhibitor   6
liposome   425
– cholesterol-modified   433
– gene expression   437
– negatively charged   433
– positively charged   433
– recognition (separation) function   425ff
– zwitterionic   433
LIPOzyme   421ff
– build-up type   431
– enzyme-like activity   428
liquid–liquid separation   454
liquid-phase direct fluorination   145ff
– industrial synthesis of perfluorinated building block   145ff
lithiation reaction   310f
– flow mode   312
luciferase reporter assay system   356
LY-248686   388
lyase   174, 322
– C–C bond forming   174

m

mandelic acid (MA)   382ff
– (R)-mandelic acid   338
– S-mandelic acid ((S)-MA)   175, 394f
(R)-2-mandelonitrile   324
manganese
– Mn(III)–salen complex   62
– Mn–salen complex   68
– Mn–salen complex/H$_2$O$_2$ system   69

– Mn–(5,10,15,20-tetrakis
  [1-hexadecylpyridium-4-yl]-21H,
  23H-porphyrin)
  (Mn–HPyP) 431
Mannich-type reaction
– enantio- and diastereoselective direct 55
– Pd-catalyzed enantioselective 118
masked silanol 103
mass transfer 424
material balance 445, 465
material of cost (MOC) 445
mathematical modeling
– individual equipment 465
membrane chip 428
membrane slurry reactor 177
3-mesyloxy-1-benzylazetidine hydrochloride 263
3-mesyloxyazetidine hydrochloride 263
metal affinity immobilized liposome chromatography (MA-ILC) 438
metal-catalyzed asymmetric epoxidation 62
metal–ONNO–tetradentate ligand-catalyzed oxidation 69
metal–salen-catalyzed oxidation 68f
metal–Schiff base-catalyzed oxidation 68
methanesulfonyl chloride 250
methanolysis 134ff
– Pd-catalyzed 138
D-methionine 185
L-methionine 183, 188
(1S,2R)-2-(methoxycarbonyl)cyclohex-4-ene-1-carboxylic acid 334
methoxyisopropyl (MIP) protection 233f
– S-1360 233
(S)-methoxyisopropylamine 332
methyl tert-butyl ether (MTBE) 199
methyl isobutyl ketone (MIBK) 193
N-methyl oxindole 308
methyl phenyl sulfide 68
2-(S)-methyl tetrahydro-4H-pyran-4-one 309
1β-methyl-carbapenem 258
1-methyl-1-methoxyethyl group 233f
(S)-(+)-N-methyl-3-(1-naphthyloxy)-3-(2-thienyl)propylamine 388
3-(methylamino)-1-(2-thienyl)propan-1-ol (MMT) 388
– (S)-MMT:(S)-MA salt 390
2-methylcyclohexanone 136
2-methylcyclohexylbenzylamine 136
2-(3,4-methylenedioxyphenyl)ethylamine (homopiperonylamine) 275
N-methylimidazole (NMI) 352
N-methylmorpholine (NMM) 250
microreactor description 311

microreactor study 308
microribonucleic acid (miRNA) 345
Mitsunobu conditions 8ff
mixed flow impeller 451
mixing 449
– equipment 449
– selection criteria 451
– utility 449
MN-447 135
model biomembrane 426
modeling
– individual equipment 465
molecular size
– evaluation 387
molecule to market 444
molecule to money 443
monoamine oxidase type B (MAO-B) inhibitor 4ff
multi-CLEA 164
multicomponent crystal 402

n
NADPH 193
D-3-(1'-naphthyl)alanine 184
L-3-(1'-naphthyl)alanine 184
D-3-(2'-naphthyl)alanine 184
L-3-(2'-naphthyl)alanine 184
(S)-naproxen 337
L-neopentylglycine 186ff
neuraminidase inhibitor 16
nickel
– π-allylnickel species 115
– catalyst 111ff
– nickel/carbene catalyst 115
niobium
– (μ-oxo)Nb–salan complex 67
– niobium–salen complex 68
nitrilase 171
nitrile hydratase (NHase) 174ff
nitroacetic acid ethyl ester 373ff
– risk assessment 374
– safety evaluation 371ff
nitrobenzene derivative 376ff
p-nitrobenzyl (1R,5R,6S)-2-(diphenylphosphoryloxy)-6-[(R)-1-hydroxyethyl]-1-methylcarbapen-2-em-3-carboxylate (MAP) 259ff
p-nitrobenzyl (PNB) ester of TBPM 259
– synthesis 265f
nitromethane method 375
nitrophenol derivative
– CHETAH calculation 378
L-norvaline (L-Nva) 188f

Novozyme 435   166f, 309, 335
nucleation   411
– crystallization   411
– primary   416

o

olefin
– metal-catalyzed asymmetric epoxidation of unfunctionalized olefin   62
– oxidative methylation   107
oligosaccharide   172
operational hazard   475
– assessment   475
optical resolution   381ff
organic synthesis in China   273
– pharmaceuticals   273ff
organo[2-(hydroxymethyl)phenyl]dimethylsilane   116f
organocatalyst
– asymmetric epoxidation   60
organosilane
– catalytic preparation   121
– halogenated   103
organosilanolate   106ff
organosilicon reagent   101ff
– Cu-catalyzed carbonyl addition reaction   119
– Rh-catalyzed 1,4-addition reaction   115
organotrimethylsilane   102
orlistat   381
oseltamivir phosphate   5, 16ff
– allylamine promoted azide-free synthesis   25
– azide-free t-butylamine-diallylamine transformation   28
– azide-free transformation   23
– commercial synthesis   22
– desymmetrization concept   32
– Diels–Alder concept   29
– meso-diester synthesis   33
– furan Diels–Alder/diphenylphosphoryl azide approach   31
– furan Diels–Alder/nitrene addition concept   30
– oxazolidinone shortcut   34
– pyridone Diels–Alder concept   29
– shikimic acid independent synthesis   27ff
– technical synthesis   18
– O-trimesylate of ethyl shikimate   35
Oslo-fluidized bed crystallizer   453
Ostwald's stage rule   402ff
oxathiane
– racemic   69
oxaziridinium salt   62
oxazolidinone moiety
– deprotection   56
oxidant   59ff
oxidation
– asymmetric, see asymmetric oxidation
– metal–ONNO–tetradentate ligand-catalyzed   69
– cis-β metal–salen-catalyzed   69
– metal–Schiff base-catalyzed   68
oxidative methylation
– N-bis(trimethylsilyl)methyl-substituted urea   107
– olefin   107
oxidoreductase   172
oxime   137
oxindole disulfide   309
L-ε-oxonorleucine acetal (L-ONA)   189
oxovanadium–salan complex   71
oxovanadium(IV)–salen complex   68
oxynitrilase   174

p

palladium   102ff, 117
– catalyst for chemoselective hydrogenation   77ff
– cationic Pd(II) species   117
– enantioselective 1,4-addition reaction   118
– enantioselective Mannich-type reaction   118
– methanolysis   138
– Pd on the activated carbon–ethylenediamine complex [Pd/C(en)]   80ff
– Pd silanolate   108
– Pd/C   86f
– Pd/C(Ph$_2$S) complex   85ff
– Pd/Fib   90ff
– Pd(OAc)$_2$ 91
– Pd-PEI   93f
– ZnX$_2$–Pd/C system   77
1-palmitoyl-2-oleoyl-sn-glycero-3-phosphatidylglycerol (POPG)   433f
1-palmitoyl-2-oleoyl-sn-glycero-3-phosphocholine (POPC)   428ff
– POPC/Ch   433
– POPC/POPG   433
PEMH$^+$ carboxylate$^-$ pair   139
penicillin   448
penicillin amidase CLEA   170
penicillin G   177
penicillin-resistant Streptococcus pneumoniae (PRSP)   257
pentafluoroanilinium triflate   41
pentafluorobenzenesulfonate anion   48

pentafluorophenylplatinum(II)(diphosphine) complex 64
PERFECT (perfluorination of esterified compounds followed by thermolysis) method 150f
perfluorinated acyl fluoride 149ff
perfluorinated building block
– industrial synthesis by liquid-phase direct fluorination 145ff
perfluorinated ketone
– PERFECT method 154
perfluorinated monomer 149
perfluorinated vinyl ether 146
perfluoro(propyl vinyl ether) (PPVE) 146ff
perfluoroacyl fluoride 150ff
perfluoroalkanesulfonic acid 154
perfluoroalkoxy copolymer (PFA) 149
peroxidase (POD) 173, 431
– POD-LIPOzyme 432
personal protective equipment (PPE) 471
pharmaceutical processing
– chemical engineering 443ff
pharmaceuticals
– organic synthesis in China 273ff
phase transfer catalyst 60
phenoxyacetic anhydride 353
(-)-phenyl-ethylamine 10
phenyl[2-(hydroxyethyl)phenyl]dimethylsilane 115ff
(RS)-1-phenyl-2-(4-methylphenyl)ethylamine (PTE) 394ff
D-phenylalanine 185
L-phenylalanine 13ff
phenylalanine aminomutase 325
phenylalanine ammonia lyase 325
phenylalanine dehydrogenase (PheDH) 329f
4-phenylbutyric acid 42
1-phenylethanol 166
(RS)-1-phenylethylamine (PEA) 382ff
– bis-PEA 384
D-phenylglycine 185
D-p-OH-phenylglycine 337
phenylsulfone 212
phospholipid
– zwitterionic 428
phosphoramidite method 346f
phytase 173
2-picoline borane (PICB) 131ff
pig liver esterase (PLE) 332f
pilot plant 446
pinacolatoboryl-substituted bromobenzene 104

pivaloyloxymethyl (POM, pivoxil) ester of C3 carboxylic acid 257
platinum
– ZnX_2–Pt/C system 77
platinum metal sulfide 77
platinum-mediated asymmetric oxidation of alkyl aryl sulfide 72
polyene macrolide RK-397 108ff
polyethylene glycol (PEG) 161
polyethyleneimine (PEI) 93
polymorph
– concomitant polymorph 409
– control 409
– disappearing 410ff
– late-appearing 402f
– process research issue 403
– screening 405
– thermodynamically stable 406
polyvinyl alcohol matrix 162
potential 424
PPARα (peroxisome proliferator-activated receptor α) agonist 473
precipitation 452
pressure vessel test 369f
pressurized filter 456
process chemistry 422
process control 475
process design 480ff
process development 469ff
process optimization 481ff
process research
– development of new drug 1ff
process safety 474ff
process safety triggers 476
prodrug esterification 267
propen-2-ylsilane 115
protease CLEA 168
protection 10, 52, 104, 151, 209, 222ff, 242, 284, 310, 459
– methoxyisopropyl (MIP) 233f
– S-1360 222
– tetrahydropyranyl (THP) 229f
– trityl (Tr) 222f
protein immobilization
– agarose beads 131
protein kinase Cβ inhibitor JTT-010 239ff
protoberberine 279
pseudopolymorph 402, 413
pyranone 310
– (S)-pyranone 309
pyridine borane (PYB) 130ff
pyridine-3-aldehyde 174
pyridoxal-5-phosphate (PLP) 331

2-pyridylboronic acid   108
γ-pyrone   43ff
pyrrolidine   253
pyruvate decarboxylase   175, 322

q
qinghaosu   285ff
qinghaosu/artemisinin   286
(-)-quinic acid   20ff
quinine   285

r
radial flow impeller   450
raltegravir   236
rate-limiting step   445
reduction
– activated C=C bond   326
– ketone   325
reductive amination
– direct   127ff
– α-keto-acid   328
– reaction solvent   138
– stereoselective   137
Relenza™   16
Residos™   311
rhodium
– enantioselective 1,2-addition reaction   116
– 1,4-addition reaction   115f
– carbonyl addition   114
– catalyst   113
– Rh/chiral diene catalyst   115
– RhCl(dppp)$_2$ 243
Riamet/Coartem   287ff
ribonucleic acid (RNA)
– interference (RNAi)   345
– oligomer   345ff
– synthesis from CEM amidite   350ff
ring construction method   228
ring modification method   228
risk assessment   364ff, 486
– AZD4619   487
– nitroacetic acid ethyl ester   374
risk matrix   484
ritonavir   413ff
RK-397   108ff
rotary filter dryer   463
rotary vacuum drum filter   456
ruthenium
– catalyst   185
– hydrosilylation of alkyne   121
– Ru(Ph$_2$pyboxazine)(pyridinedicarboxylate)   63
– Ru(pybox)(pyridinedicarboxylate) complex   63

s
S-1360 (Shionogi)   221ff
– commercial route by MIP protection   233
– discovery route   222
– process chemistry   229
S-5751   137
S/G SK-364735 (Shionogi–GlaxoSmithKline Pharmaceuticals)   236
safety evaluation
– nitroacetic acid ethyl ester   371
safety, health, and environment (SHE) consideration   469ff
safety, quality, delivery, cost, environment (SQDCE)   422
salalen (salen/salan hybrid)   62
salan (fully reduced salen)   62
salen (ethylene-1,2-bis(salicylideneiminato)   62
salt crystallization   381ff
Saquinavir™   4ff, 17
– synthetic development   13
saxagliptin   330
scale-up problem   464
SCE-2787   409ff
schizandrin C   281f
sealed-cell differential scanning calorimetry (SC-DSC)   211
seeding
– intentional   411
– unintentional   411ff
self-assembly   424
sequencing-by-synthesis method   360
D-serine   184
sertraline   381
SHE triggers model   469f
(-)-shikimic acid   22, 34ff
Shioiri's reagent   30
silica-CLEA nanocomposite   162
silicon-based carbon–carbon bond formation
– transition metal catalysis   101ff
silicon-based C–H arylation reaction   114
silicon reagent
– intramolecularly activated   104
silk-fibroin-supported chemoselective hydrogenation catalyst   90
silver-catalyzed reaction   121
silyl enol ether   110
silyl enolate   111
silylation
– aryl halide with hydrosilane   122
– C–H bond   122
– transition metal–catalyzed   122
α-silylnitrile   110f
SM-5887   207

small interfering ribonucleic acid (siRNA) 345
SOD, see superoxide dismutase
sodium triacetoxyborohydride (STAB) 127ff
solid–liquid separation 454
solid-phase synthesis
– DNA or RNA 346
sonocrystallization 416, 454
sonogashira reaction 6
L-sorbose 292
space filler 387f
spray dryer 462
standard operating procedure (SOP) 462, 466
stannyl-substituted bromobenzene 104
steel pipe test 371
cis-stilbene 94
trans-stilbene derivative 63
stoichiometric proportion 445
strategic resolution 393ff
stress-mediated bioprocess 421
stress–response function 428
Substance Avoidance Database 485
subtilisin Carlsberg 168
sulfide
– asymmetric oxidation 67f
sulfonic acid 40
– dynamic ammonium salt 40
superoxide dismutase (SOD) 426ff
– SOD-LIPOzyme 432
suspended centrifuge 458
sustainability 303
sustainable process
– enzyme 321ff
Suzuki–Miyaura C–C bond coupling reaction 131
synthesis 1ff
– development of new drug 1ff
– lab 444
synthesis with practical elegance 4
synthetic route evaluation 480ff

t

TAK-029 407ff
Tamiflu™ 4f, 16
tandem aldol reaction
– DERA-catalyzed 324
TBDPS (t-butyldiphenyl silyl) ether 85
tebipenem (TBPM) 265
– TBPM-4H$_2$O 266f
tebipenem hexetil 267
tebipenem pivoxil (TBPM-PI) 257ff
– synthesis 269
– synthetic process 265

technology transfer 477ff
telomerization 146
Tempium™ 4f
TEMPO (2,2,6,6-tetramethylpiperidine-1-oxyl) 172
TES 85
tetrabutylammonium fluoride (TBAF) 348
tetrafluoroethylene (TFE) 149
tetrahydrolipstatin 5ff
D,L-tetrahydropalmatine (THP) 277ff
– industrial synthesis 277
– primary synthetic route 280
– synthetic route 278ff
tetrahydropyranyl (THP) protective group 229f
– deprotection 230f
tetralone 208
tetraorganosilane 103
thermal stability 478
thermodynamics 445
thermogravimetry (TG) 367
Thermomyces lanuginosa lipase 165
– CLEA 165
1-(1,3-thiazolin-2-yl)azetidine-3-thio group 257
1-(1,3-thiazolin-2-yl)azetidine-3-thiol hydrochloride (TAT) 259ff
– industrial synthetic process 263
thiophene 310
TIPS 85
titanium
– (di-μ-oxo)titanium(salan) complex 65f
– di-μ-oxo-titanium–salen complex 69
– (aR,R)-di-μ-oxo titanium–salen complex 64
– hydroxo-hydroperoxo-titanium species 70
titanium/tartrate/t-butyl hydroperoxide (TBHP) system 67
p-toluenesulfonic acid 39
4-toluenesulfonyl chloride 248
2-(4-tolylsulfonyl)ethoxymethyl (TEM) 348
N-tosyl-(S)-phenylalanine (TPA) 391
– (S)-TPA 393
tosyl-D-leucine (TDL) 404
tosyl-D-valine (TDV) 403
traditional Chinese medicine (TCM) 273
ω-transaminase 331ff
transamination reaction 330
transesterification 166
– triglyceride 166

transgalactosylation 172
transition metal catalysis
– silicon-based carbon–carbon bond formation 101ff
– silylation of aryl halide with hydrosilane 122
transmetalation 101ff
– assisted by intramolecular activation 104
– organosilanolate through palladium silanolate 108
transport phenomena 425
triazole ester 228
2,3,5-tribromothiophene 305
triethylamine borane (TEAB) 129, 138
triglyceride
– transesterification 166
triisopropylsilyloxymethyl (TOM) 347
1,3,5-triketone 43ff
trimethylamine borane (TMAB) 129
4,6,9-trimethyldecan-3,5,7-trione 44
trimethylpyruvic acid (TMP) 192, 330
2-triorgano(2-pyridyl)silane 108ff
tris(dimethylamino)sulfonium difluorotrimethylsilane (TASF) 102
D-tryptophan 184
D-tyrosine 190f
L-tyrosine 190
L-*meta*-tyrosine 190
tyrosinase 173

u
ultrasonic atomization 462
unit operation 448f
– optimization and intensification 464

v
vacuum crystallizer 453
vacuum filter 456
D-valine 185
vanadium-catalyzed asymmetric oxidation of sulfide 68
vinyl butyrate 309
vinyldimethylsilanolate 108
vitamin B7 (H) 294ff
– Goldberg–Sternbach approach 295f
– synthetic method 297f
vitamin C 291ff
– Reichstein process 292

w
whole-cell biocatalyst 183ff

x
Xenical™ 4ff

y
yanhusuo 277ff

z
zanamivir 16, 323
zinc
– Friedel–Crafts alkylation 226